Handbook
of
Biochemistry
and
Molecular Biology

Handbook
of
Biochemistry
and
Molecular Biology

3rd Edition

Physical and Chemical Data
Volume I

EDITOR

Gerald D. Fasman, Ph. D.

Rosenfield Professor of Biochemistry
Graduate Department of Biochemistry
Brandeis University
Waltham, Massachusetts

CRC PRESS, Inc.
18901 Cranwood Parkway · Cleveland, Ohio 44128

Library of Congress Cataloging in Publication Data (Revised)

Main entry under title:

Handbook of biochemistry and molecular biology.

   Previous editions published under title: Handbook of
biochemistry.
   Includes bibliographies and indexes.
   CONTENTS: A. Proteins. 2. v.–B. Nucleic acids.
2. v.–C. Lipids, carbohydrates, steroids.– D. Physical and
chemical data, miscellaneous.
   1. Biological chemistry – Handbooks, manuals, etc.
2. Molecular biology – Handbooks, manuals, etc.
I. Fasman, Gerald D. II. Sober, Herbert Alexander,
1918–   [DNLM: 1. Biochemistry – Tables.
2. Molecular biology – Tables. QU16 H235]

QH345.H347 1975   574.1'92   75-29514
ISBN 0-87819-509-2 (v.1)

International Standard Book Number (ISBN)

Complete Set 0-87819-503-3
Physical and Chemical Data, Miscellaneous 0-87819-509-2

Library of Congress Card No. 75-29514

Handbook of Biochemistry and Molecular Biology

3rd Edition

Physical and Chemical Data
Volume I

Editor
Gerald D. Fasman, Ph. D.

Rosenfield Professor of Biochemistry
Graduate Department of Biochemistry
Brandeis University
Waltham, Massachusetts

The following is a list of the four major sections of the *Handbook,* each consisting of one or more volumes

Proteins — Amino Acids, Peptides, Polypeptides, and Proteins

Nucleic, Acids — Purines, Pyrimidines, Nucleotides, Oligonucleotides, tRNA, DNA, RNA

Lipids, Carbohydrates, Steroids

Physical and Chemical Data, Miscellaneous — Ion Exchange, Chromatography, Buffers, Miscellaneous, e.g., Vitamins

ADVISORY BOARD

Gerald D. Fasman
Editor

Herbert A. Sober (deceased)
Consulting Editor

MEMBERS

ADVISORY BOARD (continued)

ADVISORY BOARD (continued)

CONTRIBUTORS

Alice Adler
Graduate Department of Biochemistry
Brandeis University
Waltham, Massachusetts 02154

V. S. Ananthanarayahan
Molecular Biophysics Unit
Indian Institute of Science
Bangalore 560012
India

Norman G. Anderson
Molecular Anatomy Program
Oak Ridge National Laboratory
Oak Ridge, Tennessee 37830

Norman L. Anderson
Biology Division
Oak Ridge National Laboratory
Oak Ridge, Tennessee 37830

Roger G. Bates
Department of Chemistry
University of Florida
Gainesville, Florida 32601

H. P. J. Bloemers
Department of Biochemistry
University of Nijmegen
Nijmegen
The Netherlands

Myron K. Brakke
Crops Research Division
Agriculture Research Service
Nebraska Agricultural Experiment Station
Lincoln, Nebraska 68504

James J. Christensen
Department of Chemical Engineering
Brigham Young University
Provo, Utah 84601

Waldo E. Cohn
Biology Division
Oak Ridge National Laboratory
Oak Ridge, Tennessee 37830

Pierre Douzou
Institute National de la Sante et
    de la Recherche Medicale Groupe
U-128, B.P. 5051
34033 Montpellier Cedex
France

Murray Ettinger
Department of Biochemistry
State University of New York
Buffalo, New York 14214

Norman Good
Department of Botany and Plant
    Physiology
Michigan State University
East Lansing, Michigan 48823

R. W. Henderson
Department of Biochemistry
University of Melbourne
Parkville 3052 Victoria
Australia

G. Hui Bon Hoa
Institute National de la Sante et
    de la Recherche Medicale Groupe
U-128, B.P. 5051
34033 Montpellier Cedex
France

John O. Hutchens
Department of Physiology
University of Chicago
Chicago, Illinois 60637

Reed M. Izatt
Department of Chemistry
Brigham Young University
Provo, Utah 84601

William P. Jencks
Graduate Department of Biochemistry
Brandeis University
Waltham, Massachusetts 02154

Gordon C. Kresheck
Department of Chemistry
Northern Illinois University
DeKalb, Illinois 60115

CONTRIBUTORS (continued)

Neal Langerman
Department of Chemistry
College of Science
Utah State University
Logan, Utah 84321

Paul A. Loach
Department of Chemistry
Northwestern University
Evanston, Illinois 60201

P. Maurel
Institute National de la Sante et
de la Recherche Medicale Groupe
U-128, B.P. 5051
34033 Montpellier Cedex
France

Elemer Mihalyi
Laboratory of Biochemistry
National Heart and Lung Institute
National Institutes of Health
Bethesda, Maryland 20014

T. C. Morton
Department of Biochemistry
University of Melbourne
Parkville 3052 Victoria
Australia

Maya Paabo
Electrochemical Analysis Section
National Bureau of Standards
Washington, D.C. 20234

J. M. Regenstein
Department of Poultry Science
and Department of Food Science
Cornell University
Ithaca, New York 14853

F. Travers
Institute National de la Sante et
de la Recherche Medicale Groupe
U-128, B.P. 5051
34033 Montpellier Cedex
France

Ben A. M. van der Zeijst
Institute of Veterinary Virology
State University of Utrecht
Utrecht
The Netherlands

PREFACE

The rapid pace at which new data is currently accumulated in science presents one of the significant problems of today — the problem of rapid retrieval of information. The fields of biochemistry and molecular biology are two areas in which the information explosion is manifest. Such data is of interest in the disciplines of medicine, modern biology, genetics, immunology, biophysics, etc., to name but a few related areas. It was this need which first prompted CRC Press, with Dr. Herbert A. Sober as Editor, to publish the first two editions of a modern *Handbook of Biochemistry,* which made available unique, in depth compilations of critically evaluated data to graduate students, post-doctoral fellows, and research workers in selected areas of biochemistry.

This third edition of the *Handbook* demonstrates the wealth of new information which has become available since 1970. The title has been changed to include molecular biology; as the fields of biochemistry and molecular biology exist today, it becomes more difficult to differentiate between them. As a result of this philosophy, this edition has been greatly expanded. Also, previous data has been revised and obsolete material has been eliminated. As before, however, all areas of interest have not been covered in this edition. Elementary data, readily available elsewhere, has not been included. We have attempted to stress the areas of today's principal research frontiers and consequently certain areas of important biochemical interest are relatively neglected, but hopefully not totally ignored.

This third edition is over double the size of the second edition. Tables used from the second edition without change are so marked, but their number is small. Most of the tables from the second edition have been extensively revised, and over half of the data is new material. In addition, a far more extensive index has been compiled to facilitate the use of the Handbook. To make more facile use of the Handbook because of the increased size, it has been divided into four sections. Each section will have one or more volumes. The four sections are titled:

Proteins — Amino Acids, Peptides, Polypeptides, and Proteins
Nucleic Acids — Purines, Pyrimidines, Nucleotides, Oligonucleotides, tRNA, DNA, RNA
Lipids, Carbohydrates, Steroids
Physical and Chemical Data, Miscellaneous — Ion Exchange, Chromatography, Buffers, Miscellaneous, e.g., Vitamins

By means of this division of the data, we can continuously update the *Handbook* by publishing new data as they become available.

The Editor wishes to thank the numerous contributors, Dr. Herbert A. Sober, who assisted the Editor generously, and the Advisory Board for their counsel and cooperation. Without their efforts this edition would not have been possible. Special acknowledgments are due to the editorial staff of CRC Press, Inc., particularly Ms. Susan Cubar Benovich, Ms. Sandy Pearlman, and Mrs. Gayle Tavens, for their perspicacity and invaluable assistance in the editing of the manuscript. The editor alone, however, is responsible for the scope and the organization of the tables.

We invite comments and criticisms regarding format and selection of subject matter, as well as specific suggestions for new data (and their sources) which might be included in subsequent editions. We hope that errors and omissions in the data that appear in the Handbook will be brought to the attention of the Editor and the publisher.

Gerald D. Fasman
Editor
August 1975

PREFACE TO PHYSICAL AND CHEMICAL DATA, MISCELLANEOUS: ION EXCHANGE, CHROMATOGRAPHY, BUFFERS, MISCELLANEOUS, E.G., VITAMINS, VOLUME I

This section of the *Handbook of Biochemistry and Molecular Biology* on Physical and Chemical Data is divided into two volumes.

The first volume contains data on amino acids which consists of the coefficients of solubility in water, heat capacities, entropies of formation, and heats of combustion. Oxidation-reduction potentials of compounds of biochemical interest and of hemoproteins and metalloproteins are listed. Heats of proton ionization, pK and related thermodynamic quantities, free energies of hydrolysis and decarboxylation, and calorimetric ΔH values accompanying conformational changes of macromolecules in solution are listed. Information on the measurement of pH, buffer solutions, pH indicators, and the ionization constants of acids and bases are detailed.

Tables of refractive index of water, various solvents, salt solutions, liquids, temperature dependence, and Lorentz corrections are available.

Specific gravity of liquids, sucrose solutions, CsCl solutions, isokinetic glycerol and sucrose gradients for density gradient centrifugation, and the temperature dependence for selected compounds are included.

This collection of data, for which the editor alone is responsible, is highly selective in nature. It is hoped that this volume will be of assistance to those working in the field of biochemistry and molecular biology.

Gerald D. Fasman
Editor
March 1976

THE EDITOR

Gerald D. Fasman, Ph.D., is the Rosenfield Professor of Biochemistry, Graduate Department of Chemistry, Brandeis University, Waltham, Massachusetts.

Dr. Fasman graduated from the University of Alberta in 1948 with a B.S. Honors Degree in Chemistry, and he received his Ph.D. in Organic Chemistry in 1952 from the California Institute of Technology, Pasadena, California. Dr. Fasman did postdoctoral studies at Cambridge University, England, Eidg. Technische Hochschule, Zurich, Switzerland, and the Weizmann Institute of Science, Rehovoth, Israel. Prior to moving to Brandeis University, he spent several years at the Children's Cancer Research Foundation at the Harvard Medical School. He has been an Established Investigator of the American Heart Association, a National Science Foundation Senior Postdoctoral Fellow in Japan, and recently was a John Simon Guggenheim Fellow.

Dr. Fasman is a member of the American Chemical Society, a Fellow of the American Association for the Advancement of Science, Sigma Xi, The Biophysical Society, American Society of Biological Chemists, The Chemical Society (London), the New York Academy of Science, and a Fellow of the American Institute of Chemists. He has published 180 research papers.

The Editor and CRC Press, Inc. would like to dedicate this
third edition to the memory of Eva K. and Herbert A.
Sober. Their pioneering work on the development of the
Handbook is acknowledged with sincere appreciation.

TABLE OF CONTENTS

Nomenclature

_____

BIOCHEMICAL NOMENCLATURE

This synopsis of the recommendations of the IUPAC-IUB Commission on Biochemical Nomenclature (CBN) was prepared by Waldo E. Cohn, Director, NAS-NRC Office of Biochemical Nomenclature (OBN, located at Biology Division, Oak Ridge National Laboratory, Oak Ridge, TN 37830), from whom reprints of the CBN publications listed below and on which the synopsis is based are available.

The synopsis is divided into three sections: Abbreviations, symbols, and trivial names. Each section contains material drawn from the documents (A1 to C1, inclusive) listed below, which deal with the subjects named.

Additions consonant with the CBN Recommendations have been made by OBN throughout the synopsis.

RULES AND RECOMMENDATIONS AFFECTING BIOCHEMICAL NOMENCLATURE AND PLACES OF PUBLICATION (AS OF FEBRUARY 1975)

I. IUPAC-IUB Commission on Biochemical Nomenclature
  A1. Abbreviations and Symbols [General; Section 5 replaced by A6]
  A2. Abbreviated Designation of Amino-acid Derivatives and Peptides (1965) [Revised 1971; Expands Section 2 of A1]
  A3. Synthetic Modifications of Natural Peptides (1966) [Revised 1972]
  A4. Synthetic Polypeptides (Polymerized Amino Acids) (1967) [Revised 1971]
  A5. A One-letter Notation for Amino-acid Sequences (1968)
  A6. Nucleic Acids, Polynucleotides, and their Constituents (1970)

  B1. (Nomenclature of Vitamins, Coenzymes, and Related Compounds)
      a. Miscellaneous [A, B's, C, D's, tocols, niacins; see B2 and B3]
      b. Quinones with Isoprenoid Side-chains: E, K, Q [Revised 1973]
      c. Folic Acid and Related Compounds
      d. Corrinoids: B-12's [Revised 1973]
  B2. Vitamins B-6 and Related Compounds [Revised 1973]
  B3. Tocopherols (1973)

  C1. Nomenclature of Lipids (1967) [Amended 1970; see also II, 2]
  C2. Nomenclature of α-Amino Acids (1974) [See also II, 5]

  D1. Conformation of Polypeptide Chains (1970) [See also III, 2]

  E1. Enzyme Nomenclature (1972)[a] [Elsevier (in paperback); Replaces 1965 edition.]
  E2. Multiple Forms of Enzymes (1971) [Chapter 3 of E1]
  E3. Nomenclature of Iron-sulfur Proteins (1973) [Chapter 6.5 of E1]
  E4. Nomenclature of Peptide Hormones (1974)

II. Documents Jointly Authored by CBN and CNOC [See III]
  1. Nomenclature of Cyclitols (1968) [Revised 1973]
  2. Nomenclature of Steroids (1968) [Amended 1971; Revised 1972]
  3. Nomenclature of Carbohydrates-I (1969)
  4. Nomenclature of Carotenoids (1972) [Revised 1975]
  5. Nomenclature of α-Amino Acids (1974) [Listed under I, C2 in the following table]

III. IUPAC Commission on the Nomenclature of Organic Chemistry (CNOC)
  1. Section A (Hydrocarbons), Section B (Heterocyclics): *J. Am. Chem. Soc.,* 82, 5545;[a] Section C (Groups containing N, Hal, S, Se/Te): *Pure Appl. Chem.,* 11, Nos. 1–2[a] [A, B, and C Revised 1969:[a] Butterworth's, London (1971)]
  2. Section E (Stereochemistry):[b] *J. Org. Chem.,* 35, 2489 (1970); *Biochim. Biophys. Acta,* 208, 1 (1970); *Eur. J. Biochem.,* 18, 151 (1970) [See also I, D1]

[a]No reprints available from OBN; order from publisher.
[b]Reprints available from OBN (in addition to all in IA to ID and II).

RULES AND RECOMMENDATIONS AFFECTING BIOCHEMICAL NOMENCLATURE AND PLACES OF PUBLICATION (AS OF FEBRUARY 1975)(continued)

IV.   Physiochemical Quantities and Units (IUPAC)[a] *J. Am. Chem. Soc.*, 82, 5517 (1960) [Revised 1970: *Pure Appl. Chem.*, 21, 1 (1970)]

V.   Nomenclature of Inorganic Chemistry (IUPAC) *J. Am. Chem. Soc.*, 82, 5523[a] [Revised 1971: *Pure Appl. Chem.*, 28, No. 1 (1971)][a]

VI.   Drugs and Related Compounds or Preparations

    1.   U.S. Adopted Names (USAN) No. 10 (1972) and Supplement [U.S. Pharmacopeial Convention, Inc., 12601 Twinbrook Parkway, Rockville, Md.]

    2.   International Nonproprietary Names (INN) [WHO, Geneva]

CBN RECOMMENDATIONS APPEAR IN THE FOLLOWING PLACES[a]

	Arch. Biochem. Biophys.	Biochem. J.	Biochemistry	Biochim. Biophys. Acta	Eur. J. Biochem.	J. Biol. Chem.	Pure Appl. Chem.[b]	Biochimie (Bull. Soc.)[c]	Molek. Biol.[d]	Z. Phys. Chem.[e]
A1[f]	136,1	101,1	5,1445		1,259	241,527	40,(R)	50,3	1,872	348,245
A2(Revised)	150,1(R)	126,773(R)	11,1726(R)	263,205(R)	27,201(R)	247,977(R)	31,649(R)	49,121*	2,282*	348,256*
A3(Revised)	121,6*	104,17*	6,362*	133,1*	1,379*	242,555*	33,439(R)	49,325*	2,466*	348,262*
A4(Revised)[g]	151,597(R)	127,753(R)	11,942(R)	278,211(R)	26,301(R)	247,323(R)	31,641	51,205*	5,492(R)	349,1013*
A5	125(3),i	113,1	7,2703	168,6	5,151	243,3557	40,	50,1577	3,473	350,793
A6[h]	145,425	120,449	9,4022	247,1	15,203	245,5171			6,167	351,1055
B1*	118,505	102,15		107,1(a–c)	2,1	241,2987		49,331		348,266
B1b(Revised)	165,1(R)	147,15(R)		387,397(R)	53,15(R)		38,439			
B1d(Revised)	161(2),iii(R)	147,1(R)	13,1555(R)		45,7(R)					
B2(Revised)	162,1(R)	137,417(R)			40,325(R)	245,4229*				
B3(Revised)	165,6(R)	147,11(R)	13,1056(R)	354,155(R)	46,217(R)		33,447(R)			351,1165*
C1[f]	123,409	105,897	6,3287	152,1	2,127	242,4845		50,1363	2,784	350,279
Amendments		116(5)		202,404	12,1	245,1511				
C2			14,449		53,1					
D1[i]	145,405	121,577	9,3471	229,1	17,193	245,6489			7,289	
E2	147,1	126,769	10,4825	258,1	24,1	246,6127		54,123		
E3	160,355	135,5	12,3582	310,295	35,1	248,5907				353,852
E4		151,1	14,2559			250,3215				
II,1(Revised)	128,269*	112,17*	8,2227	165,1*	5,1*	243,5809*	37,285(R)	51,3*		350,523*
II,2[f]	136,13	113,5	10,4994	164,453	10,1	(243,5809*)	31,285(R)	51,819		351,663
Amendments	147,4	127,673		248,387	25,2					
II,3		125,673	10,3983	244,223	21,455	247,613				
II,4		127,741	10,4827	286,217	25,397	247,2633				
Amendments		151,507	14,1803							

[a] Reprints available from OBN.
[b] No reprints available from OBN; order from publisher.
[c] In French.
[d] In Russian.
[e] In German.

[f] Also in other journals.
[g] Also in *Biopolymers*, 11, 321.
[h] *J. Mol. Biol.*, 55, 299.
[i] *J. Mol. Biol.*, 52, 1.

*First, unrevised version.
(R) = revised version.

ABBREVIATIONS

Abbreviations are distinguished from **symbols** as follows (taken from Reference A1):

a. **Symbols**, for monomeric units in macromolecules, are used to make up abbreviated structural formulas (e.g., Gly-Val-Thr for the tripeptide glycylvalylthreonine) and can be made fairly systematic.

b. **Abbreviations** for semi-systematic or trivial names (e.g., ATP for adenosine triphosphate; FAD for flavinadenine dinucleotide) are generally formed of three or four capital letters, chosen for brevity rather than for system. It is the indiscriminate coining and use of such abbreviations that has aroused objections to the use of abbreviations in general.

[Abbreviations are thus distinguished from symbols in that they (a) are for semi-systematic or trivial names, (b) are brief rather than systematic, (c) are usually formed from three or four capital letters, and (d) are not used — as are symbols — as units of larger structures. ATP, FAD, etc., are abbreviations. Gly, Ser, Ado, Glc, etc., are symbols (as are Na, K, Ca, O, S, etc.); they are sometimes useful as abbreviations in figures, tables, etc., where space is limited, but are usually not permitted in text. The use of abbreviations is permitted when necessary but is never required.]

1. Nucleotides (N = A, C, G, I, O, T, U, X, ψ — see One-letter Symbols)

NMP	Nucleoside 5'-phosphate
NDP	Nucleoside 5'-di(or pyro)phosphate
NTP	Nucleoside 5'-triphosphate

Prefix d indicates deoxy.

2. Coenzymes, vitamins

CoA(or CoASH)	Coenzyme A
CoASAc	Acetyl Coenzyme A
DPN[a]	Diphosphopyridine nucleotide
FAD	Flavin-adenine dinucleotide
FMN	Riboflavin 5'-phosphate
GSH	Glutathione
GSSG	Oxidized glutathione
NAD[b]	Nicotinamide-adenine dinucleotide (cozymase, Coenzyme I, diphosphopyridine nucleotide)
NADP[b]	Nicotinamide-adenine dinucleotide phosphate (Coenzyme II, triphosphopyridine nucleotide)
NMN	Nicotinamide mononucleotide
TPN[c]	Triphosphopyridine nucleotide

3. Miscellaneous

ACTH	Adrenocorticotropin, adrenocorticotropic hormone, or corticotropin
CM-cellulose	*O*-(Carboxymethyl)cellulose
DEAE-cellulose	*O*-(Diethylaminoethyl)cellulose
DDT	1,1,1-Trichloro-2,2-bis(*p*-chlorophenyl)ethane
EDTA	Ethylenediaminetetraacetate
Hb,HbCO,HbO$_2$	Hemoglobin, carbon monoxide hemoglobin, oxyhemoglobin
P$_i$	Inorganic orthophosphate

[a]Replaced by NAD (also DPN$^+$ by NAD$^+$, DPNH by NADH).
[b]Generic term; oxidized and reduced forms are NAD$^+$, NADH (NADP$^+$, NADPH).
[c]Replaced by NADP (also TPN$^+$ by NADP$^+$, TPNH by NADPH).

PP$_i$	Inorganic pyrophosphate
TEAE-cellulose	*O*-(Triethylaminoethyl)cellulose
Tris	Tris(hydroxymethyl)aminomethan (2-amino-2-hydroxymethylpropane-1,3-diol)

4. Nucleic Acids

DNA, RNA	Deoxyribonucleic acid, ribonucleic acid (or -nucleate)
hnRNA	Heterogeneous RNA
mtDNA	Mitochondrial DNA
cRNA	Complementary RNA
mRNA	Messenger RNA
nRNA	Nuclear RNA
rRNA	Ribosomal RNA
tRNA	Transfer RNA (generic term; sRNA should not be used for this or any other purpose)
tRNA^{Ala}	Alanine tRNA; tRNA$_1^{Ala}$, tRNA$_2^{Ala}$: isoacceptor alanine tRNA's
AA-tRNA	Aminoacyl-tRNA; aminoacylated tRNA; "charged" tRNA (generic term)
Ala-tRNA or Ala-tRNA^{Ala}	Alanyl-tRNA
tRNA^{Met}	Methionine tRNA (not enzymically formylatable)
tRNA^{fMet} or tRNA$_f^{Met}$	Methionine tRNA, enzymically formylatable to . . .
fMet-tRNA	Formylmethionyl-tRNA (small f, to distinguish from fluorine F)

SYMBOLS

Symbols are distinguished from abbreviations in that they are designed to represent specific parts of larger molecules, just as the symbols for the elements are used in depicting molecules, and are thus rather systematic in construction and use. Symbols are not designed to be used as abbreviations and should not be used as such in text, but they may often serve this purpose when space is limited (as in a figure or table). Symbols are always written with a single capital letter, all subsequent letters being lower-case (e.g., Ca, Cl, Me, Ac, Gly, Rib, Ado), regardless of their position in a sequence, a sentence, or as a superscript or subscript.

Some abbreviations expressed in symbols as examples of the use of symbols:

Dimethylsulfoxide	Me$_2$SO [a]
Tetranitromethane	(NO$_2$)$_4$C [b]
Guanidine hydrochloride	Gdn · HCl [c]
Guanidinium chloride	GdmCl
Cetyltrimethylammonium bromide	CtMe$_3$NBr [d]
Ethyl methanesulfonate	MeSO$_3$Et
Methylnitronitrosoguanidine	MeN$_2$O$_3$Gdn
-nitrosourea	-Nur [e]
-nitrosamine	-Nam [f]
-fluorene	-Fln
Aminofluorene	NH$_2$Fln
Acetylaminofluorene	AcNHFln [g]
Acetoxyacetylaminofluorene	Ac(AcO)NFln
N-Acetylneuraminic acid	AcNeu [h]

[a] Replaces DMSO.
[b] Replaces TNM.
[c] Replaces Gu, Gd, and G.
[d] Replaces CTAB (similarly for other ammonium compounds).
[e] Replaces NU.
[f] Replaces NA.
[g] Replaces AAF.
[h] Not NANA.

TRIVIAL NAMES

I. Vitamins, Coenzymes, and Related Compounds
A. Vitamin A (Reference B1a)

"The term *vitamin A* should be used as the generic description for all β-ionone derivatives, other than provitamin A carotenoids, exhibiting qualitatively the biological activity of retinol. Thus, phrases such as 'vitamin A activity,' 'vitamin A deficiency,' and 'vitamin A in the form of . . .' represent preferred usage. . . ."

"The term *provitamin A carotenoid* should be used as the generic descriptor for all carotenoids exhibiting qualitatively the biological activity of β-carotene. When referring to the biological activity of the provitamin A carotenoids, the phrase 'provitamin A activity' represents preferred usage."*

*International Union of Nutritional Science (I.U.N.S.), *Nutr. Abstr. Rev.*, 40, 395 (1970).

Structure	9'Group	Name(s)
I	—CH$_2$OH	Retinol [vitamin A or A$_1$; vitamin A or A$_1$ alcohol; axerol; axerophthol] [a,b]
I	—CHO	Retinal(dehyde) [vitamin A or A$_1$ aldehyde; retinene] [b]
I	—COOH	Retinoic acid [vitamin A or A$_1$ acid] [b]
II	—CH$_2$OH	3-Dehydroretinol or dehydroretinol [vitamin A$_2$] [c,b]
II	—CHO	3-Dehydroretinal(dehyde) or dehydroretinal-(dehyde) [retinene-2] [b]
II	—COOH	3-Dehydroretinoic acid or dehydroretinoic acid

[a] 2,6,6-Trimethyl-1-(9'-hydroxy-3',7'-dimethylnona-1',3'5',7'-tetraenyl)cyclohex-1-ene.
[b] Former or alternate names.
[c] 2,6,6-trimethyl-1-(9'-hydroxy-3'-7'-dimethylnona-1',3',5',7'-tetraenyl)cyclohexa-1,3-diene.

B. Vitamin D (Reference B1a)

"The term *vitamin D* should be used as the generic descriptor for all steroids exhibiting qualitatively the biological activity of cholecalciferol.* Thus, phrases such as 'vitamin D activity' and 'vitamin D deficiency' represent preferred usage."**

Group at 17	Names
$-CH(Me)-CH_2-CH_2-CH_2-CHMe_2$	Cholecalciferol; vitamin D_3 (abbreviation: D_3)[a]
$-CH(Me)-CH=CH-CH(Me)-CHMe_2$	Ergocalciferol (calciferol), vitamin D_2 (abbreviation: D_2)[b]
$-CH(Me)-CH_2-CH_2-CH_2-C(OH)Me_2$	25-Hydroxycholecalciferol(I.U.N.S.) (abbreviation: 25(HO)D_3)

[a]9,10 Secocholesta-5,7,10(19)-trien-3β-ol; ester name substitutes -yl for -ol in trivial name (I.U.N.S.).
[b]9,10-Secoergosta-5,7,10(19), 22-tetraen-3β-ol; ester name substitutes -yl for -ol, in trivial name (I.U.N.S.).

*Replacing ergocalciferol (A.I.N.).
**International Union of Nutritional Science (I.U.N.S.), *Nutr. Abstr. Rev.,* 40, 395 (1970); American Institute of Nutrition (A.I.N.), *J. Nutr.,* 99, 244 (1969).

C. Vitamin E, Vitamin K, and Coenzyme Q (References B1a B1b, and B3)

"The term *vitamin E* should be used as the generic descriptor for all tocol and tocotrienol derivatives exhibiting qualitatively the biological activity of α-tocopherol. Thus, phrases such as 'vitamin E activity,' 'vitamin E deficiency,' and 'vitamin E in the form of . . .' represent preferred usage."*

"The term *vitamin K* should be used as the generic descriptor for 2-methyl-1,4-naphthoquinone and all derivatives exhibiting qualitatively the biological activity of phytylmenaquinone (phylloquinone). Thus, phrases such as 'vitamin K activity' and 'vitamin K deficiency' represent preferred usage."*†

*International Union of Nutritional Science (I.U.N.S.), *Nutr. Abstr. Rev.,* 40, 395 (1970).
†American Institute of Nutrition (A.I.N.), *J. Nutr.,* 99, 244 (1969).

Summary of Chemical Relationships and Nomenclature of Some Biologically Active Quinones with Isoprenoid Side-Chains, Including Vitamins E and K and Coenzyme Q (Recommended Names are in Boldface)

Aromatic nucleus	Side-chain	Trivial name(s)	Abbreviations	Cyclized form	Trivial names	Abbreviations
1,4-Naphthoquinone 2-methyl- **Menaquinone*** (Menadione)	3-(prenyl)$_n$	**Menaquinone-n** Prenylmenaquinone-n* n = 10: vitamin K$_2$(50)** n = 7: vitamin K$_2$(35)** n = 6: vitamin K$_2$(30)**	MK-n MQ-n*	2H-Naphtho[1,2b]pyran-6-ol (2,5-substituted)	Menachromenol-(n-1)	MK-n-el
	3-Phytyl	**Phylloquinone** Phytylmenaquinone* = vitamin K$_1$(20)**	**K** PMQ*	Same	Phyllochromenol	K-el
1,4-Benzoquinone 2,3-dimethoxy-5-methyl-	6-(prenyl)$_n$	**Ubiquinone-n** n = 10: coenzyme Q$_{10}$** = ubiquinone-50** n = 6 : coenzyme Q$_6$** = ubiquinone-30**	Q-n Q-10 Q-6	7,8-Dimethoxy-2H-chromen-6-ol (2,5-substituted)	Ubichromenol-(n-1) n = 10: ubichromenol-9 = ubichromenol(50)**	-Q-n-el
1,4-Benzoquinone 2,3,5-trimethyl-	6-(prenyl)$_n$	**Tocoquinone-n** n = 10: vitamin E$_2$(50)** n = 4: cf. 2,3,5-trimethyltocotrienolquinone			Tocochromanol-(n-1)	

Summary of Chemical Relationships and Nomenclature of Some Biologically Active Quinones with Isoprenoid Side-Chains, Including Vitamins E and K and Coenzyme Q (Recommended Names are in Boldface) (continued)

1,4-Benzoquinone 2,3-dimethyl-

6-(prenyl)$_n$

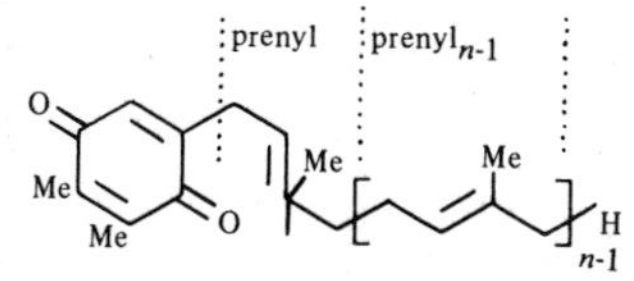

Plastoquinone-n
$n = 9$: plastoquinone**
= Kofler's quinone**
= plastochromenol-x
= solanochromene**

PQ-n
PQ-9
PQ-8-cl

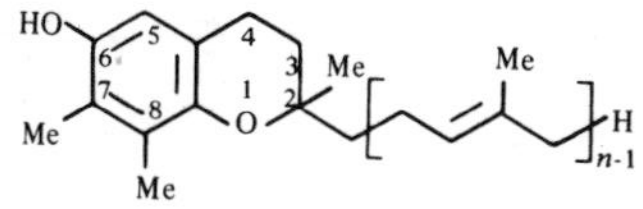

Plastochromanol-(n-1)
Plastochromanol-3
7,8-dimethyltocotrienol
= γ-tocotrienol**
= η-tocopherol**

PQ-n-al
PQ-3-al
7,8-T-3

1,4-Benzoquinone (2,3,5-substituted)

6-"phytyl"

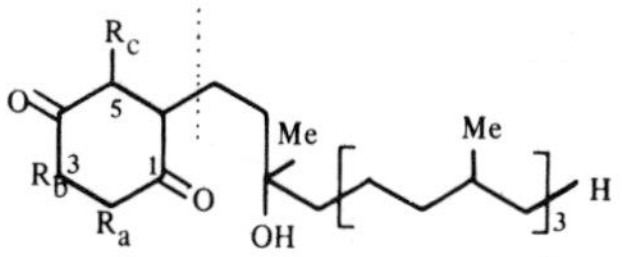

x-Tocopherolquinone(s)

x	R_a	R_b	R_c
α	Me	Me	Me (oxidized vit. E)
β	Me		Me
γ		Me	Me (cf. plastoquinone)
ζ		**Me**	

x-TQ

Cyclized, reduced form

Chroman-6-ol
(Benzopyran-6-ol)
(2,5,7,8-substituted)

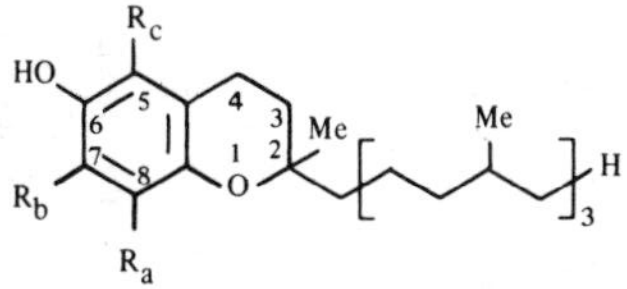

x-Tocopherol(s)
α($R_a = R_b - R_c = $Me) =
5,7,8-trimethyltocol
= vitamin E**
Tocol: $R_a = R_b = R_c = $H

x-T

1,4-Benzoquinone (2,3,5-substituted)

6-"tetraprenyl"

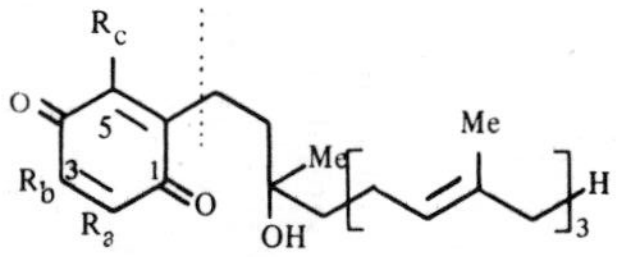

x-Tocotrienolquinone(s)

x = 2,3,5-Trimethyl-
= α** (cf. tocoquinone-4)

x = 2,5-Dimethyl-
= β**

x = 2,3-Dimethyl-
= γ** (cf. plastoquinone-4)

x = 2-Methyl-
= δ**

x-TQ-3
2,3,5-TQ-3
2,5-TQ-3
2,3-TQ-3

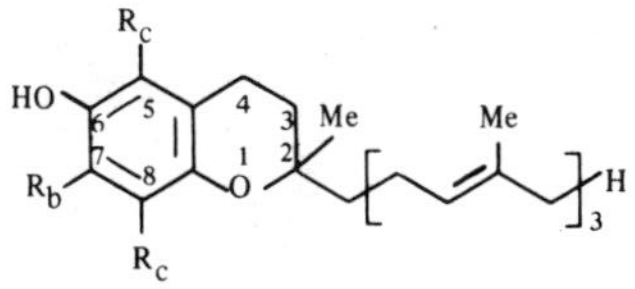

x-Tocotrienol(s)

x = **5,7,8-Trimethyl** =
tocochromanol-3
= α-tocotrienol**
= ζ-tocopherol**

x = 5,8-Dimethyl
= ϵ-tocopherol**

x = **7,8-Dimethyl** =
plastochromanol-3
= **γ-tocotrienol**
= η-tocopherol**

x-T-3
5,7,8-T-3

5,8-T-3

7,8-T-3
PQ-3-al

*Proposed by Committee on Nomenclature of IUNS [*Nutr. Abstr. Rev.*,] 40, 395, (1970), where differing from those recommended here.
**Names previously used (see first paragraph in introductory section), are not recommended.

D. Vitamin B-6 (Reference B2)

"The term *vitamin B* should be used as the generic descriptor for all 2-methylpyridine derivatives exhibiting qualitatively the biological activity of pyridoxine. This term should be used in derived terms such as vitamin B-6 deficiency, vitamin B-6 activity, vitamin B-6 antagonists."

'Pyridoxine' should not be used as a generic term synonymous with 'vitamin B-6.' Pyridoxol should not be used as a synonym for pyridoxine.

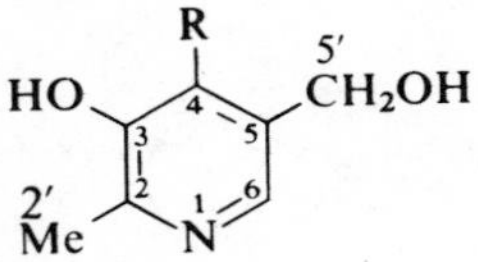

Group in position

4	5	Name	Abbreviation	Symbol[a]
$-CH_2OH$	$-CH_2OH$	Pyridoxine (not pyridoxol)	PN[b]	Pxn
$-CH_2-$	$-CH_2OH$	Pyridoxyl	—	Pxy–
$-CH=O$	$-CH_2OH$	Pyridoxal	PL[b]	Pxl
$-CH=$	$-CH_2OH$	Pyridoxylidene	—	Pxd=
$-CHNH_2$	$-CH_2OH$	Pyridoxamine	PM[b]	Pxm
$-COOH$	$-CH_2OH$	4-Pyridoxic acid	—	—
$-COOH$-(lactone)	$-CH_2OH$	4-Pyridoxolactone	—	—
$-CH_2OH$	$-CHO$	Isopyridoxal	—	—
$-CH_2OH$	$-COOH$	5-Pyridoxic acid	—	—
$-CH_2OH$-(lactone)	$-COOH$	5-Pyridoxolactone	—	—
$-CH_2OH$	$-CH_2OPO_3H_2$	Pyridoxine 5'-phosphate[c]	PNP[b]	Pxn-*P*
$-CHO$	$-CH_2OPO_3H_2$	Pyridoxal 5'-phosphate[c]	PLP[b]	Pxl-*P*
$-CHNH_2$	$-CH_2OPO_3H_2$	Pyridoxamine 5'-phosphate[c]	PMP[b]	Pxm-*P*

Group in position

4	5	2	6	Name
$-CH=O$	$-CH_3$	$-CH_3$	H	5'-Deoxypyridoxal
$-CH=O$	$-CH(CH_3)OH$	$-CH_3$	H	5'-Methylpyridoxal
$-CH=O$	$-CH_2OH$	$-H$	H	2-Demethylpyridoxal or 2-norpyridoxal
$-CH=O$	$-CH_2OH$	$-C_3H_7$	H	2-Propyl-2-norpyridoxal
$-CH=O$	$-CH_2OH$	$-CH_3$	CH_3	6-Methylpyridoxal
$-CH_3$	$-CH_2OH$	$-CH_3$	H	4-Deoxypyridoxine

[a]For use in combination with similar symbols (e.g., Lys, Me, *P*) or names, in accord with general rules of symbolism (pp. 5 *et. seq.*)
[b]Not recommended. Use full name (see Footnote c).
[c]Or -5-*P* or 5'-*P*- for 5'phosphate (or -*P* alone).

E. Vitamin B-12: Corrinoids (Reference B1d)

"The term *vitamin B–12* should be used as the generic descriptor for all corrinoids exhibiting qualitatively the biological activity of cyanocobalamin. Thus, phrases such as 'vitamin B–12 activity' and 'vitamin B–12 deficiency' represent preferred usage. The term *corrinoids* should be used as the generic descriptor for all compounds containing the corrin nucleus and thus chemically related to cyanocobalamin. The term 'corrinoid' is not synonymous with the term 'vitamin B–12'."*

"The generic name for compounds containing the corrin nucleus is corrinoid."

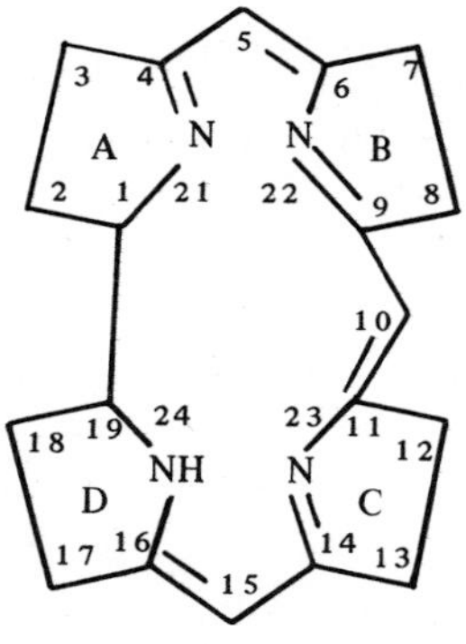

I

Corrin

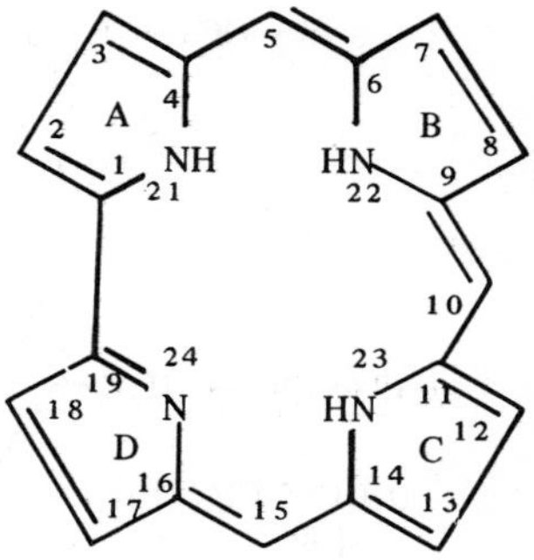

IA

Corrole

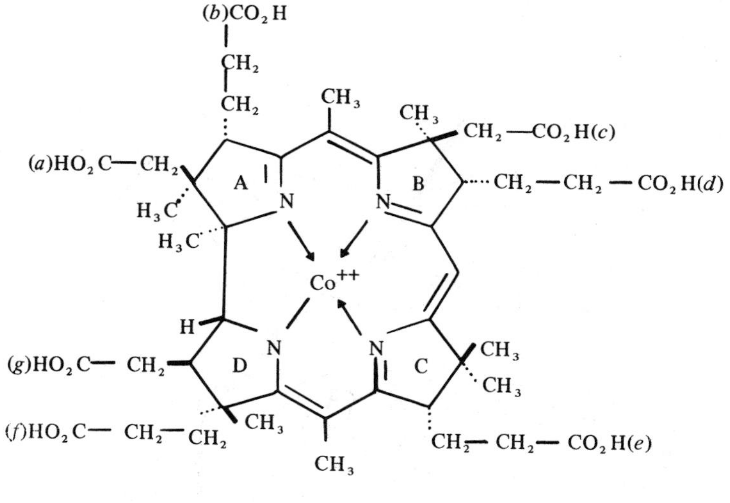

II

Cobyrinic acid

*International Union of Nutritional Science (I.U.N.S.), *Nutr. Abstr. Rev.,* 40, 395 (1970); American Institute of Nutrition (A.I.N.), *J. Nutr.,* 99, 244 (1969).

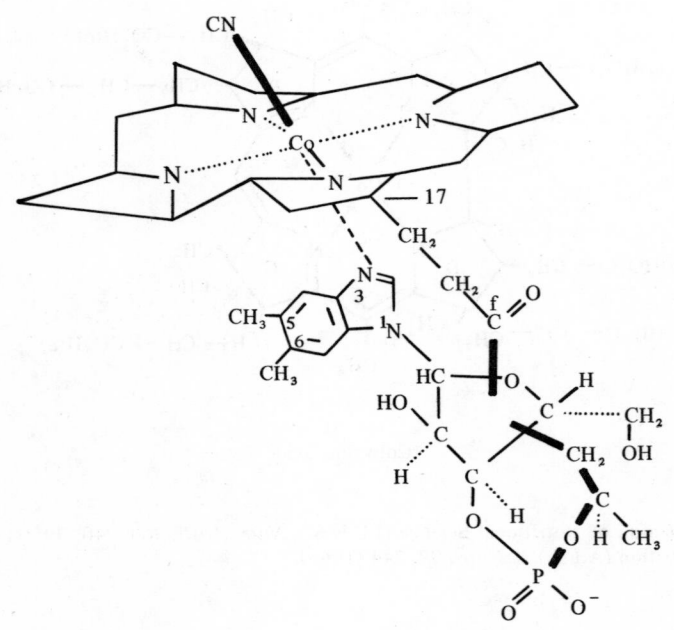

(b)CO — R

(a)R—OC—CH₂

CH₂ — CO — R(c)

CH₂—CH₂ — CO —R(d)

Co⁺⁺

(g)R— OC — H₂C

(f)OC— CH₂— H₂C

CH₂— CH₂ — CO₂ —R(e)

NH

CH₂

H► C ◄OR′

CH₃

III

Cobinic acid

IV

Cyanocobalamin (Vitamin B₁₂)

Sketch based on Hodgkin *et al.* (1957). Detail of substituents on
corrin nucleus (except side chain at C-17) is omitted for the sake of
clarity.

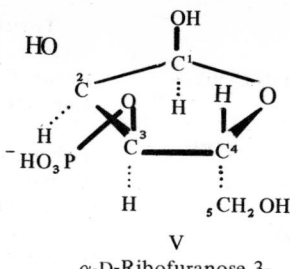

V

α-D-Ribofuranose 3-
phosphate residue.

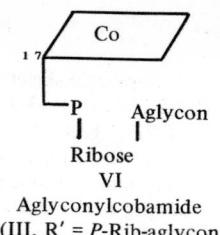

VI

Aglyconylcobamide
(III, R′ = P-Rib-aglycon)

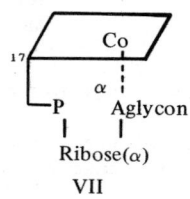

VII

Aglyconylcobamide, with
aglyconyl liganded to
cobalt (IV without CN and
with dimethylbenzimidazole
as aglycon)

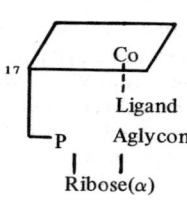

VIII

(Coα-Ligandyl)-aglyconyl-
cobamide (ligand has "displaced"
aglycon of VII)

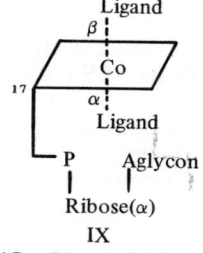

IX

(Coα-Ligandyl)-(Coβ-
ligandyl)aglyconylcobamide
(VIII with additional ligand in Coβ position)

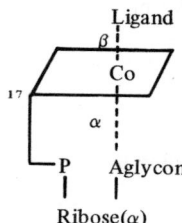

X

Xa, Aglyconyl-(Coβ-ligandyl)cobamide (VII with additional ligand
in Coβ position; IV with dimethylbenzimidazole as aglycon, CN as
Coβ ligand)

Xb, Ligandylcobalamin (if aglycon is dimethylbenzimidazole as in
IV, and with CN as Coβ ligand)

	Formula	Description	Trivial Name;[a] symbol
1	I	Tetrapyrrole skeleton (porphyrin minus its C-20)	Corrin; Crn
2	II	2,7,18-triacetic-3,8,13,17-tetrapropionic-1,2,5,7,12,12,15,17-octamethylcobaltocorrin	Cobyrinic acid
3	II	*abcdeg*-hexaamide of 2	Cobyric acid; Cby
4	III	*f*-(1-amino-2-propanol) derivative of 2 (R' = H)	Cobinic acid
5	III	*abcdeg*-hexaamide of 4	Cobinamide; Cbi
6	III-V	4 (R' = α-D-ribofuranose 3-phosphate)	Cobamic acid
7	III-V	*abcdeg*-hexaamide of 6	Cobamide; Cba
8	VII	α-(5,6-dimethylbenzimidazolyl)cobamide	Cob(II)alamin;[b] Cbl^{II}
9	IV,X	*Co*α-[α-(5,6-dimethylbenzimidazolyl)]-*Co*β-cyanocobamide	Cyanocobalamin;[c] CN-Cbl
	X	H$_2$O for CN in 9 (or in IV)	Aquacobalamin;[d] aqCbl
	X	OH for CN in 9 (or in IV)	Hydroxocobalamin;[e] HO-Cbl
	X	NO$_2$ for CN in 9 (or in IV)	Nitritocobalamin;[f] NO$_2$-Cbl
	X	5′-deoxy-5′-adenosyl for CN in 9 (or in IV)	Adenosylcobalamin;[g] AdoCbl
	X	CH$_3$ for CN in 9 (or in IV)	Methylcobalamin;[h] MeCbl

[a] Replacement of cobalt by another metal ion is indicated by ferro (or ferri), nickelo (i), hydrogeno, etc., in place of "co" (except in corrin) and symbolized as, e.g., [H]Cba, [Ni]Cbl^{II}, etc.

[b] "Vitamin B-12r" (Cob(I)alamin, as in VI, is "vitamin B-12s"). A variant with adenine as aglycon, bound to ribose by its N-7 to yield α-(7-adenyl) cobamide, has been known as "pseudovitamin B-12." The variant with a 2-methyladenyl aglycon, with CN in the *Co*β position, has been known as "Factor A."

[c] 1. Or cyanocob(III)alamin; "vitamin B-12." 2. A variant with 5-OH in place of 5,6-Me$_2$ has been known as "vitamin B-12$_{III}$." 3. The 5-OMe derivative, with methyl in the *Co*β position (cf. methylcobalamin), has been known as "Factor III$_m$."

[d] "Vitamin B-12a."

[e] "Vitamin B12b."

[f] "Vitamin B-12c."

[g] "Coenzyme B-12."

[h] Involved in several reactions, including methionine biosynthesis (see Footnote c3).

ABBREVIATIONS AND NAMES OF SOME CORRINOID DERIVATIVES AND VITAMINS B-12 (REFERENCE Bld)

(Me)aqCbi	(Methyl)aquacobinamide (methyl in α position)
(CN)MeCbi	(Cyano)methylcobinamide (methyl in β position)
(CN,aq)Cbi or (aq,CN)Cbi	Cyanoaquacobinamide (ligand location unspecified)
(aq)EtCbi	(Aqua)ethylcobinamide
(CN)(2-OAcBu)Cbi	(Cyano)(2-acetoxybutyl)cobinamide
(aq)(Bu-2)Cbi or (aq)Bu^{s}Cbi	(Aqua)(*sec*-butyl)cobinamide
(aq)3-OAcBu-2)Cby	(Aqua)(3-acetoxybut-2-yl)-cobyric acid
(aq)cHxCbi	(Aqua)cyclohexylcobinamide
(CN)(2-HOcPe)Cby	(Cyano)(2-hydroxycyclopentyl) cobyric acid
(2-MeAde/aq)MeCba [a]	*Co*α-Aqua-*Co*β-methyl(2-methyl-adenylcobamide)
(Ade/CN)CN-Cba [a]	*Co*α-Cyano-*Co*β-Cyano(adenylcobamide) or dicyanoadenyl-cobamide
(OH)MeCbl	*Co*α-Hydroxo-*Co*β-methylcobalamin or *Co*α-hydroxo-*Co*β-methyl(dimethylbenzimidazolylcobamide)
(CN)$_2$Cbl	*Co*α-Cyano-*Co*β-cyanocobalamin or dicyanocobalamin
CN-Cbl(13-epi)	Cyano(13-epi)cobalamin
CN-Cbl(13epi-*e*OH)	*Co*α-(α-5,6-Dimethylbenzimidazolyl)-*Co*β-cyano-(13-epi)co-bamic *a,b,c,d,g*-pentaamide

AdoCbl(10-Cl)	Adenosyl-10-chloro-cobalamin
(aq)AdoCbi(*e*-PhNH)	*Co*α-Aqua-*Co*β-adenosylcobinic *a,b,c,d,g*-pentaamide *e*-anilide
(CN)ClCby(8-NH-*c*-lactam)	
(CN)$_2$Cby(OMe)$_7$	Dicyanocobyrinic heptamethyl ester
(CN,aq)Cby[*a:g*-(NH$_2$)$_5$]	Cyanoaquacobyrinic acid pentaamide
(CN)$_2$Cby[10-Cl-*a:g*-(NH$_2$)$_5$]	10-Chloro derivative of the above
(Bza)Me[^{57}Co]Cba [b]	*Co*α-(α-Benzimidazolyl)-*Co*β-methyl-[^{57}Co]cobamide
(Bza)[^{14}C]MeCba [b]	*Co*α-(α-Benzimidazolyl)-*Co*β-[^{14}C]methylcobamide
([4-^{3}H]Bza)MeCba [b]	*Co*α-(α-[4-^{3}H]Benzimidazolyl)-*Co*β-methylcobamide
CN-Cbl	Cyanocob(III)alamin (vitamin B-12)
AdoCbl	Adenosylcob(III)alamin
PrCbl	*n*-Propylcob(III)alamin; methyl-, etc., similarly
(Ade)(Pr-2)Cba or (Ade)Pr^i-Cba[a]	*Co*α-[α-(Aden-9-yl)]-*Co*β-isopropylcobamide
(Bza)MeCba [b]	*Co*α-(α-Benzimidazolyl)-*Co*β-methylcobamide
2-(MeOOC)EtCbl	(2-Methoxycarbonylethyl)cob(III)alamin
(Ade-7)AdoCba [a]	*Co*α-[α-(Aden-7-yl)]-*Co*β-adenosylcobamide
(2-SHAde-7)AdoCba [a]	*Co*α-[α-(2-Thiaaden-7-yl)]-*Co*β-adenosylcobamide
(5-MeOBza)MeCba	*Co*α-(5-Methoxybenzimidazolyl)-*Co*β-methylcobamide[d]
(2-MeAde-7)CN-Cba [b]	*Co*α-[α-(2-Methyladen-7-yl)]-*Co*β-cyanocobamide[e]
(Ade)CN-Cba [a]	*Co*α-[α-(Aden-9-yl)]-*Co*β-cyanocobamide (pseudovitamin B-12)
(Ade)OH-Cba [a]	*Co*α-[α-(Aden-9-yl)]-*Co*β-hydroxocobamide (hydroxopseudovitamin B-12)
(Ade)MeCba [a]	*Co*α-[α-(Aden-9-yl)]-*Co*β-methylcobamide
[4-(Ade-9)Bu]Cbl [c]	[4-(Aden-9-yl)butyl]cob(III)alamin
(6MeSPur)AdoCba	*Co*α-(α-6-Methylthiopurinyl)-*Co*β-adenosylcobamide

[a]Ade alone represents adenine bonded to the ribosyl moiety through its 7 position (i.e., a 7-α-D-ribofuranosyladenine). Bonding to the cobalt is thus through N-9. When these positions are reversed, Ade-7 and aden-7-yl are used (i.e., the locant specifies the N linked to cobalt).
[b]Bza = benzimidazolyl.
[c]As this is a cobalamin, the adenine residue is not in the *Co*α position, but is attached (-9-yl) to a but-4-yl residue that is in turn linked to the β position of the cobalt. Named as a cobamide, it would be (Me$_2$Bza)[4-(Ade-9)Bu]Cba.
[d]Factor III$_m$.
[e]Factor A.

F. Folic Acids (Reference B1c)

"The term *folacin* should be used as a generic descriptor for all folates and related compounds exhibiting qualitatively the biological acitivity of tetrahydropteroylglutamic acid. Thus, phrases such as 'folacin activity' and 'folacin deficiency' represent preferred usage. The term *folate* may be used as the generic descriptor for the family of compounds containing the pteroic acid nucleus. Thus, the term 'folate' is not synonymous with the term 'folacin'."*

N-(2-Amino-4-hydroxypteridin-6-ylmethyl)-*p*-aminobenzoic acid. Pteroic acid (salts = pteroates; radical = pteroyl; symbol = Pte). 2-Amino-4-hydroxypteridine is known as pterin.

*American Institute of Nutrition (A.I.N.), *J. Nutr.*, 99, 244 (1969).

$$\text{Pteroyl}\left[\text{CO}\right]\text{NH}-\overset{\displaystyle\overset{\textstyle COOH}{|}}{\text{CH}}-\text{CH}_2-\text{CH}_2-\overset{\gamma}{\text{CO}}\vdots\text{NH}-\overset{\displaystyle\overset{\textstyle COOH}{|}}{\text{CH}}-\text{CH}_2-\text{CH}_2-\overset{\gamma}{\text{CO}}\vdots\text{NH}\ldots\text{etc.}$$

Pteroyl(n)glutamic acids (or -glutamates); Folic Acids (or folates)

n	H	Name[a]	Symbol[a]
1		Pteroylglutamate	PteGlu [b]
2		Pteroyldiglutamate	PteGlu$_2$
7		Pteroylheptaglutamate	PteGlu$_7$
7	7,8	7,8-Dihydropteroylheptaglutamate	7,8-H$_2$PteGlu$_7$
7	5,6,7,8	Tetrahydropteroylheptaglutamate	H$_4$PteGlu$_7$[c]; H$_4$folate [a]

[a]Folate may be substituted for pteroyl(n)glutamate or PteGlu$_n$ when the number of L-glutamate residues need not be specified. All other prefixes remain unchanged, and are indicated by standard symbols (HCNH, HCO, HOCH$_2$, CH$_3$, CH$_2$, CH).

[b]Glu signifies L-glutamic acid, as in the amino acid symbols (Al).

[c]5,6,7,8 positions for the four hydrogens are assumed and need not be specified.

Derivatives of Tetrahydropteroylglutamate (Tetrahydrofolate)

Substituent at		Prefix to tetrahydropteroylglutamate or folate	Prefix to H$_4$PteGlu or H$_4$folate[a]	Former names or symbols[b]
5	10			
–	–	–	–	PGAH$_4$; THFA; tetrahydrofolacin
CHO	–	5-Formyl-	5-CHO-	N^5-F-PGAH$_4$; "CF;" citrovorum factor; leucovorin; folinic acid; N^5-formyl-THFA
–	CHO	10-Formyl-	10-CHO	N^{10}-F-PGAH$_4$; "HLCF;" heat-labile citrovorum factor; N^{10}-formyl-THFA
CH$_3$	–	5-Methyl	5-CH$_3$-	N^5-M-PGAH$_4$; "pre-folic A;" N^5-methylfolacin; N^5-methyl-THFA
-CH$_2$-		5:10-Methylene-	5:10-CH$_2$-	–

[a]Folate is not abbreviated.

[b]All disapproved by CBN and I.U.N.S. (see Footnote a).

G. Miscellaneous Water-soluble Vitamins and Cofactors (Reference B1)

3-(4-Amino-2-methylpyrimidin-5-ylmethyl)-5-(2-hydroxethyl)-4-methylthiazolium.

Thiamin [vitamin B$_1$; aneurin(e); thiamine]

$$CH_2-(CHOH)_3-CH_2OH$$

7 , 8 - Dimethyl - 10 - (1' - D - ribityl)iso -
alloxazine. Riboflavin [vitamin B_2 ; lacto-
flavin(e); riboflavine]

"The term *niacin* should be used as the generic descriptor for pyridine 3-carboxylic acid and derivatives exhibiting qualitatively the biological activity of nicotinic acid. Thus, phrases such as "niacin activity" and "niacin deficiency" represent preferred usage."*

R = COOH
Pyridine 3-carboxylic acid
Nicotinic acid [niacin; vitamin PP]

R = CO-NH_2
Pyridine 3-carboxylic acid amide
Nicotinamide [niacinamide; nicotinic acid amide]

"The term *vitamin C* should be used as the generic descriptor for all compounds exhibiting qualitatively the biological activity of ascorbic acid. Thus phrases such as "vitamin C activity" and "vitamin C deficiency" represent preferred usage."*†

2 , 3 - didehydro - L - *threo* -
hexono - 1 , 4 - lactone.
Ascorbic acid; L-ascorbic
acid; vitamin C

*International Union of Nutritional Science (I.U.N.S.), *Nutr. Abstr. Rev.*, 40, 395 (1970).
†American Institute of Nutrition (A.I.N.), *J. Nutr.*, 99, 244, (1969).

IUPAC TENTATIVE RULES FOR THE
NOMENCLATURE OF ORGANIC CHEMISTRY
SECTION E. FUNDAMENTAL STEREOCHEMISTRY*

International Union of Pure and Applied Chemistry

INTRODUCTION

This Section of the IUPAC Rules for Nomenclature of Organic Chemistry differs from previous Sections in that it is here necessary to legislate for words that describe concepts as well as for names of compounds.

At the present time, concepts in stereochemistry (that is, chemistry in three-dimensional space) are in the process of rapid expansion, not merely in organic chemistry, but also in biochemistry, inorganic chemistry, and macromolecular chemistry. The aspects of interest for one area of chemistry often differ from those for another, even in respect to the same phenomenon. This rapid evolution and the variety of interests have led to development of specialized vocabularies and definitions that sometimes differ from one group of specialists to another, sometimes even within one area of chemistry.

The Commission on the Nomenclature of Organic Chemistry does not, however, consider it practical to cover all aspects of stereochemistry in this Section E. Instead, it has two objects in view: To prescribe, for basic concepts, terms that may provide a common language in all areas of stereochemistry; and to define the ways in which these terms may, so far as necessary, be incorporated into the names of individual compounds. The Commission recognizes that specialized nomenclatures are required for local fields; in some cases, such as carbohydrates, amino acids, peptides and proteins, and steroids, international rules already exist; for other fields, study is in progress by specialists in Commissions or Subcommittees; and further problems doubtless await identification. The Commission believes that consultations will be needed in many cases between different groups within IUPAC and IUB if the needs of the specialists are to be met without confusion and contradiction between the various groups.

The Rules in this Section deal only with Fundamental Stereochemistry, that is, the main principles. Many of these Rules do little more than codify existing practice, often of long standing; however, others extend old principles to wider fields, and yet others deal with nomenclature that is still subject to controversy.

Rule E-0

The stereochemistry of a compound is denoted by an affix or affixes to the name that does not prescribe the stereochemistry; such affixes, being additional, do not change the name or the numbering of the compound. Thus, enantiomers, diastereoisomers, and *cis–trans* isomers receive names that are distinguished only by means of different stereochemical affixes. The only exceptions are those trivial names that have stereo-chemical implications (for example, fumaric acid, cholesterol).

Note: In some cases (see Rules E-2.23 and E-3.1) stereochemical relations may be used to decide between alternative numberings that are otherwise permissible.

E-1. Types of Isomerism

E-1.1. The following nonstereochemical terms are relevant to the stereochemical nomenclature given in the Rules that follow.

---

*From *IUPAC Inf. Bull. Append. Tentative Nomencl. Sym. Units Stand.*, No. 35, August 1974, pp. 36–80. With permission.

(a) The term structure may be used in connection with any aspect of the organization of matter.

Hence: structural (adjectival)

(b) Compounds that have identical molecular formulas but differ in the nature or sequence of bonding of their atoms or in arrangement of their atoms in space are termed isomers.

Hence: isomeric (adjectival)
isomerism (phenomenological)

Examples:

$$H_3C — O — CH_3 \text{ is an isomer of } H_3C — CH_2 — OH$$

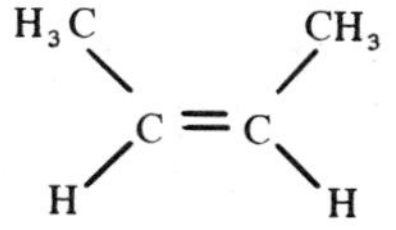

is an isomer of

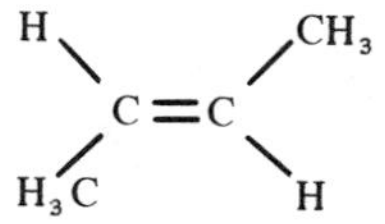

(In this and other Rules a broken line denotes a bond projecting behind the plane of the paper, and a thickened line denotes a bond projecting in front of the plane of the paper. In such cases a line of normal thickness denotes a bond lying in the plane of the paper.)

(c) The constitution of a compound of given molecular formula defines the nature and sequence of bonding of the atoms. Isomers differing in constitution are termed constitutional isomers.

Hence: constitutionally isomeric (adjectival)
constitutional isomerism (phenomenological)

Example:

$H_3C–O–CH_3$ is a constitutional isomer of $H_3C–CH_2–OH$.

Note: Use of the term "structural" with the above connotation is abandoned as insufficiently specific.

E-1.2. Isomers are termed stereoisomers when they differ only in the arrangement of their atoms in space.

Hence: stereoisomeric (adjectival)
stereoisomerism (phenomenological)

Examples:

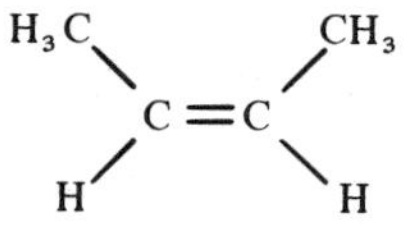

is a stereoisomer of

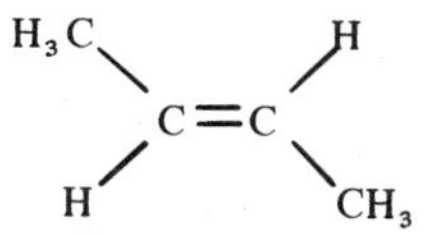

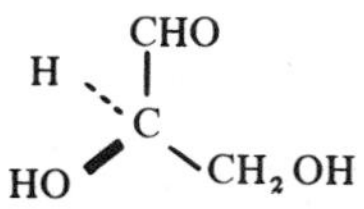

is a stereoisomer of

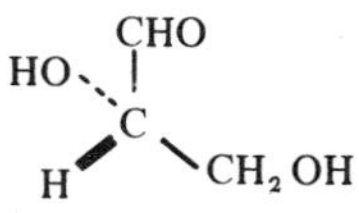

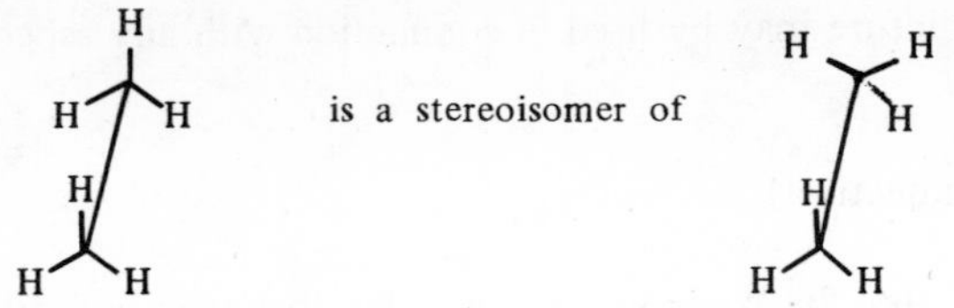

is a stereoisomer of

E-1.3. Stereoisomers are termed *cis–trans* isomers when they differ only in the positions of atoms relative to a specified plane in cases where these atoms are, or are considered as if they were, parts of a rigid structure.

Hence: *cis–trans* isomeric (adjectival)
        cis–trans isomerism (phenomenological)

Examples:

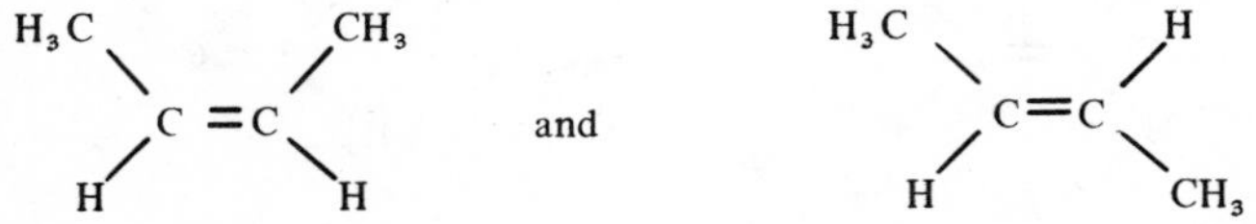

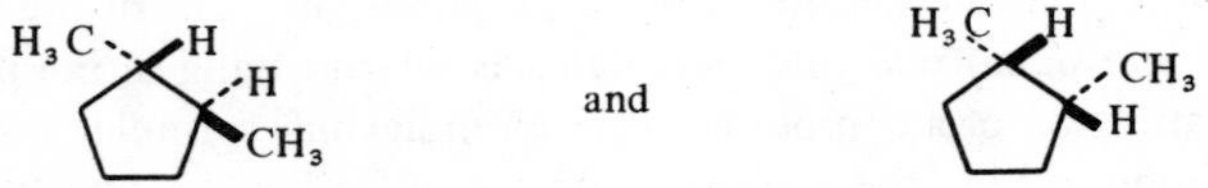

and

and

E-1.4. Various views are current regarding the precise definition of the term "configuration." (a) Classical interpretation: The configuration of a molecule of defined constitution is the arrangement of its atoms in space without regard to arrangements that differ only as after rotation about one or more single bonds. (b) This definition is now usually limited so that no regard is paid also to rotation about π bonds or bonds of partial order between one and two. (c) A third view limits the definition further so that no regard is paid to rotation about bonds of any order, including double bonds.

Molecules differing in configuration are termed configurational isomers.

Hence: configurational isomerism

Notes: (1) Contrast conformation (Rule E-1.5). (2) The phrase "differ only as after rotation" is intended to make the definition independent of any difficulty of rotation, in particular independent of steric hindrance to rotation. (3) For a brief discussion of views (a) to (c), see Appendix 1. It is hoped that a definite consensus of opinion will be established before these Rules are made "Definitive."

Examples: The following pairs of compounds differ in configuration:

(i)

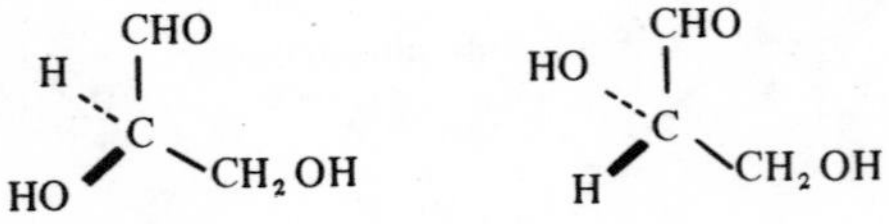

(ii)

(iii)

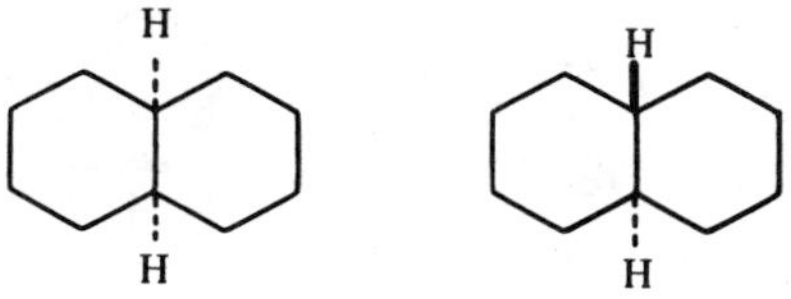

(iv)

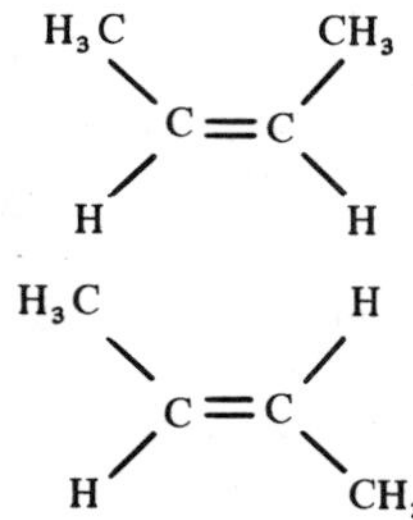

These isomers (iv) are configurational in view (a) or (b) but are conformational (see Rule E-1.5) in view (c)

E-1.5. Various views are current regarding the precise definition of the term "conformation." (a) Classical interpretation: The conformations of a molecule of defined configuration are the various arrangements of its atoms in space that differ only as after rotation about single bonds. (b) This is usually now extended to include rotation about π bonds or bonds of partial order between one and two. (c) A third view extends the definition further to include also, rotation about bonds of any order, including double bonds.

Molecules differing in conformation are termed conformational isomers.

Hence: conformational isomerism

Notes: All the Notes to Rule E-1.4 apply also to E-1.5.

Examples: Each of the following pairs of formulas represents a compound in the same configuration but in different conformations.

(a, b, c)

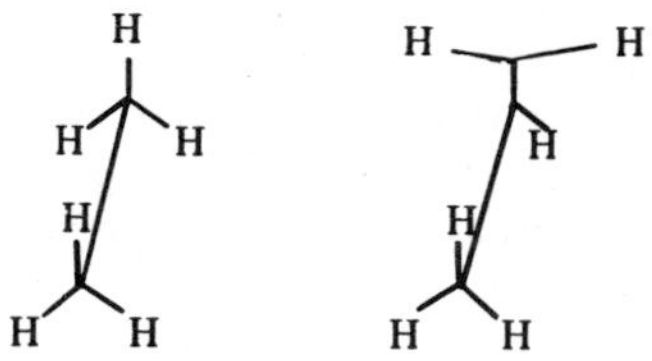

(a, b, c)

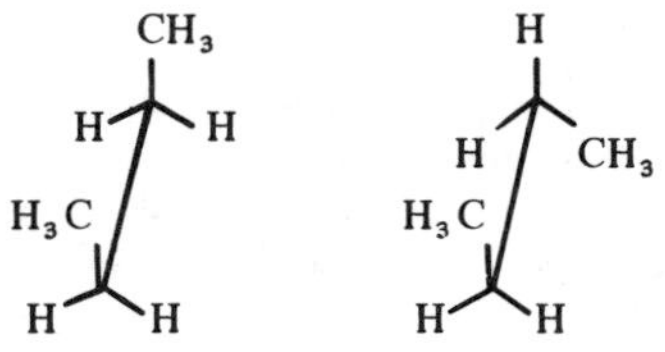

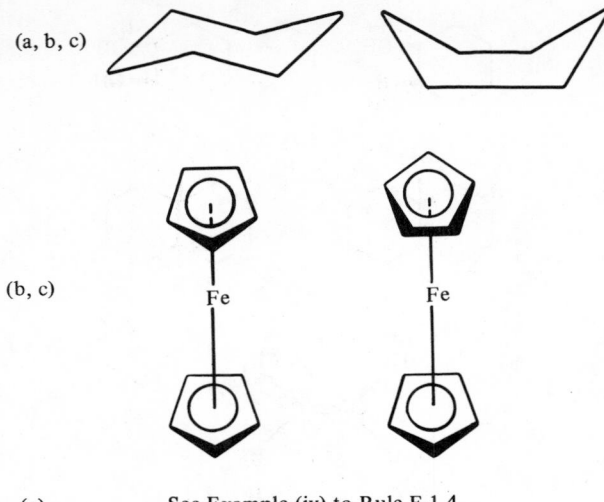

(a, b, c)

(b, c)

(c)   See Example (iv) to Rule E-1.4.

E-1.6. The terms relative stereochemistry and relative configuration are used with reference to the positions of various atoms in a compound relative to one another, especially, but not only, when the actual positions in space (absolute configuration) are unknown.

E-1.7. The terms absolute stereochemistry and absolute configuration are used with reference to the known actual positions of the atoms of a molecule in space.*

E-2. *cis−trans* Isomerism[†]

Preamble. The prefixes *cis* and *trans* have long been used for describing the relative positions of atoms or groups attached to nonterminal doubly bonded atoms of a chain or attached to a ring that is considered as planar. This practice has been codified for hydrocarbons by IUPAC.** There has, however, not been agreement on how to assign *cis* or *trans* at terminal double bonds of chains or at double bonds joining a chain to a ring. An obvious solution was to use *cis* and *trans* where doubly bonded atoms formed the backbone and were nonterminal and to enlist the sequence-rule preferences to decide other cases; however, since the two methods, when generally applied, do not always produce analogous results, it would then be necessary to use different symbols for the two procedures. A study of this combination showed that both types of symbols would often be required in one name and, moreover, it seemed wrong in principle to use two symbolisms for essentially the same phenomenon. Thus it seemed to the Commission wise to use only the sequence-rule system, since this alone was applicable to all cases. The same decision was taken independently by Chemical Abstracts Service who introduced Z and E to correspond more conveniently to *seqcis* and *seqtrans* of the sequence rule.

It is recommended in the Rules below that these designations Z and E based on the sequences rule shall be used in names of compounds, but Z and E do not always correspond to the classical *cis* and *trans* which show the steric relations of like or similar

---

*Determination of absolute configuration became possible through work by Bijvoet, J. M., Peerdeman, A. F., and van Bommel, A. J., *Nature*, 168, 271 (1951); cf. Bijvoet, J. M., *Proc. Kon. Ned. Akad. Wetensch.*, 52, 313 (1949).

[†]These Rules supersede the Tentative Rules for olefinic hydrocarbons published in the Comptes rendus of the 16th IUPAC Conference, New York, N.Y., 1951, pp. 102−103.

**Blackwood, J. E., Gladys, C. L., Loening, K. L., Petrarca, A. E., and Rush, J. E., *J. Amer. Chem. Soc.*, 90, 509 (1968); Blackwood, J. E., Gladys, C. L., Petrarca, A. E., Powell, W. H., and Rush, J. E., *J. Chem. Doc.*, 8, 30 (1968).

groups that are often the main point of interest. So the use of *Z* and *E* in names is not intended to hamper the use of *cis* and *trans* in discussions of steric relations of a generic type or of groups of particular interest in a specified case (see Rule E-2.1 and its Examples and Notes, also Rule E-5.11).

It is also not necessary to replace *cis* and *trans* for describing the stereochemistry of substituted monocycles (see Subsection E-3). For cyclic compounds the main problems are usually different from those around double bonds; for instance, steric relations of substitutents on rings can often be described either in terms of chirality (see Subsection E-5) or in terms of *cis–trans* relationships, and, further, there is usually no single relevant plane of reference in a hydrogenated polycycle. These matters are discussed in the Preambles to Subsections E-3 and E-4.

E-2.1. *Definition of cis–trans.* Atoms or groups are termed *cis* or *trans* to one another when they lie respectively on the same or on opposite sides of a reference plane identifiable as common among stereoisomers. The compounds in which such relations occur are termed *cis–trans* isomers. For compounds containing only doubly bonded atoms, the reference plane contains the doubly bonded atoms and is perpendicular to the plane containing these atoms and those directly attached to them. For cyclic compounds, the reference plane is that in which the ring skeleton lies or to which it approximates. When qualifying another word or a locant, *cis* or *trans* is followed by a hyphen. When added to a structural formula, *cis* may be abbreviated to *c*, and *trans* to *t* (see also Rule E-3.3).

Examples: (Rectangles here denote the reference planes and are considered to lie in the plane of the paper.)

The groups or atoms a,a are the pair selected for designation but are not necessarily identical; b,b are also not necessarily identical but must be different from a,a.

cis or *trans* according as a or b is taken as basis of comparison

Notes: The formulas above are drawn with the reference plane in the plane of the paper, but for doubly bonded compounds it is customary to draw the formulas so that this plane is perpendicular to that of the paper; atoms attached directly to the doubly bonded atoms then lie in the plane of the paper and the formulas appear as, for instance

$$\overset{a}{\underset{b}{\diagdown}} C = C \overset{a}{\underset{b}{\diagup}}$$

cis

Cyclic structures, however, are customarily drawn with the ring atoms in the plane of the paper, as above. However, care is needed for complex cases, such as

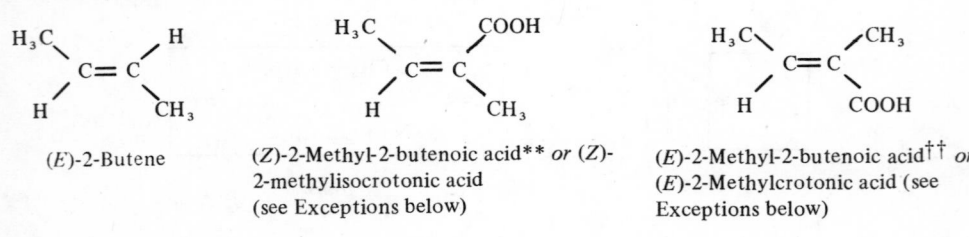

The central five-membered ring lies (approximately) in a plane perpendicular to the plane of the paper. The two a groups are *trans* to one another; so are the b groups; the outer cyclopentane rings are *cis* to one another with respect to the plane of the central ring. *cis* or *trans* (or *Z* or *E*; see Rule E-2.21) may also be used in cases involving a partial bond order when a limiting structure is of sufficient importance to impose rigidity around the bond of partial order. An example is

$$\overset{(CH_3)_2CH}{\diagdown}\underset{H}{\diagup}N - C\overset{H}{\underset{S}{\diagdown}} \quad \longleftrightarrow \quad \overset{(CH_3)_2CH}{\diagdown}\underset{H}{\diagup}N^+ = C\overset{H}{\underset{S}{\diagdown}}$$

trans (or E)

E-2.2. *cis-trans Isomerism around Double Bonds.*

E-2.21. In names of compounds steric relations around one or more double bonds are designated by affixes *Z* and/or *E*, assigned as follows. The sequence-rule-preferred* atom or group attached to one of a doubly bonded pair of atoms is compared with the sequence-rule-preferred atom or group attached to the other of that doubly bonded pair of atoms; if the selected pair are on the same side of the reference plane (see Rule 2.1) an italic capital letter *Z* prefix is used; if the selected pair are on opposite sides an italic capital letter *E* prefix is used.[†] These prefixes, placed in parentheses and followed by a hyphen, normally precede the whole name; if the molecule contains several double bonds, then each prefix is immediately preceded by the lower or less primed locant of the relevant double bond.

Examples:

$$\overset{H_3C}{\underset{H}{\diagup}} C = C \overset{H}{\underset{CH_3}{\diagdown}}$$

(E)-2-Butene

$$\overset{H_3C}{\underset{H}{\diagup}} C = C \overset{COOH}{\underset{CH_3}{\diagdown}}$$

(Z)-2-Methyl-2-butenoic acid** *or (Z)-* 2-methylisocrotonic acid (see Exceptions below)

$$\overset{H_3C}{\underset{H}{\diagup}} C = C \overset{CH_3}{\underset{COOH}{\diagdown}}$$

(E)-2-Methyl-2-butenoic acid[††] *or (E)*-2-Methylcrotonic acid (see Exceptions below)

*For sequence-rule preferences see Appendix 2.
[†]These prefixes may be rationalized as from the German *zusammen* (together) and *entgegen* (opposite).
**The name angelic acid is abandoned because it has been associated with the designation *trans* with reference to the methyl groups.
[††]The name tiglic acid is abandoned because it has been associated with the designation *cis* with reference to the methyl groups.

(Z)-3-Chloroacrylonitrile

(E)-2,3-Dichloroacrylonitrile

(Z)-1,2-Dibromo-1-chloro-2-iodoethylene
(By the sequence rule, Br is preferred to Cl,
but I to Br)

(E)-(3-Bromo-3-chloroallyl)benzene

(E)-Cyclooctene

(E)-1-sec-Butylideneindene

(Z)-1-Chloro-2-ethylidene-2H-indene

(E)-1,1'-Biindenylidene

(E)-Azobenzene

Exceptions to Rule E-2.21. The following are examples of accepted trivial names in which the stereochemistry is prescribed by the name and is not cited by a prefix.

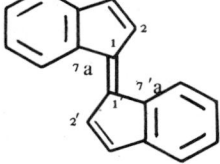

HOOCCH $\parallel$ HCCOOH	HCCOOH $\parallel$ HCCOOH	CH_3CCOOH $\parallel$ HCCOOH	HOOCCCH$_3$ $\parallel$ HCCOOH	CH$_3$CH $\parallel$ HCCOOH
Fumaric acid	Maleic acid	Citraconic acid*	Mesaconic acid*	Crotonic acid

HCCH$_3$ $\parallel$ HCCOOH	HC—(CH$_2$)$_7$—CH$_3$ $\parallel$ HC—(CH$_2$)$_7$—COOH	CH$_3$—(CH$_2$)$_7$—CH $\parallel$ HC—(CH$_2$)$_7$—COOH
Isocrotonic acid	Oleic acid	Elaidic acid

E-2.22 (*Alternative to Part of E-2.21*). (a) When more than one series of locants starting from unity is required to designate the double bonds in a molecule, or when the name consists of two words, the *Z* and *E* prefixes together with their appropriate locants may be placed before that part of the name where ambiguity is most effectively removed.

(b) [Alternative to (a)] When several *Z* or *E* prefixes are required they are arranged in

*Systematic names are recommended for derivatives of these compounds formed by substitution on carbon.

order as follows: Of the four atoms or groups attached to each doubly bonded pair of atoms, that one preferred by the sequence rule is selected; the single atoms or groups thus selected are then arranged in their sequence rule order (determined in respect of their position in the whole molecule), and the prefixes Z and/or E for the respective double bonds are placed in that order, but *without* their locants.

Note: In method (a) the final choice is left to an author or editor because of the variety of cases met and because the problems are not always the same in different languages. The presence of the locants usually eases translation from the name to a formula, but this method (a) may involve the logical difficulty explained for the third example below. Method (b) always gives a single unambiguous order and is not subject to the logical difficulty just mentioned, but translation from the name to the formula is harder than for method (a). Method (a) may be more suitable for cursive text, and method (b) for compendia. If method (b) is used it should be used whenever more than one double bond is involved, but method (a) is to be used only under the special conditions detailed in the rule.

Examples:

(a) (2E,4Z)-2,4-Hexadienoic acid
(b) (E,Z)-2,4-Hexadienoic acid

(a) (2E,4Z)-5-Chloro-2,4-hexadienoic acid
(b) (Z,E)5-Chloro-2,4-hexadienoic acid

(a) 3-[(E)-1-Chloropropenyl]-(3Z,5E)-
3,5-heptadienoic acid
(b) (E,Z,E)-3-(1-Chloropropenyl)-
3,5-heptadienoic acid

[The last example shows the disadvantages of both methods. In method (a) there is a fault of logic, namely, the 3Z,5E are not the property of the unsubstituted heptadienoic acid chain, but the 3Z arises only because of the side chain that is cited before the 3Z,5E. In method (b) it is some trouble to assign the E,Z,E to the correct double bonds.]

(a) (1Z,3E)-1,3-Cyclododecadiene
(b) (Z,E)-1,3-Cyclododecadiene

[The lower locant is assigned to the Z double bond.]

(a) 5-Chloro-4-(E-sulfomethylene)-
    (2E,5Z)-2,5-heptadienoic acid
(b) (Z,E,E)-5-Chloro-4-(sulfomethylene)-
    2,5-heptadienoic acid

[In application of the sequence rule, the relation of the SO_3H to CCl (rather than to C-3), and of the CH_3 to Cl, are decisive.]

(a) Butanone (E)-oxime*
(b) (E)-Butanone oxime

(a) 2-Chlorobenzophenone (Z)-hydrazone
(b) (Z)-2-Chlorobenzophenone hydrazone

(a) (E)-2-Péntenal (Z)-semicarbazone
(b) (Z,E)-2-Pentenal semicarbazone

(a) Benzil (Z,E)-dioxime
(b) (Z,E)-Benzil dioxime

E-2.23. When Rule C-13.1 or E-2.22(b) permits alternatives, preference for lower locants and for inclusion in the principal chain is allotted as follows, in the order stated, so far as necessary: Z over E groups; cis over trans cyclic groups; R over S groups (also r over s, etc., as in the sequence rule); if the nature of these groups is not decisive, then the lower locant for such a preferred group at the first point of difference.

Examples:

(a) (2Z,5E)-2,5-Heptadienedioic acid
(b) (E,Z)-2,5-Heptadienedioic acid

[The lower numbers are assigned to the Z double bond.]

*The terms syn, anti, and amphi are abandoned for such compounds.

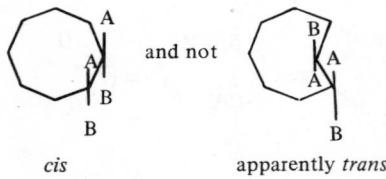

(a) 1-Chloro-3-[2-chloro-(E)-vinyl]-($1Z,3Z$)-
1,3-pentadiene
(b) (E,Z,Z)-1-Chloro-3-(2-chlorovinyl)-
1,3-pentadiene

[According to Rule C-13.1 the principal chain must
include the C=C–CH$_3$ group because this gives lower
numbers to the double bonds (1,3 rather than 1,4);
then the Cl-containing Z group is chosen for the remainder of the principal chain in accord with Rule
E-2.23.]

(a,b) (Z)-($4R$)-3-[(S)-sec-Butyl]-4-methyl-2-hexenoic acid

[The principal chain is chosen to include the (R)-group, and
the prefix Z refers to the (R)-group.]

E-3. Relative Stereochemistry of Substituents in Monocyclic Compounds[†]

Preamble. The prefixes *cis* and *trans* are commonly used to designate the positions of
substituents on rings relative to one another; when the ring is, or is considered to be,
rigidly planar or approximately so and is placed horizontally, these prefixes define which
groups are above and which below the (approximate) plane of the ring. This
differentiation is often important, so this classical terminology is retained in Subsection
E-3; since the difficulties inherent in end groups do not arise for cyclic compounds, it is
unnecessary to resort to the less immediately informative E/Z symbolism.

When the *cis–trans* designation of substituents is applied, rings are considered in their
most extended form; reentrant angles are not permitted; for example

cis   and not   apparently *trans*

The absolute stereochemistry of optically active or racemic derivatives of monocyclic
compounds is described by the sequence-rule procedure (see Rule E-5.9 and Appendix 2).
The relative stereochemistry may be described by a modification of sequence-rule
symbolism as set out in Rule E-5.10. If either of these procedures is adopted, it is then
superfluous to use also *cis* or *trans* in the names of individual compounds.

[†]Formulas in Examples to this Rule denote relative (not absolute) configurations.

E-3.1. When alternative numberings of the ring are permissible according to the Rules of Section C, that numbering is chosen which gives a *cis* attachment at the first point of difference; if that is not decisive, the criteria of Rule E-2.23 are applied. The prefixes *cis* and *trans* may be abbreviated to *c* and *t*, respectively, in names of compounds when more than one such designation is required.

Examples:

1,*c*-2,*t*-3-Trichlorocyclohexane

1-(*Z*)-Propenyl-*trans*-3-(*E*)-propenylcyclohexane

E-3.2. When one substituent and one hydrogen atom are attached at each of two positions of a monocycle, the steric relations of the two substituents are expressed as *cis* or *trans*, followed by a hyphen and placed before the name of the compound.

Examples:

cis-1,2-Dichlorocyclopentane

trans-2-Chloro-1-cyclopentanecarboxylic acid

trans-2-Chloro-4-nitro-1,
1-cyclohexanedicarboxylic acid

E-3.3. When one substituent and one hydrogen atom are attached at each of more than two positions of a monocycle, the steric relations of the substituents are expressed by adding *r* (for *reference* substituent), followed by a hyphen, before the locant of the lowest numbered of these substituents and *c* or *t* (as appropriate), followed by a hyphen, before the locants of the other substituents to express their relation to the reference substituent.

Examples:

r-1,*t*-2,*c*-4-Trichlorocyclopentant
(not *r*-1, *t*-2, *t*-4, which would follow from the
alternative direction of numbering; see Rule
E-3.1)

t-5-Chloro-*r*-1, *c*-3-cyclohexanedicarboxylic acid

E-3.4. When two different substituents are attached at the same position of a monocycle, then the lowest numbered substituent named as suffix is selected for designation as reference group in accordance with Rule E-3.2 or E-3.3; or, if none of the substituents is named as suffix, then of the lowest numbered pair that one preferred by the sequence rule is selected as reference group; and the relation of the sequence-rule preferred group at each other position, relative to the reference group, is cited as *c* or *t* (as appropriate).

Examples:

1,*t*-2-Dichloro-*r*-1-cyclopentanecarboxylic acid

r-1-Bromo-1-chloro-*t*-3-ethyl-3-methylcyclohexane
(alphabetical order of prefixes)

c-3-Bromo-3-chloro-*r*-1-cyclopentanecarboxylic acid

2-Crotonoyl-*t*-2-isocrotonoyl-*r*-1-cyclopentane-
carboxylic acid

E-4. Fused Rings

Preamble. In simple cases the relative stereochemistry of substituted fused-ring systems can be designated by the methods used for monocycles. For the absolute stereochemistry of optically active and racemic compounds the sequence-rule procedure can be used in all cases (see Rule E-5.9 and Appendix 2), and for related relative stereochemistry the procedure of Rule E-5.10 can be applied. Sequence-rule methods are, however, not descriptive of geometrical shape for other than quite simple cases. There is as yet no generally acceptable system for designating in an immediately interpretable manner the stereochemistry of polycyclic bridged ring compounds (for instance, the *endo–exo* nomenclature, which should solve one set of problems, has been used in different ways). These and related problems (e.g., cyclophanes, catenanes) will be considered in a later document.

E-4.1. Steric relations at saturated bridgeheads common to two rings are denoted by *cis* or *trans*, followed by a hyphen and placed before the name of the ring system, according to the relative positions of the exocyclic atoms or groups attached to the bridgeheads. Such rings are said to be *cis* fused or *trans* fused.

Examples:

cis-Decalin

1-Methyl-*trans*-bicyclo[8.3.1]tetradecane

E-4.2. Steric relations at more than one pair of saturated bridgeheads in a polycyclic compound are denoted by *cis* or *trans*, each followed by a hyphen and, when necessary, the corresponding locant of the lower numbered bridgehead and a second hyphen, all placed before the name of the ring system. Steric relations between the nearest atoms* of *cis*- or *trans*-bridgehead pairs may be described by affixes *cisoid* or *transoid*, followed by a hyphen and, when necessary, the corresponding locants and a second hyphen, the whole placed between the designations of the *cis*- or *trans*-ring junctions concerned. When a choice remains among nearest atoms, the pair containing the lower numbered atom is selected; *cis* and *trans* are not abbreviated in such cases. In complex cases, however, designation may be more simply effected by the sequence-rule procedure (see Appendix 2).

Examples:

cis-*cisoid*-*trans*-Perhydrophenanthrene

cis-*cisoid*-4a, 10a-*trans*-Perhydroanthracene
or *rel*-(4a*R*, 8a*S*, 9a*S*, 10a*S*)-Perhydroanthracene†

trans-3a-*cisoid*-3a, 4a-*cis*-4a-Perhydrobenz[*f*] indene
or *rel*-(3a*R*, 4a*S*, 8a*R*, 9a*R*)-Perhydrobenz[*f*] indene

E-5. Chirality

E-5.1. The property of nonidentity of an object with its mirror image is termed chirality. An object, such as a molecule in a given configuration or conformation, is termed chiral when it is not identical with its mirror image; it is termed achiral when it is identical with its mirror image.

Notes: (1) Chirality is equivalent to handedness, the term being derived from the Greek Χειρ = hand.

(2) All chiral molecules are molecules of optically active compounds, and molecules of all optically active compounds are chiral. There is a 1:1 correspondence between chirality and optical activity.

(3) In organic chemistry the discussion of chirality usually concerns the individual molecule or, more strictly, a model of the individual molecule. The chirality of an assembly of molecules may differ from that of the component molecules, as in a chiral quartz crystal or in an achiral crystal containing equal numbers of dextrorotatory and levorotatory tartaric acid molecules.

(4) The chirality of a molecule can be discussed only if the configuration or conformation of the molecule is specifically defined or is considered as defined by

*The term "nearest atoms" denotes those linked together through the smallest number of atoms, irrespective of actual separation in space. For instance, in the second Example to this Rule, the atom 4a is "nearer" to 10a than to 8a.

†For the designation *rel*, see Rule E-5.10.

common usage. In such discussions structures are treated as if they were (at least temporarily) rigid. For instance, ethane is configurationally achiral although many of its conformations, such as (A), are chiral; in fact, a configuration of a mobile molecule is chiral only if all its possible conformations are chiral; and conformations of ethane such as (B) and (C) are achiral.

(A)	(B)	(C)

Examples:

CHO	CHO
H ─ C ─ OH	HO ─ C ─ H
CH$_2$OH	CH$_2$OH
(D)	(E)

CH$_2$OH
H ─ C ─ OH
CH$_2$OH
(F)

(D) and (E) are mirror images and are not identical, not being superposable. They represent chiral molecules. They represent (D) dextrorotatory and (E) levorotatory glyceraldehyde.

(F) is identical with its mirror image. It represents an achiral molecule, namely, a molecule of *1,2,3*-propanetriol (glycerol).

E-5.2. The term asymmetry denotes absence of any symmetry. An object, such as a molecule in a given configuration or conformation, is termed asymmetric if it has no element of symmetry.

Notes: (1) All asymmetric molecules are chiral, and all compounds composed of them are therefore optically active; however, not all chiral molecules are asymmetric since some molecules having axes of rotation are chiral.

(2) Notes (3) and (4) to Rule E-5.1 apply also in discussions of asymmetry.

Examples:

CHO
H ─ C ─ OH
CH$_2$OH

has no element of symmetry and represents a molecule of an optically active compound.

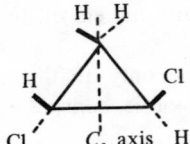

has a C_2 axis of rotation; it is chiral although not asymmetric, and is therefore a molecule of an optically active compound.

E-5.3. (a) An asymmetric atom is one that is tetrahedrally bonded to four different atoms or groups, none of the groups being the mirror image of any of the others.

(b) An asymmetric atom may be said to be at a chiral center since it lies at the center of a chiral tetrahedral structure. In a general sense, the term "chiral center" is not restricted to tetrahedral structures; the structure may, for instance, be based on an octahedron or tetragonal pyramid.

(c) When the atom by which a group is attached to the remainder of a molecule lies at a chiral center, the group may be termed a chiral group.

Notes: (1) The term "asymmetric," as applied to a carbon atom in rule E-5.3 (a), was chosen by van't Hoff because there is no plane of symmetry through a tetrahedron whose corners are occupied by four atoms or groups that differ in scalar properties. For differences of vector sense between the attached groups, see Rule E-5.8.

(2) In Subsection E-5 the word "group" is used to denote the series of atoms attached to one bond. For instance, in (i) the groups attached to C* are $-CH_3$, $-OH$, $-CH_2CH_3$, and $-COOH$; in (ii) they are $-CH_3$, $-OH$, $-COCH_2CH_2CH_2$, and $-CH_2CH_2CH_2CO$.

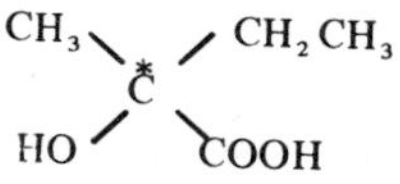

(i) 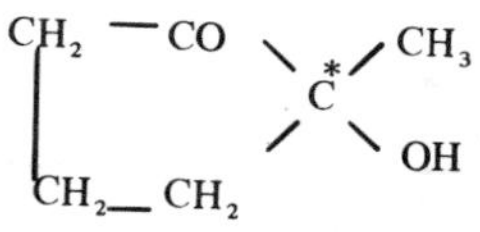   (ii)

(3) For the chiral axis and chiral plane (which are less common than the chiral center), see Appendix 2.

(4) There may be more than one chiral center in a molecule and these centers may be identical, or structurally different, or structurally identical but of opposite chirality; however, the presence of an equal number of structurally identical chiral groups of opposite chirality, and no other chiral group, leads to an achiral molecule. These statements apply also to chiral axes and chiral planes. Identification of the sites and natures of the various factors involved is essential if the overall chirality of a molecule is to be understood.

(5) Although the term "chiral group" is convenient for use in discussions it should be remembered that chirality attaches to molecules and not to groups or atoms. For instance, although the *sec*-butyl group may be termed chiral in dextrorotatory 2-*sec*-butyl-naphthalene, it is not chiral in the achiral compound $(CH_3CH_2)(CH_3)CH-CH_3$.

Examples:

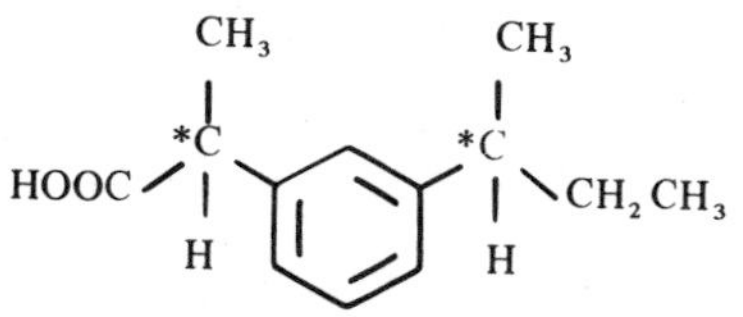

In this chiral compound there are two asymmetric carbon atoms, marked C*, each lying at a chiral center. These atoms form part of different chiral groups, namely, $-CH(CH_3)-$COOH and $-CH(CH_3)CH_2CH_3$

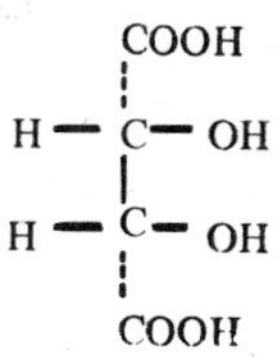

In this molecule (*meso*-tartaric acid) the two central carbon atoms are asymmetric atoms and each is part of a chiral group $-CH(OH)COOH$. These groups, however, although structurally identical, are of opposite chirality, so that the molecule is achiral.

E-5.4. Molecules that are mirror images of one another are termed enantiomers and may be said to be enantiomeric. Chiral groups that are mirror images of one another are termed enantiomeric groups.

Hence: enantiomerism (phenomenological)

Note: Although the adjective enantiomeric may be applied to groups, enantiomerism strictly applies only to molecules [see Note (5) to Rule E-5.3].

Examples: The following pairs of molecules are enantiomeric.

(i)

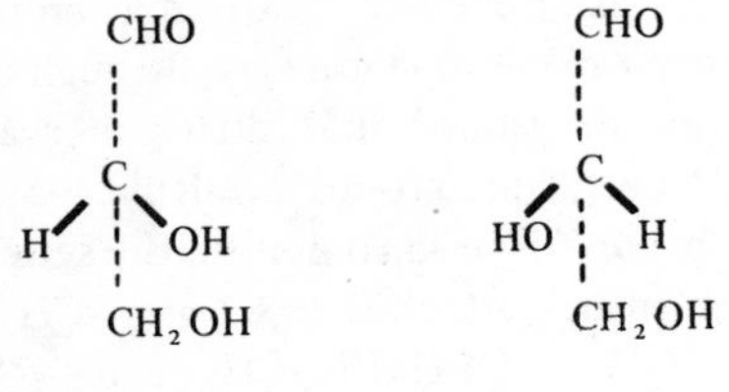

$$CHO \quad\quad\quad CHO$$

(ii)

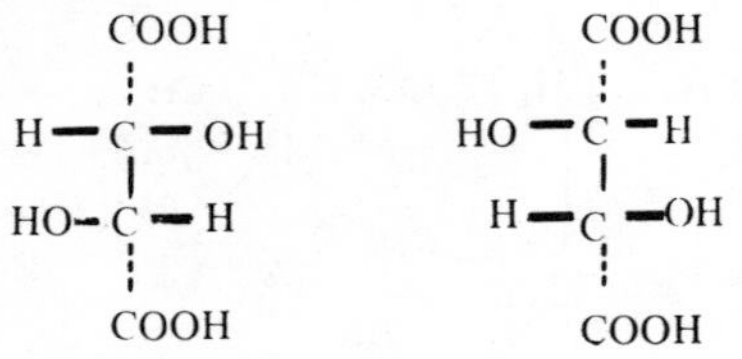

$$COOH \quad\quad\quad COOH$$

(iii)

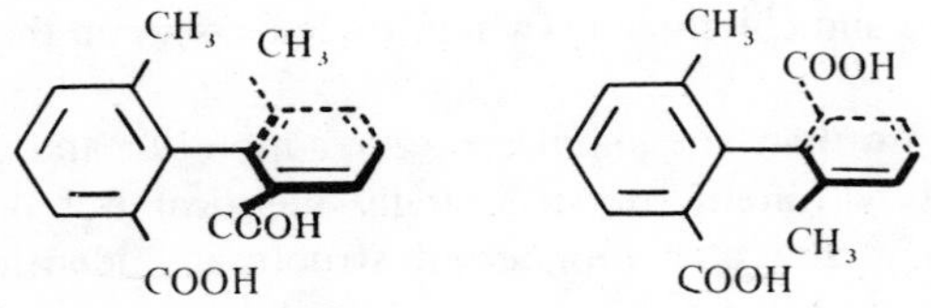

(iv)

Cyclooctene

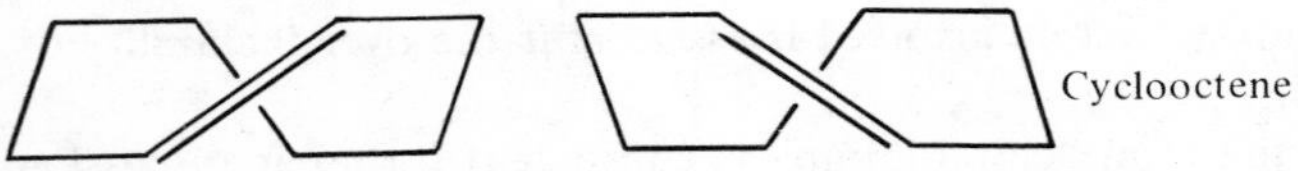

(v)

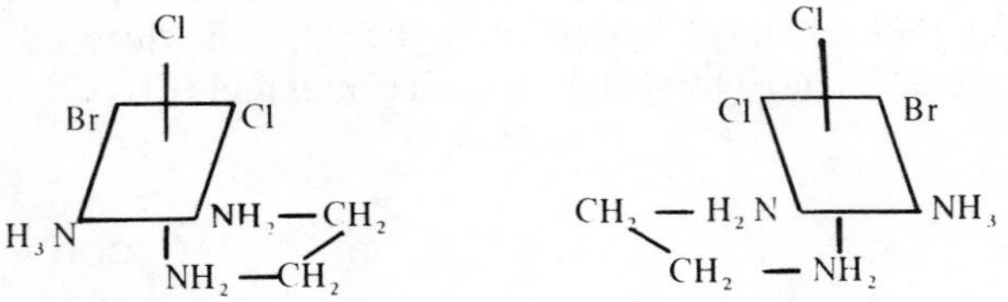

(vi)

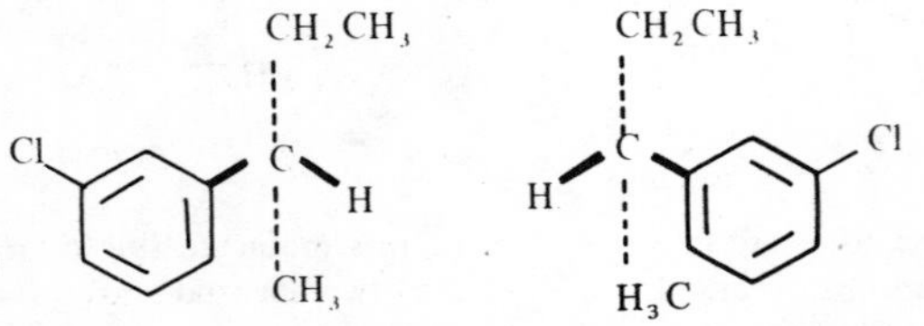

The *sec*-butyl groups in (vi) are enantiomeric.

E-5.5. When equal amounts of enantiomeric molecules are present together, the product is termed racemic, independently of whether it is crystalline, liquid, or gaseous. A homogeneous solid phase composed of equimolar amounts of enantiomeric molecules is termed a racemic compound. A mixture of equimolar amounts of enantiomeric molecules present as separate solid phases is termed a racemic mixture. Any homogeneous solid containing equimolar amounts of enantiomeric molecules is termed a racemate.

Examples: The mixture of two kinds of crystal (mirror-image forms) that separate below 28° from an aqueous solution containing equal amounts of dextrorotatory and levorotatory sodium ammonium tartrate is a racemic mixture.

The symmetrical crystals that separate from such a solution above 28°, each containing equal amounts of the two salts, provide a racemic compound.

E-5.6. Stereoisomers that are not enantiomeric are termed diastereoisomers.

Hence: diastereoisomeric (adjectival)
diastereoisomerism (phenomenological)

Note: Diastereoisomers may be chiral or achiral.
Examples:

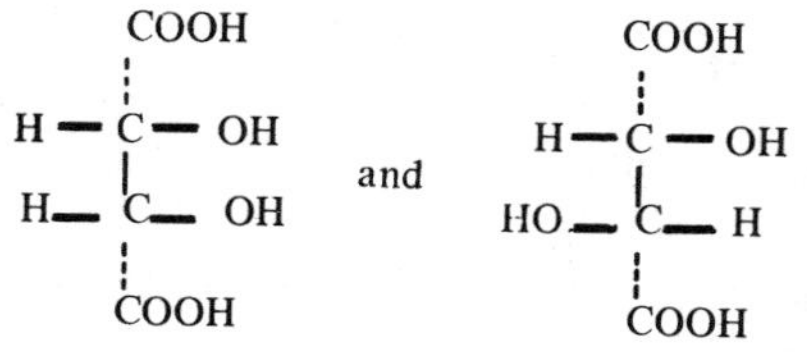

$$
\begin{array}{ccc}
\text{COOH} & & \text{COOH} \\
\text{H}-\text{C}-\text{OH} & & \text{H}-\text{C}-\text{OH} \\
\text{H}-\text{C}-\text{OH} & \text{and} & \text{HO}-\text{C}-\text{H} \\
\text{COOH} & & \text{COOH}
\end{array}
$$

are diastereoisomers; the former is achiral, and the latter is chiral.

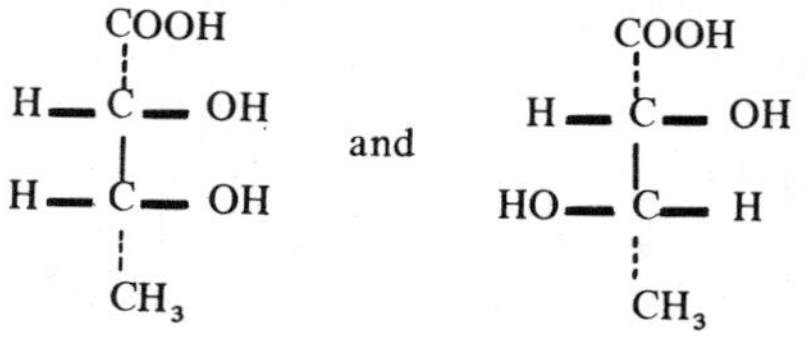

$$
\begin{array}{ccc}
\text{COOH} & & \text{COOH} \\
\text{H}-\text{C}-\text{OH} & & \text{H}-\text{C}-\text{OH} \\
\text{H}-\text{C}-\text{OH} & \text{and} & \text{HO}-\text{C}-\text{H} \\
\text{CH}_3 & & \text{CH}_3
\end{array}
$$

are diastereoisomers; both are chiral.

E-5.7. A compound whose individual molecules contain equal numbers of enantiomeric groups, identically linked, but no other chiral group, is termed a *meso* compound.
Example:

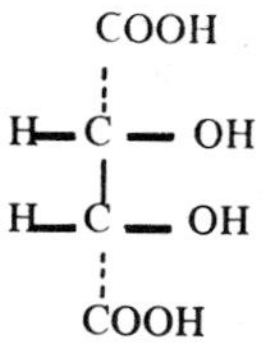

$$
\begin{array}{cc}
\text{COOH} & \text{COOH} \\
\text{H}-\text{C}-\text{OH} & \text{H}-\text{C}-\text{OH} \\
\text{H}-\text{C}-\text{OH} & \text{HO}-\text{C}-\text{H} \\
\text{COOH} & \text{HO}-\text{C}-\text{H} \\
& \text{H}-\text{C}-\text{OH} \\
& \text{COOH}
\end{array}
$$

meso-Tartaric acid 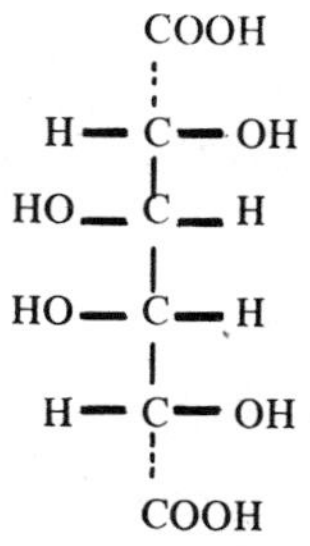   Galactaric acid

E-5.8. An atom is termed pseudoasymmetric when bonded tetrahedrally to one pair of enantiomeric groups (+)-a and (−)-a and also to two atoms or groups b and c that are different from group a, different from each other, and not enantiomeric with each other.
Examples:

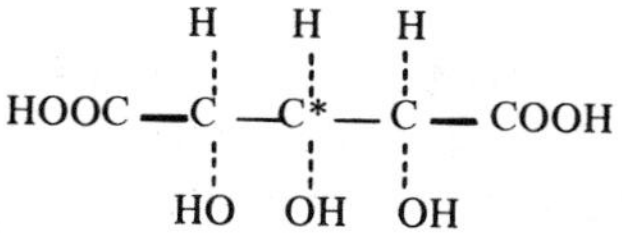

$$
\begin{array}{ccc}
& \text{H} \quad \text{H} \quad \text{H} & \\
\text{HOOC}-\text{C}-\text{C*}-\text{C}-\text{COOH} & & \text{HOOC}-\text{C}-\text{C*}-\text{C}-\text{COOH} \\
\text{HO} \quad \text{OH} \quad \text{OH} & & \text{HO} \quad \text{CH}_2 \quad \text{OH} \\
& & \text{CH}_3-\text{C}-\text{CH}_2\text{CH}_3 \\
& & \text{H}
\end{array}
$$

(A)   (B)

C* are pseudoasymmetric

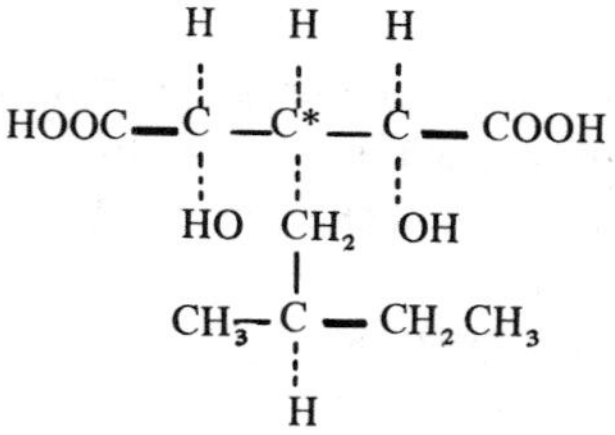

Notes: (1) The orientation, in space, of the atoms around a pseudoasymmetric atoms is not reversed on reflection; for a chiral atom (see Note to Rule E-5.3) this orientation is always reversed.

(2) Molecules containing pseudoasymmetric atoms may be achiral or chiral. If ligands b and c are both achiral, the molecule is achiral as in the first example to this Rule. If either or both of the nonenantiomeric ligands b and c are chiral, the molecule is chiral, as in the second example to this Rule, that is the molecule is not identical with its mirror image. A molecule (i) is also chiral if b and c are enantiomeric, that is, if the molecule can be symbolized as (ii), but then, by definition, it does not contain a pseudoasymmetric atom.

(i)   (ii)

(3) Compounds differing at a pseudoasymmetric atom belong to the larger class of diastereoisomers.

(4) In example (A), interchange of H and OH on C* gives a different achiral compound, which is an achiral diastereoisomer of (A) (see Rule E-5.6). In example (B), diastereoisomers are produced by inversion at C* or °C, giving in all four diastereoisomers, all chiral because of the $-CH(CH_3)CH_2CH_3$ group.

E-5.9. Names of chiral compounds whose absolute configuration is known are differentiated by prefixes *R*, *S*, etc., assigned by the sequence-rule procedure (see Appendix 2), preceded when necessary by the appropriate locants.

Examples:

(*R*)-Glyceraldehyde   (*S*)-Glyceraldehyde

(6a*S*,12*S*,5′*R*)-Rotenone   Methyl phenyl (*R*)-sulfoxide

E-5.10. (a) Names of compounds containing chiral centers, of which the relative but not the absolute configuration is known, are differentiated by prefixes *R**, *S** (spoken R star, S star), preceded when necessary by the appropriate locants, these prefixes being assigned by the sequence-rule procedure (see Appendix 2) on the arbitrary assumption that the prefix first cited is *R*.

(b) In complex cases the stars may be omitted and, instead, the whole name is prefixed by *rel* (for *relative*).

(c) When only relative configuration is known, enantiomers are distinguished by a prefix (+) or (−), referring to the direction of rotation of plane-polarized light passing through them (wavelength, temperature, solvent, and/or concentration should also be specified, particularly when known to affect the sign).

(d) When a substituent of known absolute chirality is introduced into a compound of which only the relative configuration is known, then starred symbols R^*, S^* are used and not the prefix *rel.*

Note: This Rule does not form part of the procedure formulated in the sequence-rule papers by Cahn, Ingold, and Prelog (see Appendix 2).

Examples:

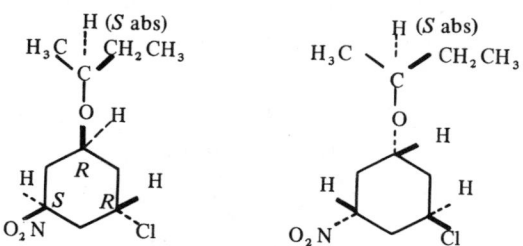

($1R^*$, $3S^*$,)-1-Bromo-3-chlorocyclohexane

rel-($1R,3R,5R$)-1-Bromo-3-chloro-5-nitrocyclohexane

($1R^*,3R^*,5S^*$)-[($1S$)-*sec*-Butoxy]-3-chloro-
5-nitrocyclohexane

E-5.11. When it is desired to express relative or absolute configuration with respect to a class of compounds, specialized local systems may be used. The sequence rule may, however, be used additionally for positions not amenable to treatment by the local system.

Examples:

gluco, arabino, etc., combined when necessary with D or L, for carbohydrates and their derivatives [see IUPAC-IUB Tentative Rules for Carbohydrate Nomenclature; see also *J. Org. Chem.,* 28, 281 (1963)].

D, L for amino acids and peptides [see Comptes rendus of the 16th IUPAC Conference, New York, N.Y., 1951., pp. 107–108; also published in *Chem. Eng. News,* 30, 4522 (1952)].

D, L, and a series of other prefixes and trivial names for cyclitols and their derivatives [see IUPAC-IUB Tentative Rules for the Nomenclature of Cyclitols, 1967, *IUPAC Inf. Bull.*, No. 32, 51 (1968); also published in *J. Biol. Chem.*, 243, 5809 (1968)].

α, β, and a series of trivial names for steroids and related compounds [see IUPAC-IUB Revised Tentative Rules for the Nomenclature of Steroids, 1967, *IUPAC Inf. Bull.*, No. 33, 23 (1968); also published in *J. Org. Chem.*, 34, 1517 (1969)].

The α, β system for steroids can be extended to other classes of compounds such as terpenes and alkaloids when their absolute configurations are known; it can also be combined with stars or the use of the prefix *rel* when only the relative configurations are known.

In spite of the Rules of Subsection E-2, *cis* and *trans* are used when the arrangement of the atoms constituting an unsaturated backbone is the most important factor, as, for instance, in polymer chemistry and for carotenoids. When a series of double bonds of the same stereochemistry occurs in a backbone, the prefix all-*cis* or all-*trans* may be used.

E-5.12. (a) An achiral object having at least one pair of features that can be distinguished only by reference to a chiral object or to a chiral reference frame is said to be prochiral, and the property of having such a pair of features is termed prochirality. A consequence is that, if one of the paired features of a prochiral object is considered to differ from the other, the resultant object is chiral.

(b) In a molecule an achiral center or atom is said to be prochiral if it would be held to be chiral when two attached atoms or groups, that taken in isolation are indistinguishable, are considered to differ.

Notes: (1) For a tetrahedrally bonded atom this requires a structure Xaabc (where none of the groups a, b, or c is the enantiomer of another).

(2) For a fuller exploration of this concept, which is of particular importance to biochemists and spectroscopists, and for its extension to axes, planes, and unsaturated compounds, see Hanson, K. R., *J. Am. Chem. Soc.*, 88, 2731 (1966).

Examples:

$$
\begin{array}{ccc}
 & & CHO \\
 & & | \\
CH_3 & & H\!-\!\overset{\displaystyle |}{C}\!-\!OH \\
| & & | \\
H\!-\!C\!-\!H & & H\!-\!\overset{\displaystyle |}{C}\!-\!H \\
| & & | \\
OH & & OH \\
 & & \\
(A) & & (B)
\end{array}
$$

In both examples (A) and (B), the methylene carbon atom is prochiral; in both cases it would be held to be at a chiral center if one of the methylene hydrogen atoms were considered to differ from the other. An actual replacement of one of these protium atoms by, say, deuterium would produce an actual chiral center at the methylene carbon atom; as a result, compound (A) would become chiral, and compound (B) would be converted into one of two diastereoisomers.

E-5.13. Of the identical pair of atoms or groups in a prochiral compound, that one which leads to an (*R*) compound when considered to be preferred to the other by the sequence rule (without change in priority with respect to other ligands) is termed *pro-R*, and the other is termed *pro-S*.

Example:

$$
\begin{array}{c}
CHO \\
| \\
H^1 \text{---} C \text{---} OH \\
| \\
H^2
\end{array}
$$

H^1 is *pro-R*.
H^2 is *pro-S*.

E-6. Conformations

E-6.1. A molecule in a conformation into which its atoms return spontaneously after small displacements is termed a conformer.

Examples:

are different conformers.

E-6.2. (a) When, in a six-membered saturated ring compound, atoms in relative positions 1, 2, 4, and 5 lie in one plane, the molecule is described as in the chair or boat conformation according as the other two atoms lie, respectively, on opposite sides or on the same side of that plane.

Examples:

Chair   Boat.

Note: These and similar representations are idealized, minor divergences being neglected.

(b) A molecule of a monounsaturated six-membered ring compound is described as being in the half-chair or half-boat conformation according as the atoms not directly bound to the doubly bonded atoms lie, respectively, on opposite sides or on the same side of the plane containing the other four (adjacent) atoms.

Examples:

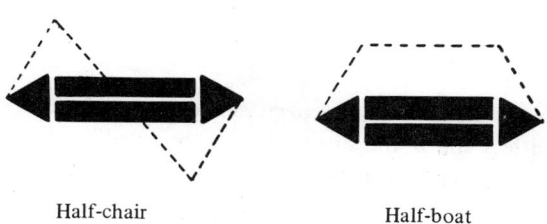

Half-chair   Half-boat

(c) A median conformation through which one boat form passes during conversion into the other boat form is termed a twist conformation. Similar twist conformations are involved in conversion of a chair into a boat form or vice versa.

Examples:

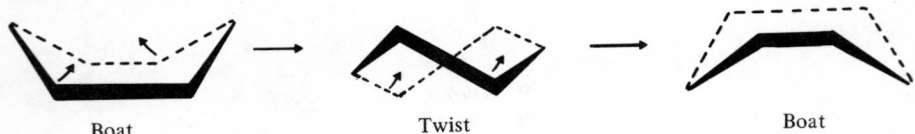

Boat   Twist   Boat

E-6.3. (a) Bonds to a tetrahedral atom in a six-membered ring are termed equatorial or axial according as they or their projections make a small or a large angle, respectively, with the plane containing a majority of the ring atoms.* Atoms or groups attached to such bonds are also said to be equatorial or axial, respectively.

Notes: (1) See, however, pseudoequatorial and pseudoaxial [Rule E-6.3(b)]. (2) The terms equatorial and axial may be abbreviated to e and a when attached to formulas; these abbreviations may also be used in names of compounds and are there placed in parentheses after the appropriate locants, for example, 1(e)-bromo-4(a)-chlorocyclohexane.

Examples:

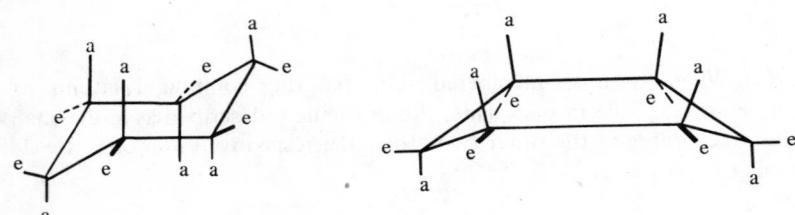

(b) Bonds from atoms directly attached to the doubly bonded atoms in a monounsaturated six-membered ring are termed pseudoequatorial or pseudoaxial according as the angles that they make with the plane containing the majority of the ring atoms approximate those made by, respectively, equatorial or axial bonds from a saturated six-membered ring. Pseudoequatorial and pseudoaxial may be abbreviated to e′ and a′, respectively, when attached to formulas; these abbreviations may also be used in names, then being placed in parentheses after the appropriate locants.

Example:

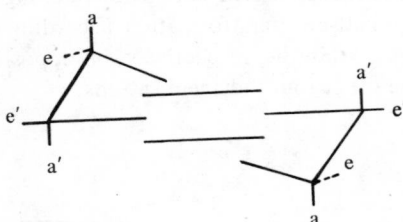

E-6.4. Torsion angle: In an assembly of attached atoms X—A—B—Y, where neither X nor Y is collinear with A and B, the smaller angle subtended by the bonds X—A and Y—B in a plane projection obtained by viewing the assembly along the axis A—B is termed the

*The terms axial, equatorial, pseudoaxial, and pseudoequatorial [see Rule E-6.3(b)] may be used also in connection with other than six-membered rings if, but only if, their interpretation is then still beyond dispute.

torsion angle (denoted by the Greek lower case letter theta θ or omega ω). The torsion angle is considered positive or negative according as the bond to the front atom X or Y requires rotation to the right or left, respectively, in order that its direction may coincide with that of the bond to the rear selected atom Y or X. The multiplicity of the bonding of the various atoms is irrelevant. A torsion angle also exists if the axis for rotation is formed by a collinear set of more than two atoms directly attached to each other.

Notes: (1) It is immaterial whether the projection be viewed from the front or the rear.

(2) For the use of torsion angles in describing molecules see Rule E-6.6.

Examples: (For construction of Newman projections, as here, see Rule E-7.2.)

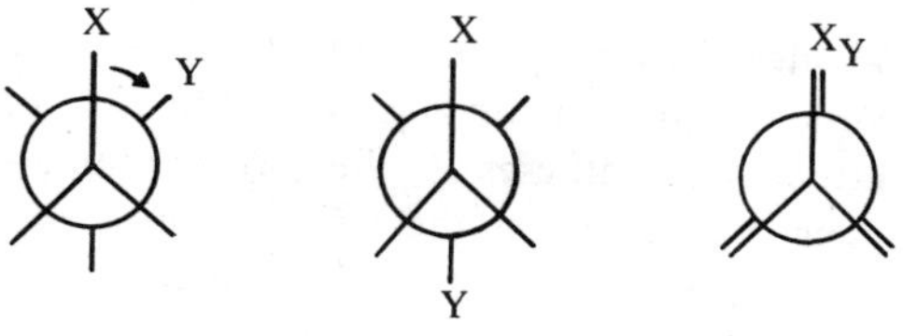

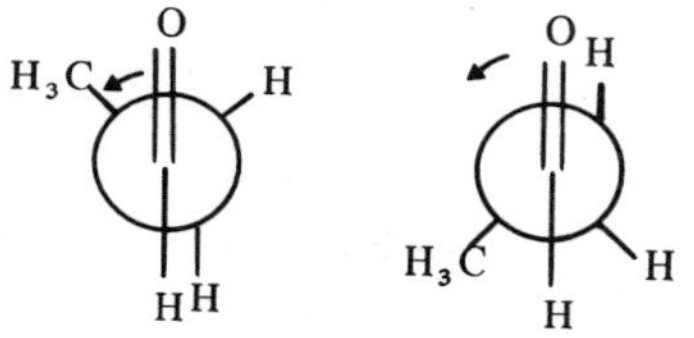

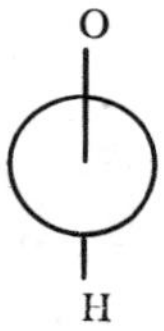

Newman projections of
propionaldehyde
$\theta = \sim -60°$ $\theta = \sim -120°$

Newman projection of
hydrogen peroxide
$\theta = \sim 180°$

E-6.5. If two atoms or groups attached at opposite ends of a bond appear one directly behind the other when the molecule is viewed along this bond, these atoms or groups are described as eclipsed, and that portion of the molecule is described as being in the eclipsed conformation. If not eclipsed, the atoms or groups and the conformation may be described as staggered.

Examples:

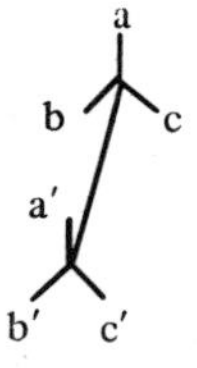

Eclipsed conformation.
The pairs a/a', b/b', and c/c' are eclipsed.

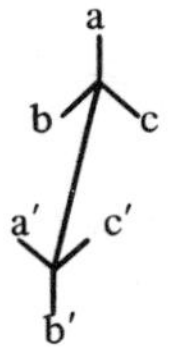

Staggered conformation.
All the attached groups are staggered.

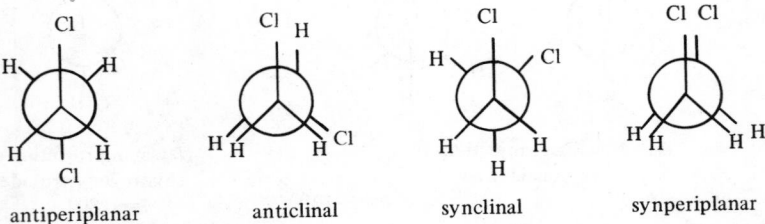

Projection of CH_3CH_2CHO.
The CH_3 and the H of the CHO are eclipsed.
The O and H's of CH_2 in CH_2CH_3 are
staggered.

E-6.6. Conformations are described as synperiplanar (*sp*), synclinal (*sc*), anticlinal (*ac*),
or antiperiplanar (*ap*) according as the torsion angle is within ±30° of 0°, ±60°, ±120°, or
±180°, respectively; the letters in parentheses are the corresponding abbreviations. Atoms
or groups are selected from each set to define the torsion angle according to the following
criteria: (1) if all the atoms or groups of a set are different, that one of each set that is
preferred by the sequence rule; (2) if one of a set is unique, that one; or (3) if all of a set
are identical, that one which provides the smallest torsion angle.
Examples:

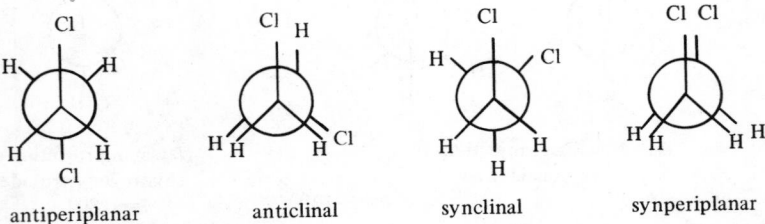

In the above conformations, all $CH_2Cl–CH_2Cl$, the two Cl atoms decide the torsion
angle.

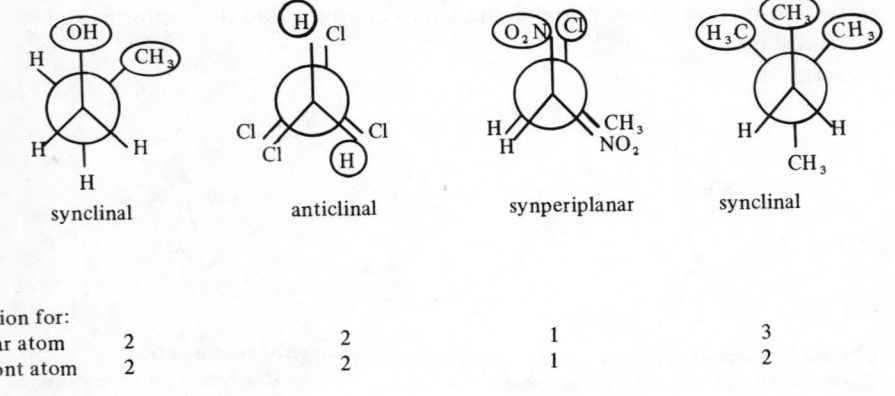

Criterion for:				
rear atom	2	2	1	3
front atom	2	2	1	2

	$(CH_3)_2 N-NH_2$ synclinal*	$CH_3 CH_2 -COCl$ anticlinal	$(CH_3)_2 CH-CONH_2$ antiperiplanar
Criterion for:			
rear atom	2	2	2
front atom	2	1	1

E-7. Stereoformulas

E-7.1. In a Fischer projection the atoms or groups attached to a tetrahedral center are projected on to the plane of the paper from such an orientation that atoms or groups appearing above or below the central atom lie behind the plane of the paper and those appearing to left and right of the central atom lie in front of the plane of the paper, and that the principal chain appears vertical with the lowest numbered chain member at the top.

Examples:

Orientation   Fischer projection

Notes: (1) The first of the two types of Fischer projection should be used whenever convenient.

(2) If a Fischer projection formula is rotated through 180° in the plane of the paper, the upward and downward bonds from the central atom still project behind the plane of the paper, and the sideways bonds project in front of that plane. If, however, the formula is rotated through 90° in the plane of the paper, the upward and downward bonds now project in front of the plane of the paper and the sideways bonds project behind that plane.

E-7.2. To prepare a Newman projection, a molecule is viewed along the bond between two atoms; a circle is used to represent these atoms with lines from outside the circle toward its center to represent bonds to other atoms; the lines that represent bonds to the nearer and the further atom end at, respectively, the center and the circumference of the circle. When two such bonds would be coincident in the projection, they are drawn at a small angle to each other.[†]

---

*The lone pair of electrons (represented by two dots) on the nitrogen atoms are the unique substituents that decide the description of the conformation (these are the "phantom atoms" of the sequence-rule symbolism).

[†]Cf. Newman, M. S., *Rec. Chem. Progr.*, 13, 111 (1952); *J. Chem. Educ.*, 33, 344 (1955); *Steric Effects in Organic Chemistry*, John Wiley & Sons, New York, 1956, 5.

Examples:

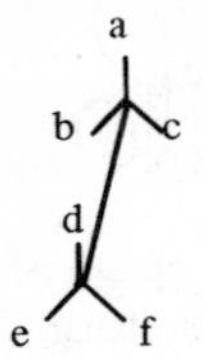

Perspective 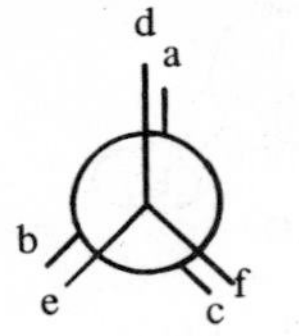   Newman projection 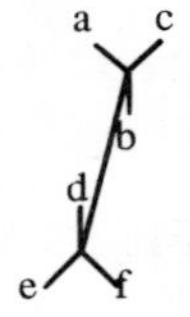   Perspective 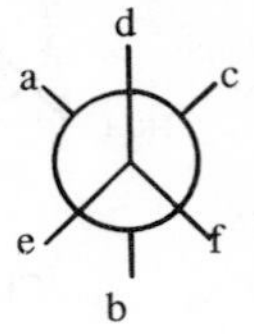   Newman projection

E-7.3. *General Note*; Formulas that display stereochemistry should be prepared with extra care so as to be unambiguous and, whenever possible, self-explanatory. It is inadvisable to try to lay down rules that will cover every case, but the following points should be borne in mind.

A thickened line (━) denotes a bond projecting from the plane of the paper toward an observer, a broken line (- - -) denotes a bond projecting away from an observer, and, when this convention is used, a full line of normal thickness (——) denotes a bond lying in the plane of the paper. A wavy line (〜〜) may be used to denote a bond whose direction cannot be specified or, if it is explained in the text, a bond whose direction it is not desired to specify in the formula. Dotted lines (· · · · · ·) should preferably not be used to denote stereochemistry, and never when they are used in the same paper to denote mesomerism, intermediate states, etc. Wedges should not be used as complement to broken lines (but see below). Single large dots have sometimes been used to denote atoms or groups attached at bridgehead positions and lying above the plane of the paper, with open circles to denote them lying below the plane of the paper, but this practice is strongly deprecated.

Hydrogen or other atoms or groups attached at sterically designated positions should never be omitted.

In chemical formulas, rings are usually drawn with lines of normal thickness, that is, as if they lay wholly in the plane of the paper even though this may be known not to be the case. In a formula such as (I) it is then clear that the H atoms attached at the A/B ring junction lie further from the observer than these bridgehead atoms, that the H atoms attached at the B/C ring junction lie nearer to the observer than those bridgehead atoms, and that X lies nearer to the observer than the neighboring atom of ring C.

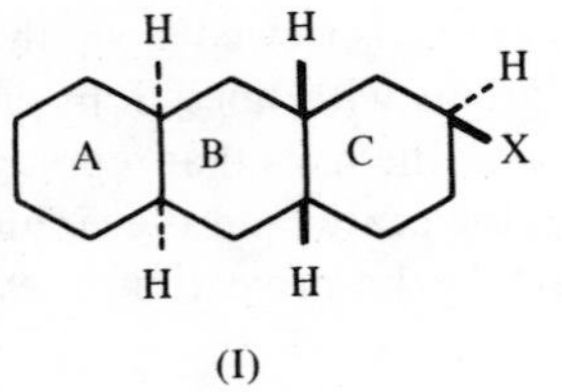

(I)

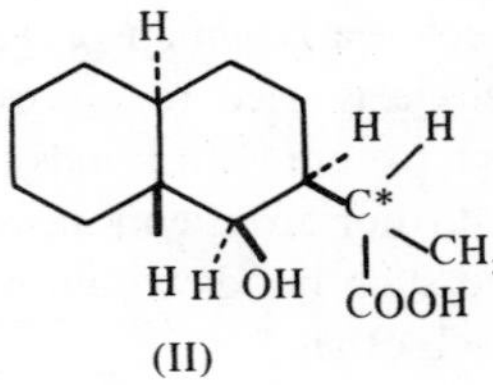

(II)

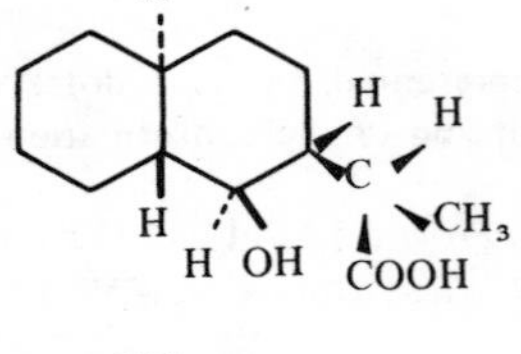

(III)

However, ambiguity can then sometimes arise, particularly when it is necessary to show stereochemistry within a group such as X attached to the rings that are drawn planar. For instance, in formula (II), the atoms O and C*, lying above the plane of the paper, are attached to ring B by thick bonds, but then, when showing the stereochemistry at C*, one finds that the bond *from* C* *to* ring B projects away from the observer and so should be a broken line. Such difficulties can be overcome by using wedges in place of lines, the broader end of the wedge being considered nearer to the observer, as in (III).

In some fields, notably for carbohydrates, rings are conveniently drawn as if they lay perpendicular to the plane of the paper, as represented in (IV); however, conventional formulas such as (V), with the lower bonds considered as the nearer to the observer, are so well established that is is rarely necessary to elaborate this to form (IV).

(IV)   (V)

By a similar convention, in drawings such as (VI) and (VII), the lower sets of bonds are considered to be nearer than the upper to the observer. In (VII), note the gaps in the rear lines to indicate that the bonds crossing them pass in front (and thus obscure sections of the rear bonds). In some cases, when atoms have to be shown as lying in several planes, the various conventions may be combined, as in (VIII). In all cases the overriding aim should be clarity.

(VI)   (VII)   (VIII)

APPENDIX 1. CONFIGURATION AND CONFORMATION

See Rules E-1.4 and E-1.5.

Various definitions have been propounded to differentiate configurations from conformations.

The original usage was to consider as conformations those arrangements of the atoms of a molecule in space that can be interconverted by rotation(s) around a single bond, and as configurations those other arrangements whose interconversion by rotation requires bonds to be broken and then re-formed differently. Interconversion of different configurations will then be associated with substantial energies of activation, and the various species will be separable, but interconversion of different conformations will normally be associated with less activation energy, and the various species, if separable, will normally be more readily interconvertible. These differences in activation energy and stability are often large.

Nevertheless, rigid differentiation on such grounds meets formidable difficulties. Differentiation by energy criteria would require an arbitrary cut in a continuous series of values. Differentiation by stability of isolated species requires arbitrary assumptions about conditions and halflives. Differentiation on the basis of rotation around single bonds meets difficulties connected both with the concept of rotation and with the selection of single bonds as requisites, and these need more detailed discussion here.

Enantiomeric biaryls are nowadays usually considered to differ in conformation, any difficulty in rotation about the $1,1'$ bond due to steric hindrance between the neighboring groups being considered to be overcome by bond bending and/or bond stretching, even though the movements required must closely approach bond breaking if these substituents are very large. Similar doubts about the possibility of rotation occur with a molecule such as (A), where rotation of the benzene ring around the oxygen-to-ring single bonds affords easy interconversion if x is large but appears to be physically impossible if x is small; and no critical size of x can be reasonably established. For reasons such as this, Rules E-1.4 and E-1.5 are so worded as to be independent of whether rotation appears physically feasible or not (see Note 2 to those Rules).

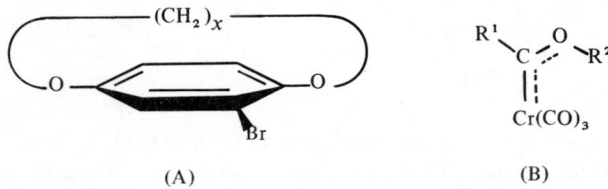

(A)   (B)

The second difficulty arises in the many cases where rotation is around a bond of fractional order between one and two, as in the helicenes, crowded aromatic molecules, metallocenes, amides, thioamides, and carbene-metal coordination compounds (such as B). The term conformation is customarily used in these cases and that appears a reasonable extension of the original conception, though it will be wise to specify the usage if the reader might be in doubt.

When interpreted in these ways, Rules E-1.4 and E-1.5 reflect the most frequent usage of the present day and provide clear distinctions in most situations. Nevertheless, difficulties remain and a number of other usages have been introduced.

It appears to some workers that once it is admitted that change of conformation may involve rotation about bonds of fractional order between one and two, it is then illogical to exclude rotation about classical double bonds because interconversion of open-chain *cis-trans* isomers depends on no fundamentally new principle and is often relatively easy, as for certain alkene derivatives such as stilbenes and for *azo* compounds, by irradiation. This extension is indeed not excluded by Rules E-1.4 and E-1.5, but if it is applied that fact should be explicitly stated.

A further interpretation is to regard a stereoisomer possessing some degree of stability (that is, one associated with an energy hollow, however shallow) as a configurational isomer, the other arrangements in space being termed conformational isomers; the term conformer (Rule E-6.1) is then superfluous. This definition, however, requires a knowledge of stability (energy relations) that is not always available.

In another view, a configurational isomer is any stereoisomer that can be isolated or (for some workers) whose existence can be established (for example, by physical methods); all other arrangements then represent conformational isomers; but it is then impossible to differentiate configuration from conformation without involving experimental efficiency or conditions of observation.

Yet another definition is to regard a conformation as a precise description of a configuration in terms of bond distances, bond angles, and dihedral angles.

In none of the above views except the last is attention paid to extension or contraction of the bond to an atom that is attached to only one other atom, such as —H or =O. Yet such changes in interatomic distance due to nonbonded interactions may be important, for instance, in hydrogen bonding, in differences due to crystal form, in association in solution, and in transition states. This area may repay further consideration.

Owing to the circumstances outlined above, the Rules E-1.4 and E-1.5 have been deliberately made imprecise, so as to permit some alternative interpretations, but they are not compatible with all the definitions mentioned above. The time does not seem ripe to legislate for other than the commoner usages or to choose finally between these. It is, however, encouraging that no definition in this field has (yet) involved atomic vibrations for which, in all cases, only time-average positions are considered.

Finally it should be noted that an important school of thought uses conformation with the connotation of "a particular geometry of the molecule, i.e., a description of atoms in space in terms of bond distances, bond angles, and dihedral angles," a definition much wider than any discussed above.

APPENDIX 2. OUTLINE OF THE SEQUENCE-RULE PROCEDURE

The sequence-rule procedure is a method of specifying the absolute molecular chirality (handedness) of a compound, that is, a method of specifying which of two enantiomeric forms each chiral element of a molecule exists. For each chiral element in the molecule it provides a symbol, usually R or S, which is independent of nomenclature and numbering. These symbols define the chirality of the specific compound considered; they may not be the same for a compound and some of its derivatives; and they are not necessarily constant for chemically similar situations within a chemical or a biogenetic class. The procedure is applied directly to a three-dimensional model of the structure, and not to any two-dimensional projection thereof.

The method has been developed to cover all compounds with ligancy up to four and with ligancy six,* and for all configurations and conformations of such compounds. The following is an outline confined to the most common situations; it is essential to study the original papers, especially the 1966 paper,[†] before using the sequence rule for other than fairly simple cases.

General Basis

The sequence rule itself is a method of arranging atoms or groups (including chains and rings) in an order of precedence, often referred to as an order of preference; for discussion this order can conveniently be generalized as a > b > c > d, where > denotes "is preferred to."

The first step, however, in considering a model is to identify the nature and position of each chiral element that it contains. There are three types of the chiral element, namely, the chiral center, the chiral axis, and the chiral plane. The chiral center, which is very much the most commonly met, is exemplified by an asymmetric carbon atom with the tetrahedral arrangement of ligands, as in (1). A chiral axis is present in, for instance, the chiral allenes such as (2) or the chiral biaryl derivatives. A chiral plane is exemplified by the plane containing the benzene ring and the bromine and oxygen atoms in the chiral compound (3), or by the underlined atoms in the cycloalkene (4). Clearly, more than one

---

*Ligancy refers to the number of bonds from an atom, independently of the nature of the bonds.
[†]Cahn, R. S., Ingold, C., and Prelog, V., *Angew. Chem. Int. Ed.,* 5, 385 (1966); errata, 5, 511 (1966); *Angew. Chem.,* 78, 413 (1966). Earlier papers: Cahn, R. S. and Ingold, C. K., *J. Chem. Soc.* (Lond.), 612 (1951); Cahn, R. S., Ingold, C., and Prelog, V., *Experientia,* 12, 81 (1956). For a partial, simplified account see Cahn, R. S., *J. Chem. Educ.,* 41, 116 (1964); errata, 41, 503 (1964).

type of chiral element may be present in one compound; for instance, group "a" in (2) migh be a *sec*-butyl group which contains a chiral center.

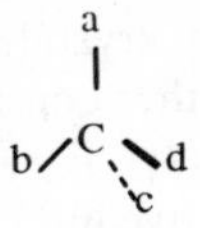

(1) 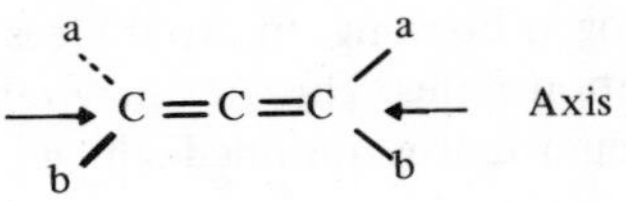   (2)

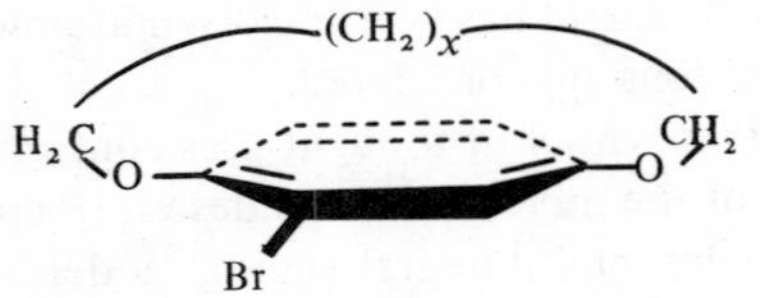

(3) 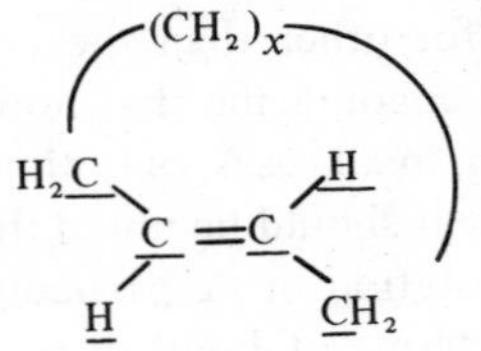   (4)

The Chiral Center

Let us consider first the simplest case, namely, a chiral center (such as carbon) with four ligands, a, b, c, and d, which are all different atoms tetrahedrally arranged as in CHFClBr. The four ligands are arranged in order of preference by means of the sequence rule; this contains five subrules, which are applied in succession so far as necessary to obtain a decision. The first subrule is all that is required in a great majority of actual cases; it states that ligands are arranged in order of decreasing atomic number, in the above case (a) Br > (b) Cl > (c) F > (d) H. There would be two (enantiomeric) forms of the compound and we can write these as (5) and (6). In the sequence-rule procedure the model is viewed from the side remote from the least-preferred ligand (d), as illustrated. Then, tracing a path from a to b to c in (5) gives a clockwise course, which is symbolized by (R) (Latin *rectus*, right; for right hand); in (6) it gives an anticlockwise course, symbolized as (S) (Latin *sinister*, left). Thus (5) would be named (R)-bromo-chlorofluoromethane, and (6) would be named (S)-bromochlorofluoromethane. Here already it may be noted that converting one enantiomer into another changes each R to S, and each S to R, always. It will be seen also that the chirality prefix is the same whether the alphabetical order is used, as above, for naming the substituents or whether this is done by the order of complexity (giving fluorochlorobromomethane).

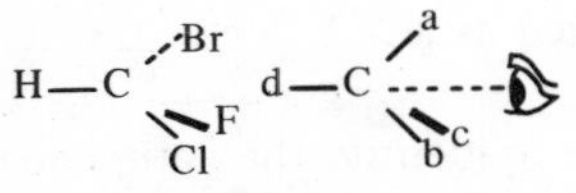

(5), (R) 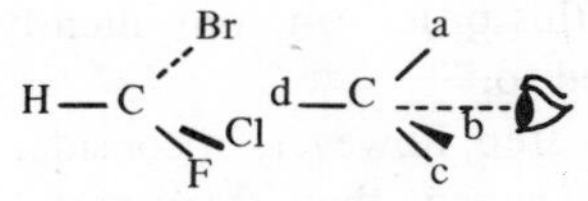   (6), (S)

Next, suppose we have $H_3C-CHClF$. We deal first with the atoms directly attached to the chiral center; so the four ligands to be considered are Cl > F > C (of CH_3) > H. Here the H's of the CH_3 are not concerned, because we do not need them in order to assign our symbol.

However, atoms directly attached to a center are often identical, as, for example, the underlined C's in $H_3\underline{C}-CHCl-\underline{C}H_2OH$. For such a compound we at once establish a preference (a) Cl > (b, c) $\underline{C},\underline{C}$ > (d) H. Then to decide between the two $\underline{C}$'s we work outward, to the atoms to which they in turn are directly attached and we then find which we can conveniently write as C(H,H,H) and C(O,H,H). We have to compare H,H,H with O,H,H, and since oxygen has a higher atomic number than hydrogen we have O > H and thence the complete order Cl > C (of CH_2OH) > C (of CH_3) > H, so that the chirality symbol can then be determined from the three-dimensional model.

$$-\underset{\underset{H}{\diagdown}}{\overset{\overset{H}{\diagup}}{C}}-H \quad \text{and} \quad -\underset{\underset{H}{\diagdown}}{\overset{\overset{O}{\diagup}}{C}}-H$$

We must next meet the first complication. Suppose that we have a molecule (7).

$$\text{(b) } H_3C-\underline{C}HCl-\overset{\overset{\textstyle Cl\ (a)}{\vdots}}{\underset{\underset{\textstyle H\ (d)}{\vdots}}{C}}-\underline{C}HF-OH\ (c)$$

$$(7)\ (S)$$

To decide between the two C's we first arrange the atoms attached to them in *their* order of preference, which gives $\underline{C}$(Cl,C,H) on the left and $\underline{C}$(F,O,H) on the right. Then we compare the preferred atom of one set (namely, Cl) with the preferred atom (F) of the other set, and as Cl > F we arrive at the preferences a > b > c > d shown in (7) and chirality (S). If, however, we had a compound (8) we should have met $\underline{C}$(Cl,C,H) and C(Cl,O,H) and, since the atoms of first preference are identical (Cl), we should have had to make the comparisons with the atoms of second preference, namely, O > C, which to the different chirality (R) as shown in (8).

$$\text{(c) } H_3C-\underline{C}HCl-\overset{\overset{\textstyle Cl\ (a)}{\vdots}}{\underset{\underset{\textstyle H\ (d)}{\vdots}}{C}}-\underline{C}HCl-OH\ (b)$$

$$(8)\ (R)$$

Branched ligands are treated similarly. Setting them out in full gives a picture that at first sight looks complex but the treatment is in fact simple. For instance, in compound (9) a first quick glance again shows (a) Cl > (b, c) $\underline{C},\underline{C}$ > (d) H: When we expand the two $\underline{C}$'s we find they are both $\underline{C}$(C,C,H), so we continue exploration. Considering first the left-hand ligand we arrange the branches and their sets of atoms in order thus: C(Cl,H,H) > C(H,H,H). On the right-hand side we have C(O,$\underline{C}$,H) > C(O,$\underline{H}$,H) (because $\underline{C}$ > $\underline{H}$). We compare first the preferred of these branches from each side and we find C(Cl,H,H) > C(O,C,H) because Cl > O, and that gives the left-hand branch preference over the right-hand branch. That is all we need to do to establish chirality (S) for this highly branched compound (9). Note that it is immaterial here that, for the lower branches, the right-hand C(O,H,H) would have been preferred to the left-hand C(H,H,H); we did not need to reach that point in our comparisons and so we are not concerned with it; but we should have reached it if the two top (preferred) branches had both been the same CH_2Cl.

Rings, when we met during outward exploration, are treated in the same way as branched chains.

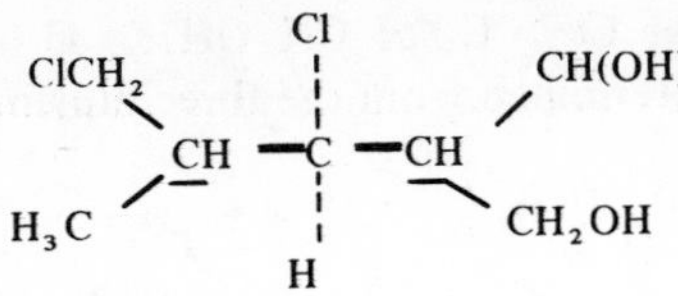

(9)

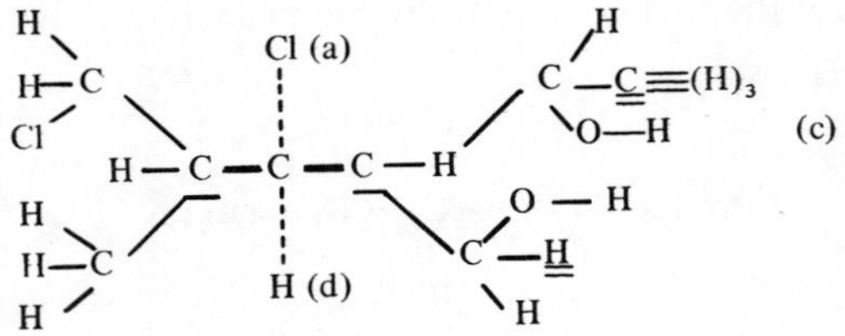

(9) (*S*)

With these simple procedures alone, quite complex structures can be handled; for instance, the analysis alongside Formula (10) for natural morphine explains why the specification is as shown. The reason for considering C-12 as C(C,C,C) is set out in the next paragraphs.

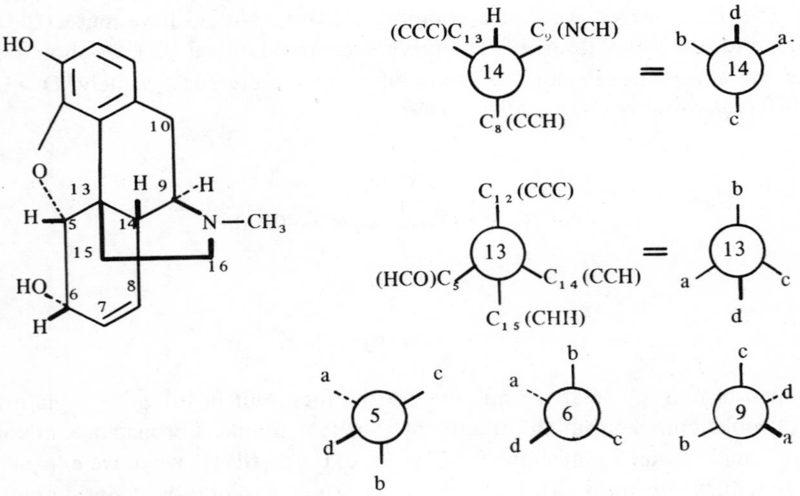

(10)   (5*R*, 6*S*, 9*R*, 13*S*, 14*R*,)-Morphine

Now, using the sequence rule depends on exploring along bonds. To avoid theoretical arguments about the nature of bonds, simple classical forms are used. Double and triple bonds are split into two and three bonds, respectively. A > C=O group is treated as (i) (below) where the (O) and the (C) are duplicate representations of the atoms at the other end of the double bond. —C≡CH is treated as (ii) and —C≡N is treated as (iii).

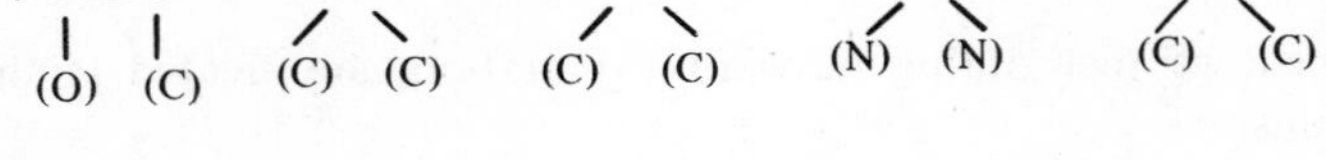

(i)   (ii)   (iii)

Thus in D-glyceraldehyde (11) the CHO group is treated as C(O,(O),H) and is thus preferred to the C(O,H,H) of the CH$_2$OH group, so that the chirality symbol is *(R)*.

D-Glyceraldehyde
(11) *(R)*

Only the doubly bonded atoms themselves are duplicated, and not the atoms or groups attached to them; the duplicated atoms may thus be considered as carrying three phantom atoms (see below) of atomic number zero. This may be important in deciding preferences in certain complicated cases.

Aromatic rings are treated as Kekulé structures. For aromatic hydrocarbon rings it is immaterial which Kekulé structure is used because "splitting" the double bonds gives the same result in all cases; for instance, for phenyl the result can be represented as (12a) where "(6)" denotes the atomic number of the duplicate representations of carbon.

For aromatic hetero rings, each duplicate is given an atomic number that is the mean of what it would have if the double bond were located at each of the possible positions. A complex case is illustrated in (13). Here C-1 is doubly bonded to one or other of the nitrogen atoms (atomic number 7) and never to carbon, so its added duplicate has atomic number 7; C-3 is doubly bonded either to C-4 (atomic number 6) or to N-2 (atomic number 7), so its added duplicate has atomic number 6½; so has that of C-8; but C-4a may be doubly bonded to C-4, C-5, or N-9, so its added duplicate has atomic number 6.33.

One last point about the chiral center may be added here. Except for hydrogen, ligancy, if not already four, is made up to four by adding "phantom atoms" which have atomic number zero and are thus always last in order of preference. This has various uses but perhaps the most interesting is where nitrogen occurs in a rigid skeleton, as, for example, in α-isosparteine (14). Here the phantom atom can be placed where the nitrogen

SOME COMMON GROUPS IN ORDER OF SEQUENCE-RULE PREFERENCE[a]

A. Alphabetical Order (Higher Number Denotes Greater Preference)

64 Acetoxy	38 Carboxyl	9 Isobutyl	55 Nitroso
36 Acetyl	74 Chloro	8 Isopentyl	6 *n*-Pentyl
48 Acetylamino	17 Cyclohexyl	20 Isopropenyl	61 Phenoxy
21 Acetylenyl	52 Diethylamino	14 Isopropyl	22 Phenyl
10 Allyl	51 Dimethylamino	69 Mercapto	47 Phenylamino
43 Amino	34 2,4-Dinitrophenyl	58 Methoxy	54 Phenylazo
44 Ammonio ⁺H₃N–	28 3,5-Dinitrophenyl	39 Methoxycarbonyl	18 Propenyl
37 Benzoyl	59 Ethoxy	2 Methyl	4 *n*-Propyl
49 Benzoylamino	40 Ethoxycarbonyl	45 Methylamino	29 1-Propynyl
65 Benzoyloxy	3 Ethyl	71 Methylsulfinyl	12 2-Propynyl
50 Benzyloxycarbonylamino	46 Ethylamino	66 Methylsulfinyloxy	73 Sulfo
13 Benzyl	68 Fluoro	72 Methylsulfonyl	25 *m*-Tolyl
60 Benzyloxy	35 Formyl	67 Methylsulfonyloxy	30 *o*-Tolyl
41 Benzyloxycarbonyl	63 Formyloxy	70 Methylthio	23 *p*-Tolyl
75 Bromo	62 Glycosyloxy	11 Neopentyl	53 Trimethylammonio
42 *ter*-Butoxycarbonyl	7 *n*-Hexyl	56 Nitro	32 Trityl
5 *n*-Butyl	1 Hydrogen	27 *m*-Nitrophenyl	15 Vinyl
16 *sec*-Butyl	57 Hydroxy	33 *o*-Nitrophenyl	31 2,6-Xylyl
19 *tert*-Butyl	76 Iodo	24 *p*-Nitrophenyl	26 3,5-Xylyl

B. Increasing Order of Sequence Rule Preference

1 Hydrogen	20 Isopropenyl	39 Methoxycarbonyl[b]	58 Methoxy
2 Methyl	21 Acetylenyl	40 Ethoxycarbonyl[b]	59 Ethoxy
3 Ethyl	22 Phenyl	41 Benzyloxycarbonyl[b]	60 Benzyloxy
4 *n*-Propyl	23 *p*-Tolyl	42 *tert*-Butoxycarbonyl[b]	61 Phenoxy
5 *n*-Butyl	24 *p*-Nitrophenyl	43 Amino	62 Glycosyloxy
6 *n*-Pentyl	25 *m*-Tolyl	44 Ammonio ⁺H₃N–	63 Formyloxy
7 *n*-Hexyl	26 3,5-Xylyl	45 Methylamino	64 Acetoxy
8 Isopentyl	27 *m*-Nitrophenyl	46 Ethylamino	65 Benzoyloxy
9 Isobutyl	28 3,5-Dinitrophenyl	47 Phenylamino	66 Methylsulfinyloxy
10 Allyl	29 1-Propynyl	48 Acetylamino	67 Methylsulfonyloxy
11 Neopentyl	30 *o*-Tolyl	49 Benzoylamino	68 Fluoro
12 2-Propynyl	31 2,6-Xylyl	50 Benzyloxycarbonylamino	69 Mercapto HS–
13 Benzyl	32 Trityl	51 Dimethylamino	70 Methylthio CH₃S–
14 Isopropyl	33 *o*-Nitrophenyl	52 Diethylamino	71 Methylsulfinyl
15 Vinyl	34 2,4-Dinitrophenyl	53 Trimethylammonio	72 Methylsulfonyl
16 *sec*-Butyl	35 Formyl	54 Phenylazo	73 Sulfo HO₃S–
17 Cyclohexyl	36 Acetyl	55 Nitroso	74 Chloro
18 1-Propenyl	37 Benzoyl	56 Nitro	75 Bromo
19 *tert*-Butyl	38 Carboxyl	57 Hydroxy	76 Iodo

[a] ANY alteration to structure, or substitution, etc., may alter the order of preference.
[b] These groups are ROC(=O)–.

lone pair of electrons is; then N-1 appears as shown alongside the formula; and the chirality (*R*) is the consequence. The same applies to N-16. Phantom atoms are similarly used when assigning chirality symbols to chiral sulfoxides (see example to Rule E-5.9).

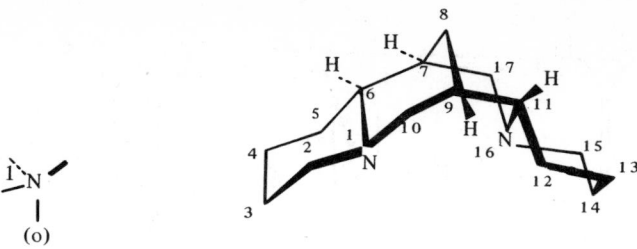

(14) (1*R*, 6*R*, 7*S*, 9*S*, 11*R*, 16*R*)-Sparteine

Symbolism

In names of compounds, the *R* and *S* symbols, together with their locants, are placed in parentheses, normally in front of the name, as shown for morphine (10) and sparteine (14), but this may be varied in indexes or in languages other than English. Positions within names are required, however, when more than a single series of numerals is used, as for esters and amines. When relative stereochemistry is more important than absolute stereochemistry, as for steroids or carbohydrates, a local system of stereochemical designation may be more useful and sequence-rule symbols need then be used only for any situations where the local system is insufficient.

Racemates containing a single center are labeled (*RS*). If there is more than one center the first is labeled (*RS*) and the others are (*RS*) or (*SR*) according to whether they are *R* or *S* when the first is *R*. For instance, the 2,4-pentanediols $CH_3-CH(OH)-CH_2-CH(OH)-CH_3$ are differentiated as

one chiral form (2*R*,4*R*)–
other chiral form (2*S*,4*S*)–
meso compound (2*R*,4*S*)–
racemic compound (2*RS*,4*RS*)–

Finally the principles by which some of the least rare of other situations are treated will be very briefly summarized.

Pseudoasymmetric Atoms

A subrule decrees that *R* groups have preference over *S* groups and this permits pseudoasymmetric atoms, as in abC(c-*R*)(c-*S*) to be treated in the same way as chiral centers, but as such a molecule is achiral (not optically active) it is given the lower case symbol *r* or *s*.

Chiral Axis

The structure is regarded as an elongated tetrahedron and viewed along the axis — it is immaterial from which end it is viewed; the nearer pair of ligands receives the first two positions in the order of preference, as shown in (15) and (16).

(15)

(*S*)
viewed from
X

or

(*S*)
viewed from
Y

(16)

Chiral Plane

The sequence-rule-preferred atom directly attached to the plane is chosen as "pilot atom." In compound (3) this is the C of the left-hand CH_2 group. Now this is attached to the left-hand oxygen atom in the plane. The sequence-rule-preferred path from this oxygen atom is then explored in the plane until a rotation is traced which is clockwise (*R*) or anticlockwise (*S*) when viewed from the pilot atom. In (3) this path is O → C → C(Br) and it is clockwise (*R*).

Other Subrules

Other subrules cater for new chirality created by isotopic labeling (higher mass number preferred to lower) and for steric differences in the ligands. Isotopic labeling rarely changes symbols allotted to other centers.

Octahedral Structures

Extensions of the sequence rule enable ligands arranged octahedrally to be placed in an order of preference, including polydentate ligands, so that a chiral structure can then always be represented as one of the enantiomeric forms (17) and (18). The face 1–2–3 is observed from the side remote from the face 4–5–6 (as marked by arrows), and the path 1 → 2 → 3 is observed; in (17) this path is clockwise (R), and in (18) it is anticlockwise (S).

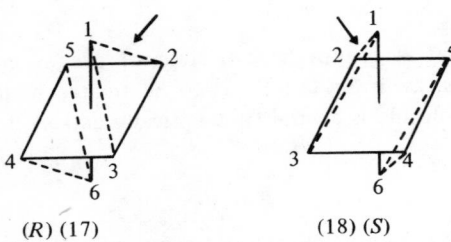

(R) (17)   (18) (S)

Conformations

The torsion angle between selected bonds from two singly bonded atoms is considered. The selected bond from each of these two atoms is that to a unique ligand, or otherwise to the ligand preferred by the sequence rule. The smaller rotation needed to make the front ligand eclipsed with the rear one is noted (this is the rotatory characteristic of a helix); if this rotation is right-handed it leads to a symbol *P* (plus); if left-handed to *M* (minus). Examples are

(M) (P) (P)

Details and Complications

For details and complicating factors the original papers should be consulted. They include treatment of compounds with high symmetry or containing repeating units (e.g., cyclitols), also π bonding (metallocenes, etc.), mesomeric compounds and mesomeric radicals, and helical and other secondary structures.

THE CITATION OF BIBLIOGRAPHIC REFERENCES IN BIOCHEMICAL JOURNALS RECOMMENDATIONS (1971)*

IUB Commission of Editors of Biochemical Journals (CEBJ)

These Recommendations were reviewed by the Commission in August 1972, when it was decided to publish them.

PREAMBLE

Two basic systems for the citation of references are used at present. The so-called Harvard System (where names of authors and the date are cited in the text, and the reference list is in alphabetical order) and the Numbering System (where numbers, but not necessarily names of authors, are cited in the text, and the reference list is in order of citation in the text). Several ways of quoting references in the list are in current use.

The Commission is of the opinion, arrived at as a result of much consultation between many senior editors, that it is unlikely that all journals would accept a recommendation to use either the Harvard or the Numbering System to the exclusion of the other. It believes, however, that most biochemists will accept the need for, and indeed welcome, a substantial degree of unification of practices, there being no strong case for the individuality of each journal on this issue. Accordingly, the Commission makes the following Recommendations to all biochemical journals; the reasons for some of them are given. The Recommendations deal first with the way in which references should be cited in the list; the proposal is suitable for journals adopting either the Harvard or the Numbering System. Secondly, there are Recommendations about the way in which each of these systems is used. Thirdly, abbreviations for titles of journals and a few other points are considered. Implementation of the Recommendations would mean that any very small differences between journals in their practices would be of the type that can be attended to at the redactory stage of preparation for press. The Commission recognizes that it cannot deal with a number of smaller problems concerning citations that arise from time to time.

RECOMMENDATIONS

1. Citations of References in the List of References Should Be as Follows

Braun, A., Brown, B. & LeBrun, C. (1971) *Journal*, 11, 111–113.

Notes: (a) This form can be used by both systems.

(b) Journals using the Numbering System should arrange the references in numerical order beside the number (which can be italicized or in brackets according to the house custom of the journal).

(c) Journals using the Harvard System should arrange the references in alphabetical order, whatever the language, except in certain situations (see Recommendation 4a below).

(d) This recommendation incorporates the following points:

  i.   Initials after surnames (full first names are not given in the list).

  ii. The use of the symbol "&" is recommended if at all possible because of its widespread usage and the fact that it is independent of the language. No comma before "&."

*From IUB Commission of Editors of Biochemical Journals (CEBJ), *J. Biol. Chem.,* 248(21), 7279–7280 (1973). With permission.

iii. Year in parentheses (this follows immediately after the authors' names because it is essential to the Harvard System).

iv. Journal title (abbreviated). This can be in italics according to house practice (see Recommendation 7 below concerning journal title abbreviations).

v. Volume number. This can be in heavy type or italics according to house practice.

vi. A few journals do not have volume numbers in which case the page numbers should follow immediately after the abbreviated journal title.

If it is necessary to quote both a volume and a part number, the reference should read: Brown, B. (1971) *Journal,* 11, pt 1, 121−123.

vii. First and last pages should be given. The Commission decided to make this Recommendation mainly on the basis of evidence that the additional information provided by quoting the last page was being required increasingly in many types of library and information retrieval services. Citation of the last page (as well as the first) has been requested for some time by the secondary and abstracting journals. Citation of both first and last pages is also an aid in the prevention of errors.

viii. The number of stops and commas is kept as small as possible.

(e) Authors' names and the abbreviated name of the journal when repeated in the next reference should be spelled out in full; ibid. and similar terms should not be used.

(f) Recommendations of the IUPAC-IUB Commission on Biochemical Nomenclature (CBN) and similar documents should be referred to as: Commission on Biochemical Nomenclature (1970) followed by a journal reference.

(g) Junior should be abbreviated to "Jr," not "jun."

2. Numbering System in the Text

The use of authors' names is permissible as authors wish; only the initial letter of the name should be in capital type. Numbers can be inserted in parentheses or as superscripts according to house custom. The printing of references at the foot of the page on which they are first quoted is considered to be helpful with the Numbering System but is not part of the Recommendation because the extra cost is generally considered to be prohibitive.

3. Harvard System in the Text

For multi-author papers, it is recommended that:

a. Not more than two authors to be named either on the first or any subsequent occasion;

b. et al. should be used for three or more authors on every occasion;

c. Each name to have the initial letter in capital type only.

Examples (Harvard System style):

Braun et al. (1969) did some work that was confirmed by LeBrun (1970).

These results (Braun et al., 1969; LeBrun, 1970) have been discussed by Brown & Braun (1971). The same Recommendation (without the year) applies when authors are quoted in the text in the Numbering System.

4. Harvard System in the List of References

a. A special problem arises in the list when there are several papers by, e.g., Green et al. in the same or over several years. While the list could be in strict alphabetical order of the full reference, the reader will find no clue in the text to the alphabetical status of the names of the second and subsequent authors (see Recommendations 3a and 3b). It is therefore recommended that all the papers by Green et al. (that is by Green and more than one co-author) should be arranged, irrespective of the names of the other authors, in chronological order (over many years if necessary) and designate tham a, b, c, etc.

Examples:

Green, G. (1970) etc.
Green, G. & Brown, B. (1971) etc.
Green, G. & White, W. (1969) etc.
Green, G., White, W. & Black, B. (1968a) etc.   sequence governed by order or date of
                                                     publication, as far as can be ascertained.
Green, G., Brown, B. & Black, B. (1968b) etc.
Green, G., White, W., Black, B. & Brown, B. (1969) etc.
Green, G., Black, B. & Brown, B. (1970) etc.

   b. Names beginning with "Mc" should be listed under "Mc" and not under "Mac," to decide alphabetical order.
   c. Names beginning with "De," "Van," or "von," etc. should be arranged under D or V/v, etc.

5. Reference to Books

These should appear in text like any reference to a journal paper. The reference in the list should read: Brown, B. & Braun, A. (1971) in *Book Title* (LeBrun, C., ed.), pp. 1–20, Publisher, Town.

Notes:
   a. If a volume number has to be quoted, this would appear before the pp. as, e.g., "vol. 2," with the number in Arabic numerals (even when Roman numerals are printed on the cover of the book).
   b. Where an author wishes to refer to a specific page within a book reference, this should be given in the text.

Example (in text): ". . . discussed on p. 21 of Braun et al.(1971)."

6. Other Forms of References

   a. *In the press.* It is recommended that (i) this should mean that the paper has been finally accepted by a journal, (ii) it is quoted in the text (both systems) just as any other paper, (iii) the year quoted should be the best estimate revised if necessary at proof stage, and (iv) the full citation in the list to read: Braun, A. & Brown, B. (1971) *Journal,* in the press.
   b. *Submitted for publication* should be used in a typescript only when it is reasonable to expect that it will be possible to alter the quotation to a final form at a stage before publication; if such alteration cannot be made then the name of the journal involved should be stated.
   c. The use of *in preparation* and *private communication* should not be allowed because they have no real value.
   d. *Personal communication* and *unpublished work* should be permitted in the text only, i.e., not in the list of references. Editors may require to see written evidence of the former.

7. Abbreviations for Journal Titles

Most biochemical journals use the *Chemical Abstract** system but a few use the World List, 4th Edition. The Commission noted that the latest information available (International List of Periodical Title Word Abbreviations prepared for the UNISIST/ICSU-AB Working Group on Bibliographical Descriptions) suggests that the abbreviations that will be recommended finally by ICSU will be very similar to those now used by *Chemical Abstracts.*

Believing that complete uniformity on this issue is highly desirable now and estimating that it may be a few more years before ICSU finally reports, the Commission recommends that all biochemical journals should now use the *Chemical Abstracts* (American Chemical Society) system. The Commission believes that any changes that will be required when ICSU eventually issues recommendations on this point will be comparatively minor ones.

8. Implementation of these Recommendations

The Commission at its meeting in Menton, May 7 to 8, 1971, has taken the view that the degree of uniformity envisaged in the Recommendations is highly desirable and therefore further recommends to all biochemical journals that the changes required should be made as soon as possible. The Commission recognizes that all journals will have to make some changes (in most cases these are minor) from their present established practices to implement these Recommendations in full. It considers that the possible objections of difficulties even for a commercial publisher with an established "house style" are outweighed by the advantage that conformity of style in the citation of references will prove to the authors, editors, and readers upon whom all journals depend for their existence.

*The journal-title abbreviations in *Biological Abstracts* are essentially the same in *Chemical Abstracts*. A *List of Serials with Title Abbreviations* is available from BioSciences Information Service of Biological Abstracts, 2100 Arch Street, Philadelphia, PA 19103.

NOMENCLATURE OF LABELED COMPOUNDS

The statement below was adopted by the IUB Commission of Editors of Biochemical Journals* (CEBJ) and appears, in the same or in similar form, in the Instructions to Authors of their journals. This system originated with the Chemical Society (London) and was subsequently adopted by the American Chemical Society (Handbook for Authors, 1967). It was adopted by CEBJ in 1971 and is the only system currently permitted in the pages of their journals.

ISOTOPICALLY LABELED COMPOUNDS

The symbol for the isotope introduced is placed in *square* brackets directly attached to the front of the name (word), as in $[^{14}C]$ urea. When more than one position in a substance is labeled by means of the same isotope and the positions are not indicated (as below), the number of labeled positions is added as a right-hand subscript, as in $[^{14}C_2]$ glycollic acid. The symbol "U" indicates uniform and "G" general labeling, e.g., $[U^{-14}C]$ glucose (where the ^{14}C is uniformly distributed among all six positions) and $[G^{-14}C]$ glucose (where the ^{14}C is distributed among all six positions, but not necessarily uniformly); in the latter case it is often sufficient to write simply "$[^{14}C]$ glucose."

The isotopic prefix precedes that part of the name to which it refers, as in sodium $[^{14}C]$ formate, iodo$[^{14}C_2]$ acetic acid, 1-amino$[^{14}C]$ methylcyclopentanol $(H_2N-^{14}CH_2-C_5H_8-OH)$, α-naphth$[^{14}C]$ oic acid $(C_{10}H-^{14}CO_2H)$, 2-acetamido-7-$[^{131}I]$ iodofluorene, fructose 1,6-$[1^{-32}P]$ diphosphate, D-$[^{14}C]$ glucose, $2H$-$[2^{-2}H]$ pyran, S-$[8^{-14}C]$ adenosyl$[^{35}S]$ methionine. Terms such as "^{131}I-labeled albumin" should not be contracted to "$[^{131}I]$ albumin" (since native albumin does not contain iodine), and "^{14}C-labeled amino acids" should similarly not be written as "$[^{14}C]$ amino acids" (since there is no carbon in the amino group).

When isotopes of more than one element are introduced, their symbols are arranged in alphabetical order, including ^2H and ^3H for deuterium and tritium, respectively.

When not sufficiently distinguished by the foregoing means, the positions of isotopic labeling are indicated by Arabic numerals, Greek letters, or prefixes (as appropriate), placed within the square brackets and before the symbol of the element concerned, to which they are attached by a hyphen; examples are $[1^{-2}H]$ ethanol $(CH_3-C^2H_2-OH)$, $[1^{-14}C]$ aniline, L-$[2^{-14}C]$ leucine (or L-$[\alpha^{-14}C]$-leucine), $[carboxy^{-14}C]$ leucine, $[Me^{-14}C]$ isoleucine, $[2,3^{-14}C]$ maleic anhydride, $[6,7^{-14}C]$ xanthopterin, $[3,4^{-13}C,^{35}S]$-methionine, $[2^{-13}C; 1^{-14}C]$ acetaldehyde, $[3^{-14}C; 2,3^{-2}H; ^{15}N]$ serine.

The same rules apply when the labeled compound is designated by a standard abbreviation or symbol, other than the atomic symbol, e.g. $[\gamma^{-32}P]$ ATP.

For simple molecules, however, it is often sufficient to indicate the labeling by writing the chemical formulae, e.g. $^{14}CO_2$, $H_2^{18}O$, ^2H_2O (not D_2O), $H_2^{35}SO_4$, with the prefix superscripts attached to the proper atomic symbols in the formulae. The square brackets are not to be used in these circumstances, nor when the isotopic symbol is attached to a word that is not a chemical name, abbreviation or symbol (e.g. ^{131}I-labeled).

---

*CEBJ consists of the Editors-in-Chief of the following journals: *Archives of Biochemistry and Biophysics, Biochemical Journal, Biochemistry, Biochimica et Biophysica Acta, Biochimie, European Journal of Biochemistry, Hoppe-Seyler's Zeitschrift für Physiologische Chemie, Journal of Biochemistry, Journal of Biological Chemistry, Journal of Molecular Biology, and Molekulyarnaya Biologiya;* corresponding members include *Proceedings of the National Academy of Sciences* (U.S.A.) and approximately 40 others.

DEFINITIVE RULES FOR THE NOMENCLATURE OF CAROTENOIDS*

**IUPAC Commission on the Nomenclature of Organic Chemistry
and IUPAC-IUB Commission on Biochemical Nomenclature**

Rule Carotenoid 1. Class of Compound

Carotenoids are a class of hydrocarbons (carotenes) and their oxygenated derivatives (xanthophylls) consisting of eight isoprenoid units joined in such a manner that the arrangement of isoprenoid units is reversed at the center of the molecule so that the two central methyl groups are in a 1,6 positional relationship and the remaining nonterminal methyl groups are in a 1,5 positional relationship. All carotenoids may be formally derived from the acyclic $C_{40}H_{56}$ structure (I below), having a long central chain of conjugated double bonds, by (i) hydrogenation, (ii) dehydrogenation, (iii) cyclization, or (iv) oxidation, or any combination of these processes.

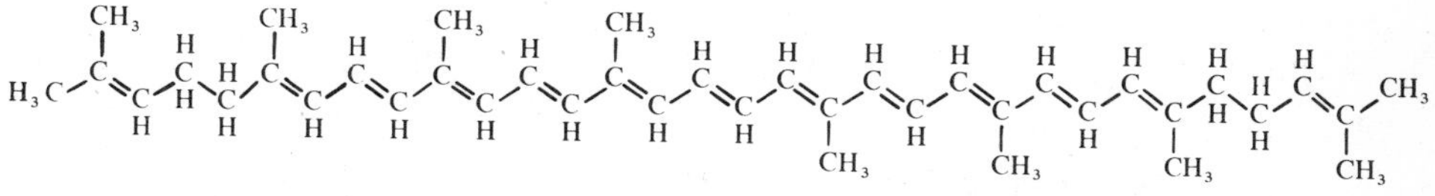

(I)

The class also includes certain compounds that arise from certain rearrangements of the carbon skeleton (I) or by the (formal) removal of part of this structure.

For convenience carotenoid formulas are often written in a shorthand form as

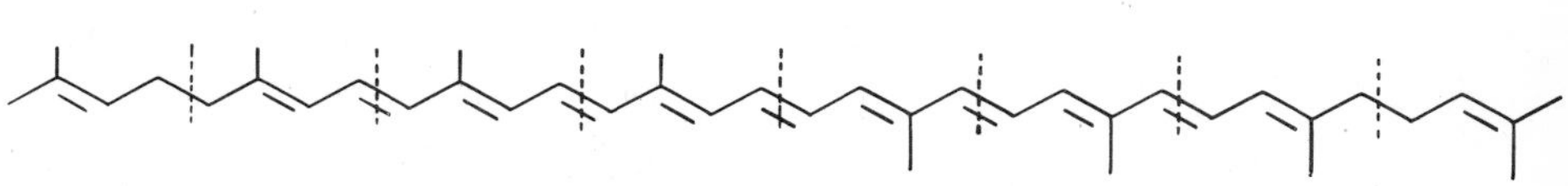

(IA)

Rule Carotenoid 2. The Stem Name

All specific names are based on the stem name "carotene," which corresponds to the structure and numbering shown in (II), where the broken lines at the two terminations are intended only to represent two "double-bond equivalents." Individual compounds may have C_9 acyclic end groups with two double bonds at positions 1,2 and 5,6 (e.g., III) or cyclic end groups (such as IV, V, VI, VII, VIII, and IX).

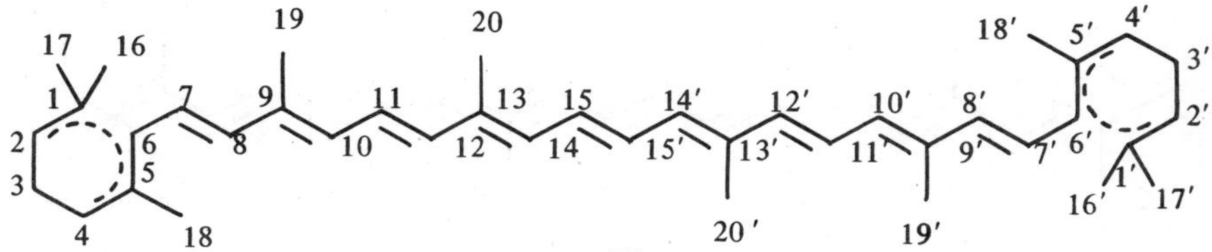

(II)

Rule Carotenoid 3. Specific Names; End-group Designations

3.1. The name of a specific carotenoid hydrocarbon is constructed by adding two Greek letters as prefixes to the stem name "carotene" (defined in Rule Carotenoid 2), these prefixes being characteristic of the two C_9 end groups.

*From IUPAC Commission on the Nomenclature of Organic Chemistry and IUPAC-IUB Commission on Biochemical Nomenclature, *Pure Appl. Chem.*, 41(3), (1975), in press. With permission.

3.2. The prefixes are

Type	Prefix	Formula	Structure
Acyclic	ψ	C_9H_{15}	III
Cyclohexene	β, ϵ	C_9H_{15}	IV, V
Methylenecyclohexane	γ	C_9H_{15}	VI
Cyclopentane	κ	C_9H_{17}	VII
Aryl	ϕ, χ	C_9H_{11}	VIII, IX

and correspond to the following end-group modifications

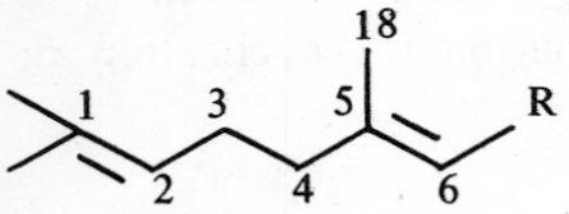

III; ψ IV; β V; ϵ VI; γ

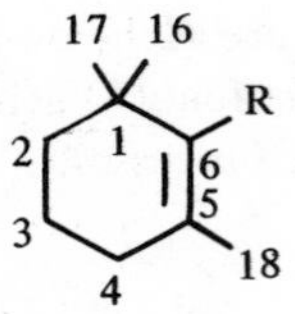

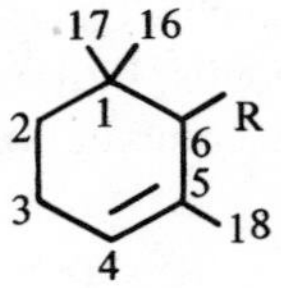

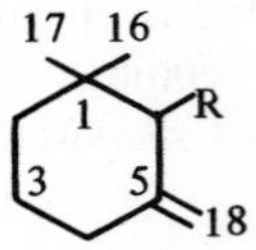

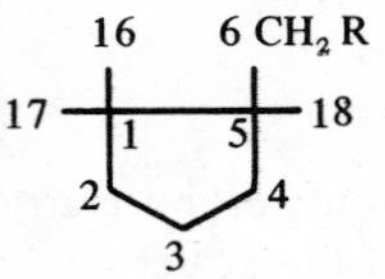

VII; κ VIII; ϕ IX; χ

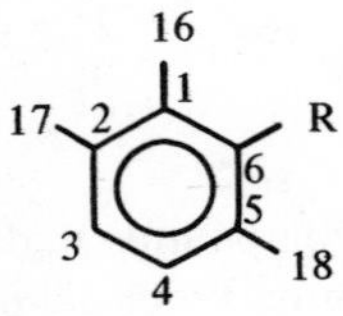

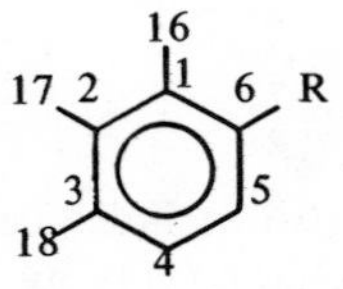

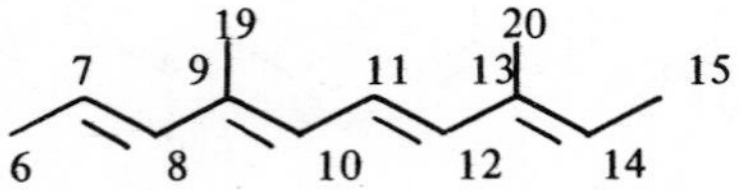

Note: The choice of locants 16 and 17 for the two methyl groups at C-1 is considered in connection with stereochemistry in Rule Carotenoid 12.4.

3.3. The Greek-letter prefixes are cited in alphabetical order; the first is separated from the second by a comma, and the second is connected to the stem name by a hyphen.

Note: The Greek-letter alphabetical order is β (beta), γ (gamma), ϵ (epsilon), κ (kappa), ϕ (phi), χ (chi), ψ (psi).

Examples:

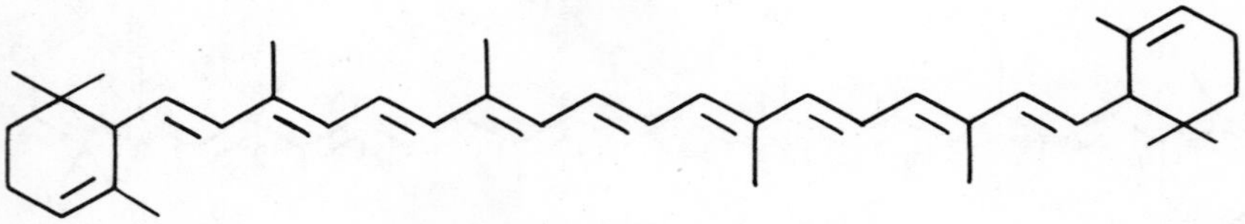

ϵ,ϵ-Carotene

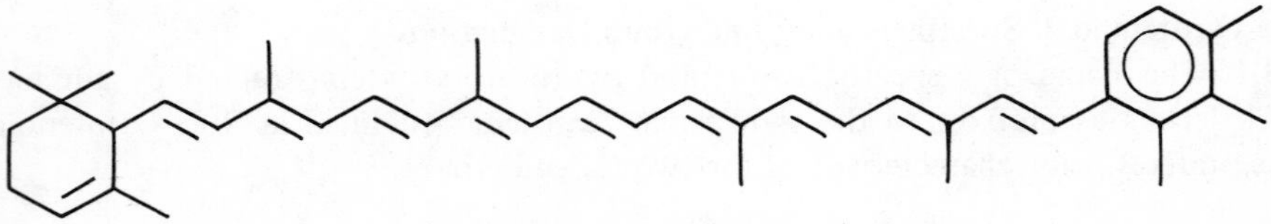

ϵ,χ-Carotene

Notes: (i) The Greek-letter prefixes are derived: β and ϵ from the symmetrical carotenoids with the trivial names "β-carotene" and ϵ-carotene," κ from the symmetrical capsorubin, γ and ψ from γ- and ψ-ionone, ϕ for phenyl, and χ, the next Greek letter after ϕ. (ii) α and δ, from the trivial names α- and δ-carotene, are not used.

3.4. When, in a modified or degraded carotenoid, an end group can be derived from more than one specific end group, the end group chosen as the basis of the name is that occurring earliest in the alphabetical order (see Rule Carotenoid 3.3).

Example:

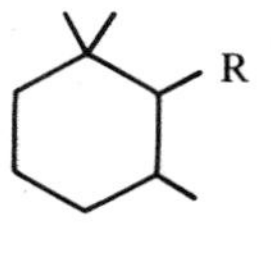

is a derivative of the β end group, not the ϵ end group.

Rule Carotenoid 4. Numbering of Carotenoid Hydrocarbons

The basic system of numbering is that shown in Structure (II) of Rule Carotenoid 2; the end groups are numbered as indicated in Rule Carotenoid 3. If the two end groups are dissimilar, lower (unprimed) numbers are given to that end of the molecule which is associated with the Greek-letter prefix cited first in the name. (All unprimed locants are cited before primed locants.)*

Examples:

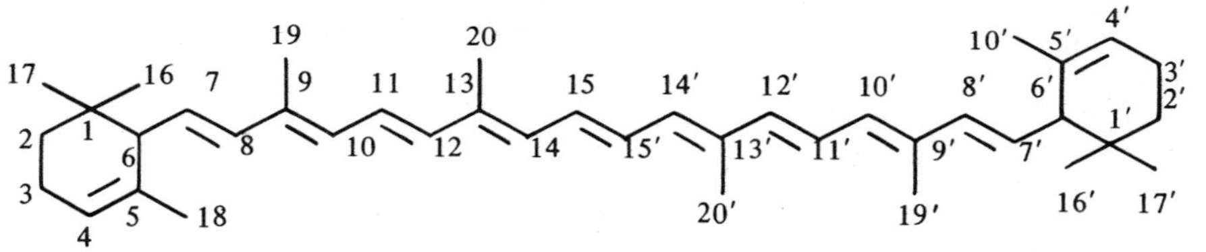

ϵ,ϵ-Carotene

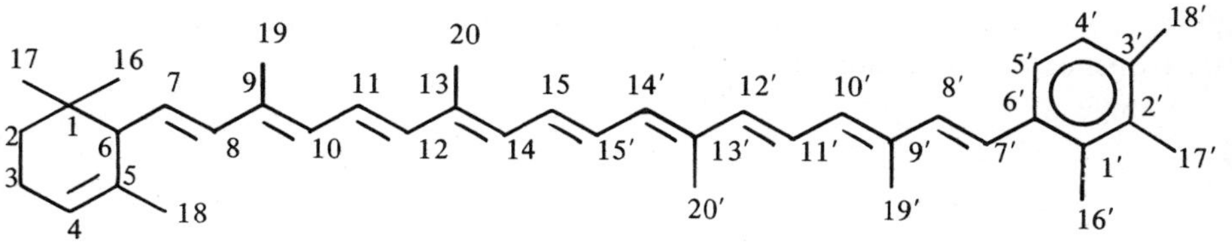

ϵ,χ-Carotene

Note: It is recommended that formulas be drawn so that unprimed numbers are on the left-hand side.

Rule Carotenoid 5. Nor Carotenoids and Seco Carotenoids

5.1. *Nor Carotenoids*. Elimination of a CH_3, CH_2, or CH group from a carotenoid is indicated by the prefix "nor," which in all cases is preceded by the locant of the carbon

---

*IUPAC Nomenclature of Organic Chemistry, [*Pure Appl. Chem.*, 11, No. 1 and 2 (1965); Sections A, B, and C published by Butterworths, London, 1971] indicates that, in ranking locants for priority, a primed numeral ranks immediately after the same numeral unprimed. In general organic nomenclature, the common practice is therefore to cite locants in the order $x,x'(x+n)$, $(x+n)'$. The established sequence in the carotenoid field is, however, to cite all unprimed numerals before any primed numerals, and this practice is followed in these rules.

atom that has been eliminated. The prefix is nondetachable. When alternatives are possible, the locant attached to nor is the lowest possible.* The basic numbering of the carotenoid is retained in the nor carotenoid.

Examples:

2,2'-Dinor-β,β-carotene

(12,13,20 formally "removed" at this point.)

12,13,20-Trinor-ββ-carotene

16,17,16',17'-Tetranor-ε,ε-carotene

5.2. *Seco Carotenoids*. Fission of the bond between two adjacent carbon atoms (other than carbon atoms 1 and 6) of a cyclic end group, with addition of one or more hydrogen atoms at each terminal group thus created, is indicated by the prefix "seco," the original carotenoid numbering being retained.

Example:

2,3-Seco-ε,ε-carotene

Rule Carotenoid 6. Changes in Hydrogenation Level

Carotenoid hydrocarbons, including carotenoid acetylenes and allenes, which differ in hydrogenation level from the corresponding carotenoid hydrocarbon defined by Rule Carotenoid 3, are named from the latter by use of the prefixes "hydro" and "dehydro" together with the locants specifying the carbon atoms at which hydrogen atoms have been added or removed.

*The contrast with steroid usage [IUPAC-IUB Definitive Rules for the Nomenclature of Steroids, *Pure Appl. Chem.*, 31, 285 (1972), Rule 2S-6, also in *Biochemistry*, 8, 2227 (1969), 10, 4994 (1971), and elsewhere] where the prefix "nor" is associated with the highest permissible number, is to be noted.

These prefixes are nondetachable[†] and immediately precede the Greek-letter prefixes denoting the end groups, and, if both occur in one name, are cited in the order: Dehydro *before* hydro (multipliers do not affect the order).

Example: Tetradehydrodihydro.

Note: Since, to maintain valency requirements, hydrogen atoms are, formally, always added or removed in pairs these prefixes will always be used with an even-number multiplier, e.g., tetrahydro, didehydro.

Examples:

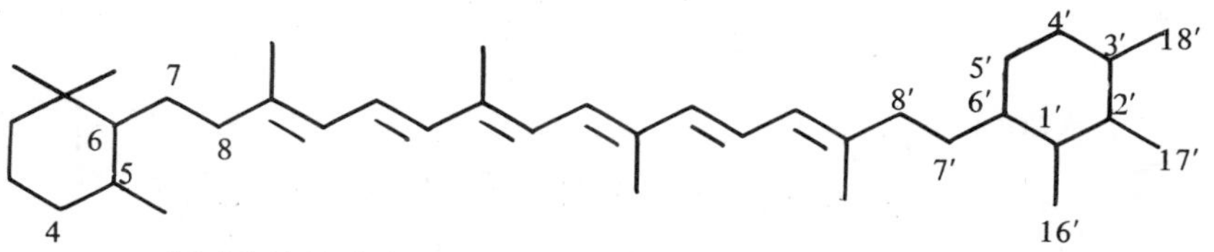

5,6,7,8,1',2',3',4',5',6',7',8'-Dodecahydro-β,χ-carotene

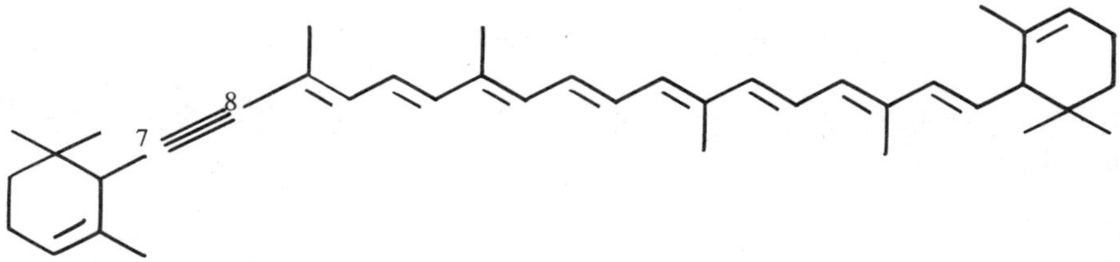

7,8-Didehydro-ε,ε-carotene

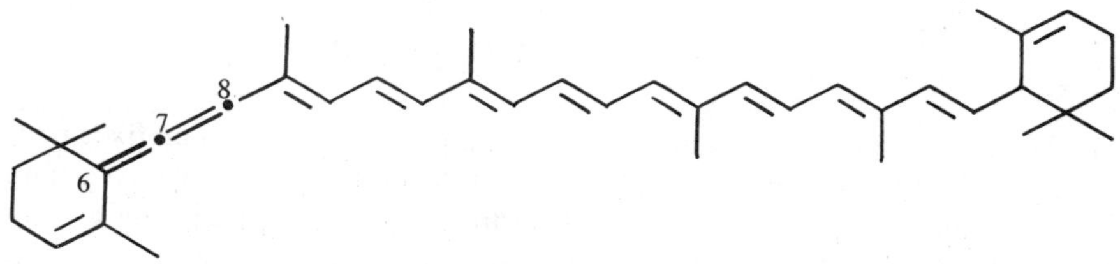

6,7-Didehydro-ε,ε-carotene

Rule Carotenoid 7. Oxygenated Derivatives

7.1. Oxygenated (and other) derivatives of carotenoid hydrocarbons are named by use of suffixes and prefixes according to the rules of general organic chemical nomenclature.[**]

Of the oxygen-containing characteristic groups present, that occurring earliest in the sequence carboxylic acid, ester of carotenoid acid, aldehyde, ketone, alcohol, and ester of carotenoid alcohol, is chosen as principal group (following IUPAC Rule C-10.3) and is cited by use of a suffix; all other groups are cited as prefixes.

[†]IUPAC Nomenclature of Organic Chemistry, *Pure Appl. Chem.,* 11, No. 1 and 2 (1965). Rule C-16.1, allows the prefix "hydro" to be detachable or nondetachable and the former has become the established usage in general organic chemistry. However, the common practice in carotenoid names is now to use this prefix as nondetachable, a practice that is followed in this set of rules.

[**]IUPAC Rules, Nomenclature of Organic Chemistry, *Pure Appl. Chem.,* 11, No. 1 and 2 (1965), Subsections C-2 to C-4.

Examples:

3-Hydroxy-β,ε-caroten-3'-one

3-Hydroxy-3'-oxo-β,ε-caroten-16-oic acid

7.2. A nonbridging ether group is named by use of the appropriate alkoxy or aryloxy prefix (Rule C-211.2).

1-Ethoxy-1,2,7',8'-tetrahydro-ψ,ψ-carotene

7.3. Oxygen bridges are indicated by use of the prefix "epoxy," this prefix is preceded by the locants of the two carbon atoms that form the bridgeheads of the oxygen bridge.

Note: The prefix "epoxy" denotes replacement, by an oxygen bridge, of a hydrogen atom at each of two carbon atoms already otherwise connected to one another. An epoxide, notionally formed by adding an oxygen atom to a double bond, is therefore an epoxydihydro derivative of the original compound.

Examples:

5,6:5',6'-Diepoxy-5,6,5',6'-tetrahydro-β,β-carotene-3,3'-diol

5,8:5',8'-Diepoxy-5,8,5'8'-tetrahydro-β,β-carotene

7.4. Compounds that may be formally derived from a carotenoid hydrocarbon by the addition of the elements of water (H, OH) or of methanol (H, OCH₃) to a double bond are named as "hydroxydihydro" or "methoxydihydro" derivatives.

Examples:

3,7'-Dihydroxy-7',8'-dihydro-β,ε-caroten-3'-one

7,7'-Dimethoxy-7,8,7',8'-tetrahydro-ε,ε-carotene-3,3'-dione

Rule Carotenoid 8. Numbering of Oxygenated Derivatives

8.1. If the two C₉ end groups of the parent carotenoid hydrocarbon are dissimilar, their oxygenated derivatives are numbered according to Rule Carotenoid 4, i.e., the end group designated by the Greek letter occurring earlier in the Greek alphabet receives unprimed locants.

Examples:

2,4-Dihydroxy-β,ε-caroten-3'-one

2,2',3'-Trimethoxy-β,ε-carotene

8.2. (a) If the two C₉ end groups of the parent carotenoid hydrocarbon are identical, then the lowest (*) locant possible is assigned to the principal group, cited as suffix.

(b) If more than one of the group chosen to be cited as suffix is present, the numbering is determined by the principle of lowest locants (**) applied to the suffixes.

(c) If no group qualifies to be cited as suffix, then the numbering is determined by the principle of lowest locants (*) for all groups cited as prefixes.

Examples:

2'-Methoxy-ε,ε-caroten-3-one
(not 2-Methoxy-ε,ε-caroten-3'-one)

ββ-caroten-19,11-olide
(The position of the lactone group is indicated following the principles of Rule C472.2)

2'-Methoxy-ε,ε-carotene-2,3'-dione [not
2-Methoxy-ε,ε-carotene,3,2'-dione; see paragraph (b)]

3,2'-Dimethoxy-1,2,5',6'-tetrahydro-ψ,ψ-carotene
[not 2,3'-Dimethoxy-1',2',5,6-tetrahydro-ψ,ψ-carotene,
1,2,3,2',5',6' having priority over 2,5,6,1',2',3'; see paragraph (c)]

Notes: (*) In the carotenoid series *all* unprimed numbers are cited *before* primed numbers and the former are therefore considered as "lower than" primed numbers, e.g., 2,6,6,1',2',6', etc. (**) When series of locants containing the same number of terms are compared term by term, that series is "lowest" which contains the lowest number on the occasion of the first difference [IUPAC Rule C-13.11(e), footnote].

Rule Carotenoid 9. *retro* Nomenclature

9.1. The prefix "*retro*" (printed in italics) and a pair of locants are used to indicate a shift, by one position, of *all* single and double bonds of the conjugated polyene system delineated by the pair of locants.

9.2. The pair of locants precedes the prefix *retro*. The first locant is that of the carbon atom that has lost a proton; the second is that of the carbon atom that has gained one.

9.3. The prefix and its accompanying locants are placed immediately before, and hyphenated to, the Greek-letter prefixes of the name defined according to Rule Carotenoid 3.

9.4. The prefix and its associated locants are not detachable from the names defined according to Rule Carotenoid 3.

Examples:

3-Hydroxy-6′,7-*retro*-β,ε-caroten-3′-one

6′,7-*retro*-ε,κ-Caroten-3-one

8′,11-*retro*-β,ψ-Caroten-11-one

Rule Carotenoid 10. Apo Nomenclature

It is often necessary to designate derivatives in which the carbon skeleton has been shortened by the formal removal of fragments from one or both ends of a carotenoid.

10.1. The unitalicized prefix "apo," preceded by a locant, is used to indicate that all of the molecule beyond the carbon atom corresponding to that locant has been replaced by hydrogen atoms. A side-chain methyl group is not considered to be "beyond" the carbon atom to which it is attached.

10.2. The prefix and its locant immediately precede the specific name (Rule Carotenoid 3) unless the locant associated with the prefix "apo" is greater than five, in which case there is no need to give a Greek-letter end-group designation for that end of the molecule.

10.3. For purposes of numbering, etc., an end that has been shortened by five or less skeletal carbon atoms is considered a ψ (acyclic) end group.

10.4. The prefix "diapo," preceded by two locants, is used to indicate removal of fragments from both ends of the molecule.

10.5. If, in a diapo compound, the two ends of the carotenoid skeleton have been shortened unequally, the lower locant associated with the prefix "diapo" is unprimed.

Examples:

2′-Apo-β,ψ-caroten-2′-al

Methyl hydrogen 6,6'-diapocarotene-6,6'-dioate (*trans*-Bixin)

Rule Carotenoid 11. Higher Carotenoids

The higher carotenoids are a class of hydrocarbons and their oxygenated derivatives consisting of *more than* eight isoprenoid units joined in a manner similar to that of the C_{40} carotenoids. They are named as mono- or disubstituted C_{40} carotenoids. The numbering of the normal carotenoid is retained.

Examples:

2-(3-Methyl-2-butenyl)-ϵ,ψ-carotene

2,2'-Bis(3-methylbutyl)-β,β-carotene-3'',3'''-diol
or
2,2'-Bis(3-hydroxy-3-methylbutyl)-β,β-carotene

Rule Carotenoid 12. Stereochemistry

12.1. Absolute Configuration at Chiral Centers. The absolute configuration at chiral centers is designated by use of the *RS* convention,* the symbols being placed, with the corresponding locants, before the carotenoid name.

Example:

(3*S*,5*R*,3'*S*,5'*R*)-3,3'-Dihydroxy-κ,κ-carotene-6,6'-dione
(Capsorubin)

---

*For a discussion on the *RS* convention and the use of thickened lines, broken lines, wavy lines, and wedges in displayed formulae, see IUPAC Tentative Rules for the Nomenclature of Organic Chemistry, Section E, Fundamental Stereochemistry, *IUPAC Inf. Bull.*, No. 35, p. 68; also published in *J. Org. Chem.*, 35, 2849 (1970), and elsewhere.

12.2. Absolute Configuration of Allenic Compounds. The absolute configuration around allene groups will be similarly designated, when known.

(3'S,5'R,6'R)-3'-Acetoxy-5,6-epoxy-3,5'-dihydroxy-6',7'-didehydro-5,6,7,8,5',6',-hexahydro-β,β-caroten-8-one (Fucoxanthin)

12.3. Geometrical Configuration around Double Bonds. The stem name "carotene" implies trans configuration about all double bonds unless the contrary is indicated. Following the designation of absolute configuration (if any), geometrical configuration is indicated by citing the double bond or bonds with a *cis* configuration.

Example:

Natural phytoene = 15-*cis*-7,8,11,12, 7',8',11',12'-
octahydro-ψ,ψ-carotene

At trisubstituted double bonds the term *cis* refers to the relative position of the two substituents forming parts of the main chain of carbon atoms.

Example:

Natural bixin = methyl hydrogen 9'-*cis*-6,6'-diapocarotene-6,6'-dioate.

In the absence of definite information on geometrical configuration, *cis* isomers may be distinguished by prefixes such as neo A, neo U, etc., (cf. Zechmeister, *Cis-Trans Isomeric Carotenoids, Vitamins A and Arylpolyenes,* Springer-Verlag, Vienna, 1962).

The stereochemical prefixes *E* and *Z* [see footnote to Rule 12.1 and *J. Am. Chem. Soc.,* 90, 509 (1968)] may be used, especially when the prefixes *cis* and *trans* might lead to ambiguity.

12.4. Numbering of gem-Dimethyl Groups at C-1. In an end group of β, γ, or ε type, the two methyl groups attached to C-1 are distinguished as follows: When the potential chirality is as shown in formula A, i.e., with the polyene chain (R) to the right of C-1, the methyl groups *below* and *above* the plane of the paper are numbered 16 and 17, respectively; if, with the polyene chain (R) to the left, the end group is as shown in B[1], then these designations are unaltered; if the end group is as shown in B[2], then they are reversed.

(A)   (B^1)   (B^2)

In an acyclic end group, the methyl groups that is *trans* to the main skeletal chain is numbered 16 and the methyl group that is *cis* is numbered 17, as shown in C.

(C)

Rule Carotenoid 13. Trivial Names

The preceding rules are designed to define precisely the structure of a given carotenoid by its name. Use of the semisystematic names derived from these rules will greatly assist communication between scientists and enable work to be more readily retrieved from the literature.

The appendix contains a list of some trival names currently in use for naturally occurring carotenoids; it also gives their semisystematic names and structures. These trival names will be of value in natural-product and biochemical work, but their use in systematic organic work should be restricted. If trivial names are used in a paper, the semisystematic name should always be given, in parentheses or in a footnote at the first mention.

While the need to coin a new trivial name must occasionally arise (e.g., because the structure of the compound is unknown), the list in the appendix should not be unnecessarily enlarged. Simple derivatives of known carotenoids should not be given new trivial names; they may, however, be named by modification of existing trivial names (cf. Steroid Rule 2S4.2, Note),* but the full systematic name according to these Rules should be given for each compound at the first mention in each paper.

*IUPAC-IUB Definitive Rules for the Nomenclature of Steroids, *Pure Appl. Chem.*, 31, 285 (1972), Rule 2S-6; also in *Biochemistry*, 8, 2227 (1969), 10, 4994 (1971), and elsewhere.

APPENDIX

SOME NATURALLY OCCURRING CAROTENOIDS HAVING TRIVIAL NAMES[a]

Hydrocarbons

Torulene

3',4'-Didehydro-β,ψ-carotene

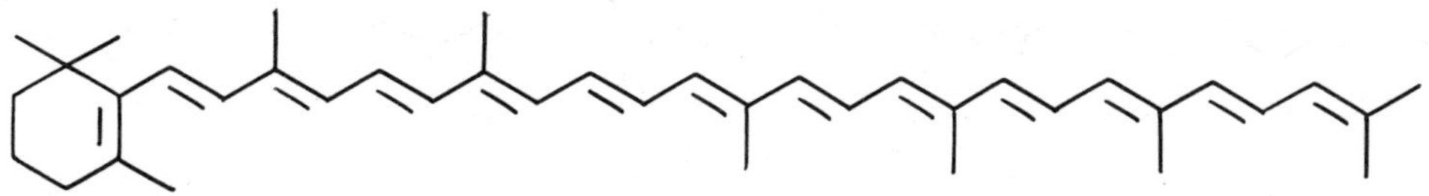

α-Zeacarotene

7',8'-Dihydro-ε,ψ-carotene

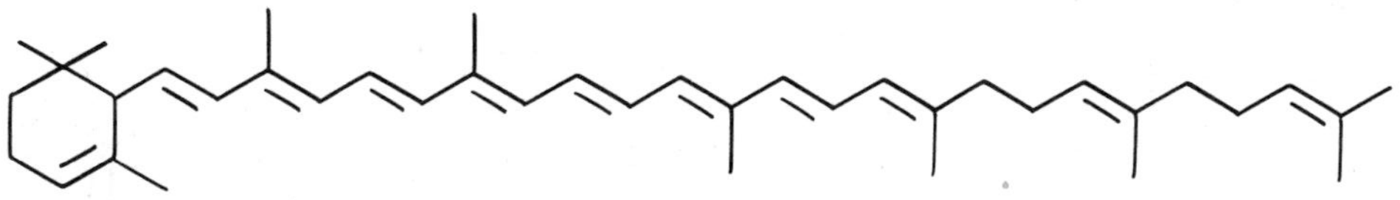

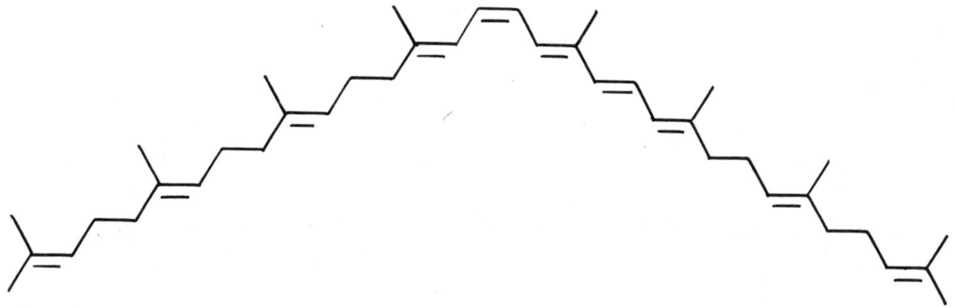

Phytofluene
Hexahydrolycopene

15-cis-7,8,11,12,7',8'-Hexahydro-ψ,ψ-carotene

[a]The examples in this appendix are based on an exhaustive list of naturally occurring carotenoids prepared by Dr. O. Straub. This list forms Chapter XII of the book, *Carotenoids*, Isler, O., Ed., Birkhäuser, Basel and Stuttgart, 1971. The two commissions are greatly indebted to Dr. Isler, Dr. Straub, and their colleagues of F. Hoffman-La Roche Co. Ltd., Basel, for providing this list of compounds. The list includes a few compounds whose natural occurrence or structure is doubtful, but whose systematic names illustrate particular points of nomenclature.

SOME NATURALLY OCCURRING CAROTENOIDS HAVING TRIVIAL NAMES (continued)

Hydrocarbons (continued)

Lycopersene

7,8,11,12,15,7',8',11',12',15'-
Decahydro-ψ,ψ-carotene

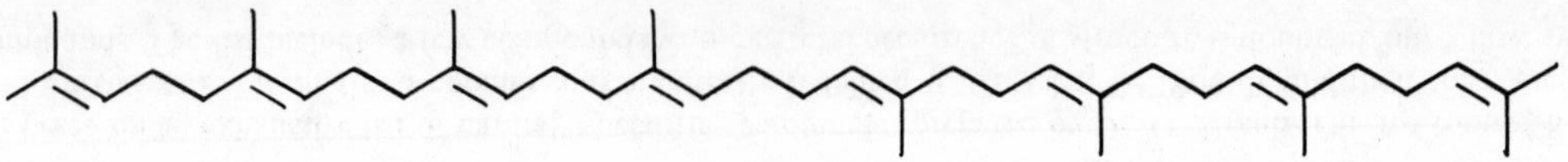

Alcohols

Gazaniaxanthin

(3R)-5'-*cis*-β,ψ-Caroten-3-ol

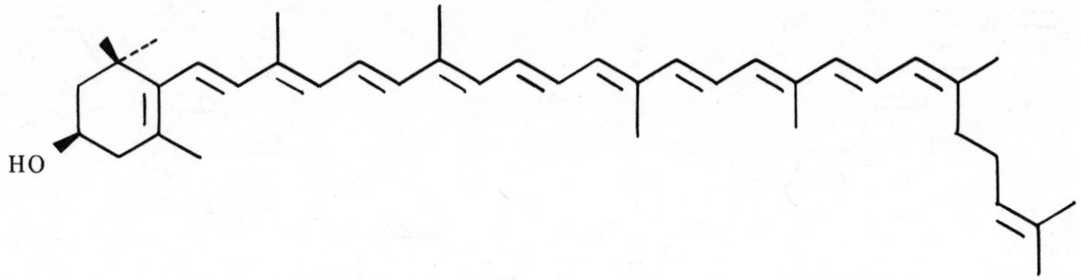

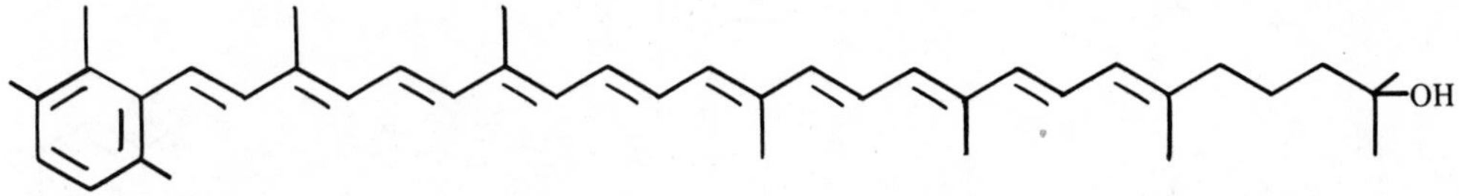

OH-Chlorobactene

1',2'-Dihydro-φ,ψ-caroten-1'-ol

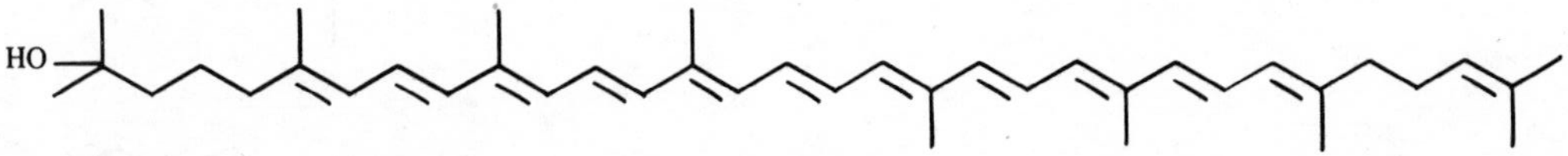

Rhodopin

1,2-Dihydro-ψ,ψ-caroten-1-ol

Lycoxanthin

ψ,ψ-Caroten-16-ol

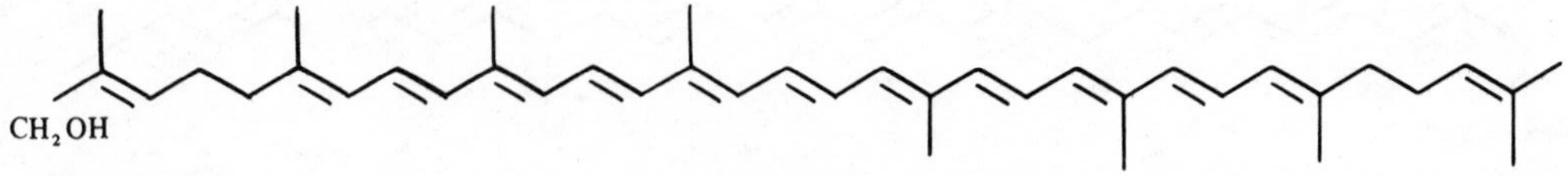

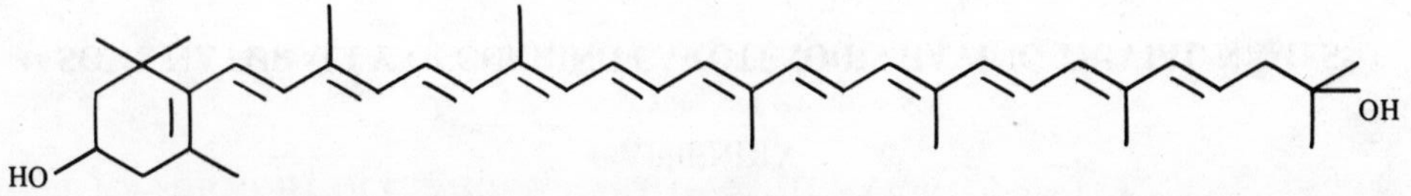

Saproxanthin

3',4'-Didehydro-1',2'-dihydro-
β,ψ-carotene-3,1'-diol

SOME NATURALLY OCCURRING CAROTENOIDS HAVING TRIVIAL NAMES (continued)

Alcohols (continued)

13-*cis*-1,2-Dihydro-ψ,ψ-carotene-1,20-diol

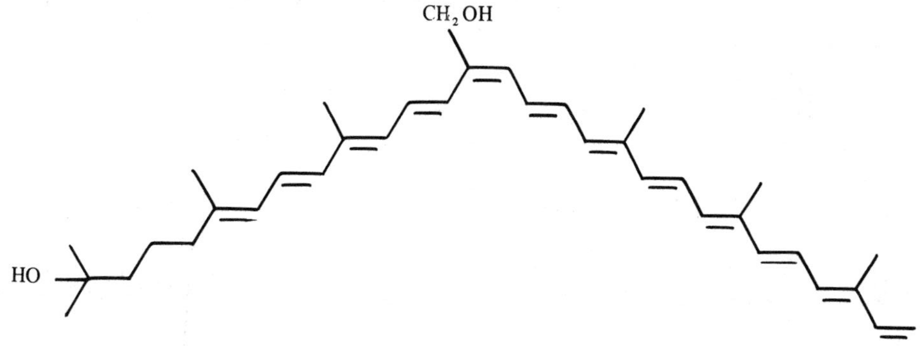

Rhodopinol
Warmingol

(3*R*,3'*R*)-7,8,7',8'-Tetradehydro-β,β-carotene-3,3'-diol

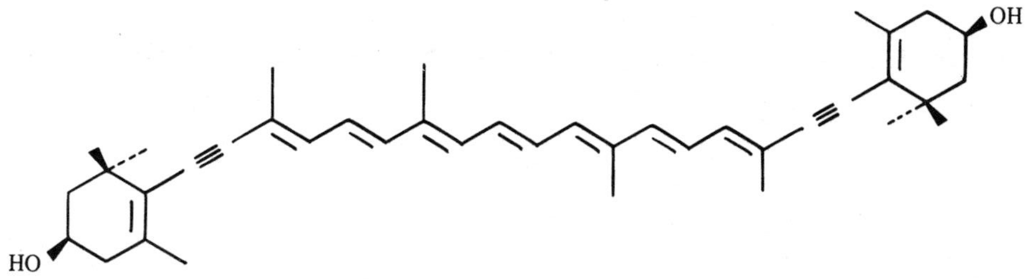

Alloxanthin
Cynthiaxanthin
Pectenoxanthin
Cryptomonaxanthin

β,β-Carotene-3,4,3',4'-tetrol

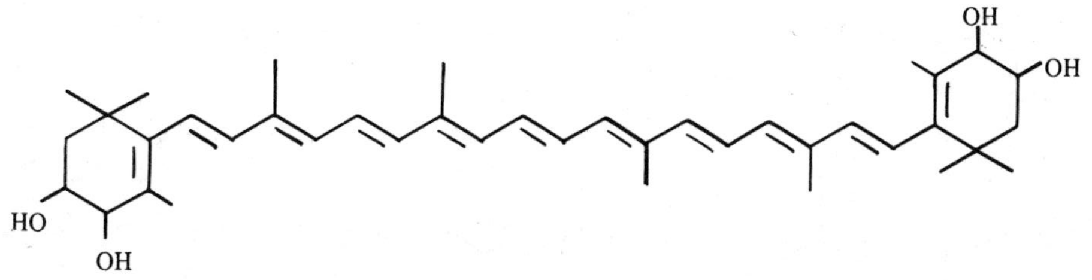

Crustaxanthin

β,ε-Carotene-3,19,3'-triol

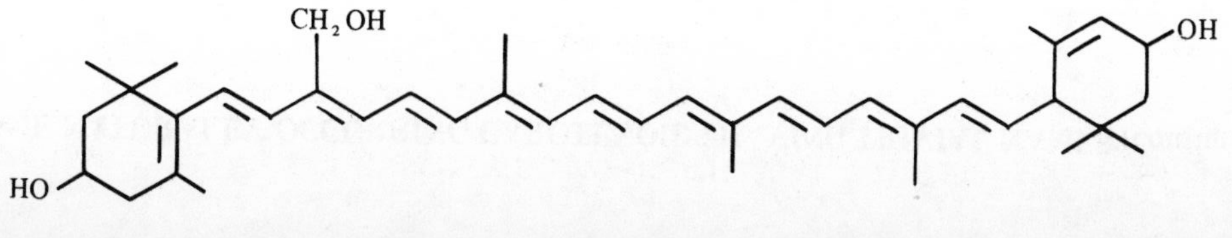

Loroxanthin

SOME NATURALLY OCCURRING CAROTENOIDS HAVING TRIVIAL NAMES (continued)

Glycosides

Phleixanthophyll

Oscillaxanthin

Ethers

Spheroidene

Rhodovibrin

Epoxides

Diadinoxanthin

1'-(β-D-Glucopyranosyloxy)-
3',4'-didehydro-1',2'-dihydro-
β,ψ-caroten-2'-ol

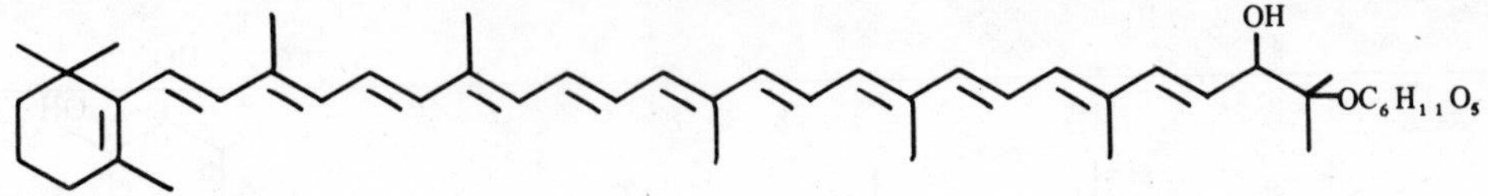

2,2'-Bis(β-L-rhamnopyranosyloxy)-
3,4,3',4'-tetradehydro-1,2,1',2'-
tetrahydro-ψ,ψ-carotene-1,1'-diol

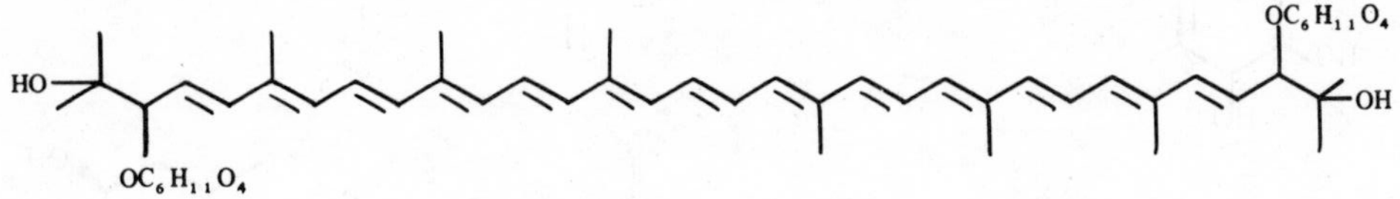

1-Methoxy-3,4-didehydro-
1,2,7',8'-tetrahydro-ψ,ψ-carotene

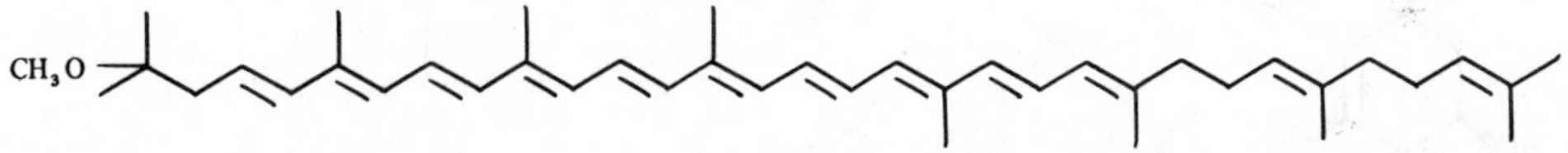

1'-Methoxy-3',4'-didehydro-
1,2,1',2'-tetrahydro-ψ,ψ-caroten-
1-ol

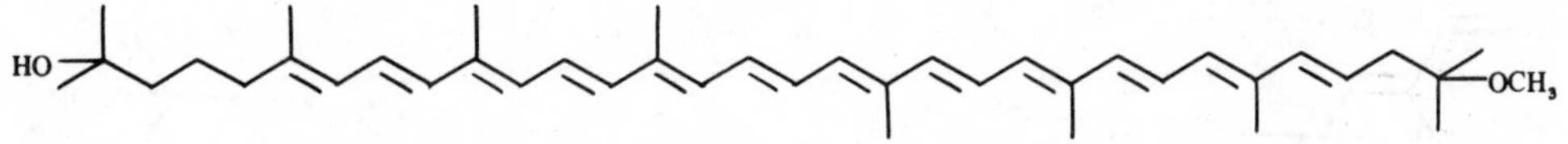

5,6-Epoxy-7',8'-didehydro-
5,6-dihydro-β,β-carotene-
3,3'-diol

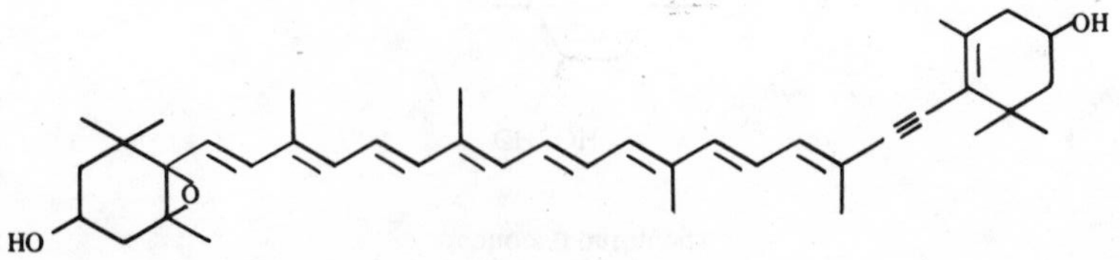

SOME NATURALLY OCCURRING CAROTENOIDS HAVING TRIVIAL NAMES (continued)

Epoxides (continued)

Vaucheriaxanthin

5',6'-Epoxy-6,7-didehydro-
5,6,5',6'-tetrahydro-β,β-caro-
tene-3,5,19,3'-tetrol

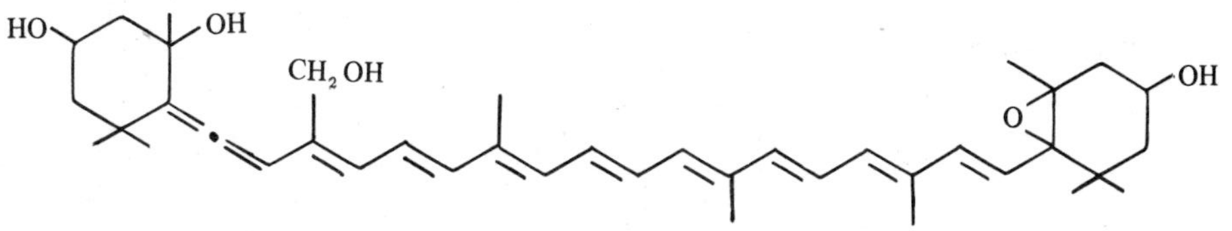

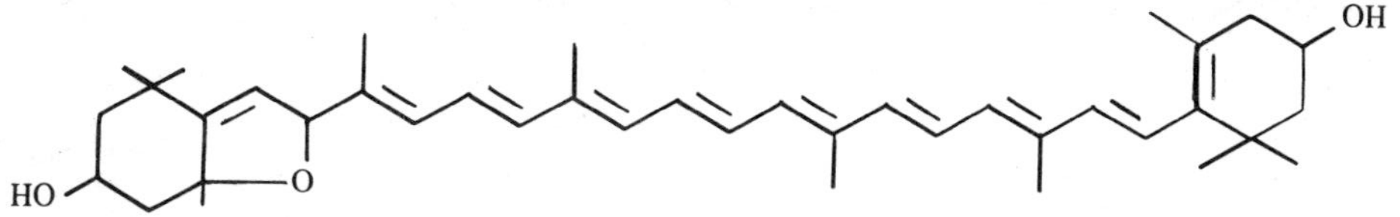

Mutatoxanthin
Citroxanthin
Zeaxanthin furanoxide

5,8-Epoxy-5,8-dihydro-β,β-caro-
tene-3,3'-diol

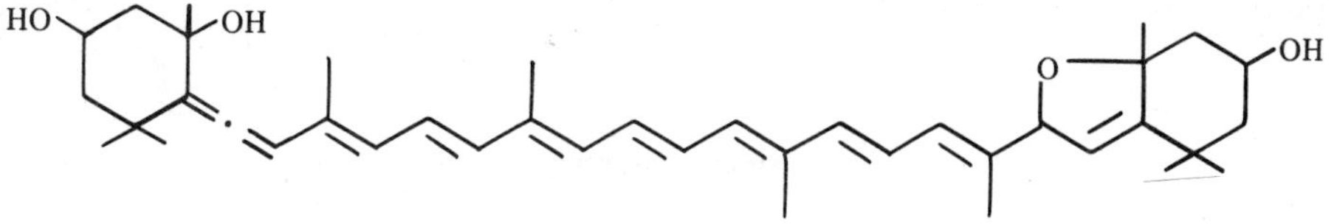

Neochrome
Foliachrome
Trollichrome

5',8'-Epoxy-6,7-didehydro-
5,6,5',8'-tetrahydro-β,β-
carotene-3,5,3'-triol

Luteoxanthin

5,6:5',8'-Diepoxy-5,6,5',8'-tet-
rahydro-β,β-carotene-3,3'-diol

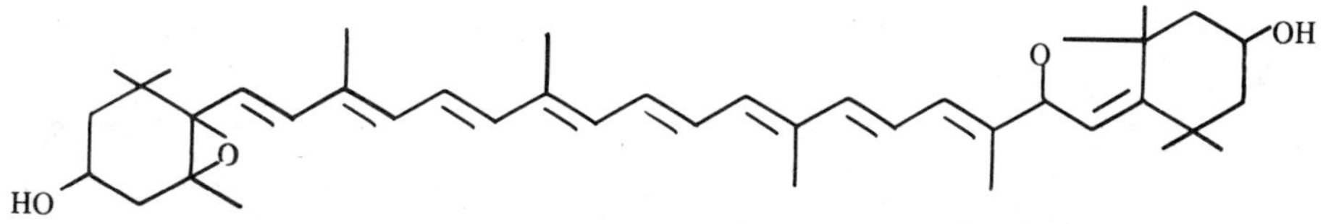

Aldehydes

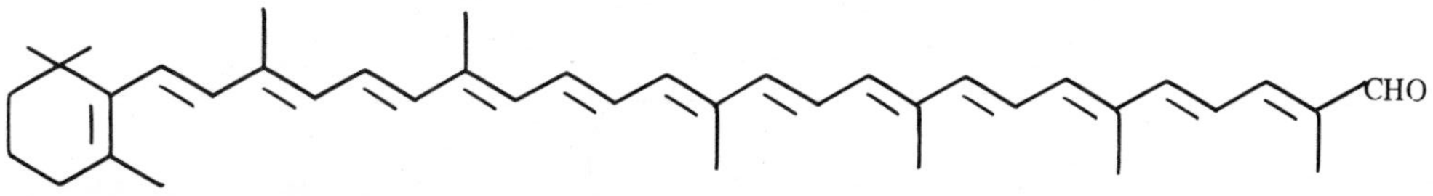

Torularhodinaldehyde

3',4'-Didehydro-β,ψ-caroten-
16-al

SOME NATURALLY OCCURRING CAROTENOIDS HAVING TRIVIAL NAMES (continued)

Aldehydes (continued)

13-*cis*-1-Hydroxy-1,2-dihydro-ψ,ψ-caroten-20-al

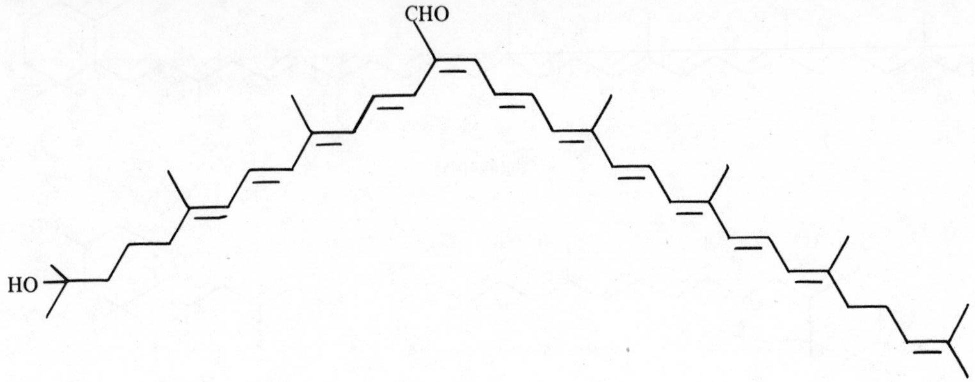

Rhodopinal
Warmingone

Acids and Acid Esters

3′,4′-Didehydro-β,ψ-caroten-16′-oic acid

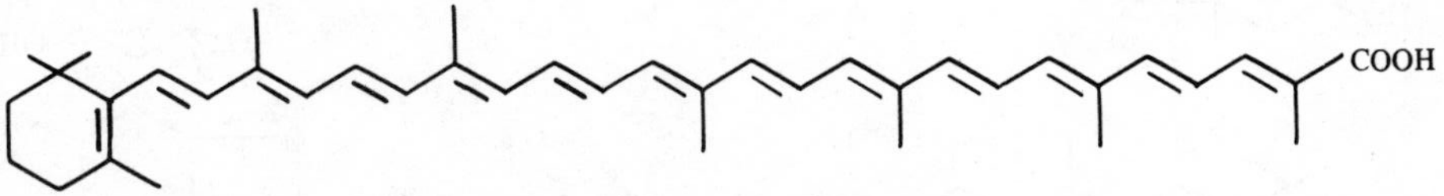

Torularhodin

Methyl 3′,4′-didehydro-β,ψ-carotein-16′-oate

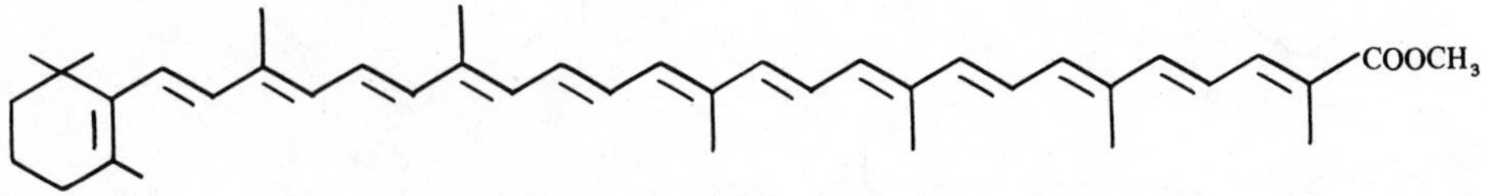

Torularhodin methyl ester

Ketones

β,ε-Caroten-4-one

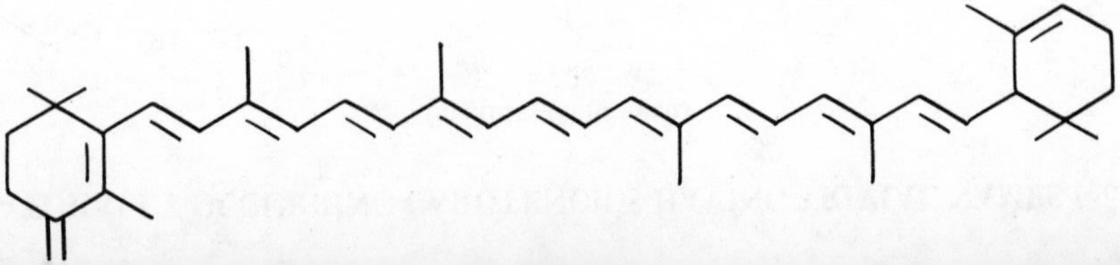

Phoenicopterone

SOME NATURALLY OCCURRING CAROTENOIDS HAVING TRIVIAL NAMES (continued)

Ketones (continued)

β,β-Carotene-4,4'-dione

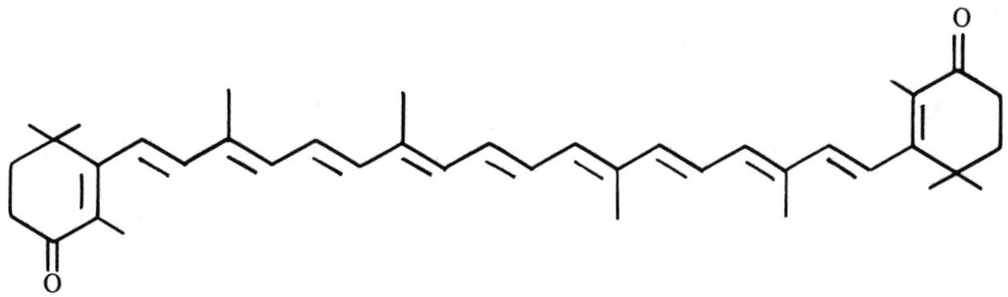

Canthaxanthin
Aphanicin
Chlorellaxanthin

3-Hydroxy-β,ψ-caroten-4'-one

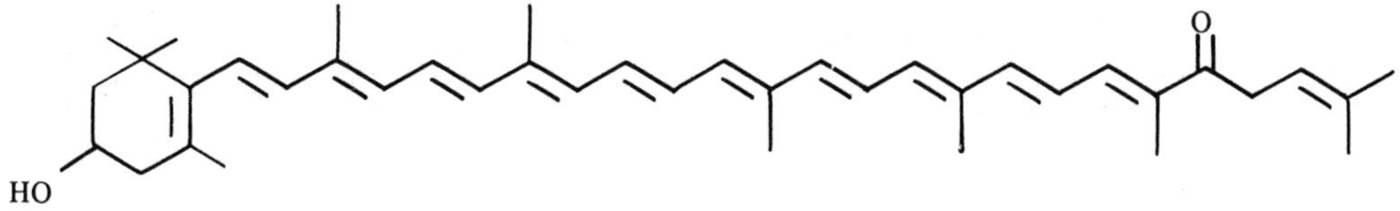

Rubixanthone

1'-Methoxy-1',2'-dihydro-
χ,ψ-caroten-4'-one

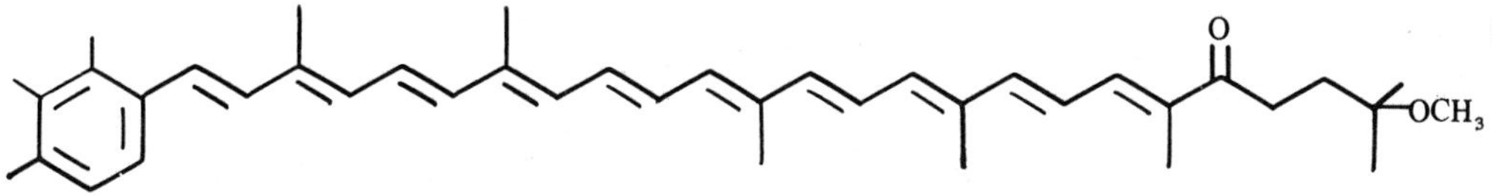

Okenone

1'-Hydroxy-1-methoxy-
3,4-didehydro-1,2-
1',2',7',8'-hexahydro-ψ,ψ-
caroten-2-one

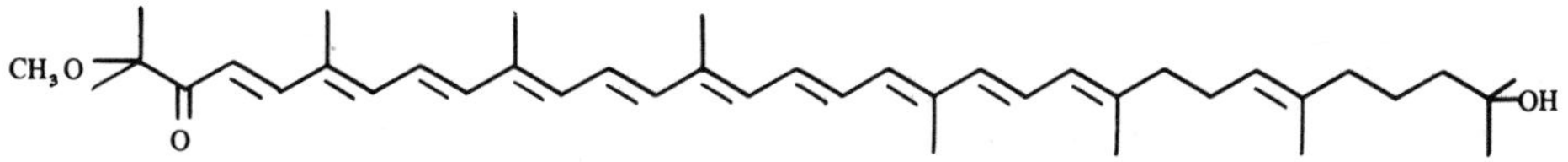

Hydroxyspheriodenone

1,1'-Dimethoxy-3,4,3',4'-
tetradehydro-1,2,1',2'-tetrahydro-
ψ,ψ-carotene-2,2'-dione

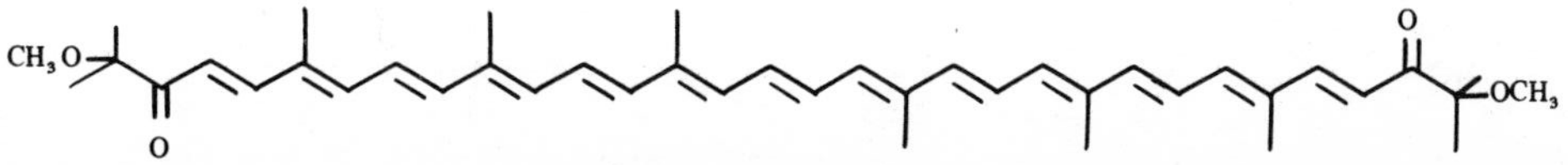

2,2'-Diketospirilloxanthin

SOME NATURALLY OCCURRING CAROTENOIDS HAVING TRIVIAL NAMES (continued)

Ketones (continued)

Siphonaxanthin

3,19,3'-Trihydroxy-7,8-
dihydro-β,ε-caroten-8-one

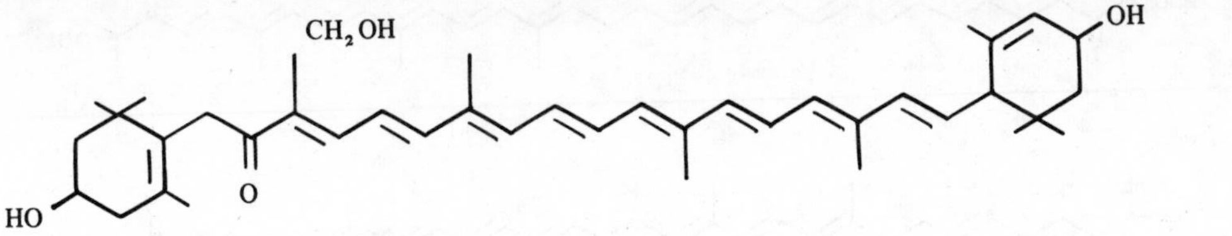

Flexixanthin

3,1'-Dihydroxy-3',4'-didehydro-
1',2'-dihydro-β,ψ-caroten-4-one

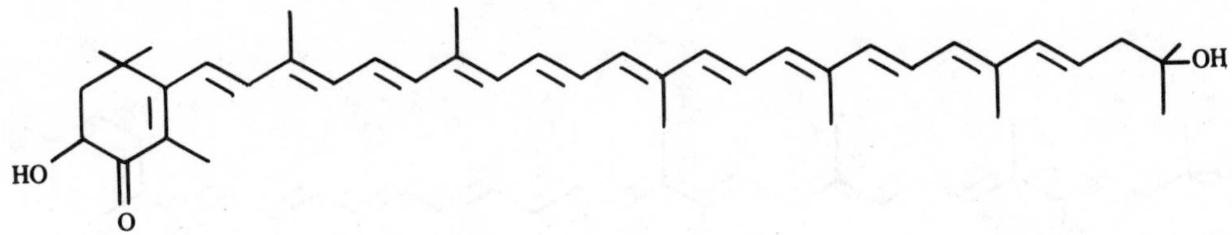

Cryptocapsin

(3'S,5'R)-3'-Hydroxy-β,κ-
caroten-6'-one

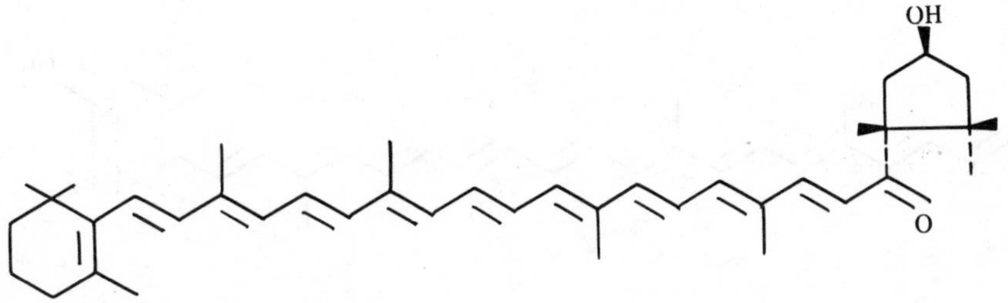

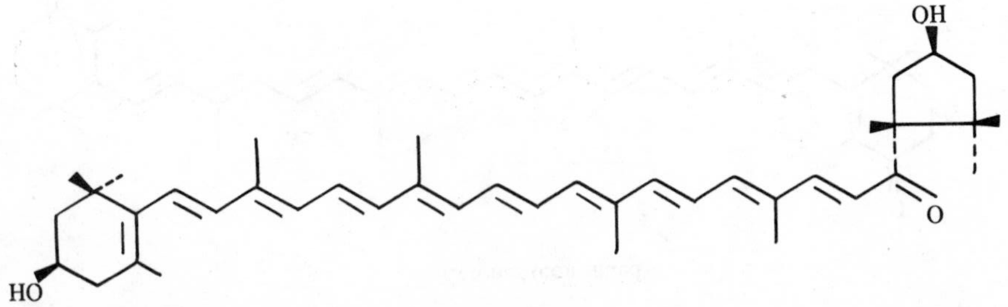

Capsanthin

(3R,3'S,5'R)-3,3'-Dihydroxy-
β,κ-caroten-6'-one

SOME NATURALLY OCCURRING CAROTENOIDS HAVING TRIVIAL NAMES (continued)

Ketones (continued)

Capsorubin

(3S,5R,3'S,5'R)-3,3'-Dihydroxy-
κ,κ-carotene-6,6'-dione

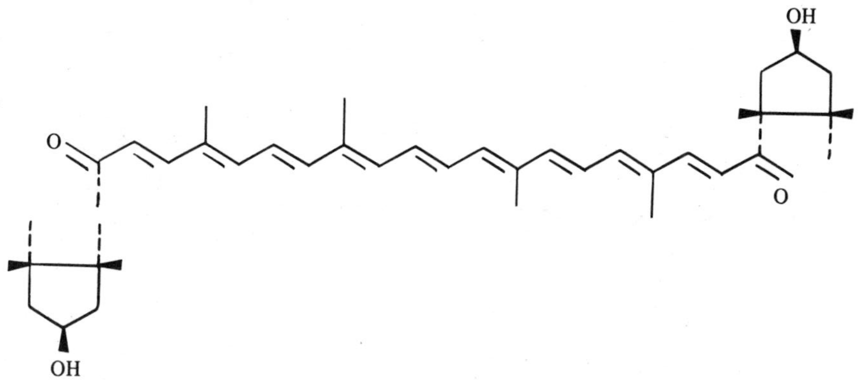

Pectenolone

3,3'-Dihydroxy-7',8'-didehydro-
β,β-caroten-4-one

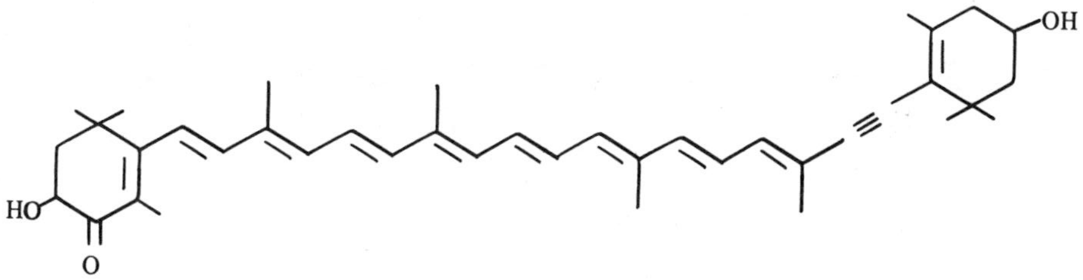

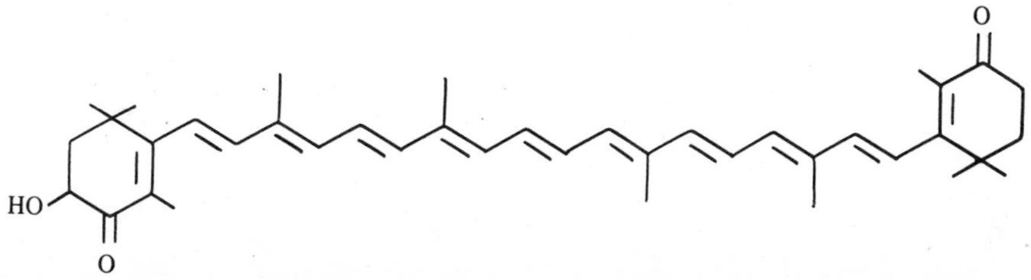

Adonirubin
Phoenicoxanthin
3-OH-Canthaxanthin

3-Hydroxy-β,β-carotene-4,4'-dione

SOME NATURALLY OCCURRING CAROTENOIDS HAVING TRIVIAL NAMES (continued)

Ketones (continued)

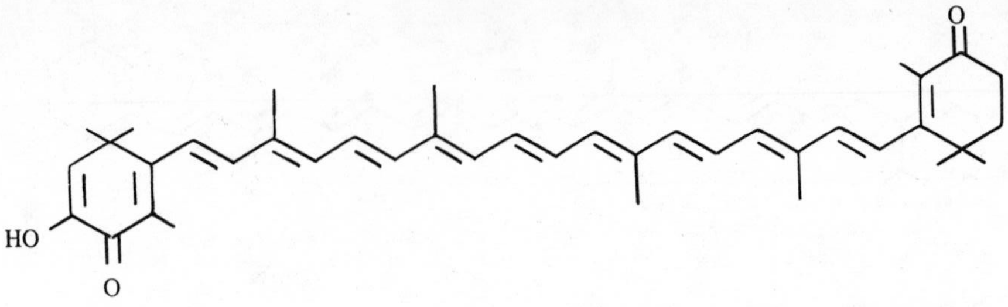

Phoeniconone
Dehydroadonirubin

3-Hydroxy-2,3-didehydro-β,β-
carotene-4,4'-dione

Esters of Alcohols

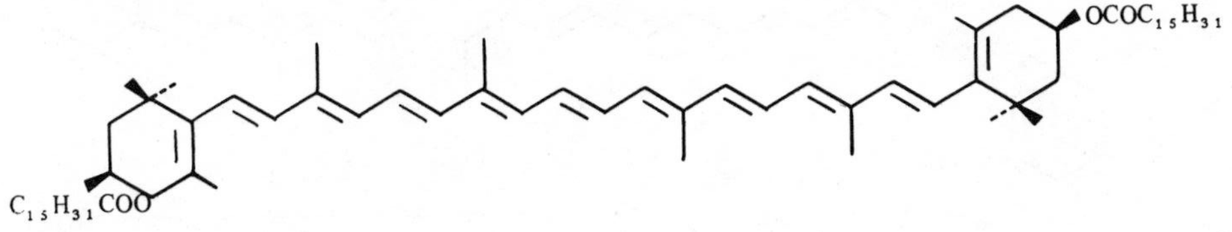

Zeaxanthin dipalmitate
Physalien

(3R,3'R)-3,3'-Bispalmitoyloxy-
β,β-carotene or (3R,3'R)-β,β-
carotene-3,3'-diol dipalmitate

Astacein

3,3'-Bispalmitoyloxy-2,3,2',3'-
tetradehydro-β,β-carotene-
4,4'-dione or 3,3'-dihydroxy-
2,3,2',3'-tetradehydro-β,β-caro-
tene-4,4'-dione dipalmitate

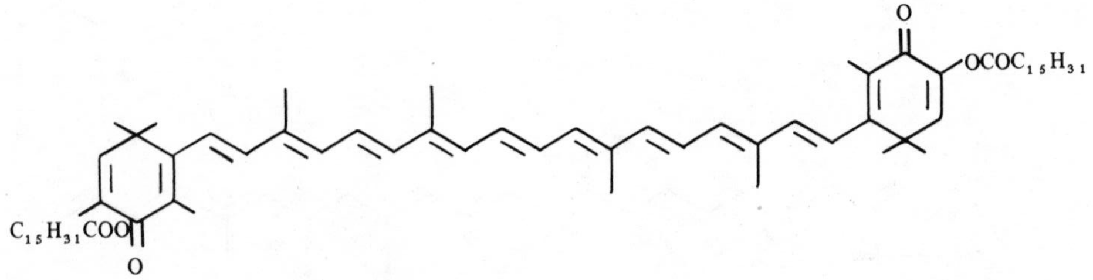

Siphonein

3,3'-Dihydroxy-19-lauroyloxy-
7,8-dihydro-β,ε-caroten-8-one
or 3,19,3'-trihydroxy-7,8-
dihydro-β,ε-caroten-8-one 19-
laurate

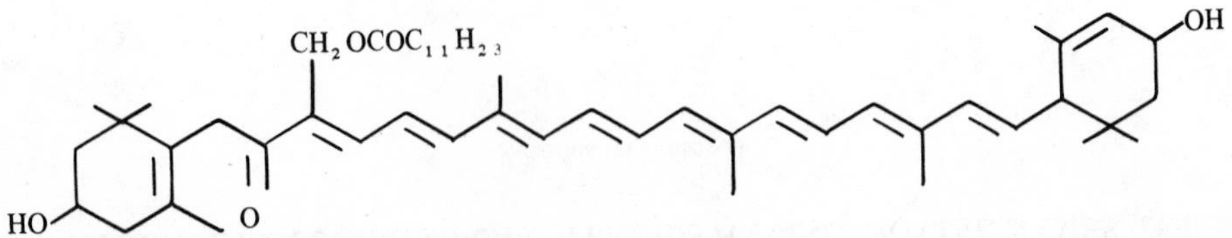

SOME NATURALLY OCCURRING CAROTENOIDS HAVING TRIVIAL NAMES (continued)

Esters of Alcohols (continued)

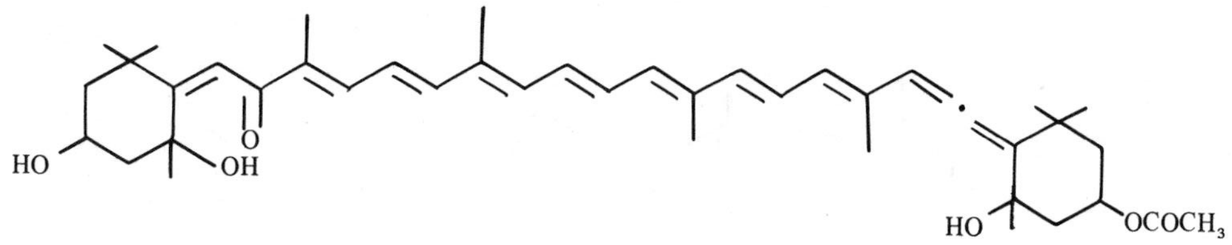

Isofucoxanthin

3'-Acetoxy-3,5,5'-trihydroxy-
6',7'-didehydro-5,8,5',6'-
tetrahydro-β,β-caroten-8-one

Fucoxanthin

3'-Acetoxy-5,6-epoxy-3,5'-
dihydroxy-6',7'-didehydro-
5,6,7,8,5',6'-hexahydro-β,β-
caroten-8-one

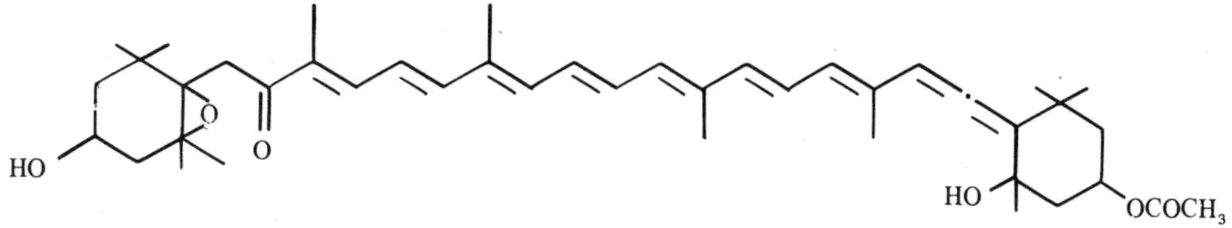

Apo Carotenoids

β-Apo-2'-carotenal

3',4'-Didehydro-2'-apo-β-
caroten-2'-al

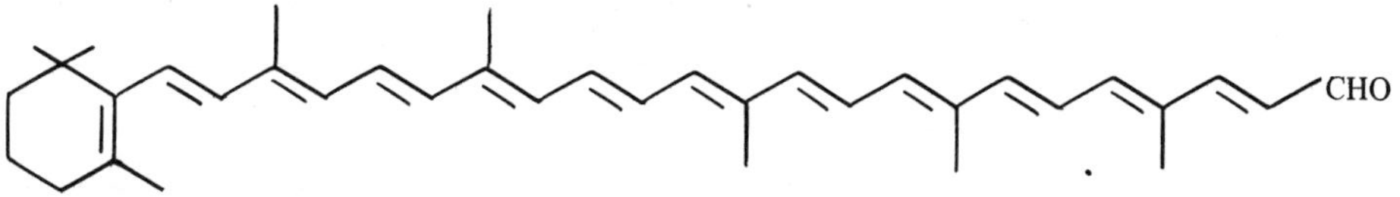

Azafrinaldehyde

5,6-Dihydroxy-5,6-dihydro-10'-
apo-β-caroten-10'-al

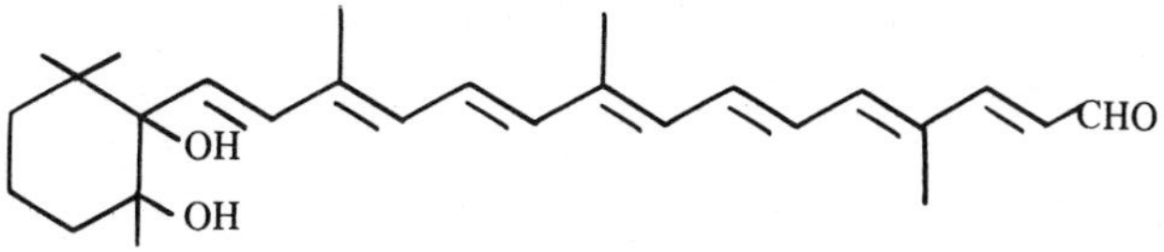

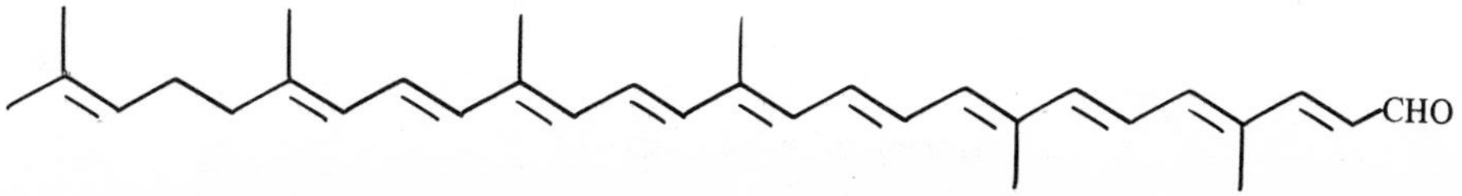

Apo-2-lycopenal
Apo-6'-lycopenal

6'-Apo-ψ-caroten-6'-al

SOME NATURALLY OCCURRING CAROTENOIDS HAVING TRIVIAL NAMES (continued)

Apo Carotenoids (continued)

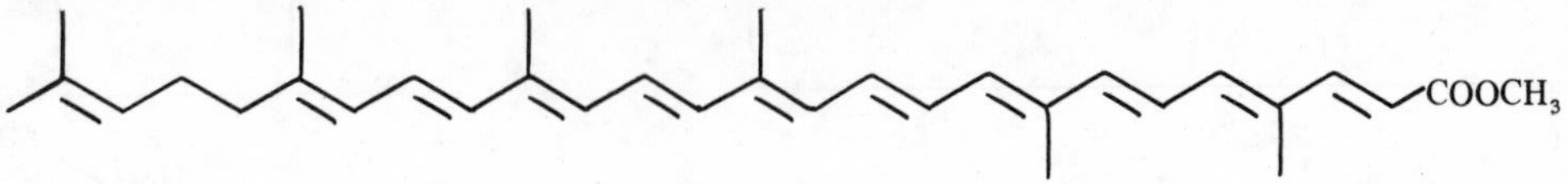

Methyl apo-6'-lycopenoate

Citranaxanthin

Sintaxanthin

Paracentrone

Hopkinsiaxanthin

Methyl 6'-apo-ψ-caroten-6'-oate

5',6'-Dihydro-5'-apo-β-caroten-6'-one or 5',6'-dihydro-5'-apo-18'-nor-β-caroten-6'-one or 6'-methyl-6'-apo-β-caroten-6'-one

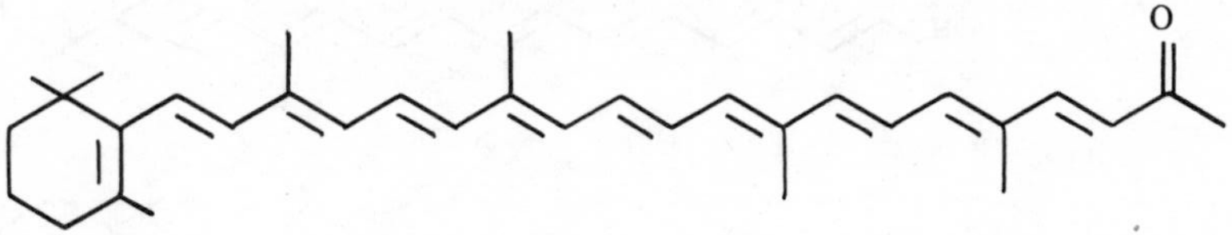

7',8'-Dihydro-7'-apo-β-caroten-8'-one or 8'-methyl-8'-apo-β-caroten-8'-one

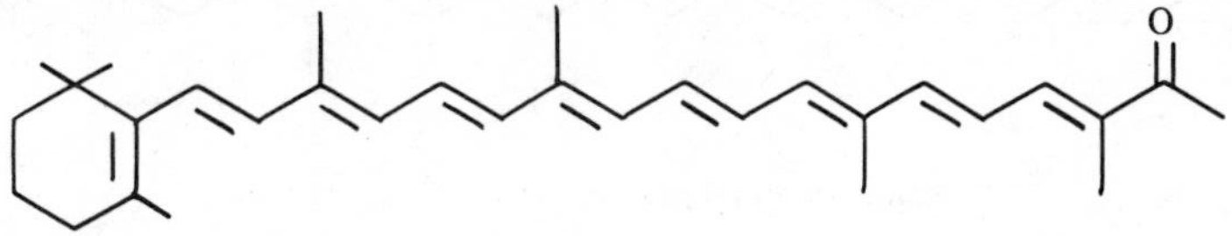

3,5-Dihydroxy-6,7-didehydro-5,6,7',8'-tetrahydro-7'-apo-β-caroten-8'-one or 3,5-dihydroxy-8'-methyl-6,7-didehydro-5,6-dihydro-8'-apo-β-caroten-8'-one

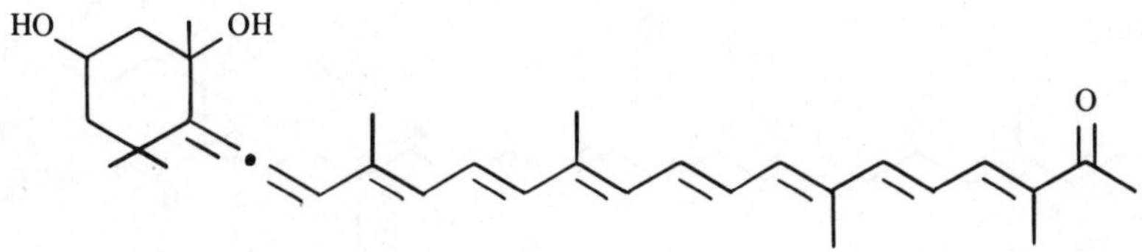

3-Hydroxy-7,8-didehydro-7',8'-dihydro-7'-apo-β-carotene-4,8'-dione or 3-hydroxy-8'-methyl-7,8-didehydro-8'-apo-β-carotene-4,8'-dione

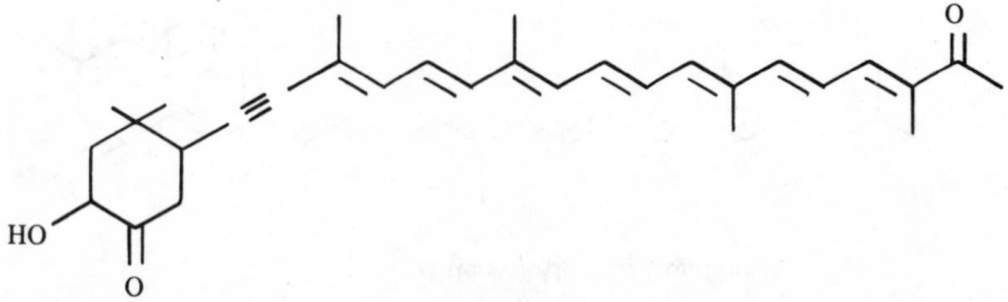

SOME NATURALLY OCCURRING CAROTENOIDS HAVING TRIVIAL NAMES (continued)

Apo Carotenoids (continued)

Bixin

6'-Methyl hydrogen 9'-*cis*-6,6'-diapocarotene-6,6'-dioate

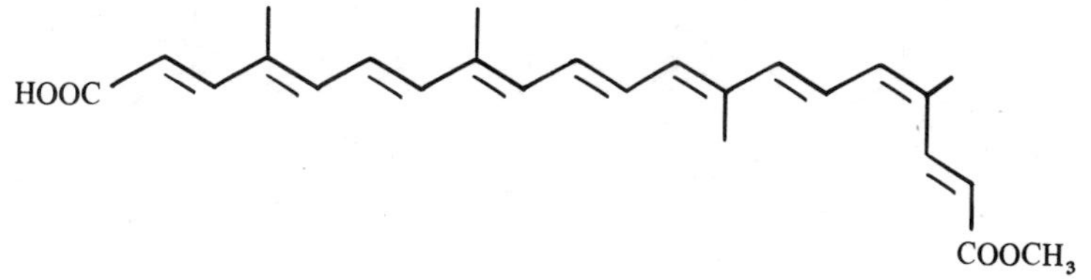

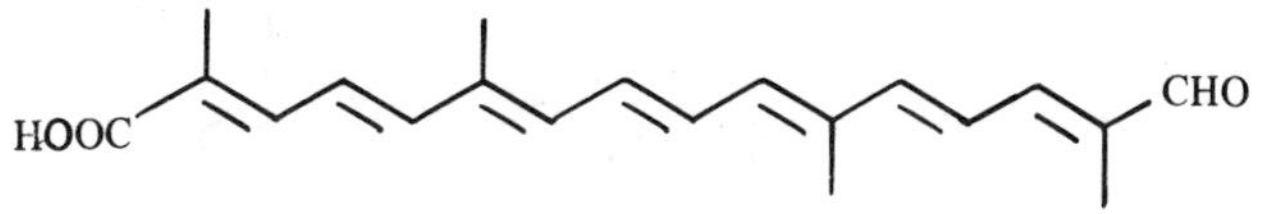

Crocetinsemialdehyde

8'-Oxo-8,8'-diapo-8-carotenoic acid

Crocetin

8,8'-Diapo-8,8'-carotenedioic acid

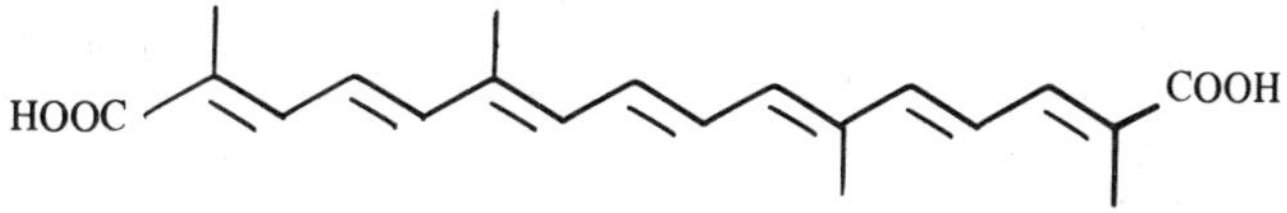

Crocin

Digentiobiosyl 8,8'-diapo-8,8'-carotenedioate

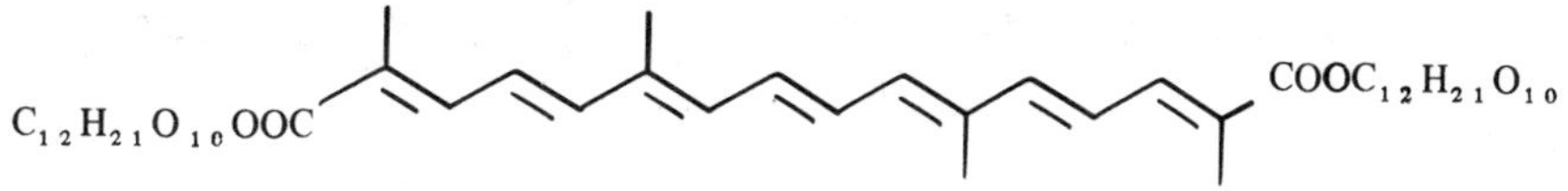

Nor and Seco Carotenoids

Actinioerythrin

3,3'-Bisacyloxy-2,2'-dinor-β,β-carotene-4,4'-dione

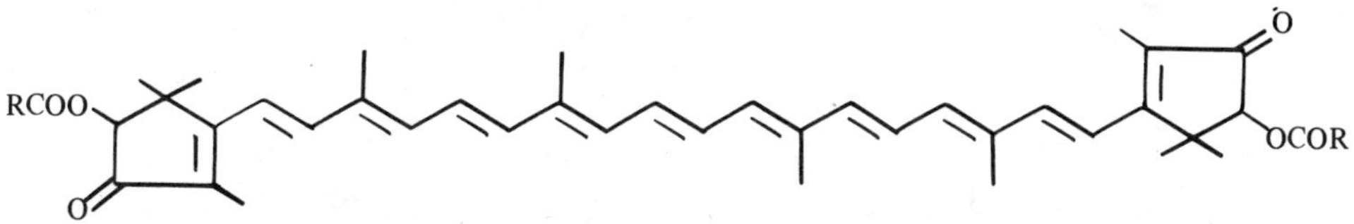

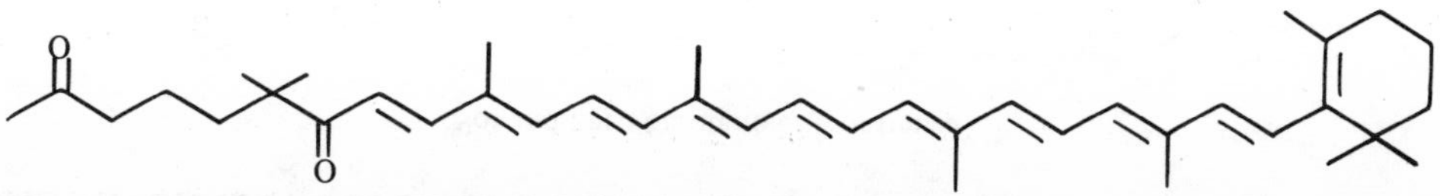

Semi-β-carotenone

5,6-Seco-β,β-carotene-5,6-dione or 5',6'-seco-β,β-carotene-5',6'-dione

SOME NATURALLY OCCURRING CAROTENOIDS HAVING TRIVIAL NAMES (continued)

Nor and Seco Carotenoids (continued)

Semi-α-carotenone

5,6-Seco-β,ε-carotene-5,6-dione

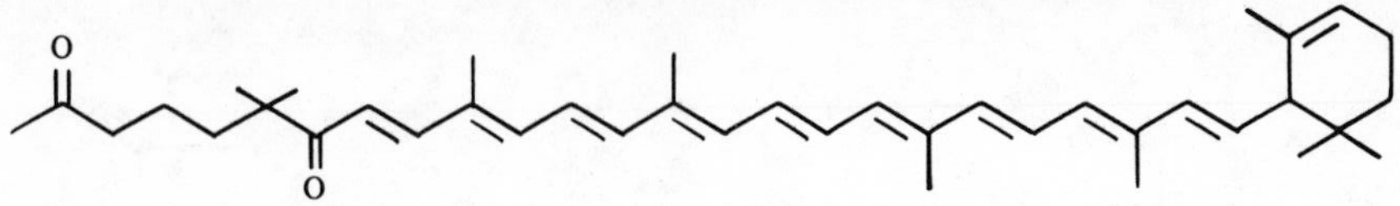

β-Carotenone

5,6:5',6'-Diseco-β,β-carotene-5,6,5',6'-tetrone

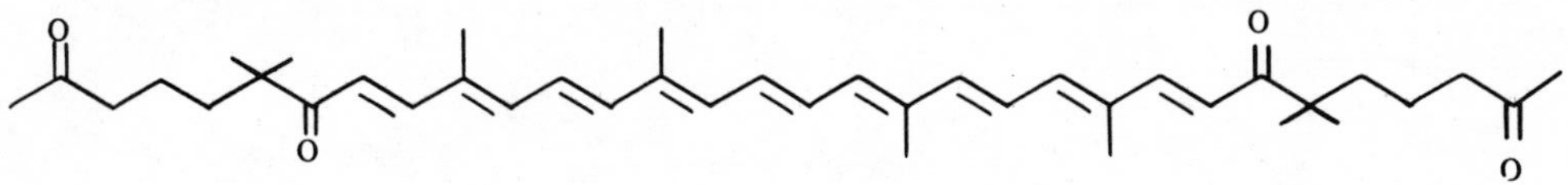

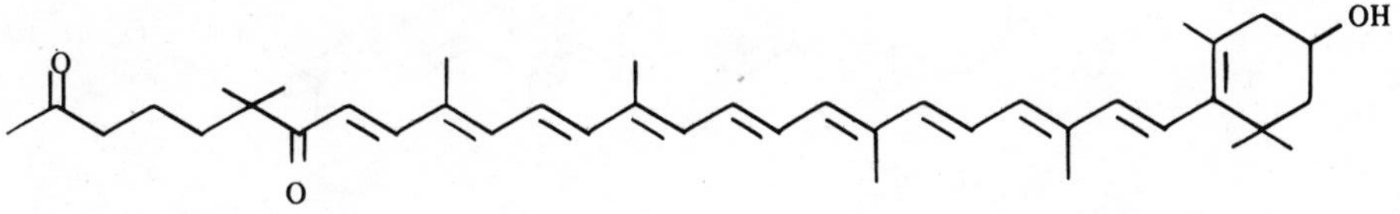

Triphasiaxanthin
3-Hydroxysemi-β-carotenone

3'-Hydroxy-5,6-seco-β,β-carotene-5,6-dione or 3-hydroxy-5',6'-seco-β,β-carotene-5',6'-dione

Peridinin

3'-Acetoxy-5,6-epoxy-3,5'-dihydroxy-6',7'-didehydro-5,6,5',6'-tetrahydro-12',13',20'-trinor-β,β-caroten-19,11-olide

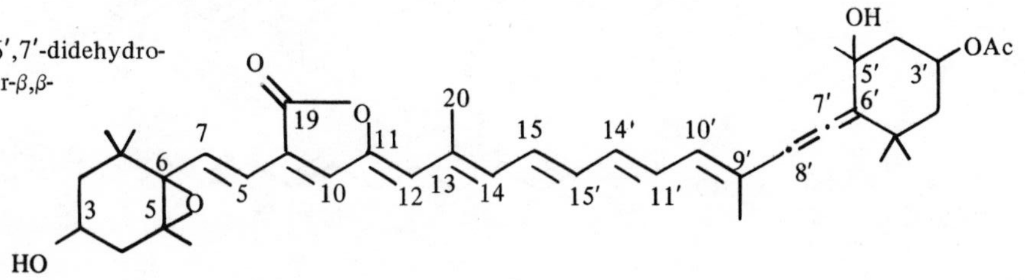

Pyrrhoxanthininol

5,6-Epoxy-3,3'-dihydroxy-7',8'-didehydro-5,6-dihydro-12',13',20'-trinor-β,β-caroten-19,11-olide

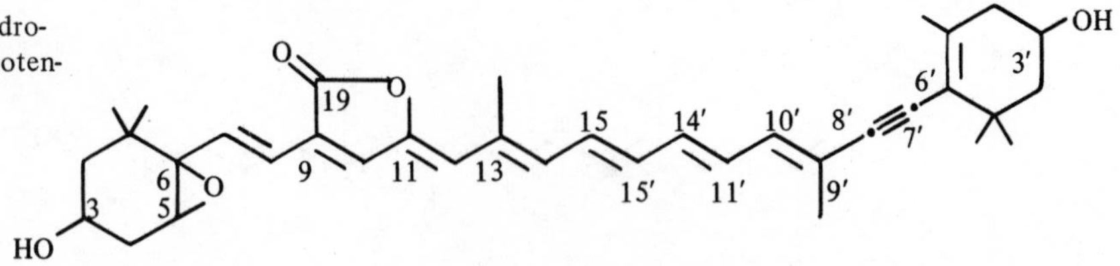

SOME NATURALLY OCCURRING CAROTENOIDS HAVING TRIVIAL NAMES (continued)

retro Carotenoids and *retro* Apo Carotenoids

4',5'-Didehydro-4,5'-*retro*-β,β-carotene-3,3'-diol

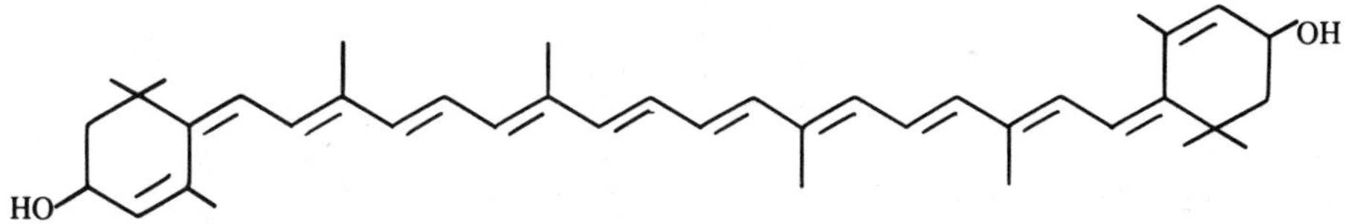

Eschscholtzxanthin

3'-Hydroxy-4',5'-didehydro-4,5'-*retro*-β,β-caroten-3-one

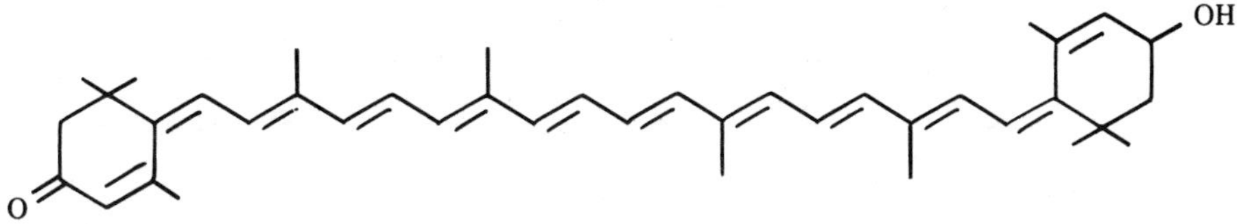

Eschscholtzxanthone

4',5'-Didehydro-4,5'-*retro*-β,β-carotene-3,3'-dione

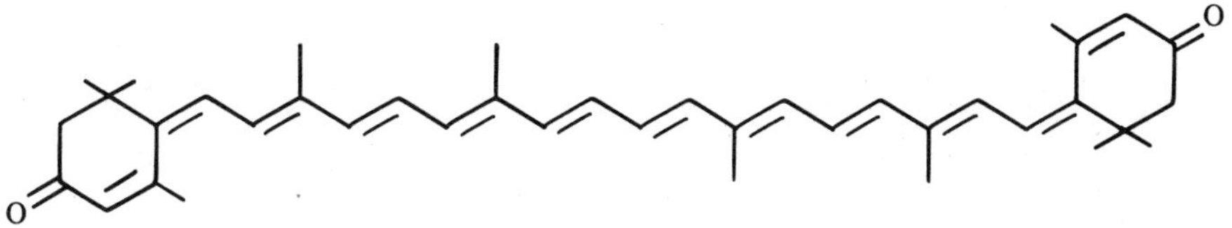

Rhodoxanthin

7,8-Dihydro-8,7'-*retro*-β,β-carotene or 7,7'-dihydro-β,β-carotene

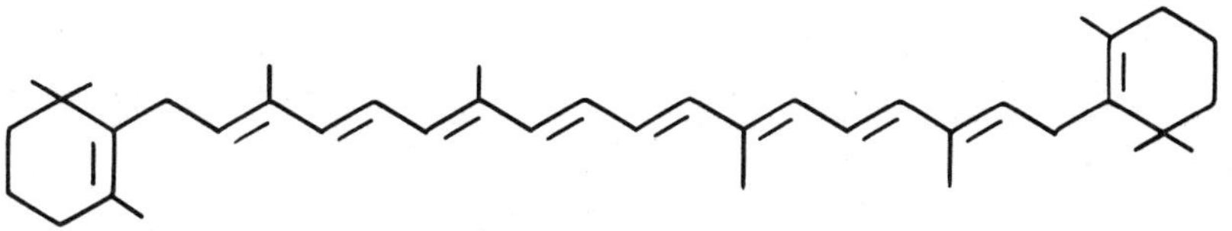

No trivial name

3-Hydroxy-5'-methyl-4,5'-*retro*-5'-apo-β-caroten-5'-one or 3-hydroxy-4,5'-*retro*-5'-apo-β-caroten-5'-one

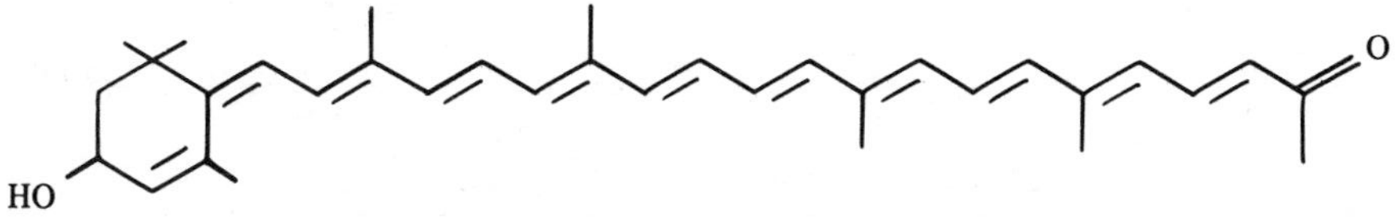

Tangeraxanthin

SOME NATURALLY OCCURRING CAROTENOIDS HAVING TRIVIAL NAMES (continued)

Higher Carotenoids

Nonaprenoxanthin

2-(4-Hydroxy-3-methyl-2-butenyl)-
7',8',11',12'-tetrahydro-ε,ψ-
carotene

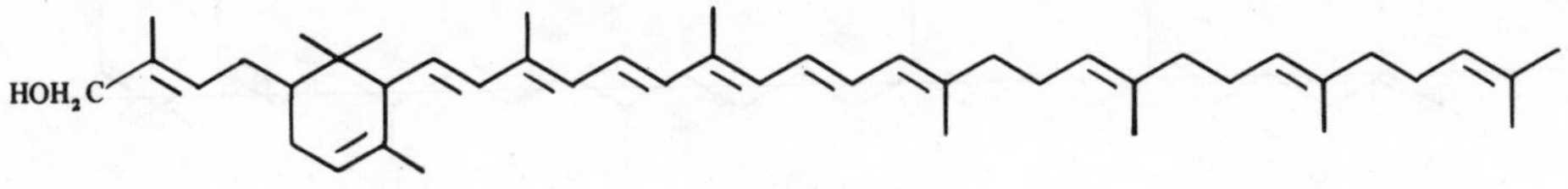

Decaprenoxanthin

2,2'-Bis(4-hydroxy-3-methyl-2-
butenyl)-ε,ε-carotene

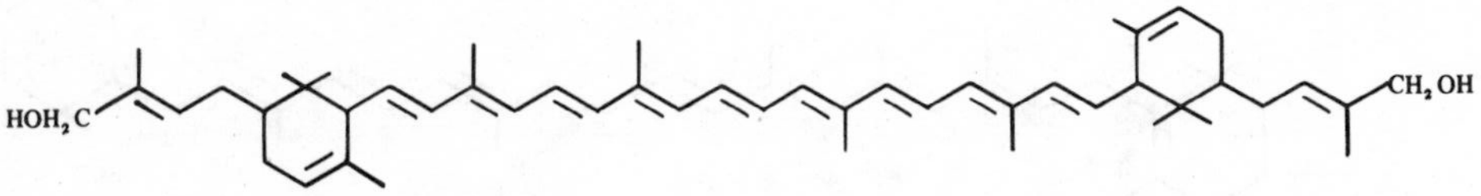

C.p. 450

2-[4-Hydroxy-3-(hydroxymethyl)-2-
butenyl]-2'-(3-methyl-2-butenyl)-
β,β-carotene

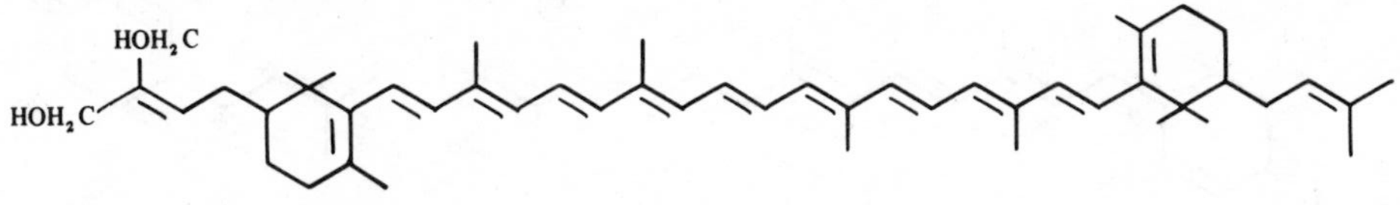

C.p. 473

2'-(4-Hydroxy-3-methyl-2-butenyl)-
2-(3-methyl-2-butenyl)-3',4'-
didehydro-1',2'-dihydro-β,ψ-
caroten-1'-ol

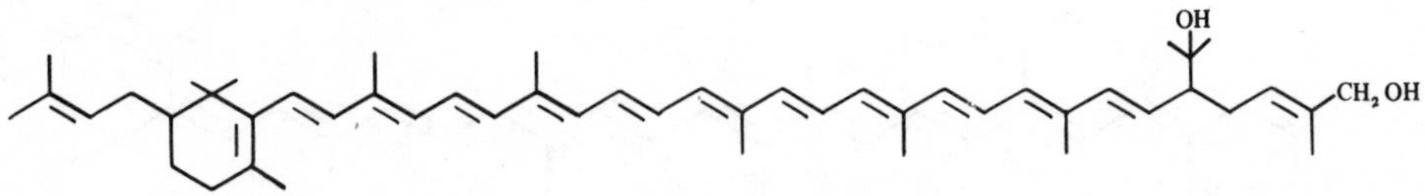

Bacterioruberin

2,2'-Bis(3-hydroxy-3-methylbutyl)-
3,4,3'-4'-tetradehydro-1,2,1',2'-
tetrahydro-ψ,ψ-carotene-1,1'-diol

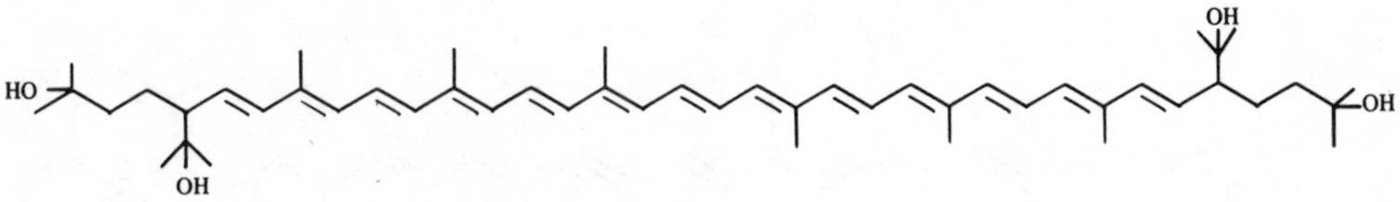

Physical and Chemical Data

RECOMMENDATIONS FOR MEASUREMENT AND PRESENTATION OF BIOCHEMICAL EQUILIBRIUM DATA

Prepared by the Interunion Commission on Biothermodynamics*

Equilibrium data are of importance in describing and understanding biochemical systems. At present there is marked variation among different investigators in the choice of experimental conditions for the study of similar or identical reactions and in the manner of reporting the data. In many cases the description of the system does not provide all the essential information that would permit reproduction of the experiments. This can lead to confusion and difficulties in correlating the results of different workers.

Equilibrium studies in biochemistry involve special problems that are not encountered in general chemistry. The attainment of biochemical equilibria commonly involves the addition of a specific enzyme to the system to catalyze the reaction studied; sometimes two or more enzymes must be added. In addition the enzyme may require the presence of certain cofactors, such as metal ions. The reactants, or products, or both, may bind or release protons or other ions during the reaction under study. Thus the experimental system must be described with particular care.

The acquisition and reporting of meaningful thermodynamic data relating to biochemical systems, as well as to other complex reacting systems, involves two fundamental problems:

1. It is not always possible to define the reacting species precisely.
2. Even in cases where the reacting species can be defined, their thermodynamic activities frequently cannot be determined.

We offer in this report several recommendations with the aim of increasing the usefulness of biochemical equilibrium data and coordinating the results of different investigators. These recommendations include a set of standard conditions which would facilitate the attainment of a common body of knowledge of a wide range of biochemical equilibria. This does not preclude the choice of special experimental conditions that may be appropriate for certain reactions; but whenever possible these reactions should also be studied under the recommended standard conditions. To avoid confusion in interpretation we also recommend standardization of terminology, symbols, and units in the presentation of such data.

For other discussions of the presentation of numerical data and of thermodynamic data derived from experiments, we call the attention of the reader to guides prepared by CODATA[1] and IUPAC.[2]

PART I. STANDARD CONDITIONS FOR EQUILIBRIUM MEASUREMENTS

True thermodynamic equilibrium constants are defined in terms of activities of the

*H. Gutfreund (Bristol, U.K.), P. Privalov (Poustchino, U.S.S.R.), representing International Union of Biochemistry (IUB). J.T. Edsall (Cambridge, Mass., U.S.A.) and W.P. Jencks (Waltham, Mass., U S.A.) representing International Union of Pure and Applied Biophysics (IUPAB). G. T. Armstrong (Washington, D.C., U.S.A.) and I. Wadsö (chairman; Lund, Sweden) representing International Union of Pure and Applied Chemistry (IUPAC). R.L. Biltonen (associated member; Charlottesville, Va., U.S.A.)

reactants and products. In many systems of biochemical interest it is not possible to evaluate the activities of all components. It is, therefore, frequently necessary to calculate equilibrium constants in terms of concentrations. The proper quotient of equilibrium concentrations is acceptably constant for many purposes, and will be referred to in this document as the concentration equilibrium constant, with the symbol K_c. However, it should be recognized that values for such equilibrium constants, K_c, and corresponding Gibbs energy changes, ΔG_c°, may not be truly constant as the composition of the system is changed.* It is essential that reported values for such quantities be accompanied by a complete and accurate description of the composition of the reacting system and the methods by which the composition was established. References to prior work will in many cases provide adequate descriptions of these methods.

Experimental Conditions

It is recommended that measurements be made at 25° (or 37°C) and an ionic strength brought to 0.10 mol · dm^{-3} with potassium chloride,** using the lowest effective buffer concentration. If these conditions are not practicable, conditions should be chosen that are well-defined and can be maintained constant throughout a series of experiments. Reagents should be as pure as practicable, and the method of purification, estimated purity, and source of each reagent should be given. The existence of a true equilibrium condition should be demonstrated by approaching the equilibrium state from both directions and by making determinations at several different concentrations of the reactants. It is desirable also to study the effect of variation of the ionic strength and the presence of specific salt effects. Complete or representative experimental data from which equilibrium constants are derived should be reported.

Temperature

A standard temperature of 25°C (298.15 K) is preferred to permit comparison with most available chemical data. The numerical values of the equilibrium constant at both 25°C and 37°C should be reported when possible.

It is desirable to measure equilibrium constants at several temperatures. The standard enthalpy and entropy changes for a specific reaction can then be calculated from the temperature dependence of the equilibrium constant. However, when such experiments are performed, it is essential to take account of changes in the dissociation constants of reactants and buffers with changing temperature and the resulting changes in pH and in the concentrations of dissociating species. It may be noted that calorimetry is generally a more accurate technique for determination of the enthalpy change of a reaction. The calorimetrically determined enthalpy change must be corrected for the enthalpy change from any reaction of a buffer involving protons that are taken up or given off in the reaction.

Buffer and pH

If only a limited number of measurements are to be made, they should be carried out at pH = 7.0 and, if possible, also at a pH value at which the apparent equilibrium constant K_c' has little or no dependence on pH. (K_c' is defined in a later section.) If direct measurements at pH 7.0 are not practicable, the calculated values for this pH should be reported. The procedure used in making these calculations must be carefully described. Care should be taken that the solution is adequately buffered so that the pH is

---

*The standard Gibbs energy change, ΔG_c°, is now the correct term for what is frequently called the standard Gibbs free energy change or simply the free energy change and often given the symbol ΔF°.
**The recommended unit of concentration is mol·dm^{-3}, commonly denoted by M in the biochemical literature.

well-defined throughout the experiment. It is desirable to determine the effect of varying the nature and concentration of the buffer in order to identify buffer effects. Buffers that are known to interact with reactants (including macromolecules) or salts, such as phosphate or pyrophosphate in the presence of divalent metal ions, should be avoided.

Complexation with Ions

The equilibrium constants for complexation of ions with reactants, such as magnesium ion with phosphate compounds, should be determined under the conditions of the experiments and the concentrations of reactants should be corrected for complexation. If this is not practicable, there are three alternatives:

a. Metal ions that are necessary for activation of enzymes may be added in trace amounts that are sufficient to permit attainment of equilibrium but are too small to change the concentrations of reactants significantly by complexation.

b. Measurements may be carried out at a concentration of metal ions such that essentially all of one (or more) of the reactants exists in the complexed form and the degree of complexation of other reactants is not significant; i.e., the equilibrium constant and standard Gibbs energy change are determined for the reaction involving the complexed species.

c. The measurements may be carried out under physiological conditions to obtain equilibria and standard Gibbs energy changes that are directly applicable to biological systems.

In any case, the concentrations of metal ions should be specified and, when possible, equilibrium constants should be reported corresponding to the three situations above, i.e., uncomplexed reactants, fully complexed reactants (e.g., Mg·ATP), and physiological conditions. In view of the limited availability and occasional unreliability of equilibrium constants for ion complexation, it is highly desirable to determine directly the effect of variation of the concentration of metal ions under the experimental conditions employed. In some cases it may be necessary to correct for complexation of reactants by appropriate extrapolation procedures or to use special experimental conditions to avoid complexation of the reactants with components of the reaction medium. For example, tetramethyl-ammonium chloride may be used instead of potassium chloride for measurements with polyphosphates or other polyanions that complex with alkali ions.

General Considerations in Reporting Equilibrium Data

Consider the following reaction

$$aA + bB \rightleftharpoons cC + dD \tag{1}$$

where a, b, c, and d, respectively, are the stoichiometric coefficients of components, A, B, C, and D. In this case the concentration equilibrium constant is

$$K_c = \frac{[C]^c \cdot [D]^d}{[A]^a \cdot [B]^b} \tag{2}$$

and the corresponding standard Gibbs energy* change is

$$\Delta G_c^\circ = -RT \ln K_c \tag{3}$$

where [A], [B], etc. are the concentrations of distinct molecular components involved in the chemical reaction, R is the gas constant and T is the thermodynamic temperature. Water as a reactant in *dilute aqueous* solutions provides a special problem. In order to conform to the general principles of dilute solutions, the value for the factor [H_2O] in equilibrium expressions should be taken as unity in such cases (see the last paragraph of part I).

In some cases it is known that the designated concentration of a component includes several distinct molecular species (e.g., [C] is [CH_3COOH] + [CH_3COO^-]) but insufficient information exists to allow a further description of the equilibrium system. For example, in a buffered system it may not be possible to evaluate the participation of the H^+ ion in all of the possible equilibria. However, if the [H^+] may be maintained constant it is possible to write an apparent equilibrium constant, K_c', in which the [H^+] does not appear explicitly. Such an apparent equilibrium constant will be pH dependent and will generally have a different value from K_c (Equation 2) which will be pH independent. The K_c' is written in terms of the total concentrations of the measurable components

$$K_c' \text{ (pH} = x, \text{ etc.)} = \frac{[C_1 + C_2 + \ldots]^c \, [D_1 + D_2 + \ldots]^d}{[A_1 + A_2 + \ldots]^a \, [B_1 + B_2 + \ldots]^b} \tag{4}$$

and an apparent standard Gibbs energy change

$$\Delta G_c^{\circ\prime} \text{ (pH} = x, \text{ etc.)} = -RT \ln K_c' \tag{5}$$

The value of K_c' is subject to certain constraints. The value of pH, and perhaps that of the concentration of certain metal ions, or other factors must be fixed and specified in order to obtain a definite K_c' value. The values of these constraints must be stated as indicated above, but these constraining factors do not explicitly appear among the concentrations of components on the right hand side of Equation 4. Usually in biochemistry subscript c in K_c can be deleted. However, it is recommended that the notation "apparent" should never be left out in connection with K_c' and $\Delta G_c^{\circ\prime}$.

A simple example should help clarify the situation. Consider the reaction:

$$R_1COOR_2 + H_2O \overset{K_I}{\rightleftharpoons} R_1COOH + HOR_2 \overset{K_{II}}{\rightleftharpoons} R_1COO^- + HOR_2 + H^+ \tag{6}$$

---

*Equation 3 implies that the equilibrium in question is for a "dilute-ideal" solution, in which the activity coefficients of all reactants and products are independent of concentration over the concentration range under consideration. Although this condition cannot be expected to hold rigorously in any actual system, in practice ΔG_c° is frequently found to be constant within the experimental error of the measurements in biochemical systems. Therefore, we shall refer to ΔG_c° as a "standard Gibbs energy change," although it should be recognized that this usage is an approximation.

It is also to be noted that the argument of a logarithm (e.g., K_c in Equation 3) must be dimensionless. For concentration equilibrium constants this can be achieved by using "relative concentrations," that is, concentration divided by a standard concentration = 1. This operation is of importance in principle, but can be ignored in practice. It is necessary, however, that the standard state concentration units be explicitly described when reporting such values.

The concentration equilibrium constants and standard Gibbs energy changes for steps I and II are

$$K_I = \frac{[R_1COOH][HOR_2]}{[R_1COOR_2][H_2O]};$$

(7)

Since the factor for water in eq. 7 is taken as unity in *dilute aqueous solutions*, it may be omitted in writing the equations for equilibria in such systems.

$$\Delta G_I^\circ = -RT \ln K_I$$

(8)

$$K_{II} = \frac{[R_1COO^-][HOR_2][H^+]}{[R_1COOH][HOR_2]} = \frac{[R_1COO^-][H^+]}{[R_1COOH]};$$

(9)

$$\Delta G_{II}^\circ = -RT \ln K_{II}$$

(10)

The concentration equilibrium constant and standard Gibbs energy change for the overall reaction are

$$K = \frac{[R_1COO^-][HOR_2][H^+]}{[R_1COOR_2]} = K_I K_{II}$$

(11)

$$\Delta G^\circ = -RT \ln K$$

(12)

However, an apparent equilibrium constant at some pH = x may be written in which the concentrations of R_1COO^- and R_1COOH are summed and do not have to be differentiated.

$$K'(pH = x) = \frac{([R_1COO^-] + [R_1COOH])[HOR_2]}{[R_1COOR_2]}$$

(13)

The corresponding standard Gibbs energy change is

$$\Delta G^{\circ\prime}(pH = x) = -RT \ln K'$$

(14)

In cases where the details of the chemical equilibria are known, the interconversion between K and K' is generally straightforward. For the example just described

$$[R_1COOH] = \frac{[R_1COO^-][H^+]}{K_{II}}$$

(15)

and

$$K'(pH = x) = \frac{[R_1COO^-](1 + [H^+]/K_{II})[HOR_2]}{[R_1COOR_2]} = K\left(1 + \frac{[H^+]}{K_{II}}\right) \cdot \frac{1}{[H^+]}$$

(16)

The well known problem of the standard Gibbs energy change of ATP hydrolysis to ADP and inorganic phosphate (P_i) furnishes a more complex example of concentration and apparent equilibrium constants. The overall reaction may be written:

$$\text{Total ATP} + H_2O \rightleftharpoons \text{total ADP} + \text{total } P_i$$

Thus, the apparent equilibrium constant and $\Delta G^{o\prime}$ values at some specified pH are

$$K' = \frac{[\text{total ADP}]\,[\text{total P}_i]}{[\text{total ATP}]} \tag{17}$$

$$\Delta G^{o\prime} = -RT \ln \frac{[\text{total ADP}]\,[\text{total P}_i]}{[\text{total ATP}]} \tag{18}$$

Data are available to convert these apparent constants into concentration constants. In order to deal with these phosphate compounds in a medium related to their biological environment, it is necessary to consider their interactions with protons and with magnesium ions. At pH values near 7 and at physiological $[\text{Mg}^{2+}]$ values, it is found that each of the three phosphates can be described by one acidic dissociation, and each acid and its conjugate base may bind one Mg^{2+} ion. Thus the total concentrations in K' and $\Delta G^{o\prime}$ may be resolved as follows:

$$[\text{Total ATP}] = [\text{ATP}^{4-}] + [\text{ATP}^{3-}] + [\text{ATPMg}^{2-}] + [\text{ATPMg}^-] \tag{19}$$

$$[\text{Total ADP}] = [\text{ADP}^{3-}] + [\text{ADP}^{2-}] + [\text{ADPMg}^-] + [\text{ADPMg}] \tag{20}$$

$$[\text{Total P}_i] = [\text{P}_i^{2-}] + [\text{P}_i^-] + [\text{P}_i\text{Mg}] + [\text{P}_i\text{Mg}^+] \tag{21}$$

In fact $[\text{P}_i\text{Mg}^+]$ is always so small as to be negligible, and $[\text{P}_i\text{Mg}]$ is of minor importance.

The three pKa values involved — one for each of the compounds ATP, ADP, and P_i — are known, as are the association constants of the various species for Mg^{2+} ion. We can then formulate the reaction in terms of the relations between particular species of reactants and products. For instance, in the limiting case of high pH (>8) and $[\text{Mg}^{2+}] = 0$, the predominant reaction becomes:

$$\text{ATP}^{4-} + \text{H}_2\text{O} \rightleftharpoons \text{ADP}^{3-} + \text{P}_i^{2-} + \text{H}^+ \tag{22}$$

It would be equally valid to formulate the reaction in terms of the equation:

$$\text{ATP}^{3-} + \text{H}_2\text{O} \rightleftharpoons \text{ADP}^{2-} + \text{P}_i^- \tag{23}$$

and other choices are obviously possible, provided that the equations balance with respect to total charge and stoichiometry. We note that the equilibrium constants of all such reactions must be independent of pH, whether or not $[\text{H}^+]$ enters explicitly into the reaction. Formulation of such equilibrium constants implies nothing as to the mechanism of the reaction in this or any other case.

If we choose to formulate the reaction in terms of Equation 22, we have:

$$K_{\text{ATP}^{4-}} = \frac{[\text{ADP}^{3-}]\,[\text{P}_i^{2-}]\,[\text{H}^+]}{[\text{ATP}^{4-}]} \tag{24}$$

In the limiting case where $[\text{Mg}^{2+}] = 0$, we can then write for the relation between K' (of Equation 17) and $K_{\text{ATP}^{4-}}$ (Equation 24):

$$K' (\text{pH} = x, [\text{Mg}^{2+}] = 0) = \frac{([\text{ADP}^{3-}] + [\text{ADP}^{2-}])([\text{P}_i^{2-}] + [\text{P}_i^{-}])}{([\text{ATP}^{4-}] + [\text{ATP}^{3-}])} = K_{\text{ATP}^{4-}} \frac{\left(1 + \frac{[\text{H}^+]}{K_\beta}\right)\left(1 + \frac{[\text{H}^+]}{K_\gamma}\right)}{\left(1 + \frac{[\text{H}^+]}{K_\alpha}\right)[\text{H}^+]}$$

$$(25)$$

where K_α, K_β, and K_γ are, respectively, the acidic dissociation constants of ATP^{3-}, ADP^{2-}, and P_i^{-}.

Since all the constants in Equation 25 are known, the apparent equilibrium constant (Equation 17) is thus explicitly related to the pH-independent constant $K_{\text{ATP}^{4-}}$. In the presence of Mg^{2+} ion the analysis is in principle exactly similar but is much more complicated in detail. A detailed discussion is given for instance by Alberty.[3]

We note that the usefulness of Equation 17 depends upon the fact that it is possible to determine analytically the *total* concentrations of ATP, ADP, and P_i without distinguishing between the different forms listed in Equations 19, 20, and 21.

We recommend that concentration equilibrium constants and corresponding standard Gibbs energy changes be reported when the reaction can be completely and accurately defined. In cases where this is impossible corresponding apparent quantities may be reported, if possible over a range of pH values and other constraining variables. In any case the values must be accompanied by a precise definition of the reaction scheme assumed.

Quantities

It is recommended that values of equilibrium constants and standard Gibbs energy changes should normally be reported based on concentrations (amount of substance divided by volume, with units of moles per cubic decimetre) except for the proton factor, for water in aqueous systems in which water is involved as a reactant or product, for gases, or for solids. For special purposes it may be desirable to report additional values that are based on molalities, mole fractions, or partial pressures; in any case the unit employed should be clearly identified. The use of equilibrium constants based on concentrations expressed in such units as millimoles or micromoles per cubic decimetre (litre) should be avoided.

It is recommended that for reactions in dilute aqueous solutions the factor for H_2O be taken as unity in the absence of a measurement of its relative activity. An actual concentration such as the conventional 55.5 $\text{mol} \cdot \text{dm}^{-3}$ should only be used for liquid water for well defined reasons. It is recommended that $10^{-\text{pH}}$ be used for the proton factor. The conventions chosen for the liquid water and the proton factors must always be clearly and explicitly stated. For gases the partial pressure in atmospheres or (if the partial pressure is expressed in kilopascals) the ratio of the partial pressure to 101.325 kPa should be used.

Gibbs energies should be reported in joules per mole ($\text{J} \cdot \text{mol}^{-1}$). Until this unit comes into general use, results may in addition be reported in thermochemical calories per mole (1 cal_{th} = 4.184 j).

PART II. SYMBOLS, UNITS, AND TERMINOLOGY

General

The precision of statement, the universal intelligibility, and the enduring quality of a thermodynamic document will be enhanced if the symbols, units, and terminology used in it are those recommended by an international standardizing body. The following information summarizes recommendations of international scientific bodies in the physical and measurement sciences, selected for particular relevance to the thermodynamics of biological processes. The author is referred to recommendations of the IUPAC[4],* for more detailed lists of recommended symbols and terminology for physicochemical quantities, and for a discussion of the International System of Units (SI). The SI is also recommended by the International Organization for Standardization (ISO), which provides rules for their use.[5] Other descriptions of the SI are provided by the International Bureau of Weights and Measures.[6]

In adhering to these recommendations for the presentation of scientific data, authors are urged:

1. To use the recommended name of each physical quantity and the preferred symbol, as far as possible.

2. To use SI units and their internationally accepted symbols.[4-6]

3. To use functional expressions where feasible, rather than subscripts or superscripts, for specifying substance, physical state, and temperature to which the quantity refers; e.g., for a process such as reaction (1)

$$\Delta G_1^{\circ\prime} \ (\text{pH} = 7.0, I_c = 0.10 \ \text{mol}\cdot\text{dm}^{-3}, \text{at } 25°C)$$

or for a dissolved substance

$$C_p \ (C_2H_5OH, \text{in } H_2O, 25°C)$$

4. To indicate both numerical values and units in referring to physical quantities in the text, and in labels for axes of graphs and rows and columns of tables.

The preferred convention for statements in the text is

physical quantity = numerical value $\times$ unit

e.g., $\Delta G^{\circ\prime} \ (\text{pH} = 7.0, I_c = 0.10 \ \text{mol}\cdot\text{dm}^{-3}, \text{at } 25°C) = -8.7 \ \text{kJ}\cdot\text{mol}^{-1}$

The preferred label for tables and graphs, obtained by rearrangement of the equation above is

physical quantity/unit = numerical value

that is, in the equation $\Delta G/\text{kJ}\cdot\text{mol}^{-1} = -8.7$, the left hand side is the label and the right hand side is the numerical quantity to be tabulated. If a power of ten is used in addition to or in place of a prefix on the units the power of ten should be shown correctly with

*Reference 4 gives a summary of the International System of Units (SI) and recommendations for the usage of these units including reference to usage of other commonly found non-SI units. It also lists recommended names and symbols for quantities relating to space and time, mechanical properties, molecular masses and concentrations, thermodynamics, electricity and magnetism, electrochemistry, light and electromagnetic radiation, and transport properties.

Recommendations are also given in Reference 4 for subscripts and superscripts and for format of presenting formulas, symbols, units, numerical values, mathematical symbols, and relationships. Special chapters discuss notation and conventions for electromotive force and electrode processes, pH measurements, and reaction rates and related quantities. Also of interest to the biochemist and biothermodynamicist is an appendix which gives recommendations concerning the definitions of activities and related properties in aqueous solutions as well as other phases.

the units (thus $\Delta G/\text{J}\cdot\text{mol}^{-1} = -0.087 \times 10^5$, or $\Delta G/10^5 \text{ J}\cdot\text{mol}^{-1} = -0.087$, or any other algebraically correct form).

Values other than SI units may be included in parentheses when it is felt this will improve communication between the author and his readers. This may be considered desirable when there is still widespread current usage of a non-SI unit for a certain quantity; such as the calorie or the atmosphere. When a non-SI unit is used, the author should define it in terms of SI units.

The symbols for physical quantities should be printed in italic (sloping) type or underlined in typescript and the symbols for units should be printed in roman (upright) type.

The usage of these recommendations is illustrated below for certain specific applications in thermodynamic measurements.

No attempt will be made here to discuss all of the names and symbols for physical quantities, which are adequately described in the references given. Our attention will be directed toward those for which the International Recommendations form a distinct variation from much established practice.

Temperature

Thermodynamic temperatures and temperature differences are expressed in kelvins, symbol K, not degrees Kelvin or °K. Celsius temperatures and temperature differences are expressed in degrees Celsius, symbol °C. The degree Celsius, which is identical to the kelvin, is sometimes improperly called the degree "centigrade."

Pressure

The unit of pressure is the pascal, which is one newton per square metre ($1 \text{ Pa} = 1 \text{ N}\cdot\text{m}^{-2}$). A convenient unit for many pressure measurements is the kilopascal (symbol kPa). The biological thermodynamicist should recognize that the common thermodynamic term, pV, is an energy term, and if p is in pascals and V is in cubic metres, then the $p\cdot V$ product is obtained directly in joules. No combination of customary non-SI units avoids the use of a conversion factor; and such awkward energy units as litre-atmosphere are avoided by the recommended usage.

The commonly found units mmHg or Torr, and their submultiples, should be avoided.

The standard atmosphere is defined as 101.325 pascals ($1 \text{ atm} = 101.325 \text{ kPa}$). This is a non-SI unit. In calculating equilibrium constants and standard thermodynamic functions based on pressure measurements it should be recognized that the accepted standard pressure is 101.325 kPa, which is often called one atmosphere in this context.

Energy

Energy measurements, including all thermal measurements, should be reported in joules (J), kilojoules (kJ), or millijoules (mJ) as appropriate.

In cases where energies are also expressed in thermochemical calories (cal_{th}) or a multiple or submultiple, the author should state the conversion factor used ($1 \text{ cal}_{th} = 4.184 \text{ J}$). The "nutritional" or "large" calorie (sometimes abbreviated "Cal" and equal to one kilocalorie) has little use outside of its special area, and its use should be avoided. Other definitions of calorie have little or no use in biochemistry or in other thermochemical work.

The author should realize that the use of calories is on the decline because of the acceptance of the SI units, so that data reported in calories will probably require conversion relatively soon.

For other units of measure, the author is referred to IUPAC and BIPM recommendations.[4-6]

Composition of Solutions

For thermodynamic applications, the composition of a solution is commonly described in terms of concentration, molality, or mole fraction. Other descriptors, such as mass fraction or volume fraction, may be found convenient for special applications. All of these should be clearly distinguished, and the solutions used should be unambiguously described using proper units.

The concentration of a solute substance B is the amount of B (moles) divided by the volume of the solution. Accepted symbols are c_B and [B]. A convenient unit of concentration is $mol \cdot dm^{-3}$. Concentration is sometimes called "molarity." A solution with a concentration of $0.1 \ mol \cdot dm^{-3}$ is often called a 0.1 molar solution or a 0.1 M solution. Because the term molarity and the unit M are liable to be confused with molality the term concentration and the unit $mol \cdot dm^{-3}$ are preferred.

The mass concentration of a solute substance B is the mass of B divided by the volume of the solution. An accepted symbol is ρ_B and an appropriate unit is $kg \cdot m^{-3}$.

The molality of solute substance B is the amount of B (moles) divided by the mass of solvent. An accepted symbol is m_B and an appropriate unit is $mol \cdot kg^{-1}$. A solution having a molality equal to $0.1 \ mol \cdot kg^{-1}$ is sometimes called a 0.1 molal solution or a 0.1 m solution. Because of possible confusion of "molal" with "molar" and because m is the symbol for the metre, the unit $mol \cdot kg^{-1}$ is preferred.

The mole fraction of substance B is the amount of B (moles) in a solution divided by the total amount (moles) of all substances in the solution. Accepted symbols are x_B and y_B, where x_B or $y_B = n_B / \sum_i n_i$. The mole fraction is a dimensionless number.

State Functions

In stating energy quantities associated with the state functions of a substance, care should be taken to use the precisely defined functions and the recommended symbols, internal energy (U), enthalpy (H), Helmholtz energy (A), Gibbs energy (G), entropy (S), or heat capacity (C).

The term heat should be avoided in reference to state thermodynamic functions. Thus, enthalpy is preferable to heat content and enthalpy of formation is preferable to heat of formation. An exception to this restriction is heat capacity for which no alternative has been accepted. The term specific heat capacity should refer only to heat capacity per unit mass. The term specific heat should be avoided. The corresponding molar quantity is molar heat capacity.

The term free energy is ambiguous and should not be used. Proper terms are Helmholtz energy or Helmholtz function and Gibbs energy or Gibbs function, symbols A and G, respectively, and referring to $U - TS$ or $H - TS$, respectively, The extensive quantities have the unit joule (J), The specific quantities have the unit joule per kilogram ($J \cdot kg^{-1}$). The molar quantities have the unit joule per mole ($J \cdot mol^{-1}$). Entropy, symbol S, an extensive quantity, has the unit joule per kelvin ($J \cdot K^{-1}$); the specific entropy has the unit joule per kilogram kelvin ($J \cdot kg^{-1} \cdot K^{-1}$); the molar entropy has the unit joule per mole kelvin ($J \cdot mol^{-1} \cdot K^{-1}$).

Equilibrium Constants

Ideally the thermodynamic equilibrium constant, based on activities of the reacting components, should be reported. In the case of most biochemical studies, however, this is not practicable. Therefore, the recommendations made in the preceding section of this document regarding the calculation and reporting of equilibrium constants and Gibbs energy changes should be observed. In accordance with the general recommendation given above, that symbols for physical quantities be differentiated from symbols for units, the symbol K should be in italic (sloping) type in printed documents, or underlined in typewritten documents, to differentiate it from the symbol for kelvin, K, which should be in roman (upright) type.

Electrode Measurements

Conventions concerning the signs of electric potential differences, electromotive forces, and electrode potentials, are given in Reference 4 (Chapter 9) and are sometimes called the "Stockholm Convention" of 1953. The author should strictly adhere to them.

By these conventions, the direction of a written chemical reaction, the order of representation of the corresponding electrolytic cell elements, and the assignment of sign to the electromotive force, E, of the cell obey the following consistency relationships.

1. The cell reaction and the cell diagram are written from left to right in such a way that *when* reaction occurs in this direction positive electric charge flows in the cell from left to right. The electric potential difference is equal in sign and magnitude to the electric potential of the electrode on the right minus that of the electrode on the left.

2. The electrode potential of an electrode (half cell) is the electromotive force of a cell in which the electrode on the right is the electrode in question and the electrode on the left is a standard hydrogen electrode. A more positive electromotive force indicates a greater oxidizing potential.

Summary Tables

Table 1 lists the thermodynamic functions found in the recommendations of the main text, gives their units and a few relationships involving them. It also summarizes some of the principal recommended conditions and practices. Table 2 gives some SI units and their symbols.

Table 1a
THERMODYNAMIC TERMS USED IN THESE RECOMMENDATIONS

$[A]$	Concentration of A
y_B	Activity coefficient of substance B (concentration basis)
a_B	Relative activity of substance B (*Note:* $a_B = y_B \cdot [B]$)
I	Ionic strength ($I_c = 1/2 \sum\limits_1^S c_i \cdot z_i^2$; $I_m = 1/2 \sum\limits_1^S m_i \cdot z_i^2$)
	where c_i is the concentration of the ith ion, m_i is the molality of the ith ion, and z_i is the charge number of the ith ion. S is the number of ion types present
K_c	pH-independent proper product of equilibrium concentrations (concentration equilibrium constant)
K	Thermodynamic equilibrium constant
$K'(\text{pH} = x; \text{etc.})$	pH-dependent apparent proper product of summed equilibrium concentrations, constrained with respect to concentrations of stated species (apparent equilibrium constant)
$K_{\exp}$	See K'
K_{app}	See K'
K_{obsd}	See K'
ΔG_c°	Standard Gibbs energy change corresponding to the pH-independent product of equilibrium concentrations ($\Delta G_c^\circ = -RT \ln K_c$)
$\Delta G_c^{\circ\prime} \ (\text{pH} = x)$	Apparent standard Gibbs energy change corresponding to the apparent product of equilibrium concentrations at fixed pH in a buffered solution ($\Delta G_c^{\circ\prime} = -RT \ln K_c'$)
$\Delta G_{\exp}^\circ$	See $\Delta G^{\circ\prime}$
ΔG_{app}°	See $\Delta G^{\circ\prime}$
ΔF	Formerly common symbol for free energy change; now more properly called Gibbs energy change and given the symbol ΔG
ΔF°	See ΔF, ΔG°, $\Delta G^{\circ\prime}$
ΔS	Entropy change
ΔH	Enthalpy change
C_p	Heat capacity at constant pressure

The units of molar enthalpy change and molar Gibbs energy change are generally expressed in joules per mole ($J \cdot mol^{-1}$) or kilojoules per mole ($kJ \cdot mol^{-1}$) of a reaction shown, but the units should be stated in every case. For particular purposes specific enthalpy change or specific Gibbs energy change may be given per unit mass of a given reactant or product (joules per kilogram) ($J \cdot kg^{-1}$).

The units of entropy change are generally expressed in joules per mole kelvin ($J \cdot mol^{-1} \cdot K^{-1}$) for a particular process, but the units should be stated in every case. For particular purposes specific entropy change may be given per unit mass of a given reactant or product, joule per kilogram kelvin, ($J \cdot kg^{-1} \cdot K^{-1}$) or the total entropy change for a stated process in which case the units are joules per kelvin ($J \cdot K^{-1}$).

The units of absolute entropy and heat capacity are the same as those for entropy change ($J \cdot mol^{-1} \cdot K^{-1}$, $J \cdot kg^{-1} \cdot K^{-1}$ or $J \cdot K^{-1}$) but refer to a particular substance or aggregate of substances rather than to a process.

Table 1a (continued)
THERMODYNAMIC TERMS USED IN THESE RECOMMENDATIONS

Some thermodynamic relations: The units for these quantities are joules, J, for the extensive quantity total energy, joules per mole ($J \cdot mol^{-1}$) for molar energy, and joule per kilogram ($J \cdot kg^{-1}$) for specific energy.

$\Delta U = Q + W$ — First law of thermodynamics — the increase of internal energy of a system is the sum of heat supplied to the system and work done on the system.

$\Delta H = \Delta U + p\Delta V$ — For a constant pressure system an increase in enthalpy of the system (ΔH) results in an increase in internal energy (ΔU) and work done by the system ($p\Delta V$).

$\Delta H = H(T_2) - H(T_1) = \int C_p dT$ — For a system at constant pressure to which energy is supplied the increase of enthalpy is the integral of C_p over the range of the temperature change it causes.

$\Delta G = \Delta H - T\Delta S$ — The increase in Gibbs energy of a system (ΔG) at constant temperature is the increase in enthalpy (ΔH) minus the increase in $T\Delta S$.

Table 1b
SUMMARY OF THERMODYNAMIC SYMBOLS
AND OTHER RECOMMENDATIONS

Quantity	Apparent value at pH = constant (x)	pH-independent value
Equilibrium constant*	K'	K
Standard molar Gibbs energy change**	$\Delta G°'$ (pH = x)	$\Delta G°$
Concentration of reactant A	$[A]_{total}$	$[A]$

Recommended Conventions Concerning Factors in Expressions for Equilibrium Constants

In cases where water occurs as a reactant or product, state whether the factor for water is taken as unity or as 55.5 or other number.

State whether 10^{-pH} or some other measure is used for the hydrogen ion factor. The pH is not uncontested as an accurate measure of concentration or activity of the hydrogen ion.

Recommended Measurement Conditions

	Primary conditions	Secondary conditions
Temperature $t/°C$ (or T/K)	25°C (298.15 K) (also, vary t)	37°C (310.15 K) (also, vary t)
Ionic strength, $I/mol \cdot dm^{-3}$	0.1 (made up with KCl)	0.1 (made up with KCl)
Hydrogen ion concentration	pH = 7	
Buffer concentration	Lowest effective	

*Approximately constant proper quotient of equilibrium concentrations.
**Formally calculated from the equilibrium constant.

Table 2

PHYSICAL QUANTITIES, SI UNITS, AND THEIR SYMBOLS

Name of unit   Symbol for unit

(a) Base Units

Length	metre	m
Mass	kilogram	kg
Time	second	s
Electric current	ampere	A
Temperature	kelvin	K
Amount of substance	mole	mol
Luminous intensity	candela	cd

(b) Derived Units (Examples)

Force	newton	N	$(kg \cdot m \cdot s^{-2})$
Pressure	pascal	Pa	$(N \cdot m^{-2})$
Energy	joule	J	$(kg \cdot m^2 \cdot s^{-2})$
Power	watt	W	$(J \cdot s^{-1})$
Electric charge	coulomb	C	$(A \cdot s)$
Electric potential difference	volt	V	$(J \cdot A^{-1} \cdot s^{-1})$
Electric resistance	ohm	Ω	$(V \cdot A^{-1})$
Frequency	hertz	Hz	(s^{-1})
Area	square metre	m^2	
Volume	cubic metre	m^3	
Density	kilogram per cubic metre	$kg \cdot m^{-3}$	

(c) SI Prefixes

Fraction	Prefix	Symbol	Multiple	Prefix	Symbol
10^{-1}	deci	d	10	deca	da
10^{-2}	centi	c	10^2	hecto	h
10^{-3}	milli	m	10^3	kilo	k
10^{-6}	micro	μ	10^6	mega	M
10^{-9}	nano	n	10^9	giga	G
10^{-12}	pico	p	10^{12}	tera	T
10^{-15}	femto	f	10^{15}	peta	P
10^{-18}	atto	a	10^{18}	exa	E

REFERENCES

1. Guide for the presentation in the primary literature of numerical data derived from experiments, *CODATA Bull.*, No. 9, December 1973.
2. A guide to procedures for the publication of thermodynamic data, *J. Chem. Thermodynamics*, 4, 511 (1972); *Pure Appl. Chem.*, 29, 395 (1972).
3. **Alberty,** *J. Biol. Chem.*, 244, 3290 (1969).
4. **McGlashan,** *Manual of Symbols and Terminology for Physiochemical Quantities and Units*, Butterworth and Co., London, 1970; also in *Pure Appl. Chem.*, 21, No. 1 (1970).
5. International Organization for Standardization (ISO), SI units and recommendations for the use of their multiples and of certain other units, International Standard ISO-1000, 1st ed., 1973-02-01, American National Standards Institute, New York. See also, for more detail: International Standard ISO-31/0-1974 and ISO-31/I-XII, 1965–1975 which deal with quantities, units, symbols, conversion factors, and conversion tables for various branches of science and technology. Copies are obtained through the ISO-member national standards organizations of various countries.
6. Bureau International des Poids et Mesures (BIPM), *Le Système International d'Unités*, OFFILIB, Paris, France, 1970. Authorized English translations are available: *The International System of Units (SI)*, National Bureau of Standards Publication 330, U.S. Gov. Print. Off. Washington, D.C., 1971, or *The International System of Units (SI)*, Her Majesty's Stationery Office, London.

COEFFICIENTS OF SOLUBILITY EQUATIONS OF CERTAIN AMINO ACIDS IN WATER[a]

Amino acid	a_1	$b_1 \times 10^2$	$c_1 \times 10^5$	a_2	a_3	$b_3 \times 10^2$	$c_3 \times 10^5$	a_4	$b_4 \times 10^2$	$c_4 \times 10^5$
L-Alanine	2.1048	0.4669	—	0.1551	−2.5792	1.075	—	−6.5150	1.037	—
D,L-Alanine	2.0830	0.5608	—	0.1333	−3.2199	1.291	—	−7.1317	1.245	—
L-Asparagine H$_2$O	0.9289	2.311	−4.981	−1.2475	−25.9584	1.059	−11.47	−30.2463	11.79	−11.84
L-Aspartic acid	0.3194	1.519	—	−1.8047	−13.7113	3.499	—	−17.7370	3.502	—
D,L-Aspartic acid	0.4181	2.016	−4.999	−1.7060	−25.1918	10.93	−11.51	−29.2797	10.98	−11.61
L-Cystine	−1.299	1.357	—	−3.680	−18.643	3.125	—	−21.023	3.125	—
D,L-Cystine[c]	−1.7959	0.8013	27.89	−4.1766	19.4912	−23.66	47.61	15.4747	−23.66	47.61
D,L-Cystine[c]	−2.1087	3.367	−22.56	−4.4894	−36.9568	14.48	−16.80	−40.9733	14.48	−16.80
Meso-cystine[c]	−1.7190	0.4514	27.39	−4.0997	34.7268	−33.38	63.01	30.7013	−33.38	63.01
Meso-cystine[c]	−2.6034	5.890	−49.41	−4.9841	−133.4125	75.4125	−113.8	−137.429	−75.73	−113.8
L-Dibromotyrosine (hydrated)	0.0839	1.627	—	−2.445	−15.881	3.753	—	−19.894	3.752	—
L-Dibromotyrosine (anhydrous)	0.188	0.9884	—	−2.343	−11.610	2.276	—	−15.537	2.247	—
L-Dichlorotyrosine	0.0065	1.038	4.648	−2.392	−4.058	−3.450	1.069	−8.426	−3.215	1.030
L-Diiodotyrosine	−0.690	1.92	—	−3.326	−19.745	4.42	—	−23.761	4.43	—
D,L-Diiodotyrosine	−0.827	1.43	—	−3.464	−16.989	3.30	—	−21.006	3.30	—
D-Glutamic acid	0.5331	1.613	—	−1.6345	−13.9054	3.714	—	−17.9095	3.709	—
D,L-Glutamic acid	0.9317	1.523	—	−1.2359	−12.4244	3.507	—	−16.4071	3.495	—
Glycine	2.1516	1.087	−4.114	0.2762	−13.2619	7.676	−9.473	−17.8976	8.171	−10.50
L-Hydroxyproline	2.4603	0.3891	—	0.3428	−1.6575	0.8959	—	−5.5906	0.8514	—
L-Isoleucine	1.5787	0.07682	2.594	−0.5389	2.7190	−3.081	5.972	−1.3913	−3.020	5.866
D,L-Isoleucine	1.2616	0.2512	3.794	−0.8560	2.9651	−4.193	8.736	−1.1373	−4.134	8.632
L-Leucine[d]	1.3561	0.02233	3.727	−0.7615	4.5073	−4.683	8.582	−0.7252	−3.814	7.198
D,L-Leucine	0.9013	0.2635	4.591	−1.2163	3.4260	−5.167	10.57	−0.6258	−5.143	10.53
D,L-Methionine	1.2597	1.108	−1.221	−0.9140	−11.1682	4.086	−2.811	−15.2099	4.111	−2.871
D,L-Norleucine	0.9258	0.4524	3.402	−1.1918	0.2523	−3.236	7.833	−4.2067	−2.941	7.340
L-Phenylalanine	1.2974	0.6982	—	−0.9204	−6.510	1.608	—	−10.5103	1.601	—
D,L-Phenylalanine	0.9986	0.5252	3.140	−1.2192	−0.7184	−2.739	7.229	−5.0876	−2.495	6.808
L-Proline	3.1050	0.4206	—	1.0441	−0.2407	0.9686	—	−3.8586	0.7586	—
D,L-Serine	1.3432	1.520	−3.548	−0.6782	−17.2153	7.963	−8.169	−21.4529	8.134	−8.504
Taurine	1.5945	1.916	−8.500	−0.5029	−27.8015	15.10	−19.57	−32.1283	15.35	−20.07
L-Tryptophan	0.9156	0.4834	2.988	−1.3942	−11.3824	4.872	6.881	−15.3928	4.869	−6.879

[a] Solubility equations:

   a. $\text{Log } S = a_1 + b_1 t + c_1 t^2$ (grams per 1,000 gms water)

   b. $\text{Log } m = a_2 + b_1 t + c_1 t^2$ (moles per 1,000 gms water)

   c. $\ln m = a_3 + b_3 T + c_3 T^2$

   d. $\ln N_2 = a_4 + b_4 T + c_4 T^2$

[c] The first set of values refer to 273.1° to 303.1° and the second set to 298.1° to 323.5° absolute.

[d] Values are not strictly accurate due to contamination of the leucine with a small amount of methionine.

COEFFICIENTS OF SOLUBILITY EQUATIONS OF CERTAIN AMINO ACIDS IN WATER (continued)

Amino acid	a_1	$b_1 \times 10^2$	$c_1 \times 10^5$	a_2	a_3	$b_3 \times 10^2$	$c_3 \times 10^5$	a_4	$b_4 \times 10^2$	$c_4 \times 10^5$
L-Tyrosine	−0.708	1.46	—	−2.966	−10.799	3.36	—	−20.062	3.37	—
D,L-Tyrosine	−0.833	1.51	—	−3.091	−16.562	3.46	—	−20.577	3.46	—
L-Valine[b]	1.9211	8.1515	8.589	−0.1456	0.6274	−8.927	1.978	−3.4570	−8.528	1.9058
L-Valine[b]	1.6675	97.75	80.22	−0.4011	−20.8468	123.4	18.47	−24.2293	119.4	17.85
L-Valine[b]	1.8847	11.42	4.799	−0.1839	−0.3175	−3.406	1.105	−3.6570	−8.125	1.910
L-Valine[b]	1.9227			−0.1459	−0.3359			−4.3653		
L-Valine[b]	1.8836			−0.1850	−0.4260			−4.4542		
D,L-Valine	1.7749	0.2389	2.607	−0.2966	−2.2921	−2.729	6.003	−1.7417	−2.705	5.928

[b]The five sets of values which are given under L-valine refer to the various crystal forms.

Reprinted with slight modification from *The Chemistry of the Amino Acids and Proteins*, C. L. A. Schmidt, Ed. Charles C Thomas, Publisher, Springfield, Ill., 1944, 845. Courtesy of the publisher.

HEAT CAPACITIES, ABSOLUTE ENTROPIES, AND ENTROPIES OF FORMATION OF AMINO ACIDS AND RELATED COMPOUNDS

John O. Hutchens

Heat capacity (C_p°), absolute entropy (S°) and the entropy change for formation from the elements (ΔSf°) are given for a temperature of 298.15°K. The units are cal $\deg^{-1} \text{mole}^{-1}$ for all entries except for proteins where they are cal $\deg^{-1} g^{-1}$. 1 cal = 4.1840 absolute joules and 0°C−273.15°K. International Atomic Weights of 1959 are employed. In calculating ΔSf° the entropies of the elements used are (in cal $\deg^{-1} \text{mole}^{-1}$) C (graphite), 1.3609; H_2 (gas), 31.211; O_2 (gas), 49.003; N_2 (gas), 45.767; Cl_2 (gas), 53.31; and S (rhombic), 7.62.[1] S° for D_2 (gas) is 34.620 cal $\deg^{-1} \text{mole}^{-1}$.[2]

The references cited give heat capacities at other temperatures, and, in most cases, the additional thermodynamic functions $(H^\circ - H_0^\circ)$, $(H^\circ - H_0^\circ)/T$, and $-(F^\circ - H_0^\circ)/T$. None of the stated values is more accurate than ± 0.2%. Values obtained by extrapolating below 90°K are uncertain by ± 1–2 cal $\deg^{-1} \text{mole}^{-1}$. Entries are listed in alphabetical order under (1) amino acids, (2) peptides, (3) proteins, and (4) miscellaneous related substances.

HEAT CAPACITIES, ENTROPIES AND ENTROPIES OF FORMATION

Compound	C_p°	S°	$-\Delta Sf^\circ$	Reference	Remarks
AMINO ACIDS					
L-Alanine	29.22	30.88	154.33	3, 4	—
D-Arginine	55.8[a]	59.9	307.3[a]	5	Extrapolated below 90°K
L-Arginine · HCl	26.37	68.43	341.01	6	
L-Asparagine	38.3[a]	41.7	207.9[a]	7	—
L-Asparagine · H_2O	49.69	50.10	255.17	8	—
L-Aspartic Acid	37.09	40.66	194.91	8	—
DL-Citrulline	55.2[a]	60.8	292.4[a]	9	Extrapolated below 90°K
L-Cysteine	38.8[a]	40.6	152.2[a]	10	Extrapolated below 90°K
L-Cystine	62.60	67.06	287.38	11	
L-Glutamic Acid	41.84	44.98	223.16	8	—
D-Glutamic Acid · HCl	50.0[a]	59.3	251.1[a]	12	Extrapolated below 90°K
L-Glutamine	44.02	46.62	235.51	8	—
Glycine	23.71	24.74	127.90	3, 4	—
L-Histidine · HCl	59.64	65.99	242.54	6	—
L-Hydroxyproline	36.79	41.19	202.45	13	Transitions at 21.9 and 31.5°K
L-Hydroxyproline (deuterated)	39.40	43.04	205.72	14	Transitions at 25.9 and 28.9°K. From 99.6% D_2O. Carboxyl, hydroxyl, and imido hydrogens presumably replaced
L-Isoleucine	45.00	49.71	233.21	4	—
L-Leucine	48.03	50.62	232.30	4	—
L-Lysine · HCl	57.10	63.21	300.46	6	—
L-Methionine	69.32	55.32	202.65	11	Transition at 305.5°K
DL-Ornithine	45.6[a]	46.2	242.6[a]	9	Extrapolated below 90°K
L-Phenylalanine	48.52	51.06	204.74	15	—
L-Proline	36.13	39.21	179.93	15	—
L-Serine	32.40	35.65	174.06	16	—
L-Threonine	35.2	36.5	205.8	13	Several samples tested. Difficult to dry. Anomalous behavior 250–300°K
L-Tryptophan	56.92	60.00	237.01	15	—
L-Tyrosine	51.73	51.15	229.15	15	—
L-Valine	40.35	42.75	207.60	4	—

[a]Calculated by compiler, not in reference cited.

HEAT CAPACITIES, ENTROPIES AND ENTROPIES OF FORMATION (continued)

Compound	C_p°	S°	$-\Delta Sf^\circ$	Reference	Remarks
PEPTIDES					
DL-Alanylglycine	43.6[a]	51.0	231.1[a]	18	Extrapolated below 90°K. MW not stated in reference. C_p° given per g in reference
Glycylglycine	39.19	43.09	206.47	13	
DL-Leucylglycine	61.3[a]	67.2	312.6[a]	17	Extrapolated below 90°K. MW not stated in reference. C_p° given per g in reference
PROTEINS					
Bovine Serum Albumin	—	—	—	18	
Native Hydrated	0.3161	0.3249	—	—	2.14% H_2O. No heat of fusion noted
Native Anhydrous	0.3049	0.3175	—	—	—
Denatured Anhydrous	0.3096	0.3205	—	—	—
Bovine Zinc Insulin	—	—	—	13	
Native Hydrated	0.3155	0.3252	—	—	4.0% H_2O. No heat of fusion noted
Native Anhydrous	0.2996	0.3144	—	—	
Collagen	—	—	—	14	Bovine serosal collagen
Native Hydrated	0.3834	0.3589	—	—	13.53% H_2O. No heat of fusion noted
Native Anhydrous	0.2921	0.3081	—	—	
α-Chymotrypsinogen	—	—	—	18	Bovine source
Native Hydrated	0.3834	0.3635	—	—	10.7% H_2O. No heat of fusion noted
Native Anhydrous	0.3090	0.3227	—	—	—
MISCELLANEOUS RELATED SUBSTANCES					
Creatine	41.1[a]	45.3	218.2[a]	19	—
Creatinine	33.2[a]	40.0	167.8[a]	19	—
Urea	22.26	25.00	109.06	20	—

REFERENCES

1. National Bureau of Standards Circular 500 (1961), Selected Values of Chemical Thermodynamic Properties, U.S. Gov. Print. Off., Washington, D.C.
2. National Bureau of Standards Report No. 8504 (1964), Preliminary Report on the Thermodynamic Properties of Selected Light-Element and Some Related Compounds, U.S. Gov. Print. Off., Washington, D.C.
3. Hutchens, Cole, and Stout, *J. Am. Chem. Soc.*, 82, 4813 (1960).
4. Hutchens, Cole, and Stout, *J. Phys. Chem.*, 67, 1128 (1963).
5. Huffman and Ellis, *J. Am. Chem. Soc.*, 59, 2150 (1937).
6. Cole, Hutchens, and Stout, *J. Phys. Chem.*, 67, 2245 (1963).
7. Huffman and Borsook, *J. Am. Chem. Soc.*, 54, 4297 (1932).
8. Hutchens, Cole, Robie, and Stout, *J. Biol. Chem.*, 238, 2407 (1963).
9. Huffman and Fox, *J. Am. Chem. Soc.*, 63, 3464 (1940).
10. Huffman and Ellis, *J. Am. Chem. Soc.*, 57, 46 (1935).
11. Hutchens, Cole, and Stout, *J. Biol. Chem.*, 239, 591 (1964).
12. Huffman, Ellis, and Borsook, *J. Am. Chem. Soc.*, 62, 297 (1940).
13. Hutchens, J. O., Cole, A. G., and Stout, J. W., unpublished data.
14. Hutchens, J. O., Kim, S., and Stout, J. W., unpublished data.
15. Cole, Hutchens, and Stout, *J. Phys. Chem.*, 67, 1852 (1963).
16. Hutchens, Cole, and Stout, *J. Phys. Chem.*, 239, 4194 (1964).
17. Huffman, *J. Am. Chem. Soc.*, 63, 688 (1941).
18. Hutchens, J. O., unpublished data.
19. Huffman and Borsook, *J. Am. Chem. Soc.*, 54, 4297 (1932).
20. Ruerhwein and Huffman, *J. Am. Chem. Soc.*, 68, 1759 (1946).

This table originally appeared in Sober, Ed., *Handbook of Biochemistry and selected data for Molecular Biology*, 2nd ed., Chemical Rubber Co., Cleveland, 1970.

HEAT OF COMBUSTION, ENTHALPY AND FREE ENERGY OF FORMATION OF AMINO ACIDS AND RELATED COMPOUNDS

John O. Hutchens

Heat of combustion (ΔH_c°) is given as the enthalpy change for the reaction of burning the compound at constant pressure to produce CO_2 (gas), H_2O (liquid), N_2 (gas) and S (rhombic). The enthalpy of formation (ΔHf°) is given for formation of the compound from C (graphite), H_2 (gas), O_2 (gas), N_2 (gas) and S (rhombic). In calculating ΔHf°, the enthalpies of formation of CO_2 (gas, $p = 0$) = 94.0518 Kcal mole^{-1} and of H_2O (liquid) = 68.3174 Kcal mole^{-1} were employed.[1] Most of the heats of combustion were originally reported as the enthalpy change for the combustion reaction with all gases at $p = 1$ atmosphere. No correction has been made for further expansion of the gases. However, all heats of combustion have been corrected so that International Atomic Weights of 1959 apply. The listings are for a temperature of 298.15°K. 0°C = 273.15°K and 1 cal = 4.1840 absolute joules. The units for all entries are Kcal mole^{-1}.

In calculating ΔGf° the free energy change for formation of the compound from C (graphite), H_2 (gas, f = 1); O_2 (gas, f = 1), N_2 (gas, f = 1) and S (rhombic) the entropies of formation (ΔSf°) of amino acids and peptides listed in table on page B-6 were employed. $\Delta Gf^{\circ} = \Delta Hf^{\circ} - T\Delta Sf^{\circ}$. Where available ΔSf° for the L-isomer of an amino acid was used in calculating ΔGf° for D- and DL-form on the assumption that this entropy value was more reliable. Available entropy data on D- and DL-forms depend on extrapolation of heat capacity data below 90°K while those for the L-amino acids in most cases do not.

The table is intended to be comprehensive rather than selective. All heats of combustion which have come to the compiler's attention have been included regardless of degree of reliability. Where more than one heat of combustion has been reported for a compound, ΔHf° and ΔGf° have been calculated only from the apparently more reliable data. Where no reasonable choice could be made between apparently equally reliable recent heats of combustion (since 1930) ΔHf° and ΔGf° have been calculated from both.

Compounds are listed alphabetically under (1) Amino Acids, (2) Peptides, and (3) Miscellaneous related compounds.

HEAT OF COMBUSTION, ENTHALPY AND FREE ENERGY OF FORMATION

Compound	$-\Delta H_c^{\circ}$	Reference	$-\Delta Hf^{\circ}$	$-\Delta Gf^{\circ}$	Remarks
AMINO ACIDS					
D-Alanine	387.1	2	134.2	88.2	—
	387.6	3	—	—	—
L-Alanine	386.8[a]	4, 5	134.5	88.4	—
DL-Alanine	386.6	6	134.7	88.7	—
	387.6	3	—	—	—
	387.8	7	—	—	—
	389.5	8	—	—	—
D-Arginine	893.5	6	149.0	57.4	—
L-Asparagine	460.8	2	188.7	126.7	—
	463.5	9	—	—	—
L-Asparagine · H_2O	458.1	2	259.7	183.6	—
	459.8	10	—	—	—
L-Aspartic Acid	382.6	2	232.6	174.5	—
	383.2	10	—	—	—
	385.0	11	—	—	—
	385.0	12	—	—	—
	385.7	8	—	—	—

[a]Authors give only ΔE for bomb process. ΔH_c° calculated by compiler.

HEAT OF COMBUSTION, ENTHALPY AND FREE ENERGY OF FORMATION (continued)

Compound	$-\Delta H_c^\circ$	Reference	$-\Delta Hf^\circ$	$-\Delta Gf^\circ$	Remarks
AMINO ACIDS (continued)					
L-Cysteine	394.6	13	126.7	81.3	—
L-Cystine	724.6	13, 14	249.6	163.9	—
D-Glutamic Acid	537.5	2	240.2	173.7	—
L-Glutamic Acid	536.4[a]	15, 16	241.3	174.8	—
	536.9	12	—	—	—
	542.6	8	—	—	—
L-Glutamine	614.3[a]	15, 16	197.5	127.3	—
Glycine	230.5[a]	4, 5	128.4	90.3	—
	232.6	6	126.3	88.2	—
	233.4	3	—	—	—
	234.5	9	—	—	—
L-Isoleucine	855.8[a]	4, 5	152.5	83.0	—
D-Leucine	856.0	6	152.4	83.1	—
L-Leucine	856.0	6	152.4	83.1	—
	853.7[a]	15	154.6	85.4	—
DL-Leucine	855.2	6	153.2	83.9	—
	856.0	9	—	—	—
L-Methionine	664.8[a]	4, 14	181.2	120.9	—
L-Phenylalanine	1110.6	4	111.6	50.6	—
DL-Phenylalanine	1111.9	8	110.9	49.9	—
	1112.3	17	—	—	—
L-Serine	347.7	18	173.6	121.6	Calculated from structural factor -OH for -H on L-Alanine
Isoserine	343.8	3	177.5	125.5	Secondary alcohol. $-\Delta H_c^\circ$ lower than Serine expected
L-Threonine	490.7[a]	4	192.9	131.5	—
L-Tryptophan	1345.2[a]	4	99.2	28.5	—
L-Tyrosine	1061.7	12	160.5	92.2	Agrees with structural change -OH for -H on L-Phenylalanine
	1058.3	6	—	—	—
	1070.8	10	—	—	—
L-Valine	698.3[a]	15, 5	147.7	85.8	—
DL-Valine	701.1	3	144.9	83.0	—
PEPTIDES					
DL-Alanylglycine	625.9	19	186.0	117.1	—
DL-Alanyl-DL-Phenylalanine	1505.4	20	169.8	77.4[b]	—
DL-Alanyl-DL-Phenylalanyl-glycine	1742.7	21	223.0	107.2[b]	—
Glycyl-DL-Alanyl-DL-phenylalanine	1744.2	20	221.5	105.7	—
Glycylglycine	471.4	19	178.1	116.6	—
	470.8	8	—	—	—
di-Glycylglycine	710.0	3	230.1	145.1[b]	—
tri-Glycylglycine	946.9	3	283.7	175.3[b]	—
Glycyl-DL-Phenylalanine	1349.2	20	163.6	79.1[b]	—
Glycyl-DL-Tryptophan	1586.2	21	148.8	54.7[b]	—
Glycyl-DL-Valine	937.0	20	199.6	115.3[b]	—
DL-Leucylglycine	1093.4	19	205.6	112.4	—
	1095.8	3	—	—	—
DL-Leucylglycylglycine	1333.7	8	255.8	139.7[b]	—
DL-Serylserine	692.7	20	281.5	192.4[b]	—
DL-Valyl-DL-Phenylalanine	1816.8	21	183.1	74.9[b]	—
MISCELLANEOUS					
Creatine	555.1	2	128.4	63.3	—
Creatinine	558.1	2	57.0	7.0	—
Urea	151.0	22	79.6	47.1	—

[b]Calculated assuming $\Delta S_{298.15} = 10.3$ cal deg^{-1} mole^{-1} for the reaction: Amino Acid (solid) + Amino Acid (solid) → Dipeptide (solid) + H_2O (liquid).

HEAT OF COMBUSTION, ENTHALPY AND FREE ENERGY OF FORMATION (continued)

REFERENCES

1. National Bureau of Standards Circular 500 (1961), *Selected Values of Chemical Thermodynamic Properties,* U.S. Gov. Print. Off., Washington, D.C.
2. Huffman, Ellis, and Fox, *J. Am. Chem. Soc.,* 58, 1728 (1936).
3. Wrede, *Z. Phys. Chem.,* 75, 81 (1910).
4. Tsuzuki, Harper, and Hunt, *J. Phys. Chem.,* 62, 1594 (1958).
5. Hutchens, Cole, and Stout, *J. Phys. Chem.,* 67, 1128 (1963).
6. Huffman, Fox, and Ellis, *J. Am. Chem. Soc.,* 59, 2144 (1937).
7. Landrieu, *Compt. Rend.,* 142, 580 (1906).
8. Fischer and Wrede, *Sitz. Kgl. Preuss. Akad. Wiss.,* (1904) p 687.
9. Stohmann and Langbein, *J. Prakt. Chem.,* 44, 336 (1891).
10. Emery and Benedict, *Am. J. Physiol.,* 28, 301 (1911).
11. Stohmann, *Z. Phys. Chem.,* 10, 410 (1892).
12. Oka, *Nippon Seirigaku Zasshi,* 9, 365 (1944).
13. Huffman and Ellis, *J. Am. Chem. Soc.,* 57, 41 (1935).
14. Hutchens, Cole, and Stout, *J. Biol. Chem.,* 239, 591 (1964).
15. Tsuzuki and Hunt, *J. Phys. Chem.,* 61, 1668 (1957).
16. Hutchens, Cole, Robie, and Stout, *J. Biol. Chem.,* 238, 2407 (1963).
17. Breitenbach, Derkosch, and Wessely, *Nature,* 169, 922 (1952).
18. Hutchens, Cole, and Stout, *J. Phys. Chem.,* 67, 1852 (1963).
19. Huffman, *J. Phys. Chem.,* 46, 885 (1952).
20. Ponomarev, Alekseeva, and Akimova, *Russ. J. Phys. Chem.* (Engl. Transl.), 36, 457 (1963).
21. Alekseeva and Ponomarev, *Russ. J. Phys. Chem.* (Engl. Transl.), 38, 731 (1964).
22. Huffman, *J. Am. Chem. Soc.,* 62, 1009 (1940).

This table originally appeared in Sober, Ed., *Handbook of Biochemistry and selected data for Molecular Biology,* 2nd ed., Chemical Rubber Co., Cleveland, 1970.

SOLUBILITIES OF AMINO ACIDS IN WATER AT VARIOUS TEMPERATURES

John O. Hutchens

The table gives the solubilities of the amino acids in grams per kilogram of water at various temperatures. None of the experimental observations were made at temperatures above 70°C. In most cases the experimental values were obtained both from a colder originally unsaturated solution and a warmer originally saturated solution. Even where four significant figures are given, many of the values probably are not reliable to better than 1 to 2% as judged by agreement between investigators. Lack of a value at a given temperature indicates that, in the opinion of the compiler, extrapolation of the curve from lower temperatures could not be justified.

SOLUBILITIES OF AMINO ACIDS IN WATER AT VARIOUS TEMPERATURES (g/kg)

Substance	0°	10°	20°	30°	40°	50°	60°	70°	80°	90°	100°	References
L-Alanine	127.3	141.7	157.8	175.7	195.7	217.9	242.6	270.2	300.8	335.0	373.0	1, 2
DL-Alanine	121.1	137.8	156.7	178.3	202.9	230.9	262.7	299.0	340.1	387.0	440.4	1, 2, 3
L-Arginine · HCL	400.0	553.0	718.0	931.0	1240.0	—	—	—	—	—	—	2
L-Asparagine · H₂O	8.49	14.29	23.5	37.79	59.37	91.18	136.8	200.6	287.7	403.0	551.7	4
L-Aspartic Acid	1.72	2.82	4.18	5.94	8.38	11.99	17.01	24.14	34.25	48.59	68.93	1, 3
DL-Aspartic Acid	2.62	4.12	6.33	9.50	14.0	20.0	28.0	38.4	51.4	67.3	85.9	1, 3
L-Cystine	0.0502	0.0686	0.0938	0.1281	0.1751	0.2394	0.3272	0.4472	0.612	0.836	1.142	4
L-Diiodotyrosine	0.204	0.318	0.494	0.769	1.197	1.862	2.897	4.508	7.015	10.91	16.98	1
L-Glutamic Acid	3.29	4.98	7.20	10.19	14.70	—	—	—	—	—	—	1
DL-Glutamic Acid	8.55	12.13	17.22	24.47	34.75	49.34	70.06	99.50	141.3	200.8	284.9	2
Glycine	141.8	180.4	225.2	275.9	331.6	391.0	452.6	513.9	572.7	626.2	671.7	1, 3
L-Hydroxyproline	288.6	315.6	345.2	377.6	413.0	451.7	494.1	540.4	591.0	646.5	706.9	1, 3
L-Isoleucine	—[a]	32.0	33.6	35.4	37.5	—	—	—	—	—	—	5
DL-Isoleucine	18.26	19.52	21.23	23.50	26.47	30.3	35.39	42.01	50.75	62.37	78.02	2
L-Leucine	22.70	23.01	23.74	24.90	26.58	28.87	31.89	35.84	40.98	47.65	56.38	1, 3
DL-Leucine	7.97	8.56	9.39	10.51	12.03	14.06	16.78	20.46	25.46	32.38	42.06	1
L-Lysine · HCL	462.0	556.0	666.0	799.0	965.0	—	—	—	—	—	—	1, 3
L-Methionine	36.8	43.7	51.4	60.4	70.9	—	—	—	—	—	—	2
DL-Methionine	18.19	23.41	29.95	38.12	48.23	60.70	75.95	94.52	116.9	143.9	176.0	2
DL-Norleucine	8.43	9.43	10.71	12.36	14.49	17.27	20.88	25.7	32.02	40.60	52.29	4
L-Phenylalanine	19.83	23.29	27.35	32.13	37.73	44.31	52.04	61.11	71.78	84.29	99.00	1, 3
DL-Phenylalanine	9.97	11.33	13.07	15.29	18.15	21.87	26.71	33.12	41.66	53.16	68.8	4
L-Proline	1274.0	1403.0	1545.0	1703.0	1876.0	2066.0	2277.0	2508.0	2764.0	3045.0	3355.0	1, 3
L-Serine	133.0	247.0	362.0	476.0	592.0	—	—	—	—	—	—	5
DL-Serine	22.04	31.03	42.95	58.52	78.42	103.4	134.1	171.1	214.8	265.4	322.4	5
Taurine	39.31	59.92	87.84	123.8	167.8	218.8	274.2	330.5	383.1	427.0	457.6	2
L-Tryptophan	8.23	9.27	10.57	12.23	14.35	17.06	20.57	25.14	31.16	39.14	49.87	4
L-Tyrosine	0.196	0.274	0.384	0.537	0.752	1.052	1.473	2.1	2.884	4.036	5.7	4
DL-Tyrosine	0.147	0.208	0.294	0.417	0.743	0.836	1.2	1.7	2.4	3.4	4.8	4
L-Valine	—[b]	53.7	56.5	59.7	63.6	—	—	—	—	—	—	6
DL-Valine	59.6	63.3	68.1	74.2	81.7	91.1	102.8	117.4	135.8	158.9	188.1	2, 7

[a] Experimental value of 33 g/kg H₂O at 1°C.

[b] Experimental value of 53.4 g/kg H₂O at 1°C.

REFERENCES

1. Dalton and Schmidt, J. Biol. Chem., 103, 549 (1933).
2. Hade, Thesis, University of Chicago, December 1962.
3. Dunn, Ross, and Read, J. Biol. Chem., 103, 579 (1933).
4. Dalton and Schmidt, J. Biol. Chem, 109, 241 (1935).
5. Tomiyama and Schmidt, J. Gen. Physiol., 19, 379 (1935).
6. Winnick and Schmidt, J. Gen. Physiol., 18, 889 (1935).
7. Dalton and Schmidt, J. Gen. Physiol., 19, 767 (1936).

This table originally appeared in Sober, Ed., *Handbook of Biochemistry and selected data for Molecular Biology*, 2nd ed., Chemical Rubber Co., Cleveland, 1970.

HEATS OF SOLUTION OF AMINO ACIDS IN AQUEOUS SOLUTION AT 25°C

John O. Hutchens

Enthalpy changes are given for: (1) Solution of the solid crystalline amino acid into the infinitely dilute solution ($\bar{H}_2^\circ - H_2^-$), (2) Solution of the solid crystalline substance into saturated solution ($\bar{H}_2$ sat $- H_2^-$), and (3) Dilution of the saturated solution to infinite dilution ($\bar{H}_2^\circ - \bar{H}_2$ sat). All entries are Kcal mole^{-1} and are for 25°C. Columns headed "cal" indicate direct calorimetric measurement. Columns headed "soly" indicate that the value was calculated from solubility and activity data. Generally speaking the two methods agree within the limits of error of either method.

The data suffer from multiple defects. The purity of many of the amino acids used in older work[1] is subject to question. Calculations from solubility data are handicapped by poor knowledge of activity coefficients, particularly for sparingly soluble amino acids. Some amino acids form hydrates, and it is not always clear that the solubility data are for the crystalline anhydrous form or that equilibrium has been reached as regards either crystalline form or solubility.

HEATS OF SOLUTION OF AMINO ACIDS IN AQUEOUS SOLUTION AT 25°C

Compound	$\bar{H}_2^\circ - H_2^-$		$\bar{H}_2(sat) - H_2^-$		$\bar{H}_2^\circ - \bar{H}_2(sat)$	Reference	Remarks
	cal	soly	cal	soly			
D-Alanine	—	1.83	—	1.83	0	1	Agrees with (2). Assumed $\partial \ln \gamma_m / \partial \ln m = 0$
L-Alanine	—	1.83	—	2.0	−0.2	2	Assumes $\partial \ln \gamma_N / \partial \ln N = 0.09$
DL-Alanine	2.0	2.2	2.2	2.2	−0.2	1	—
D-Arginine	1.5	—	1.1	—	0.4	1	
L-Arginine · HCl	—	7.9	—	7.0	0.9	2	Assumes $\partial \ln \gamma_N / \partial \ln N = -0.245$
L-Asparagine	5.8	—	5.8	—	0	1	—
L-Asparagine · H$_2$O	8.0	8.4	8.0	—	0	1	—
	—	—	—	7.7	—	2, 3	—
L-Aspartic Acid	6.0	6.2	5.8	5.6	0.2?	1	Questionable assumption that $\partial \ln \gamma_m / \partial \ln m = 0.549$
	—	6.3	—	6.3	0	2, 3	Assumes $\partial \ln \gamma_N / \partial \ln N = 0$
DL-Aspartic Acid	7.1	7.2	6.9	6.5	0.2?	1	Questionable assumption that $\partial \ln \gamma_m / \partial \ln m = 0.549$
L-Cystine	—	5.5	—	—	—	1	Too sparsely soluble for reliable activity measurements
L-Diiodotyrosine	—	7.8	—	7.8	0	1	
D-Glutamic Acid	—	6.5	6.3	6.0	—	1	Questionable assumption that $\partial \ln \gamma_m / \partial \ln m = 0.539$
DL-Glutamic Acid	6.5	6.2	6.3	5.7	0.2?	1	Questionable assumption that $\partial \ln \gamma_m / \partial \ln m = 0.539$
Glycine	3.8	3.4	3.4	3.4	0.4	1	$\partial \ln \gamma_m / \partial \ln m = 0.06$
L-Histidine	3.3	—	3.2	—	0.1	1	$\bar{H}_2^\circ - \bar{H}_2$ (sat) from dilution of 0.5 M solution
L-Histidine · HCl	—	10.2	—	—	—	2	Activity data needed $\gamma_\pm \neq 1$
L-Hydroxyproline	1.4	1.4	1.4	—	<0.1	1	$\bar{H}_2^\circ - \bar{H}_2$ (sat) from dilution of 2 M solution
D-Isoleucine	—	0.8	—	—	—	1	—
L-Isoleucine	—	0.9	—	0.9	<0.1	2	—
DL-Isoleucine	—	1.8	—	1.8	—	1	—
L-Leucine	—	0.8	—	0.8	—	1	Sample contained methionine
	—	1.0	—	1.0	<0.1	2, 3	—
DL-Leucine	—	2.0	—	2.1	−0.1?	1	Questionable assumption that $\partial \ln \gamma_m / \partial \ln m = 0.38$
D-Lysine	−4.0	—	−3.5	—	−0.5	1	$\bar{H}_2^\circ - \bar{H}_2$ (sat) from dilution of 1 M solution

HEATS OF SOLUTION OF AMINO ACIDS IN AQUEOUS SOLUTION AT 25°C (continued)

Compound	$\overline{H}_2^\circ-\overline{H}_2$		$\overline{H}_2(sat)-\overline{H}_2$		$\overline{H}_2^\circ-\overline{H}_2(sat)$	Reference	Remarks
L-Lysine·HCl	—	5.0	—	7.0	−2.0	2	Assumes $\partial \ln \gamma_N/\partial \ln N = 0.44$
L-Methionine	—	2.8	—	2.8	<0.1	2	—
DL-Norleucine	—	2.5	—	2.5	—	1	—
L-Phenylalanine	—	2.8	—	—	—	1	—
	—	2.7	—	2.7	—	2, 3	Forms hydrate
DL-Phenylalanine	—	2.8	—	2.8	—	1	—
L-Proline	−0.8	1.3	>0.3	—	−1.0	1	$\overline{H}_2^\circ-\overline{H}_2$ (sat) from dilution of 8 M solution
L-Serine·H$_2$O	—	4.6	—	3.7	0.9	2	
L-Serine	2.8	—	—	—	—	4	$\overline{H}_2^\circ-\overline{H}_2$ includes heat of hydration
DL-Serine	5.2	5.4	—	5.0	0.1	1	Must form hydrate in solution
L-Tryptophan	—	1.4	—	—	—	1	—
	—	—	—	2.5	—	3	Possibly forms hydrate
L-Tyrosine	—	6.0	—	6.0	—	1	—
D-Valine	—	0.5	—	—	—	1	—
L-Valine	—	0.9	—	1.0	−0.1	2, 3	—
DL-Valine	1.4	1.5	1.7	1.6	−0.3	1	Assumes $\partial \ln \gamma_m/\partial \ln m = -0.549$

REFERENCES

1. **Huffman and Borsook,** in *Chemistry of the Amino Acids and Proteins,* Schmidt, Ed., C C Thomas, Springfield, Ill., 1938. A compilation, original references are cited.
2. **Hade,** Thesis, University of Chicago, December 1962.
3. **Hade, E. P. K.,** personal communication.
4. **Hutchens, J. O. and Hade, E. P. K.,** unpublished data.

This table originally appeared in Sober, Ed., *Handbook of Biochemistry and selected data for Molecular Biology,* 2nd ed., Chemical Rubber Co., Cleveland, 1970.

FREE ENERGIES OF SOLUTION AND STANDARD FREE ENERGY OF FORMATION OF AMINO ACIDS IN AQUEOUS SOLUTION AT 25°C

John O. Hutchens

The table lists the molality (m, moles per Kg of H_2O) of the saturated solution at 25°C (298.15°K); the appropriate molal activity coefficient (γ_m or $\gamma_{\underline{m}}^{\pm}$); the free energy change for transporting one mole of the solute from the saturated solution to a hypothetical aqueous solution at an activity of 1 molal (ΔG soln); and the free energy change for formation of the substance in hypothetical 1 molal solution from the elements ($\Delta \widetilde{G}f°$). The units of ΔG soln and $\Delta \widetilde{G}f°$ are Kcal mole^{-1}. The free energies of formation of the solid crystalline amino acids are given elsewhere in this handbook.

FREE ENERGIES OF SOLUTION AND STANDARD FREE ENERGY OF FORMATION OF AMINO ACIDS IN AQUEOUS SOLUTION AT 25°C SATURATED SOLUTION

Substance	Saturated Solution m	Reference	γ_m	Reference	ΔG soln	$-\Delta \widetilde{G}f°$	Remarks
	mole/kg				Kcal mole^{-1}	Kcal mole^{-1}	
L-Alanine	1.862	1	1.045	2	−0.368	88.8	—
L-Arginine·HCl	4.061	1	0.587[a]	3	−1.03	—	Assumes 1 : 1 electrolyte ΔG solution = 2RT ln $(m_{\pm}\gamma_m^{\pm})$
L-Asparagine·H$_2$O	0.190	1	1.0	4	0.983	182.6	—
L-Aspartic Acid	0.0375	1	1.0	4	2.06	172.4	—
L-Cystine	4.57×10^{-4}	5	1.0	—	4.53	159.4	Activity coefficient assumed
L-Glutamic Acid	0.0586	1	1.0	4	1.77	173.0	—
L-Glutamine	0.291	6	1.0	4	0.731	126.6	—
Glycine	3.33	5	0.729	7	−0.525	$\begin{cases}90.8\\88.7\end{cases}$	Heats of combustion disagree
L-Hydroxyproline	2.75	5	1.05	8	−0.629	—	—
L-Isoleucine	0.263	1	1.0	4	0.791	82.2	—
L-Leucine	0.165	1	1.0	4	1.07	$\begin{cases}82.0\\84.3\end{cases}$	Heats of combustion disagree
L-Methionine	0.377	1	0.875	4	0.656	120.2	—
L-Phenylalanine	0.167	1	1.0	4	1.06	49.5	—
L-Proline	14.1	5	3.13	8	−2.24	—	Activity coefficient from extrapolation above m = 7.3
L-Serine	4.02	1	0.602	3	−0.524	122.1	—
L-Tryptophan	0.0666	1	1.0	4	1.60	26.9	—
L-Tyrosine	2.51×10^{-3}	5	1.0	—	3.55	88.6	Activity coefficient assumed
L-Valine	0.496	1	0.923	4	0.461	85.3	—

[a]Activity coefficient is $\gamma_m^{\pm}$.

REFERENCES

1. **Hade,** Thesis, University of Chicago, December 1962.
2. **Smith and Smith,** *J. Biol. Chem.,* 121, 607 (1937).
3. **Hutchens, Figlio and Granito,** *J. Biol. Chem.,* 238, 1419 (1963).
4. **Hutchens, J. O. and Nancy Norton,** unpublished data.
5. **Borsook and Huffman,** in *The Chemistry of the Amino Acids and Proteins,* Schmidt, Ed., C C Thomas, Springfield, Ill., 1938.
6. **Weast, Ed.,** *Handbook of Chemistry and Physics,* 38th ed., The Chemical Rubber Co., Cleveland, Ohio, 1956.
7. **Smith and Smith,** *J. Biol. Chem.,* 117, 209 (1937).
8. **Smith and Smith,** *J. Biol. Chem.,* 132, 57 (1940).

This table originally appeared in Sober, Ed., *Handbook of Biochemistry and selected data for Molecular Biology,* 2nd ed., Chemical Rubber Co., Cleveland, 1970.

ACTIVITIES OF AMINO ACIDS AND PEPTIDES AT 25°C

John O. Hutchens

The negative logarithms ($- \log_{10}$) of the molal activity coefficients (γ_m) are listed as a function of molality of the compound. γ_m = a/m. All of the data were obtained by a method involving isopiestic solutions.

The following amino acids of limited solubility have been shown to have essentially unit activity up to saturation, i.e. $-\log \gamma_m$ = 0.00 ± 0.01.[1]

L-Aspartic Acid	(m = 0.038)
L-Glutamic Acid	(m = 0.058)
L-Isoleucine	(m = 0.263)
L-Leucine	(m = 0.166)
L-Phenylalanine	(m = 0.167)
L-Tryptophan	(m = 0.069)

MOLAL ACTIVITY COEFFICIENTS ($-\log_{10}$)

Substance	0.2 m	0.3 m	0.5 m	0.7 m	1.0 m	1.5 m	2.0 m	2.5 m	3.0 m	3.5 m	4.0 m	Reference
Alanine	−0.002	−0.003	−0.005	−0.007	−0.010	−0.015[a]	—	—	—	—	—	2
Alanylalanine	0.008	0.010	0.006	−0.001	−0.015	—	—	—	—	—	—	3
Alanylglycine	0.031	0.042	0.061	0.064	0.068	—	—	—	—	—	—	3
α-amino *n*-butyric acid	−0.0046	−0.0070	−0.0122	−0.0177	−0.0275	−0.0421	−0.0664	—	—	—	—	5
Arginine · HCl $(-\log_{10})\gamma^{\pm}_{m}$	0.202	0.243	0.303	0.346	0.398	0.456	0.499	0.529	0.551	0.570	0.585	4
Glycine	0.0176	0.0232	0.0421	0.0489	0.0579	0.0868	0.1041	0.1181	0.1297[b]	—	—	5
Glycylalanine	0.029	0.040	0.054	0.061	0.068	—	—	—	—	—	—	3
Glycylglycine	0.0404	0.0568	0.0844	0.1069	0.1532	0.1616	—	—	—	—	—	5
Tri-glycine	0.070	0.095	—	—	—	—	—	—	—	—	—	3
Hydroxyproline	0.000	0.000	−0.001	−0.001	−0.003	−0.006	−0.011	—	—	—	—	6
Methionine	−0.008	−0.012	−0.020	−0.029	−0.040	−0.060	−0.081	−0.103	−0.126	−0.148	−0.174[c]	1
Proline	0.042	0.052	—	—	—	—	—	—	—	—	—	6
Serine	0.022	0.032	0.052	0.070	0.094	0.127	0.152	0.174	0.193	0.207	0.220	4
Threonine	0.005	0.007	0.011	0.015	0.018	0.022	0.025	—	—	—	—	6
Valine	−0.013	−0.019	−0.032	—	—	—	—	—	—	—	—	2

[a] Also −0.019 at 1.86 m.
[b] Also 0.1321 at 3.114 m.
[c] Also −0.224 at 5.0 m, −0.262 at 6.0 m, −0.296 at 7.0 m, and −0.302 at 7.3 m.

REFERENCES

1. **Hutchens, J. O. and Nancy Norton**, unpublished data.
2. **Smith and Smith**, *J. Biol. Chem.*, 121, 607 (1937).
3. **Smith and Smith**, *J. Biol. Chem.*, 135, 273 (1940).
4. **Hutchens, Figlio, and Granito**, *J. Biol. Chem.*, 238, 1419 (1963)
5. **Ellerton, Reinfelds, Mulcathy, and Dunlop**, *J. Phys. Chem.*, 68, 398 (1964).
6. **Smith and Smith**, *J. Biol. Chem.*, 132, 57 (1940).

This table originally appeared in Sober, Ed., *Handbook of Biochemistry and selected data for Molecular Biology*, 2nd ed., Chemical Rubber Co., Cleveland, 1970.

ENTHALPY, ENTROPY, AND FREE ENERGY VALUES FOR BIOCHEMICAL REDOX REACTIONS

(Data are reported for pH 7.0 and 298 K)

Oxidation half reaction	$E^{\circ\prime}$ volts	ΔG kJ mole^{-1}	ΔH kJ mole^{-1}	ΔS J mole^{-1} deg^{-1}	References (enthalpy data)
Non protein reactions					
$2\,Cys \rightleftharpoons (Cys)_2 + 2H^+ + 2e^-$	0.32	−61.5	40.2	341	1, 2
$2GSH \rightleftharpoons (GS)_2 + 2H^+ + 2e^-$	0.23	−44.4	25.1	233	1, 2
$2HOC_2H_4SH \rightleftharpoons (HOC_2H_4S)_2 + 2H^+ + 2e^-$	−	−	38.1	−	1, 2
$L(+)Lactate \rightleftharpoons pyruvate + 2H^+ + 2e^-$	0.18	−34.7	78.2	379	1, 2
$H_4Folate \rightleftharpoons H_2\,folate + 2H^+ + 2e$	0.18	−34.7	211.7	828	3
$Ascorbate \rightleftharpoons dehydroascorbate + 2H^+ + 2e^-$	−0.06	11.7	77.8	222	−
$FMNH_2 \rightleftharpoons FMN + 2H^+ + 2e^-$	−0.22	42.3	56.1	46.0	4
$Hydroquinone \rightleftharpoons benzoquinone + 2H^+ + 2e^-$	−0.29	55.6	177.4	408	1, 2
$NADH \rightleftharpoons NAD^+ + H^+ + 2e^-$	−0.32	61.6	29.2	−108	2, 5, 6
$NADPH \rightleftharpoons NADP^+ + H^+ + 2e^-$	−0.324	62.4	25.3	−124	2, 5, 7
$Fe(CN)_6 \rightleftharpoons Fe(CN)_6^{-3} + e^-$	−0.36	69.4	111.7	142	8[a]
Protein reactions					
$Fe(II)hemerythrin \rightleftharpoons Fe(III)hemerythrin + e^-$	−	−	102.5	−	9
$Ferrocytochrome\ c \rightleftharpoons Ferricytochrome\ c + e^-$ (mammalian)	−0.26	25	59	114	1
$Ferrocytochrome\ c \rightleftharpoons Ferricytochrome\ c + e^-$ (bacterial)	−	−	79.5	−	1

Compiled by Neal Langerman.

[a]This reaction is the commonly used reference reaction.

REFERENCES

1. Watt and Sturtevant, personal communication.
2. Schott and Sturtevant, personal communication.
3. Rothman, Kisliuk, and Langerman, *J. Biol. Chem.*, 248, 7845 (1973).
4. Beaudette and Langerman, *Arch. Biochem. Biophys.*, 161, 125 (1974).
5. Burton, *Biochem. J.*, 143, 365 (1974).
6. Poe, Gutfreund, and Estabrook, *Arch. Biochem. Biophys.*, 122, 204 (1967).
7. Engel and Dalziel, *Biochem. J.*, 105, 691 (1967).
8. Hanania, Irvine, Eaton, George, *J. Phys. Chem.*, 71, 2022 (1967).
9. Langerman and Sturtevant, *Biochemistry*, 10, 2809 (1971).

OXIDATION-REDUCTION POTENTIALS, ABSORBANCE BANDS AND MOLAR ABSORBANCE OF COMPOUNDS USED IN BIOCHEMICAL STUDIES

Paul A. Loach

In addition to the references[1-5,7,8] in the table, other generally used sources of oxidation-reduction data are: *Biochemist's Handbook,* D. Van Nostrand Co., Princeton, N.J. (1961); *The Encyclopedia of Electrochemistry,* C. A. Hampel, Ed., Reinhold Publishing Corp. New York (1964); *Oxidation-Reduction Potentials in Bacteriology and Biochemistry,* sixth ed., L. F. Hewitt, McCorquodale and Co. Ltd., London, (1950); *Biochemisches Taschenbuch* part II, Springer-Verlag, New York, (1964).·

The oxidation-reduction couples are listed according to decreasing values of $E°$ or $E°'$. When both values are available, the order is according to $E°'$. Unless otherwise indicated, $E°'$ is the mid-point potential for a particular couple at pH7. Temperatures are not listed; most of the data are relevant to room temperature (20°C to 30°C). When more exact conditions are desired (ionic strength, concentration, temperature, nature of data used to derive $E°$ or $E°'$) the reader should consult the reference listed.

OXIDATION-REDUCTION POTENTIALS, ABSORBANCE BANDS AND MOLAR ABSORBANCE OF COMPOUNDS USED IN BIOCHEMICAL STUDIES

System	$E°$	$E°'$	λ_{max}	E_{mM}	Reference
1. $F_2(gas)/F^-$	2.87	—	—	—	1
2. $H_2N_2O_2/N_2$ (gas)	2.65	—	—	—	1
3. $S_2O_8^{2-}/SO_4^{2-}$	2.0	—	—	—	1
4. H_2O_2/H_2O	1.77	—	—	—	1
5. MnO_4^-/MnO_2	1.69	—	—	—	1
6. $HClO_2/HClO$	1.64	—	—	—	1
7. H_5IO_6/IO_3^-	1.6	—	—	—	1
8. MnO_4^-/Mn^{2+}	1.51	—	—	—	1
9. Mn^{3+}/Mn^{2+}	1.4	—	—	—	2
10. $Cl_2(gas)/Cl^-$	1.359	—	—	—	1
11. $ClO_2(gas)/HClO_2$	1.27	—	—	—	1
12. MnO_2/Mn^{2+}	1.23	—	—	—	1
13. $[Mn^{3+}(PO_4)_2]^{3-}/[Mn^{2+}(PO_4)_2]^{4-}$	1.22	—	—	—	2
14. Pt^{2+}/Pt	1.2	—	—	—	1
15. IO_3^-/I_2	1.19	—	—	—	1
16. ClO_4^-/ClO_3^-	1.19	—	—	—	1
17. $ClO_3^-/ClO_2(gas)$	1.15	—	—	—	1
18. $[Cu^{3+}(IO_6)_2]^{7-}/[Cu^{2+}(IO_6)_2]^{8-}$, pH 8	—	1.1			2
pH 12		0.7			
19. Br_2/Br^-	1.087	—	—	—	1
20. $N_2O_4(gas)/HNO_2$	1.07	—	—	—	1
21. Fe^{3+}/Fe^{2+} O-phenanthroline	1.06	—	—	—	3
22. $[IrCl_6]^{2-}/[IrCl_6]^{3-}$	1.05	—	—	—	1
23. VO_3^-/VO^{2+}	1.0	—	—	—	2
24. IO_4^-/IO_3^-	1.375	0.96	—	—	2
25. $HNO_2/NO(gas)$	0.99	—	—	—	1
26. p-Toluene-sulphochloramide, Na salt (Chloramine-T)	1.52	0.90	—	—	2
27. Nitrosoguanidine/Nitroguanidine	0.85	—	—	—	4
28. O_2/H_2O	1.229	0.816	—	—	1
29. $NO_3^-/N_2O_4(gas)$	0.80	—	—	—	1
30. Ag^+/Ag	0.7994	—	—	—	1
31. 1,2-Benzoquinone	0.792	—	—	—	5
32. Hg_2^{2+}/Hg	0.792	—	—	—	1
33. Zn Octaethylporphyrin (methanol)	—	0.78	—	—	71
34. Fe^{3+}/Fe^{2+}	0.771	—	—	—	1
35. $[Mo^{3+}(CN)_6]^{3-}/[Mo^{2+}(CN)_6]^{4-}$	0.73	—	—	—	5
36. Porphyrexide	—	0.725	—	—	5
37. Pyrogallol	0.713	—	—	—	5
38. $NO(gas)/H_2N_2O_2$	0.71	—	—	—	1
39. Hg^{2+}/Hg_2^{2+}	0.91	—	—	—	1
40. 1,2-Naphthoquinone-4-sulfonate	0.628	—	—	—	5
41. Mn^{4+}/Mn^{3+} Hematoporphyrin IX, pH 9.9	—	0.626	400 (ox)	70	6
42. MnO_4^-/MnO_4^{2-}	0.6	—	—	—	1
43. $S_2O_6^{2-}/H_2SO_3$	0.6	—	—	—	1
44. $[W(CN)_8]^{3-}/[W(CN)_8]^{4-}$	0.57	—	—	—	5
45. Porphyrindin	—	0.565	—	—	5
46. NH_2OH/NH_4	—	0.562	—	—	7
47. $H_3AsO_4/HAsO_2$	0.56	—	—	—	1
48. o-Tolidine	—	0.55	—	—	5
49. Cu^{2+}/Cu^+ Hemocyanin	—	0.540	350		8, 9
			600		
50. I_2/I^-	0.536	—	—	—	1
51. Cu^+/Cu	0.521	—	—	—	1
52. Bacteriochlorophyll a (methanol)	—	0.52	—	—	71
53. $S_2O_3^{2-}/S$	0.5	—	—	—	1
54. $S_2O_4^{2-}/S_2O_3^{2-}$	1.03	0.484	—	—	1, 7
55. MoO_2^{2+}/MoO^{3+}	0.48	—	—	—	1
56. Phenylhydrazine sulfonate	0.437	—	—	—	5
57. 2-Methyl-1,4-naphthoquinone (Menadione-Vitamin K_3)	0.422	—	—	—	5
58. P_{700}	—	0.43	—	—	10

OXIDATION-REDUCTION POTENTIALS, ABSORBANCE BANDS AND MOLAR ABSORBANCE OF COMPOUNDS USED IN BIOCHEMICAL STUDIES (continued)

System	$E°$	$E°'$	λ_{max}	E_{mM}	Reference
59. $P_{890}(P_{0.44})$	—	0.44	—	⌐	11, 12 13, 14
60. NO_3^-/NO_2^-	0.94	0.421	—	—	1, 7
61. $H_2SO_3/S_2O_3^{2-}$	0.40	—	—	—	1
62. 2,5-Dihydroxy-1,4-benzoquinone		0.38	—	—	5
63. Adrenalin	0.809	0.380	—	—	4, 5
64. *p*-Amino-dimethyl aniline	—	0.38	—	—	5
65. Fe^{3+}/Fe^{2+} Cytochrome f	—	0.365	413 (ox) 423(red) 525(red) 555(red)	—	66
66. $[Fe(CN)_6]^{3-}/[Fe(CN)_6]^{4-}$	—	0.36	— 420(ox)	— 1.000	5 15
67. *o*-Quinone/Diphenol	—	0.35	—	—	5
68. Fe^{3+}/Fe^{2+} Cytochrome c_{550} (*R. rubrum*)	—	0.338	409(ox) 416(red) 521(red) 550(red)	—	16
69. Cu^{2+}/Cu	0.337	—	—	—	1
70. Fe^{3+}/Fe^{2+} Acetate, pH 5	—	0.34	—	—	5
71. Fe^{3+}/Fe^{2+} Cytochrome c_5 (*Azotobacter*)	—	0.32	420(red) 526(red) 555(red)	—	17
72. $As5^+/As^{3+}$	—	0.316	—	—	2
73. *p*-Aminophenol	0.779	0.314	—	—	5
74. $O_2(gas)/H_2O_2$	0.69	0.295	—	—	7
75. Fe^{3+}/Fe^{2+} Cyt. c_4 (*Azotobacter*)	—	0.30	411(ox) 416(red) 522(red) 551(red)	115.8 157.2 17.6 23.8	17
76. Fe^{3+}/Fe^{2+} Cyt c_{552} (*Pseudomonas*)	—	0.300	409(ox) 416(red) 520(red) 552(red)	—	36
77. 1,4-Benzoquinone	0.699	0.293	—	—	5
78. Fe^{3+}/Fe^{2+} Cyt a	—	0.29	—	—	18
79. *p*-Quinone/Hydroquinone	—	0.28	—	—	5
80. 2,6-Dibromo-2'-SO_3H indophenol	—	0.273	—	—	5
81. Fe^{3+}/Fe^{2+} Malonate, pH 4	—	0.26	—	—	5
82. 2,5-Dihydroxyphenylacetic acid (Homogentisic acid)	0.687	0.260	—	—	5
83. Fe^{3+}/Fe^{2+} Salicylate, pH 4	—	0.26	—	—	5
84. Fe^{3+}/Fe^{2+} Cyt c	—	0.254	407(ox) 415(red) 521(red) 550(red)	— 125 15.9 27.7	19, 20
85. 2,6,2'-Trichloroindophenol	—	0.254	—	—	5
86. Fe^{3+}/Fe^{2+} Chlorocruorin(pyridine)$_2$	—	0.246	434(red) 544(red) 562(red)	—	29
87. Indophenol	—	0.228	—	—	5
88. *o*-Toluidine Blue	0.677	0.224	—	—	5
89. Phenol Blue	—	0.224	—	—	5
90. Fe^{3+}/Fe^{2+} Cyt c_1	—	0.22	410(ox) 418(red) 524(red) 554(red)	116 11.6 24.1	37
91. Fe^{3+}/Fe^{2+} Mesoporphyrin poly D,L-(lysine-phenylalanine), pH 4	—	0.22	—	—	21
92. Fe^{3+}/Fe^{2+} Cyt b_2 (yeast)	—	0.219	—	—	22

OXIDATION-REDUCTION POTENTIALS, ABSORBANCE BANDS AND MOLAR ABSORBANCE OF COMPOUNDS USED IN BIOCHEMICAL STUDIES (continued)

System	$E°$	$E°'$	λ_{max}	E_{mM}	Reference
93. 2,6-Dichlorophenolindophenol (DCPIP)	—	0.217	600	20.6	5, 23
94. 2,6-Dibromoindophenol	—	0.216	—	—	5
95. Janus Green	—	0.21	—	—	5
96. 3-Amino thiazine	—	0.208	—	—	5
97. Butyryl-Co A dehydrogenase $FAD^+/FADH_2$ (Cu present)	—	0.187	—	—	24
98. Fe^{3+}/Fe^{2+} Hemoglobin (H 6.0),	—	0.17	500(ox)	9.0	25, 26
			630(ox)	4.0	27
pH 7.0	—	0.144	—	—	26, 28
99. SO_4^{2-}/H_2SO_3	0.17	—	—	—	1
100. 2,6-Dibromo-2'-methoxy-indophenol	—	0.161	—	—	5
101. Sn^{4+}/Sn^{2+}	—	0.15	—	—	2
102. Adrenodoxin	—	0.15	414(ox)	5.7	9
103. 2,6-Dimethylindophenol	—	0.148	—	—	5
104. 1,2-Naphthoquinone	0.547	0.143	—	—	4
105. Fe^{3+}/Fe^{2+} PPIX(pyridine)$_2$,	—	0.137	419(red)	192	29
			525(red)	17.5	30
			557(red)	34.4	
pH 9	—	0.09	—	—	5
106. 1-Naphthol-2-sulfonate indophenol	—	0.123	—	—	5
107. Fe^{3+}/Fe^{2+} Cyt$_{553}$ (*R. spheroides*)	—	0.120	412(ox)		31
			418(red)		
			523(red)		
			553(red)		
108. Toluylene Blue	—	0.115	—	—	5
109. Fe^{3+}/Fe^{2+} Cyt$_{552}$ (*Chromatium*)	—	0.100	410(ox)		16
			417(red)		
			525(red)		
			552(red)		
110. Ubiquinone/Ubihydroquinone (in 95% ethanol)	0.542	0.10	275	15	32
111. TiO^{2+}/Ti^{3+}	—	0.10	—	—	3
112. $S_4O_6^{2-}/S_2O_3^{2-}$	0.08	—	—	—	1
113. Dehydroascorbic acid/ascorbic acid,	—	0.058	—	—	5
pH 4	—	0.166	—	—	7
pH 8.7	—	−0.012	—	—	2
114. *N*-Methylphenazinium methosulfate (PMS)	—	0.08	387(ox)	23.8	33
			388(semi-		5
			450 (quinone)		
115. Fe^{3+}/Fe^{2+} Cyt b (mitochondrial)	—	0.077	—	—	34
		0.050	429	114	
			532		
			561	21	
		−0.040			18
116. $[W^{5+}(OH^-)_4(CN^-)_4]^{3-}/ [W^{4+}/OH^-)_4(CN^-)_4]^{4-}$	—	0.07	—	—	35
117. Thionine	0.563	0.064	—	—	5
118. Thioindigo-tetrasulfonate	0.409	0.063	—	—	5
119. Phenazine ethosulphate	—	0.055	—	—	5
120. Cresyl Blue	0.583	0.047	632		5
121. Fe^{3+}/Fe^{2+} Myoglobin	—	0.046	500(ox)	9.1	28, 27
			630(ox)	3.5	
122. Fe^{3+}/Fe^{2+} Cyt b$_3$ (plants)	—	0.040	560(red)	—	39
			529(red)		
123. 1,4-Naphthoquinone	0.470	0.036	—	—	5
124. Toluidine Blue	0.534	0.034	—	—	5
125. Fumaric/Succinate	—	0.031	—	—	7
126. $[Ni(C_{10}H_{10})]^+/Ni(C_{10}H_{10})$	—	0.03	—	—	40
127. Thiazine Blue	—	0.027	—	—	5
128. Gallocyanine	—	0.021	—	—	5

OXIDATION-REDUCTION POTENTIALS, ABSORBANCE BANDS AND MOLAR ABSORBANCE
OF COMPOUNDS USED IN BIOCHEMICAL STUDIES (continued)

System	$E°$	$E°'$	λ_{max}	E_{mM}	Reference
129. Fe^{3+}/Fe^{2+} Cyt b_5 (microsomal)	—	0.02	413(ox)	117	41
			423(red)	170	
			526(red)	13	
			555(red)	26	
130. Thioindigo disulfonate	0.347	0.014	—	—	5
131. Methylene Blue	0.532	0.011	688(ox)	—	5
132. Fe^{3+}/Fe^{2+} Methylated heme undecapeptide of Cyt c (pyridine)	—	0.008			68
133. Fe^{3+}/Fe^{2+} Hematoporphyrin (pyridine)$_2$	—	+0.004	519(red)	—	8, 29
			545(red)		
134. Fe^{3+}/Fe^{2+} Oxalate	—	0.002			5
135. 3-Methyl-9-phenyl isoalloxazine	—	−0.002			5
136. Fe^{3+}/Fe^{2+} Cytochromoid c (*Chromatium*)	—	−0.040	406(ox)		42
			418(red)		
			525(red)		
			552(red)		
137. Fe^{3+}/Fe^{2+} Cytochromoid c (*R. rubrum*)	—	−0.008	390(ox)		70
			424(red)		
			568(red)		
138. Crotonyl-CoA/Butyryl-CoA	—	−0.015		—	38
139. Pyocyanine	0.235	−0.034	690(ox)	4.5	33
		−0.038	370		5
140. Indigo-tetrasulfonate	0.365	−0.046	—	—	5
141. 2-Methyl-3-phytyl-1,4-naphthoquinone (Vitamin K_1/Dihydro-Vitamin K_1)	0.363	−0.05	—	—	5
142. Luciferin	—	−0.05		—	4
			490(ox)	8.85	
			380(ox)	10.8	
143. Fe^{3+}/Fe^{2+} Rubredoxin		−0.057	333(ox)	6.3	69
			311(red)	10.8	
144. Fe^{3+}/Fe^{2+} Cyt b_6 (Chloroplasts)	—	−0.06	563(red)	—	43
145. Methyl Capri Blue	0.477	−0.061			5
146. Fe^{3+}/Fe^{2+} Mesoporphyrin (pyridine)$_2$	—	−0.063			8
147. $H_2SO_3/HS_2O_4^-$	−0.08	—			1
148. Fe^{3+}/Fe^{2+} Mesoporphyrin poly-D,L-(glu-phe), pH 9	—	−0.07			21
149. Xanthine oxidase	—	−0.08	—	—	5
150. Indigo-trisulfonate	0.332	−0.081	—	—	5
151. Fe^{3+}/Fe^{2+} 1, 3, 5, 8-Tetramethyl porphyrin-6,7-dipropionic acid methyl ester-2,4-disulfonic acid	—	−0.09			44
152. Thiohistidine	—	−0.09	—	—	4
153. $[V(C_{10}H_{10})]^{2+}/[V(C_{10}H_{10})]^+$	—	−0.08			40
154. Glyoxylate/Glycollate	—	−0.090			7
155. Fe^{3+}/Fe^{2+} Heme undecapeptide from Cyt c (pyridine)	—	−0.092	403(ox)	117	68
			413(red)	155	
			521(red)		
			551(red)		
156. 6,8,9-Trimethyl isoalloxazine	—	−0.109	—	—	5
157. Chloraphine	0.274	−0.115	—	—	5
158. CO_2(gas)/CO(gas)	−0.12	—	—	—	1
159. Yellow enzyme $FMN/FMNH_2$	—	−0.122	—	—	45
160. Indigo-disulfonate	0.291	−0.125	—	—	5
161. 9-Phenyl isoalloxazine	—	−0.126	—	—	4
162. Vitamin K reductase	—	−0.127	—	—	46
163. Fe^{3+}/Fe^{2+} PPIX (histidine)$_2$, pH9.5	—	−0.138	—	—	5
164. 2-OH-1,4-Naphthoquinone	—	−0.139	—	—	5
165. Thioglycolic acid	—	−0.14	—	—	4

OXIDATION-REDUCTION POTENTIALS, ABSORBANCE BANDS AND MOLAR ABSORBANCE OF COMPOUNDS USED IN BIOCHEMICAL STUDIES (continued)

System	$E°$	$E°'$	λ_{max}	E_{mM}	Reference
166. Fe^{3+}/Fe^{2+} (Pyrophosphate)	—	−0.14	—	—	5
167. 2-Amino-N-methyl phenazine methosulfate	—	−0.145	—	—	5
168. Indigo-monosulfonate	0.262	−0.157	—	—	5
169. Hydroxypyruvate/Glycerate	—	−0.158	—	—	7
170. Oxaloacetate/Malate	—	−0.166	—	—	47
171. Brilliant Alizarin Blue	—	−0.173	—	—	5
172. Alloxazine	—	−0.170	—	—	5
173. Mn^{3+}/Mn^{2+} Methyl pheophorbide a	—	−0.180	370(ox)	39	48
			425(ox)	31	
			475(ox)	13	
			665(ox)	17	
			418(red)	120	
			647(red)	24	
174. 2-Methyl-3-hydroxy-1,4-Naphthoquinone (Phthiocol)	—	−0.180	—	—	5
175. 9-Methyl-isoalloxazine	—	−0.183	—	—	4
176. Fe^{3+}/Fe^{2+} PPIX $(CN^-)_2$, pH9.9	—	−0.183	—	—	5
177. Anthraquinone-2,6-disulfonate	0.228	−0.184	—	—	5
178. Pyruvate/Lactate	—	−0.185	—	—	4
	—	−0.190			7
179. Fe^{3+}/Fe^{2+} Protoporphyrin IX (borate buffer), pH 8.2	—	−0.188	—	—	4, 49
180. Neutral Blue	0.17	−0.19	—	—	5
181. Dihydroxy acetone-P/ α-Glycero-P	—	−0.19	—	—	4
182. Acetaldehyde/Ethanol	—	−0.192	—	—	7
183. $[Ti(C_{10}H_{10})]^{2+}/[Ti(C_{10}H_{10})]^+$	—	−0.197	—	—	4, 7
184. $SO_4^{2-}/S_2O_6^{2-}$	−0.2	−0.20	—	—	40
185. Fe^{3+}/Fe^{2+} Heme undecapeptide, Cyt c (imidazole)	—	−0.201	—	—	1
186. Riboflavin	—	−0.208	260	27.7	68
			375(ox)	10.6	4
			450(ox)	12.2	
187. Fe^{3+}/Fe^{2+} Cyt c_3 (*Desulforibro desulfuricans*)	—	−0.205	410(ox)	—	50
			419(red)		51
			525(red)		
			553(red)	4.2	
188. Fe^{3+}/Fe^{2+} Heme octapeptide from Cyt c	—	−0.205	397(ox)	140	52
			414(red)	128	
			520(red)	6	
			550(red)	10	
189. $[Ru(NH_3)_6]^{3+}/[Ru(NH_3)_6]^{2+}$	—	−0.214	—	—	53
190. Fe^{3+}/Fe^{2+} Heme octapeptide from Cyt c (imidazole)	—	−0.217	405(ox)	122	68
			416(red)	162	
			520(red)		
			550(red)		
191. Anthraquinone-1-sulfonate	0.195	−0.218	—	—	5
192. FMN/FMNH$_2$, pH 7.09	—	−0.219	260	27.1	54
		−0.211	375(ox)	10.4	50
			450(ox)	12.2	
193. FAD/FADH$_2$	—	−0.219	260	37	54
			375(ox)	9.3	
			450(ox)	11.3	
194. 6,7,9-Trimethyl-isoalloxazine (Lumiflavin)	—	−0.223	—	—	54
195. Janus Green B	—	−0.225	—	—	5
196. Fe^{3+}/Fe^{2+} Protoporphyrin IX (phosphate buffer), pH 8.2	—	−0.226	395(ox)	55	4
			495(ox)	7	30
			620(ox)	6	
197. Glutathione	—	−0.23	—	—	7, 5
		−0.34			
198. Acetoacetyl CoA/B-OH-Butyryl CoA	—	−0.238	—	—	7

OXIDATION-REDUCTION POTENTIALS, ABSORBANCE BANDS AND MOLAR ABSORBANCE
OF COMPOUNDS USED IN BIOCHEMICAL STUDIES (continued)

System	$E°$	$E°'$	λ_{max}	E_{mM}	Reference
199. S(rhombic)/H_2S	0.14	−0.243	—	—	1, 7
200. Acetylmethyl carbinol/butane-2,3-diol	—	−0.244			7
201. Fe^{3+}/Fe^{2+} Copoporphyrin $(CN^-)_2$, pH 9.6	—	−0.247			5
202. 3-Acetylpyridine-NAD	—	−0.248			55
203. Phenosafranine	0.280	−0.252	—	—	5
204. V^{3+}/V^{2+}	−0.255	—			5
205. Co^{3+}/Co^{2+} Mesoporphyrin (pyridine)$_2$	—	−0·265			5
206. Mn^{3+}/Mn^{2+} Hematoporphyrin IX dimethyl ester	—	−0.268			6
207. Fe^{3+}/Fe^{2+} Peroxidase (horseradish)	—	−0.271	415(ox)	60	56, 27, 57
			500(ox)	10.0	
			640(ox)	3.0	
208. Fructose-sorbitol	—	−0.272			5
209. H_3PO_4/H_3PO_3	−0.276	—			1
210. Rosindulin 2G	0.139	−0.281			5
211. Thionicotinamide-NAD	—	−0.285	400(red)	—	58
212. Acetone/Isoproponol	—	−0.281	—	—	5
		−0.286			
213. Safranine T	0.235	−0.289			5
214. Lipoic acid	—	−0.29			5
215. Indulin Scarlet	0.047	−0.299			5
216. Thiophenol	—	−0.30			4
217. 4-Aminoacridine	—	−0.301			59
218. Acridine	—	−0.313	—	—	59
219. $NAD^+/NADH$	−0.105	−0.320	259(ox)	18	7, 50
			259(red)	15	
			339(red)	6.2	
220. $NADP^+/NADPH$	—	−0.324	259(ox)	18	7, 50
			259(red)	15	
			339(red)	6.2	
221. Neutral Red	0.240	−0.325			5
222. Cystine/Cysteine	—	−0.340	240(ox) (shoulder)	0.050	7
223. Lipoyl dehydrogenase	—	−0.34	—	—	60
224. NAD^+/α-NADH	—	−0.341	259(ox)	17	61
			346(red)		
225. Mn^{3+}/Mn^{2+} Hematoporphyrin IX	—	−0.342	370(ox)	79	6
			460(ox)	50	
			545(ox)	12	
			770(ox)	1.3	
			416(red)	175	
			545(red)	18	
226. Acetoacetate/β-hydroxybutyrate	—	−0.346	—	—	7
227. Uric aid/Xanthine	—	−0.36	—	—	7
228. Benzyl viologen	—	−0.36			5
229. Gluconolactone/Glucose	—	−0.364			7
230. 3-Aminoacridine	—	−0.369	—	—	59
231. Xanthine/Hypoxanthine	—	−0.371	248.5(ox)	10.2	7, 50
			278　(ox)	8.9	
232. Mn^{3+}/Mn^{2+} Mesoporphyrin (pyridine)$_2$	—	−0.387	—	—	5
233. 1-Aminoacridine	—	−0.394			59
234. Cr^{3+}/Cr^{2+}	−0.40	—	—	—	1, 2
235. N-Methyl nicotinamide	—	−0.419			5
236. CO_2/Formate	−0.20	−0.42			1
237. Fe^{3+}/Fe^{2+} Ferredoxin (*Clostridium*)	—	−0.413	300(ox) 390(ox)	6	65
238. H^+/H_2	0.000	−0.421	—	—	5
239. Fe^{3+}/Fe^{2+} Ferredoxin (spinach)	—	−0.432	325 420(ox) 463(ox)	—	65
240. Methyl violgoen	—	−0.44	—	—	5
241. Xanthine oxidase	—	−0.45	—	—	63

OXIDATION-REDUCTION POTENTIALS, ABSORBANCE BANDS AND MOLAR ABSORBANCE OF COMPOUNDS USED IN BIOCHEMICAL STUDIES (continued)

System	$E°$	$E°'$	λ_{max}	E_{mM}	Reference
242. SO_4^{2-}/SO_3^{2-}	—	−0.454	—	—	7
243. Gluconate/Glucose	—	−0.44	—	—	62
		−0.47			7
244. 2-Aminoacridine	—	−0.486		—	59
245. Oxallate/Glyoxalate	—	−0.50	—	—	7
246. H_3PO_3/H_3PO_2	−0.50	—	—	—	1
247. $SO_3^{2-}/S_2O_4^{2-}$	—	−0.527	—	—	7
		−0.471			
248. Acetate/acetaldehyde	—	−0.581	—	—	7
		−0.589			
249. 2,8-Diaminoacridine	—	−0.731	—	—	59
250. SiO_2/Si	−0.86	—	—	—	1
251. 5-Aminoacridine	—	−0.916	—	—	59
252. $N_2(gas)/H_3NOH^+$	−1.87	—	—	—	1
253. Formamidine sulfinic acid	—	−1.5	—	—	64

Compiled by Paul A. Loach.

REFERENCES

1. **Latimer,** in *The Oxidation States of the Elements and Their Potentials in Aqueous Solution,* 2nd ed., Prentice-Hall, New York, 1952.
2. **Berka, Vulterin, and Zyka,** in *Newer Redox Titrants,* Pergamon Press, New York, 1965.
3. **Koltoff, Belcher, Stenger, and Matsuyama,** in *Volumetric Analysis.* Vol. III. Interscience, New York, 1957.
4. **Lardy,** in *Respiratory Enzymes,* Burgess, Minneapolis, 1949.
5. **Clark,** in *Oxidation-Reduction Potentials of Organic Systems,* Williams & Wilkins, Baltimore, 1960.
6. **Loach and Calvin,** *Biochemistry,* 2, 361 (1963).
7. **Burton,** *Ergeb. Physiol.,* 49, 275 (1957).
8. **Martell and Calvin,** in *Chemistry of Metal Chelate Complexes,* Prentice-Hall, New York, 1958.
9. **Klotz and Klotz,** *Science,* 121, 477 (1955).
10. **Kok,** *Biochim. Biophys. Acta,* 48, 527 (1961).
11. **Goodheer,** *Biochim. Biophys. Acta,* 38, 389 (1960).
12. **Clayton,** *Photochem. Photobiol.,* 1, 201 (1962).
13. **Loach, Androes, Maksim, and Calvin,** *Photochem. Photobiol.,* 2, 443 (1963).
14. **Kuntz, Loach, and Calvin,** *Biophys. J.,* 4, 277 (1964).
15. **Minakami, Ringler, and Singer,** *J. Biol. Chem.,* 237, 569 (1962).
16. **Kamen and Vernon,** *Biochim. Biophys. Acta,* 17, 10 (1955).
17. **Tissieres,** *Biochem. J.,* 64, 582 (1956).
18. **Ball,** *Biochem. Z.,* 295, 262 (1938).
19. **Rodkey and Ball,** *J. Biol. Chem.,* 182, 17 (1950).
20. **Theorell and Åkeson,** *J. Am. Chem. Soc.,* 63, 1804 (1941).
21. **Lautsch, Brouer, and Becker,** *Z. Electrochem.,* 61, 174 (1957).
22. **Cutolo,** *Arzneimittelforsch,* 8, 581 (1958).
23. **Armstrong,** *Biochim. Biophys. Acta,* 86, 194 (1964).
24. **Green, Mii, Mahler, and Bock,** *J. Biol. Chem.,* 206, 1 (1954).
25. **Havemann,** *Biochem. Z.,* 314, 118 (1943).
26. **Taylor and Hastings,** *J. Biol. Chem.,* 131, 649 (1939).
27. **George, Beetleston, and Griffith,** in *Symposia on Hematin Enzymes,* Canberra, 1959.
28. **Taylor and Morgan,** *J. Biol. Chem.,* 144, 15 (1942).
29. **Falk,** in *Porphyrins and Metalloporphyrins,* Elsevier, New York, 1964.
30. **Shack and Clark,** *J. Biol. Chem.,* 171, 143 (1947).
31. **Orlando,** *Biochim. Biophys. Acta,* 57, 373 (1962).
32. **Morton, Gloor, Schindler, Wilson, Chopard-dit-Jean, Hemming, Isler, Leat, Pennock, Ruegg, Schwieter, and Wiss,** *Helv. Chim. Acta,* 41, 2343 (1958).
33. **Jagendorf and Marguiliea,** *Arch. Biochem. Biophys.,* 90, 184 (1960).

34.   Holton and Colpa-Boonstra, *Biochem. J.,* 76, 179 (1960).
35.   Mikhalevich and Litvinchuk, *Zh. Neorgan. Khim.,* 9, 2391 (1964).
36.   Kamen and Lakeda, *Biochim. Biophys. Acta,* 21, 518 (1956).
37.   Green, Jarnefelt, and Tisdale, *Biochim. Biophys. Acta,* 31, 34 (1959).
38.   Hauge, *J. Am. Chem. Soc.,* 78, 5266 (1956).
39.   Hartree, *Advanc. Enzymol.,* 18, 1 (1957).
40.   Pauson, *Quart. Rev.,* 9, 391 (1955).
41.   Velick and Strittmatter, *J. Biol. Chem.,* 221, 265 (1956).
42.   Newton and Kamen, *Biochim. Biophys. Acta,* 21, 71 (1956).
43.   Hill, *Nature,* 174, 501 (1954).
44.   Walter, *J. Biol. Chem.,* 196, 151 (1952).
45.   Vestling, *Acta Chem. Scand.,* 9, 1600 (1955).
46.   Martius and Marki, *Biochem. Z.,* 333, 111 (1960).
47.   Burton and Wilson, *Biochem. J.,* 54, 86 (1953).
48.   Loach and Calvin, *Nature,* 22, 343 (1964).
49.   Cowgill and Clark, *J. Biol. Chem.,* 198, 33 (1952).
50.   Weber, in *Biochemist's Handbook,* Long, Ed., Spon, London, 1961, 81.
51.   Postgate, *Biochim. Biophys. Acta,* 18, 427 (1955).
52.   Harbury and Loach, *J. Biol. Chem.,* 235, 3640 (1960).
53.   Endicott, and Taube, *Inorg. Chem.,* 4, 437 (1965).
54.   Lowe and Clark, *J. Biol. Chem.,* 221, 983 (1956).
55.   Rodkey, *J. Biol. Chem.,* 234, 188 (1959).
56.   Harbury, *J. Biol. Chem.,* 225, 1009 (1957).
57.   Theorell, *Enzymologia,* 10, 3 (1942).
58.   Anderson and Kaplan, *J. Biol. Chem.,* 234, 1226 (1959).
59.   Breyer, Buchanan, and Duewell, *J. Chem. Soc.,* 360 (1944).
60.   Searls and Sanadi, *Proc. Natl. Acad. Sci. (USA),* 45, 697 (1957).
61.   Kaplan, in *The Enzymes,* 2nd, ed., Vol. 3, Boyer, Lardy, and Myrbäck, Eds., Academic Press, New York, 1960, 105.
62.   Strecker and Korkes, *J. Biol. Chem.,* 196, 769 (1952).
63.   Mackler, Mahler, and Green, *J. Biol. Chem.,* 210, 149 (1954).
64.   Shashova, *Biochemistry,* 3, 1719 (1964).
65.   Tagawa and Arnon, *Nature,* 195, 537 (1962).
66.   Davenport and Hill, *Proc. R. Soc. B (England),* 139, 327 (1952).
67.   Wateri and Kimura, *Biochem. Biophys. Res. Commun.,* 24, 106 (1966).
68.   Harbury and Loach, *J. Biol. Chem.,* 235, 3646 (1960).
69.   Lovenberg and Sobel, *Proc. Natl. Acad. Sci. (USA),* 54, 193 (1965).
70.   Bartsch and Kamen, *J. Biol. Chem.,* 230, 41 (1958).
71.   Fuhrhop and Mauzerall, *J. Am. Chem. Soc.,* 91, 4174 (1969).

OXIDATION–REDUCTION POTENTIALS OF HEMOPROTEINS AND METALLOPORPHYRINS

R. W. Henderson and T. C. Morton

The potentials quoted in the following Table, except for Nos. 276–282, were referred to the standard hydrogen electrode; the sign convention used is that the couple which accepts electrons from another couple is the more positive and is the oxidizing couple. Oxidation-reduction potentials (E_m) should be quoted with the pH value and the temperature of the determination; unfortunately this information was not always available. For discussions on errors of measurement, sign conventions, techniques, etc., see *Oxidation-Reduction Potentials of Organic Systems* by W. Mansfield Clark.[1] E_m values of hemoproteins may be considerably influenced by (i) method of preparation (ii) degree of purity (iii) degree of modification; for these reasons widely different values for the one pigment may be found in some cases. Classification of the cytochromes follows the recommendations of *Enzyme Nomenclature*.[2] Many of the systems listed have E_m v. pH relationships; where such information is available indication is made in the Notes. Preliminary attempts at correlation of E_m values with ligand types and heme peripheral substituents have been made by several authors, see e.g., References 3 and 4. The section *Metalloporphyrins with and without simple ligands* is an extension of the table compiled by Clark.[1]

OXIDATION-REDUCTION POTENTIALS OF HEMOPROTEINS AND METALLOPORPHYRINS

No.	Type	Source	E_m(Volt)[a],[b] HEMOPROTEINS	pH	Temperature °C	Method of measurement[e]	Notes*	Reference
	Cytochromes[d]							
1	a	Heart-muscle	0.29	7.4	20	C	i	5, 7
2	a	Heart-muscle	0.278	7.5	10	C	i, ii	6, 7
3	"cyt. oxidase"	Heart-muscle	0.285	7.4	25	C	i	8
4	a_2 (see cyt. d)	—	—	—	—	—	—	–
5	a_3	—	[0.55][c]	—	—	—	iii	9
6	a (CO exposed)	Heart-muscle	0.25	8	—	C	—	140
7	"cyt. oxidase" (a) [a – 1] (b) [a – 2] (c) [a – 3] (d) (CO-exposed)	Heart-muscle	0.360 0.300 0.208 0.269	7	20	C	i, lxxxv	148
8	b	Heart-muscle	–0.04	7.4	20	C	—	5
9	b	Heart-muscle	–0.05	—	—	P	—	10
10	b	Heart-muscle	0.06 to 0.09	7	25	C	—	11
11	b	Heart-muscle	–0.34	7	30	C	iv	12
12	b	Heart-muscle	0.06 to 0.08	7	25	P	—	13
13	b	Heart-muscle	0.06	7	25	C	v	14
14	b	Heart-muscle	0.073	7	25	C	lxxxvi	149
15	b	Wheat roots	0.14 to 0.17	—	—	C	vi	15
16	[dh][c]	Wheat roots	0.02 to 0.05	—	—	C	vi	16
17	b_1	*Escherichia coli* crude prep. purified prep.	–0.01 –0.34	7	25	C	—	17
18	b_1	*Escherichia coli*	–0.02	7	25	C	—	18
19	b (562)	*Escherichia coli*	0.113	7	21	C	—	19, 142

*Notes appear on pages 144 to 147.

[a] E_m = potential (Volt) referred to the standard hydrogen electrode when $S_o = S_r$; also known as E'_o; may be written to include pH e.g. E_m^7.

[b] Two E_m values listed at one pH indicates that values were obtained in the one report within the range given.

[c] Square brackets indicate a very doubtful E_m value or classification.

[d] Where the official names of cytochromes are not yet final, α-peak positions (mμ) of the reduced forms are listed where possible.

[e] Methods of potential measurement are broadly divided into potentiometric, P, see Reference 1 and C, comparative; in these cases, various systems are used to poise the potential at a known value and the % oxidation/reduction of the species under investigation is determined, usually spectrophotometrically. See, e.g., References 29, 32, and 55.

OXIDATION-REDUCTION POTENTIALS OF HEMOPROTEINS AND METALLOPORPHYRINS (continued)

Hemoproteins (Continued)

No.	Type	Source	E_m (Volt)[a,b]	pH	Temperature °C	Method of measurement[e]	Notes	Reference
	Cytochromes[d] (continued)							
20	b (562)	*Bacterium anitratum*	0.120 to 0.140	7	20 to 25	C	—	20
21	b_2	Yeast	0.22	7	20 to 25	P	vii	21
22	b_2	Yeast	0.12	7	25	C	viii	22, 23
23	b_2 (modified)	Yeast	0.24 to 0.29	7	—	C	ix	24
24	b_2 (Trypsin-treated)	Yeast	−0.028	7	30	C	—	146
25	b_3	Plant tissue (pea leaves)	0.04	7	—	C	—	25
26	b_4	See halotolerant cyt. c types.	—	—	—		—	—
27	$[b_5]^c$	Liver	−0.13	7.5	15	C	x	26
28	b_5	Liver	−0.12	7	23	C	xi	27
29	b_5	Liver	0.02	7	26	C	xii	29
30	b_5	Liver	0.014 / −0.098	6.5 / 8	—	C	—	30
31	b_5	Liver	−0.14	7	23	C	xiii	31
32	b_5 (modified)	Liver	0.01	7	23	C	xiii	31
33	b_6	Plant tissue (chloroplasts)	−0.06	7	20	C	—	32
34	b_7	Plant tissue (spadices of *Arum maculatum*)	−0.03	7	—	C	—	25
35	b (555)	Plant tissue (Mung bean)	−0.03	7	—	C	—	33
36	b (561)	Plant tissue (Mung bean)	0.002	7	—	C	—	33
37	[b] (559)	Plant tissue (Pea chloroplasts)	0.37	(6.5–7.5)	25	C	xcii	157
38	b (559)	Liver	−0.02	7	20	C	xiv	34
39	b (562) (helicorubin)	*Helix pomatia*	0.20	7	24	C	—	35
40	b (562)	Alga (*Monostroma nitidum*)	0.06 to 0.25	—	—	C	—	36
41	b (563)	*Musca domestica* (larva)	0.011	7	25	C	—	141
42	b (555)	*Musca domestica* (larva)	0.006	7	12	C	—	151
43	[b]	*Rhodospirillum rubrum*	0	7	—		—	158
44	c	Heart-muscle	0.27	7.4	30	C	xv	5
45	c	Heart-muscle	0.256	7.2	30	P	xvi	37
46	c	Heart-muscle	0.255	6.4	25	P	xvii	38
47	c (Trichloroacetic acid treated)	Heart-muscle	0.310	6.4	25	P	—	39

OXIDATION-REDUCTION POTENTIALS OF HEMOPROTEINS AND METALLOPORPHYRINS (continued)

Hemoproteins (Continued)

No.	Type	Source	E_m(Volt)[a,b]	pH	Temperature °C	Method of measurement[e]	Notes	Reference
	Cytochromes[d] (continued)							
48	c	Heart-muscle	0.26	7	30	P	—	40
49	c (Guanidated)	Heart-muscle	[Slightly]e <0.26	—	—	C	—	41
50	c (Acetylated)	Heart-muscle	−0.3	7	—	C	—	42
	c (Proteinase-treated) (Nos. 46, 47 pepsin treated; Nos. 48–51 pepsin and trypsin treated)							
51	Undecapeptide + imidazole	Heart-muscle	−0.195	7	30	P	xviii	43, 44, 45
52	Undecapeptide + pyridine	Heart-muscle	−0.099	7.3	30	P	xviii	43, 44, 45
53	Octapeptide	Heart-muscle	−0.207	7.04	30	P	xviii	43, 44, 45
54	Octapeptide + imidazole	Heart-muscle	−0.221	7.12	30	P	xviii	43, 44, 45
55	Octapeptide + NH₃	Heart-muscle	−0.240	10.0	30	P	xviii	43, 44, 45
56	Octapeptide + N-acetyl-DL-methionine	Heart-muscle	−0.05	7	30	P	—	147
57	c	Calf thymus nuclei	0.272	7	—	C	—	159
58	c	Calf thymus mitochondria	0.265	7	—	C	—	159
59	c	Yeast	0.26	7	about 10	C	—	46
60	c	Yeast	0.28	6.4	25	P	—	47
61	c	Tuna	0.250	7	30	P	—	143
62	c	Tuna	0.247	7	30	P	—	143
	c (Guanidated)							
63	c_1	Heart-muscle	0.220	7	30	C	xix	48, 49
64	c_2	Rhodospirillum rubrum	0.338	7	—	C	xx	50, 51
65	c_2	Rhodospirillum rubrum	0.320	7.4	20	P	—	148
66	c_3	Desulfovibrio desulfuricans	−0.14 to −0.25	7	—	C	—	52
67	c_3	Desulfovibrio desulfuricans	−0.204	7	30	P	—	53
68	c_3	Desulfovibrio gigas	−0.22	—	30	C	—	54
69	c_6 (formerly f)	Plant tissue (chloroplasts)	0.365	7	30	C	xxi	55,157

OXIDATION-REDUCTION POTENTIALS OF HEMOPROTEINS AND METALLOPORPHYRINS (continued)

Cytochromes[d] (continued) — Hemoproteins (Continued)

No.	Type	Source	E_m (Volt)[a,b]	pH	Temperature °C	Method of measurement[e]	Notes	Reference
70	c (556) (formerly h)	Helix pomatia	0.20	7	24	C	xxii	35
71	[c][c]	Cryptophalus aspersus	0.126 / 0.138	6.5 / 4.8	—	C	—	56
72	c (552)	Red alga (Grateloupia sp.)	0.30	7	20	C	—	58
73	c (553)	Red alga (Porphyra tenera)	0.34	7	20	C	xxiii	58, 59
74	c (553)	Red alga (Porphyra yezoensis)	0.342	7	20	C	—	162
75	c (553)	Red alga (Porphyra pseudolinealis)	0.340	7	20	C	—	162
76	c (553)	Red alga (Nemalion vermiculare)	0.363	7	20	C	—	162
77	c (553)	Red alga (Gliopeltis complanata)	0.340	7	20	C	—	162
78	c (552)	Green alga (Ulva sp.)	0.30	7	20	C	—	58
79	c (552)	Green alga (Monostroma nitidum)	0.31	7	20	C	—	58
80	c (553)	Green alga (Enteromorpha prolifera)	0.364	7	20	C	—	162
81	c (553)	Green alga (Cladophora sp.)	0.378	7	20	C	—	162
82	c (553)	Green alga (Chaetomorpha spiralis)	0.382	7	20	C	—	162
83	c (553)	Green alga (Chaetomorpha crassa)	0.381	7	20	C	—	162
84	c (553)	Green alga (Bryopsis sp.)	0.385	7	20	C	—	162
85	c (553)	Green alga (Caulerpa brachypus)	0.385	7	20	C	—	162
86	c (555)	Green alga (Codium latum)	0.390	7	20	C	—	162
87	c (553)	Brown alga (Undaria pinnatifida)	0.34	7	20	C	—	58
88	c (553)	Brown alga (Endarachne binghamiae)	0.361	7	20	C	—	162
89	c (553)	Brown alga (Petalonia fascia)	0.360	7	20	C	—	162
90	c (553)	Blue-green alga (Tolypothrix tenuis)	0.30	7	20	C	—	58
91	c (549)	Blue-green alga (Anacystis nidulans)	−0.26	7	—	C	—	60, 154
92	c (554)	Blue-green alga (Anacystis nidulans)	0.35	7	—	C	—	60, 154
93	c (554)	Diatom (Navicula pelliculosa)	0.34	7	22	C	lxxxviii	61
94	c (552)	Euglena gracilis	0.36	7	25	C	xxiv	62
95	c (552)	Euglena wild type	0.37	7	—	C	xxv	63
96	c (556)	Euglena wild type	0.31	7	—	C	xxvi	63
97	c (554)	Chlorella vulgaris	0.35	7.2	30	C	—	160
98	c (553)	Chlamydomonas reinhardi	0.37	7	—	C	—	161
99	c (553)	Tetrahymena pyriformis	0.245	7	—	C	—	155
100	c (552)	Pseudomonas denitrificans	0.32	7	—	C	xxvii	64
101	c (552)	Pseudomonas denitrificans	0.30	7	—	C	—	65
102	c (551)	Pseudomonas aeruginosa	0.250	7	—	C	xxviii	66
103	c (551)	Pseudomonas aeruginosa	0.286	6.5	20	P	xxix	67
104	c (552)	Pseudomonas aeruginosa	0.30	7	—	C	xxx	68
105	c (554)	Pseudomonas aeruginosa	0.225	7	—	—	xxviii	66

OXIDATION-REDUCTION POTENTIALS OF HEMOPROTEINS AND METALLOPORPHYRINS (continued)

No.	Type	Source	E_m(Volt)[a,b]	pH	Temperature °C	Method of measurement[e]	Notes	Reference
	Cytochromes[d] (continued)							
		Hemoproteins (Continued)						
106	c (550)	*Pseudomonas saccharophila*	0	7.8	—	C	xxxi	69
107	c (552)	*Pseudomonas saccharophila*	0.237	7.0	25	C	—	70
108	c (553)	*Pseudomonas fluorescens*	0.235	6.9	—	—	xxxii	71
109	c (551)	*Azotobacter vinelandii*	0.30	7	—	C	xxxii	72
110	c (555)	*Azotobacter vinelandii*	0.32	7	—	—	xxxiii	72
111	c (550)	*Micrococcus denitrificans*	0.25	7	—	C	—	64
112	c (551)	*Halotolerant micrococcus*	0.249	7	25	C	—	73
113	c (554, 548)	*Halotolerant micrococcus*	0.113	7	25	C	xxxiv	74
114	c (552–548)	*Halotolerant micrococcus*	0.223	7.2	—	C	xciii	150
115	c (554)	*Halotolerant micrococcus*	0.180	7	25	C	—	74, 150
116	c (553)	*Chlorobium thiosulfatophilum*	0.163	7	—	P	—	75
117	c (554)	*Chlorobium thiosulfatophilum*	0.140	7	—	P	—	75
118	c (551)	*Chlorobium thiosulfatophilum*	0.135	7	20	C	lxxxix	156
119	c (553)	*Chlorobium thiosulfatophilum*	0.098	7	20	C	xc	156
120	c (555)	*Chlorobium thiosulfatophilum*	0.145	7	20	C	xci	156
121	c (550)	*Thiobacillus thioparus*	−0.14 to −0.15	7	—	—	—	76
122	c (552)	*Thiobacillus denitrificans*	0.270	6.8	23	C	xxxv	77, 78
123	c (550)	*Thiobacillus X*	0.200	7	30	C	—	79
124	c (554)	*Thiobacillus X*	0.210	7	30	C	—	79
125	c (557)	*Thiobacillus X*	0.155	7	30	C	—	79
126	c (551)	*Rhodospirillum rubrum*	0.350	7	21	C	xxxvi	80
127	c (552)	*Rhodospirillum rubrum* (see Cyt. c_2)	—	—	—	—	—	—
128	c (550)	*Rhodopseudomonas capsulatus*	0.33	7	—	C	xxxvii	64, 81
129	c (552)	*Rhodopseudomonas palustris*	0.31	7	—	C	xxxviii	81
130	c (552)	*Rhodopseudomonas palustris*	0.33	7	20	C	—	82
131	c (552)	*Rhodopseudomonas palustris* (van Niel strain 2137)	0.35	7	—	C	—	83
132	c (550)	*Rhodopseudomonas spheroides* soluble / particle bound	0.34 / [0.25][c]	7	30	C	—	144, 153
133	c (553)	*Rhodopseudomonas spheroides*	0.120	7	—	C	—	84
134	c (552)	*Chromatium* (strain D)	−0.04	7	—	C	xxxix	85
135	c (552)	*Chromatium* (strain D)	0.01	7	—	C	—	86
136	c (550)	*Rhodomicrobium vannielii*	0.304	7	25	C	—	87
137	c (550)	*Bacillus megaterium*	0.250	7	—	P	—	88
138	c (550)	*Bordetella* spp.	0.259	7	—	P	—	89
139	c (553)	*Bordetella* spp.	0.192	7	—	P	—	89
140	c (552)	*Escherichia coli*	−0.2 (approx.)	7	—	C	—	90

OXIDATION-REDUCTION POTENTIALS OF HEMOPROTEINS AND METALLOPORPHYRINS (continued)

No.	Type	Source	E_m (Volt)[a,b]	pH	Temperature °C	Method of measurement[e]	Notes	Reference
		Hemoproteins (Continued)						
	Cytochromes[d] (Continued)							
141	c (552)	*Escherichia coli*	−0.252	7	—	C	xciv	163
142	c (552)	*Escherichia coli*	0.150	7	20	C	lxxxvii	152
143	c (552)	Iron oxidizing bacteria	0.31 / 0.38	7 } 2.9	—	C	—	91
144	c (550)	A Pseudomonad (unidentified)	0.26	7	—	C	—	92
145	c (553)	A Pseudomonad (unidentified)	0.2 (approx.)	7	—	C	—	92
146	c (550)	*Rhizobium japonicum*	0.28	7	20	C	—	93
147	c (552)	*Rhizobium japonicum*	0.19	7	20	C	xl	93
	$(cc')_1$ Also called RHP or cytochromoid c.							
148	$(cc')_1$	*Rhodospirillum rubrum*	−0.008	7	30	C	—	94
149	c_1^1 (formerly called $(cc^1)_1$)	*Rhodopseudomonas palustris*	0.250	7	—	—	—	95
150	c_2^1 (formerly called $(cc^1)_2$)	*Rhodopseudomonas palustris* (van Niel strain 2137)	0.105	7	—	C	—	83
151	$(cc')_1$	*Chromatium* (strain D)	−0.005	7	—	C	—	86
152	(cc')	*Pseudomonas denitrificans*	0.09	6.8	—	C	—	138
	(Also called crypto-cytochrome)							
153	o (559)	"Colourless" blue-green alga (*Vitreoscilla*)	0.10	7	—	C	—	145
154	o (553)	"Colourless" blue-green alga (*Vitreoscilla*)	−0.09	7	—	C	—	145
155	d (Formerly called a_2)	—	[0.30]c	—	—	—	xli	96
	Hemoglobins and Myoglobins							
	Hemoglobin/methemoglobin (Hb)							
156	Hb	Mammals	Early values	—	—	—	—	1, p. 458
157	Hb	Beef	0.095	7	38	P	xlii	97
158	Hb	Horse	0.139	7.0	30	P	xliii	98
159	Hb	Beef and Horse	0.15	7.0	30	P	xliv	99
160	Hb	Horse	0.132	7.0	30	P	xlv	100
161	Hb	Horse	Wide range (see note)	7.9	1	P	xlvi	101
162	Hb	Horse	0.184 (without correction for mediator)	7.1	—	—	xlvii	102

OXIDATION-REDUCTION POTENTIALS OF HEMOPROTEINS AND METALLOPORPHYRINS (continued)

No.	Type	Source	E_m (Volt)^{a,b}	pH	Temperature °C	Method of measurement^e	Notes	Reference
	Hemoglobins and Myoglobins (Continued)		**Hemoproteins (Continued)**					
163	Hb	Dog	0.132 0.154	7.1 5.9	30	P	—	103
164	Hb	Human	0.150	7	30	P	xlviii	104
165	Hb	*Chironomus plumosus* (larva)	0.035	7	30	P	xlix	100
166	Hb	Horse	0.110	6.94	30	P	l	103
167	Treated with 4M urea Hb	Dog	0.106	6.92	30	P	li	103
168	Treated with 4M urea Hb	Horse	0.116	7	30	P	lii	105
169	Treated with 2M urea Hb	Horse	0.105 0.071	7 7.6	30	P	liii	105
170	Treated with 4M urea Hb	Horse	0.11	7	30	P	liv	106
171	Guanidated Hb	Horse	0.07	7	30	P	lv	100
172	Treated with *N*-ethylmaleimide Hb	Human	0.10	7	30	P	lvi	107
173	Treated with 4M urea (See note re treatment with other reagents). Hb	Human						
	Treated with—							
	(a) carboxypeptidase A		0.057	7	30	P	lvii	108
	(b) carboxypeptidase B		0.095	7	30	P	lvii	108
	(c) carboxypeptidases A and B		0.042	7.1	30	P	lvii	108
174	Treated with Bromthymol blue Hb	Human	0.164	7	30	P	lviii	109
	Myoglobin/metmyoglobin (Mb)							
175	Treated with Hb	Human						
	(a) Iodoacetamide		{0.149 0.021	{6 9	30	P & C	—	164
	(b) Cystine		{0.150 0.032	{6 9	30	P & C	—	164
	(c) Cystamine		{0.156 0.037	{6 9	30	P & C	—	164
	(d) *p*-mercuribenzoate		{0.163 0.037	{6 9	30	P & C	—	164

OXIDATION-REDUCTION POTENTIALS OF HEMOPROTEINS AND METALLOPORPHYRINS (continued)

No.	Type	Source	E_m (Volt)[a,b]	pH	Temperature °C	Method of measurement[e]	Notes	Reference
	Hemoglobins and Myoglobins (Continued)		**Hemoproteins (Continued)**					
176	Hb	Human						
	In combination with haptoglobin (Hp) Molar ratio Hp:Hb							
	2:1		⎰0.076 ⎱0.012	⎱7 ⎰9	30	P	xcviii	165
177	Hb	Human						
	1:1		⎰0.103 ⎱0.013	⎱7 ⎰9				
	(a) α-chain		0.052	7	5	P & C	xcv	166
	(b) β-chain		0.11	7	5	P & C	xcvi	166
	(c) β-chain combined with p-mercuribenzoate		0.11	7	5	P & C	xcvii	166
178	Mb	Horse	0.046	7.0	30	P	lix	110
179	Mb	Horse	0.06	7.0	30	P	lx	100
180	Mb Guanidated	Horse	See note	See note	30	P	lxi	107
181	Mb In presence of Potassium chloroiridate	Horse	0.9	7	20.4	C	lxii	111
182	Mb	Whale	⎰0.061 ⎱0.024	⎱6 ⎰9	30	P & C	—	164
	Miscellaneous metalloporphyrin-protein complexes							
183	Manganese-protoporphyrin-globin	Globin from human Hb	0.048	7.0	30	P	lxiii	112
184	Chlorocruorin		[0.20][c]	—	—	—	lxiv	96
185	Leghemoglobin (Lb)	Soybean	0.240	7.0	25	P	—	115
186	Peroxidase	Horse-radish	-0.271	7.0	30	P	lxv	113
187	Chlorophyll	Chlorella						
	a		[0.65][c]					
	b		[0.43][c]					
	Bacteriochlorophyll	Photosynthetic bacteria	[0.55][c]	—	—	—	lxvi	116, 117

METALLOPORPHYRINS WITH AND WITHOUT SIMPLE LIGANDS

No.	Metal	Porphyrin	S^f ($M \times 10^5$)	Ligand^h-molarity	Solutions^i	$E_m\,9^{a,g}$	$E_m\,11^{a,g}$	pH range investigated	Slopes $\frac{dE_h}{-dpH}$	pH of inflexions	Notes*	Reference
188	Fe	Etio. I	4.76	Pyr.	1.09	A75	—	-0.1	11–12.3	0.06	—	—	120
189	Fe	Copro. I	4.2	Pyr.	1.05	W	0.002	-0.118	8.4–12.5	0, 0.06	9.94	—	120
190	Fe	Copro. I	5.6	Nic.	0.22	W	-0.002	-0.123	8.6–12.3	0, 0.06	8.92	—	120
191	Fe	Copro. I	5.6	CN⁻	0.02	W	-0.249	-0.249	10–12.4	0	—	—	120
192	Fe	Hemato. IX	11.0	Pyr.	1.13	W	—	-0.076	10–11.7	0.06	—	—	121
193	Fe	Hemato. IX	13.2	Pic.	0.93	W	—	-0.179	10.05	0.06	—	—	121
194	Fe	Hemato. IX	17.2	CN⁻	0.022 to 0.008	W	—	-0.200	9.7–11.4.	0	—	—	121
195	Fe	Meso. IX	34	None		A23	-0.278	-0.398	8.5–10.4	0.06	—	lxvii	122
196	Fe	Meso. IX	14.6	Nic.	0.24	A47.5	—	-0.146	11.5	0.06	—	lxvii	121
197	Fe	Meso. IX	14.6	Pyr.	0.456	A47.5	—	-0.144	9.1–12.4	0.06	—	lxvii	121
198	Fe	Meso. IX	12.5	CN⁻	0.00815	W	-0.229	-0.229	10–11.4	0	—	—	121
199	Fe	Meso. IX	34	Pila.	0.1	A23	-0.354	-0.354	8.7–9.9	0	—	lxviii	122
200	Fe	Meso. IX	10	Pyr.	2.5	W	0.029	0.020 (E_m10)	8.5–10.5	0, 0.03	9.5	lxviii	105
201	Fe	Meso. IX	10	Pyr.	2.5	U	0.013	0.002 (E_m10)	8.7–10.4	0, 0.03	9.6	lxviii	105
202	Fe	Proto. IX	38	None		W	-0.22	—	—	—	—	22°	123
203	Fe	Proto. IX	7.6	None		W	-0.16	—	—	—	—	22°	123
204	Fe	Proto. IX	7.6	Pyr.	0.113	W	0.056	—	—	—	—	22°	123
205	Fe	Proto. IX	38	CN⁻	0.04	W	-0.16	—	—	—	—	22°	123
206	Fe	Proto. IX	10	None		W Bo.	-0.28	-0.40	7–12.6	0.06	—	lxix	124
207	Fe	Proto. IX	10	None		W Ph.	-0.24	-0.36	—	—	—	lxix	124
208	Fe	Proto. IX	46	None		A47.5	-0.235	-0.355	8.3–11.1	0.06	—	—	125
209	Fe	Proto. IX	14.4	Nic.	0.353	W	-0.010	-0.053	10.2–11.2	0.06	—	lxx	121
210	Fe	Proto. IX	10	Nic.	0.04	W	—	-0.130	9.1–11.9	0.06	—	—	124
211	Fe	Proto. IX	47.6	Nic.	0.29	W	—	0.07	10.2–11.8	0.06	—	—	121
212	Fe	Proto. IX	10	Pyr.	1.0	W	—	-0.09	9.5–12	0.06	—	—	124
213	Fe	Proto. IX	10	Pyr.	2.5	W	0.09	-0.01	7.4–12.8	0, 0.06	9.04	—	125
214	Fe	Proto. IX	10	Pic.	1	W	-0.008	-0.128	7.9–11.9	0.06	—	—	124
215	Fe	Proto. IX	10	His.	0.07	W	-0.132	—	7–9.6	0.015	—	—	124
216	Fe	Proto. IX	20	Pila.	0.0206	W	-0.161	—	7.5–9.5	0.015	—	—	124

*Notes appear on pages 144 to 147.

^{f} $S = S_0 + S_r$, where S_0 and S_r = sums of concentrations (molarities) of all species of oxidant and reductant respectively.

^{g} E_m values obtained potentiometrically (but see Note lxxix) and at 30° unless otherwise indicated under Notes.

^{h} Ligands: Pyr. = pyridine; Nic. = nicotine; Pic. = α-picoline; Pil. = pilocarpine; Pila. = pilocarpate (pilocarpinate), doubt exists as to whether some investigators used pilocarpine or pilocarpate, see Reference 122; His. = histidine; His. his. = histidyl histidine; DMF = dimethylformamide; Imid = imidazole.

^{i} Solutions: W, aqueous buffers; A + number = aqueous buffer with indicated per cent ethanol; U = 4M urea in aqueous buffer; Bo. = borate; Ph. = phosphate; Gly. = glycine buffer; Cit. = citrate buffer; NaD. S. = sodium dodecyl sulphate.

METALLOPORPHYRINS WITH AND WITHOUT SIMPLE LIGANDS (continued)

No.	Metal	Porphyrin	S^f ($M \times 10^5$)	Ligand[h]-molarity	Solutions[i]	$E_m 9^a{}_g$	$E_m 11^a{}_g$	pH range investigated	Slopes $-\dfrac{dE_h}{dpH}$	pH of inflexions	Notes	Reference
217	Fe	Proto. IX	10	CN^-	0.05 to 0.08	W	-0.185	-0.185	8–12.3	0	—	—	124
218	Fe	Proto. IX	10	Pic.	1.5	W	-0.07 (E_m8)	—	7.8–9.2	0.05	—	lviii	105
219	Fe	Proto. IX	10	Pic.	1.5	U	-0.09 (E_m8)	—	8.1–9.2	0.05	—	lxviii	105
220	Fe	Proto. IX	10	His.	0.1	W	-0.165 (E_m8)	—	7.3–8.7	0.06, 0.04	8.1	lxviii	105
221	Fe	Proto. IX	10	His.	0.1	U	-0.183 (E_m8)	—	7.4–8.6	0.06, approx. 0 and 0.04	7.8, 8.2 approx.	lxviii	105
222	Fe	Proto. IX	10	His.his.	0.04	W	-0.210 (E_m8)	—	7.4–8.4	0.06, 0.01	7.8	lxviii	105
223	Fe	Proto. IX	10	His.his.	0.04	U	-0.219 (E_m8)	—	7.5–8.4	0.06, 0.01	7.8	lxviii	105
224	Fe	Proto. IX	10	Pyr.	2.5	W	0.103 (E_m8)	0.038 (E_m10.5)	7.6–10.5	0, 0.06	9.4	lxviii	105
225	Fe	Proto. IX	10	Pyr.	2.5	U	0.091 (E_m8)	0.015 (E_m10.5)	7.6–10.7	0, 0.06	9.2	lxviii	105
226	Fe	Proto. IX	10	CN^-	0.1	W	-0.175 (E_m9.3)	—	9.3–10.5	0	—	lxviii	105
227	Fe	Proto. IX	10	CN^-	0.1	U	-0.175 (E_m9.3)	—	9.3–10.5	0	—	lxviii	105
228	Fe	Proto. IX	10	Pil.	0.04	W	-0.150 (E_m8)	—	7.4–8.3	0.01	—	lxviii	105
229	Fe	Proto. IX	10	Pil.	0.04	U	-0.170 (E_m8)	—	7.4–8.2	0.01	—	lxviii	105
230	Fe	Proto. IX	10	Pila.	0.08	W	-0.230 (E_m8)	—	8.2–9.0	0.01	—	lxxi	105
231	Fe	Proto. IX	10	Pila.	0.08	U	-0.215 (E_m8)	—	8.0–9.0	0.01	—	lxxi	105
232	Fe	Proto. IX	50	None	—	W	0.005 (E_m4.5)	-0.054 (E_m7.5)	4.5–9.5	0, 0.06	6.5	lxxii	100
233	Fe	Proto. IX	50	DMF	—	DMF/H_2O 4:1(v/v)	-0.030 (E_m4.5)	-0.050 (E_m7.5)	3.5–7.5	See note	See note	lxxii	100
234	Fe	Spirog.	10	None	—	W	-0.194	—	7.2–9.6	0.059	—	lxxiii	127
235	Fe	Spirog.	10	Pic.	—	W	-0.01	—	9.63	—	—	lxxiii	127
236	Fe	Spirog.	10	Pila.	—	W	-0.067 (E_m9.63)	—	9.63	—	—	lxxiii	127

METALLOPORPHYRINS WITH AND WITHOUT SIMPLE LIGANDS (continued)

No.	Metal	Porphyrin	S^f ($M \times 10^5$)	Ligand^h-molarity	Solutions^j	$E_m 9^{a,g}$	$E_m 11^{a,g}$	pH range investigated	Slopes $\frac{dE_h}{-dpH}$	pH of inflexions	Notes	Reference
237	Fe	Spirog.	10	CN$^-$	—	W	-0.113 (E_m 9.95)	—	9.95	—	—	lxxiii	127
238	Fe	Spirog.	—	Pyr	2.5	W	0.205	—	—	—	—	lxxiii	128
239	Fe	*a*	—	Pyr.	2.5	W	0.233	—	—	—	—	lxxiv 25°	128
240	Fe	*a*	1–3	His.	0.3	W	-0.17	—	7.0–10.3	0.058, 0.035	8.75	lxxiv 25°	139
241	Fe	*a*	1–3	His.	0.3	W + NaD.S.	-0.10	—	7.0–10.4	0.034, 0, 0.034	8.5 8.9	lxxiv	139
242	Fe	*a*	1–3	Pyr.	2.5	W	0.278 (E_m7)	—	7	—	—	lxxiv	139
243	Fe	*a*	1–3	Imid.	0.3	W	0	—	7.4–10.0	0.018, 0.044, 0.080	7.9 8.5	lxxxiv	139
244	Fe	*a*	1–3	Imid.	0.3	W	0 $\begin{bmatrix}-0.04\ (E_m7.1)\end{bmatrix}^c$	—	7.0–10.2	0.018, 0.09	8.8	lxxxiv	139
245	Fe	*a*	1–3	Imid.	0.3	W + NaD.S.	—	—	—	—	—	lxxxiv	139
246	Fe	Deutero-diester disulfonate	26	None	—	W	-0.19	-0.31	1.6–14	0.03, 0.06	6.8	lxxv 25°	131
247	Fe	Deutero-diester disulfonate	26	Pyr.	1.1	W	0.16	0.04	6–14	0, 0.06	8.8	lxxv 25°	131
248	Fe	Deutero-diester disulfonate	26	CN$^-$	0.02	W	-0.06	-0.06	7.5–14	0	—	lxxv 25°	131
249	Fe	Pheophorbide *a*	10	None	—	W	-0.17	-0.30	8.0–12.6	0.06	—	lxxvi	132
250	Fe	Pheophorbide *a*	10	CN$^-$	0.1	W	-0.124	-0.127	9.1–10.3	0	—	lxxvi	132
251	Fe	Pheophorbide *a*	10	Pyr.	1.0	W	0.125 (E_m9.6)	0.035	9.6–11	0.06	—	lxxvi	132
252	Fe	Pheophorbide *a*	10	Pila.	0.2	W	-0.08	—	7.2–9.1	0.03	—	lxxvi	132
253	Fe	Pheophorbide *a*	10	Pic.	1.0	W	0.09	0 (E_m10.6)	8.9–10.6	0.06	—	lxxvi	132
254	Fe	Pheophorbide *b*	10	None	—	W	-0.15	-0.28	8–12.6	0.06	—	lxxvi	132
255	Fe	Pheophorbide *b*	10	CN$^-$	0.1	W	-0.058 (E_m9.7)	—	—	—	—	lxxvi	132
256	Fe	Pheophorbide *b*	10	Pyr.	1.0	W	0.124 (E_m9.4)	—	—	—	—	lxxvi	132
257	Fe	Pheophorbide *b*	10	Pila.	0.2	W	-0.025 (E_m9.3)	—	—	—	—	lxxvi	132

METALLOPORPHYRINS WITH AND WITHOUT SIMPLE LIGANDS (continued)

No.	Metal	Porphyrin	S^f ($M \times 10^5$)	Ligand[h]	molarity	Solutions[i]	$E_m\,9^{a \cdot g}$	$E_m\,11^{a \cdot g}$	pH range investigated	Slopes $\dfrac{d\,E_h}{-d\,pH}$	pH of inflexions	Notes	Reference
258	Fe	Pheophorbide b	10	Pic.	1.0	W	0.074 ($E_m\,9.4$)	—	—	—	—	lxxvi	132
259	Co	Meso. IX	25	Pic.	0.9	W	−0.18	−0.18	7.5–11.7	0	—	lxxvii	133
260	Co	Meso. IX	25	Nic.	0.9	W	−0.19	−0.19	8.6–12.4	0	—	lxxvii	133
261	Co	Meso. IX	25	Pyr.	0.9	W	−0.27	−0.27	8.3–12.5	0	—	lxxvii	133
262	Co	Proto. IX	50	None		W	0.06 ($E_m\,7$)	−0.07 ($E_m\,9$)	7–9.4	0.06	—	—	134
263	Mn (III/II)	Hemato. IX	1.8	None	—	W	−0.357	−0.357	6.7–13.6	0.03, 0.02	7.7, 11.4	lxxviii 22.5°	135
264	Mn (III/II)	Hemato. IX dimethyl ester	—	None		A20	−0.268	—	6.9–9.6	0	—	22.5°	135
265	Mn (IV/III)	Hemato. IX	10	None		W	0.635 ($E_m\,9.9$)	0.509	9.9–13.6	0.12 and approx. 0.06	approx. 12	lxxix 22.5°	135
266	Mn (III/II)	Meso. IX	25	Nic.	0.9	W	−0.30	−0.30	10–13.5	0	—	lxxx	133
267	Mn (III/II)	Meso. IX	25	Pic.	0.9	W	−0.32	−0.32	7.6–11.9	0	—	lxxx	133
268	Mn (III/II)	Meso. IX	25	Pyr.	0.9	W	−0.39	−0.39	7.4–12.5	0	—	lxxxi	133
269	Mn (III/II)	Meso. IX	25	None		W	−0.483	—	12.35	—	—	lxxxi	133
270	Mn (III/II)	Meso. IX	25	Pyr.	—	W	−0.33	—	12.35	0	—	lxxxi	133
271	Mn (III/II)	Proto. IX	50	None		W	0.094 ($E_m\,7$)	−0.043 ($E_m\,10$)	6–10	See note	—	lxxxii	134
272	Mn (III/II)	Phaeophorbide a di-methyl ester	4.4	None		A20	−0.180	—	7.9–10.1	0	—	22.5°	136
273	Fe	Meso. IX-di-L-Glu	10	None		W	−0.24	−0.31	8–11	0.03	—	lxxxiii	137
274	Fe	Meso. IX-poly-DL-(Phe-Glu)	5	None		W	−0.08	−0.20	9–11	0.06	—	lxxxiii 20°	137
275	Fe	Meso. IX-poly-DL-(Lys-Phe)	5	None		W	0.24 ($E_m\,4$)	—	1–4	0.03	—	lxxxiii 20°	137

MID-POINT OXIDATION POTENTIALS OF METALLOPORPHYRINS (REFERRED TO CALOMEL ELECTRODE)

No.	Metal	Porphyrin or compound	Solvent	Oxidant	E_m(cal.) Volt	Temperature °C	Notes	Reference
276	Mg	Octaethylporphyrin	Methanol	$Fe(ClO_4)_3$	0.427	23	xcix	167
277	Zn	Octaethylporphyrin	Methanol	$Fe(ClO_4)_3$	0.525	23	xcix	167
278	Cu	Octaethylporphyrin	Chloroform: methanol, 4:1	$Fe(ClO_4)_3$	0.601	23	xcix	167
279	Ni	Octaethylporphyrin	Chloroform: methanol, 4:1	$Fe(ClO_4)_3$	0.636	23	xcix	167
280	Pd	Octaethylporphyrin	Chloroform: methanol, 4:1	$Fe(ClO_4)_3$	0.726	23	xcix	167
281	Mg	Chlorin (Chlorophyll)	Methanol	$Fe(ClO_4)_3$	0.550	23	xcix, c	167
282	Mg	Bacteriochlorophyll	Methanol	I_2	0.270	23	xcix	167

NOTES

i   Here cyt. *a* equals cyt. oxidase, see Reference 6 for relation to cyt. a_3.

ii E_m calculated on assumption that E_m cyt. *c* at 10° and 30° are equal.

iii   Method of determination not known.

iv E_m quoted for purified cytochrome. On addition of mitochondrial protein, E_m changes to approximately 0 Volt. See Reference 14 for criticism of this determination.

v E_m determined pH 6.4–8.2; slope, 0.06.

vi E_m calculated on assumption that wheat root cyt. *c* has the same E_m as mammalian cyt. *c*.

vii E_m determined pH 5.4–7.6; slope, 0.

viii E_m determined pH 6.4–9; slopes, 0 and 0.06, inflexion at pH 7.8. See Reference 23 for comments.

ix   Treatment with *p*-mercuriphenylsulfonate produces an apparent *c* type cytochrome.

x   Crude preparation, α-band approximately 560 mμ, compare type b_5, α-band = 556 mμ.

xi   Particle bound preparation. See Reference 28 for relevance of E_m values of particulate preparations.

xii   Purified preparation, non-particulate. E_m determined pH 5.2–7.5; slopes approximately 0 and 0.01, inflexion approximately pH 6.3.

xiii   Particulate preparation, E_m 7 = –0.14 Volt, purified (modified) by pancreatic lipase, E_m 7 = 0.01 Volt, on addition to latter of microsomal lipid, E_m 7 = –0.12 Volt.

xiv   Microsomal bound form called by authors P450, when solubilized P420; similar to cyt. b_1, but combines with CO.

xv   For early cyt. *c* E_m values see Reference 57.

xvi   Preparation Fe content 0.29 per cent. E_m determined pH 0.4–10.0; slopes, 0.12, 0 and 0.06, inflexions at pH 1.75 and 7.8.

xvii   Preparation resin-purified. Fe content 0.44 per cent.

xviii   Undecapeptide: +imidazole. E_m determined pH 4–13; +pyridine. E_m determined pH 3.08–12.8. Octapeptide: alone, E_m determined pH 5.47–12.49; +imidazole. E_m determined pH 4.6-12.8; +NH$_3$, E_m determined 10.0-12.4. Generally cyt. *c* type E_m v. pH curves obtained, but with considerable modification in some cases. Nitrous acid treated undecapeptide gave E_m values similar to octapeptide.

xix   Small error in E_m value reported in Reference 49 corrected in Reference 48.

xx E_m determined pH 5-9.5; slopes, 0.03 and 0, inflexion at pH 8.

xxi E_m determined pH 5.0–10.5; slopes, 0 and 0.06, inflexion at pH 8.4.

xxii   Acid-acetone treatment indicated this cytochrome intermediate in type between *b* and *c*.

xxiii E_m determined pH 5.0–11.0; slope, 0.014.

Notes (continued)

xxiv E_m determined pH 6.2–7.0; slope, 0.

xxv E_m determined pH 5.0–10.4; slopes, 0 and 0.06, inflexion at pH 8.6.

xxvi E_m determined pH 5.0–10.8; slope, 0.

xxvii E_m determined pH 6.0–8.0; slope, 0.

xxviii   Method uncertain; quoted value may be low as author also reported E_m 7 = 0.243 Volt for heart-muscle cyt. c, compare 41, 42.

xxix   Crystalline sample.

xxx E_m determined pH 5.0–9.0; slopes, 0.02 and approximately 0, inflexion at approximately pH 8. See Reference 67 for possible presence of two cytochromes in this preparation.

xxxi E_m quoted as 0.08 mV.

xxxii E_m determined pH 6.0–7.5; slope, 0. Author named c (551) as c_4 and c(555) as c_5.

xxxiii E_m 7 same for this pigment from both anaerobic and aerobic cultures. E_m of pigment from anaerobic culture only, determined pH 7.

xxxiv E_m determined pH 6–8; slopes, 0 and approximately 0.06, inflexion at pH 7. Preparation showed two peaks in vicinity of 550 mμ.

xxxv   Quoted value may be high as authors also reported E_m 6.8 = 0.284 Volt for beef heart-muscle cyt. c, compare Nos. 45, 46.

xxxvi   Authors suggest this pigment is particle bound cyt. c_2.

xxxvii E_m determined at pH 5.0–10.0; slope, 0.

xxxviii E_m determined at pH 5.0–9.0; slope, 0.

xxxix E_m determined at pH 6–8; some evidence for an inflexion at approximately pH 7.

xl   An atypical c type as it combined readily with CO.

xli   Authors gave no detail.

xlii E_m determined at pH 4.7–8.2; slopes, 0 and 0.06, inflexion at approximately pH 5.5.

xliii E_m determined at pH 5.08–9.18; slopes, 0 and 0.06, inflexion at pH 6.7.

xliv E_m determined at pH approximately 5–9 for temperatures 10–45°. Slopes in each case, 0 and 0.05, inflexions varied from pH 7.3–6.2 as temperature increased. See also Reference 1, pp. 458–461.

xlv E_m determined at pH 5.7–10.6; slopes, 0 and 0.067 (calculated from published graph), inflexion at pH 6.8.

xlvi   Variations in E_m found to be dependent on time of dialysis and on type of titration i.e., reductive or oxidative.

xlvii E_m determined at pH 6.6, 7.1 and 9.0. Corrections were made for presence of mediator; calculated values due to stepwise reaction of heme groups also included.

xlviii E_m determined at pH 5.9–9.4; slopes, 0 and 0.06, inflexion at approximately 6.6.

xlix E_m determined approximately 5–9; authors consider results show inflexions at pH 6.05, 7.0 and 7.35. E_m values unchanged on addition of N-ethylmaleimide.

l E_m determined at pH 6.2–7.7, decreases as pH increases beyond 6, insufficient data for slope and inflexion calculations.

li E_m determined at pH 6.2, 6.9 and 7.3, same remarks apply as for 1.

lii E_m determined at 6.6–7.5; slope, approximately 0.05, lessens as pH becomes lower. E_m horse Hb/MetHb determined at pH 6.6-8.6, similar results to References 98 and 99.

liii E_m values estimated from published graph.

liv E_m determined at pH 5.8-7.5; slope, 0.015.

lv E_m determined at pH 6.3-10.5; complex curve obtained which is compared with that for the untreated Hb.

lvi   Reversibility of urea effect on E_m 6–9 and temperature dependence of this effect on E_m 7 shown with human Hb. Lowering of E_m of Hb by 0.5 M guanidine hydrochloride (irreversible), 4M nicotinamide, 10 mM potassium cyanate, formaldehyde, formaldehyde + 5 mM N-ethylmaleimide approximately 6–9; extraction of accurate values from published curves difficult due to some lack of detail. Also plotted is urea conc. (1 to 8 M) against deviation from the untreated value of E_m 7 for human, horse, sheep, dog and Chironomus Hb.

Notes (continued)

lvii — The E_m values listed were taken from authors' titration curves at pH stated. E_m determined (a) pH 6.0-9.3, maximum at pH 6.4, slopes, 0.054 and 0.06, weak inflexion at pH 8.5; (b) pH range 5.7-9.5, maximum at pH 6.2, slopes, 0.054 and 0.06, weak inflexion at pH 8.5; (c) pH 6.0-9.0, slopes, approximately 0 and 0.05, inflexion at pH 7.1. Oxidative Bohr effect discussed for normal and treated Hb in terms of above results.

lviii — E_m determined at pH 6.0-10.0; slope, 0 and 0.05, inflexion at pH 6.7.

lix — E_m determined at pH 5.9-7.5; slope, 0.007.

lx — E_m determined at pH 5.5-10.5; authors consider inflexions at pH 6.9, 7.9 and 8.9. No change in E_m on addition of 5 mM N-ethylmaleimide.

lxi — E_m determined at pH 6-9 in presence of 0.5 M guanidine hydrochloride. E_m lowered by approximately 15 mV, only in region pH 6-7.

lxii — E_m determined at pH 3-12; curve contains weak inflexions at pH 6.3 and 8.5.

lxiii — E_m determined at pH 5.2-9.7; slopes, 0 to approximately pH 7 then a series of slopes which correspond to inflexions at pH 7.1, 7.8 and 9.3. E_m unaffected by addition of 5 mM N-ethylmaleimide.

lxiv — Authors gave no detail.

lxv — E_m determined at pH 4.0-11.3; slopes, 0.06, 0 and approximately 0.06, inflexions at pH 7.4 and 10.6. Calculated values for the cyanide complex, E_m 6 = −0.40 Volt, E_m 9 = −0.48 Volt. Reference 114 reports E_m determinations on complex systems of horse-radish peroxidase in the presence of hydrogen peroxide and various electron donor systems.

lxvi — E_m determinations on chlorophylls in aqueous methanol are reported in References 116 and 117. Similar studies on aqueous suspensions of chloroplast fragments are reported in Reference 118. There are doubts as to the thermodynamic reversibilities of these systems. E_m values obtained by a polarographic procedure are reported in Reference 119.

lxvii — See Reference 121 for ionic strength effect.

lxviii — Values estimated from published graphs.

lxix — Difference between borate and phosphate buffers not confirmed in Reference 121 but see Reference 131.

lxx — See Reference 121 for n (of Nernst electrode equation) = 1 in 47.5% ethanol; n = 2 in water.

lxxi — Values estimated from published graphs. E_m in this case shown to be increased by increase in dielectric strength of the medium.

lxxii — Values estimated from published graphs. E_m of No. 232 determined in presence of 6.8×10^{-5} M polyvinyl pyrrolidone; this prevented precipitation of hemin at low pH and overcame a significant time-dependence of E_m. See also Reference 126.

lxxiii — Spirographis porphyrin is 1,3,5,8-tetramethyl-2-formyl-4-vinyl-porphin-6,7-dipropionic acid.

lxxiv — Porphyrin a structural details see References 129 and 130.

lxxv — This porphyrin is 1,3,5,8-tetramethyl-porphin-2,4 disulfonic acid-6,7-dipropionic acid methyl ester. E_m values estimated from published graph. Changes in slopes at pH > 13 discussed.

lxxvi — Values estimated from published graphs. Note errors of ordinate labeling in each graph: Figure 1, E_h should read $E_{calomel}$; Figure 2, from +0.140 Volt mark upwards, potential drops 0.080 Volt/division.

lxxvii — E_m values estimated from published graph.

lxxviii — Slopes and inflexions estimated from published graph.

lxxix — E_m estimated by a spectrophotometric method involving comparison with a poising system the potential of which was determined potentiometrically.

lxxx — E_m values estimated from published graph. Influence of ionic strength on oxidation-reduction potential discussed.

lxxxi — See lxxx and also Chapter 8, Reference 1.

lxxxii — Slope gradually increases until pH 8.5 after which it remains at 0.06.

lxxxiii — Glutamic acid was in peptide link with the propionic acid residues of the heme, in the other cases these residues were hydrazine linked to polypeptide carboxyl groups.

lxxxiv — The effects of dextrins and also of mixed ligands on E_m are discussed. Two forms of the imidazole hemochrome occur.

Notes (continued)

lxxxv	Titration curves indicate 3 steps involved in complete reduction, values quoted represent half-reduction potentials for each step. $[a-1]$, $n = 1.7$; $[a-2]$, $n = 1.7$; $[a-3]$, $n = 1.5$; CO-complex, $n = 3-4$.
lxxxvi	Particulate preparation. E_m determined pH 6–8.5; slopes 0 and 0.06, inflexion at approximately pH 6.8; $n = 0.55$.
lxxxvii	Cells were grown aerobically. Some conditions pigment modified, then combined with CO, but $E_m 7$ unaltered.
lxxxviii	E_m determined pH 6.6–9.4; slope, 0.04 calculated from authors' results.
lxxxix	E_m determined pH 6–8; slope, 0; 2 hemes/molecule.
xc	E_m determined pH 6–8; slope, 0; contains at least one flavin/molecule; authors' state c (553) (No. 116) probably a modified form.
xci	Authors' state E_m decreases from 0.145 Volt at pH 6 to 0.114 Volt at pH 8 and give $E_m 7 = 0.145$ Volt; in this case slope = 0, pH 6–7 and 0.03, pH 7–8; a uniform slope would require $E_m 7 = 0.13$ Volt. This pigment appears identical to No. 117.
xcii	Determination *in situ* in chloroplasts.
xciii	Pigment exhibits a split α-band
lciv	Pigment from anaerobically grown cells.
xcv	Solution in $1M$ glycine. E_m determined pH 6–8; slope, 0; $n = 1$; system stable.
xcvi	Solution in $1M$ glycine. E_m determined pH 6–8; slopes, 0 and 0.06; inflexion at pH 7.0; some doubt re values as β-chain not stable on reduction.
xcvii	Results almost identical with those for uncombined β-chain, system stable.
xcviii	Values in table are selected from a number given in pH range 4.86–9.15. Results indicate – (i) genetic type of haptoglobin no effect, (ii) Hp:Hb ratio little effect at pH9, but considerable effect near neutrality, (iii) two oxidation–reduction systems present in equimolar amounts, each of $n = 1$ and with $E_m 7$ values different by 0.056 Volt; this consistent with α- and β-chains acting as separate entities. Compare No. 177.
xcix	Fully reversible, one electron oxidation; LiCl, LiClO$_4$ usually added to buffer salt effects during titration. Solution in concentration range 10^{-4}–$10^{-5} M$.
c	Chlorophyll could not be titrated directly. After initial oxidation a chlorin was produced which could be reversibly oxidized and reduced.

Compiled by R. W. Henderson and T. C. Morton.

REFERENCES

1. Clark, in *Oxidation-Reduction Potentials of Organic Systems,* Williams and Wikins, Baltimore, 1960.
2. *Enzyme Nomenclature,* Elsvier, New York, 1965.
3. Williams, in *Haematin Enzymes,* Falk, Lemberg, and Morton, Eds., Pergamon Press, 1961, 41.
4. Falk, in *Porphyrins and Metalloporphyrins,* Elsevier, New York, 1964, 67.
5. Ball, *Biochem. Z.,* 295, 262 (1938).
6. Slater, van Gelder, and Minnaert, in *Oxidases and Related Redox Systems,* King, Mason, and Morrison, Eds., John Wiley & Sons, New York, 1965, 667.
7. Minnaert, *Biochim. Biophys. Acta,* 110, 42 (1965).
8. Wainio, *J. Biol. Chem.,* 216, 593 (1955).
9. Straub, *Biochemie. Verlag Ungar. Akad. Wiss. (Budapest),* 182 (1960), quoted by T. Yonetani, in *The Enzymes,* 2nd ed., Boyer, Lardy and Myrbäck, Eds., Academic Press, New York, 1963, 60.
10. Sezuku and Okunuki, *J. Biochem.* (Tokyo), 43, 107 (1956).
11. Holton and Colpa-Boonstra, *Biochem. J.,* 76, 179 (1960).
12. Goldberger, Pumphrey, and Smith, *Biochim. Biophys. Acta,* 58, 307 (1962).
13. Feldman and Wainio, *J. Biol. Chem.,* 235, 3635 (1960).
14. Straub and Colpa-Boonstra, *Biochim. Biophys. Acta,* 60, 650 (1962).
15. Lundegårdh, *Arkiv. Kemi.,* 5, 97 (1953).
16. Lundegårdh, *Nature,* 173, 939 (1954).
17. Deeb and Hager, *J. Biol. Chem.,* 239, 1024 (1964).
18. Fujita, Itagaki, and Sato, *J. Biochem.* (Tokyo), 53, 282 (1963).
19. Itagaki, E. and Hager, L. P., personal communication, January 1966.
20. Hauge, *Arch. Biochem. Biophys.,* 94, 308 (1961).
21. Boeri and Cutolo, *Boll. Soc. Ital. Biol. Sper.,* 33, 1711 (1957).
22. Hasegawa and Ogura, in *Haematin Enzymes,* Falk, Lemberg, and Morton, Eds., Pergamon Press, New York, 1961, 534.
23. Hiromi and Sturtevant, *J. Biol. Chem.,* 240, 4662 (1965).
24. Morton and Shepley, *Biochim. Biophys. Acta,* 96, 349 (1965).
25. Bendall and Hill, *New Phytol.,* 55, 206 (1956).
26. Yoshikawa, *J. Biochem.* (Tokyo), 38, 1 (1951).
27. Strittmatter and Ball, *Proc. Natl. Acad. Sci. U.S.A.,* 38, 19 (1952).
28. Discussion, *Haematin Enzymes,* Falk, Lemberg, and Morton, Eds., Pergamon Press, New York, 1961, 477.
29. Velick and Strittmatter, *J. Biol. Chem.,* 221, 265 (1956).
30. Raw, Molinari, do Amaral, and Mahler, *J. Biol. Chem.,* 233, 225 (1958).
31. Kawai, Yoneyama, and Yoshikawa, *Biochim. Biophys. Acta,* 67, 522 (1963).
32. Hill, *Nature,* 174, 501 (1954).
33. Shichi and Hackett, *J. Biol. Chem.,* 237, 2959 (1962).
34. Omura and Sato, *J. Biol. Chem.,* 239, 2379 (1964).
35. Keilin, *Nature,* 180, 427 (1957).
36. Katoh and San Pietro, *Biochem. Biophys. Res. Commun.,* 20, 406 (1965).
37. Rodkey and Ball, *J. Biol. Chem.,* 182, 17 (1950).
38. Henderson and Rawlinson, *Biochem. J.,* 62, 21 (1956).
39. Henderson and Rawlinson, *Nature,* 177, 1180 (1956).
40. Hettinger and Harbury, *Proc. Natl. Acad. Sci. U.S.A.,* 52, 1469 (1964).
41. Minakami, Titani, and Ishikura, *J. Biochem.* (Tokyo), 45, 341 (1958).
42. Minakami, Titani, and Ishikura, *J. Biochem.* (Tokyo), 44, 535 (1957).
43. Harbury and Loach, *Proc. Natl. Acad. Sci. U.S.A.,* 45, 1344 (1959).
44. Harbury and Loach, *J. Biol. Chem.,* 235, 3640 (1960).
45. Harbury and Loach, *J. Biol. Chem.,* 235, 3646 (1960).
46. Minakami, *J. Biochem.* (Tokyo), 42, 749 (1955).
47. Henderson and Rawlinson, in *Haematin Enzymes,* Falk, Lemberg, and Morton, Eds., Pergamon Press, New York, 1961, 370.
48. Green, Järnefelt, and Tisdale, *Biochem. Biophys. Acta,* 38, 160 (1960).
49. Green, Järnefelt, and Tisdale, *Biochim. Biophys. Acta,* 31, 34 (1959).
50. Vernon and Kamen, *J. Biol. Chem.,* 211, 643 (1954).
51. Kamen and Vernon, *Biochim. Biophys. Acta,* 17, 10 (1955).
52. Ishimoto, Koyama, and Nagai, *J. Biochem.* (Tokyo), 41, 763 (1954).
53. Postgate, in *Haematin Enzymes,* Falk, Lemberg, and Morton, Eds., Pergamon Press. New York, 1961, 407.
54. le Gall, Mazza, and Dragoni, *Biochim. Biophys. Acta,* 99, 385 (1965).
55. Davenport and Hill, *Proc. Roy. Soc.* (London), B139, 327 (1952).

56. Tosi and Ghiretti, *Atti. Acad. Nazl. Lincei. Rend. Classe. Sci. Fis. Mat. Nat.*, 29, 104 (1960). See *Chem. Abstr.*, 55, 10731b (1961).
57. Margoliash and Schejter, *Advanc. Prot. Chem.*, 21, 114 (1966).
58. Katoh, *J. Biochem.* (Tokyo), 46, 629 (1959).
59. Katoh, *Plant Cell Physiol.* (Tokyo), 1, 91 (1960).
60. Holton and Myers, *Science,* 142, 234 (1963).
61. Yamanaka and Kamen, *Biochem. Biophys. Res. Commun.*, 19, 751 (1965).
62. Nishimura, *J. Biochem.* (Tokyo), 46, 219 (1959).
63. Perini, Kamen, and Schiff, *Biochim. Biophys. Acta,* 88, 74 (1964).
64. Kamen and Vernon, *Biochim. Biophys. Acta,* 17, 10 (1955).
65. Iwasaki, *J. Biochem.* (Tokyo), 47, 174 (1960).
66. Horio, *J. Biochem.* (Tokyo), 45, 267 (1958).
67. Horio, Higashi, Sasagawa, Kusai, Nakai, and Okunuki, *Biochem. J.,* 77, 194 (1960).
68. Kamen and Takeda, *Biochim. Biophys. Acta,* 21, 518 (1956).
69. Bone, *Nature,* 197, 517 (1963).
70. Yamanaka, Miki, and Okunuki, *Biochim. Biophys. Acta,* 77, 654 (1963).
71. Kogut, *Biochem. J.,* 65, 35P (1957).
72. Tissières, *Biochem. J.,* 64, 582 (1956).
73. Hori, *J. Biochem.* (Tokyo), 50, 481 (1961).
74. Hori, *J. Biochem.* (Tokyo), 50, 440 (1961).
75. Gibson, *Biochem. J.,* 79, 151 (1961).
76. Skarzyński, Klimek, and Szczepkowski, *Bull. Acad. Polon. Sci. Classe II,* 4, 299 (1956). See *Chem. Abstr.*, 53, 15191d (1959).
77. Aubert, Milhaud, Moncel, and Millet, *Acad. Sci. (Paris) Compt. Rend.,* 246, 1616 (1958).
78. Aubert, Millet, and Milhaud, *Ann. Inst. Pasteur,* 96, 559 (1959).
79. Trudinger, *Biochem. J.,* 78, 673 (1961).
80. Taniguchi and Kamen, *Biochim. Biophys. Acta,* 96, 395 (1965).
81. Kamen, *Bacteriol. Rev.,* 19, 250 (1955).
82. Morita, *J. Biochem.* (Tokyo), 48, 870 (1960).
83. De Klerk, Bartsch, and Kamen, *Biochim. Biophys. Acta,* 97, 275 (1965).
84. Orlando, *Biochim. Biophys. Acta,* 57, 373 (1962).
85. Newton and Kamen, *Biochim. Biophys. Acta,* 21, 71 (1956).
86. Bartsch and Kamen, *J. Biol. Chem.,* 235, 825 (1960).
87. Morita and Conti, *Arch. Biochem. Biophys.,* 100, 302 (1963).
88. Vernon and Mangum, *Arch. Biochem. Biophys.,* 90, 103 (1960).
89. Sutherland, *Biochim. Biophys. Acta,* 73, 162 (1963).
90. Fujita and Sato, *Biochim. Biophys. Acta,* 77, 690 (1963).
91. Vernon, Mangum, Beck, and Shafia, *Arch. Biochem. Biophys.,* 88, 227 (1960).
92. Vernon, *J. Biol. Chem.,* 222, 1045 (1956).
93. Appleby, C. A., personal communication. February 1966.
94. Bartsch and Kamen, *J. Biol. Chem.,* 230, 41 (1958).
95. Kamen, in *Bacterial Photosynthesis,* Gest, San Pietro, and Vernon, Eds., Antioch Press, 1963, 61.
96. James, Lyons, and Williams, *Biochemistry,* 1, 379 (1962).
97. Havemann and Wolff, *Biochem. Z.,* 293, 399 (1937).
98. Taylor and Hastings, *J. Biol. Chem.,* 131, 649 (1939).
99. Havemann, *Biochem. Z.,* 314, 118 (1943).
100. Behlke and Scheler, *Acta Biol. Med. Ger.,* 8, 88 (1962).
101. Abel, *Arch. Biochem. Biophys.,* 101, 286 (1963).
102. Stopp, Ristau, and Jung, *Acta Biol. Med. Ger.,* 11, 751 (1963).
103. Taylor, *J. Biol. Chem.,* 144, 7 (1942).
104. Antonini, Wyman, Brunori, Taylor, Rossi-Fanelli, and Caputo, *J. Biol. Chem.,* 239, 907 (1964).
105. Banerjee, *J. Chim. Phys.,* 57, 615 (1960).
106. Banerjee, *J. Chim. Phys.,* 57, 627 (1960).
107. Behlke and Scheler, *Acta Biol. Med. Ger.,* 12, 629 (1964).
108. Brunori, Antonini, Wyman, Zito, Taylor, and Rossi-Fanelli, *J. Biol. Chem.,* 239, 2340 (1964).
109. Antonini, Wyman, Brunori, and Rossi-Fanelli, *Biochim. Biophys. Acta,* 94, 188 (1965).
110. Taylor and Morgan, *J. Biol. Chem.,* 144, 15 (1942).
111. George and Irvine, *Biochem. J.,* 60, 596 (1954).
112. Thiele, Behlke, and Scheler, *Acta Biol. Med. Ger.,* 12, 19 (1964).
113. Harbury, *J. Biol. Chem.,* 225, 1009 (1957).
114. Matkovics, Kovács, Földeák, and Gy. Sipos, *Enzymologia,* 21, 96 (1959/60).

115. **Henderson, R. W. and Appleby, C. A.,** to be published.
116. **Goedheer,** *Biochim. Biophys. Acta,* 27, 478 (1958).
117. **Goedheer, Horreus de Haas, and Schuller,** *Biochim. Biophys. Acta,* 28, 278 (1958).
118. **Kok,** *Biochim. Biophys. Acta,* 48, 527 (1961).
119. **Stanienda,** *Naturwissenschaften,* 50, 731 (1963).
120. **Vestling,** *J. Biol. Chem.,* 135, 623 (1940).
121. **Davies,** *J. Biol. Chem.,* 135, 597 (1940).
122. **Cowgill and Clark,** *J. Biol. Chem.,* 198, 33 (1952).
123. **Conant and Tongberg,** *J. Biol. Chem.,* 86, 733 (1930).
124. **Barron,** *J. Biol. Chem.,* 121, 285 (1937).
125. **Shack and Clark,** *J. Biol. Chem.,* 171, 143 (1947).
126. **Scheler and Rahmel,** *Acta Biol. Med. Ger.,* 10, 218 (1963).
127. **Barron,** *J. Biol. Chem.,* 133, 51 (1940).
128. **Henderson, R. W.,** Ph.D. thesis, University of Melbourne, 1960.
129. **Lemberg,** *Rev. Pure Appl. Chem.,* 15, 125 (1965).
130. **Grassl, Coy, and Lynen,** *Biochem. Z.,* 338, 771 (1963).
131. **Walter,** *J. Biol. Chem.,* 196, 151 (1952).
132. **Kikuchi and Barron,** *J. Am. Chem. Soc.,* 81, 3990 (1959).
133. **Taylor,** *J. Biol. Chem.,* 135, 569 (1940).
134. **Thiele, Behlke, and Scheler,** *Acta Biol. Med. Ger.,* 11, 767 (1963).
135. **Loach and Calvin,** *Biochemistry,* 2, 361 (1963).
136. **Loach and Calvin,** *Nature,* 202, 343 (1964).
137. **Lautsch, Broser, and Becker,** *Z. Elektrochem.,* 61, 174 (1957).
138. **Suzuki and Iwasaki,** *J. Biochem.* (Tokyo), 52, 193 (1962).
139. **Vanderkooi and Stotz,** *J. Biol. Chem.,* 241, 3316 (1966).
140. **Tzagoloff and Wharton,** *J. Biol. Chem.,* 240, 2628 (1965).
141. **Ohnishi,** *J. Biochem.* (Tokyo), 59, 17 (1966).
142. **Itagaki and Hager,** *J. Biol. Chem.,* 241, 3687 (1966).
143. **Hettinger and Harbury,** *Biochemistry,* 4, 2585 (1965).
144. **Motokawa and Kikuchi,** *Biochim. Biophys. Acta,* 120, 274 (1966).
145. **Webster and Hackett,** *J. Biol. Chem.,* 241, 3308 (1966).
146. **Labeyrie, Groudinsky, Jacquot-Armand, and Naslin,** *Biochim. Biophys. Acta,* 128, 492 (1966).
147. **Harbury, Cronin, Fanger, Hettinger, Murphy, Meyer, and Vinogradov,** *Proc. Natl. Acad. Sci. U.S.A.,* 54, 1658 (1965).
148. **Horio and Ohkawa,** *J. Biochem.* (Tokyo), 64, 393 (1968).
149. **Urban and Klingenberg,** *Eur. J. Biochem.,* 9, 519 (1969).
150. **Mori and Hirai,** in *Structure and Function of Cytochromes,* Okunuki, Kamen, and Sekuzu, Eds., University of Tokyo Press and University Park Press, 1968, 681.
151. **Okada and Okunuki,** *J. Biochem.* (Tokyo), 65, 581 (1969).
152. **Barrett and Sinclair,** *Biochim. Biophys. Acta,* 143, 279 (1967).
153. **Orlando,** *Biochim. Biophys. Acta,* 143, 634 (1967).
154. **Holton and Myers,** *Biochim. Biophys. Acta,* 131, 375 (1967).
155. **Yamanaka, Nagata, and Okunuki,** *J. Biochem.* (Tokyo), 63, 753 (1968).
156. **Meyer, Bartsch, Cusanovitch, and Mathewson,** *Biochim. Biophys. Acta,* 153, 854 (1968).
157. **Bendall,** *Biochem. J.,* 109, 46p (1968).
158. **Horio and Kitahara,** unpublished, quoted by Horio and Yamashita in *Bacterial Photosynthesis,* Gest, San Pietro, and Vernon, Eds., Antioch Press, 1963, 275.
159. **Yamagata and Sato,** *J. Biochem.* (Tokyo), 64, 549 (1968).
160. **Honda, Baker, and Muenster,** *Plant Cell Physiol.,* 2, 151 (1961).
161. **Gorman and Levine,** *Plant Physiol.,* 41, 1643 (1966).
162. **Sugimura, Toda, Murata, and Yakushiji,** in *Structure and Function of Cytochromes,* Okunuki, Kamen, and Sekuzu, Eds., University of Tokyo Press and University Park Press, 1968, 452.
163. **Fujita,** *J. Biochem.* (Tokyo), 60, 204 (1966).
164. **Brunori, Taylor, Antonini, Wyman, and Rossi-Fanelli,** *J. Biol. Chem.,* 242, 2295 (1967).
165. **Brunori, Alfsen, Saggese, Antonini, and Wyman,** *J. Biol. Chem.,* 243, 2950 (1968).
166. **Banerjee and Cassaly,** *J. Mol. Biol.,* 42, 337 (1969).
167. **Fuhrhop and Mauzerall,** *J. Am. Chem. Soc.,* 91, 4174 (1969).

These tables originally appeared in Sober, Ed., *Handbook of Biochemistry and selected data for Molecular Biology,* 2nd ed., Chemical Rubber Co., Cleveland, 1970.

HEATS OF PROTON IONIZATION, pK, AND RELATED THERMODYNAMIC QUANTITIES

Reed M. Izatt and James J. Christensen

A. Compound

1. The nomenclature follows that in the 46th Edition of the *Handbook of Chemistry and Physics.* The Chemical Rubber Co., 1965–66.

B. Formula, Synonyms

1. Formulas are given as necessary. The protonated form of each species is given. Extensive cross-referencing is included as an aid in the location of compounds.

C. ΔH

1. The first criterion for construction of the Table was to include all reliable calorimetric ΔH data. ΔH values derived from temperature variation studies are included when calorimetric data are not available and where the temperature variation data extend over a wider temperature range than the calorimetric data. When two or more sets of temperature variation data of apparently equal reliabilities exist, the one giving the widest temperature range is chosen. Generally, ΔH values calculated from temperature variation data are less reliable than calorimetric values and care should be exercised in their use.

2. Uncertainties of the ΔH values are not reported; however, the significant figures are given in each case as reported in the original article.

3. Values of ΔH listed first are those which are valid at zero ionic strength. Values reported as a function of ionic strength and solvent composition are then reported in that order. In each category, calorimetric results are listed first, followed by temperature variation data.

D. pK

1. The pK value is valid at the temperature listed in the $T, °C$ column, unless otherwise specified in parentheses following the pK value.

E. ΔS

1. The ΔS value is valid at the temperature listed in the $T, °C$ column, unless otherwise specified in parentheses following the pK value.

F. $T, °C$

1. ΔH values are not reported for temperature intervals less than five degrees, except for 38°C.

G. Method

1. The method of ΔH determination is designated by C (calorimetric) of T (temperature variation).

H. Conditions

1. The conditions under which ΔH was determined are given. μ refers to the ionic strength, and C to the concentration of the compound. Supporting electrolyte, when present, is identified by its composition and molarity (M). Composition of non-aqueous solvents, when present, are indicated in the *Conditions* column with the non-aqueous component listed in the *Remarks* column.

2. The ionic strength dependence of ΔH is given whenever data have been reported at widely varying ionic strengths.

3. Values are given for more than one ionic strength only if reported at widely varying ionic strengths.

I. References

1. References in this column are those from which the data were taken.

2. The data in certain references are known by the compilers to be questionable. This has been indicated by a superscript on the reference, i.e., 40^a, 54^b, 88^c, and 107^d. The superscript letters have the following significance:

a. Some data in this reference appear to be unreliable (private communication from author to compilers).

b. These data appear unreliable.

c. These data are questionable; see Reference 106.

d. These data should be disregarded; see Reference 106.

J. Other References

1. These references also contain ΔH values, but were not included in the Table.

K. Remarks

1. In those cases in which ΔH values appeared unusual, an appropriate comment to this effect is made.

2. Solvents other than water are designated with the weight percent non-aqueous solvent being given in the *Conditions* column.

3. In some cases, the reliability of the ΔH data is uncertain. This is indicated under *Remarks* by the designation 3. followed by a., b., c., d. where these letters have the following meanings.

a. ΔH is calculated from pK values determined at only two temperatures.

b. ΔH values are average values over the indicated temperature range, resulting in a combination of thermodynamic data valid at different temperatures.

c. No correction has been made for competing ions prior to calculation of ΔH.

d. The ionic strength is unknown or not specified.

4. Reactions involving microspecies are given where these have been reported.

HEATS OF PROTON IONIZATION, pK, AND RELATED THERMODYNAMIC QUANTITIES

Compound	Formula, Synonyms	ΔH (kcal/mole)	pK	ΔS cal/deg-mole	T°C	Method	Conditions	Reference	Other References	Remarks
Acetic acid	CH$_3$COOH Ethanoic acid	0.47	4.762	−20.2	10	C	$\mu = 0$	324	6–13, 325, 326	—
		−0.02	4.756	−21.9	25	C	$\mu = 0$	1	389	—
		−0.50	4.769	−23.4	40	C	$\mu = 0$	324	—	—
		0.714	4.781	−19.2	0	T	$\mu = 0$	2	—	—
		0.552	4.770	−19.7	5	—	—	—	—	—
		0.389	4.762	−20.3	10				—	—
		0.223	4.758	−20.9	15				—	—
		0.057	4.756	−21.5	20				—	—
		−0.112	4.756	−22.1	25				—	—
		−0.282	4.757	−22.6	30				—	—
		−0.455	4.762	−23.2	35				—	—
		−0.628	4.769	−23.8	40				—	—
		−0.804	4.777	−24.3	45				—	—
		−0.982	4.787	−24.8	50				—	—
		−1.161	4.799	−25.4	55				—	—
		−1.342	4.812	−26.0	60				—	—
		0.09	—	—	25	C	$\mu = 0.1$	3	—	—
		0.18	—	—	25	C	$\mu = 0.2$	4	—	—
		0.721	—	—	25	C	$\mu = 2.0$ (NaClO$_4$)	328	—	—
		0.09	4.754	−22.1	25	T	0	5	—	Conditions refer to wt. % dioxane
		−0.05	5.290	−24.4	—	—	20	—	—	—
		−0.44	6.318	−30.3	—	—	45	—	—	—
		−0.61	8.318	−40.1	—	—	70	—	—	—
		−1.34	10.135	−50.8	—	—	82	—	—	—
		−0.17	4.93	−23.0	25	C	10.0	327	—	Conditions refer to wt. % ethanol
		−0.10	5.06	−23.5	25	C	17.0	—	—	—
		−0.11	5.11	−23.8	25	C	20.0	—	—	—
		−0.10	5.25	−24.4	25	C	26.0	—	—	—
		−0.10	5.32	−24.7	25	C	30.0	—	—	—
		−0.13	5.43	−25.3	25	C	35.0	—	—	—
		−0.15	5.55	−25.9	25	C	40.0	—	—	—
		−0.40	5.80	−27.9	25	C	50.0	—	—	—
		−0.50	6.10	−29.6	25	C	60.0	—	—	—
		−0.72	6.45	−31.9	25	C	70.0	—	—	—
		−1.24	6.82	−35.4	25	C	80.0	—	—	—
		−2.12	—	—	25	C	—	328	—	95% v/vCH$_3$OH

HEATS OF PROTON IONIZATION, pK, AND RELATED THERMODYNAMIC QUANTITIES (continued)

Acetic acid—(Continued)

Compound	Formula, Synonyms	ΔH (kcal/mole)	pK	ΔS cal/deg-mole	T°C	Method	Conditions	Reference	Other References	Remarks
—, d_1	CH_3COOD	1.192	5.3463	−20.2	0	T	$\mu = 0$	14	—	—
		0.275	5.3130	−23.4	25	—	—	—	—	—
		−0.730	5.3245	−26.6	40	—	—	—	—	—
—, d_3	CD_3COOH	0.735	4.795	−19.2	0	T	$\mu = 0$	7	—	—
		0.069	4.771	−22.1	25	—	—	—	—	—
		−0.945	4.799	−24.9	50	—	—	—	—	—
—, d_4	CD_3COOD	1.174	5.360 (5°)	−20.3	0	T	$\mu = 0$	15	—	Solvent: D_2O
		0.279	5.325	−23.4	25	—	—	—	—	—
		−0.695	5.336	−26.6	50	—	—	—	—	—
—, bromo-	—	−1.10	2.902	−17.0	25	C	$\mu = 0$	324	6, 326	—
		−1.43	2.93	−19.9	40	C	$\mu = 0$	—	—	—
		−1.054	2.8877	−16.8	20	T	$\mu = 0$	16	—	—
		−1.239	2.9021	−17.4	25	—	—	—	—	—
		−1.435	2.9180	−18.1	30	—	—	—	—	—
—, [(carboxy methyl-imino)- bis(ethylenenitrilo)] tetra-	CH_2COOH $(HOOCCH_2)_2\overset{+}{N}H(CH_2)_2NH(CH_2)_2\overset{+}{N}H(CH_2COOH)_2$ Diethylenetriaminepentaacetic acid	1.73	4.26 (pK_6)	−13.6	20	C	$0.1\,M\,KNO_3$	17	18	—
		4.32	7.09 (pK_7)	−17.7	20	C	$0.1\,M\,KNO_3$	17	18	—
		7.96	10.57 (pK_8)	−21.2	20	C	$0.1\,M\,KNO_3$	324	326	—
—, chloro-	—	−0.46	2.827	−15.0	10	C	$\mu = 0$	324	6, 10	—
		−0.99	2.858	−16.4	25	C	$\mu = 0$	—	—	—
		−1.53	2.910	−18.2	40	C	$\mu = 0$	—	—	—
		−0.149	2.816	−13.4	0	T	$\mu = 0$	19	11, 16	—
		−0.593	2.827	−15.0	10	—	—	—	—	—
		−0.914	2.833	−16.1	18	—	—	—	—	—
		−1.170	2.858	−17.0	25	—	—	—	—	—
		−1.402	2.883	−17.9	32	—	—	—	—	—
		−1.639	2.910	−18.5	40	—	—	—	—	—
—, dibromo-	—	−0.5	1.39	−8	25	—	—	329	—	3.d.
—, dichloro-	—	0.17	1.30	−6.5	25	C	$\mu = 0$	326	—	—
—, diphenyl-(hydroxy)-	Benzilic acid $(C_6H_5)_2COHCOOH$	−0.168	3.049 (35°)	−14	16.8	T	—	12	—	3.a.b.d.
—, [ethylenebis (oxyethyl- enenitrilo)] tetra-	$(HOOCCH_2)_2NH(CH_2)_2O(CH_2)_2O(CH_2)_2NH(CH_2COOH)_2$ Di (2-aminoethoxy)ethanetetraacetic acid	5.76	8.84 (pK_5)	−20.8	20	C	$0.1\,M\,KNO_3$	20	—	—
		5.40	8.95 (pK_5)	−22.5	20	T	$\mu = 0.1$	21	—	—
		4.88	8.88 (pK_5)	−24.2	25	T	$\mu = 0.1$	21	—	—

HEATS OF PROTON IONIZATION, pK, AND RELATED THERMODYNAMIC QUANTITIES (continued)

Compound	Formula, Synonyms	ΔH (kcal/mole)	pK	ΔS cal/deg-mole	T °C	Method	Conditions	Reference	Other References	Remarks
Acetic acid—(Continued)										
	EGTA	5.84	9.44 (pK$_6$)	−23.3	20	C	0.1 M KNO$_3$	20	—	—
		6.10	9.60 (pK$_6$)	−23.1	20	T	$\mu = 0.1$	21	—	—
		6.33	9.53 (pK$_6$)	−22.4	25	T	$\mu = 0.1$	21	—	—
		−0.680	2.586	−14.1	25	C	$\mu = 0$	326	—	—
		−1.232	2.5711	−16.0	20	T	$\mu = 0$	16	6	—
		−1.390	2.5864	−16.5	25	—	$\mu = 0$	—	—	—
		−1.558	2.6038	−17.1	30	—	—	—	—	—
—, (hexamethylene-dinitrilo) tetra-	(HOOCH$_2$)$_2$NH$^+$(CH$_2$)$_6$NH$^+$(CH$_2$COOH)$_2$ Hexamethylenedinitrilotetraacetic acid HDTA	6.24	9.79	−23.5	20	C	0.1 M KNO$_3$	20	—	—
		7.91	10.84	−22.5	20	C	0.1 M KNO$_3$	20	—	—
—, hydroxy-	HOCH$_2$COOH Glycolic acid	0.53	3.848	−15.8	10	C	$\mu = 0$	324	—	—
		0.11	3.831	−17.2	25	C	$\mu = 0$	1	6	—
		−0.39	3.835	−18.8	40	C	$\mu = 0$	324	—	—
		1.020	3.875	−14.0	0.3	T	$\mu = 0$	22	—	—
		0.660	3.844	−15.29	12.5	—	—	—	—	—
		0.210	3.831	−16.83	25	—	—	—	—	—
		−0.320	3.833	−18.59	37.5	—	—	—	—	—
		−0.920	3.849	−20.55	50	—	—	—	—	—
		0.56	—	—	25	C	$\mu = 2.0$ (NaClO$_4$)	328	—	—
—, hydroxy-(phenyl)-	C$_6$H$_5$CHOHCOOH Mandelic acid Phenylglycolic acid	−0.067	3.372 (35°)	−16	16.8	T	—	12	—	3.a.b.d.
—, iminodi-	H$_2$N(CH$_2$COOH)$_2$ Diglycolamidic acid	6.98	9.89 (pK$_3$)	−21.8	25	C	$\mu = 0.02$	330	—	—
		8.15	9.44 (pK$_3$)	−15.4	20	C	0.1 M KNO$_3$	20	—	—
—, iminotri-	H$^+$N(CH$_2$COOH)$_3$ Nitrilotriacetic acid Triglycolamic acid	2.0	1.687 (pK$_2$)	−2.1	0	T	$\mu = 0$	23	25–27	—
		0.25	1.65	−7.3	10	—	—	—	—	—
		−0.25	1.65	−8.4	20	—	—	—	—	—
		−0.75	1.66	−9.5	30	—	—	—	—	—
		−1.75	1.686	−12.3	40	—	—	—	—	—
		0.4	2.953 (pK$_3$)	−12.1	0	T	$\mu = 0$	23	—	—
		0.3	2.948	−12.5	10	—	—	—	—	—
		0	2.940	−13.5	20	—	—	—	—	—
		−0.8	2.956	−16.1	30	—	—	—	—	—
		−1.1	2.978	−17.2	40	—	—	—	—	—
		5.20	10.594 (pK$_4$)	−29.6	0	T	$\mu = 0$	23	—	—
		4.80	10.454	−30.9	10	—	—	—	—	—
		4.40	10.334	−32.4	20	—	—	—	—	—

HEATS OF PROTON IONIZATION, pK, AND RELATED THERMODYNAMIC QUANTITIES (continued)

Compound	Formula, Synonyms	ΔH (kcal/mole)	pK	ΔS cal/deg-mole	T°C	Method	Conditions	Reference	Other References	Remarks
Acetic acid—(Continued)										
—, iodo	—	4.05	10.230	−33.6	30	—	—	330	—	—
		4.53	10.28 (pK_4)	−31.9	25	C	$\mu = 0.04$	24	6, 326	—
		4.56	9.73 (pK_4)	−29	20	C	$\mu = 0.1$	324	—	—
		−0.75	3.124	−16.9	10	C	$\mu = 0$	—	—	—
		−1.21	3.175	−18.6	25	C	$\mu = 0$	—	6	—
		−1.51	3.211	−19.5	40	T	$\mu = 0$	16	—	—
		−1.256	3.1582	−18.7	20	—	$\mu = 0$	—	—	—
		−1.416	3.1752	−19.3	25	—	—	—	—	—
		−1.585	3.1934	−19.8	30	—	—	—	—	—
—, mercapto-	$HSCH_2COOH$ Thioglycolic acid	6.2	10.56 (pK_2)	−27.6	25	C	$\mu = 0$	28	29	—
—, 2-mercaptophenyl-	$C_6H_5SCH_2COOH$ Phenylthioglycolic acid Thioglycolic acid *S*-phenyl	−0.600	3.56 (25°)	−18.3	0–25	T	—	30	—	3.a.b.d.
—, methoxy-	CH_3OCH_2COOH Methoxyacetic acid Methylglycolic acid	−0.35	3.544	−17.4	10	C	$\mu = 0$	324	—	—
		−0.94	3.570	−19.4	25	C	$\mu = 0$	—	—	—
		−1.148	3.613	−22.3	40	C	$\mu = 0$	—	—	—
		−0.960	3.5704	−19.6	25	T	$\mu = 0$	31	—	—
—, *N*-methyl-*N*-nitrilodi-	$CH_3\overset{+}{N}H(CH_2COOH)_2$ *N*-Methylaminediacetic acid	0	2.150 (pK_2)	−9.5	30	T	$\mu = 0$	32	—	—
		6.8	9.920 (pK_3)	−23	30	T	$\mu = 0$	32	—	—
		6.94	9.65 (pK_3)	−20.5	20	C	$0.1\,M\,KNO_3$	33	—	—
—, (octamethylene-dinitrilo) tetra-	$(CH_2COOH)_2\overset{+}{N}H(CH_2)_8\overset{+}{N}H(CH_2COOH)_2$ ODTA	5.79	9.89	−25.5	20	C	0.1	20	—	—
		8.09	10.84	−22.0	20	C	0.1	20	—	—
—, oxo-	$HCOCO_2H$ Glyoxylic acid Aldehydoformic acid	0.53	3.46	−14	25	C	$\mu = 0$	34	—	—
		0.641	3.46	−13.68	25	C	$\mu = 0.05$	—	—	—
—, [oxybis(ethylene-nitrilo)] tetra-	$(HOOCCH_2)_2\overset{+}{N}H(CH_2)_2O(-CH_2)_2\overset{+}{N}H-(CH_2COOH)_2$ β,β′-Diaminodiethylether-*N,N*′-tetraacetic acid BATA	7.25	8.83 (pK_5)	−15.7	20	C	$0.1\,M\,KNO_3$	20	—	—
		6.23	9.47 (pK_6)	−22.1	20	C	$0.1\,M\,KNO_3$	20	—	—
—, (pentamethyl-enedinitrilo)tetra-	$(CH_2COOH)_2\overset{+}{N}H(CH_2)_5\overset{+}{N}H(CH_2COOH)_2$ PETA	6.3	9.52 (pK_5)	−22.1	20	C	$0.1\,M\,KNO_3$	20	—	—
		7.5	10.68 (pK_6)	−23.3	20	C	$0.1\,M\,KNO_3$	20	—	—
—, phenyl-	Phenylethanoic α-Toluic	−0.311	4.296 (35°)	−21	16.8	T	—	12	—	3.a.b.d.

HEATS OF PROTON IONIZATION, pK, AND RELATED THERMODYNAMIC QUANTITIES (continued)

Compound	Formula, Synonyms	ΔH (kcal/mole)	pK	ΔS cal/deg-mole	T °C	Method	Conditions	Reference	Other References	Remarks
Acetic acid—(Continued)										
—, phenylsulfonyl-	$C_6H_5SO_2CH_2COOH$	−1.300	2.44 (25°)	−15.5	0–25	T	—	30	—	3.a.b.d.
—, phenylsulfoxy-	$C_6H_5SOCH_2COOH$	−1.000	2.66 (25°)	−15.5	0–25	T	—	30	—	3.a.b.d.
—, (propylenedinitrilo) tetra-	$(CH_2COOH)_2NH(CH_2)_3NH(CH_2COOH)_2$ TMTA	4.43	8.62	−21.6	20	C	0.1 M KNO$_3$	20	—	—
		5.16	10.45	−30.2	20	C	0.1 M KNO$_3$	20	—	—
—, (tetramethylene dinitrilo)tetra-	$(CH_2COOH)_2NH(CH_2)_4NH(CH_2COOH)_2$ TETA	5.81	9.05 (pK$_5$)	−21.6	20	C	0.1 M KNO$_3$	20	—	—
		6.68	10.66 (pK$_6$)	−26.0	20	C	0.1 M KNO$_3$	20	—	—
—, [thio bis(ethylene nitrilo)]tetra-	$(HOOCCH_2)_2NH(CH_2)_2S(CH_2)_2NH$ $(CH_2COOH)_2$ β,β'-Diaminodiethylsulfide-N,N'-tetraacetic acid BSTA	6.59	8.47 (pK$_5$)	−16.3	20	C	0.1 M KNO$_3$	20	—	—
		6.69	9.42 (pK$_6$)	−20.3	20	C	0.1 M KNO$_3$	20	—	—
—, thiolo-	CH_3COSH Ethanethiolic acid	0.56	3.62	−14.7	25	C	$\mu = 0.015$	29	—	—
—, tribromo-	—	−0.8	−0.147	−2	25	—	—	329	—	3.d.
—, trichloro-	—	1.5	1.54	−2	25	—	—	329	—	3.d.
—, trifluoro-	—	0	0.22	−1	25	—	—	329	—	3.d.
—, trimethyl-	—	−0.724	5.031	−25.45	25	T	$\mu = 0$	13	—	—
Acetonic acid	See: Propanoic acid, 2-hydroxy-2-methyl-									
Acetophenone	(structural formula: benzene ring with —C(=O)—CH$_3$ and OH)									
—, 2-hydroxy-		8.6	10.8	−21	25	T	$\mu = 3.0$	332	—	—

β-Acetopropionic acid	See: Pentanoic acid, 4-oxo-
m-Acetoxybenzoic acid	See: Benzoic acid, 3-ethoxy-
N-Acetyl-α-alanine	See: α-Alanine, N-acetyl-
N-Acetyl-β-alanine	See: β-Alanine, N-acetyl-
Acetylacetone	See: 2,4-Pentanedione
N-Acetyl-α-amino-n-butyric acid	See: Butanoic Acid, 2-amino-N-acetyl-
β-Acetylaminoethanethiol	See: Ethanethiol, 2-acetylamino-
Acetylformic acid	See: Propanoic acid, 2-oxo-
Acetylmethylcarbinol	See: Butanoic acid, 3-hydroxy-
2-Acetylpyridine	See: Pyridine, 2-acetyl-
3-Acetylpyridine	See: Pyridine, 3-acetyl-
4-Acetylpyridine	See: Pyridine, 4-acetyl-
Acetylsalicylic acid	See: Benzoic acid, 2-hydroxyacetate

HEATS OF PROTON IONIZATION, pK, AND RELATED THERMODYNAMIC QUANTITIES (continued)

Compound	Formula, Synonyms	ΔH (kcal/mole)	pK	ΔS cal/deg-mole	T°C	Method	Conditions	Reference	Other References	Remarks
Acridine		8.32	5.65	-3.01	15	T	—	36	—	3.d.
	2,3,5,6-Dibenzopyridine	3.480	4.07	-6.95	25	C	—	35	—	3.d., 50 wt. % ethanol
—, 1-hydroxy-		6.71	5.72	-0.56	15	T	—	36	—	3.d.
—, 2-hydroxy-		7.69	5.62	0.54	15	T	—	36	—	3.d.
—, 3-hydroxy-		5.10	5.30	-1.64	15	T	—	36	—	3.d.
Adenine	6-Aminopurine, Purine, 6-amino-	4.9	4.2 (pK$_1$)	-2.7	25	C	$\mu = 0$	38	41–43	—
		4.2	4.12 (pK$_1$)	-4.7	25	T	$\mu = 0.1$	39	—	—
		9.1	9.8 (pK$_2$)	-14	25	C	$\mu = 0$	38	—	—
		9.5	9.72 (pK$_2$)	-12.4	25	T	$\mu = 0.1$	39	—	—
—, 9-β-D-xylo-furanosyl-		8.4	12.34	-28.3	25	C	$\mu = 0$	44	—	—
Adenosine	9-β-D-Ribofuranosyladenine	3.1	3.5 (pK$_1$)	-5.7	25	C	$\mu = 0$	38	40[a], 42	—
		3.8	3.55 (pK$_1$)	-3.4	25	T	$\mu = 0.1$	39	—	—
		9.7	12.35 (pK$_2$)	-23.9	25	C	$\mu = 0$	45	—	—

HEATS OF PROTON IONIZATION, pK, AND RELATED THERMODYNAMIC QUANTITIES (continued)

Compound	Formula, Synonyms	ΔH (kcal/mole)	pK	ΔS cal/deg-mole	T°C	Method	Conditions	Reference	Other References	Remarks
Adenosine—(Continued)										
—, 2'-deoxy-		3.870	3.8 (pK$_1$)	−4.4	25	C	$\mu = 0.1$	40a	—	—
—, 2'-deoxy-5'-monophosphoric acid	 d AMP 5'-Deoxyadenylic acid Deoxyadenosine-, 5 phosphoric acid	−1.04	6.65 (pK$_3$)	−33.8	25	T	$\mu = 0$	46	40a	98% pure AMP
—, 5'-di-phosphoric acid	 ADP	4.1 −1.3 −1.37	4.2 (pK$_3$) 7.0 (pK$_4$) 7.20 (pK$_4$)	−5.4 −36 −37.5	25 25 25	C C T	$\mu = 0$ $\mu = 0$ $\mu = 0$	38 38 47	46 333 —	— — —

HEATS OF PROTON IONIZATION, pK, AND RELATED THERMODYNAMIC QUANTITIES (continued)

Compound	Formula, Synonyms	ΔH (kcal/mole)	pK	ΔS cal/deg-mole	T °C	Method	Conditions	Reference	Other References	Remarks
Adenosine—(Continued)										
—, 5'-monophosphoric acid		4.2	3.7 (pK₁)	−3.0	25	C	$\mu = 0$	38	46	—
		−1.8	6.4 (pK₂)	−36	25	C	$\mu = 0$	38	—	(pK₂ value valid at $\mu = 0.15$)
	AMP 5'-Adenylic acid	−0.85	6.67 (pK₂)	−33.4	25	T	$\mu = 0$	333	—	—
		10.9	13.06 (pK₃)	−23	25	C	$\mu = 0$	45	—	—
—, 5'-triphosphoric acid	ATP	3.7	4.0 (pK₄)	−5.7	25	C	$\mu = 0$	38	46, 47	—
		4.1	4.05 (pK₄)	−4.5	25	T	$\mu = 0.1$	48	333	—
		−1.2	7.0 (pK₅)	−36.0	25	C	$\mu = 0$	38	—	—
		0.5	6.52 (pK₅)	−27.8	25	T	$\mu = 0.1$	48	—	—
5'-Adenylic acid	See: Adenosine, 5'-monophosphoric acid									
Adipic acid	See: Hexanedioic acid									
ADP	See: Adenosine, 5'-diphosphoric acid									
α-Alanine (DL)	CH₃-CH-NH₃⁺-COOH α-Aminopropanoic acid	1.25	2.392 (pK₁)	−6.5	10	C	$\mu = 0$	324	10, 51–54	—
		0.75	2.348	−8.2	25	C	$\mu = 0$	i	10, 51–54	—
		0.615	2.35	−8.0	25	C	$\mu = 0$	49	—	—
		0.23	2.328	−9.9	40	C	$\mu = 0$	324	—	—
		1.510	2.426	−5.6	1	T	$\mu = 0$	50	—	—
		1.210	2.383	−6.6	12.5	—	—	—	—	—
		0.800	2.348	−8.0	25	—	—	—	—	—
		0.320	2.330	−9.6	37.5	—	—	—	—	—
		−0.250	2.332	−11.4	50	—	—	—	—	—
		10.838	9.83 (pK₂)	−8.0	25	C	$\mu = 0$	49	—	—
		11.68	10.29	−5.8	10	C	$\mu = 0$	51	—	—
		10.91	9.87	−8.6	25	C		330	—	—
		10.26	9.51	−10.8	40	—	—	—	—	—
		10.990	10.586	−8.3	1.0	T	$\mu = 0$	50	—	—
		11.080	10.225	−8.0	12.5	—	—	—	—	—
		11.040	9.866	−8.0	25	—	—	—	—	—

HEATS OF PROTON IONIZATION, pK, AND RELATED THERMODYNAMIC QUANTITIES (continued)

Compound	Formula, Synonyms	ΔH (kcal/mole)	pK	ΔS cal/deg-mole	T°C	Method	Conditions	Reference	Other References	Remarks
α-Alanine (DL)—(Continued)										
—, N-acetyl-	$CH_3CH(NHCOCH_3)COOH$	10.890	9.548	−8.6	37.5	—	—	—	—	—
		10.580	9.256	−9.6	50	—	—	55	—	—
		−0.631	3.7152 (pK_1)	−19.12	25	T	μ = 0	56	—	—
α-Alanine(L)										
—, alanyl(L)-	$CH_3CH\overset{+}{N}H_3CONHCH(CH_3)COOH$	−0.134	3.342 (pK_1)	−15.7	25	T	μ = 0.1	56	—	—
		2.291	3.447	−7.9	37	—	—	—	—	—
		8.371	8.17 (pK_2)	−9.4	25	T	μ = 0.1	56	—	—
		5.930	8.28	−17.5	37	—	—	—	—	—
α-Alanine(D)										
—, alanyl(L)-	$H_3\overset{+}{N}CH(CH_3)CONHCH(CH_3)CO_2H$	−1.023	3.119 (pK_1)	−17.3	25	T	μ = 0.10	57	—	—
		−0.157	3.138	−14.7	37	—	—	—	—	—
		9.599	8.296 (pK_2)	−5.8	25	T	μ = 0.10	57	—	—
		9.829	8.021	−5.1	37	—	—	—	—	—
—, alanyllysyl-		10.4	8.23 (25°, pK_1)	−2.76	10–35	T	μ = 2	189	—	3.b.
		13.0	10.81 (25°, pK_2)	−5.84	10–35	T	μ = 2	189	—	3.b.
—, carbamoyl-	$\overset{+}{N}H_3CONHCH(CH_3)COOH$	−0.232	3.8924 (pK_1)	−18.59	25	T	μ = 0	58	—	—
—, glycyl-	$\overset{+}{N}H_3CH_2CONHCH(CH_3)COOH$	−0.563	3.1532 (pK_1)	−16.32	25	T	μ = 0	6	53	—
—, 3-methoxy-	β-Methoxy-DL-alanine	0.820	2.037 (pK_1)	−6.56	25	T	μ = 0	31	—	—
		10.150	9.176 (pK_2)	−7.95	25	T	μ = 0		—	—
β-Alanine	$\overset{+}{H_3N}CH_2CH_2COOH$ β-Aminopropionic acid	0.62	3.606 (pK_1)	−14.1	10	C	μ = 0	324	59	—
		1.08	3.551	−12.6	25	C	μ = 0	1	—	—
		0.38	3.517	−9.2	40	C	μ = 0	324	—	—
		11.30	10.238 (pK_2)	−9.0	25	C	μ = 0	330	325, 334	—
—, N-acetyl-		0.255	4.4452 (pK_1)	−19.48	25	T	μ = 0	55	—	—
—, carbamoyl-	Carbamoyl-β-alanine	0.194	4.4873	−19.88	25	T	μ = 0	58	—	—
L-Alanyl-D-alanine	See: α-Alanine(D), N-alanyl(L)-									
L-Alanyl-L-alanine	See: α-Alanine(L), alanyl(L)-									
Alanyllysyl-α-alanine	See: α-Alanine, alanyllysyl-									
Albumin										
—, Bovine serum	(phenolic)	11.5	10–10.3	−8	25	T	μ = 0.15	60	—	—
Alanyllysylalanine	See: α-Alanine, alanyllysyl-									
Allobarbitone	See: Barbituric acid, 5,5-diallyl-									
DL-Allothreonine	See: Threonine, allo-									
Allylamine	See: Propene, 3-amino-									

HEATS OF PROTON IONIZATION, pK, AND RELATED THERMODYNAMICS QUANTITIES (continued)

Compound	Formula, Synonyms	ΔH (kcal/mole)	pK	ΔS cal/deg-mole	T°C	Method	Conditions	Reference	Other References	Remarks
α-N-Allylamine oxime	See: 2-Butanone, 3-(allylamine)-3-methyl-, oxime									
5-Allylbarbituric acid	See: Barbituric acid, 5-allyl-									
5-Allyl-5-(α-methylbutyl)-barbituric acid	See: Barbituric acid, 5-allyl-5-(α-methyl-butyl)-									
Aluminum										
—, EDTA complex (EDTA) = Y = ethylenediamine-tetraacetic acid	$Al(H_2O)HY \rightarrow Al(H_2O)Y^- + H^+$	3.9	2.96 (pK_1)	0	20	C	$\mu = 0.1$	61	—	—
—	$Al(H_2O)Y \rightarrow Al(OH)Y^{2-} + H^+$	5.0	6.12 (pK_2)	−10.9	20	C	$\mu = 0.1$	61	—	—
Amine										
—, 2-[2(2-aminoethoxy)ethyl-thio]-ethyl-	$NH_3^+CH_2CH_2SCH_2CH_2OCH_2CH_2NH_3^+$	12.0	8.54 (30°, pK_1)	−1	10–40	T	$\mu = 0$	62	—	3.b.
		12.0	9.46 (30°, pK_2)	−4	—	—	—	—	—	—
1,8-Diamino-3-oxa-6-thiaoctane								63		
1,8-Oxa-6-thiaoctane-1, 8-diamine										
—, butylmethyl-	$CH_3CH_2CH_2CH_2NH_2CH_3$ n-Butylmethylamine	12.57	10.90	−7.7	25	C	$\mu = 0$	330	—	—
		13.1	10.6	−5	25	C	$\mu = 0.02$	63	—	—
		11.67	9.29	−3.4	25	C	$\mu = 0.02$	330	—	—
—, diallyl-	$(CH_2 = CHCH_2)_2NH_2^+$	11.33	8.52	−1.0	25	C	$\mu = 0$	330	—	—
—, dibenzyl-	$(C_6H_5CH_2)_2NH_2^+$	13.66	11.25	−5.7	25	C	$\mu = 0$	330	—	—
—, di-n-butyl-	$(CH_3CH_2CH_2CH_2)_2NH_2^+$									
—, di-sec-butyl-		14.03	10.91	−2.8	25	C	$\mu = 0$	330	—	—
—, dicyclohexyl-	$(C_6H_{11})_2NH_2^+$	14.21	11.25	−3.8	25	C	$\mu = 0$	330	—	—
—, dicyclopentyl-	$(C_5H_9)_2NH_2^+$	14.17	10.93	−2.5	25	C	$\mu = 0$	330	—	—
—, diethyl-	$(C_2H_5)_2NH_2^+$	12.73	10.93	−7.3	25	C	$\mu = 0$	330	—	—
		12.77	10.489 (40°)	−7.2	20–40	T	$\mu = 0$	64	—	3.b.
		12.83	10.489 (40°)	−7.0	20–40	T	$\mu = 0.05$	64	65, 66	3.b.
		12.85	10.503 (40°)	−7.0	20–40	T	$\mu = 0.10$	64	—	3.b.
		12.98	10.507 (40°)	−6.6	20–40	T	$\mu = 0.20$	64	—	3.b.
—, 2,2'-diaminodiethyl-	$(^+NH_3CH_2CH_2)_2NH_2^+$ Diethylenetriamine	7.20	4.25 (pK_1)	4.7	25	C	0.1M KCl	67	145	3.a.b.
		8.2	4.59 (pK_1)	5.2	30–40	T	0.1M BaCl$_2$	176	—	—
							0.1M HNO$_3$			3.a.b.
		11.95	8.98 (pK_2)	−1.0	25	C	0.1M KCl	67	—	—
		12.6	8.94 (pK_2)	−0.7	30–40	T	0.1M BaCl$_2$	176	—	—
							0.1M HNO$_3$			3.a.b.
		11.20	9.78 (pK_3)	−7.2	25	C	0.1M KCl	67	—	—
		11.7	9.68 (pK_3)	−6.9	30–40	T	0.1M BaCl$_2$	176	—	—
							0.1M HNO$_3$			3.a.b.

HEATS OF PROTON IONIZATION, pK, AND RELATED THERMODYNAMIC QUANTITIES (continued)

Compound	Formula, Synonyms	ΔH (kcal/mole)	pK	ΔS cal/deg-mole	T°C	Method	Conditions	Reference	Other References	Remarks
Amine—(Continued)										
—, 3,3'-diaminodipropyl-	$(H_3^+NCH_2CH_2CH_2)_2NH_2$	10.47	7.72 (pK₁)	-0.2	25	C	0.1 M KCl	68	—	—
		12.99	9.56 (pK₂)	-0.2	25	C	0.1 M KCl	68	—	—
		12.26	10.65 (pK₃)	-7.5	25	C	0.1 M KCl	68	—	—
—, diethyl-2,2'-dihydroxy-	$(C_2H_5OH)_2NH_2^+$ Diethanolamine	10.01	8.88	-7.1	25	C	$\mu = 0$	330	69	3.b.
—, diethyl-2-hydroxy-		11.401	9.961 (25°)	-7.3	0–60	T	$\mu = 0$	335	—	—
—, diisobutyl-	$(CH(CH_3)_2(CH_2)_2NH_2^+$	13.38	10.82	-4.5	25	C	$\mu = 0$	330	—	—
—, N,N-diethyl-2-hydroxyethyl-		9.719	9.804 (25°)	-12.3	0–60	T	$\mu = 0$	335	—	—
—, diisopentyl-		13.39	10.92	-5.0	25	C	$\mu = 0$	330	—	—
—, diisopropyl-		13.55	11.20	-5.9	25	C	$\mu = 0$	330	—	—
—, dimethyl-	$(CH_3)_2NH_2^+$	12.04	10.732	-8.7	25	C	$\mu = 0$	330	70, 335	—
		11.88	10.77	-9.5	25	T	$\mu = 0$	73	—	—
		12.2	10.8	-8	25	C	$\mu = 0.002$	63	—	—
		11.46	11.39	-10.9	5	T	$\mu = 0$	71	—	—
		11.65	11.08	-10.2	15	—	—	—	—	—
		11.85	10.78	-9.6	25	—	—	—	—	—
		12.05	10.49	-8.9	35	—	—	—	—	—
		12.27	10.22	-8.2	45	—	—	—	—	—
		12.030	9.828	-4.6	25	T	$\mu = 0.10$	72	—	60 wt. % methanol
—, ethyl(2-hydroxy)-isopropyl(2-hydroxy)-	$COH(CH_3)_2NH_2^+CH_2CH_2OH$ Ethanolisopropanolamine	9.8	8.87	-7	25	T	$\mu = 0$	74	—	—
—, 2-hydroxy-1,1-dimethylethyl-		12.894	9.694 (25°)	-1.1	0–60	T	$\mu = 0$	335	—	—
—, methyl-	$CH_3NH_3^+$ Monomethylamine	12.91	11.31	-5.3	5	T	$\mu = 0$	71	—	—
		13.00	10.96	-5.0	15	—	—	—	—	—
		13.09	10.63	-4.7	25	—	—	—	—	—
		13.18	10.32	-4.4	35	—	—	—	—	—
		13.27	10.02	-4.1	45	—	—	—	—	—
		12.790	9.752	-1.73	25	T	$\mu = 0$	72	—	60 wt. % methanol
—, N,N-di(3-aminopropyl)-N-methyl-	N-Methyl di(3-aminopropyl)amine	9.96	6.32 (30°, pK₁)	4	10–40	T	$\mu = 0$	75	—	3.b.
		12.0	9.19 (30°, pK₂)	-2						
		11.0	10.33 (30°, pK₃)	-11	—	T	$\mu = 0$	74	—	—
—, N,N-diethanol-N-methyl-	N-Methyldiethanolamine	9.2	8.52	-8	25	T	$\mu = 0$	74	—	—

HEATS OF PROTON IONIZATION, pK, AND RELATED THERMODYNAMIC QUANTITIES (continued)

Compound	Formula, Synonyms	ΔH (kcal/mole)	pK	ΔS cal/deg-mole	T°C	Method	Conditions	Reference	Other References	Remarks
Amine—(Continued)										
—, dipropyl-		13.17	11.00	-6.2	25	C	$\mu = 0$	330	—	—
—, triallyl-	$(CH_2 = CHCH_2)_3NH^+$	8.83	8.31	-8.5	25	C	$\mu = 0$	330	—	—
—, 2,2',2''-triamino-tri-ethylamine-	$N(CH_2CH_2NH_3)_3^{3+}$	12.15	8.41 (pK$_1$)	2.3	25	C	—	76	—	3.d.
		12.85	9.42 (pK$_2$)	-0.2	25	C	—	76	—	3.d.
		11.70	10.14 (pK$_3$)	-7.0	25	C	—	76	—	3.d.
—, triethyl-	—	10.32	10.715	-14.4	25	C	$\mu = 0$	330	65, 77 79, 336	—
—, triethyl-2,2',2''-trihydroxy-	$(HOCH_2CH_2)_3N$ Triethanolamine	10.38	10.75	-14.4	25	C	0.1 M KCl	78	—	—
		8.16	7.762	-8.2	25	C	$\mu = 0$	330	—	—
		7.696	8.290	-9.75	0	T	$\mu = 0$	80	—	—
		7.753	8.173	-9.54	5	—	—	—	—	—
		7.813	8.067	-9.32	10	—	—	—	—	—
		7.873	7.963	-9.13	15	—	—	—	—	—
		7.932	7.861	-8.91	20	—	—	—	—	—
		7.994	7.762	-8.70	25	—	—	—	—	—
		8.057	7.666	-8.48	30	—	—	—	—	—
		8.121	7.570	-8.29	35	—	—	—	—	—
		8.186	7.477	-8.08	40	—	—	—	—	—
		8.253	7.387	-7.86	45	—	—	—	—	—
		8.296	7.299	-7.65	50	—	—	—	—	—
—, trimethyl-	—	8.80	9.752	-15.1	25	C	$\mu = 0$	330	—	—
		7.97	10.24	-18.2	5	T	$\mu = 0$	71	70, 73	—
		8.37	10.02	-16.8	15	—	—	—	—	—
		8.79	9.80	-15.3	25	—	—	—	—	—
		9.22	9.59	-13.9	35	—	—	—	—	—
		9.67	9.38	-12.5	45	—	—	—	—	—
		8.86	9.79	-15.1	25	C	0.1 M KCl	78	—	—
		9.970	8.744	-6.6	25	T	$\mu = 0.10$	72	—	60 wt. % methanol
—, tripropyl-	$(CH_3CHOHCH_2)_3\overset{+}{N}H$ Triisopropanol amine	10.5	10.66	-14.7	25	C	$\mu = 0$	330	—	—
—, tripropyl-2,2',2''-tri-hydroxy-		8.9	7.86	-6	25	T	$\mu = 0$	74	—	—
Amine oxime	See: 2-Butanone, 3-amino-3-methyl-, oxime									
Aminoacetic acid	See: Glycine									
4-Aminobenzophenone	See: Benzophenone, 4-amino-									
DL-α-Amino-n-butyric acid	See: Butanoic acid, 2-amino (d)-									
Aminobenzene	See: Aniline									
3-Aminobenzo(d,e,f)phenanthrene	See: Pyrene, 3-amino-									

HEATS OF PROTON IONIZATION, pK, AND RELATED THERMODYNAMIC QUANTITIES (continued)

Compound	Formula, Synonyms	ΔH (kcal/mole)	pK	ΔS cal/deg-mole	T °C	Method	Conditions	Reference	Other References	Remarks
2-Aminobenzoic acid	See: Benzoic acid, 2-amino-									
3-Aminobenzoic acid	See: Benzoic acid, 3-amino-									
4-Aminobenzoic acid	See: Benzoic acid, 4-amino-									
γ-Aminobutyric acid	See: Butanoic acid, 4-amino-									
α-Amino-n-caproic acid	See: Hexanoic acid, 2-amino-									
ε-Aminocaproic acid	See: Hexanoic acid, 6-amino-									
β-Aminoethanethiol	See: Ethanethiol, 2-amino-									
2-Aminoethyl-2′-hydroxyethyl sulfide	See: Sulfide, 2-aminoethyl-2′-hydroxyethyl-									
4-Amino-1,2-dihydro-1,3-diazine-2-one	See: Cytosine									
o-Aminodiphenyl	See: Biphenyl, 2-amino-									
m-Aminodiphenyl	See: Biphenyl, 3-amino-									
p-Aminodiphenyl	See: Biphenyl, 4-amino-									
2-Aminoethane-1-phosphoric acid	See: Ethane, 2-amino-1-phosphoric acid-									
2-Aminoethanesulfonic acid	See: Taurine									
2-Aminoethanol-1-phosphoric acid	See: Ethanol, 2-amino-1-phosphoric acid-									
bis-2(2-aminoethylether)	See: Ether, di(2-aminoethyl)-									
β-Aminoethylglyoxaline	See: Histamine									
2-Aminoethylmethyl sulfide	See: Ethane, 1-amino-2-(methylthio)-									
2-Aminoethylphosphate	See: Ethanephosphonic acid, 2-amino-									
2,2′-Aminoethylpiperidine	See: Piperidine, 2-(2′-aminoethyl)-									
2,2′-Aminoethylpyridine	See: Pyridine, 2-(2′-aminoethyl)-									
2-Aminoethylsulfate	See: Sulfuric acid, 2-aminoethyl-									
α-Aminoglutaric acid	See: Glutamic acid									
DL-2-Amino-5-guanidopentanoic acid	See: Arginine									
2 Aminohydroxanthine	See: Guanine									
2-Amino-3-hydroxybutyric acid	See: Threonine									
2-Amino-2′-hydroxydiethylsulfide	See: Sulfide, 2-amino-2′-hydroxydiethyl-									
2-Amino-2-(hydroxymethyl)-1,3-propanediol	See: 1,3-Propanediol-2-amino-2(hydroxymethyl)-									
2-Amino-3(4-hydroxyphenyl)-propanoic acid	See: Tyrosine									

HEATS OF PROTON IONIZATION, pK, AND RELATED THERMODYNAMIC QUANTITIES (continued)

Compound	Formula, Synonyms	ΔH (kcal/mole)	pK	ΔS cal/deg-mole	T °C	Method	Conditions	Reference	Other References	Remarks
2-Amino-3-hydroxypropanoic acid	See: Serine									
α-Amino-β-indolepropionic acid	See: Tryptophan									
α-Aminoisobutyric acid	See: Propanoic acid, 2-amino-2-methyl-									
D-α-Aminoisocaproic acid	See: Leucine (DL)									
α-Aminoisovaleric acid	See: Valine									
1-Aminoisoquinoline	See: Isoquinoline, 1-amino-									
3-Aminoisoquinoline	See: Isoquinoline, 3-amino-									
4-Amino-3-methylbenzene sulfonic acid	See: Benzenesulfonic acid, 4-amino-3-methyl-									
2-Aminomethylpyridine	See: Pyridine, 2-aminomethyl-									
α-Amino-β-methylvaleric acid	See: Pentanoic acid, 2-amino-3-methyl-									
2-Aminopentanedioic acid	See: Glutamic acid									
α-Aminopropanoic acid	See: α-Alanine (DL)									
β-Aminopropanoic acid	See: β-Alanine									
6-Aminopurine	See: Adenine									
2-Aminopyridine	See: Pyridine, 2-amino-									
3-Aminopyridine	See: Pyridine, 3-amino-									
4-Aminopyridine	See: Pyridine, 4-amino-									
4-Amino-2(1)pyrimidone	See: Cytosine									
2-Aminoquinoline	See: Quinoline, 2-amino-									
3-Aminoquinoline	See: Quinoline, 3-amino-									
4-Aminoquinoline	See: Quinoline, 4-amino-									
5-Aminoquinoline	See: Quinoline, 5-amino-									
6-Aminoquinoline	See: Quinoline, 6-amino-									
8-Aminoquinoline	See: Quinoline, 8-amino-									
6-Aminoribofuranosidepurine	See: Adenosine									
4-Aminosalicylamide	See: Benzoic acid, 4-amino-2-hydroxy-, amide-									
5-Aminosalicylic acid	See: Benzoic acid, 5-amino-2-hydroxy-, amide-									
Aminosuccinic acid	See: Aspartic acid									
4-Amino-2-thiabutane	See: 2-Thiabutane, 4-amino-									
3-Amino-4-toluenesulfonic acid	See: 4-Toluenesulfonic acid, 3-amino-									
4-Amino-3-toluenesulfonic acid	See: 3-Toluenesulfonic acid, 4-amino-									
Aminourea	See: Semicarbazide									
α-Amino-n-valeric acid	See: Pentanoic acid, 2-amino-									

HEATS OF PROTON IONIZATION, pK, AND RELATED THERMODYNAMIC QUANTITIES (continued)

Compound	Formula, Synonyms	ΔH (kcal/mole)	pK	ΔS cal/deg-mole	T°C	Method	Conditions	Reference	Other References	Remarks
δ-Aminovaleric acid	See: Pentanoic acid, 5-amino-									
Ammonia	NH_4^+	12.51	9.242	−0.3	25	C	$\mu = 0$	330	70, 81, 82	—
		12.54	9.90	−0.21	5	T	$\mu = 0$	71	83	—
		12.50	9.57	−0.34	15	—	—	—	—	—
		12.48	9.25	−0.45	25	—	—	—	—	—
		12.44	8.95	−0.56	35	—	—	—	—	—
		12.41	8.67	−0.67	45	—	—	—	—	—
		12.43	9.29	−0.8	25	C	0.1 M KCl	78	—	—
		1.240	4.66	−17.6	18	C	$\mu = 0$	337	—	Reaction: $NH_3 + H_2O = NH_4OH$
		0.875	4.75	−18.8	25	C	$\mu = 0.5$ NaNO$_3$	—	—	—
		0.600	4.82	−19.7	30	C	$\mu = 0.5$ NaNO$_3$	—	—	—
		0.330	4.90	−20.6	35	C	$\mu = 0.5$ NaNO$_3$	—	—	—
		0.080	4.97	−21.4	40	C	$\mu = 0.5$ NaNO$_3$	—	—	—
		1.192	4.33	−16.1	18	C	$\mu = 0.5$ NaNO$_3$	—	—	—
		0.850	4.41	−17.3	25	C	$\mu = 0.5$ NaNO$_3$	—	—	—
		0.600	4.47	−18.1	30	C	$\mu = 0.5$ NaNO$_3$	—	—	—
		0.350	4.55	−18.9	35	C	$\mu = 0.5$ NaNO$_3$	—	—	—
		0.081	4.62	−19.8	40	C	$\mu = 0.5$ NaNO$_3$	—	—	—
		1.020	4.27	−16.5	18	C	$\mu = 1.0$ NaNO$_3$	—	—	—
		0.700	4.36	−17.6	25	C	$\mu = 1.0$ NaNO$_3$	—	—	—
		0.462	4.42	−18.4	30	C	$\mu = 1.0$ NaNO$_3$	—	—	—
		0.222	4.49	−19.2	35	C	$\mu = 1.0$ NaNO$_3$	—	—	—
		−0.025	4.57	−20.0	40	C	$\mu = 1.0$ NaNO$_3$	—	—	—
		0.361	4.19	−18.4	18	C	$\mu = 3.0$ NaNO$_3$	—	—	—

HEATS OF PROTON IONIZATION, pK, AND RELATED THERMODYNAMIC QUANTITIES (continued)

Compound	Formula, Synonyms	ΔH (kcal/mole)	pK	ΔS cal/deg-mole	T°C	Method	Conditions	Reference	Other References	Remarks
Ammonia—(Continued)										
		0.109	4.28	−19.2	25	C	$\mu = 3.0$ NaNO$_3$	—	—	—
		−0.075	4.36	−19.8	30	C	$\mu = 3.0$ NaNO$_3$	—	—	—
		−0.266	4.43	−20.5	35	C	$\mu = 3.0$ NaNO$_3$	—	—	—
		12.560	8.626	2.65	25	T	$\mu = 0$	72	—	60 wt. % methanol
AMP	See: Adenosine, 5'-monophosphoric acid									
dAMP	See: Adenosine, 2'-deoxy-5'-monophosphoric acid									
***n*-Amylamine**	See: Propane, 1-amino-									
Angiotensin II	(Tyrosyl)	6.3	10.37	−27	25	T	$\mu = 0$	84	—	—
***o*-Anisic acid**	See: Benzoic acid, 2-methoxy-									
Aniline		7.377	4.596 (25°)	3.72	10–50	T	$\mu = 0$	340	70, 87–89, 339	3.b.
	Aminobenzene	7.24	4.60	3.3	25	C	0.095M HCl	338	—	—
	Phenylamine	5.9	4.19 (20°)	1.0	0–20	T	C = 0.01	85	—	3.b. 50 wt. % ethanol
		7.490	—	—	10	—	—	—	—	—
		7.370	—	—	20	—	—	—	—	—
		7.276	4.60	3.4	25	C	C = 0.06 M	86	87	—
		7.180	—	—	30	C	—	—	—	—
—, 2-bromo-		4.99	2.53	5.2	25	C	$\mu = 0$	330	87	—
—, 3-bromo-		6.23	3.53	4.7	25	C	$\mu = 0$	330	87, 341	—
—, 4-bromo-		6.72	3.88	4.8	25	C	$\mu = 0$	330	87	—
—, 2-chloro-	*o*-Chloroaniline	6.0	2.63	8.1	25	T	$\mu = 0$	70	89	—
		6.002	2.632	8.1	25	T	$\mu = 0$	2	—	3.d.
—, 3-chloro-		6.268	3.521	4.91	25	T	$\mu = 0$	341	87	—
—, 3-hydroxy-*N,N,N*-trimethyl-	*m*-Hydroxyphenyltrimethylammonia	6.09	8.06	−16.4	25	C	$\mu = 0$	90	—	—
—, 4-hydroxy-*N,N,N*-trimethyl-	*p*-Hydroxyphenyltrimethylammonia	5.46	8.35	−19.9	25	C	$\mu = 0$	90	—	—
—, 3-iodo-		6.326	3.583	4.81	25	T	$\mu = 0$	341	87	—
—, 3-methoxy-		7.013	4.204	4.29	25	T	$\mu = 0$	341	87	—
—, 3-nitro-		4.5	—	—	25	C	$\mu = 0$	338	87	—
—, 3-nitro-		4.980	2.460	5.45	25	T	$\mu = 0$	341	—	—
—, 2-sulfonic acid-		2.43	2.458	−3.1	25	C	$\mu = 0.02$	330	—	—

HEATS OF PROTON IONIZATION, pK, AND RELATED THERMODYNAMIC QUANTITIES (continued)

Compound	Formula, Synonyms	ΔH (kcal/mole)	pK	ΔS cal/deg-mole	T°C	Method	Conditions	Reference	Other References	Remarks
Aniline—(Continued)										
—, 3-sulfonic acid-										
—, 4-sulfonic acid-										
p-Anisic acid	See: Benzoic acid, 4-methoxy-	5.02	3.738	−0.3	25	C	μ = 0.012	330	—	—
m-Anisic acid	See: Benzoic acid, 3-methoxy-	4.44	3.232	0.1	25	C	μ = 0.02	330	—	—
Anserine	$^+NH_3CH_2CH_2CONHCH(COOH)CH_2C=CH$ $N—C$ CH_3 H	7.5	6.8 (37°, pK$_2$)	−6.9	22–37	T	μ = 0.3	91	—	3.d.
Anthracene, 1-amino-	1-Anthrylamine (structure)	5.2	3.22 (20°)	2.9	0–20	T	C = 0.005	85	—	3.b. 50 wt. % ethanol
Anthranilic acid	See: Benzoic acid, 2-amino-									
1-Anthrylamine	See: Anthracene, 1-amino- (imino group)									
Aquocobalamine complex										
Arabinose	(structure)	10.5	10.25	−12	25	T	μ = 0	92	—	—
		8.3	12.54	−29.6	25	C	μ = 0	28	—	—
Arginine	$H_3NC(:NH)NH(CH_2)_3$-$CHNH_2CO_2H$ DL-2-Amino-5-guanidopentanoic acid	1.63	1.914 (pK$_1$)	−2.8	0	T	μ = 0	93	54, 95	—
		1.50	1.8847	−3.3	5				342	—
		1.37	1.8697	−3.7	10					—
		1.25	1.8488	−4.2	15					—
		1.12	1.8374	−4.6	20					—
		0.98	1.8217	−5.0	25					—
		0.85	1.8138	−5.5	30					—
		0.71	1.8013	−6.0	35					—
		0.57	1.7995	−6.4	40					—
		0.43	1.7852	−6.8	45					—
		0.28	1.787	−7.3	50					—
		10.95	9.7178 (pK$_2$)	−4.4	0	T	μ = 0	93		—
		10.91	9.5626	−4.5	5					—
		10.86	9.4073	−4.7	10					—

HEATS OF PROTON IONIZATION, pK, AND RELATED THERMODYNAMIC QUANTITIES (continued)

Compound	Formula, Synonyms	ΔH (kcal/mole)	pK	ΔS cal/deg-mole	T°C	Method	Conditions	Reference	Other References	Remarks
Arginine—(Continued)										
		10.82	9.2669	−4.9	15	—	—	—	—	—
		10.77	9.1226	−5.0	20	—	—	—	—	—
		10.73	8.9936	−5.2	25	—	—	—	—	—
		10.68	8.8589	−5.3	30	—	—	—	—	—
		10.64	8.7390	−5.5	35	—	—	—	—	—
		10.59	8.614	−5.6	40	—	—	—	—	—
		10.54	8.5040	−5.8	45	—	—	—	—	—
		10.49	8.3852	−5.9	50	—	—	—	—	—
		10.44	8.2824	−6.1	55	—	—	—	—	—
		10.25	9.00	6.7	25	T	$\mu = 0.0237$	342	—	—
		12.400	12.48 (pK$_3$)	−15.5	0–25	T	$\mu = 0.01$	94	—	3.a.b.
—, phenylalanyl-	(carboxyl)	0.450	2.66 (25°)	−10.7	0–25	T	$\mu = 0.01$	94	—	3.a.b.
	(amino)	10.150	7.57 (25°)	−0.6	0–25	T	$\mu = 0.01$	94	—	3.a.b.
	(guanidine)	11.950	12.40 (25°)	−5.0	0–25	T	$\mu = 0.01$	94	—	3.a.b.
—, tyrosyl-	(carboxyl)	0.0	2.65	−12.1	0–25	T	$\mu = 0.01$	94	—	3.a.b.
	(amino)	10.5	7.39	1.41	0–25	T	$\mu = 0.01$	94	—	3.a.b.
	(oxyphenol)	6.0	9.36	−22.7	0–25	T	$\mu = 0.01$	94	—	3.a.b.
	(guanidine)	13.0	11.62	−9.6	0–25	T	$\mu = 0.1$	94	—	—
Arsenic acid	H$_3$AsO$_4$	−1.69	2.25 (pK$_1$)	−16.0	25	C	$\mu = 0.1$–0.3	96	—	—
		0.77	6.77 (pK$_2$)	−28.4	25	C	$\mu = 0.1$–0.3	96	—	—
—, ortho-		4.35	11.53 (pK$_3$)	−38.5	25	C	$\mu = 0.1$–0.3	96	—	—
Arsenous acid	H$_3$AsO$_3$	6.58	9.23 (pK$_1$)	−20.2	25	C	$\mu = 0.1$–0.3	96	—	—
Asparacemic acid	See: Aspartic acid									
Asparagine	H$_3^+$NCOCH$_2$CH(NH$_3^+$)COOH	0.133	2.9420 (pK$_1$)	−13.01	25	T	$\mu = 0$	6	—	—
—, glycyl-	$^+$H$_3$HCH$_2$CONHCHCH$_2$CO(NH$_3^+$)COOH	1.85	2.05 (pK$_1$)	−3.2	25	C	$\mu = 0$	1	—	—
Aspartic acid	HOOCCH$_2$CH(NH$_3^+$)COOH	2.324	2.122 (pK$_1$)	−1.3	1	T	$\mu = 0$	97	—	—
	Aminosuccinic acid	2.094	2.054	−2.1	12.5	—	—	—	—	—
	Asparacemic acid	1.783	1.990	−3.1	25	—	—	—	—	—
		1.373	1.945	−4.5	37.5	—	—	—	—	—
		0.889	1.907	−6.0	50	—	—	—	—	—
		0.96	3.87 (pK$_2$)	−14.5	25	C	$\mu = 0$	1	—	—
		1.761	4.006 (pK$_2$)	−11.9	1	T	$\mu = 0$	97	—	—
		1.448	3.944	−13.0	12.5	—	—	—	—	—
		1.110	3.900	−14.2	25	—	—	—	—	—
		0.619	3.878	−15.8	37.5	—	—	—	—	—
		0.057	3.870	−17.8	50	—	—	—	—	—
		9.280	10.604 (pK$_3$)	−14.6	1	T	$\mu = 0$	97	—	—

HEATS OF PROTON IONIZATION, pK, AND RELATED THERMODYNAMIC QUANTITIES (continued)

Compound	Formula, Synonyms	ΔH (kcal/mole)	pK	ΔS cal/deg-mole	T°C	Method	Conditions	Reference	Other References	Remarks
Aspartic acid—(Continued)										
	Asparacemic acid—(Continued)	9.220	10.304	−14.8	12.5	—	—	—	—	—
		9.025	10.002	−15.5	25	—	—	—	—	—
		8.695	9.742	−16.6	37.5	—	—	—	—	—
		8.220	9.512	−18.1	50	—	—	—	—	—
Aspartyltyrosine	See: Tyrosine, aspartyl-									
Aspirin	See: Benzoic acid, 2-hydroxyacetate									
ATP	See: Adenosine, 5'-triphosphoric acid									
1-Azacyclopentane	See: Pyrrolidine									
1-Aza-cyclohexane										
—, N-methyl-		11.5	8.99 (25°)	−2.6	15–35	T	—	99	—	60 wt. % methanol 3.b.d.
1-Aza-cyclononane	$[^+NH_2 \; CH_2(CH_2)_6CH_2]$	12.9	9.66 (25°)	−0.93	25	T	—	99	—	60 wt. % methanol 3.b.d.
1-Aza-cyclooctadecane	$[^+NH_2 \; CH_2(CH_2)_{15}CH_2]$	12.0	9.41 (25°)	−2.81	25	T	—	99	—	60 wt. % methanol 3.b.d.
Azimidobenzene	See: 1,2,3-Benzotriazole									
Azine	See: Pyridine									
Azirane	Dihydroazirine Ethyleneimine	9.06	8.04	−6.4	25	C	μ = 0	330	100, 343	—
Azoimide	See: Hydroazoic acid									
Azophenylene	See: Phenazine, 2-hydroxy-									
Barbital	See: Barbituric acid, 5,5-diethyl-									
Barbitone	See: Barbituric acid, 5,5-diethyl-									
Barbituric acid	Pyrimidinetroine	0.06	4.04	−18.8	25	C	μ = 0	101	—	—

HEATS OF PROTON IONIZATION, pK, AND RELATED THERMODYNAMIC QUANTITIES (continued)

Compound	Formula. Synonyms	ΔH (kcal/mole)	pK	ΔS cal/deg-mole	T°C	Method	Conditions	Reference	Other References	Remarks
Barbituric acid—(Continued)										
—, 5-allyl-	—	−1.89	4.78	−28.2	25	C	μ = 0	101	—	—
—, 5-allyl-5(α-methylbutyl)-	—	5.00	8.08	−20.2	25	C	μ = 0	101	—	—
—, 5,5-diallyl-	Allobarbitone Curral Dial	4.92	7.78	−19.1	25	C	μ = 0	101	—	—
—, 5,5-diethyl-	Barbital Barbitone Veronal	5.81	7.98	−17.0	25	C	μ = 0	101	102	—
—, 1,3-dimethyl-	—	−0.05	4.68	−21.6	25	C	μ = 0	101	—	—
—, 5-ethyl-5-(α-methylbutyl)-	—	5.81	8.11	−17.6	25	C	μ = 0	101	—	—
—, 5-ethyl-5-pentyl-	5-Ethyl-5-n-amylbarbituric acid	5.22	7.95	−18.6	25	C	μ = 0	101	—	—
—, 5-ethyl-5-phenyl-	Luminal Phenobarbital	4.60	7.44	−18.6	25	C	μ = 0	101	—	—
—, 5-methyl-	—	−1.28	4.40	−24.4	25	C	μ = 0	101	—	—
BATA	See: Acetic acid, [oxy *bis* (ethylenenitrilo)]-tetra-									
Benzaldehyde	CHO									
—, 4-dimethylamino-		1.800	1.647	−1.5	25	T	μ = 0	103	—	—
—, 2-hydroxy-	Salicylaldehyde o-Hydroxybenzaldehyde	5.15	8.37	−21.0	25	C	μ = 0	104	—	—
		6.3	8.80	−20	25	T	μ = 3	35	—	—
—, 3-hydroxy-	m-Hydroxybenzaldehyde	5.17	9.02	−23.9	25	C	μ = 0	104	344	—
—, 4-hydroxy-	p-Hydroxybenzaldehyde	4.992	8.982 (25°)	−24.36	5–60	T	μ = 0.04	345	344	3.b.
		4.26	7.62	−20.6	25	C	μ = 0	104	—	—
—, 2-hydroxy-4-amino-		11.0	13.74 (25°)	−27	15–35	T	3.0M NaClO₄	332	—	3.b.
—, 2-hydroxy-3-methoxy-	o-Vanillin	4.0	7.4	−2	25	T	μ = 0.10	346	—	—
—, 3-hydroxy-4-methoxy-	Isovanillin	4.13	7.91	−22.3	25	C	μ = 0	104	—	—
—, 4-hydroxy-3-methoxy-	Vanillin	4.62	8.89	−25.2	25	C	μ = 0	104	—	—
		3.75	7.40	−21.3	25	C	μ = 0	104	—	—
1-Benzazine	See: Quinoline									
2-Benzazine	See: Isoquinoline									
Benzene										
—, 1-amino-2,3-dimethyl-	2,3-Xylidine	8.79	3.35	8.59	25	C	μ = 0	88	—	—

HEATS OF PROTON IONIZATION, pK, AND RELATED THERMODYNAMIC QUANTITIES (continued)

Compound	Formula, Synonyms	ΔH (kcal/mole)	pK	ΔS cal/deg-mole	T°C	Method	Conditions	Reference	Other References	Remarks
Benzene—(Continued)										
—, 1-amino-2,4-dimethyl-	2,4-Xylidine	7.90	3.7	3.63	25	C	$\mu = 0$	88	—	—
—, 1-amino-3,5-dimethyl-	3,5-Xylidine	9.07	3.48	8.75	25	C	$\mu = 0$	88	—	—
—, 2-amino-1,3-dimethyl-	2,6-Xylidine	5.24	3.08	-1.63	25	C	$\mu = 0$	88	—	—
—, 2-amino-1,4-dimethyl-	2,5-Xylidine	6.53	3.4	0.87	25	C	$\mu = 0$	88	—	—
—, 4-amino-1,2-dimethyl-	3,4-Xylidine	8.04	3.77	3.42	25	C	$\mu = 0$	88	—	—
—, 1,3-dichloro 2,5-dihydroxy-	2,6-Dichlorohydroquinone	2.3	7.30 (26.1°, pK_1)	-26	13–25	T	$\mu = 0.65$	105	—	3.b.
		5.1	9.99 (25.1°, pK_2)	-29	13–25	T	$\mu = 0.65$	105	—	3.b.
1,4-dihydroxy-	Hydroquinone / Quinol	3.8	9.85 (25°, pK_1)	-32	13–25	T	$\mu = 0.65$	105	—	—
		3.0	11.39 (25.9°, pK_2)	-42	13–25	T	$\mu = 0.65$	105	—	—
1,4-dihydroxy-2-methyl-	Toluhydroquinone	5.0	10.05 (25.1°, pK_1)	-29	13–25	T	$\mu = 0.65$	105	—	—
		3.3	11.64 (25.9°, pK_2)	-42	13–25	T	$\mu = 0.65$	105	—	—
1,4-dihydroxy-2,3,5,6-tetramethyl-	Durohydroquinone	7.8	11.25 (25°, pK_1)	-25	12–30	T	$\mu = 0.65$	105	—	—
		12.6	12.70 (29.8°, pK_2)	-17	12–30	T	$\mu = 0.65$	105	—	—
1,2-dimethyl-3-hydroxy-	2,3-Xylenol	5.70	10.544	-29.1	25	T	$\mu = 0$	106	107d	—
-1,2-dimethyl-4-hydroxy-	3,4-Xylenol	5.37	10.356	-29.4	25	T	$\mu = 0$	106	107d	—
1-3-dimethyl-2-hydroxy-	2,6-Xylenol	5.46	10.615	-30.3	25	T	$\mu = 0$	106	107d	—
-1,3-dimethyl-4-hydroxy-	2,4-Xylenol	5.76	10.595	-29.2	25	T	$\mu = 0$	106	107d	—
-1,4-dimethyl-2-hydroxy-	2,5-Xylenol	5.58	10.404	-28.9	25	T	$\mu = 0$	106	107d	—
-1,5-dimethyl-3-hydroxy-	3,5-Xylenol	5.34	10.203	-28.8	25	T	$\mu = 0$	106	107d	—
—, 1-hydroxy-2,3,4-trimethyl-		5.80	10.59	-29.0	25	—	—	329	—	—
—, 1-hydroxy-2,3,5-trimethyl-	2,3,5-Trimethylphenol	6.00	10.67	-28.7	25	C	$\mu = 0$	90	—	—
—, 1-hydroxy-2,4,5-trimethyl-	2,4,5-Trimethylphenol / Pseudoemenol	6.40	10.57	-26.9	25	C	$\mu = 0$	90	—	—
—, 1-hydroxy-2,4,6-trimethyl-	2,4,6-Trimethylphenol	5.44	10.89	-31.6	25	C	$\mu = 0$	90	—	—
—, 1-hydroxy-3,4,5-trimethyl-	3,4,5-Trimethylphenol	5.68	10.25	-27.9	25	C	$\mu = 0$	90	—	—
—. 3-hydroxy-1-ethyl-		5.294	10.068 (25°)	-28.31	5–60	T	$\mu = 0.04$	345	—	3.b.
Benzene boronic acid	$B(OH)_2$; Phenylboric acid	1.9	8.84 (25°)	-34	20–38.5	T	$\mu = 0$	108	—	3.b.
Benzenecarboxylic acid	See: Benzoic acid									

HEATS OF PROTON IONIZATION, pK, AND RELATED THERMODYNAMIC QUANTITIES (continued)

-1,2-Benzenedicarboxylic acid

Compound	Formula, Synonyms	ΔH (kcal/mole)	pK	ΔS cal/deg-mole	T °C	Method	Conditions	Reference	Other References	Remarks
Phthalic acid		-0.122	2.925 (pK$_1$)	-13.8	0	T	$\mu = 0$	109	—	—
		-0.221	2.927	-14.2	5	—	—	—	—	—
		-0.323	2.931	-14.6	10	—	—	—	—	—
		-0.426	2.937	-14.7	15	—	—	—	—	—
		-0.530	2.943	-15.3	20	—	—	—	—	—
		-0.637	2.950	-15.6	25	—	—	—	—	—
		-0.746	2.958	-16.0	30	—	—	—	—	—
		-0.856	2.967	-16.4	35	—	—	—	—	—
		-0.968	2.978	-16.7	40	—	—	—	—	—
		-1.082	2.988	-17.3	45	—	—	—	—	—
		-1.197	3.001	-17.4	50	—	—	—	—	—
		-1.315	3.014	-17.8	55	—	—	—	—	—
		-1.434	3.028	-18.2	60	—	—	—	—	—
		1.183	5.432 (pK$_2$)	-20.5	0	T	$\mu = 0$	110	—	—
		0.859	5.418	-21.7	5	—	—	—	—	—
		0.529	5.410	-22.9	10	—	—	—	—	—
		0.194	5.405	-24.1	15	—	—	—	—	—
		-0.148	5.405	-25.2	20	—	—	—	—	—
		-0.496	5.408	-26.4	25	—	—	—	—	—
		-0.849	5.416	-27.6	30	—	—	—	—	—
		-1.208	5.427	-28.8	35	—	—	—	—	—
		-1.574	5.442	-29.9	40	—	—	—	—	—
		-1.945	5.462	-31.1	45	—	—	—	—	—
		-2.322	5.485	-32.3	50	—	—	—	—	—
		-2.705	5.512	-33.5	55	—	—	—	—	—
		-3.093	5.541	-34.6	60	—	—	—	—	—
		-0.469	4.59 (25°)	-21.0	5-45	T	0	347	—	Conditions refer to wt. % ethylene glycol in water. 3.b.
		-0.591	5.63 (25°)	-27.6	5-45	T	10	347	—	3.b.
		-1.053	6.05 (25°)	-31.2	5-45	T	30	347	—	3.b.
		-1.674	6.53 (25°)	-35.5	5-45	T	50	347	—	3.b.
		-1.735	7.28 (25°)	-32.9	5-45	T	70	347	—	3.b.
		-1.844	8.34 (25°)	-44.4	5-45	T	90	347	—	3.b.

HEATS OF PROTON IONIZATION, pK, AND RELATED THERMODYNAMIC QUANTITIES (continued)

Compound	Formula, Synonyms	ΔH (kcal/mole)	pK	ΔS cal/deg·mole	T°C	Method	Conditions	Reference	Other References	Remarks
1,3-Benzenedicarboxylic acid										
—, mononitrile-	COOH ... CN 3-Cyanobenzoic acid	−0.04	3.598	−16.6	25	T	$\mu = 0$	89	—	—
1,4-Benzenedicarboxylic acid										
—, mononitrile-	COOH ... CN 4-Cyanobenzoic acid	0.03	3.551	−16.1	25	T	$\mu = 0$	89	—	
Benzenesulfonic acid	SO_3H									
—, 2-amino-	NH_3^+ ... SO_3H Orthanilic acid	2.374	2.4580	−3.28	25	T	$\mu = 0$	111	112	—
—, 3-amino-	Metanilic acid	4.64	3.808	−1.9	25	C	$\mu = 0$	348	111	—
—, 3-amino-4-methyl-		5.05	3.64	0.3	25	T	$\mu = 0$	329	—	—
—, 4-amino-	Sulfanilic acid	4.2	3.23	−0.7	25	C	$\mu = 0$	113	111, 114	—
—, 4-amino-3-methyl-	4-Amino-3-methylbenzene sulfonic acid	4.489	3.126 (pK$_2$)	0.75	25	T	$\mu = 0$	115	—	—
—, 4-hydroxy-	p-Phenolsulfonic acid	4.036	9.055	−27.9	25	T	$\mu = 0$	116	—	—
Benzenol	See: Phenol									
Benzilic acid	See: Acetic acid, diphenyl-(hydroxy)-									

HEATS OF PROTON IONIZATION, pK, AND RELATED THERMODYNAMIC QUANTITIES (continued)

Compound	Formula, Synonyms	ΔH (kcal/mole)	pK	ΔS cal/deg-mole	T°C	Method	Conditions	Reference	Other References	Remarks
Benzimidazole										
—, 2-ethyl-		9.3	6.27	2.5	4-35	T	μ = 0.16	117	—	3.b.
	Benzoglioxaline									
	1,3-Benzodiazole									
—, 2-methyl-	2-Methylbenzimidazole	9.8	6.29	4.1	4-35	T	μ = 0.16	117	—	3.b.
—, 2-(2-pyridyl)-	2-(2-pyridyl)benzimidazole	8.7	5.58	−6	25	T	μ = 0.16	117	—	3.b.
		3.8	3.44 (25°)	−3	0-40	T	μ = 0.005	118	—	3.b. 50 wt. % dioxane
Benzoate, methyl-*m*-amino-	See: Benzoic acid, 3-aminomethyl ester-									
Benzoate, methyl-*o*-amino-	See: Benzoic acid, 2-aminomethyl ester-									
Benzoate, methyl-*p*-amino-	See: Benzoic acid, 4-aminomethyl ester-									
Benzohydroquinone,2,6-dichloro-	See: Benzene, 1,3-dichloro-2,5-dihydroxy-									
Benzohydroquinone,2-methyl-	See: Benzene, 1,4-dihydroxy-2-methyl-									
Benzohydroquinone, 2,3,5,6-tetramethyl-	See: Benzene, 1,4-dihydroxy-2,3,5,6-tetramethyl-									
Benzoic acid	COOH	7.85	—	—	10	C	μ = 0	119	8, 9	—
		0.15	4.201	−18.7	25	C	μ = 0	1	12, 89	—
		0.105	4.20	−18.9	25	C	μ = 0	119	120−123	—
		0.10	4.19	−18.8	25	C	μ = 0	327	—	—
		−0.465	—	—	40	C	μ = 0	119	—	—
	Benzenecarboxylic acid	−0.06	4.50	−20.8	25	C	10.0	327	—	—
		−0.15	4.68	−21.9	25	C	17.0	327	—	—
		−0.15	4.76	−22.3	25	C	20.0	327	—	—
		0.05	4.94	−22.4	25	C	26.0	327	—	—
		0.47	5.04	−21.5	25	C	30.0	327	—	—
		0.77	5.20	−21.2	25	C	35.0	327	—	—
		1.05	5.35	−21.0	25	C	40.0	327	—	—
		1.02	5.66	−22.5	25	C	50.0	327	—	—
		0.86	5.98	−24.5	25	C	60.0	327	—	—
		0.32	6.32	−27.8	25	C	70.0	327	—	—
		−0.08	6.74	−31.1	25	C	80.0	327	—	—
—, m-acetoxy-	See: Benzoic acid, 3-ethoxy-									

HEATS OF PROTON IONIZATION, pK, AND RELATED THERMODYNAMIC QUANTITIES (continued)

Compound	Formula, Synonyms	ΔH (kcal/mole)	pK	ΔS cal/deg-mole	T°C	Method	Conditions	Reference	Other References	Remarks
Benzoic acid—(Continued)										
—, 4-amino-2-hydroxyamide-	$CONH_2$, OH, NH_2 4-Aminosalicylamide	5.1	9.16	−25	25	T	$\mu = 3$	332	35	—
—, 5-amino-2-hydroxy-amide,	5-Aminosalicylic acid	11.0	13.74	−27	25	T	$\mu = 3$	35		—
—, 2-amino-	Anthranilic acid 2-Aminobenzoic acid	3.79	2.09 (pK$_1$)	3.0	25	C	$\mu = 0$	124		—
		2.79	4.79 (pK$_2$)	−13.4	25	C	$\mu = 0$	124		—
		6.091	5.125 (pK$_2$)	−2.64	20	T	$\mu = 0$	125	125	—
		2.242	5.025	−15.63	30					—
		−1.793	5.030	−28.74	40					—
		−6.243	5.100	−41.85	50					—
		−10.23	5.200	−54.87	60					—
—, 2-amino-N,N-dimethyl-		0.94	1.63 (pK$_1$)	−4.4	25	C	$\mu = 0.023$	330		—
		5.18	8.42 (pK$_2$)	−21.1	25	C	$\mu = 0$	330		—
—, 2-aminomethyl ester	Methyl-o-aminobenzoate	4.57	2.36	4.5	25	C	$\mu = 0$	124		—
—, 3-amino-	3-Aminobenzoic acid	2.56	3.07 (pK$_1$)	−5.4	25	C	$\mu = 0$	124	125	—
		4.17	4.79 (pK$_2$)	−7.9	25	C	$\mu = 0$	124		—
		0.0	3.30	−15.1	25	C		124		—
		6.5	3.45	6.0	—					
		6.5	4.56	0.9	—					
		0.1	4.41	−19.8	—					
		−6.5	−0.15	−21.1	—					
—, 3-aminomethyl ester	Methyl-m-aminobenzoate	6.50	3.58	5.4	25	C	$\mu = 0$	124		—
—, 4-amino-	4-Aminobenzoic acid	4.97	2.41 (pK$_1$)	5.6	25	C	$\mu = 0$	124		—
		0.70	4.85 (pK$_2$)	−19.8	25	C	$\mu = 0$	124	125	—
		0.1	3.49	−15.6	25	C		124		—
		5.1	2.45	5.9	—					
		5.1	3.79	−0.2	—					
		0.4	4.83	−20.8	—					
		−5.0	1.04	−21.5	—					

Equations for the 3-amino micro species group:

$$^+H_3NC_6H_4COOH = {}^+H_3NC_6H_4COO^- + H^+$$
$$^+H_3NC_6H_4COOH = H_2NC_6H_4COOH + H^+$$
$$^+H_3NC_6H_4COO^- = H_2NC_6H_4COO^- + H^+$$
$$H_2NC_6H_4COOH = H_2NC_6H_4COO^- + H^+$$
$$H_2NC_6H_4COOH = {}^+H_3NC_6H_4COO^-$$

Above equations refer to micro species

Equations for the 4-amino micro species group:

$$^+H_3NC_6H_4COOH = {}^+H_3NC_6H_4COO^- + H^+$$
$$^+H_3NC_6H_4COOH = H_2NC_6H_4COOH + H^+$$
$$^+H_3NC_6H_4COO^- = H_2NC_6H_4COO^- + H^+$$
$$H_2NC_6H_4COOH = H_2NC_6H_4COO^- + H^+$$
$$H_2NC_6H_4COOH = H_3^+NC_6H_4COO^-$$

Above equations refer to micro species

HEATS OF PROTON IONIZATION, pK, AND RELATED THERMODYNAMIC QUANTITIES (continued)

Compound	Formula, Synonyms	ΔH (kcal/mole)	pK	ΔS cal/deg-mole	T°C	Method	Conditions	Reference	Other References	Remarks
Benzoic acid—(Continued)										
—, 4-aminomethyl ester	Methyl-p-aminobenzoate	5.09	2.45	5.9	25	C	$\mu = 0$	124	—	—
—, 3-bromo-	3-Bromobenzoic acid	−0.06	3.809	−17.6	25	T	$\mu = 0$	89	12	—
—, 4-bromo-	4-Bromobenzoic acid	0.11	4.002	−17.9	25	T	$\mu = 0$	89	—	—
—, 2-chloro-	2-Chlorobenzoic acid	−0.673	2.924	−11	16.8	T	—	12	8	3.a.b.d.
—, 3-chloro-	3-Chlorobenzoic acid	0.019	3.83	−17.4	25	T	—	126	12, 89	3.d.
—, 4-chloro-	4-Chlorobenzoic acid	0.226	3.98	−17.5	25	T	—	126	12, 89	3.d.
—, 2,4-dihydroxy-	2,4-Dihydroxybenzoic acid β-Resorcylic acid	1.360	3.280 (35°)	−20	16.8	T	—	12	—	3.a.b.d.
—, 2,5-dihydroxy-	2,5-Dihydroxybenzoic acid Gentisic acid	1.029	2.889 (35°)	−10	16.8	T	—	12	—	3.a.b.d.
—, 2,4-dihydroxyethyl ester	p-Hydroxyethylsalicylate	1.5	8.43	−33.5	25	T	$\mu = 0$	120	—	—
—, 2,6-dihydroxyethyl ester	o-Hydroxyethylsalicylate	7.1	10.0	−22.1	25	T	$\mu = 0$	120	—	—
—, 3,5-dinitro-	3,5-Dinitrobenzoic acid	0.965	2.785 (35°)	−10	16.8	T	—	12	127	3.a.b.d.
—, 3-ethoxy-	m-Acetoxybenzoic acid	0.495	3.879 (35°)	−16	16.8	T	—	12	—	3.a.b.d.
—, 2-hydroxy-	Salicylic acid	0.73	2.973 (pK$_1$)	−11.2	25	C	$\mu = 0$	128	8, 12	—
	2-Hydroxybenzoic acid	0.888	2.80	−9.8	25	C	$\mu = 0.05$	350	35	—
		0.925	2.68	−9.2	25	C	$\mu = 0.1$	350	120, 129	—
		1.159	2.34	−6.8	25	C	$\mu = 0.5$	350	332, 350	—
		0.461	3.32	−9.2	40	C	$\mu = 0.5$	349	—	—
		0.273	2.32	−9.9	45	C	$\mu = 0.5$	349	—	—
		0.046	2.32	−10.5	50	C	$\mu = 0.5$	349	—	—
		−0.182	2.32	−11.2	55	C	$\mu = 0.5$	349	—	—
		1.375	2.43	−6.5	25	C	$\mu = 1.0$	350	—	—
		2.178	1.88	−1.3	25	C	$\mu = 3.0$	350	—	—
		0.500	3.000 (25°, pK$_1$)	−12.0	30	T	0	130	—	(Conditions refer to wt. % ethanol)
		−0.200	3.180 (25°, pK$_1$)	−15.2	—	—	16.2	—	—	
		−0.300	3.585 (25°, pK$_1$)	−17.4	—	—	33.2	—	—	
		−0.400	4.046 (25°, pK$_1$)	−19.8	—	—	52.0	—	—	
		−1.200	4.745 (25°, pK$_1$)	−25.0	—	—	73.5	—	—	
		−2.200	5.346 (25°, pK$_1$)	−31.8	—	—	85.7	—	—	
		−1.700	6.398 (25°, pK$_1$)	−35.0	—	—	95.0	—	—	
		0.800	—	—	—	—	100.0	—	—	
		8.51	13.596 (pK$_2$)	−33.7	25	C	$\mu = 0$	128	35	—
		9.465	13.31	−27.4	10	C	$\mu = 0.5$ (NaNO$_3$)	349	—	—
		9.229	13.15	−28.4	18	C	$\mu = 0.5$ (NaNO$_3$)	349	—	—

HEATS OF PROTON IONIZATION, pK, AND RELATED THERMODYNAMIC QUANTITIES (continued)

Compound	Formula, Synonyms	ΔH (kcal/mole)	pK	ΔS cal/deg-mole	T°C	Method	Conditions	Reference	Other References	Remarks
Benzoic acid—(Continued)										
—, 2-hydroxy- (Continued)	2-Hydroxybenzoic acid (Continued)	8.965	12.94	−29.1	25	C	$\mu = 0.5$ ($NaNO_3$)	349	—	—
		8.960	12.94	−29.0	25	C	$\mu = 0.5$	350	—	—
		8.390	12.64	−31.0	40	C	$\mu = 0.5$ ($NaNO_3$)	349	—	—
		8.215	12.45	−31.6	50	C	$\mu = 0.5$ ($NaNO_3$)	349	—	—
		9.070	12.83	−28.2	25	C	$\mu = 1.0$	350	—	—
		9.390	12.80	−27.5	25	C	$\mu = 2.0$	350	—	—
		9.660	12.80	−26.0	25	C	$\mu = 3.0$	350	—	—
		7.1	10.04	−22	25	T	0.01M Buffer	120	—	—
—, 3-hydroxy-	3-Hydroxybenzoic acid	0.348	—	—	20	C	$C = 1.1 \times 10^{-4}$	122	8, 12	—
		0.159	4.075	−18.1	25		$\mu = 0$			—
		−0.029	—	—	30					—
Benzoic acid, 3-hydroxy-nitrile-	See: Phenol, 3-cyano-	—	—	—	—			—	—	—
Benzoic acid, 4-hydroxy-nitrile-	See: Phenol, 4-cyano-	—	—	—	—			—	—	—
—, 4-hydroxy-	4-Hydroxybenzoic acid	0.945	—	—	10	C	$\mu = 0$	122	12, 89	—
		0.561	4.595 (pK_1)	—	20			131		—
		0.363	9.2 (pK_2)	−9.8	25					—
		0.167	—	—	30					—
—, 2-hydroxyacetate-	Acetylsalicylic acid	3.4	—	−30.8	25	T	$\mu = 0$	120	12	—
		−0.835	3.506	−18.75	35	C	$\mu = 0.005$	132		—
—, 2-hydroxyamide-	Salicylamide	6.7	8.66 (35°)	−18	15–35	T	$\mu = 3$	332	35	3.b.
—, 4-hydroxyethyl ester-	—	1.5	8.43	−34	25	T	0.01M Buffer	120	—	—
—, 4-hydroxy-3-methoxy-	Vanillic acid / 4-Hydroxy-3-methoxybenzoic acid	0.145	4.355	−19.50	25	T	$\mu = 0$	133	12	—
—, 2-hydroxyethyl ester-	Methylsalicylate	7.5	10.19	−22	25	T	$\mu = 3$	332	35	—
—, 2-hydroxy-5-sulfo-	2-Hydroxy-5-sulfobenzoic acid / 5-Sulfosalicylic acid	7.1	11.9	−30	25	T	$\mu = 3$	332	35	—

HEATS OF PROTON IONIZATION, pK, AND RELATED THERMODYNAMIC QUANTITIES (continued)

Compound	Formula, Synonyms	ΔH (kcal/mole)	pK	ΔS cal/deg-mole	T°C	Method	Conditions	Reference	Other References	Remarks
—, 2-iodo-	2-Iodobenzoic acid	-3.250	2.860	-24.0	25	T	—	8	—	3.d. This ΔH value appears to be an abnormally large negative number for this type of ionization.
Benzoic acid—(Continued)										
—, 3-iodo-	3-Iodobenzoic acid	0.190	3.86	-17.0	25	T	—	126	8	3.b.
—, 2-mercapto-	2-Mercaptobenzoic acid	5.72	8.88 (pK$_2$)	-21.4	25	C	$\mu = 0.10$	29	—	—
	Thiosalicylic acid	—	—	—	—	—	—	—	—	—
—, *m*-methoxy-	See: Benzoic acid-3-methoxy-									
—, *o*-methoxy-	See: Benzoic acid-2-methoxy-									
—, 2-methoxy-	*o*-Anisic acid	-1.60	4.09	-24.1	25	C	$\mu = 0$	123	—	—
	2-Methoxybenzoic acid									
—, 3-methoxy-	3-Methoxybenzoic acid	0.06	4.09	-18.5	25	C	$\mu = 0$	123	8, 12	—
—, 4 methoxy-	4-Methoxybenzoic acid	0.57	4.47	-18.5	25	C	$\mu = 0$	123	8, 12	—
	Anisic acid									
—, 2-methyl-	2-Methylbenzoic acid	-1.50	3.91	-22.9	25	C	$\mu = 0$	123	8, 12	—
	o-Toluic acid									
—, 3-methyl-	3-Methylbenzoic acid	0.07	4.24	-19.2	25	C	$\mu = 0$	123	8, 12	—
	m-Toluic acid									
—, 4-methyl-	4-Methylbenzoic acid	0.24	4.34	-19.0	25	C	$\mu = 0$	123	8, 12	—
	p-Toluic acid									
—, 2-nitro-	2-Nitrobenzoic acid	-3.355	2.184	-21.3	25	T	—	8	—	3.d. This ΔH value appears to be an abnormally large negative number for this type of ionization.
—, 3-nitro-	3-Nitrobenzoic acid	0.320	3.447	-14.7	25	T	—	8	12	3.b.
—, 4-nitro-	4-Nitrobenzoic acid	0.03	3.442	-15.6	25	T	$\mu = 0$	89	12	—
—, 3,4,5-trihydroxy-	3,4,5-Trihydroxybenzoic acid Gallic acid	0.733	4.404 (35°)	-18	16.8	T	—	12	—	3.a.b.d.

HEATS OF PROTON IONIZATION, pK, AND RELATED THERMODYNAMIC QUANTITIES (continued)

Compound	Formula, Synonyms	ΔH (kcal/mole)	pK	ΔS cal/deg-mole	T°C	Method	Conditions	Reference	Other References	Remarks
Benzoic acid—(Continued)										
—, 3,4,5-trimethoxy-	3,4,5-Trimethoxybenzoic acid	−1.83	4.19	−25.33	20	T	—	134	—	3.d. The variation of these ΔH values with temperature appears to be abnormally large
		−2.98	4.30	−29.19	30	—	—	—	—	
		−4.17	4.38	−33.06	40	—	—	—	—	
		−5.39	4.53	−36.93	50	—	—	—	—	
		−6.66	4.65	−40.79	60	—	—	—	—	
Benzophenone										
—, 4-amino-		4.541	2.15 (25°)	5.39	10-40	T	—	135	—	3.d.
Benzo(b)pyridine	See: Quinoline									
Benzo(c)pyridine	See: Isoquinoline									
5,6-Benzoquinoline		4.54	5.00	−6.45	20	T	—	136	—	3.d.
7,8-Benzoquinoline		6.09	4.15	2.7	20	T	—	136	—	3.d.
1,2,3-Benzotriazole		7.47	8.38	−13.3	25	C	$\mu = 0$	352	—	
Benzoxazole										
—, 2(2'-pyridyl)-		12.2	9.27 (25°)	−1	15-40	T	$\mu = 0.005$	118	—	3.b.
		12.5	8.98 (25°)	0.8	0-40	T	$\mu = 0.005$	118	—	3.b. 50 wt. % dioxane
Benzylamine	See: Toluene, α-amino-									

HEATS OF PROTON IONIZATION, pK, AND RELATED THERMODYNAMIC QUANTITIES (continued)

Compound	Formula, Synonyms	ΔH (kcal/mole)	pK	ΔS cal/deg-mole	T°C	Method	Conditions	Reference	Other References	Remarks
Benzyldiethylamine	See: Toluene,α-amino-N,N-diethyl-									
Betaine	$(CH_3)_3\overset{+}{N}CH_2COOH$ / Lycine / Oxyneurine	−0.08	1.832	−8.7	25	C	$\mu = 0$	137	—	—
2-2-Biimidazolyl		6.08	5.01	−2.5	25	C	$0.3M$ $NaClO_4$	353	—	—
Biphenyl										
—, 2-amino-	NH_2 (structure, positions 2,3,4,5,6; 2',3',4',5',6')	5.7	3.03 (25°)	5.2	0–20	T	$C = 0.01$	85	—	3.b. 50 wt. % ethanol
—, 3-amino-	2-Aminodiphenyl / 3-Aminodiphenyl	5.2	3.82 (20°)	0.2	0–20	T	$C = 0.01$	85	—	3.b. 50 wt. % ethanol
—, 4-amino-	4-Aminodiphenyl / Xenylamine	5.6	3.81 (20°)	1.5	0–20	T	$C = 0.01$	85	—	3.b. 50 wt. % ethanol
2,2′-Bipyridyl	2.2′-Bipyridine (structure, $\overset{+}{N}H$ / N)	3.37	4.352	−8.62	25	T	$\mu = 0$	77	118, 136, 139–141	—
		3.66	4.52	−8.2	20	C	$0.1\,M\ NaNO_3$	138	—	—
		4.02	4.62	−7.88	30.3	C	$1\,M\ KNO_3$	56	354	—
		2.7	3.33 (25°)	−6	0–40	T	$\mu = 0.05$	118	—	3.b. 50 wt. % dioxane
Boric acid										
—, **meta-**	HBO_2 / meta-Boric acid	3.920	9.380	−29.1	10	T	$\mu = 0$	142	11	—
		3.750	9.327	−29.7	15	—				—
		3.570	9.280	−30.3	20	—				—
		3.360	9.236	30.0	25	—				—
		3.140	9.197	−31.7	30	—				—
		2.900	—		35	—				—
		2.630	9.132	−33.4	40	—				—
		2.350	—		45	—				—
		2.040	9.080	−35.2	50	—				—
—, **ortho-**	H_3BO_3 / ortho-Boric acid	3.920	9.380	−29.1	10	T		142	—	3.d.
		3.750	9.327	−29.6	15	—				—
		3.570	9.280	−30.2	20	—				—
		3.360	9.236	−31.0	25	—				—
		3.140	9.197	−31.7	30	—				—
		2.900	—		35	—				—
		2.630	9.132	−33.4	40	—				—

HEATS OF PROTON IONIZATION, pK, AND RELATED THERMODYNAMIC QUANTITIES (continued)

Compound	Formula, Synonyms	ΔH (kcal/mole)	pK	ΔS cal/deg-mole	T°C	Method	Conditions	Reference	Other References	Remarks
Boric acid—(Continued)										
—, *ortho*—(Continued)	*ortho*-Boric acid—(Continued)									
		2.350	—	—	45	—	—	—	—	—
		2.040	9.080	-35.2	50	—	—	—	—	—
Bovine albumin	(Tyrosyl)	3.3	9.22	-31.1	25	C	$C = 0.01\,M$	355	—	—
		11	10.3	-10	25	T	$\mu = 0.5$	143	—	—
α-Bromobutyric acid	See: Butanoic acid, 2-bromo-									
3-Bromo-4-dimethylamino-pyridine	See: Pyridine, 3-bromo-4-(dimethylamino)-									
3-Bromo-4-methylamino-pyridine	See: Pyridine, 3-bromo-4-amino,N-methyl-									
α-Bromopropionic acid	See: Propanoic acid, 2-bromo-									
β-Bromopropionic acid	See: Propanoic acid, 3-bromo-									
3-Bromopyridine	See: Pyridine, 3-bromo-									
BSTA	See: Acetic acid, [thiobis(ethylenenitrilo)]-tetra-									
Butane										
—, 1-amino-	$CH_3CH_2CH_2CH_2\overset{+}{N}H_3$ n-Butylamine	13.98	10.640	-1.8	25	C	$\mu = 0$	330		
		13.9	10.640	-2.1	25	T	$\mu = 0$	77	66	
		14.07	10.107 (40)	-1.3	20-40	T	$\mu = 0$	64		3.b.
		14.09	10.114	-1.3			$\mu = 0.05$			
		14.04	10.127	-1.5			$\mu = 0.10$			
		14.09	10.141	-1.4			$\mu = 0.20$			
		13.97				C	$\mu = 0.2$	4		
—, 2-amino-		13.9	10.95 (25)	-3.46	0-25	T	$\mu = 2$	189	336, 356	3.b.
—, 1-amino-3-methyl-		14.03	10.56	-1.3	25	C	$\mu = 0$	330		
		14.03	10.64	-1.6	25	C	$\mu = 0$	330		
—, 1,2-diamino-	$H_3\overset{+}{N}CH_2CH(CH_2CH_3)\overset{+}{N}H_3$ 1,2-Diamino-1-ethylethane	9.92	6.399 (pK_1)	4.0	25	C	$\mu = 0$	144		
		11.45	9.388 (pK_2)	-4.5	25	C	$\mu = 0$	144		
—, 1,4-diamino-	$H_3\overset{+}{N}CH_2CH_2CH_2CH_2\overset{+}{N}H_3$ Tetramethylenediamine Putrescine	13.20	9.20 (pK_1)	2.2	25	C	$\mu = 0.02$	330		
		12.0	9.04 (30°, pK_1)	-2	10-40	T	$\mu = 0$	145		3.b.
		13.58	10.65 (pK_2)	-3.2	25	C	$\mu = 0.02$	330		
		12.0	10.50 (30°, pK_2)	-8	10-40	T	$\mu = 0$	145		
—, *meso*-2,3-diamino-	$CH_3CH(\overset{+}{N}H_3)CH(\overset{+}{N}H_3)CH_3$ meso-2,3-Diaminobutane	9.4	6.92 (25°, pK_1)	0.13	0-25	T	$0.5\,M\,KNO_3$	146		3.b.
		9.8	9.97 (25°, pK_2)	-12.7	0-25	T	$0.5\,M\,KNO_3$	146		3.a.b.
—, *rac*-2,3-diamino-	rac-2,3-Diaminobutane	10.3	6.91 (25°, pK_1)	2.93	0-25	T	$0.5\,M\,KNO_3$	146		3.a.b.
		10.3	10.00 (25°, pK_2)	-11.2	0-25	T	$0.5\,M\,KNO_3$	146		3.a.b.
—, 2,3-diamino-2,3-dimethyl-	2,3-Diamino-2,3-dimethylbutane	9.2	6.56 (25°, pK_1)	1.03	0-25	T	$0.5\,M\,KNO_3$	146		3.a.b.
		8.9	10.13 (25°, pK_2)	-16.50	0-25	T	$0.5\,M\,KNO_3$	146		3.a.b.

HEATS OF PROTON IONIZATION, pK, AND RELATED THERMODYNAMIC QUANTITIES (continued)

Compound	Formula, Synonyms	ΔH (kcal/mole)	pK	ΔS cal/deg-mole	T°C	Method	Conditions	Reference	Other References	Remarks
Butane—(Continued)										
—, 1,1-dinitro-		3.618	5.90	-12.33	20	T	—	147	—	3.d.
Butanedioic acid	See: Succinic acid									
Butanoic acid	$CH_3CH_2CH_2COOH$ Butyric acid	-0.242	—	—	10	C	μ = 0	122	—	—
		-0.542	—	—	20					—
		-0.64	4.817	-24.2	25	C	μ = 0	324	1, 12, 149	—
		-0.698	4.818	-24.4	25	—		122	—	—
		-0.853	—	—	30	—			—	—
		-1.12	4.854	-25.8	40	C	μ = 0	324	—	—
		0.273	4.806	-21.0	0	T	μ = 0	148	—	—
		0.106	4.804	-21.6	5	—			—	—
		0	4.803	-22.0	8					—
		-0.073	4.803	-22.2	10					—
		-0.266	4.805	-22.9	15					—
		-0.472	4.810	-23.6	20					—
		-0.691	4.817	-24.4	25					—
		-0.926	4.827	-25.1	30					—
		-1.174	4.840	-26.0	35					—
		-1.437	4.854	-26.8	40					—
		-1.714	4.871	-27.7	45					—
—, 2-amino(d)-	DL-α-Amino-n-butyric acid	0.77	2.315	-7.7	10	C	μ = 0	324	—	—
		0.38	2.286	-9.2	25	C	μ = 0	324	—	—
		0.04	2.289	-10.4	40	C	μ = 0	324	—	—
		1.090	2.334	-6.7	1	T	μ = 0	50	150	—
		0.750	2.310	-7.9	12.5				325	—
		0.310	2.286	-9.4	25					—
		-0.220	2.288	-11.2	37.5					—
		-0.830	2.297	-13.1	50					—
		10.86	9.830 (pK₂)	-8.6	25	C	μ = 0	330	—	—
		10.750	10.530	-9.0	1.0	T	μ = 0	50	—	—
		10.820	10.180	-8.7	12.5					—
		10.770	9.830	-8.8	25					—
		10.580	9.518	-9.4	37.5					—
		10.260	9.234	-10.5	50					—
—, 4-amino-	γ-Aminobutyric acid Piperidinic acid	0.75	4.057 (pK₁)	-15.9	10	C	μ = 0	324	59	—
		0.39	4.031	-17.2	25	C	μ = 0	324	151	—
		0.01	4.027	-18.5	40	C	μ = 0	324	325	—
		0.405	4.0312	-17.0	25	T	μ = 0	151	59	—
		12.45	10.556 (pK₂)	-6.5	25	C	μ = 0	330	151, 325	—
		12.070	10.5557	-7.8	25	T	μ = 0	151	—	—

HEATS OF PROTON IONIZATION, pK, AND RELATED THERMODYNAMIC QUANTITIES (continued)

Compound	Formula, Synonyms	ΔH (kcal/mole)	pK	ΔS cal/deg-mole	T°C	Method	Conditions	Reference	Other References	Remarks
Butanoic acid—(Continued)										
—, 2-amino-N-acetyl-	$CH_3CH_2CH(CH_3CONH)COOH$ N-Acetyl-α-amino-n-butyric acid	-0.773	3.7158	-19.59	25	T	μ = 0	55	—	—
—, 2-amino carbamoyl-	$\overset{+}{N}H_3CONHCH(CH_3CH_2)COOH$ Carbamoyl-α-amino-n-butyric acid	-0.493	3.8856 (pK₁)	-19.43	25	T	μ = 0	58	—	—
—, 4-amino carbamoyl-	$H_3\overset{+}{N}CONH(CH_2)_3COOH$ Carbamoyl-γ-aminobutyric acid	-0.115	4.6831	-21.83	25	T	μ = 0	58	—	—
—, 2-amino-3-methyl-	See: Valine									
—, 2-amino-4-methyl-	$(CH_3)_2-CH-CH(NH_2)-COOH$	10.640	9.744 (pK₂)	-8.72	25	C	—	325	—	3.d.
—, 2-amino-4(methylthio)-	See: Methionine									
—, 2-bromo-	α-Bromobutyric acid	-1.263	2.939 (35°)	-18	16.8	T	—	12	13	3.a.b.d.
—, 2-ethyl-	$(CH_3CH_2)_2CHCOOH$ Diethylacetic acid	-1.86	4.74	-27.6	25	C	μ = 0	324	—	—
—, glycyl-2-amino-	$\overset{+}{N}H_3CH_2CONHCHCH_2CH_3COOH$ Glycyl-α-amino-n-butyric acid	-0.672	3.1546 (pK₁)	-16.69	25	T	μ = 0	6	—	—
—, 3-hydroxy-	Acetylmethylcarbinol	2.022	0.663	3.48	20	T	μ = 0	152	—	—
		-0.767	—	—	10	C	μ = 0	122	12, 13	—
—, 3-methyl-	$(CH_3)_2CHCH_2COOH$ Isovaleric acid	-1.039	—	—	20	—	—	—	122	—
		-1.15	—	-25.9	25	C	μ = 0	324	—	—
		-1.168	4.770	-25.7	25	—	—	—	—	—
		-1.302	4.770	—	30	—	—	—	—	—
—, 2-oxo-	α-Ketobutyric acid	2.82	2.50	-2.0	25	C	μ = 0	34	—	—
		2.932	2.50	-1.61	25	C	μ = 0.05	357	—	—
1-Butanol										
—, 2-amino-	—	12.460	9.516 (25°)	-1.8	0-60	T	μ = 0	335	—	3.b.
—, 2-amino-2-ethyl-	—	9.66	9.82	-12.6	25	C	μ = 0	330	—	—
2-Butanone,3-(allylamine)-3-methyl, oxime	$\underset{\text{α-N-Allylamine oxime}}{\overset{\overset{\displaystyle CH_3 \quad \overset{\text{NOH}}{\|}}{H_3C-C-C-CH_3}}{H_2^+N-CH_2-CH=CH_2}}$	10.8	—	—	25	C	μ = 0	63	—	—
—, 3-amino-3-methyl-, oxime	$\underset{\text{Amine oxime}}{\overset{\overset{\displaystyle CH_3 \quad \overset{\text{NOH}}{\|}}{H_3C-C-C-CH_3}}{^+NH_3}}$	12.7	9.09	1	25	C	μ = 0	63	—	—

HEATS OF PROTON IONIZATION, pK, AND RELATED THERMODYNAMIC QUANTITIES (continued)

Compound	Formula, Synonyms	ΔH (kcal/mole)	pK	ΔS cal/deg-mole	T °C	Method	Conditions	Reference	Other References	Remarks
2-Butanone—(Continued)										
—, 3-n-butylamine-3-methyl, oxime	$H_3C-C-C-CH_3$ with CH₃, NOH; $^+H_2N-CH_2-CH_2-CH_2-CH_3$ N-n-Butylamine oxime	11.0	9.09	−5	25	C	$\mu = 0$	63	—	—
—, 3-(ethylamino)-3-methyl, oxime	$H_3C-C-C-CH_3$ with CH₃, NOH; $H_2^+N-CH_2-CH_3$ α-N-Ethylamine oxime	10.6	9.23	−7	25	C	$\mu = 0$	63	—	—
—, 3-(isopropylamino)-3-methyl, oxime	$H_3C-C-C-CH_3$ with CH₃, NOH; CH₃; H_2^+N-CH with CH₃ α-N-Isopropylamine oxime	10.6	9.09	−6	25	C	$\mu = 0$	63	—	—
—, 3-(methylamino)-3-methyl, oxime	$H_3C-C-C-CH_3$ with CH₃, NOH; $H_2^+N-CH_3$ α-N-Methylamine oxime	10.3	9.23	−8	25	C	$\mu = 0.002$	63	—	—
—, 3-(n-propylamine)-3-methyl, oxime	$H_3C-C-C-CH_3$ with CH₃, NOH; $H_2^+N-CH_2-CH_2-CH_3$ α-N-n-Propylamine oxime	10.8	9.09	−5	25	C	$\mu = 0$	63	—	—
cis-Butenedioic acid	See: Maleic acid									
trans-Butenedioic acid	See: Fumaric acid									
2-Butenoic acid (*trans*)	$CH_3-CH=CH-C-OH$ with =O Crotonic acid	0.280	4.676 (35°)	−20	16.8	T	—	12	—	3.a.b.d.
sec-Butylacetic acid	See: Pentanoic acid, 3-methyl-									

HEATS OF PROTON IONIZATION, pK, AND RELATED THERMODYNAMIC QUANTITIES (continued)

Compound	Formula, Synonyms	ΔH (kcal/mole)	pK	ΔS cal/deg-mole	T°C	Method	Conditions	Reference	Other References	Remarks
Butanoic acid—(Continued)										
n-Butylamine	See: Butane, 1-amino-									
sec-Butylamine	See: Butane, 2-amino-									
t-Butylamine	See: Propane, 2-amino-2-methyl-									
N-*n*-Butylamine oxime	See: 2-Butanone, 3-*n*-butylamine-3-methyl, oxime									
N,*N'*-di-*n*-Butylethylene-diamine	See: Ethane, 1,2-diamino-*N*,*N'*-di-*n*-butyl-									
n-Butylmethylamine	See: Amine, butylmethyl-									
t-Butylthiol	See: 1-Ethanethiol, 1,1-dimethyl-									
Butyric acid	See: Butanoic acid									
Camphoric acid		-0.550	4.595 (35°)	-23	16.8	T	—	12	—	3.a.b.d.
n-Caproic acid	See: Hexanoic acid									
Caprylic acid	See: Octanoic acid									
Carbamoyl-α-alanine	See: Alanine, carbamoyl-									
Carbamoyl-β-alanine	See: β-Alanine, carbamoyl-									
Carbamoyl-α-amino-*n*-butyric acid	See: Butanoic acid, 2-aminocarbamoyl-									
Carbamoyl-γ-aminobutyric acid	See: Butanoic acid, 4-aminocarbamoyl-									
Carbamoyl-α-aminoiso-butyric acid	See: Propanoic acid, carbamoyl-α-amino-2-methyl-									
Carbamoylglycine	See: Glycine, carbamoyl-									
N-Carbamoylglycine	See: Glycine, carbamoyl-									
Carbamoylhydrazine	See: Semicarbazide									
Carbinol	See: Methanol									
Carbolic acid	See: Phenol									
Carbonic acid	H_2CO_3	0.580	3.807	-15.3	0	T	$\mu = 0$	153	91	$H_2CO_3 \rightarrow$ H^+ +
		0.0	3.754	-17.2	17	—	—	—	155–7	HCO_3^-, see
		-0.412	3.764	-18.5	27	—	—	—	—	ref 322
		-0.880	3.798	-20.2	37	—	—	—	—	—

HEATS OF PROTON IONIZATION, pK, AND RELATED THERMODYNAMIC QUANTITIES (continued)

Compound	Formula, Synonyms	ΔH (kcal/mole)	pK	ΔS cal/deg-mole	T°C	Method	Conditions	Reference	Other References	Remarks
Carbonic acid—(Continued)										
		4.484	6.583	−13.7	0	T	$\mu = 0$	**154**	—	$(CO_2 + H_2O \rightarrow$
		2.952	6.429	−19.1	15	—	—	—	—	$H^+ +$
		2.075	6.365	−22.2	25	—	—	—	—	HCO_3^-), see
		1.109	7.317	−29.9	38	—	—	—	—	ref 322
		3.500	10.24	−35.2	25	C	$\mu = 0$	**81**	—	3.d.
—, trithio-	H_2CS_3	2.86	2.64	−2.51	20	T	—	**158**	—	
cis-Caronic acid	See: 1,2 Cyclopropanedicarboxylic acid, 3,3-dimethyl-(*cis*)-									
trans-Caronic acid	See: 1,2-Cyclopropanedicarboxylic acid, 3,3-dimethyl-(*trans*)-									
Carubinose	See: Mannose (D)									
4-Chloro-4'-aminodiphenyl-sulfone	See: Sulfone, 4-chloro-4'-aminodiphenyl-									
o-Chloroaniline	See: Aniline, 2-chloro-									
α-Chloropropionic acid	See: Propanoic acid, 2-chloro-									
β-Chloropropionic acid	See: Propanoic acid, 3-chloro-									
Chlorous acid	$HClO_2$	−4.100	2.021	−22.7	25	C	$\mu = 0$	**81**		
Cinnamic acid	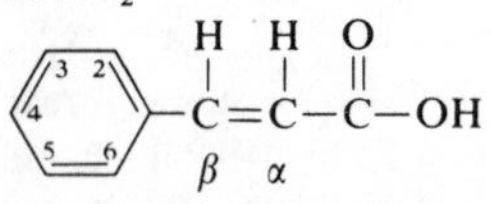 (structure shown) *cis-β-Phenylacrylic acid*	0.600	4.404	−18.1	25	T	—	**8**	**12**	3.d. *cis* or *trans* not specified
—, α-bromo-	—	−2.990	1.977	−19.1	25	T	—	**8**	—	3.d. *cis* or *trans* not specified
—, 2-hydroxy-	o-Coumaric acid	0.181	4.616 (35°)	−20	16.8	T	—	**12**	—	3.a.b.d. *cis* or *trans* not specified
—, o-hydroxy-	See: Cinnamic acid, 2-hydroxy-									
Citric acid	$HOC(CH_2CO_2H)_2CO_2H$ 2-Hydroxy-1,2,3-propanetricarboxylic acid	1.760	3.220 (pK$_1$)	−8.3	0	T	$\mu = 0$	**159**	—	—
		1.612	3.200	−8.8	5	—	—	—	—	—
		1.462	3.176	−9.4	10	—	—	—	—	—
		1.310	3.160	−9.9	15	—	—	—	—	—
		1.155	3.142	−10.4	20	—	—	—	—	—
		0.997	3.128	−11.0	25	—	—	—	—	—
		0.836	3.116	−11.5	30	—	—	—	—	—
		0.673	3.109	−12.0	35	—	—	—	—	—
		0.507	3.099	−12.6	40	—	—	—	—	—

Structure (Cinnamic acid):

$$\text{benzene ring} - \underset{\beta}{C}H = \underset{\alpha}{C}H - \underset{\overset{\|}{O}}{C} - OH$$

HEATS OF PROTON IONIZATION, pK, AND RELATED THERMODYNAMIC QUANTITIES (continued)

Compound	Formula, Synonyms	ΔH (kcal/mole)	pK	ΔS cal/deg-mole	T°C	Method	Conditions	Reference	Other References	Remarks
Citric acid—(Continued)										
		0.338	3.097	−13.1	45	—	—	—	—	—
		0.167	3.095	−13.6	50	—	—	—	—	—
		1.654	4.837 (pK$_2$)	−16.1	0	T	$\mu = 0$	159	—	—
		1.447	4.813	−16.8	5	—	—	—	—	—
		1.237	4.797	−17.6	10	—	—	—	—	—
		1.022	4.782	−18.3	15	—	—	—	—	—
		0.804	4.769	−19.1	20	—	—	—	—	—
		0.582	4.761	−19.8	25	—	—	—	—	—
		0.357	4.755	−20.6	30	—	—	—	—	—
		0.128	4.751	−21.3	35	—	—	—	—	—
		−0.105	4.750	−22.1	40	—	—	—	—	—
		−0.342	4.754	−22.8	45	—	—	—	—	—
		−0.583	4.757	−23.6	50	—	—	—	—	—
		0.660	6.393 (pK$_3$)	−26.8	0	T	$\mu = 0$	159	—	—
		0.378	6.386	−27.9	5	—	—	—	—	—
		0.090	6.383	−28.9	10	—	—	—	—	—
		−0.202	6.384	−29.9	15	—	—	—	—	—
		−0.500	6.388	−30.9	20	—	—	—	—	—
		−0.803	6.396	−32.0	25	—	—	—	—	—
		−1.111	6.406	−33.0	30	—	—	—	—	—
		−1.424	6.423	−34.0	35	—	—	—	—	—
		−1.742	6.439	−35.0	40	—	—	—	—	—
		−2.065	6.462	−36.0	45	—	—	—	—	—
		−2.394	6.484	−37.1	50	—	—	—	—	—
CMP	See: Cytidine,5'-monophosphoric acid									
dCMP	See: Cytidine, 5'-monophosphoric acid, 2'-deoxy-									
Colamine	See: Ethanol, 2-amino-									
o-Coumaric acid	See: Cinnamic acid, 2-hydroxy-									
m-Cresol	See: Toluene, 3-hydroxy-									
o-Cresol	See: Toluene, 2-hydroxy-									
p-Cresol	See: Toluene, 4-hydroxy-									
Crotonic acid	See: 2-Butenoic acid (*trans*)									
CTP	See: Cytidine, 5'-triphosphoric acid									
Cyanic acid	HOCN	1	3.58 (25°)	−10	20–33	T	$\mu = 0$	216	—	3.b.
—, thio-	HSCN	13	0.95 (25°)	40	20–33	T	$\mu = 0$	216	—	3.b.
Cyanoacetic acid	See: Malonic acid, mononitrile-									
3-Cyanobenzoic acid	See: 1,3-Benzenedicarboxylic acid, mononitrile-									

HEATS OF PROTON IONIZATION, pK, AND RELATED THERMODYNAMIC QUANTITIES (continued)

Compound	Formula, Synonyms	ΔH (kcal/mole)	pK	ΔS cal/deg-mole	T°C	Method	Conditions	Reference	Other References	Remarks
4-Cyanobenzoic acid	See: 1,4-Benzenedicarboxylic acid, mononitrile-									
3-Cyanophenol	See: Phenol, 3-cyano-									
4-Cyanophenol	See: Phenol, 4-cyano-									
Cyclohexane										
—, amino-	NH₂	14.38	10.64	−0.5	25	C	$\mu = 0$	330	—	—
—, carboxylic acid	Hexahydrobenzoic acid	−0.92	4.903	−25.5	25	C	$\mu = 0$	1	—	—
—, amino-N,N-dimethyl-		10.11	10.72	−15.2	25	C	$\mu = 0$	330	—	—
—, 1,2-diamino (*cis*)-		10.1	5.94 (30°, pK_1)	6	10–40	T	$\mu = 0$	145	—	3.b.
		11.5	9.66 (30°, pK_2)	−6	10–40	T	$\mu = 0$	145	—	3.b.
—, 1,2-diamino (*trans*)-		10.2	6.20 (30°, pK_1)	5	10–40	T	$\mu = 0$	145	—	—
		11.7	9.60 (30°, pK_2)	−6	10–40	T	$\mu = 0$	145	—	—
—, 1,2-diamino-N,N,N',N'-tetraacetic acid (*trans*)-	NH⁺(CH₂COOH)₂	2.06	6.15 (pK_5)	−21.1	20	C	0.1 M KNO₃	160	—	—
		1.34	6.12 (pK_5)	−23.51	25	T	0.1 M KNO₃	161	—	—
	DCTA NH⁺(CH₂COOH)₂	6.65	12.35 (pK_6)	−33.9	20	C	0.1 M KNO₃	160	—	—
		7.35	11.58 (pK_6)	−28.44	25	T	0.1 M KNO₃	161	—	—
—, 1,2-dicarboxylic acid (*cis*)-	Hexahydrophthalic acid	1.1	4.34 (pK_1)	−16	25	C	$\mu = 0$	59	—	—
—, 1,2-dicarboxylic acid (*cis*)-		−0.30	6.76 (pK_2)	−31	25	C	$\mu = 0$	59	—	—
—, 1,2-dicarboxylic acid (*trans*)-		−1.9	4.18 (pK_1)	−25	25	C	$\mu = 0$	59	—	—
		−0.24	5.93 (pK_2)	−27	25	C	$\mu = 0$	59	—	—
cis-cyclohexane-1,2-dicarboxylic acid	See: Cyclohexane, 1,2-dicarboxylic acid (*cis*)-									
trans-cyclohexane-1,2-dicarboxylic acid	See: Cyclohexane, 1,2-dicarboxylic acid (*trans*)-									
Cyclopentane										
—, amino-		14.30	10.65	−0.8	25	C	$\mu = 0$	330	—	—
Cyclopropane										
—, amino-		11.72	9.10	−2.3	25	C	$\mu = 0$	330	—	—
1,2-Cyclopropanedicarboxylic acid		−1	2.34 (pK_1)	−14	25	T	$\mu = 0$	59	—	—
—, 3,3-dimethyl-(*cis*)-	CH₃—C—CH₃	−1	8.31 (pK_2)	−41	25	T	$\mu = 0$	59	—	—
—, 3,3-dimethyl-(*trans*)-	HOOC—CH—CH—COOH *cis*-Caronic acid *trans*-Caronic acid	−2	3.92 (pK_1)	−25	25	T	$\mu = 0$	59	—	—
		−2	5.32 (pK_2)	−31	25	T	$\mu = 0$	59	—	—

HEATS OF PROTON IONIZATION, pK, AND RELATED THERMODYNAMIC QUANTITIES (continued)

Compound	Formula, Synonyms	ΔH (kcal/mole)	pK	ΔS cal/deg-mole	T°C	Method	Conditions	Reference	Other References	Remarks
Cysteine (I)	HSĊH₂CH(NH₃⁺)CO₂H L-β-Mercaptoalanine	8.6	8.39 (pK₂)	−9.5	25	C	$\mu = 0$	28	—	For a correction involving this article, see ref. 322
		8.0	7.98 (45°)	−11.4	50	T	$\mu = 0$	360	—	—
		7.6	7.52	−12.5	75	T	$\mu = 0$	—	—	3.b.
		8.1	10.76 (pK₃)	−21.9	25	C	$\mu = 0$	28	—	—
		7.0	10.34 (45°)	−25.3	50	T	$\mu = 0$	360	—	3.b.
		5.8	9.93	−29.1	75	T	$\mu = 0$	—	—	—
L-Cysteine, S-methyl- Cysteine	H₃CSCH₂CH(NH₃⁺)COOH	10.1	8.97 (pK₂)	−7.2	25	C	$\mu = 0$	28	—	—
—, diacetyl ester- —,N,N'-diformyl-	[HOOC—CH₂—CH₂S]₂ HNCHO	5.800	8.53	−20	25	T	$\mu = 0.03\text{–}0.15$	162	—	—
		6.300	9.50	−23	25	T	$\mu = 0.03\text{–}0.15$	162	—	—
5'Cytidylic acid **Cytidine**	See: Cytidine-5'-monophosphoric acid 3(α-Ribosido) cytosine	4.47	4.08 (pK₁)	−3.9	25	C	$\mu = 0$	163	164	—
		4.4	4.22 (pK₁)	−5.0	25	T	$\mu = 0.1$	39	—	—
		10.7	12.24 (pK₂)	−20.2	25	C	$\mu = 0$	163	—	—
—, 2'-deoxy-	CMP	4.300	4.3	−5.3	25	C	$\mu = 0.1$	40	—	—
—, 5'-diphosphoric acid	5'Cytidylic acid	−1.34	7.18 (pK₄)	−37.4	25	T	$\mu = 0$	46	—	—
—, 5'-monophosphoric acid	d CMP	−1.35	6.62 (pK₃)	−34.8	25	C	$\mu = 0$	46	—	—
—, 5'-monophosphoric acid-2'-deoxy-		4.280	4.4	−5.8	25	C	$\mu = 0.1$	40	—	—
—, 5'-triphosphoric acid-	CTP	−1.75	7.65 (pK₅)	−40.8	25	T	$\mu = 0$	46	—	—
Cytosine	4-Amino-1,2-dihydro-1,3-diazin-2-one 4-Amino-2(1)pyrimidone	5.14	4.58 (pK₁)	−3.7	25	C	$\mu = 0$	163	40, 164	—
		5.0	4.5 (pK₁)	−3.7	25	T	$\mu = 0.1$	39	—	—
		11.5	12.15 (pK₂)	−17.0	25	C	$\mu = 0$	163	—	—
		11.0	11.82 (pK₂)	−17.1	25	T	$\mu = 0.1$	39	—	—

Cytidine structural formula:

HOCH₂—C—C—C—C—N with ribose (H OH OH H, O ring) and cytosine base (OH, NH⁺, NH₂)

HEATS OF PROTON IONIZATION, pK, AND RELATED THERMODYNAMIC QUANTITIES (continued)

Compound	Formula. Synonyms	ΔH (kcal/mole)	pK	ΔS cal/deg-mole	T°C	Method	Conditions	Reference	Other References	Remarks
DCTA	See: Cyclohexane, 1,2-diamino-N,N,N',N'-tetraacetic acid(*trans*)-									
Decane										
—, 1,1-dinitro-	$C_9H_{19}CH(NO_2)_2$	2.50	3.597	−16.46	5–60	T	—	165	—	3.b.d.
2′-Deoxyadenosine	See: Adenosine, 2′-deoxy-									
Deoxyadenosine-5′-phosphoric acid	See: Adenosine, 2′-deoxy-5′-monophosphoric acid-									
5′-Deoxyadenylic acid	See: Adenosine, 2′-deoxy-5′-monophosphoric acid-									
2-Deoxyglucose	See: Glucose, 2-deoxy-									
2-Deoxyribose	See: Ribose, 2-deoxy-									
Deuteriophosphoric acid	See: Phosphoric acid, deuterio-									
Dextrose	See: Glucose									
Deuterium oxide	See: Water, deuterated									
Diacetylcystine ester	See: Cystine, diacetyl ester									
Diacetylmethane	See: 2,4-Pentanedione									
5,5-Diallylbarbituric acid	See: Barbituric acid, 5,5-diallyl-									
1,3-Diamino-2-aminomethylethylpropane	See: Propane, 1,3-diamino-2-aminomethylethyl-									
***meso*-2,3-Diaminobutane**	See: Butane, *meso*-2,3-diamino-									
***rac*-2,3-Diaminobutane**	See: Butane, *rac*-2,3-diamino-									
L-α,ε-Diaminocaproic acid	See: Lysine									
***trans*-1,2-Diaminocyclohexane-N,N,N',N'-tetraacetic acid**	See: Cyclohexane, 1,2-diamino-N,N,N',N'-tetraacetic acid-									
2,2′-Diaminodiethylamine	See: Amine, diethyl-2,2′-diamino-									
β,β′-Diaminodiethylether-N,N'-tetraacetic acid	See: Acetic acid, [oxy-*bis*-(ethylenenitrilo)] tetra-									
2,3-Diamino-2,3-dimethylbutane	See: Butane, 2,3-diamino-2,3-dimethyl-									
1,2-Diamino-1,1-dimethylethane	See: Propane, 1,2-diamino-2-methyl-									
3,3′-Diaminodipropylamine	See: Amine, 3,3′-diamino-dipropyl-									
1,8-Diamino-3,6-dithiaoctane	See: Octane, 1,8-diamino-3,6-dithia-									
β,β′-Diaminodiethylsulfide-N,N'-tetraacetic acid	See: Acetic acid [thio-*bis*-(ethylenenitrilo)] tetra-									
1,2-Diamino-1-ethylethane	See: Butane, 1,2-diamino-									
Di(2-aminoethyl)ether	See: Ether, di(2-aminoethyl)-									

HEATS OF PROTON IONIZATION, pK, AND RELATED THERMODYNAMIC QUANTITIES (continued)

Compound	Formula, Synonyms	ΔH (kcal/mole)	pK	ΔS cal/deg-mole	T°C	Method	Conditions	Reference	Other References	Remarks
Di(2-aminoethoxy)ethane-tetraacetic acid	See: Acetic acid [ethylene bis-(oxyethylenenitrilo) tetra-									
1,2-Diaminoisobutane	See: Propane, 1,2-diamino-2-methyl-									
1,8-Diamino-3-oxa-6-thiaoctane	See: Amine, 2-[2(2-aminoethoxy) ethyl-thio]-ethyl-									
1,2 Diamino-1-phenylethane	See: Ethane, 1,2-diamino-1-phenyl-									
1,2-Diaminopropane	See: Propane, 1,2-diamino-									
1,3-Diaminopropane-2-ol	See: Propane, 1,3-diamino-2-hydroxyl-									
N,N-Di(3-aminopropyl)-N-methylamine	See: Amine, N,N-di(3-aminopropyl)-N-methylamine-									
1,5-Diazaphenanthrene	See: 1,5-Phenanthroline									
4,5-Diazaphenanthrene	See: 4,5-Phenanthroline									
1,3-Diazole	See: Imidazole									
2,3,5,6-Dibenzopyridine	See: Acridine									
N,N'-Di-n-butylethylene-diamine	See: Ethane, 1,2-diamino-N,N'-di-n-butyl-									
2,6-Dichlorohydroquinone	See: Benzene, 1,3-dichloro-2,5-dihydroxy-									
Dicyanomethane	See: Malonic acid, dinitrile-									
Diethylamine	See: Amine, diethyl-									
Diethanolamine	See: Amine, diethyl-2,2'-dihydroxy-									
N,N'-Diethylethylene-methylamine	See: Ethane, 1,2-diamino-N,N'-diethyl-									
Diethylacetic acid	See: Butanoic acid, 2-ethyl-									
2-Diethylaminoethanol	See: Butanol, 2-amino-2-ethyl-									
5,5-Diethylbarbituric acid	See: Barbituric acid, 5,5-diethyl-									
Diethyleneimide oxide	See: Morpholine									
N,N'-Diethylethylene-diamine	See: Ethane, 1,2-diamino-N,N'-diethyl-									
Diethylenediamine	See: Piperazine									
Diethylenetriamine	See: Propane, 1,3-diamino-2-aminomethylethyl-propane-									
Diethylenetriaminepenta-acetic acid	See: Acetic acid [(carboxymethylimino) bis-(ethylenenitrilo)] tetra-									
Diglycine	See: Glycine, N-glycyl-									
Diglycolamidic acid	See: Acetic acid, iminodi-									
Dihydroazrine	See: Azirane									
Dihydroxybenzoic acid	See: Benzoic acid, dihydroxy-									
2,3-Dihydroxybutanedioic acid	See: Tartaric acid									

HEATS OF PROTON IONIZATION, pK, AND RELATED THERMODYNAMIC QUANTITIES (continued)

Compound	Formula, Synonyms	ΔH (kcal/mole)	pK	ΔS cal/deg-mole	T °C	Method	Conditions	Reference	Other References	Remarks
N,N-Di(2-hydroxyethyl)-glycine	See: Glycine, *N,N*-di(2-hydroxyethyl)-									
2,4-(Dihydroxymethyl)-phenol	See: Phenol, 2,4-(dihydroxymethyl)-									
2,3-Dihydroxysuccinic acid	See: Tartaric acid									
3,5-Diiodotyrosine	See: Tyrosine, 3,5-diiodo-									
2,7-Dimethyl-1,8-phenanthroline	See: 1,8-Phenanthroline, 4,5-dimethoxy-2,7-dimethyl-									
Dimethylamine	See: Amine, dimethyl-									
Dimethylaminoacetic acid	See: Glycine, *N,N*-dimethyl-									
4-Dimethylaminobenzal-dehyde	See: Benzaldehyde, 4-dimethylamino-									
2-Dimethylaminoethanol	See: Propanol, 2-amino-2-methyl-									
4-Dimethylaminopyridine	See: Pyridine, 4(*N,N*-dimethylamino)-									
N,N-Dimethylanthranilic acid	See: Benzoic, acid, 2-amino-*N,N*-dimethyl-									
1,3-Dimethylbarbituric acid	See: Barbituric acid, 1,3-dimethyl-									
N,N-Dimethylcyclohexyl-amine	See: Cyclohexane, amino-*N,N*-dimethyl-									
3,5-Dimethyl-4-dimethyl-aminopyridine	See: Pyridine, 3,5-dimethyl-4(*N,N*-dimethyl-amino)-									
N,N'-Dimethylethylene-diamine	See: Ethane, 1,2-diamino-*N,N'*-dimethyl-									
N,N'-Dimethylethylene-diamine-*N,N'*-diacetic acid	See: Ethane, 1,2-diamino-*N,N'*-dimethyl-*N,N'*-diacetic acid-									
β,β-Dimethylglutaric acid	See: Pentanedoic acid, 3,3-dimethyl-									
N,N-Dimethylglycine	See: Glycine, *N,N*-dimethyl-									
3,5-Dimethyl-4-methyl-aminopyridine	See: Pyridine, 3,5-dimethyl-4(*N*-methylamino)-									
2,7-Dimethyl-1,8-phenan-throline	See: 1,8-Phenanthroline, 2,7-dimethyl-									
3,5-Dimethylpyridine	See: Pyridine, 3,5-dimethyl-									
1,1-Dinitropropane	See: Propane, 1,1-dinitro-									
3,6-Dioxaoctane-1,8-diamine	See: 3,6-Octanedione, 1,8-diamino-									
meso-1,2-Diphenylethylene-diamine	See: Ethane, 1,2-diamino-1,2-diphenyl (*meso*)-									
rac-1,2-Diphenylethylene-diamine	See: Ethane, 1,2-diamino-1,2-diphenyl-(*racemic*)-									

HEATS OF PROTON IONIZATION, pK, AND RELATED THERMODYNAMIC QUANTITIES (continued)

Compound	Formula, Synonyms	ΔH (kcal/mole)	pK	ΔS cal/deg-mole	T°C	Method	Conditions	Reference	Other References	Remarks
Diphosphate	See: Phosphoric acid. pyro-									
N,N'-Di-i-propylethylenediamine	See: Ethane, 1,2-diamino-N-N'-di-i-propyl-									
N,N'-Di-n-propylethylenediamine	See: Ethane, 1,2-diamino-N,N'-di-n-propyl-									
Disulfide										
—, 3,3'-dicarboxyl-dipropyl-	$[S(CH_2)_3COOH]_2$ Dithiodipropionic acid	6.300	10.35	−26	25	T	$\mu = 0.03\text{--}0.15$	162	—	—
Dithiodipropionic acid	See: Disulfide, 3,3'-dicarboxyldipropyl-									
Dodecaborane										
—, 1,12-dicarboxy-	$[1,12\text{-}B_{12}H_{10}(COOH)_2]^{-2}$	2.15	9.07 (pK$_1$)	−34.4	25	C	$\mu = 0$	361	—	—
		2.30	10.23 (pK$_2$)	−39.1	25	C	$\mu = 0$	361	—	—
Durohydroquinone	See: Benzene, 1,4-dihydroxy-2,3,5,6-tetramethyl-									
EDTA	See: Ethane,1,2-diamino-N,N,N',N'-tetraacetic acid									
EGTA	See: Acetic acid [ethylene bis-(oxyethylene-nitrilo)]tetra-									
Ephedrine(-) enantiomorph	CH(OH)CH(CH$_3$)NH$_2^+$CH$_3$	10.837	9.39	−6.57	25	C	$\mu = 0$	362	166–167	—
Pseudo-ephedrine (+) enantiomorph										
Ethane	C$_2$H$_6$	11.037	9.53	−6.57	25	C	$\mu = 0$	362	—	—
—, amino-	Ethylamine	13.71	10.63	−2.7	25	C	$\mu = 0$	330	65	—
		13.58	10.158 (40°)	−2.4	20–40	T	$\mu = 0.00$	64	66, 168	3.b.
		13.60	10.153 (40°)	−2.9			$\mu = 0.05$		363	—
		13.63	10.150 (40°)	−3.0			$\mu = 0.10$			—
		13.76	10.134 (40°)	−3.1			$\mu = 0.20$			—
—, 1-amino-2-methoxy-	CH$_3$OCH$_2$CH$_2$NH$_3^+$ 2-Methoxyethylamine	11.7	9.28 (30°)	−4	10–40	T	$\mu = 0$	62	—	3.b.
—, 1-amino-2(methylthio)-	NH$_3^+$CH$_2$CH$_2$SCH$_3$ 2(Methylthio)ethylamine 2-Aminoethylmethyl sulfide	12.3	9.18 (30°)	1	10–40	T	$\mu = 0$	169	—	3.b.
—, 1,2-diamino-	1,2-Ethanediamine Ethylenediamine	11.03	7.13 (pK$_1$)	4.4	25	C	$\mu = 0$	171	144	—
		10.940	—	—	25	C	0.1 M KCl	172	145	—
		10.89	7.32 (pK$_1$)	3.0	25	C	0.3 M NaClO$_4$	353	330	—

HEATS OF PROTON IONIZATION, pK, AND RELATED THERMODYNAMIC QUANTITIES (continued)

Compound	Formula. Synonyms	ΔH (kcal/mole)	pK	ΔS cal/deg-mole	T°C	Method	Conditions	Reference	Other References	Remarks
Ethane—(Continued)										
		10.60	7.44 (pK₁)	1.5	25	C	1 M KCl	67	169	—
		11.94	9.910 (pK₂)	−5.3	25	C	μ = 0	171	173	—
		11.910	—		25	C	0.1 M KCl	172	174	—
		12.00	9.90 (pK₂)	−5.0	25	C	0.3M NaClO₄	353	—	—
—, 1,2-diamino-N,N'-di-n-butyl-	N,N'-di-n-Butylethylenediamine	12.20	10.19 (pK₂)	−5.7	25	C	1 M KCl	67	—	—
		10.7	7.46 (25°, pK₁)	1.74	0–25	T	0.5 M KNO₃	146	—	3.a.b.
—, 1,2-diamino-N,N'-diethyl-	N,N'-Diethylethylenediamine	11.0	10.19 (25°, pK₂)	−9.73	0–25	T	0.5 M KNO₃	146	—	3.a.b.
		12.4	7.70 (25°, pK₁)	6.37	0–25	T	0.5 M KNO₃	146	—	3.a.b.
		8.9	10.46 (25°, pK₂)	−18.0	0–25	T	0.5 M KNO₃	146	—	3.a.b.
—, 1,2-diamino-N,N'-dimethyl-	N,N'-Dimethylethylenediamine	12.32	7.47 (pK₁)	7.17	25°	T	μ = 0	364	—	—
		12.4	7.47 (25°, pK₁)	7.41	0–25	T	0.5 M KNO₃	146	—	3.a.b.
		8.9	10.29 (25°, pK₂)	−17.24	0–25	T	0.5 M KNO	146	—	3.a.b.
—, 1,2-diamino-N,N'-dimethyl-N,N'-diacetic acid-	N,N'-Dimethylethylenediamine-N,N'-diacetic acid	3.68	6.294 (pK₃)	−15.3	0	T	μ = 0	32	—	—
		3.68	6.169	−15.2	10				—	—
		3.69	6.047	−15.1	20				—	—
		3.70	5.926	−14.9	30				—	—
		3.72	5.803	−14.7	40				—	—
		6.78	10.446 (pK₄)	−23.1	0	T	μ = 0	32	—	—
		6.82	10.268	−22.9	10				—	—
		6.83	10.068	−22.8	20				—	—
		6.85	9.882	−22.6	30				—	—
		6.88	9.684	−22.3	40				—	—
—, 1,2-diamino-1,2-diphenyl-(meso)-	meso-1,2-Diphenylethylenediamine	11.6	4.78 (25°, pK₁)	17.10	0–25	T	0.005 M Ba(ClO₄)₂	146	—	3.a.b. 50 wt % dioxane
		11.0	7.85 (25°, pK₂)	1.01	0–25	T	0.005 M Ba(ClO₄)₂	146	—	—
—, 1,2-diamino-1,2-diphenyl-(racemic)-	rac-1,2-Diphenylethylenediamine	9.7	3.95 (25°, pK₁)	14.46	0–25	T	0.005 M Ba(ClO₄)₂	146	—	3.a.b. 50 wt % dioxane
		11.3	8.09 (25°, pK₂)	0.872	0–25	T	0.005 M Ba(ClO₄)₂	146	—	—
—, 1,2-diamino-N,N'-di-i-propyl-	N,N'-di-i-Propylethylenediamine	10.0	7.59 (25°, pK₁)	−1.21	0–25	T	0.5 M KNO₃	146	—	3.a.b.
		10.7	10.40 (25°, pK₂)	−11.74	0–25	T	0.5 M KNO₃	146	—	3.a.b.
—, 1,2-diamino-N,N'-di-n-propyl-	N,N'-di-n-Propylethylenediamine	9.1	7.53 (25°, pK₁)	−4.02	0–25	T	0.5 M KNO₃	146	—	3.a.b.
		10.4	10.27 (25°, pK₂)	−12.07	0–25	T	0.5 M KNO₃	146	176	3.a.b.
—, 1,2-diamino-N,N'-di(2-aminoethyl)-	⁺NH₃CH₂CH₂NH₂CH₂CH₂NH₂CH₂CH₂NH₃⁺ Triethylenetetramine	6.83	3.24 (pK₁)	8.1	25	C	0.1 M KCl	175	—	—
		9.53	6.54 (pK₂)	2.0	25	C	0.1 M KCl	175	—	—
		11.27	9.06 (pK₃)	−3.7	25	C	0.1 M KCl	175	—	—
		11.01	9.75 (pK₄)	−7.8	25	C	0.1 M KCl	175	—	—

HEATS OF PROTON IONIZATION, pK, AND RELATED THERMODYNAMIC QUANTITIES (continued)

Compound	Formula, Synonyms	ΔH (kcal/mole)	pK	ΔS cal/deg-mole	T°C	Method	Conditions	Reference	Other References	Remarks
Ethane—(Continued)										
—, 1,2-diamino-N-hydroxy-N,N',N'-triacetic acid	—	-1.12	2.63 (pK$_3$)	-15.8	25	C	$\mu = 0.1$ (KCl)	365	—	—
		3.09	5.37 (pK$_4$)	-14.2	25	C	$\mu = 0.1$ (KCl)	365	—	—
		6.65	9.80 (pK$_5$)	-22.5	25	C	$\mu = 0.1$ (KCl)	365	—	—
—, 1,2-diamino-N-hydroxyethyl-N,N',N'-triacetic acid	(CH$_2$COOH)$_2$NHCH$_2$CH$_2$NH$^+$(CH$_2$CH$_2$OH)(CH$_2$COOH) N-Hydroxyethylethylenediamine-N,N',N'-triacetic acid	-0.34	2.39 (25°, pK$_4$)	-12.1	15–40	T	$\mu = 0.1$	177	—	3.b.
—, 1,2-diamino-N-methyl-	NH$_3^+$CH$_2$CH$_2$NH$_2^+$CH$_3$ N-Methylethylenediamine	2.83	5.37 (25°, pK$_5$)	-15.1	15–40	T	$\mu = 0.1$	177	—	3.b.
		7.29	9.93 (25°, pK$_6$)	-21.0	15–40	T	$\mu = 0.1$	177	—	3.b.
—, 1,2-diamino-1-phenyl-	NH$_3^+$CH$_2$CH(C$_6$H$_5$)NH$_3^+$	10.34	7.26 (pK$_1$)	1.5	25	C	$\mu = 0.5$	174	169	—
		11.25	10.14 (pK$_2$)	-8.7	25	C	$\mu = 0.5$	174	—	—
		10.62	5.757 (pK$_1$)	9.3	25	C	$\mu = 0$	144	—	—
—, 1,2-diamino-N,N,N',N'-tetraacetic acid	(CH$_2$COOH)$_2$NHCH$_2$CH$_2$NH$^+$(CH$_2$COOH)$_2$ Ethylenediaminetetraacetic acid EDTA	11.10	8.55 (pK$_2$)	-1.9	25	C	$\mu = 0$	144	—	—
		-0.490	1.55 (pK$_2$)	-8.8	20	C	$\mu = 0.1$	178	160	—
		-0.180	1.99 (pK$_3$)	-9.8	20	C	$\mu = 0.1$	178	179	—
		-0.180	(pK$_3$)	—	25	C	$\mu = 0.44$	367	180	—
		-1.430	2.67 (pK$_4$)	-17.1	20	C	$\mu = 0.1$	178	366, 367	—
		-1.43	(pK$_4$)	—	25	C	$\mu = 0.44$	367	—	—
		3.69	6.236 (pK$_5$)	-16.4	30	T	$\mu = 0$	32	—	—
		4.390	6.16 (pK$_5$)	-13.2	20	C	$\mu = 0.1$	178	—	—
		3.61	6.273 (pK$_5$)	-16.7	25	C	$\mu = 0.03$–0.10	330	—	—
		5.34	10.883 (pK$_6$)	-32.2	30	T	$\mu = 0$	32	—	—
		5.690	10.26 (pK$_6$)	-27.5	20	C	$\mu = 0.1$	178	—	—
		4.35	10.948 (pK$_6$)	-35.6	25	C	$\mu = 0.04$–0.17	330	—	—
—, 1,2-diamino-N,N,N',N'-tetrakis-(2-aminoethyl)-	[(H$_3^+$NCH$_2$CH$_2$)$_2$NHCH$_2$]$_2$ N,N,N',N'-Tetrakis(2-aminoethyl)-ethylenediamine	4.5	1.81 (pK$_1$)	9.0	25	C	$\mu = 0.1$ (KCl)	76	—	—
		12.00	11.47 (pK$_2$)	1.8	25	C	$\mu = 0.1$ (KCl)	368	—	3.d.
		13.15	12.23 (pK$_3$)	3.1	25	C	$\mu = 0.1$ (KCl)	368	—	3.d.
		11.45	13.04 (pK$_4$)	-5.3	25	C	$\mu = 0.1$ (KCl)	368	—	3.d.
		11.30	13.72 (pK$_5$)	-8.1	25	C	$\mu = 0.1$ (KCl)	368	—	3.d.
—, 1,1-dinitro-	See: Ethane, 1,2-diamino-N,N'-di(2-amino-ethyl)-	3.540	5.21	-11.82	20	T	—	147	—	3.d.
Ethane,1,2-bis(2-amino-ethylamino)-	See: Ethane, 1,2-diamino-...									
1,2-Ethanediamine	See: Ethane, 1,2-diamino-Ethylenediamine									

HEATS OF PROTON IONIZATION, pK, AND RELATED THERMODYNAMIC QUANTITIES (continued)

Compound	Formula, Synonyms	ΔH (kcal/mole)	pK	ΔS cal/deg-mole	T°C	Method	Conditions	Reference	Other References	Remarks
Ethanedioic acid	See: Oxalic acid									
Ethanephosphonic acid										
—, 2-amino-	$^+NH_3CH_2CH_2PO_3H_2$ 2-Aminoethylphosphate	−0.452 11.04	5.84 (pK₂) 10.64 (pK₃)	−28.2 −11.7	25 25	T T	$\mu=0$ $\mu=0$	181 181	— —	3.d. 3.d.
Ethanethiol	CH_3CH_2SH Ethyl hydrosulfide Ethyl mercaptan Ethyl thioalcohol	6.42	10.61	−27.0	25	C	$\mu=0.015$	29	—	—
—, 2-acetylamino-	$CH_3CONH_2^+CH_2CH_2SH$ β-Acetylaminoethanethiol	6.26	9.92	−24.4	25	C	$\mu=0.015$	29	—	—
—, 2-amino-	β-Aminoethanethiol	7.43	8.23 (pK₁)	−12.7	25	C	$\mu=0.01$	29	—	—
—, 1,1-dimethyl-	t-Butylthiol	5.30	11.22	−33.6	25	C	$\mu=0.01$–0.10	29	—	—
—, 2-hydroxy-	β-Hydroxyethanethiol	6.21	9.72	−23.6	25	C	$\mu=0.015$	29	—	—
—, 1-methyl-	$(CH_3)_2CHSH$ Isopropylthiol	5.38	10.86	−31.6	25	C	$\mu=0.10$	29	—	—
Ethanethiolic acid	See: Acetic acid, thiolo-									
Ethanoic acid	See: Acetic acid									
Ethanol	CH_3CH_2OH	12.04	9.498	−3.1	25	C	$\mu=0$	330	356, 363	3.d.
Ethanol, 2-amino-	2-Hydroxyethylamine Ethanolamine Colamine	11.965	—	—	10	C	$C=0.06$	86	62	—
		12.050	—	—	20					
		12.070	9.45	−2.7	25					
		12.090	—	—	30					
		12.107	10.306	−2.84	0	T	$\mu=0$	182		
		12.102	10.132	−2.86	5					
		12.096	9.964	−2.88	10					
		12.090	9.803	−2.90	15					
		12.085	9.646	−2.92	20					
		12.079	9.495	−2.94	25					
		12.073	9.349	−2.96	30					
		12.068	9.208	−2.98	35					
		12.061	9.071	−3.00	40					
		12.055	8.939	−3.02	45					
		12.049	8.811	−3.04	50					
—, 2-amino-1-phosphoric acid-	—	0.335 0.143 −0.052	5.836 (pK₂) 5.832 5.832	−25.5 −26.2 −26.9	5 10 15	T	$\mu=0$	170	—	—

HEATS OF PROTON IONIZATION, pK, AND RELATED THERMODYNAMIC QUANTITIES (continued)

Compound	Formula, Synonyms	ΔH (kcal/mole)	pK	ΔS cal/deg-mole	T °C	Method	Conditions	Reference	Other References	Remarks
Ethanol-2-amino—(Continued)		−0.250	5.834	−27.5	20	—	—	—	—	—
		−0.452	5.838	−28.2	25	—	—	—	—	—
		−0.658	5.845	−28.9	30	—	—	—	—	—
		−0.866	5.854	−29.6	35	—	—	—	—	—
		−0.951	5.858	−29.9	37	—	—	—	—	—
		−1.079	5.865	−30.3	40	—	—	—	—	—
		−1.294	5.878	−31.0	45	—	—	—	—	—
		−1.513	5.893	−31.6	50	—	—	—	—	—
—, 2-dimethylamino-	See: Ethanol, 2-amino-									
—, 2-methylamino-	See: Amine, ethyl(2-hydroxy)-isopropyl-(2-hydroxy)-									
Ethanolamine		8.74	9.26	−13.1	25	C	$\mu = 0$	330	—	—
Ethanolisopropanol amine		11.06	9.88	−8.1	25	C	$\mu = 0$	330	—	—
Ethanthiolic acid	See: Acetic acid, thiolo-									
Ether										
—, di(2-aminoethyl)-	$(\overset{+}{N}H_3CH_2CH_2)_2O$ *bis*-2(2-aminoethylether)	13.2	8.62 (30°, pK_1)	4	10–40	T	$\mu = 0$	62	—	3.b.
		11.7	9.59 (30°, pK_2)	−5	10–40	T	$\mu = 0$	62	—	3.b.
Ethylamine	See: Ethane, amino-									
3-Ethyl-4-aminopyridine	See: Pyridine, 4-amino-3-ethyl-									
α-N-Ethylamine oxime	See: 2-Butanone, 3-(ethylamino)-3-methyl-, oxime									
5-Ethyl-5-n-amylbarbituric acid	See: Barbituric acid, 5-ethyl-5-pentyl-									
3-Ethyl-4-dimethylamine-pyridine	See: Pyridine, 4(N,N-dimethylamino)-3-ethyl-									
Ethylenediamine	See: Ethane, 1,2-diamino-									
Ethylenediaminetetraacetic acid	See: Ethane, 1,2-diamino-N,N,N',N'-tetraacetic acid-									
Ethyl glycinate	See: Glycine, ethylester-									
Ethyl hydrosulfide	See: Ethanethiol									
Ethylenimine	See: Azirane									
Ethylisoamylmalonic acid	See: Hexanoic acid, 2-carboxyl-5-methyl-									
Ethyl mercaptan	See: Ethanethiol									
3-Ethyl-4-methylamino-pyridine	See: Pyridine, 3-ethyl-4-methylamino-									
5-Ethyl-5(α-methylbutyl)-barbituric acid	See: Barbituric acid, 5-ethyl-5-(α-methylbutyl)-									

HEATS OF PROTON IONIZATION, pK, AND RELATED THERMODYNAMIC QUANTITIES (continued)

Compound	Formula, Synonyms	ΔH (kcal/mole)	pK	ΔS cal/deg-mole	T°C	Method	Conditions	Reference	Other References	Remarks
5-Ethyl-5-phenylbarbituric acid	See: Barbituric acid, 5-ethyl-5-phenyl-									
3-Ethylpyridine	See: Pyridine, 3-ethyl-									
Ethyl thioalcohol	See: Ethanethiol									
Factor B-histamine complex	(amino)	4.7	4.49	−5	25	T	$\mu = 0.11$	92	—	—
	(amino)	11.7	11.00	−11	25	T	$\mu = 0.11$	92	—	—
Factor B-imidazole complex		12.1	11.39	−12	25	T	$\mu = 0.16$	92	—	—
Ferrihemoproteins	$Fe^+ \cdot OH_2 = FeOH + H^+$ where Fe = hematin Fe(III)									
	Soya-bean-nodule pigments									
Slow component		2.20	8.49	−31.5	25	T	$\mu = 0$	183		
Fast component		2.60	8.57	−30.5	25	T	$\mu = 0$	183		
Horse blood		3.91	8.81	−27.2	25	T	$\mu = 0$	183		
ferrihemoglobin		5.75	8.97	−21.6	25	T	$\mu = 0$	183		
Chironomus blood ferrihemoglobin		3.8	8.19	−25	25	T	$\mu = 0$	183		
Ferrimyoglobin complex		11.4	10.34	−9	25	T	$\mu = 0$	92		—
	[structure: ring system with NH_3^+, positions 1–9]	5.7	4.21 (20°)	0.3	0–20	T	C = 0.005	85	—	3.b. 50 wt.% ethanol
Fluorophenol	See: Phenol, fluoro-									
Formic acid	HCOOH	0.01	3.751	−17.1	25	C	$\mu = 0$	1	9–11	—
	Methanoic acid	0.931	3.786	−13.9	0	T	$\mu = 0$	184	13, 185	—
		0.755	3.772	−14.5	5				389	—
		0.573	3.762	−15.2	10					—
		0.384	3.757	−15.8	15					—
		0.189	3.753	−16.5	20					—
		−0.013	3.751	−17.2	25					—
		−0.221	3.752	−17.9	30					—
		−0.436	3.758	−18.5	35					—
		−0.657	3.766	−19.3	40					—
		−0.884	3.773	−20.0	45					—
		−1.118	3.782	−20.8	50					—
		−1.358	3.793	−21.5	55					—
		−1.605	3.809	−22.2	60					—
		−0.02	—	−17.6	25	T	0	5	—	Conditions refer to wt.% dioxane
		−0.36	—	−17.9			20			—
		−1.07	—	−19.7			45			—
		−1.47	—	−27.1			70			—
		−2.51	—	−31.8			82			—

HEATS OF PROTON IONIZATION, pK, AND RELATED THERMODYNAMIC QUANTITIES (continued)

Compound	Formula, Synonyms	ΔH (kcal/mole)	pK	ΔS cal/deg-mole	T°C	Method	Conditions	Reference	Other References	Remarks
Formonitrile	See: Hydrocyanic acid									
Fructose	—									
Fumaric acid	HOOC—CH=CH—COOH *trans*-Butendioic acid	8.2	12.27	−28.6	25	C	$\mu = 0$	186		—
		0.11	3.095 (pK$_1$)	−13.8	25	C	$\mu = 0$	1	59	—
		−0.68	4.602 (pK$_2$)	−23.3	25	C	$\mu = 0$	1		—
2-Furancarboxylic acid	α-Furoic acid Pyromucic acid	−2.0	3.164	−21.1	25	T	$\mu = 0$	187		—
		−2.0	3.200	−21.3	30	—	—			—
		−2.1	3.216	−21.5	35	—	—			—
		−2.2	3.239	−21.7	40	—	—			—
α-Furoic acid	See: 2-Furancarboxylic acid									
2-Furoic acid	See: 2-Furancarboxylic acid									
Galactose	(Fischer projection)	9.0	12.48	−26.9	25	C	$\mu = 0$	186		—
Gallic acid	See: Benzoic acid, 3,4,5-trihydroxy-									
Glucose	Dextrose (Fischer projection)	7.7	12.46 (pK$_3$)	−31.3	25	C	$\mu = 0$	186		—
—, 2-amino-	α-glucosamine	10.000	7.526	−1.375	25	T	$\mu = 0.05$	369	188	—
—, 3-amino-	β-glucosamine	10.050	7.703	0.504	25	T	$\mu = 0.05$	369		—
—, 2-deoxy-	(Fischer projection)	8.2	12.52	−29.7	25	C	$\mu = 0$	186		—
—, 1-phosphate		0.474	6.506	−28.1	5	T	$\mu = 0$	190		—
		0.254	6.500	−28.85	10	—	—			—
		0.029	6.499	−29.6	15	—	—			—
		−0.199	6.500	−30.4	20	—	—			—
		−0.431	6.504	−31.2	25	—	—			—
		−0.668	6.510	−32.0	30	—	—			—
		−0.908	6.519	−32.8	35	—	—			—
		−1.005	6.524	−33.1	37	—	—			—
		−1.152	6.531	−33.6	40	—	—			—
		−1.400	6.545	−34.35	45	—	—			—
		−1.652	6.561	−35.1	50	—	—			—
—, 6-phosphate		8.4	11.71	−25.0	25	C	$\mu = 0$	186		—

HEATS OF PROTON IONIZATION, pK, AND RELATED THERMODYNAMIC QUANTITIES (continued)

Compound	Formula, Synonyms		ΔH (kcal/mole)	pK	ΔS cal/deg-mole	T °C	Method	Conditions	Reference	Other References	Remarks
Glutamic acid	HOOCCH($\overset{+}{N}H_3$)CH$_2$CH$_2$COOH		0.947	2.186 (pK$_1$)	−6.6	5	T	$\mu = 0$	191	54, 98	—
	α-Aminoglutaric acid		−0.064	2.162	−10.1	25				192	—
	2-Aminopentanedioic acid		−1.427	2.204	−14.5	50					—
			1.083	4.326 (pK$_2$)	−15.9	5	T	$\mu = 0$	191		—
			0.373	4.272	−18.3	25					—
			−0.584	4.282	−21.4	50					—
			9.578	9.358 (pK$_3$)	−10.7	25					—
—, lysyl-		(carboxyl)	0.750	2.93 (25°, pK$_1$)	−10.9		T	$\mu = 0$	191		—
		(carboxyl)	0.0	4.47 (25°, pK$_2$)	−20.40		T	$\mu = 0.01$	94		3.a.b.
		(amino)	10.50	7.75 (25°, pK$_3$)	−0.2						—
		(amino)	11.95	10.50 (25°, pK$_4$)	−8.0						—
Glutaric acid	See: Pentanedioic acid										
Glutaric acid, 3-methyl-	See: Pentanedioic acid, 3-methyl-										
Glycerol	HOCH$_2$CHOHCH$_2$OH										
—, 1-phosphoric acid	Glycerol-1-phosphate		0.185	6.642	−29.7	5	T	$\mu = 0$	193		—
			−0.045	6.641	−30.5	10					—
			−0.274	6.643	−31.4	15					—
			−0.510	6.648	−32.2	20					—
			−0.749	6.656	−33.0	25					—
			−0.993	6.666	−33.8	30					—
			−1.241	6.679	−34.6	35					—
			−1.341	6.685	−34.9	37	T		193		—
			−1.493	6.695	−35.4	40					—
			−1.749	6.713	−36.2	45					—
			−2.009	6.733	−37.0	50					—
—, 2-phosphoric acid	Glycerol-2-phosphate		−1.396	1.223 (pK$_1$)	−10.6	5	T	$\mu = 0$	194		—
			−1.760	1.245	−11.9	10					—
			−2.132	1.271	−13.2	15					—
			−2.509	1.301	−14.5	20					—
			−2.893	1.335	−15.8	25					—
			−3.284	1.372	−17.1	30					—
			−3.681	1.413	−18.4	35					—
			−3.842	1.430	−19.0	37					—
			−4.085	1.457	−19.7	40					—
			−4.495	1.504	−21.0	45					—
			−4.912	1.554	−22.3	50					—
			0.630	6.657 (pK$_2$)	−28.2	5	T	$\mu = 0$	194		—
			0.376	6.650	−29.1	10					—
			0.118	6.646	−30.0	15					—
			−0.145	6.646	−30.9	20					—

HEATS OF PROTON IONIZATION, pK, AND RELATED THERMODYNAMIC QUANTITIES (continued)

Compound	Formula, Synonyms	ΔH (kcal/mole)	pK	ΔS cal/deg-mole	T°C	Method	Conditions	Reference	Other References	Remarks
Glycerol—(Continued)										
		−0.412	6.650	−31.8	25	—	—	—	—	—
		−0.684	6.657	−32.7	30	—	—	—	—	—
		−0.961	6.666	−33.6	35	T	$\mu = 0$	194	—	—
		−1.073	6.71	−34.0	37	—	—	—	—	—
		−1.242	6.679	−34.5	40	—	—	—	—	—
		−1.528	6.694	−35.4	45	—	—	—	—	—
		−1.818	6.712	−36.3	50	—	—	—	—	—
		1.41	2.397	−5.8	10	C	$\mu = 0$	324	6, 8	—
Glycine	$H_3\overset{+}{N}CH_2COOH$ Aminoacetic acid Glycocoll	0.98	2.351	−7.5	25	C	$\mu = 0$	1	10, 11 53, 54 71, 171 197, 348	—
		0.47	2.327	−9.2	40	C	$\mu = 0$	324	198, 199, 200, 201, 202, 203, 205	—
		1.670	—	—	10	T	$\mu = 0$	10		—
		1.180	—	—	25	—	—	—	—	
		0.630		—	40	—	—	—	—	
		0.958	2.350 (pK$_1$)	−7.54	25	T	0.000	196	—	Conditions refer to $\mu = 0$ for glycine and indicated M for NaCl
		0.977	2.353	−7.49	—	—	0.100	—	—	
		1.005	2.359	−7.43	—	—	0.300	—	—	
		1.083	2.385	−7.28	—	—	0.725	—	—	
		1.201	2.430	−7.09	—	—	1.250	—	—	
		1.388	2.501	−6.79	—	—	2.000	—	—	
		1.635	2.605	−6.44	—	—	3.000	—	—	
		1.16	2.358 (pK$_1$)	−6.9	25	T	0	5	—	Conditions refer to wt. % dioxane
		1.18	2.635	−8.1	—	—	20	—	—	
		0.99	3.108	−10.9	—	—	45	—	—	
		0.83	3.974	−15.4	—	—	70	—	—	
		10.640	9.779 (pK$_2$)	−8.90	25	C	—	325	—	3.d.
		10.76	9.77	−8.6	25	—	—	—	—	
		10.22	9.46	−10.7	40	—	—	—	—	

HEATS OF PROTON IONIZATION, pK, AND RELATED THERMODYNAMIC QUANTITIES (continued)

Compound	Formula. Synonyms	ΔH (kcal/mole)	pK	ΔS cal/deg-mole	T°C	Method	Conditions	Reference	Other References	Remarks
Glycine—(Continued)										
—, N-acetyl-	CH$_3$CONHCH$_2$COOH	10.550	9.777 (pK$_2$)	−9.4	25	T	0	198	—	Conditions refer to $\mu = 0$ for glycine and indicated M for NaCl
		10.749	9.603	−7.9	—	—	0.1	—	—	
		10.879	9.577	−7.3	—	—	0.3	—	—	3.d.
—, carbamoyl-	NH$_2^+$CONHCH$_2$COOH	−0.150	3.6698	−17.3	25	T	$\mu = 0$	6	—	
		−0.210	3.68	−17.48	25	C	—	325	—	
—, N,N-di(2-hydroxyethyl)-	H$_3$N$^+$CH$_2$COOH	0.290	3.8758 (pK$_1$)	−16.76	25	T	$\mu = 0$	58	—	
		6.278	8.3335 (pK$_2$)	−17.08	25	T	$\mu = 0$	206	—	
—, N,N-dimethyl-	Dimethylaminoacetic acid	−0.155	2.146 (pK$_1$)	−9.3	25	T	$\mu = 0$	35	—	
		7.38	10.34 (pK$_2$)	−20.8	5	T	$\mu = 0$	71	—	
		7.52	10.14	−20.3	15						
		7.66	9.94	−19.8	25						
		7.80	9.76	−19.3	35						
		7.95	9.59	−18.9	45						
—, ethyl ester	Ethyl glycinate	7.60	7.75	−9.96	25	T	$\mu = 0$	207	6, 53, 94	
—, N-glycyl-	Glycylglycine Diglycine	0.032	3.08 (pK$_1$)	−13.9	25	C	$\mu = 0.1$	358	199, 208	
		1.190	3.201	−10.4	1	T	$\mu = 0$	97		
		0.862	3.166	−11.5	12.5						
		0.391	3.126	−12.9	25						
		−0.128	3.141	−14.8	37.5						
		−0.736	3.159	−16.7	50						
		10.6	8.08 (pK$_2$)	−1.5	25	C	$\mu = 0.1$	358	97	
		10.660	8.594	−2.0	12.5						
		10.600	8.252	−2.0	25						
		10.400	7.948	−2.8	37.5						
		10.060	7.668	−4.0	50						
—, N-glycyl-copper complex	—	6.9	4.06	4.5	25	C	$\mu = 0.1$	358	—	Reaction: CuGG$^+$ = CuA + H$^+$
—, glycylglycyl-	Triglycine	0.2	3.18 (pK$_1$)	−13.9	25	C	$\mu = 0.1$	358	199	
		10.1	7.86 (pK$_2$)	−2.1	25	C	$\mu = 0.1$	358	—	
—, glycylglycyl-copper complex	—	7.5	5.06	2.0	25	C	$\mu = 0.1$	358	—	Reaction: CuGGG$^+$ → CuA + H$^+$

HEATS OF PROTON IONIZATION, pK, AND RELATED THERMODYNAMIC QUANTITIES (continued)

Compound	Formula, Synonyms	ΔH (kcal/mole)	pK	ΔS cal/deg-mole	T°C	Method	Conditions	Reference	Other References	Remarks
Glycine—(Continued)										
—, glycylglycyl-copper complex	—	7.4	6.78	−6.2	25	C	$\mu = 0.1$	358	—	CuA→CuB⁻ + H⁺; A = GGG; −H
—, N-glycylglycylglycyl-	—	0.18 / 10.40 / 7.5	3.17 (pK₁) / 7.87 (pK₂) / 5.40	−13.9 / −1.2 / 0.4	25 / 25 / 25	C / C / C	$\mu = 0.10$	410 / 410 / 410	—	Reaction: CuGGG⁺ →CuA + H⁺
—, N-glycylglycylglycyl-copper complex	—	6.6	6.80	−9.0	25	C	$\mu = 0.10$	410	—	Reaction: CuA→ CuB⁻ + H⁺
		8.9	9.15	−12.0	25	C	$\mu = 0.10$	410	—	Reaction: CuB⁻→ CuC²⁻ + H⁺
—, histidyl-	(carboxyl) (imidazole) (amino)	0.300 / 7.50 / 10.8	2.40 / 5.80 / 7.82	−9.98 / −1.4 / 0.5	0–25 / 0–25 / 0–25	T / T / T	$\mu = 0.01$	94 / 94 / 94	—	3.a.b. 3.a.b. 3.a.b.
—, phenylalanyl-	(carboxyl) (amino)	0.68 / 10.0	3.10 (25°) / 7.71 (25°)	−11.9 / −1.7	0–25 / 0–25	T / T	$\mu = 0.01$	94	—	3.a.b. 3.a.b.
—, N-propionyl-	CH₃CH₂CONHCH₂COOH	−0.140	3.717	−17.5	25	T	$\mu = 0$	55	—	—
Glycocoll	See: Glycine									
Glycolic acid	See: Acetic acid, hydroxy-									
Glycylalanine	See: Alanine, glycyl-									
Glycylasparagine	See: Asparagine, glycyl-									
Glycylglycine	See: Glycine, N-glycyl-									
Glycylserine	See: Serine, glycyl-									
Glyoxaline	See: Imidazole									
Glyoxylic acid	See: Acetic acid, oxo-									
Glysyl-α-amino-n-butyric acid	See: Butanoic acid, glycyl-2-amino-									
Guaiacol	See: Phenol, 2-methoxy-									
Guanine	2-Aminohydroxanthine 6-Hydroxy-2-aminopurine	10.1	9.42	−9.1	25	T	$\mu = 0.1$	39	—	—

HEATS OF PROTON IONIZATION, pK, AND RELATED THERMODYNAMIC QUANTITIES (continued)

Compound	Formula, Synonyms	ΔH (kcal/mole)	pK	ΔS cal/deg-mole	T°C	Method	Conditions	Reference	Other References	Remarks
Guanosine		1.0	1.6	−4.0	25	T	$\mu = 0.1$	39	40[a], 209	—
		3.6	9.24	−13.0	25	T	$\mu = 0.1$	39	—	—
	9-D-Ribosidoguanine Vernine									
—, 2'-deoxy-	—	1.910	2.5	−5.0	25	C	$\mu = 0.1$	40[a]	—	—
—, 5'-diphosphate	—	−1.48	7.19 (pK$_4$)	−37.7	25	T	$\mu = 0$	46	—	—
—, 5'-monophosphate	—	−1.45	6.66 (pK$_3$)	−35.3	25	T	$\mu = 0$	46	—	—
—, 5'-monophosphate-2'-deoxy-	—	0.140	2.9	−12.8	25	C	$\mu = 0.1$	40[a]	—	—
—, 5'-triphosphate	—	−1.75	7.65 (pK$_5$)	−40.8	25	T	$\mu = 0$	46	—	—
		3.85	8.55	−27.0	25	T	—	370	—	3.d.
Haemin-water (in metmyglobin)										
Haemoglobin A (human)	—	3.400	8.81	−28.7	20	T	$\mu = 0$	371	—	
Haemoglobin C (human)	—	4.880	8.66 (20°)	−23.0	20	T	$\mu = 0$	371	—	
Haemoglobin S (human)	—	4.090	8.74	−26.0	20	T	$\mu = 0$	371	—	
HDTA	See: Acetic acid (hexamethylenedinitrilo)tetra-									
H$_5$DTPA	See: Acetic acid [(carboxymethylimino)tetra-(ethylenenitrilo)]tetra-									
HEDTA	See: Ethane, 1,2-diamino-N-hydroxy-N,N',N'-triacetic acid									
Helianthin	See: Methyl orange									
Heptanedioic acid	HOOC(CH$_2$)$_5$COOH Pimelic acid	−0.33	4.484 (pK$_1$)	−21.6	25	C	$\mu = 0$	1	—	
		−0.93	5.424 (pK$_2$)	−27.9	25	C	$\mu = 0$	1	—	
2,4-Heptanedione	—	4.1	8.43 (25°)	−25	5–45	T	—	372	—	3.b.d. Enol form
		2.5	9.15 (25°)	−33	5–45	T	—	—	—	3.b.d. Keto form
4,6-Heptanedione,3-methyl-	—	3.1	8.52 (25°)	−29	5–45	T	—	372	—	3.b.d. Enol form
		1.5	9.10 (25°)	−36	5–45	T	—	—	—	3.b.d. Keto form

HEATS OF PROTON IONIZATION, pK, AND RELATED THERMODYNAMIC QUANTITIES (continued)

Compound	Formula, Synonyms	ΔH (kcal/mole)	pK	ΔS cal/deg-mole	T°C	Method	Conditions	Reference	Other References	Remarks
Heptanoic acid										
—, 7-amino-	ω-Aminoheptanoic acid	−0.47	4.502	−22.2	25	C	$\mu = 0$	324	—	—
Hexahydroaniline	See: Cyclohexane, amino-									
Hexahydrobenzoic acid	See: Cyclohexane, carboxylic acid-									
Hexahydrophthalic acid	See: 1,2-Dicarboxylic acid (cis)-									
Hexahydropyrazine	See: Piperazine									
Hexamethylenediamine	See: Hexane, 1,6-diamino-									
Hexamethylenedinitrilo-tetraacetic acid	See: Acetic acid, hexamethylenedinitrilo-tetra-									
2,3,4,5,6,7-Hexamethyl-1,8-phenanthroline	See: 1,8-Phenanthroline, 2,3,4,5,6,7-hexa-methyl-									
Hexane										
—, 1,6-diamino-	$CH_3(CH_2)_4CH_3$ Hexamethylenediamine	13.820	9.830 (pK₁)	1.3	25	T	$\mu = 0$	173	—	—
		11.360	(pK₁)	—	0	T	$\mu = 0$	173	—	—
		13.910	10.930 (pK₂)	−3.3	25	T	$\mu = 0$	173	—	—
		11.410	(pK₂)	—	0	T	$\mu = 0$	173	—	—
—, 1,1-dinitro-		4.02	5.386 (25°)	−11.08	5-60	T	—	165	—	3.b.d.
Hexanedioic acid	$HOOC(CH_2)_4COOH$ Adipic acid	−0.30	4.418 (pK₁)	−21.5	25	C	$\mu = 0$	1	—	—
		−0.64	5.412 (pK₂)	−26.9	25	C	$\mu = 0$	1	210	—
2,4-Hexanedione		4.8	8.49 (25°)	−22	5-45	T	—	372	—	3.b.d. Enol form
		3.1	9.32 (25°)	−31	5-45	T	—	372	—	3.b.d. Keto form
3,5-Hexanedione										
—, 2,2-dimethyl-	$(CH_3)_3CCOCH_2COCH_3$	3.5	10.01	−55	25	T	—	211	—	3.b.d.
—, 2-methyl-	$(CH_3)_2CHCOCH_2COCH_3$	2.6	9.44	−50	25	T	—	211	—	3.b.d.
		3.8	8.66 (25°)	−26	5-45	T	—	372	—	3.b.d. Enol form
		2.0	9.31 (25°)	−35	5-45	T	—	372	—	3.b.d. Keto form
Hexanoic acid	$CH_3(CH_2)_4COOH$ n-Caproic acid	−0.197	—	—	10	C	—	122	13	—
		−0.493	—	—	20			—	—	—
		−0.644	4.872	−24.5	25		$\mu = 0$	—	—	—
		−0.793	—	—	30			—	—	—
		−0.700	4.856	−24.6	25	T	$\mu = 0$	6	—	—
—, 2-amino-	$CH_3CH_2CH_2CH_2CH(\overset{+}{N}H_3)COOH$ DL-Norleucine α-Amino-n-caproic acid	0.92	2.363	−7.6	10	C	$\mu = 0$	324	—	—
		0.43	2.335	−9.2	25	C	$\mu = 0$	324	—	—
		−0.7	2.324	−10.7	40	C	$\mu = 0$	324	—	—
		1.300	2.394 (pK₁)	−6.3	1.0	T	$\mu = 0$	50	—	—

HEATS OF PROTON IONIZATION, pK, AND RELATED THERMODYNAMIC QUANTITIES (continued)

Compound	Formula. Synonyms	ΔH (kcal/mole)	pK	ΔS cal/deg-mole	T°C	Method	Conditions	Reference	Other References	Remarks
Hexanoic acid—(Continued)										
		0.980	2.356	−7.3	12.5	—	—	—	—	—
		0.560	2.335	−8.8	25	—	—	—	—	—
		0.060	2.324	−10.4	37.5	—	—	—	—	—
		−0.540	2.328	−12.3	50	—	—	—	—	—
—, 6-amino-	ε-Aminocaproic acid	10.950	10.546 (pK$_2$)	−8.3	1.0	T	$\mu = 0$	50	—	—
		11.030	10.190	−8.0	12.5	—	—	—	—	—
		11.050	9.834	−7.9	25	—	—	—	—	—
		10.840	9.513	−8.6	37.5	—	—	—	—	—
		10.540	9.224	−9.6	50	—	—	—	—	—
		0.10	4.392	−19.8	10	C	$\mu = 0$	324	—	—
		−0.32	4.373	−21.1	25	C	$\mu = 0$	—	—	—
		−0.79	4.388	−22.6	40	C	$\mu = 0$	—	—	—
		0.818	4.420 (pK$_1$)	−17.2	1	T	$\mu = 0$	97	—	—
		0.459	4.387	−18.2	12.5	—	—	—	—	—
		−0.008	4.373	−20.0	25	—	—	—	—	—
		−0.561	4.384	−21.9	37.5	—	—	—	—	—
		−1.204	4.410	−23.9	50	—	—	—	—	—
		13.120	11.666 (pK$_2$)	−5.6	1	T	$\mu = 0$	97	—	—
		13.390	11.244	−4.5	12.5	—	—	—	—	—
		13.560	10.804	−4.0	25	—	—	—	—	—
		13.620	10.406	−3.8	37.5	—	—	—	—	—
		13.550	10.036	−3.9	50	—	—	—	—	—
—, 2-carboxyl-5-methyl-	Ethylisoamylmalonic acid $CH_3CH_2C(COOH)_2(CH_2)_2CH(CH_3)CH_3$	−1.31	2.50 (pK$_1$)	−14.2	25	C	$\mu = 0$	1	—	—
		−0.36	7.31 (pK$_2$)	−34.7	25	C	$\mu = 0$	1	—	—
Hexoic acid	See: Hexanoic acid									
Histamine		10.1	5.87 (30°, pK$_1$)	7	10–40	T	$\mu = 0$	212	168	3.b.
		6.9	5.784 (pK$_1$)	−3	25	T	$\mu = 0$	373	—	—
		7.4	5.595	−2	37	T	$\mu = 0$	—	—	—
		8.1	5.358	0	50	T	$\mu = 0$	—	—	—
		13.4	9.70 (30°, pK$_2$)	0	10–40	T	$\mu = 0$	212	—	3.b.
		11.0	9.756 (pK$_2$)	−8	25	T	$\mu = 0$	373	—	—
		11.9	9.386	−5	37	T	$\mu = 0$	—	—	—
		12.9	9.047	−1	50	T	$\mu = 0$	—	—	—

Histamine structure:

$$HC{-}N{=}CH,\quad C{-}NH,\quad CH_2,\quad CH_2,\quad NH_3^+$$

4-Imidazoleethylamine
β-Aminoethylglyoxaline

HEATS OF PROTON IONIZATION, pK, AND RELATED THERMODYNAMIC QUANTITIES (continued)

Compound	Formula, Synonyms	ΔH (kcal/mole)	pK	ΔS cal/deg-mole	T°C	Method	Conditions	Reference	Other References	Remarks
Histidine	$CH_2CH(NH_3^+)COOH$ (imidazole ring)	7.14	6.00 (pK$_1$)	3.1	25	C	$\mu = 0$	330	94, 374 375	—
		10.43	9.16 (pK$_2$)	−6.5	25	C	$\mu = 0$	330	—	—
		4.0	6.3 (20°)	−15.3	10–32	T	$\mu = 0.2$ (NaCl)	376	—	3.b. Values valid for histidine in 12 position in ribonuclease.
		21.1	6.1 (41°)	41.0	32–41	T	$\mu = 0.2$ (NaCl)	376	—	3.b. Values valid for histidine in 12 position in ribonuclease.
		8.2	6.7 (20°)	−3.5	10–32	T	$\mu = 0.2$ (NaCl)	—	—	3.b. Values valid for histidine in 105 position in ribonuclease.
		8.2	6.8 (41°)	−3.5	32–41	T	$\mu = 0.2$ (NaCl)	—	—	3.b. Values valid for histidine in 105 position in ribonuclease.
		6.3	5.8 (20°)	−5.5	10–32	T	$\mu = 0.2$ (NaCl)	—	—	3.b. Values valid for histidine in 119 position in ribonuclease.
		11.4	5.7 (41°)	11.5	32–41	T	$\mu = 0.2$ (NaCl)	—	—	3.b. Values valid for histidine in 119 position in ribonuclease.

HEATS OF PROTON IONIZATION, pK, AND RELATED THERMODYNAMIC QUANTITIES (continued)

Compound	Formula, Synonyms	ΔH (kcal/mole)	pK	ΔS cal/deg-mole	T°C	Method	Conditions	Reference	Other References	Remarks
Histidine—(Continued)										
—, 1-methyl-	—	6.9	6.3 (37°)	-6.1	22–37	T	$\mu = 0.3$	91	—	3.b.
Histidylglycine	See: Glycine, histidyl-									
Human plasma albumin	—	-9.91	3.438	-48.9	55	T	—	214	—	3.b.d.
Hydrantoic acid	See: Glycine, carbamoyl-									
Hydrazine	$NH_3^+NH_3^+$	8.9	-0.67 (pK$_1$)	33	25	C	$\mu = 0.42$	330	—	—
		9.97	7.956 (pK$_2$)	-3.0	25	C	$\mu = 0$	—	70	—
		15.1	8.263	14.6	15	T	$\mu = 0$	215	—	—
		12.9	8.142	6.9	20	—	—	—	—	—
		10.6	7.968	-0.7	25	—	—	—	—	—
		6.0	7.753	-16.1	35	—	—	—	—	—
Hydroazoic acid	HN_3	3.60	4.72	-7.8	25	T	$\mu = 0$	70	216	—
Hydrocinnamic acid	See: Propanoic acid, 3-phenyl-									
Hydrocyanic acid	HCN Hydrogen cyanide Formonitrile	10.4	9.21	-3.9	25	C	$\mu = 0$	217	216, 377	—
Hydrogen cyanide	See: Hydrocyanic acid									
Hydrogen peroxide	—	8.2	11.75	-25.7	20	T	$\mu = 0$	378	379	—
Hydroquinone	See: Benzene, 1,4-dihydroxy-									
Hydroquinone monomethyl ether	See: Phenol, 4-methoxy-									
2-Hydroxyacetophenone	See: Acetophenone, 2-hydroxy-									
Hydroxyacridine	See: Acridine, hydroxy-									
6-Hydroxy-2-aminopurine	See: Guanine									
Hydroxybenzaldehyde	See: Benzaldehyde, hydroxy-									
Hydroxybenzene	See: Phenol									
Hydroxybenzoic acid	See: Benzoic acid, hydroxy-									
β-Hydroxyethanethiol	See: Ethanethiol, 2-hydroxy-									
2-Hydroxyethylamine	See: Ethanol, 2-amino-									
o-Hydroxyethylbenzoate	See: Benzoic acid, 2-hydroxy, ethyl ester									
p-Hydroxyethylbenzoate	See: Benzoic acid, 4-hydroxy, ethyl ester									
Hydroxyethylenediaminetriacetic acid	See: Ethane, 1,2-diamino-N-hydroxy-N,N',N'-triacetic acid									
N-Hydroxyethylethylenediamine-N,N,N'-triacetic acid	See: Ethane, 1,2-diamino-N-hydroxyethyl-N,N,N'-triacetic acid									

HEATS OF PROTON IONIZATION, pK, AND RELATED THERMODYNAMIC QUANTITIES (continued)

Compound	Formula, Synonyms	ΔH (kcal/mole)	pK	ΔS cal/deg-mole	T°C	Method	Conditions	Reference	Other References	Remarks
N-N-Di(2-hydroxyethyl)-glycine	See: Glycine, *N,N*-di(2-hydroxyethyl)-									
o-Hydroxyethyl salicylate	See: Benzoic acid, 2,6-dihydroxyethyl ester-									
p-Hydroxyethyl salicylate	See: Benzoic acid, 2,4-dihydroxyethyl ester-									
Hydroxyisobutyric acid	See: Propanoic acid, 2-hydroxy-2-methyl-									
Hydroxylamine	NH_3^+OH	9.5	6.186	4.6	15	T	$\mu = 0$	70	—	—
		9.4	6.063	4.2	20	—	—	—	—	—
		9.3	5.948	3.8	25	—	—	—	—	—
		9.0	5.730	3.0	35	—	—	—	—	—
8-Hydroxylepidine	See: Quinoline, 8-hydroxy-4-methyl-									
4-Hydroxy-3-methoxy-benzoic acid	See: Benzoic acid, 4-hydroxy-3-methoxy-									
Hydroxymethylphenol	See: Phenol, hydroxymethyl-									
3-Hydroxyphenazine	See: Phenazine, 3-hydroxy-									
m-Hydroxyphenyltrimethyl-ammonia	See: Aniline, 3-hydroxy-*N,N,N*-trimethyl-									
p-Hydroxyphenyltrimethyl-ammonia	See: Aniline, 4-hydroxy-*N,N,N*-trimethyl-									
Hydroxyproline	See: Proline, 4-hydroxy-									
2-Hydroxy-1,2,3-propane-tricarboxylic acid	See: Citric acid									
8-Hydroxyguinalidine	See: Quinoline, 8-hydroxy-2-methyl-									
8-Hydroxyquinoline-5-sulfonic acid	See: 5-Quinolinesulfonic acid, 8-hydroxy-									
2-Hydroxy-5-sulfobenzoic acid	See: Benzoic acid, 2-hydroxy-5-sulfo-									
3'-Hydroxy-5'-sulfo-phenylazo-2-naphthol	See: Naphthalene, 2,2'-dihydroxy-5'-sulfophenylazo-									
Hypochlorous acid	HOCl	3.320	7.424	−22.8	25	T	$\mu = 0$	81		
		7.2	8.8	−16.1	25	T	$\mu = 0.01$	39	321	—
Hypoxanthine	6(1)Purinone Sarcine See: Inosine									
Hypoxanthine riboside										

HEATS OF PROTON IONIZATION, pK, AND RELATED THERMODYNAMIC QUANTITIES (continued)

Compound	Formula, Synonyms	ΔH (kcal/mole)	pK	ΔS cal/deg-mole	T°C	Method	Conditions	Reference	Other References	Remarks
Hypoxanthosine	See: Inosine									
IDP	See: Inosine, 5'-diphosphoric acid-									
Imidazole		8.78	6.99 (pK₁)	-2.5	25	C	$\mu = 0$	330	168	—
		8.66	7.581	-2.99	0	T	$\mu = 0$	218	219–221	—
		8.70	7.467	-2.82	5	—		—	381, 382	—
		8.74	7.334	-2.69	10	—		—	—	—
		8.77	7.216	-2.59	15	—		—	—	—
		8.79	7.103	-2.53	20	—		—	—	—
		8.79	6.993	-2.51	25	—		—	—	—
		8.79	6.887	-2.52	30	—		—	—	—
		8.77	6.784	-2.57	35	—		—	—	—
		8.75	6.685	-2.66	40	—		—	—	—
		8.70	6.589	-2.79	45	—		—	—	—
		8.65	6.497	-2.95	50	—		—	—	—
		8.78	6.986 (pK₁)	-2.5	—	C	$\mu = 0.011$	348	—	—
		8.79			—	C	$\mu = 0.2$	4	—	—
		9.03	7.06	-2.0	25	C	$0.3M$ NaClO₄	353	—	—
		17.6	14.44 (pK₂)	-7	25	T	$\mu = 0$	92	—	—
		11.4	7.58 (25°)	-9	15–35	T	pH = 9–12	381	—	3.b. Imidazole complex with sperm whale ferri-myoglobin. Reaction: Fe±ImH→ Fe— Im+H⁺
—, 4(5)-2'-aminoethyl-	—	10.28	9.82 (pK₁)	-10.4	25	C	$0.3M$ NaClO₄	353	—	—
		9.25	6.09 (pK₂)	3.1	25	C	$0.3M$ NaClO₄	353	—	—
—, 4(5)-aminomethyl-	—	9.73	9.15 (pK₁)	-9.3	25	C	$0.3M$ NaClO₄	353	—	—
		7.43	4.77 (pK₂)	3.1	25	C	$0.3M$ NaClO₄	353	—	—
—, 2,4-dimethyl-	—	9.2	8.38	-7.4	25	T	$\mu = 0$	220	—	—
—, 2-hydroxymethyl-	—	9.99	7.660	-1.54	25	C	$\mu = 3.0(\text{ClO}_4^-)$	383	—	—
—, 4-hydroxymethyl-	—	9.32	7.419	-2.69	25	C	$\mu = 3.0(\text{ClO}_4^-)$	383	—	—
—, 4-methyl-	—	8.6	7.55	-5.7	25	T	$\mu = 0$	220	—	—
Imidazoline										
—, 2-(2'-pyridyl)-		9.2	8.98 (25°)	-10	0–40	T	$\mu = 0.005$	118	—	3.b.
		10.01	8.54 (25°)	-5	0–40	T	$\mu = 0.05$	118	—	3.b. 50 wt % dioxane
		13.7	9.01	4.73	25	T	$\mu = 0.044$	222	—	—

HEATS OF PROTON IONIZATION, pK, AND RELATED THERMODYNAMIC QUANTITIES (continued)

Compound	Formula, Synonyms	ΔH (kcal/mole)	pK	ΔS cal/deg-mole	T°C	Method	Conditions	Reference	Other References	Remarks
7-Imidazol[4,5-d]pyrimidine	See: Purine									
4-Imidazoleethylamine	See: Histamine									
Iminazole	See: Imidazole									
Iminodiacetic acid	See: Acetic acid, iminodi-									
Iminotriacetic acid	See: Acetic acid, iminotri-									
IMP	See: Inosine, 5'-monophosphoric acid-									
ITP	See: Inosine, 5'-triphosphoric acid-									
inosine		7.2	8.9	−16.4	25	T	$\mu = 0.1$	39	—	—
	Hypoxanthosine									
	Hypoxanthine riboside									
—, 5'-diphosphoric acid	IDP	−1.34	7.18 (pK$_4$)	−37.4	25	T	$\mu = 0$	46	—	—
—, 5'-monophosphoric acid	IMP	−1.43	6.66 (pK$_3$)	−35.3	25	T	$\mu = 0$	46	—	—
—, 5'-triphosphoric acid	ITP	−1.61	7.68 (pK$_5$)	−40.5	25	T	$\mu = 0$	46	—	3.b.
Iodinated insulin	(ε-amino)	13.4	11.12 (25°)	−5.95	0–35	T	$\mu = 2$	189	—	3.b.
(beef)	(quanidinium)	14.3	12.50 (25°)	−9.21	0–35	T	$\mu = 2$	189	—	
2-Iodobenzoic acid	See: Benzoic acid, 2-iodo-									
3-Iodobenzoic acid	See: Benzoic acid, 3-iodo-									
Iodogorgoic acid	See: Tyrosine, 3,5-diiodo-									
7-Iodo-8-hydroxyquinoline-5-sulfonic acid	See: 5-Quinolinesulfonic acid, 8-hydroxy-7-iodo-									
β-Iodopropionic acid	See: Propanoic acid, 3-iodo-									
Iron (II)										
—, *bis*(pyridine-2,6-dial-doxime)-	—	0	7.40	−34	25	T	$\mu = 0$	223	—	—
—, 2(2'-pyridyl)imidazoline	—	6.9	6.46	−6.42	25	T	$\mu = 0.107$	226	—	—
Isobutylacetic acid	See: Pentanoic acid, 4-methyl-									
Isobutylamine	See: Propane, 1-amino-2-methyl-									
Isobutyric acid	See: Propanoic acid, 2-methyl-									
Isobutyric acid, α-amino-	See: Propanoic acid, 2-amino-2-methyl-									

HEATS OF PROTON IONIZATION, pK, AND RELATED THERMODYNAMIC QUANTITIES (continued)

Compound	Formula, Synonyms	ΔH (kcal/mole)	pK	ΔS cal/deg-mole	T°C	Method	Conditions	Reference	Other References	Remarks
Isobutyric acid, hydroxy-	See: Propanoic acid, 2-hydroxy-2-methyl-									
Isocaproic acid	See: Pentanoic acid, 4-methyl-									
Isohexanoic acid	See: Pentanoic acid, 4-methyl-									
DL-Isoleucine	See: Pentanoic acid, 2-amino-3-methyl-									
α-N-Isopropylamine oxime	See: 2-Butanone, 3-(isopropylamino)-3-methyl-,oxime									
Isonicotinic acid	See: 4-Pyridinecarboxylic acid									
Isopentylamine	See: Butane, 1-amino-3-methyl-									
Isopropylamine	See: Propane, 2-amino-									
α-N-Isopropylamine oxime	See: 2-Butanone,3-(isopropylamino)-3-methyl-, oxime									
3-Isopropyl-4-dimethyl-aminopyridine	See: Pyridine, 4(N,N-dimethylamino)-3-isopropyl-									
3-Isopropyl-4-aminopyridine	See: Pyridine, 4-amino-3-isopropyl-									
3-Isopropyl-4-di-methylaminopyridine	See: Pyridine, 4(N,N-dimethylamino)-3-isopropyl-									
β-Isopropylglutaric acid	See: Pentanedioic acid, 3-isopropyl-									
3-Isopropyl-4-methyl-aminopyridine	See: Pyridine, 3-isopropyl-4-(N-methylamino)-									
3-Isopropylpyridine	See: Pyridine, 3-isopropyl-									
Isopropylthiol	See: Ethanethiol, 1-methyl-									
Isoquinoline	2-Benzazine / Leucoline / Benzo(c)pyridine	5.925	5.07	−3.30	25	C	—	37	—	3.d.
—, 1-amino-	—	10.5	7.62 (20°)	0.1	5-35	T	C = 0.01 M	85	—	3.b.
—, 3-amino-	—	5.0	5.05 (20°)	−6.0	5-35	T	C = 0.005 M	85	—	3.b.
Isovaleric acid	See: Butanoic acid, 3-methyl-									
Isovaleric acid, α-amino-	See: Valine									
Isovanillin	See: Benzaldehyde-3-hydroxy-4-methoxy-									
α-Ketobutyric acid	See: Butanoic acid, 2-oxo-									
Lactic acid	See: Propanoic acid, 2-hydroxy-									
Leucine(DL)	(CH$_3$)$_2$CHCH$_2$CHNH$_3^+$COOH, D-α-Aminoisocaproic acid	1.180	2.383 (pK$_1$)	−6.5	1.0	T	$\mu = 0$	50	—	—
		0.860	2.348	−7.7	12.5	—	—	—	—	—
		0.420	2.328	−9.2	25	—	—	—	—	—

HEATS OF PROTON IONIZATION, pK, AND RELATED THERMODYNAMIC QUANTITIES (continued)

Compound	Formula, Synonyms	ΔH (kcal/mole)	pK	ΔS cal/deg-mole	T°C	Method	Conditions	Reference	Other References	Remarks
Leucine(DL)—(Continued)		-0.090	2.327	-10.9	37.5	—	—	—	—	—
		-0.700	2.333	-12.8	50	—	—	—	—	—
		10.870	10.454 (pK₂)	-8.2	1.0	T	μ = 0	50	—	—
		10.940	10.095	-7.9	12.5	—	—	—	—	—
		10.900	9.744	-8.0	25	—	—	—	—	—
		10.730	9.434	-8.6	37.5	—	—	—	—	—
		10.410	9.142	-9.6	50	—	—	—	—	—
Leucine										
—, glycyl-		-0.752	3.1800 (pK₁)	-17.07	25	T	μ = 0	6	—	—
Leucoline	See: Isoquinoline									
Levulinic acid	See: Pentanoic acid, 4-oxo-									
Loretine	See: Quinoline, 8-hydroxy-7-iodo-									
Luminal	See: Barbituric acid, 5-ethyl-5-phenyl-									
2,4-Lutidine	See: Pyridine, 2,4-dimethyl-									
2,6-Lutidine	See: Pyridine, 2,6-dimethyl-									
3,5-Lutidine	See: Pyridine, 3,5-dimethyl-									
Lycine	See: Betaine									
Lysine	H₃N⁺(CH₂)₄CH(NH₃⁺)COOH, L-α,ε-Diaminocaproic acid	0.30	2.18 (25°, pK₁)	-9.0	0–25	T	μ = 0.01	94	—	3.a.b.
		12.80	8.95 (25°, pK₂)	2.0	0–25	T	μ = 0.01	94	—	3.a.b.
		12.3	8.95 (pK₂)	0	25	T	0.1 M KCl	374	—	3.b.
		11.0	9.53 (25°, pK₂)	-6.70	0–25	T	μ = 2	189	—	3.a.b.
		11.60	10.53 (25°, pK₃)	-9.30	0–25	T	μ = 0.01	94	—	
		11.6	10.53 (pK₃)	-9	25	T	0.1 M KCl	374	—	3.b.
		13.3	10.94 (25°, pK₃)	-5.43	0–35	T	μ = 2	189	—	3.a.b.
—, lysyl-		2.00	1.95 (25°, pK₁)	-2.2	0–25	T	μ = 0.01	94	—	3.b.
		12.7	8.17 (25°, pK₂)	5.4	—	—	—	—	—	3.a.b.
		11.35	9.45 (25°, pK₃)	-5.3	—	—	—	—	—	—
		13.30	10.63 (25°, pK₄)	-4.0	—	—	—	—	—	—
Lysylglutamic acid	See: Glutamic acid, lysyl-									
Lysyllysine	See: Lysine, lysyl-									
Lyxose	CHO, HO—C—H, HO—C—H, H—C—OH, CH₂OH	8.0	12.22	-28.7	25	C	μ = 0	186	—	—

HEATS OF PROTON IONIZATION, pK, AND RELATED THERMODYNAMIC QUANTITIES (continued)

Compound	Formula. Synonyms	ΔH (kcal/mole)	pK	ΔS cal/deg-mole	T°C	Method	Conditions	Reference	Other References	Remarks
Maleic acid	HOOCCH:CHCOOH *cis*-Butenedioic acid	0.08	1.910 (pK$_1$)	-8.5	25	C	$\mu = 0$	1	59	—
		-0.83	6.332 (pK$_2$)	-31.8	25	C	$\mu = 0$	1	—	—
D,L-Malic acid	See: Butanedioic acid (D,L),2-hydroxy-									
Malonic acid	CH$_2$(COOH)$_2$ Propanedioic acid	0.29	2.826 (pK$_1$)	-12.0	25	C	$\mu = 0$	1	320	—
		-0.92	5.696 (pK$_2$)	-29.2	25	C	$\mu = 0$	1	—	—
		0.322	5.669 (pK$_2$)	-24.8	0	T	$\mu = 0$	224	—	—
		0.034	5.665	-25.8	5	—	—	—	—	—
		-0.259	5.668	-26.8	10	—	—	—	—	—
		-0.556	5.673	-27.9	15	—	—	—	—	—
		-0.856	5.683	-28.9	20	—	—	—	—	—
		-1.161	5.696	-30.0	25	—	—	—	—	—
		-1.469	5.710	-31.0	30	—	—	—	—	—
		-1.781	5.730	-32.0	35	—	—	—	—	—
		-2.096	5.750	-33.0	40	—	—	—	—	—
		-2.414	5.777	-34.0	45	—	—	—	—	—
		-2.736	5.803	-35.0	50	—	—	—	—	—
		-3.060	5.833	-36.0	55	—	—	—	—	—
		-3.387	5.866	-37.0	60	—	—	—	—	—
—, 2-amino-2(hydroxymethyl)-	—	11.35	8.07	-1.12	25	C	$\mu = 0$	356	—	—
—, 2-amino-2-methyl-	—	11.93	8.80	-0.26	25	C	$\mu = 0$	356	—	—
—, diethyl-	3,3-Pentanedicarboxylic acid	-1.25	2.211 (pK$_1$)	-14.3	25	C	$\mu = 0$	1	—	—
		-0.82	7.292 (pK$_2$)	-36.1	25	C	$\mu = 0$	1	—	—
—, dinitrile-	CH$_2$(CN)$_2$ Malononitrile	13.4	11.20 (pK$_1$)	-6.4	25	C	$\mu = 0$	225	—	—
—, mononitrile-	Cyanoacetic acid	-0.032	2.4447	-11.3	5	T	$\mu = 0$	226	6, 12	—
		-0.283	2.4467	-12.2	10					384	—
		-0.502	2.4523	-13.0	15						—
		-0.705	2.4597	-13.7	20						—
		-0.888	2.4600	-14.3	25						—
		-1.074	2.4818	-14.9	30						—
		-1.255	2.4955	-15.5	35						—
		-1.465	2.5107	-16.2	40						—
		-1.705	2.5281	-16.9	45						—
—, mononitrile-di-isopropyl-	—	-2.783	2.39	-20.95	5	T	$C = 10^{-4} M$	329	—	—
		-2.941	2.43	-21.51	10	T	$C = 10^{-4} M$			—
		-3.098	2.47	-22.06	15	T	$C = 10^{-4} M$			—
		-3.252	2.51	-22.59	20	T	$C = 10^{-4} M$			—
		-3.402	2.56	-23.10	25	T	$C = 10^{-4} M$			—
		-3.548	2.60	-23.58	30	T	$C = 10^{-4} M$			—
		-3.689	2.64	-24.05	35	T	$C = 10^{-4} M$			—

HEATS OF PROTON IONIZATION, pK, AND RELATED THERMODYNAMIC QUANTITIES (continued)

Compound	Formula, Synonyms	ΔH (kcal/mole)	pK	ΔS cal/deg-mole	T°C	Method	Conditions	Reference	Other References	Remarks
Malonic acid—(Continued)										
		−3.824	2.68	−24.48	40	T	$C = 10^{-4} M$	—	—	—
		−3.952	2.72	−24.89	50	T	$C = 10^{-4} M$	—	—	—
Malononitrile	See: Malonic acid, dinitrile-									
Mandelic acid	See: Acetic acid, hydroxy-(phenyl)-									—
Mannose (D)	CH₂OH	7.9	12.08 (pK$_1$)	−28.9	25	C	$\mu = 0$	186	—	—
	Carubinose									
	Seminose									
Mercaptoacetic acid	See: Acetic acid, mercapto-									
L-β-Mercaptoalanine	See: Cysteine (L)									
2-Mercaptobenzoic acid	See: Benzoic acid, 2-mercapto-									
β-Mercaptopropionic acid	See: Propanoic acid, 3-mercapto-									
Metanilic acid	See: Benzenesulfonic acid, 3-amino-									
Methemoglobins	Mouse 7.14		8.000	−12.2	20	T	$\mu = 0.051$	227	229	—
	Rat 3.34		8.036	−25.3			—	—	385	—
	Baboon 4.66		8.112	−21.2			—	—	—	—
	Patas monkey 5.18		8.130	−19.5			—	—	—	—
	Mona monkey 4.93		8.163	−20.5			—	—	—	—
	Pig 7.87		8.211	−10.7			—	—	—	—
	Dog 6.09		8.266	−17.0			—	—	—	—
	Hyena 5.50		8.295	−19.2			—	—	—	—
	Tantalus monkey 5.65		8.300	−18.7			—	—	—	—
	Cat 4.35		8.301	−23.1			—	—	—	—
	Pigeon 9.16		8.290	−6.68			—	—	—	—
	Horse 3.66		8.405	−26.0			—	—	—	—
	Human 3.40		8.136	−25.6			—	—	—	—
	Human 4.09		8.189	−23.5			—	—	—	—
	Human 4.88		8.146	−20.6			—	—	—	—
	Guinea pig 5.860		8.213	−17.6	20	T	$\mu = 0.0506$	228	—	—
	Lizard 4.120		8.260	−23.7			—	—	—	—

HEATS OF PROTON IONIZATION, pK, AND RELATED THERMODYNAMIC QUANTITIES (continued)

Compound	Formula, Synonyms	ΔH (kcal/mole)	pK	ΔS cal/deg-mole	T °C	Method	Conditions	Reference	Other References	Remarks
Methemoglobins—(Continued)										
		Bat 8.340	8.288	−9.45	—	—	—	—	—	—
		Shrew 10.220	8.280	−3.00	—	—	—	—	—	—
		Duck 6.550	8.260	−15.45	—	—	—	—	—	—
		Turkey 8.110	8.204	−9.86	—	—	—	—	—	—
		Chicken 6.650	8.247	−15.0	—	—	—	—	—	—
		Guinea fowl 6.830	8.244	−14.4	—	—	—	—	—	—
		Cow 1.570	8.230	−32.3	—	—	—	—	—	—
		Mangabey monkey 4.120	8.090	−23.0	—	—	—	—	—	—
		Rabbit 1.760	8.020	−30.7	—	—	—	—	—	—
Methane	$CH_3NH_3^+$	13.29	10.641	−4.1	25	C	$\mu = 0$	330	63, 70, 73	—
—, amino-	Methylamine									
—, dinitro-	Nitroform	2.220	3.60	−9.22	20	T	—	147	171, 230	3.d.
—, trinitro-	—	1.592	0.17 (20°)	4.64	5–60	T	$\mu = 0$	231	—	3.b.
Methanol	CH_3OH	12.5	12.22 (25°)	−14.0	18–37	T	—	232	—	3.b.d.
	Carbinol	12.0	16.66	−36.0	25	T	0.1 M MeOH	233	—	—
	Methyl alcohol									
Methanoic acid	See: Formic acid									
Methionine	$CH_3SCH_2CH_2CH(\overset{+}{N}H_3)COOH$	0.140	2.125 (25°, pK$_1$)	−9.3	10–40	T	$\mu = 0.05$	234	95	3.b.
		0.14	2.12 (pK$_1$)	−9.3	25	T	0	386	—	Conditions refer to wt.% dioxane
		3.5	3.09	−2.3	25	T	44.6	386	—	—
		0.65	3.51	−13.9	25	T	59.7	386	—	—
		3.05	3.88	−7.5	25	T	69	386	—	—
		10.4	9.28 (25°, pK$_2$)	−7.5	10–40	T	$\mu = 0.05$	234	—	—
		10.40	9.28	−7.5	25	T	0	386	—	3.b. Conditions refer to wt.% dioxane
		9.3	9.75	−13.2	25	T	44.6	386	—	—
		10.10	10.19	−12.8	25	T	59.7	386	—	—
		9.00	10.83	−19.4	25	T	69	386	—	—
Methoxyacetic acid	See: Acetic acid, methoxy-									
β-Methoxy-DL-alanine	See: DL-Alanine, 3-methoxy-									

HEATS OF PROTON IONIZATION, pK, AND RELATED THERMODYNAMIC QUANTITIES (continued)

Compound	Formula, Synonyms	ΔH (kcal/mole)	pK	ΔS cal/deg-mole	T°C	Method	Conditions	Reference	Other References	Remarks
2-Methoxybenzoic acid	See: Benzoic acid, 2-methoxy-									
3-Methoxybenzoic acid	See: Benzoic acid, 3-methoxy-									
4-Methoxybenzoic acid	See: Benzoic acid, 4-methoxy-									
2-Methoxyethylamine	See: Ethane, 1-amino-2-methoxy-									
***m*-Methoxyphenol**	See: Phenol, 3-methoxy-									
***o*-Methoxyphenol**	See: Phenol, 2-methoxy-									
***p*-Methoxyphenol**	See: Phenol, 4-methoxy-									
Methyl alcohol	See: Methanol									
Methylamine	See: Methane, amino-									
***N*-Methylaminediacetic acid**	See: Acetic acid, *N*-methyl-*N*-nitrilodi-									
α-*N*-Methylamine oxime	See: 2-Butanone, 3-(methylamino)-3-methyl-, oxime									
2-Methylaminoethanol	See: Ethanol, 2-amino-2-methyl-									
2-(2-Methylaminoethyl)-pyridine	See: Pyridine, 2-(2-methylaminoethyl)-									
2-(Methylaminomethyl)-pyridine	See: Pyridine, 2-(methylaminomethyl)-									
4-Methylaminopyridine	See: Pyridine, 4(*N*-methylamino)-									
3-Methyl-4-aminopyridine	See: Pyridine, 4-amino-3-methyl-									
Methyl-*m*-aminobenzoate	See: Benzoic acid, 3-aminomethyl ester-									
Methyl-*o*-aminobenzoate	See: Benzoic acid, 2-aminomethyl ester-									
Methyl-*p*-aminobenzoate	See: Benzoic acid, 4-aminomethyl ester-									
***o*-Methylaniline**	See: Toluene, 2-amino-									
3-Methylaniline	See: Toluene, 3-amino-									
4-Methylaniline	See: Toluene, 4-amino-									
5-Methylbarbituric acid	See: Barbituric acid, 5-methyl-									
2-Methylbenzimidazole	See: Benzimidazole, 2-methyl-									
S-Methyl-L-cysteine	See: L-Cysteine, S-methyl-									
***N*-Methyl-di(3-aminopropyl)-amine**	See: Amine, *N*,*N*-di(3-aminopropyl)-*N*-methylamine-									
Methyldicyanoacetate	See: Propanoic acid, 2,2-dinitrile-									
***N*-Methyldiethanolamine**	See: Amine, *N*,*N*-diethanol-*N*-methyl-									
3-Methyl-4-dimethylamino-pyridine	See: Pyridine, 3-methyl-4(*N*,*N*-dimethylamino)-									
***N*-Methylethylenediamine**	See: Ethane, 1,2-diamino-*N*-methyl-									
***N*-Methylglycine**	See: Sarcosine									
***N*-Methyliminodiacetic acid**	See: Acetic acid, *N*-methyliminodi-									
3-Methyl-4-methylamino-pyridine	See: Pyridine, 3-methyl-4(*N*-methylamino)-									
Methylene cyanide	See: Malonic acid, dinitrile-									

HEATS OF PROTON IONIZATION, pK, AND RELATED THERMODYNAMIC QUANTITIES (continued)

Compound	Formula, Synonyms	ΔH (kcal/mole)	pK	ΔS cal/deg-mole	T°C	Method	Conditions	Reference	Other References	Remarks
N-Methylethylenediamine	See: Ethane, 1,2-diamino-*N*-methyl-									
β-Methylglutaric acid	See: Pentanedioic acid, 3-methyl-									
N-Methylglycine	See: Sarcosine									
Methylglycolic acid	See: Acetic acid, methoxy-									
Methyl orange	Acid → base form	3.35	3.43 (25°)	−4.4	20–50	T	—	235	—	3.b., In Tartrate Buffer
	Sodium 4'-dimethylaminoazobenzene-4-sulfonate									
	Helianthin									
	Orange II									
3-Methyl-2,4-pentanedione	See: 2,4-Pentanedione, 3-methyl-									
2-Methylphenol	See: Toluene, 2-hydroxy-									
3-Methylphenol	See: Toluene, 3-hydroxy-									
4-Methylphenol	See: Toluene, 4-hydroxy-									
6-Methyl-2-picolylamine	See: Pyridine, 2-(methylaminomethyl)-6-methyl-									
6-Methyl-2-picolylmethylamine	See: Pyridine, 2-methyl-6-methylamino-methyl-									
3-Methylpyridine	See: Pyridine, 3-methyl-									
Methyl-2-pyridylketone	See: Pyridine, 2-acetyl-									
Methyl-3-pyridylketone	See: Pyridine, 3-acetyl-									
2(Methylthio)ethylamine	See: Ethane, 1-amino-2-(methylthio)-									
Methylsalicylate	See: Benzoic acid, 2-hydroxymethyl ester-									
Molonitrile	See: Malonic acid, dinitrile-									
Mono-2-hydroxyethylamine	See: Ethanol, 2-amino-									
Monomethylamine	See: Amine, methyl-									
Morpholine	Diethylenimide oxide Tetrahydro-1, 4-isoxazine	9.33	8.492	7.58	25	T	$\mu = 0$	236	—	—
—, 4(2-aminoethyl)-	—	−2.9	4.06 (30°, pK$_1$)	−9	10–40	T	$\mu = 0$	237	—	3.b.
		−9.0	9.15 (30°, pK$_2$)	−12						
Myoglobin, sperm whale	—	(Carboxyl) 1.6	4.40	−14.8	25	C	$\mu = 0.15$	238	—	—
		(Histidine) 7.1	6.62	−6.48	—	—	—	—	—	—
		(α-amino) 11	7.8	1.21	—	—	—	—	—	—
		(Hemic acid) 6	8.9	−20.6	—	—	—	—	—	—
		(Tyrosine) 6.1	10.0	−25.3	—	—	—	—	—	—
		(Lysine) 12.7	10.0	−3.17	—	—	—	—	—	—

HEATS OF PROTON IONIZATION, pK, AND RELATED THERMODYNAMIC QUANTITIES (continued)

Compound	Formula, Synonyms	ΔH (kcal/mole)	pK	ΔS cal/deg-mole	T°C	Method	Conditions	Reference	Other References	Remarks
Naphthionic acid	See: 1-Naphthalenesulfonic acid, 4-amino-									
Naphthalene										
—, 1-amino-	1-Naphthylamine, α-Naphthylamine	5.9	3.40 (20°)	4.6	0–20	T	C = 0.01 M	85	—	3.b. 50 wt % EtOH, EtOH
—, 2-amino-	2-Naphthylamine, β-Naphthylamine	4.8	3.77 (20°)	−0.9	0–20	T	C = 0.01 M	85	—	3.b. 50 wt % EtOH, EtOH
—, benzene-azo-1-amino-6,8-disulfonic acid	[structure: N=N–N–H, HSO₃, SO₃H]	7.1	3.56 (40°)	3.2	40–80	T	$\mu = 0$	239	—	3.b.
—, 3-benzene carboxaldehyde-azo-1-amino-6,8-disulfonic acid-	[structure: N=N–N, HSO₃, SO₃H, COH]	8.3	3.96 (40°)	4.2	40–80	T	$\mu = 0$	239	—	3.b.
—, 2-benzenesulfonic acid-azo-1-amino-6,8-disulfonic acid-	—	−2.6	3.37 (40°)	−12.1	40–80	T	$\mu = 0$	239	—	3.b.
—, 3-benzenesulfonic acid-azo-1-amino-6,8-disulfonic acid-	—	3.4	3.29 (40°)	−2.3	40–80	T	$\mu = 0$	239	—	3.b.
—, 4-benzenesulfonic acid-azo-1-amino-6,8-disulfonic acid-	—	3.5	3.31 (40°)	−2.0	40–80	T	$\mu = 0$	239	—	3.b.
—, 2-benzoic acid-azo-1-amino-6,8-disulfonic acid-	—	3.8	6.46 (40°)	−20.9	40–80	T	$\mu = 0$	239	—	3.b.
—, 3-benzoic acid-azo-1-amino-6,8-disulfonic acid-	—	5.8	3.70 (40°)	0.7	40–80	T	$\mu = 0$	239	—	3.b.

HEATS OF PROTON IONIZATION, pK, AND RELATED THERMODYNAMIC QUANTITIES (continued)

Compound	Formula. Synonyms	ΔH (kcal/mole)	pK	ΔS cal/deg-mole	T°C	Method	Conditions	Reference	Other References	Remarks
Naphthalene—(Continued)										
—,4-benzoic acid-azo-1-amino-6,8-sulfonic acid-	—	3.2	3.69 (40°)	-3.3	40–80	T	$\mu = 0$	239	—	3.b.
—,2-chlorobenzene-azo-1-amino-6,8-disulfonic acid-	—	5.8	3.17 (40°)	2.1	40–80	T	$\mu = 0$	239	—	3.b.
—,3-chlorobenzene-azo-1-amino-6,8-disulfonic acid-	—	8.7	3.84 (40°)	5.2	40–80	T	$\mu = 0$	239	—	3.b.
—,4-chlorobenzene-azo-1-amino-6,8-disulfonic acid-	—	8.1	3.75 (40°)	4.5	40–80	T	$\mu = 0$	239	—	3.b.
—,1,4-dihydroxy-	1,4-Naphthohydroquinone 1,4-Naphthalenediol	6.2	9.37 (26.4°)	-22	13	T	$\mu = 0.65$	105	—	3.b.
		5.5	10.93 (26.6°)	-32	25	T	$\mu = 0.65$	105	—	3.b.
—,2,2'-dihydroxy-5-sulfophenylazo-		3.4	6.94 (30°, pK$_2$)	-22	25	T	$\mu = 0.1$	149	—	3.b.
		2.2	6.83 pK$_2$	-26	50	T				
		10.8	12.78 (30°, pK$_3$)	-25	25	T	$\mu = 0.1$	149	—	3.b.
		8.2	12.36 (pK$_3$)	-33	50	T				
3-Hydroxy-5-sulfophenylazo-2-naphthol Solochrome violet R		—	—	—				—	—	
—,2,4-dimethylbenzene-azo-1-amino-6,8-disulfonic acid-	—	6.4	3.59 (40°)	2.0	40–80	T	$\mu = 0$	239	—	3.b.
—,2,5-dimethylbenzene-azo-1-amino-6,8-disulfonic acid-	—	5.40	3.64 (40°)	0.4	40–80	T	$\mu = 0$	239	—	3.b.
—,2,6-dimethylbenzene-azo-1-amino-6,8-disulfonic acid-	—	-1.2	3.56 (40°)	-10.0	40–80	T	$\mu = 0$	239	—	3.b.
—,4-hydroxybenzene-azo-1-amino-6,8-disulfonic acid-	—	6.7	3.42 (40°)	2.9	40–80	T	$\mu = 0$	239	—	3.b.
—,2-methoxybenzene-azo-1-amino-6,8-disulfonic acid-	—	6.3	3.59 (40°)	1.7	40–80	T	$\mu = 0$	239	—	3.b.
—,3-methoxybenzene-azo-1-amino-6,8-disulfonic acid-	—	9.60	3.88 (40°)	4.5	40–80	T	$\mu = 0$	239	—	3.b.

HEATS OF PROTON IONIZATION, pK, AND RELATED THERMODYNAMIC QUANTITIES (continued)

Compound	Formula, Synonyms	ΔH (kcal/mole)	pK	ΔS cal/deg-mole	T°C	Method	Conditions	Reference	Other References	Remarks
Naphthalene—(Continued)										
—, 4-methoxybenzene-azo-1-amino-6,8-disulfonic acid-	—	8.5	3.57 (40°)	5.4	40-80	T	$\mu = 0$	239	—	3.b.
—, 2-methylbenzene-azo-1-amino-6,8-disulfonic acid-		1.7	3.44 (40°)	-5.1	40-80	T	$\mu = 0$	239	—	3.b.
—, 3-methylbenzene-azo-1-amino-6,8-disulfonic acid-		9.4	3.71 (40°)	6.4	40-80	T	$\mu = 0$	239	—	3.b.
—, 4-methylbenzene-azo-1-amino-6,8-disulfonic acid-		8.8	3.58 (40°)	5.9	40-80	T	$\mu = 0$	239	—	3.b.
—, 3-nitrobenzene-azo-1-amino-6,8-disulfonic acid-		4.2	3.60 (40°)	-1.5	40-80	T	$\mu = 0$	239	—	3.b.
—, 4-nitrobenzene-azo-1-amino-6,8-disulfonic acid-		4.2	3.91 (40°)	-2.5	40-80	T	$\mu = 0$	239	—	3.b.
1,4-Naphthalenediol	See: Naphthalene, 1,4-dihydroxy-									
1-Naphthalenesulfonic acid	SO_3H (naphthalene, positions 1–8)									
—, 4-amino-	Naphthionic acid	4.101	2.559 (35°)	2	16.8	0-35	—	12	—	3.a.d.
Naphthionic acid	Naphthionic acid		—	—						
1,4-Naphthohydroquinone	See: Naphthalene, 1,4-dihydroxy-									
α-Naphthylamine	See: Naphthalene, 1-amino-									
1-Naphthylamine	See: Naphthalene, 1-amino-									
β-Naphthylamine	See: Naphthalene, 2-amino-									
2-Naphthylamine	See: Naphthalene, 2-amino-									
Nicotinic acid	See: 3-Pyridinecarboxylic acid									
Nitric acid	HNO_3	-3.30	-1.44	-4.46	25	T	$\mu = 0$	70	—	
Nitrilotriacetic acid	See: Acetic acid, iminotri-									
Nitrilo-2,2',2''-triethanol	See: Amine, triethyl-2,2',2''-hydroxy-									
Nitroform	See: Methane, trinitro-									
7-Nitro-8-hydroxyquinoline-5-sulfonic acid	See: 5-Quinolinesulfonic acid, 8-hydroxy-7-nitro-									
o-Nitrophenol	See: Phenol, 2-nitro-									
m-Nitrophenol	See: Phenol, 3-nitro-									
p-Nitrophenol	See: Phenol, 4-nitro-									

HEATS OF PROTON IONIZATION, pK, AND RELATED THERMODYNAMIC QUANTITIES (continued)

Compound	Formula, Synonyms	ΔH (kcal/mole)	pK	ΔS cal/deg-mole	T°C	Method	Conditions	Reference	Other References	Remarks
7-(4-Nitrophenylazo)-8-hydroxyquinoline-5-sulfonic acid	See: 5-Quinolinesulfonic acid. 7-(4-nitrophenylazo)-8-hydroxy-									
Nitrous acid	HNO_2	3.8	3.224	−1.6	15	T	$\mu = 0$	387	70	—
		3.0	3.177	−4.4	20	T	$\mu = 0$	—	—	—
		2.2	3.138	−7.1	25	T	$\mu = 0$	—	—	—
		0.47	3.100	−12.7	35	T	$\mu = 0$	—	—	—
—, hypo-	$H_2O_2N_2$	3.0	6.92 (pK$_1$)	−22	25	T	$\mu = 0.1$	388	70	—
		8.0	9.90 (pK$_2$)	−19	25	T	$\mu = 0.1$	388	—	—
		11.10	18.05 (pK$_1$ · pK$_2$)	−9.9	25	C	$\mu = 0$	240	—	—
DL-Norleucine	See: Hexanoic acid, 2-amino-									
Norvaline	See: Pentanoic acid, 2-amino-									
Octamethylenedinitrilotetraacetic acid	See: Acetic acid, octamethylenedinitrilotetra-									
Octane	$CH_3(CH_2)_6CH_3$	11.0	8.60 (30°, pK$_1$)	−3	10–40	T	$\mu = 0.002$	62	—	3.b.
		11.4	9.57 (30°, pK$_2$)	−6	10–40	T	$\mu = 0.002$	62	—	3.b.
—, 1,8-diamino-3,6-dioxo-		12.2	8.43 (30°, pK$_1$)	−2	10–40	T	$\mu = 0$	169	—	3.b.
		12.2	9.31 (30°, pK$_2$)	2	10–40	T	$\mu = 0$	169	—	3.b.
—, 1,8-diamino-3,6-dithia-		2.97	2.90 (pK$_1$)	−3.6	25	C	$\mu = 0$	241	—	—
		3.01	2.97 (pK$_1$)	−3.5	25	C	0.1 M KCl	78	—	—
—, 1,4-diazabicyclo[2.2.2]-	Triethylenediamine	7.29	8.60 (pK$_2$)	−15.2	25	C	0.1 M KCl	241	—	—
		7.30	8.82 (pK$_2$)	−15.9	25	C	$\mu = 0$	78	—	—
Octanedioic acid	HOOC(CH$_2$)$_6$COOH	−0.39	4.512 (pK$_1$)	−21.9	25	C	$\mu = 0$	1	—	—
Suberic acid		−0.64	5.404 (pK$_2$)	−26.9	25	C	$\mu = 0$	62	—	3.b.
3,6-Octanedione-,1,8-diamino-	H$_3$NCH$_2$CH$_2$CO(CH$_2$)$_2$COCH$_2$CH$_2$NH$_3^+$	11.0	8.60 (30°, pK$_1$)	−3	10–40	T	$\mu = 0$	62	—	3.b.
		11.4	9.57 (30°, pK$_2$)	−6	10–40	T	$\mu = 0$	12	—	3.b.
Octanoic acid	CH$_3$(CH$_2$)$_6$COOH	−0.077	4.910 (35°)	−23	16.8	T	—	—	—	3.a.b.d.
	Caprylic acid									
ODTA	See: Octamethylenedinitrilotetraacetic acid									
Orange II	See: Methyl orange									
Orthanilic acid	See: Benzenesulfonic acid, 2-amino-									
Oxalic acid	(COOH)$_2$	−1.02	1.271 (pK$_1$)	−9.2	25	C	$\mu = 0$	1	243	—
	Ethanedioic acid	−1.50	4.266 (pK$_2$)	−24.6	25	C	$\mu = 0$	1	389	—
		−0.346	4.201 (pK$_2$)	−20.5	0	T	$\mu = 0$	242	—	—
		−0.600	4.207	−21.4	5			—	—	—
		−0.857	4.218	−22.3	10			—	—	—

HEATS OF PROTON IONIZATION, pK, AND RELATED THERMODYNAMIC QUANTITIES (continued)

Compound	Formula, Synonyms	ΔH (kcal/mole)	pK	ΔS cal/deg-mole	T°C	Method	Conditions	Reference	Other References	Remarks
Oxalic acid—(Continued)										
		−1.120	4.231	−23.2	15	—	—	—	—	—
		−1.387	4.247	−24.2	20	—	—	—	—	—
		−1.659	4.266	−25.1	25	—	—	—	—	—
		−1.936	4.287	−26.0	30	—	—	—	—	—
		−2.217	4.312	−26.9	35	—	—	—	—	—
		−2.502	4.338	−27.8	40	—	—	—	—	—
		−2.793	4.369	−28.8	45	—	—	—	—	—
		−3.088	4.399	−29.7	50	—	—	—	—	—
Oxine	See: Quinoline, 8-hydroxy-									
Oxine-5-sulfonic acid	See: Quinolinesulfonic acid, 8-hydroxy-									
Oxine, 2-methyl-	See: Quinoline, 8-hydroxy-2-methyl-									
Oxine, 4-methyl-	See: Quinoline, 8-hydroxy-4-methyl-									
Oxine, thio, 2-methyl-	See: Quinoline, 2-methyl-8 mercapto-									
Oxoacetic acid	See: Acetic acid, oxo-									
Oxyneurine	See: Betaine									
6-Oxypurine	See: Hypoxanthine									
β-Parvoline	See: Pyridine, 2,3,5,6-tetramethyl-									
Parvuline	See: Pyridine, 2,3,5,6-tetramethyl-									
Pentamethylenedinitrilo-tetraacetic acid	See: Acetic acid, pentamethylenedinitrilo-tetra-									
β,β-pentamethyleneglutaric acid	See: Pentanedioic acid, 3,3-pentamethylene-									
Pentamethyleneimine	See: Piperidine									
Pentane	$CH_3(CH_2)_3CH_3$									
—, 1-amino-	—	13.98	10.64	−1.8	25	C	$\mu = 0$	330	—	—
—, 3-aza-1,5-diamino-N,N',N'',N'''-pentaacetic acid-	—	2.4	8.50 (pK$_4$)	−30.9	27	C	$\mu = 0.1(KNO_3)$	390	—	—
		8.4	10.41 (pK$_5$)	−19.7	27	C	$\mu = 0.1(KNO_3)$	390	—	—
—, 3,3-Dicarboxylic acid	See: Malonic acid, diethyl-									
—, 1,1-dinitro-	$CH_3(CH_2)_3CH(NO_2)_2$, 1.1-Dinitropentane	3.86	5.337 (25°)	−11.44	5–60	T	—	165	—	3.b.
Pentanedioic acid	$HOOC(CH_2)_3COOH$ Glutaric acid	−0.12	4.344 (pK$_1$)	−20.3	25	C	$\mu = 0$	1	59	—
		−0.58	5.420 (pK$_2$)	−26.7	25	C	$\mu = 0$	1	—	—
—, 2-amino-	See: Glutamic acid									
—, 3,3-dimethyl-	$HOOCCH_2C(CH_3)_2CH_2COOH$ β,β-Dimethylglutaric acid	−3	3.70 (pK$_1$)	−27	25	T	$\mu = 0$	59	—	—
		−2.5	6.34 (pK$_2$)	−37	25	T	$\mu = 0$	59	—	—
—, 3-isopropyl-	$(CH_3)_2(CH_2)_2(CH_2COOH)_2$ β-Isopropylglutaric acid	−1	4.30 (pK$_1$)	−23	25	T	$\mu = 0$	59	—	—
		−1.5	5.51 (pK$_2$)	−30	25	T	$\mu = 0$	59	—	—
—, 3-methyl-	$HOOCCH_2CH(CH_3)CH_2COOH$ β-Methylglutaric acid	−0.3	4.25 (pK$_1$)	−20	25	T	$\mu = 0$	59	—	—
		−1.0	5.41 (pK$_2$)	−28	25	T	$\mu = 0$	59	—	—

HEATS OF PROTON IONIZATION, pK, AND RELATED THERMODYNAMIC QUANTITIES (continued)

Compound	Formula, Synonyms	ΔH (kcal/mole)	pK	ΔS cal/deg-mole	T°C	Method	Conditions	Reference	Other References	Remarks
Pentanedioic acid—(Continued)										
—,3,3-pentamethylene-	$\overset{\displaystyle CH_2}{(H_2C)_2}\diagdown\!\!\diagup(CH_2)_2$	−2.5	3.49 (pK₁)	−24	25	T	μ = 0	59	—	—
		−1.5	6.96 (pK₂)	−36	25	T	μ = 0	59	—	—
	$HOOC-CH_2-C-CH_2-COOH$ β,β-Pentamethyleneglutaric acid									
2,4-Pentanedione	$CH_3COCH_2COCH_3$	2.8	9.02 (20°)	−32	10–40	T	μ = 0	244	211	3.b.
	Acetylacetone	5.3	8.24 (25°)	−20	5–45	T	—	372	—	Enol form
	Pentane-2,4-dione	2.2	8.93 (25°)	−35	5–45	T	—	—	—	Keto form
	Diacetylmethane	3.21	8.82	−29.6	25	C	μ = 0.1 (NaClO₄)	403	—	Solvent: 50 vol. % dioxane in water
		2.6	9.12	−32.2	0	T	0.0	391	—	Conditions refer to wt. fraction methanol in H₂O
		2.5	9.02	−32.8	25	T	0.0			
		2.5	8.93	−32.6	40	T	0.0			
		2.5	9.28	−33.3	0	T	0.098			
		2.6	9.16	−33.3	25	T	0.099			
		2.5	9.00	−33.2	40	T	0.099			
		2.4	9.44	−33.4	0	T	0.220			
		2.4	9.31	−34.6	25	T	0.222			
		2.4	9.21	−34.4	40	T	0.224			
		2.4	9.52	−34.7	0	T	0.289			
		2.2	9.46	−36.1	25	T	0.295			
		2.2	9.35	−35.8	40	T	0.229			
		2.3	9.68	−37.0	0	T	0.374			
		2.0	9.60	−37.0	25	T	0.383			
		2.0	9.49	−36.6	40	T	0.387			
		2.3	9.84	−38.6	0	T	0.474			
		1.8	9.75	−38.3	25	T	0.485			
		1.8	9.70	−38.8	40	T	0.489			
		2.1	10.08	−40.1	0	T	0.590			
		1.7	10.04		25	T	0.610			
		1.7	9.98	−40.2	40	T	0.621			

HEATS OF PROTON IONIZATION, pK, AND RELATED THERMODYNAMIC QUANTITIES (continued)

Compound	Formula. Synonyms	ΔH (kcal/mole)	pK	ΔS cal/deg-mole	T°C	Method	Conditions	Reference	Other References	Remarks
2,4-Pentanedione—(Continued)										
		2.4	9.12	−33.1	0	T	0.00	—	—	Conditions refer to wt. fraction 1-propanol in water.
		2.3	9.02	−33.4	25	T	0.00	—	—	
		2.2	8.93	−33.6	40	T	0.00	—	—	
		2.4	9.28	−34.8	0	T	0.056	—	—	
		2.2	9.16	−34.5	25	T	0.056	—	—	
		2.2	9.07	−34.3	40	T	0.056	391	—	
		2.1	9.52	−36.0	0	T	0.132	391	—	
		2.0	9.38	−36.4	25	T	0.134	391	—	
		2.0	9.35	−36.3	40	T	0.134	391	—	
		2.1	9.76	−36.9	0	T	0.183	391	—	
		1.9	9.60	−37.8	25	T	0.188	391	—	
		1.8	9.56	−38.1	40	T	0.188	391	—	
		1.9	10.00	−38.7	0	T	0.251	391	—	
		1.8	9.90	−39.4	25	T	0.256	391	—	
		1.6	9.77	−40.0	40	T	0.256	391	—	
		1.8	10.32	−40.6	0	T	0.334	391	—	
		1.3	10.26	−42.4	25	T	0.346	391	—	
		1.2	10.12	−42.3	40	T	0.346	391	—	
		1.5	10.72	−43.7	0	T	0.450	391	—	
		0.8	10.70	−46.0	25	T	0.473	391	—	
		0.7	10.61	−46.4	40	T	0.473	391	—	
—, 3-methyl-	$CH_3COCH(CH_3)COCH_3$ 3-Methyl-2,4-pentanedione	5	10.87 (25°)	−63	10–38	T	—	211	—	3.b.d.
Pentanoic acid	$CH_3(CH_2)_3COOH$ Valeric acid Valerianic acid Propylacetic acid	−0.720	4.842	−24.57	25	T	$\mu = 0$	13	—	—
—, 2-amino-	$CH_3(CH_2)_2CHNH_3^+COOH$ Norvaline α-Amino-n-valeric acid	0.83	2.347 (pK$_1$)	−8.0	10	C	$\mu = 0$	324	—	—
		0.39	2.318	−9.3	25	C	$\mu = 0$	324	—	—
		0.06	2.309	−10.7	40	C	—	324	—	—
		1.290	2.376 (pK$_1$)	−6.1	1	T	$\mu = 0$	50	—	—
		0.970	2.340	−7.3	12.5		—	—	—	—
		0.550	2.318	−8.8	25		—	—	—	—
		0.050	2.309	−10.4	37.5		—	—	—	—
		−0.540	2.313	−12.3	50		—	—	—	—
		10.850	9.808 (pK$_2$)	−8.32	25	C	$\mu = 0$	325	—	3.d.
		10.840	10.508	−8.5	1	T	$\mu = 0$	50	—	—
		10.910	10.154	−8.2	12.5		—	—	—	—
		10.880	9.808	−8.4	25		—	—	—	—
		10.700	9.490	−9.0	37.5		—	—	—	—

HEATS OF PROTON IONIZATION, pK, AND RELATED THERMODYNAMIC QUANTITIES (continued)

Compound	Formula, Synonyms	ΔH (kcal/mole)	pK	ΔS cal/dg-mole	T °C	Method	Conditions	Reference	Other References	Remarks
Pentanoic acid—(Continued)										
—, 5-amino-	$^+NH_3(CH_2)_4COOH$ δ-Aminovaleric acid	10.380	9.198	−9.9	50	C	μ = 0	324	—	—
		0.28	4.26 (pK₁)	−18.5	10	C	μ = 0	324	—	—
		−0.22	4.20	−19.9	25	C	μ = 0	324	—	—
		−0.61	4.25	−21.4	40	C	μ = 0	324	—	—
		0.160	4.229	−18.8	25	T	μ = 0	6	—	—
		12.970	10.766 (pK₂)	−5.57	25	C	μ = 0	325	—	3.d.
—, 2-amino-2-methyl-	See: Leucine									
—, 2-amino-3-methyl-	CH₃CH₂CH(CH₃)CH(NH₃⁺)COOH α-Amino-β-methyl- n-Valeric acid DL-Isoleucine	1.080	2.365 (pK₁)	−6.9	1.0	T	μ = 0	50	—	—
		0.740	2.338	−8.1	12.5				—	—
		0.300	2.318	−9.6	25				—	—
		−0.220	2.317	−11.3	37.5				—	—
		−0.840	2.332	−13.3	50				—	—
		10.800	10.460 (pK₂)	−8.4	1.0	T	μ = 0	50	—	—
		10.870	10.100	−8.1	12.5				—	—
		10.830	9.758	−8.3	25				—	—
		10.650	9.439	−8.9	37.5				—	—
		10.330	9.157	−9.9	50				—	—
—, 4-methyl-	(CH₃)₂CH(CH₂)₂COOH Isohexanoic acid Isobutylacetic acid Isocaproic acid	−0.61	4.845	−24.2	25	C	μ = 0	324	13	3.a.b.d.
—, 4-oxo-	CH₃COCH₂CH₂COOH Levulinic acid	0.534	4.609 (35°)	−19	16.8	T		12	—	—
3-Pentene-1,2,3,4,4-pentanitrile-2-phenyl-	β-Acetopropionic acid p-(Tricyanovinyl)- Phenyldicyanomethane	−0.4	0.75	−4.7	25	T	μ = 0	225	—	—
Pepsinogen		6.3	10.97	−24	25	T	μ = 0.15	245	—	3.a.
PETA	See: Acetic acid. (pentamethylenedinitrilo)-tetra-									
Phenanthrene										
—, 1-amino-	1-Phenanthrylamine	5.6	3.23 (20°)	4.2	0–20	T	C = 0.0025	85	—	3.b. 50 wt. % ethanol
—, 2-amino-	2-Phenanthrylamine	4.4	3.60 (20°)	−1.3	0–20	T	C = 0.005	85	—	3.b. 50 wt. % ethanol
—, 3-amino-	3-Phenanthrylamine	4.4	3.59 (20°)	−1.3	0–20	T	C = 0.005	85	—	3.b. 50 wt. % ethanol

HEATS OF PROTON IONIZATION, pK, AND RELATED THERMODYNAMIC QUANTITIES (continued)

Compound	Formula, Synonyms	ΔH (kcal/mole)	pK	ΔS cal/deg-mole	T°C	Method	Conditions	Reference	Other References	Remarks
Phenanthrene—(Continued)										
—, 9-amino-	9-Phenanthrylamine	5.6	3.19 (20°)	4.3	0–20	T	C = 0.0025	85	—	3.b. 50 wt. % ethanol
m-Phenanthroline	See: 1,5-Phenanthroline									
1,5-Phenanthroline		2.17	4.30	−11.4	20	T	$\mu = 0$	136	—	3.d.
1,8-Phenanthroline	—	4.97	7.21	−15.1	20	T	—	136	—	3.d.
—, 4,5-dimethoxy-2,7-dimethyl-		—	—		—					
—, 2,7-dimethyl-	—	6.50 / 5.28	2.56 (pK$_1$) / 5.32 (pK$_2$)	11.4 / −5.4	20 / 20	T / T	$\mu = 0$ / $\mu = 0$	136 / 136	—	—
—, 2,3,4,5,6,7-hexamethyl-	—	4.97	7.26	−15.3	20	T	$\mu = 0$	136	—	—
—, 2,4,5,7-tetramethyl-	—	6.53 / 4.33	3.38 (pK$_1$) / 5.35 (pK$_2$)	7.74 / −8.87	20 / 20	T / T	$\mu = 0$ / $\mu = 0$	136 / 136	—	—
1,10-Phenanthroline		3.5 / 3.95	4.857 (25°) / 4.96	−10.2 / −9.2	0–50 / 20	T / C	$\mu = 0$ / 0.1 M NaNO$_3$	246 / 190	35	3.b.
—, 5-nitro-	—	2.26	3.23	−7.0	25	T	—	392	—	3.d. Sodium Acetate Buffer 0.2M
4,5-Phenanthroline	4,5-Diazaphenanthrene o-Phenanthroline	3.0 / 3.47 / 4.07	3.95 / 4.85 / 5.06 (25°)	−8 / −9.41 / −9.5	25 / 20 / 25–45	C / T / T	$\mu = 1.0$ / — / —	354 / 136 / 139	—	— / 3.d. / 3.b.d.
—, 10-methyl-	—	4.40	5.11	−8.6	25	T	$\mu = 0$	393	—	—
—, 10-nitro-	—	2.26	3.23	−7.2	25	T	$\mu = 0$	394	—	—

HEATS OF PROTON IONIZATION, pK, AND RELATED THERMODYNAMIC QUANTITIES (continued)

Compound	Formula, Synonyms	ΔH (kcal/mole)	pK	ΔS cal/deg-mole	T °C	Method	Conditions	Reference	Other References	Remarks
***o*-Phenanthroline**	See: 4,5-Phenanthroline									
Phenanthrylamine	See: Phenanthrene, amino-									
Phenazine	(structure)	3.35	2.67	0.09	15	T	—	36	—	3.d.
	3-Hydroxyphenazine (structure, OH)									
Phenobarbital	See: Barbituric acid, 5-ethyl-5-phenyl-									
Phenol	OH (structure) Hydroxybenzene Benzenol	5.650	9.979	−26.7	25	C	$\mu = 0$	274	87, 90, 106, 107, 120, 248, 249	—
		5.47	9.95	−27.2	5–60	T	$\mu = 0.04$	395		3.b.
		6.1	9.8	−24	25	T	$\mu = 0.10$	346		—
—, 2 bromo-		4.40	8.452 (25°)	−23.9	5–50	T	$\mu = 0$	396	—	3.b.
—, 3 bromo-		5.35	9.031 (25°)	−23.4	10–60	T	$\mu = 0$	397	—	3.b.
—, 4 bromo-		5.74	9.34 (25°)	−23.5	5–60	T	$\mu = 0$	395	—	3.b.
—, 2-chloro-		5.101	—	—	10	C	$\mu = 0$	122	106, 247, 250	—
		4.795	—	—	20	—				
		4.590	8.487	−23.5	25	T				
		4.387	—	—	30					
—, 3-chloro-		4.26	8.555 (25°)	−24.9	5–50	T	$\mu = 0$	396	—	3.b.
		4.7	6.3	−22	25	T	$\mu = 0.10$	346	—	—
—, 4-chloro-		5.286	9.08	−23.8	25	C	$\mu = 0$	250	397	—
		6.1	8.7	−19	25	C	$\mu = 0.10$	346	—	—
—, 2-cyano-		5.800	9.379	−23.5	25	T	$\mu = 0$	247	250	—
		6.1	9.1	−21	25	T	$\mu = 0.10$	346	—	—
—, 3-cyano-	Benzoic acid. 3-hydroxy-, nitrile *m*-Cyanophenol 3-Cyanophenol	4.0	6.8	−18	25	C	$\mu = 0$	346	395	—
		5.20	8.57	−21.8	25	T	$\mu = 0.10$	90	—	—
		5.3	8.3	−20	25	C	$\mu = 0.10$	346	—	—
—, 4-cyano-	Benzoic acid. 4-hydroxy-, nitrile 4-Cyanophenol *p*-Cyanophenol	4.92	7.97	−20.0	25	C	$\mu = 0$	90	—	—
		4.8	7.6	−19	25	T	$\mu = 0.10$	346	—	—

HEATS OF PROTON IONIZATION, pK, AND RELATED THERMODYNAMIC QUANTITIES (continued)

Compound	Formula, Synonyms	ΔH (kcal/mole)	pK	ΔS cal/deg-mole	T °C	Method	Conditions	Reference	Other References	Remarks
Phenol—(Continued)										
—, 2,4-dihydroxymethyl-	[structure: OH, CH₂OH, CH₂OH]	4.9	9.79	−28.4	25	T	$\mu = 0$	248	—	—
—, 2,3-dimethyl-	See: Benzene, 1,2-dimethyl-3-hydroxy-									
—, 2,4-dimethyl-	See: Benzene, 1,3-dimethyl-4-hydroxy-									
—, 2,5-dimethyl-	See: Benzene, 1,4-dimethyl-2-hydroxy-									
—, 2,6-dimethyl-	See: Benzene, 1,3-dimethyl-2-hydroxy-									
—, 3,4-dimethyl-	See: Benzene, 1,2-dimethyl-4-hydroxy-									
—, 3,6-dimethyl-	See: Benzene, 1,5-dimethyl-3-hydroxy-									
—, 2-ethoxy-	See: Benzene, 3-hydroxy-1-ethyl-									
Phenol, 3-ethyl-		6.18	10.109 (25°)	−25.5	5–20	T	$\mu = 0.04$	396	—	3.b.
—, 2-fluoro-	—	4.66	8.73	−24.3	25	C	$\mu = 0$	251	—	—
—, 3-fluoro-	—	5.52	9.29	−24.0	25	C	$\mu = 0$	251	—	—
—, 4-fluoro-	—	5.93	9.89	−25.4	25	C	$\mu = 0$	251	—	—
—, 2-hydroxymethyl-	[structure: OH, CH₂OH]	5.4	9.95	−27.4	25	T	$\mu = 0$	248	—	—
—, 4-hydroxymethyl-	[structure: OH, CH₂OH]	4.5	9.84	−30.0	25	T	$\mu = 0$	248	—	—
—, 2-iodo-	—	4.27	8.513 (25°)	−24.6	5–50	T	$\mu = 0.04$	396	—	3.b.
—, 3-iodo-	—	5.53	9.023 (25°)	−22.7	10–60	T	$\mu = 0.4$	397	—	3.b.
—, 4-iodo-	—	5.38	9.32 (25°)	−24.6	5–60	T	$\mu = 0.04$	395	—	3.b.
—, 2-methoxy-	Guaiacol, Pyrocatechol monomethyl ether, o-Methoxyphenol	5.74	9.99	−26.5	25	C	$\mu = 0$	249	345	—
—, 3-methoxy-	Resorcinol monomethyl ether, m-Methoxyphenol	5.26	9.65	−26.5	25	C	$\mu = 0$	249	—	—
		—	—	—	—			—	—	—

HEATS OF PROTON IONIZATION, pK, AND RELATED THERMODYNAMIC QUANTITIES (continued)

Compound	Formula. Synonyms	ΔH (kcal/mole)	pK	ΔS cal/deg-mole	T°C	Method	Conditions	Reference	Other References	Remarks
Phenol, 3-ethyl—(Continued)										
—, 4-methoxy-	p-Methoxyphenol	5.70	10.20	-27.6	25	C	μ = 0	249	—	—
	Hydroquinone monomethyl ether	—	—	—	—	—	—	—	—	—
—, 2-nitro-	o-Nitrophenol	4.655	7.221	-17.4	25	C	μ = 0	247	—	3.b.
		4.59	7.230 (25°)	-17.67	5-60	T	μ = 0.1	398	247, 252	—
		5.145	—	—	10	C	μ = 0	122	345	—
—, 3-nitro-	m-Nitrophenol	4.833	—	—	20	—	—	—	—	—
		4.705	8.267	-22.1	25	—	—	—	—	—
		4.575	—	—	30	—	—	—	—	—
		4.8	8.1	-21	25	T	μ = 0.10	346	346	—
—, 4-nitro-	p-Nitrophenol	4.700	7.143	-16.9	25	C	μ = 0	247	253	—
		4.895	—	—	10	C	μ = 0	122	—	—
		4.669	—	—	20	—	—	—	—	—
		4.570	7.150	-17.4	25	—	—	—	—	—
		4.466	—	—	30	—	—	—	—	—
		4.71	7.156 (25°)	-16.96	5-60	T	μ = 0.1	398	—	3.b.
		4.5	9.56	-28.6	25	T	μ = 0	248	—	—
—, 2,4,6-trihydroxymethyl- —, 2,4,6-trinitro-	Picric acid	-1.483	0.19 (20°)	-6.03	5-60	T	μ = 0	231	254	3.b.
cis-β-Phenylacrylic acid	See: Cinnamic acid	—	—	—	—	—	—	—	—	—
Phenylalanine	$CH_2CH(\overset{+}{N}H_3)COOH$	11.40	9.75 (pK$_2$)	-4.3	10	C	μ = 0	197	—	—
		10.67	9.31	-6.7	25						
		10.53	8.96	-7.4	40						
Phenylalanylarginine	See: Arginine, phenylalanyl-									
7-Phenylazo-8-hydroxy-quinoline-5-sulfonic acid	See: Quinolinesulfonic acid. 7-phenylazo-8-hydroxy-									
Phenylboric acid	See: Benzeneboronic acid									
Phenylethanoic acid	See: Acetic acid, phenyl-									
Phenylglycolic acid	See: Acetic acid, hydroxyphenyl-									
Phenylpropiolic acid	See: Propynoic acid, phenyl-									
p-Phenolsulfonic acid	See: Benzenesulfonic acid, 4-hydroxy-									
Phenylsulfonylacetic acid	See: Acetic acid, phenylsulfonyl-									
Phenylsulfoxyacetic acid	See: Acetic acid, phenylsulfoxy-									
Phenylthioglycolic acid	See: Acetic acid, 2-mercaptophenyl-									
Phosphoramidic acid		~0	3.08 (25°, pK$_1$)	-14	18-39	T	μ = 0	399	—	3.b.
		5.2	8.63 (25°, pK$_2$)	-22	18-39	T	μ = 0	399	—	3.b.
Phosphoric acid		2.034	7.2810 (5°, pK$_2$)	-26.0	0	T	μ = 0	255	—	—
		0.987	7.2005	-29.6	25						
		-0.423	7.1828	-34.2	30						
—, deuterio-	Deuteriophosphoric acid	2.504	7.885 (5°, pK$_2$)	-27.1	0	T	μ = 0	255	—	Solvent: D$_2$O

HEATS OF PROTON IONIZATION, pK, AND RELATED THERMODYNAMIC QUANTITIES (continued)

Phosphoric acid—(Continued)

Compound	Formula, Synonyms	ΔH (kcal/mole)	pK	ΔS cal/deg-mole	T°C	Method	Conditions	Reference	Other References	Remarks
—, monofluoro-	—									
—, ortho-	H_3PO_4	1.376	7.780	−31.0	25	—	—	—	—	—
		−0.144	7.744	−35.9	50	—	—	—	—	—
		−1.90	4.79 (pK$_2$)	−28.2	25	—	—	329	—	—
		−1.880	2.124 (pK$_1$)	−16.0	25	C	$\mu = 0$	81	10, 46	—
		−1.466	2.107 (pK$_1$)	−14.7	15	T	$\mu = 0$	256	81, 259−261	—
		−1.646	2.127	−15.3	20				333	—
		−1.828	2.148	−16.0	25				389	—
		−2.014	2.171	−16.6	30					—
		−2.203	2.196	−17.2	35					—
		−2.394	2.224	−17.8	40					—
		−2.589	2.251	−18.4	45					—
		0.90	7.20 (pK$_2$)	−30	25	C	$\mu = 0$	38		—
		2.284	7.3139 (pK$_2$)	−25.1	0	T	$\mu = 0$	257		—
		2.034	7.2806	−26.0	5					—
		1.779	7.2533	−26.9	10					—
		1.520	7.2305	−27.8	15					—
		1.256	7.2129	−28.7	20					—
		0.987	7.1972	−29.6	25					—
		0.714	7.1909	−30.5	30					—
		0.437	7.1856	−31.4	35					—
		0.155	7.1822	−32.4	40					—
		−0.132	7.1818	−33.3	45					—
		−0.432	7.1851	−34.2	50					—
		−0.718	7.1893	−35.1	55					—
		−1.019	7.1981	−36.0	60					—
		6.000	6.76 (pK$_2$)	−9.4	5	T	$\mu = 0.7$	91		—
		2.900	6.68	−20.5	15					—
		−0.200	6.64	−31.0	25					—
		−4.200	6.68	−44.0	37					—
—, pyro-	$PO_3H_2OPO_3H_2$ Pyrophosphoric acid	4.20	12.40 (pK$_3$)	−43	25	C	$\mu = 0$	258		—
		−0.3	6.7 (pK$_3$)	−32	25	C	$\mu = 0$	38	46, 262	—
		0.1	6.76 (pK$_3$)	−31	25	C	$\mu = 0.13$	259		—
		−1.7	9.4 (pK$_4$)	−49	25	C	$\mu = 0$	38		—
		0.4	9.41 (pK$_4$)	−42	25	C	$\mu = 0.22$	259		—
—, tripoly-	$PO_3H_2PO_4HPO_3H_2$ Tripolyphosphoric acid	−1.3	6.2 (pK$_4$)	−33	25	C	$\mu = 0$	38		—
		−1.5	6.50 (pK$_4$)	−31	25	C	$\mu = 0.4$	259		—
		−3.8	8.9 (pK$_5$)	−53	25	C	$\mu = 0$	38		—
		0.10	8.82 (pK$_5$)	−40	20	C	0.1 M (CH$_3$)$_4$ NOH	33		—
		0.1	9.24 (pK$_5$)	−42	25	C	$\mu = 0.65$	259		—

HEATS OF PROTON IONIZATION, pK, AND RELATED THERMODYNAMIC QUANTITIES (continued)

Compound	Formula, Synonyms	ΔH (kcal/mole)	pK	ΔS cal/deg-mole	T °C	Method	Conditions	Reference	Other References	Remarks
Phthalic acid	See: 1,2-Benzene-dicarboxylic acid									
α-Picoline	See: Pyridine, 2-methyl-									
2-Picoline	See: Pyridine, 2-methyl-									
β-Picoline	See: Pyridine, 3-methyl-									
3-Picoline	See: Pyridine, 3-methyl-									
γ-Picoline	See: Pyridine, 4-methyl-									
4-Picoline	See: Pyridine, 4-methyl-									
Picolinic acid	See: 2-Pyridinecarboxylic acid									
2-Picolylamine	See: Pyridine, 2-(methylamino)-									
3-Picolylamine	See: Pyridine, 3-(methylamino)-									
2-Picolylmethylamine	See: Pyridine, 2-(methylaminomethyl)-									
Picric acid	See: Phenol, 2,4,6-trinitro-									
Pimelic acid	See: Heptanedioic acid									
2-Pipecolylamine	See: Piperidine, 2-(aminomethyl)-									
Piperazine	$H-N_4^{\ 6\ 5}\!\!^{\ \ \ }{}_{3\ \ 2}^{\ 1}N-H$	7.42	5.333 (25°, pK$_1$)	0.53	0–50	T	$\mu = 0$	364	263	3.b.
		7.12	5.60 (30°, pK$_1$)	−1.8	25	C	0.1 M KCl	175	—	—
		10.24	9.731 (25°, pK$_2$)	−10.17	0–50	T	$\mu = 0$	364	263	3.b.
		10.17	9.72 (pK$_2$)	−10.3	25	C	0.1 M KCl	175	—	—
	Diethylenediamine									
	Hexahydropyrazine									
Piperidine		12.26	11.963	−9.85	0	T	$\mu = 0$	264	35, 236	—
		12.36	11.786	−9.51	5	—	—	—	—	—
		12.45	11.613	−9.15	10	—	—	—	—	—
		12.55	11.443	−8.80	15	—	—	—	—	—
		12.66	11.280	−8.46	20	—	—	—	—	—
		12.76	11.123	−8.10	25	—	—	—	—	—
		12.86	10.974	−7.74	30	—	—	—	—	—
		12.97	10.818	−7.41	35	—	—	—	—	—
		13.08	10.670	−7.05	40	—	—	—	—	—
		13.20	10.526	−6.69	45	—	—	—	—	—
		13.31	10.384	−6.33	50	—	—	—	—	—
		12.190	11.13	−10.0	25	C	$\mu = 0$	37	—	3.d.
		12.9	11.06	−7.34	25	T	$\mu = 0$	99	—	10 wt. % CH$_3$OH
		12.4	10.96	−8.56	25	T	$\mu = 0$	—	—	60 wt. % CH$_3$OH
	Pentamethyleneimine									
		13.3	10.09	−1.56	25	T	$\mu = 0$	264	—	—
—, 2-(2-aminoethyl)-	2,2'-Aminoethylpiperidine	12	8.79 (10°, pK$_1$)	2 (10°)	10–40	T	$\mu = 0$	265	—	3.b.
		11	10.78 (10°, pK$_2$)	−9 (10°)	10–40	T	$\mu = 0$	265	—	3.b.

HEATS OF PROTON IONIZATION, pK, AND RELATED THERMODYNAMIC QUANTITIES (continued)

Compound	Formula, Synonyms	ΔH (kcal/mole)	pK	ΔS cal/deg-mole	T°C	Method	Conditions	Reference	Other References	Remarks
Piperidine—(Continued)										
—, 2-(aminomethyl)-	[ring structure, positions 2–6, N$^+$H, CH$_2$NH$_3^+$]	10	6.54 (10°, pK$_1$)	4 (10°)	10–40	T	$\mu = 0$	265	—	3.b.
		11	9.97 (10°, pK$_2$)	−6 (10°)	10–40	T	$\mu = 0$	265	—	3.b.
Proline	[ring structure, N$^+$H, COOH]	1.149	2.011 (pK$_1$)	−5.0	1					—
		0.780	1.964	−6.3	12.5					—
		0.342	1.952	−7.8	25	T	$\mu = 0$	266	150	—
		−0.181	1.950	−9.5	37.5					—
		−0.746	1.958	−11.3	50					—
		10.360	11.296 (pK$_2$)	−13.9	1					—
		10.400	10.972	−13.8	12.5					—
		10.310	10.640	−14.1	25	T	$\mu = 0$	266	150	—
		10.085	10.342	−14.7	37.5					—
		9.715	10.064	−16.0	50					—
—, 4-hydroxy-	Hydroxyproline	1.602	1.900 (pK$_1$)	−2.8	1					—
		1.310	1.850	−3.9	12.5					—
		0.918	1.818	−5.2	25	T	$\mu = 0$	266	150	—
		0.446	1.798	−6.8	37.5					—
		−0.115	1.796	−8.5	50					—
		9.585	10.274 (pK$_2$)	−12.0	1					—
		9.543	9.958	−12.2	12.5					—
		9.385	9.662	−12.7	25	T	$\mu = 0$	266	150	—
		9.080	9.394	−13.7	37.5					—
		8.635	9.138	−15.1	50					—
Propane										
—, 1-amino-	n-Propylamine $H_3CCH_2CH_2NH_3^+$	13.84	10.568	−2.0	25	C	$\mu = 0$	330	—	—
		13.7	10.568	−2.5	25	T	$\mu = 0$	77	66	—
		13.95	10.086 (40°)	−1.6	20–40	T	$\mu = 0.20$	64	336, 356	3.b.
		13.94	10.062 (40°)	−1.5			$\mu = 0.10$			—
		13.85	10.060 (40°)	−1.8			$\mu = 0.05$			—
		13.85	10.047 (40°)	−1.7						—
—, 2-amino-	—	13.97	10.67	−2.0	25	C	$\mu = 0$	330	—	—
—, 1-amino-2-methyl-	—	13.92	10.48	−1.3	25	C	$\mu = 0$	330	356	—
—, 2-amino-2-methyl-	t-Butylamine	14.43	10.685	−0.5	25	C	$\mu = 0$	330	—	—
		14.29	11.439	−0.98	5	T	$\mu = 0$	267	—	—
		14.30	11.240	−0.91	10					—
		14.32	11.048	−0.86	15					—

HEATS OF PROTON IONIZATION, pK, AND RELATED THERMODYNAMIC QUANTITIES (continued)

Compound	Formula, Synonyms	ΔH (kcal/mole)	pK	ΔS cal/deg-mole	T °C	Method	Conditions	Reference	Other References	Remarks
Propane—(Continued)										
—, 1,2-diamino-	1,2-Propanediamine	14.34	10.862	-0.79	20	—	—	—	—	—
		14.36	10.685	-0.74	25	—	—	—	—	—
		14.37	10.511	-0.67	30	—	—	—	—	—
		14.39	10.341	-0.62	35	—	—	—	—	—
		9.65	6.607 (pK$_1$)	2.1	25	C	$\mu = 0$	144	—	—
		10.1	7.13 (25°, pK$_2$)	1.25	0-25	T	0.50 M KNO$_3$	146	174	3.a.b.
		11.92	9.720 (pK$_2$)	-4.5	25	C	$\mu = 0$	146	—	—
		11.3	10.00 (25°, pK$_2$)	-7.85	0-25	T	0.50 M KNO$_3$	146	—	3.a.b.
—, 1,3-diamino-	Trimethylenediamine	12.44	8.49 (pK$_1$)	2.9	25	C	$\mu = 0$	330	145	—
		12.97	9.13 (pK$_1$)	1.7	25	C	0.3 M NaClO$_4$	353	174	—
		13.19	10.47 (pK$_2$)	-3.7	25	C	$\mu = 0$	330	—	3.b.
		13.3	10.23 (30°, pK$_2$)	-3	10-40	T	$\mu = 0$	145	—	—
		13.9	10.62 (pK$_2$)	-4.6	25	C	0.3 M NaClO$_4$	353	—	—
—, 1,3-diamino-2-hydroxyl-	1,3-Diaminopropane-2-ol	13.08	10.55 (pK$_2$)	-4.4	25	C	$\mu = 0.5$	174	—	3.b.
		10.4	7.68 (30°, pK$_1$)	-1	10-40	T	$\mu = 0$	145	—	3.b.
—, 1,2-diamino-2-methyl-	11.2	11.2	9.42 (30°, pK$_2$)	-6	10-40	T	$\mu = 0$	145	146	—
	$NH_3^+CH_2C(CH_3)_2^+NH_3$ 1,2-Diamino-1,1-dimethylethane 1,2-Diaminoisobutane	9.66	6.178 (pK$_1$)	4.1	25	C	$\mu = 0$	144	—	—
		11.78	9.420 (pK$_2$)	-3.6	25	C	$\mu = 0$	144	—	—
—, 1,1-dinitro-	1,1-Dinitropropane	3.675	5.5	-12.83	20	T	—	147	—	3.b.
Propanedioic acid	See: Malonic acid									
1,3-Propanediol,-2-amino-2-(hydroxymethyl)-	$(CH_2OH)_3CNH_3^+$	11.347	8.030	—	25	C	$\mu = 0$	417	77, 270	—
		11.39	—	1.5	25	C	$\mu = 0.010$	348	271	—
	THAM Tris(hydroxymethyl) Aminomethane TRIS	11.95	8.6792	3.24	5	T	$\mu = 0$	268	—	—
		11.60	8.3602	1.99	15	—	—	—	—	—
		11.33	8.0686	1.08	25	—	—	—	—	—
		11.16	7.8006	0.52	35	—	—	—	—	—
		11.10	7.5515	0.31	45	—	—	—	—	—
		11.14	7.3204	0.46	55	—	—	—	—	—
		11.36	—	—	25	—	—	—	—	—
		11.70	8.38	0.90	25	C	$\mu = 0.1$	3	—	—
		11.43	7.818	2.55	25	T	$\mu = 0.65$	91	269	50 wt. % methanol
—, 2-amino-2-methyl-	$CH_3CNH_3^+(CH_2OH)_2$ 2-Amino-2-methyl-1,3-propanediol	12.17	9.6116	0.60	0	T	$\mu = 0$	272	—	—
		12.12	9.4328	0.41	5	—	—	—	—	—
		12.07	9.2658	0.24	10	—	—	—	—	—

HEATS OF PROTON IONIZATION, pK, AND RELATED THERMODYNAMIC QUANTITIES (continued)

Compound	Formula, Synonyms	ΔH (kcal/mole)	pK	ΔS cal/deg-mole	T°C	Method	Conditions	Reference	Other References	Remarks
1,3-Propanediol—(Continued)										
		12.02	9.1044	0.05	15	—	—	—	—	—
		11.97	8.9508	−0.12	20	—	—	—	—	—
		11.92	8.8013	−0.31	25	—	—	—	—	—
		11.86	8.6588	−0.48	30	—	—	—	—	—
		11.81	8.5193	−0.67	35	—	—	—	—	—
		11.75	8.3854	−0.84	40	—	—	—	—	—
		11.70	8.2569	−1.03	45	—	—	—	—	—
		11.64	8.1322	−1.20	50	—	—	—	—	—
Propanoic acid	CH$_3$CH$_2$COOH Propionic acid	−0.14	4.874	−22.8	25	C	μ = 0	1	5, 9, 10	—
		0.737	4.895	−19.7	0	T	μ = 0	273	12, 13	—
		0.562	4.884	−20.3	5	—	—	—	326	—
		0.384	4.877	−21.0	10	—	—	—	—	—
		0.203	4.874	−21.6	15	—	—	—	—	—
		0.019	4.873	−22.2	20	—	—	—	—	—
		−0.168	4.874	−22.9	25	—	—	—	—	—
		−0.358	4.877	−23.5	30	—	—	—	—	—
		−0.551	4.883	−24.1	35	—	—	—	—	—
		−0.746	4.891	−24.8	40	—	—	—	—	—
		−0.945	4.901	−25.4	45	—	—	—	—	—
		−1.147	4.910	−26.0	50	—	—	—	—	—
		−1.351	4.923	−26.6	55	—	—	—	—	—
		−1.559	4.936	−27.3	60	—	—	—	—	—
		−0.163	4.874	−22.8	25	T	0	185	—	—
		−0.148	5.467	−25.5	25	T	20	—	—	Conditions refer to wt. % dioxane
		−0.206	6.554	−30.7	25	T	45	—	—	
		−0.201	8.614	−40.1	25	T	70	—	—	
		−1.064	10.415	−51.2	25	T	82	274	—	Conditions refer to wt. % iso-propyl-alcohol
		0.6721	4.9924	−20.38	0	T	5	—	—	
		−0.2424	4.9766	−23.58	25	T	5	—	—	
		−0.8294	4.9953	−25.50	40	T	5	—	—	Conditions refer to wt. % iso-propyl-alcohol
		0.8227	5.1064	−20.35	0	T	10	274	—	

HEATS OF PROTON IONIZATION, pK, AND RELATED THERMODYNAMIC QUANTITIES (continued)

Compound	Formula, Synonyms	ΔH (kcal/mole)	pK	ΔS cal/deg-mole	T°C	Method	Conditions	Reference	Other References	Remarks
Propanoic acid—(Continued)		-0.2493	5.0857	-24.10	25	T	10	—	—	—
		-0.9376	5.1062	-26.35	40	T	10	274	—	Conditions refer to wt. % iso-propyl-alcohol
		1.1199	5.3633	-20.44	0	T	20	—	—	—
—, 2-amino-		-0.2168	5.3310	-25.12	25	T	20	—	—	—
—, 2-amino-2-methyl-	α-Aminoisobutyric acid	-1.0750	5.3533	-27.92	40	T	20	—	—	3.d.
		10.750	9.866 (pK$_2$)	-8.93	0	C	—	325	—	
		1.41	2.36 (pK$_1$)	-6	25	C	μ = 0	201	150	
		1.300	2.419 (pK$_1$)	-6.2	1	T	μ = 0	50	—	
		0.980	2.380	-7.1	12.5	—	—	—	—	
		0.560	2.357	-8.9	25	—	—	—	—	
		0.060	2.351	-10.6	37.5	—	—	—	—	
		-0.540	2.356	-12.4	50	—	—	—	—	
		11.57	10.205 (pK$_2$)	-7.9	25	C	μ = 0	330	201	
		11.480	10.960 (pK$_2$)	-8.2	1	T	μ = 0	50	—	
		11.600	10.580	-7.8	12.5	—	—	—	—	
		11.610	10.205	-7.7	25	—	—	—	—	
		11.510	9.872	-8.1	37.5	—	—	—	—	
		11.260	9.561	-8.9	50	—	—	—	—	
		11.090	10.205	-9.29	25	C	μ = 0	325	326	
—, 2-amino-3-methyl-								1	326	3.d.
—, 2-bromo-	α-Bromopropionic acid	-1.31	2.971	-18.0	25	C	μ = 0	324	1, 12, 326	
—, 3-bromo-	β-Bromopropionic acid	-0.17	3.992	-18.9	25	C	μ = 0	326	326	
—, carbamoyl-α-amino-2-methyl-	$\overset{+}{N}H_3CONHC(CH_3)_2COOH$ Carbamoyl-α-amino-isobutyric acid	0.217	4.4627 (pK$_1$)	-19.69	25	T	μ = 0	58	1, 326	
—, 2-chloro-	α-Chloropropionic acid	-0.51	3.992	-20.0	25	C	μ = 0	324	122, 326	
—, 3-chloro-	β-Chloropropionic acid	-0.32	3.992	-19.3	25	C	μ = 0	1	—	
—, 2,3-dibromo-		-1.89	2.33	-17.0	25	C	μ = 0	326	—	
—, 2,2-dichloro-		-0.45	2.06	-10.9	25	C	μ = 0	326	—	
—, 2,3-dichloro-		-0.17	2.85	-13.6	25	C	μ = 0	326	—	
—, 2,2-dinitrile-	Methyldicyanoacetate	-4.5	-2.8	-2	25	T	μ = 0	225	—	
—, 3,3'-dithio-di-	See: Dipropyl-3,3'-dicarboxydisulfide									
—, 2-hydroxy-	Lactic acid Sarcolactic acid	0.768	3.880	-14.9	0	T	μ = 0	275	276	
		0.619	3.873	-15.5	5	—	—	—	—	
		0.458	3.868	-16.1	10	—	—	—	—	
		0.285	3.861	-16.7	15	—	—	—	—	
		0.098	3.857	-17.3	20	—	—	—	—	

HEATS OF PROTON IONIZATION, pK, AND RELATED THERMODYNAMIC QUANTITIES (continued)

Propanoic acid—(Continued)

Compound	Formula, Synonyms	ΔH (kcal/mole)	pK	ΔS cal/deg-mole	T°C	Method	Conditions	Reference	Other References	Remarks
—, 2-hydroxy-2-methyl-	Hydroxyisobutyric acid	-0.102	3.858	-18.0	25	—	—	—	—	—
		-0.315	3.861	-18.7	30	—	—	—	—	—
		-0.543	3.867	-19.45	35	—	—	—	—	—
		-0.686	3.870	-19.9	38	—	—	—	—	—
		-0.785	3.873	-20.2	40	—	—	—	—	—
		-1.041	3.883	-21.0	45	—	—	—	—	—
		-1.313	3.895	-21.9	50	—	—	—	—	—
	Acetonic acid	0.518	3.971 (35°)	-16	16.8	T	—	12	—	3.a.b.d.
—, 3-iodo-	β-Iodopropionic acid	-1.38	4.08	-23.2	25	C	μ = 0	326	12	—
—, 3-mercapto-	β-Mercaptopropionic	-0.450	4.031 (35°)	-20	16.8	T	—	12	—	3.a.b.d.
—, 2-methyl-	Isobutyric acid	6.10	10.84 (pK_2)	-29.1	25	C	μ = 0.05	29	—	—
		-0.345	—	—	10	C	μ = 0	122	9, 12	—
		-0.623	—	—	20	—	—	—	13	—
		-0.775	4.853	-24.8	25	—	—	—	324	—
		-0.921	—	—	30	—	—	—	—	—
—, 2-oxo-	$CH_3COCOOH$ Pyruvic acid Acetylformic acid	2.90	2.49	-1.7	25	C	μ = 0	34	—	—
		3.011	2.60	-1.80	25	C	μ = 0.05	357	—	—
—, 3-phenyl-	Hydrocinnamic acid	-0.066	4.664 (35°)	-22	16.8	T	—	12	—	3.a.b.d.
1-Propanol	$CH_3CH_2CH_2OH$									
—, 1-amino-		14.4	9.96	3	25	T	μ = 0	74	—	—
—, 2-amino-		12.042	9.469 (25°)	-2.9	0-60	T	μ = 0	335	—	3.b.
—, 3-amino-		12.74	9.96	-2.9	25	C	μ = 0	330	356	3.b.
—, 2-amino-2-methyl-	2-Amino-2-methyl-1-propanol	12.93	9.69	-0.98	25	C	μ = 0	356	267	—
Propene										
—, 3-amino-		13.06	9.52	0.2	25	C	μ = 0	330	—	—
		12.89	9.691	-1.1	25	T	μ = 0	267	—	—

Propionic acid — See: Propanoic acid

—, 3-iodo- — See: Propanoic acid, 3-iodo-

N-Propionylglycine — See: Glycine, N-propionyl-

Propylacetic acid — See: Pentanoic acid

Propylamine — See: Propane, 1-amino-

α-N-n-Propylamine oxime — See: 2-Butanone, 3-(n-propylamine)-3-methyl-, oxime

Propylenediamine — See: Propane-1,2-diamino-

Propylenedinitrilotetra-acetic acid — See: Acetic acid, propylenedinitrilotetra-

HEATS OF PROTON IONIZATION, pK, AND RELATED THERMODYNAMIC QUANTITIES (continued)

Compound	Formula, Synonyms	ΔH (kcal/mole)	pK	ΔS cal/deg-mole	T°C	Method	Conditions	Reference	Other References	Remarks
Propynoic acid —, phenyl-	⬡—C:CCOOH	−0.792	2.269	−13	16.8	T	—	12	—	3.a.b.d.
Pseudoemenol	Phenylpropiolic acid See: Benzene, 1-hydroxy-2.4.5-trimethyl-									
Pseudoephedrine · HCl	See: Ephedrine, pseudo·HCl									
Purine		8.2	8.74 (30°, pK$_2$)	−14.0	20–50	T	$\mu = 0$	43	—	3.b.
—, 6-amino-	See: Adenine									
Putrescine	See: Butane, 1,4-diamino-									
Pyrene										
—, 3-amino-	3-Aminobenzo(d.e.f)phenanthrene	3.9	2.91 (20°)	−0.1	0–20	T	C = 0.005	85	—	3.b. 50 wt. % ethanol
Pyridine	Azine	4.92	5.17	−7.2	25	C	$\mu = 0$	330	37	
		4.130	—	—	10	C	C = 0.05	86	277, 278	3.d.
		4.640	—	—	20				280, 285	3.d.
		4.721	5.30	−8.4	25				348, 400	
		4.800	—	—	30					
		5.230	4.38	−2.50	25	C		37		3.d. 50 wt. % ethanol
—, 2-aldehyde-2'-Pyridylhydrazone-		5.3	2.91 (25°)	4	5–60	T	$\mu = 0$	401		3.b.
		6.9	5.62 (25°)	−3	5–60	T	$\mu = 0$	401		3.b.
—, 2,4-dimethyl-	2,4-Lutidine	7.165	6.79	−7.00	25	C	$\mu = 0$	37		3.d.
—, 2-acetyl-	Methyl, 2-pyridylketone	3.073	2.64	−1.7	25	C	$\mu = 0$	277		
—, 3-acetyl-	2-Acetylpyridine Methyl, 3-pyridylketone	3.127	3.26	−4.4	25	C	$\mu = 0$	277		
—, 4-acetyl-	3-Acetylpyridine	3.168	3.51	−3.4	25	C	$\mu = 0$	277		
—, 2-aldehyde-	4-Acetylpyridine	5.0	4.13	−0.7	5	T	C = 10^{-4} M	281		3.d.
		5.7	4.00	1.7	15					
		6.5	3.84	4.1	25					
		6.8	3.76	5.3	30					

HEATS OF PROTON IONIZATION, pK, AND RELATED THERMODYNAMIC QUANTITIES (continued)

Compound	Formula, Synonyms	ΔH (kcal/mole)	pK	ΔS cal/deg-mole	T°C	Method	Conditions	Reference	Other References	Remarks
Pyridine—(Continued)										
—, 2-aldoxime-		7.6	3.57	7.7	40	—	—	—	—	—
		8.3	3.42	10.1	50	—	—	—	—	—
		9.1	3.25	12.6	60	—	—	—	—	—
		3.32	3.42 (pK$_1$)	−4.5	25	T	$\mu = 0$	282	—	—
		7.7	3.88 (pK$_1$)	9.9	5	T	$C = 10^{-4}\,M$	281	—	3.d.
		6.3	3.70	4.9	15	—	—	—	—	—
		4.8	3.56	0.0	25	—	—	—	—	—
		4.1	3.51	−2.5	30	—	—	—	—	—
		2.6	3.42	−7.5	40	—	—	—	—	—
		1.0	3.39	−12.5	50	—	—	—	—	—
		−0.7	3.38	−17.5	60	—	—	—	—	—
		6.8	10.22 (pK$_2$)	−24.0	25	T	$\mu = 0$	282	—	—
		0.7	10.25 (pK$_2$)	−44.3	5	T	$C = 10^{-4}$	281	—	3.d.
		1.4	10.21	−41.8	15	—	—	—	—	—
		2.1	10.17	−39.3	25	—	—	—	—	—
		2.5	10.13	−38.1	30	—	—	—	—	—
		3.3	10.08	−35.7	40	—	—	—	—	—
		4.0	10.00	−33.2	50	—	—	—	—	—
		4.9	9.91	−30.7	60	—	—	—	—	—
—, 2-amino-	—	8.4	6.86 (20°)	−2.9	5–35	T	$C = 0.02\,M$	85	—	3.a.d.
—, 3-amino-	—	5.1	5.98 (20°)	−7.4	5–35	T	$C = 0.02\,M$	85	—	3.b.d.
—, 4-amino-	—	11.338	9.8731	−3.65	0	T	$\mu = 0$	283	85, 285	—
		11.321	9.7043	−3.70	5	—	—	—	—	—
		11.305	9.5486	−3.78	10	—	—	—	—	—
		11.288	9.3979	−3.82	15	—	—	—	—	—
		11.271	9.2524	−3.90	20	—	—	—	—	—
		11.254	9.1141	−3.94	25	—	—	—	—	—
		11.235	8.9783	−4.02	30	—	—	—	—	—
		11.216	8.8455	−4.06	35	—	—	—	—	—
		11.200	8.7170	−4.13	40	—	—	—	—	—
		11.180	8.5941	−4.18	45	—	—	—	—	—
		11.161	8.4768	−4.25	50	—	—	—	—	—
		11.133	8.942	−1.67	10	T	$\mu = 0$	284	—	50 wt. % methanol
—, 4-amino-3-bromo-	—	10.739	8.520	−2.96	25	—	—	285	—	—
		10.346	8.149	−4.24	40	—	—	285	—	—
—, 4-amino-3,5-dimethyl-	—	7.69	7.04	−5.97	20	T	$\mu = 0$	285	—	—
—, 4-amino-3-ethyl-	—	10.48	9.54	−7.88	20	T	$\mu = 0$	285	—	—
		10.89	9.51	−6.34	20	T	$\mu = 0$	285	—	—

HEATS OF PROTON IONIZATION, pK, AND RELATED THERMODYNAMIC QUANTITIES (continued)

Compound	Formula, Synonyms	ΔH (kcal/mole)	pK	ΔS cal/deg-mole	T°C	Method	Conditions	Reference	Other References	Remarks
Pyridine—(Continued)										
—, 4-amino-3-isopropyl-	—	11.28	9.54	−5.15	20	T	$\mu = 0$	285	—	—
—, 4-amino-3-methyl-	—	11.66	9.43	−3.34	20	T	$\mu = 0$	285	—	—
—, 4-amino-2,3-5,6-tetra-methyl-		10.35	10.58	−13.10	20	T	$\mu = 0$	285	—	—
—, 2(2-aminoethyl)-	CH₂CH₂NH₃⁺ ring; 2-(2-Aminoethylpyridine)	6.27	3.96 (pK₁)	−2.9	25	C	0.3 M NaClO₄	353	286	—
		11.60	9.46 (pK₂)	−4.4	25	C	0.3 M NaClO₄	—		
—, 2(aminomethyl)-	CH₂NH₃⁺ ring	6.31	2.33 (pK₁)	−10.5	25	C	0.3 M NaClO₄	353	286	—
		10.57	8.55 (pK₂)	−3.7	25	C	0.3 M NaClO₄	—		
—, 2-(aminomethyl)-6-methyl-	2-Picolylamine 6-Methyl-2-picolylamine	9.3	8.70 (30°, pK₂)	−9.1	0–40	T	—	287	—	3.b.d.
—, 3-bromo-		1.85	2.91	−7.03	20	T	$\mu = 0$	285	—	—
—, 3-bromo-4-amino-N-methyl-	3-Bromo-4-methylaminopyridine	9.01	7.47	−3.45	20	T	$\mu = 0$	285	—	—
—, 3-bromo-4-dimethyl-aminopyridine		6.36	6.52	−8.15	20	T	$\mu = 0$	285	—	—
4(dimethylamino)-										
—, 2-carboxylic acid		0.52	1.01 (pK₁)	−2.9	25	C	$\mu = 0.13–0.27$	330	—	—
		2.35	5.32 (pK₂)	−16.5	25	C	$\mu = 0$	—	—	—
		2.49	(pK₂)	—	25	C	$\mu = 0.019$	—	—	—
		2.51	(pK₂)	—	25	C	$\mu = 0.036$	—	—	—
		2.72	(pK₂)	—	25	C	$\mu = 0.190$	—	—	—
—, 3-carboxylic acid		0.75	2.07 (pK₁)	−6.9	25	C	$\mu = 0.01–0.09$	330	—	—
		2.71	4.81 (pK₂)	−12.9	25	C	$\mu = 0$	—	—	—
		2.76	(pK₂)	—	25	C	$\mu = 0.008$	—	—	—
		2.81	(pK₂)	—	25	C	$\mu = 0.017$	—	—	—
		2.97	(pK₂)	—	25	C	$\mu = 0.168$	—	—	—
—, 4-carboxylic acid		0.48	1.84 (pK₁)	−6.8	25	C	$\mu = 0.03–0.14$	330	—	—
		3.02	4.86 (pK₂)	−12.1	25	C	$\mu = 0$	—	—	—
		2.99	(pK₂)	—	25	C	$\mu = 0.008$	—	—	—
		3.05	(pK₂)	—	25	C	$\mu = 0.017$	—	—	—
		3.19	(pK₂)	—	25	C	$\mu = 0.026$	—	—	—
		3.20	(pK₂)	—	25	C	$\mu = 0.168$	—	—	—

HEATS OF PROTON IONIZATION, pK, AND RELATED THERMODYNAMIC QUANTITIES (continued)

Compound	Formula. Synonyms	ΔH (kcal/mole)	pK	ΔS cal/deg-mole	T°C	Method	Conditions	Reference	Other References	Remarks
Pyridine—(Continued)										
-2,6-diacetyldioxime		5.3	10.08 (pK$_1$)	−28	25	T	$\mu = 0$	223	—	3.b.
		6.7	10.88 (pK$_2$)	−27	25	T	$\mu = 0$	223	—	3.b.
Pyridine-2,6-dialdoxime	See: Pyridine, 2,6-diacetyldioxime									
—, 2,4-dimethyl-	2,4-Lutidine	7.165	6.79	−7.00	25	C	—	37	—	3.d.
—, 2,6-dimethyl-	2,6-Lutidine	7.235	6.75	−6.60	25	C	—	37	—	3.d.
—, 3,5-dimethyl-	3,5-Lutidine	6.365	6.18	−6.95	25	C	—	37	—	3.d.
—, 3,5-dimethyl-4(N,N′-dimethylamino)-	—	9.83	8.15	−3.75	20	T	$\mu = 0$	285	285	—
—, 3,5-dimethyl-4-(N-methylamino)-	—	12.06	9.43	−2.01	20	T	$\mu = 0$	285	—	—
—, 4(N,N-dimethylamino)-	—	10.75	9.71	−7.71	20	T	$\mu = 0$	285	—	—
—, 4(N,N′-dimethylamino)-3-isopropyl-	—	8.76	8.27	−7.95	20	T	$\mu = 0$	285	—	—
—, 3-ethyl-	—	5.30	5.80	−8.43	20	T	$\mu = 0$	285	—	—
—, 4(N,N-dimethylamino)-3-ethyl-	—	9.15	8.66	−8.43	20	T	$\mu = 0$	285	—	—
—, 4(N,N-dimethylamino)-3-ethyl-	—	9.02	8.68	−8.94	20	T	$\mu = 0$	285	—	—
—, 3-ethyl-4-methyl-amino-	—	11.54	9.90	−5.90	20	T	$\mu = 0$	285	—	—
—, 3-isopropyl-	—	5.57	5.88	−7.91	20	T	$\mu = 0$	285	—	—
—, 3-isopropyl-4(N-methylamino)-	—	11.93	9.96	−4.88	20	T	$\mu = 0$	285	—	—
—, 2-methyl-	2-Picoline	5.990	5.97	−7.20	25	C	C = 0.5 M	37	280	3.d.
	α-Picoline	5.930	—	—	10	C		86	—	3.d.
	—	6.075	—	—	20	—		—	—	—
	—	6.095	6.02	−7.0	25	—		—	—	—
—, 3-methyl-	β-Picoline	6.115	—	—	30	—		—	—	—
	3-Picoline	5.640	5.68	−7.05	25	C		37	285	3.d.
—, 4-methyl-	4-Picoline	6.020	6.02	−7.35	25	C		37	—	3.d.
	γ-Picoline									
—, 3-(methylamino)-	3-Picolylamine	9.3	8.70 (30°, pK$_2$)	−9.1	10–40	T	$\mu = 0$	287	—	3.b.
—, 4(N-methylamino)-	—	11.02	9.66 (30°, pK$_2$)	−6.62	20	T	$\mu = 0$	285	—	—

HEATS OF PROTON IONIZATION, pK, AND RELATED THERMODYNAMIC QUANTITIES (continued)

Compound	Formula. Synonyms	ΔH (kcal/mole)	pK	ΔS cal/deg-mole	T°C	Method	Conditions	Reference	Other References	Remarks
Pyridine-2,6-dialdoxime—(Continued)										
—, 2-(2-methylaminoethyl)-		2.4 10.1	3.58 (30°, pK$_1$) 9.65 (30°, pK$_2$)	−8 −11	10–40 10–40	T T	$\mu = 0$ $\mu = 0$	289 289	— —	3.b. 3.b.
—, 2-methylamino-methyl-	2-Picolylmethylamine	9.43	9.30 (10°, pK$_2$)	−9 (10°)	10–40	T	$\mu = 0$	286	—	3.b.
—, 3-methyl-4 (N-methyl-amino)-		9.43 10.87	8.82 (30°, pK$_2$) 9.83	−9 (10°) −7.88	10–40 20	T T	$\mu = 0$ $\mu = 0$	286 285	— —	3.b. —
—, 2-methyl-6-methyl-aminomethyl-	$CH_3-NH_2^+-CH_2$	4.21	8.79 (30°, pK$_2$)	−3.0 (30°)	10–40	T	$\mu = 0$	289	—	3.b.
	6-Methyl-2-picolylmethylamine									
—, 4(N-methylamino)-2,3,5,6-tetramethyl-		9.96	10.06 (20°)	−12.08	5–35	T	$\mu = 0$	285	—	—
—, 2,3,5,6-tetramethyl-	β-Parvoline Parvuline	8.08	7.88 (20°)	−8.49	5–35	T	$\mu = 0$	285	—	—
2-Pyridinecarboxylic acid	COOH	2.45	5.29 (pK$_2$)	−16.0	25	C	$\mu = 0$	290	—	—
3-Pyridinecarboxylic acid	COOH	2.57	4.77 (pK$_2$)	−13.2	25	C	$\mu = 0$	290	—	—
4-Pyridinecarboxylic acid	COOH Nicotinic acid	3.00	4.84	−12.1	25	C	$\mu = 0$	290	—	—
	Isonicotinic acid									
2-(2-Pyridyl)-benzimidazole	See: Benzimidazole, 2-(2-pyridyl)-									

HEATS OF PROTON IONIZATION, pK, AND RELATED THERMODYNAMIC QUANTITIES (continued)

Compound	Formula, Synonyms	ΔH (kcal/mole)	pK	ΔS cal/deg-mole	T°C	Method	Conditions	Reference	Other References	Remarks
2-(2-Pyridyl)-benzoxazole	See: Benzoxazole, 2(2'-pyridyl)-									
2-(2-Pyridyl)-imidazole	See: Imidazole, 2-(2'-pyridyl)-									
2,2'-Pyridylimidazoline-iron(II)	See: Iron II-2(2'-pyridyl)imidazoline-									
Pyrimidinetroine	See: Barbituric acid									
Pyrocatechol monomethyl ether	See: Phenol, 2-methoxy-									
Pyromucic acid	See: 2-Furancarboxylic acid									
Pyrrole	HNCHC(COOH)CHCH									
—, 3-carboxylic acid		0.45	4.453		25	C	$\mu = 0$	330	—	—
		0.47	—		25	C	$\mu = 0.002$	—	—	—
		0.53			25	C	$\mu = 0.017$	—	—	3.d.
		0.68			25	C	$\mu = 0.172$	—	—	—
Pyrrolidine	[pyrrolidine ring] N—H	12.370	11.11	—	25	C	—	37	236	—
		12.63	12.17	-9.30	0	T	$\mu = 0$	402	—	—
		12.71	11.98	-9.42	5	T	$\mu = 0$	—	—	—
		12.78	11.81	-9.15	10	T	$\mu = 0$	—	—	—
		12.86	11.63	-8.87	15	T	$\mu = 0$	—	—	—
		12.94	11.43	-8.60	20	T	$\mu = 0$	—	—	—
	1-Azacyclopentane	13.02	11.305	-8.34	25	T	$\mu = 0$	—	—	—
	Tetrahydropyrrole	13.10	11.15	-8.05	30	T	$\mu = 0$	—	—	—
	Tetramethylenimine	13.18	10.99	-7.79	35	T	$\mu = 0$	—	—	—
		13.27	10.84	-7.53	40	T	$\mu = 0$	—	—	—
		13.35	10.70	-7.24	45	T	$\mu = 0$	—	—	—
		13.44	10.56	-6.98	50	T	$\mu = 0$	—	—	—
				-6.72						
D-2-Pyrrolidinecarboxylic acid	See: Proline									
Pyrophosphoric acid	See: Phosphoric acid, pyro-									
Pyruvic acid	See: Propanoic acid, 2-oxo-									
Quinol	See: Benzene, 1,4-dihydroxy-									
Quinoline	[quinoline structure] 1-Benzazine, Benzo(b)-pyridine	5.365	4.80	-3.95	25	C	—	37	—	3.d.
—, 2-amino-	Quinolylamine	9.8	7.34 (20°)	-0.5	5–35	T	C = 0.01	85	—	3.b.
—, 3-amino-		5.3	4.95 (20°)	-4.6	5–35	T	C = 0.01	85	—	3.b.

HEATS OF PROTON IONIZATION, pK, AND RELATED THERMODYNAMIC QUANTITIES (continued)

Compound	Formula. Synonyms	ΔH (kcal/mole)	pK	ΔS cal/deg-mole	T°C	Method	Conditions	Reference	Other References	Remarks
Quinoline—(Continued)										
—, 4-amino-	β-Quinolylamine	11.9	9.17 (20°)	−1.2	5–35	T	C = 0.01	85	—	3.b.
—, 5-amino-	5-Quinolylamine	5.6	5.46 (20°)	−6.2	5–35	T	C = 0.01	85	—	3.b.
—, 6-amino-	6-Quinolylamine	6.1	5.63 (20°)	−5.3	5–35	T	C = 0.01	85	—	3.b.
—, 8-amino-	8-Quinolylamine	5.0	3.99 (20°)	1.3	5–35	T	C = 0.01	85	—	3.b.
—, 8-hydroxy-	—	5.83	4.13	−0.7	25	C	$\mu = 0.1$ (NaClO$_4$)	403	—	Solvent = 50 v/v° dioxane in water
		6.41	10.95	−28.6	25	C	$\mu = 0.1$ (NaClO$_4$)	—	—	Solvent = 50 v/v° dioxane in water
—, 2-methyl-	—	6.96	4.72 (pK$_1$)	−1.7	25	C	$\mu = 0.1$ (NaClO$_4$)	403	—	Solvent = 50 v/v° dioxane in water
		6.92	11.31 (pK$_2$)	−28.5	25	C	$\mu = 0.1$ (NaClO$_4$)	—	—	Solvent = 50 v/v° dioxane in water
—, 4-methyl-	—	6.74	4.83	−0.5	25	C	$\mu = 0.1$ (NaClO$_4$)	403	—	Solvent = 50 v/v° dioxane in water
		6.18	11.10	−30.0	25	C	$\mu = 0.1$ (NaClO$_4$)	—	—	Solvent = 50 v/v° dioxane in water
—, 8-mercapto-	—	1.73	0.77	2.3	25	C	$\mu = 0.1$	403	—	Solvent = 50 v/v° dioxane in water
		3.02	4.80	−32.1	25	C	$\mu = 0.1$	—	—	Solvent = 50 v/v° dioxane in water
—, 2-methyl-8-hydro-	—	7.5	11.71	−28	25	T	$\mu = 0.005$	404	—	—
		6.9	4.58	2	25	T	$\mu = 0.005$	—	—	—
—, 4-methyl-8-hydro-	—	9.1	11.62	−23	25	T	$\mu = 0.005$	404	—	—
		7.1	4.67	2	25	T	$\mu = 0.005$	—	—	—

HEATS OF PROTON IONIZATION, pK, AND RELATED THERMODYNAMIC QUANTITIES (continued)

Compound	Formula, Synonyms	ΔH (kcal/mole)	pK	ΔS cal/deg-mole	T °C	Method	Conditions	Reference	Other References	Remarks
Quinoline—(Continued)										
—, 2-methyl-8-mercapto-	—	2.3	1.96	1	25	C	$\mu = 0.1$ (NaClO$_4$)	403	—	Solvent = 50 v/v° dioxane in water
		3.58	9.76	−33	25	C	$\mu = 0.1$ (NaClO$_4$)	—	—	Solvent = 50 v/v° dioxane in water
5-Quinolinesulfonic acid										
—, 8-hydroxy-		4.2	4.405 (pK$_2$)	−4.8	0	T	$\mu = 0$	291	292	—
		4.5	4.104	−3.8	25	—	—	—	—	—
		5.2	3.813	−2.0	50	—	—	—	—	—
		4.37	3.88 (pK$_2$)	3.1	25	C	$\mu = 0.1$ (NaClO$_4$)	403	—	—
		4.04	3.56 (pK$_2$)	2.7	25	C	$\mu = 0.1$ (NaClO$_4$)	403	—	Solvent = 50 v/v° dioxane in water
		5.9	9.125 (pK$_3$)	−20	0	T	$\mu = 0$	291	—	—
		5.0	8.750	−23	25	—	—	—	—	—
		3.8	8.491	−27	50	—	—	—	—	—
		4.00	8.43 (pK$_3$)	25.2	25	C	$\mu = 0.1$ (NaClO$_4$)	403	—	—
		4.95	9.85 (pK$_3$)	28.5	25	C	$\mu = 0.1$ (NaClO$_4$)	403	—	Solvent = 50 v/v° dioxane in water
—, 8-hydroxy-7-iodo-	Loretine	1.3	2.61 (pK$_2$)	−7.2	0	T	$\mu = 0$	291	—	—
		1.5	2.51	−6.4	25	—	—	—	—	—
		1.8	2.42	−6.0	50	—	—	—	—	—
		6.1	7.768 (pK$_3$)	−13.5	0	T	$\mu = 0$	291	—	—
		4.1	7.417	−21	25	—	—	—	—	—
		1.0	7.267	−29	50	—	—	—	—	—
—, 8-hydroxy-7-nitro-	—	−0.2	1.93 (pK$_2$)	−9.3	0	T	$\mu = 0$	291	—	—
		−0.2	1.94	−9.6	25	—	—	—	—	—
		−0.4	1.94	−10.0	50	—	—	—	—	—
		2.5	5.919 (pK$_3$)	−18	0	T	$\mu = 0$	291	—	—
		2.6	5.750	−18	25	—	—	—	—	—
		2.6	5.605	−18	50	—	—	—	—	—

HEATS OF PROTON IONIZATION, pK, AND RELATED THERMODYNAMIC QUANTITIES (continued)

Compound	Formula, Synonyms	ΔH (kcal/mole)	pK	ΔS cal/deg-mole	T °C	Method	Conditions	Reference	Other References	Remarks
5-Quinolinesulfonic acid—(Continued)										
—, 7(4-nitrophenylazo)-8-hydroxy-		4.5	3.41 (pK$_2$)	0.9	0	T	$\mu = 0$	291	—	—
		3.5	3.14	−2.7	25	—	—	—	—	—
		2.1	2.98	−7.2	50	—	—	—	—	—
		4.9	7.811 (pK$_3$)	−18	0	T	$\mu = 0$	291	—	—
		4.5	7.495	−19	25	—	—	—	—	—
		3.7	7.262	−22	50	—	—	—	—	—
—, 7-phenylazo-8-hydroxy-	—	4.0	3.65 (pK$_2$)	−2.1	0	T	$\mu = 0$	291	—	—
		3.1	3.41	−5.2	25	—	—	—	—	—
		1.1	3.27	−10	50	—	—	—	—	—
		5.2	8.183 (pK$_3$)	−18	0	T	$\mu = 0$	291	—	—
		4.6	7.850	−21	25	—	—	—	—	—
		3.4	7.620	−25	50	—	—	—	—	—
8-Quinolinol	See: Quinoline, 8-hydroxy-									
α-Quinolylamine	See: Quinoline, 2-amino-									
β-Quinolylamine	See: Quinoline, 4-amino-									
5-Quinolylamine	See: Quinoline, 5-amino-									
6-Quinolylamine	See: Quinoline, 6-amino-									
8-Quinolylamine	See: Quinoline, 8-amino-									
Racephedrine	See: Ephedrine(DL)									
Resorcinol monomethyl ether	See: Phenol, 3-methoxy-									
3-(D)-Ribofuranosidouracil	See: Uridine									
9-β-D-Ribofuranosyladenine	See: Adenosine									
Ribonuclease	(phenolic dissociation)	7.0	9.92 (25°)	−22	6–25	T	$\mu = 0.15$	293	—	3.a.b.
Ribose		8.1	12.22	−28.7	25	C	$\mu = 0$	186	—	—
—, 2-deoxy-		7.7	12.67	−32.1	25	C	$\mu = 0$	186	—	—

HEATS OF PROTON IONIZATION, pK, AND RELATED THERMODYNAMIC QUANTITIES (continued)

Compound	Formula, Synonyms	ΔH (kcal/mole)	pK	ΔS cal/deg-mole	T°C	Method	Conditions	Reference	Other References	Remarks
Ribose—(Continued)										
—, 5-phosphate-	2-Deoxyribose [structure: $HO-C_2-CH_2$, $HO-C_3-C^4-H$, $^5CH_2OPO_3H_2$]	−2.7	6.70 (pK$_2$)	−40	25	C	$\mu = 0$	38	—	—
		6.1	13.05 (pK$_3$)	−39.4	25	C	$\mu = 0$	45	—	—
3(α-Ribosido)cytosine	See: Cytidine									
9-D-Ribosidoguanine	See: Guanosine									
β-Resorcylic acid	See: Benzoic acid, 2,4-dihydroxy-									
Riboflavin	Lactoflavin, Vitamin B$_2$, Vitamin G	8.3	9.69 (25°)	−16	10–40	T	$\mu = 0.01$	294	—	3.b.
Salicylaldehyde	See: Benzaldehyde, 2-hydroxy-									
Salicylamide	See: Benzoic acid, 2-hydroxyamide-									
Salicylic acid	See: Benzoic acid, 2-hydroxy-									
Salicylic acid-acetyl	See: Benzoic acid, 2-hydroxy acetate-									
Sarcine	See: Hypoxanthine									
Sarcosine	CH$_3$$\overset{+}{N}H_2CH_2$COOH, N-Methylglycine	1.72	2.12 (pK$_1$)	−4	25	C	$\mu = 0$	201	325	—
		9.75	10.20 (pK$_2$)	−13	25	C	$\mu = 0$	201		—
		9.74	10.71 (pK$_2$)	−14.0	5	T	$\mu = 0$	71		—
		9.71	10.45	−14.1	15	—	—	—		—
		9.68	10.20	−14.2	25	—	—	—		—
		9.65	9.97	−14.3	35	—	—	—		—
		9.62	9.76	−14.4	45	—	—	—		—
SDA	See: Semicarbazide,1,1-diacetic acid									
Selenic acid	H$_2$SeO$_4$	−5.57	1.65 (pK$_2$)	−26.3	25	T	$\mu = 0$	295	—	—
Semicarbazide	Aminourea carbamylhydrazine, NH$_2$NHCONH$_2$	6.08	3.53	3.9	25	C	$\mu = 0.1$	405	—	—
—, 1,1-diacetic acid-	SDA	−0.2	2.96 (pK$_1$)	−14	30	C	$\mu = 0.1$ (KNO$_3$)	406	—	—
		−0.6	4.04 (pK$_2$)	−20	30	C	$\mu = 0.1$ (KNO$_3$)	406	—	—
—, seleno-	NH$_2$NHCSeNH$_2$	4.5	0.8	11.2	25	C	$\mu = 0.1$	405	—	—
—, 3-thio-		4.53	1.5	8.1	25	C	$\mu = 0.1$	405	—	—
—, 3-thio-1,1-diacetic acid-	TSDA	−0.04	2.94 (pK$_1$)	−14	30	C	$\mu = 0.1$ (KNO$_3$)	406	—	—
		−0.2	4.07 (pK$_2$)	−19	30	C	$\mu = 0.1$ (KNO$_3$)	406	—	—

HEATS OF PROTON IONIZATION, pK, AND RELATED THERMODYNAMIC QUANTITIES (continued)

Compound	Formula, Synonyms	ΔH (kcal/mole)	pK	ΔS cal/deg-mole	T°C	Method	Conditions	Reference	Other References	Remarks
Seminose	See: Mannose(D)									
Serine	HOCH$_2$CH(NH$_3^+$)COOH 2-Amino-3-hydroxypropanoic acid	1.981	2.296 (pK$_1$)	−3.3	1	T	$\mu = 0$	266	31	—
		1.721	2.232	−4.2	12.5	—	—	—		—
		1.366	2.186	−5.4	25.0	—	—	—		—
		0.932	2.154	−6.8	37.5	—	—	—		—
		0.411	2.132	−8.5	50.0	—	—	—		—
		10.450	9.880 (pK$_2$)	−7.0	1	T	$\mu = 0$	266		—
		10.490	9.542	−6.9	12.5	—	—	—		—
		10.405	9.208	−7.2	25.0	—	—	—		—
		10.200	8.904	−7.9	37.5	—	—	—		—
		9.840	8.628	−9.1	50.0	—	—	—		—
—, glycyl-	$^+$NH$_3$CH$_2$CONHCH(CH$_2$OH)COOH Glycylserine	0.189	2.9808	−13.0	25	T	$\mu = 0$	6		—
Sodium 4-dimethylamino-azobenzene-4-sulfonate	See: Methyl orange									
Solochrome violet R	See: Naphthalene, 2,2′-dihydroxy-5′-sulfophenylazo-									
Suberic acid	See: Octanedioic acid									
Succinic acid	HOOCCH$_2$CH$_2$COOH Butanedioic acid	0.80	4.207 (pK$_1$)	−16.6	25	C	$\mu = 0$	1	298	—
		1.526	4.2845 (pK$_1$)	−14.0	0	T	$\mu = 0$	296		—
		1.378	4.2631	−14.6	5	—	—	—		—
		1.228	4.2449	−15.1	10	—	—	—		—
		1.075	4.2316	−15.6	15	—	—	—		—
		0.920	4.2176	−16.2	20	—	—	—		—
		0.761	4.2066	−16.7	25	—	—	—		—
		0.601	4.1980	−17.2	30	—	—	—		—
		0.437	4.1914	−17.8	35	—	—	—		—
		0.270	4.1878	−18.3	40	—	—	—		—
		0.101	4.1869	−18.8	45	—	—	—		—
		−0.070	4.1863	−19.4	50	—	—	—		—
		0.06	5.635 (pK$_2$)	−25.6	25	C	$\mu = 0$	1		—
		1.144	5.674 (pK$_2$)	−21.8	0	T	$\mu = 0$	297		—
		0.902	5.660	−22.6	5	—	—	—		—
		0.656	5.649	−23.5	10	—	—	—		—
		0.406	5.642	−24.4	15	—	—	—		—
		0.151	5.639	−25.3	20	—	—	—		—

HEATS OF PROTON IONIZATION, pK, AND RELATED THERMODYNAMIC QUANTITIES (continued)

Compound	Formula, Synonyms	ΔH (kcal/mole)	pK	ΔS cal/deg-mole	T°C	Method	Conditions	Reference	Other References	Remarks
Succinic acid—(Continued)										
		−0.107	5.635	−26.1	25	—	—	—	—	—
		−0.371	5.641	−27.0	30	—	—	—	—	—
		−0.639	5.647	−27.9	35	—	—	—	—	—
		−0.911	5.654	−28.8	40	—	—	—	—	—
		−1.188	5.669	−29.6	45	—	—	—	—	—
		−1.469	5.680	−30.5	50	—	—	—	—	—
Succinic acid(D,L), 2-hydroxy-	HOOCCH(OH)CH$_2$COOH, Butanedioic acid(D,L), —, 2-hydroxy-, D.L-Malic acid	1.592	3.537 (pK$_1$)	−10.4	0	T	$\mu = 0$	299	—	—
		1.421	3.520	−11.0	5	—	—	—	—	—
		1.247	3.494	−11.6	10	—	—	—	—	—
		1.070	3.482	−12.2	15	—	—	—	—	—
		0.890	3.472	−12.8	20	—	—	—	—	—
		0.706	3.458	−13.4	25	—	—	—	—	—
		0.520	3.452	−14.1	30	—	—	—	—	—
		0.330	3.446	−14.7	35	—	—	—	—	—
		0.138	3.444	−15.3	40	—	—	—	—	—
		−0.058	3.446	−15.9	45	—	—	—	—	—
		−0.257	3.445	−16.6	50	—	—	—	—	—
		0.983	5.119 (pK$_2$)	−19.8	0	T	$\mu = 0$	299	—	—
		0.738	5.108	−20.7	5	—	—	—	—	—
		0.490	5.098	−21.6	10	—	—	—	—	—
		0.237	5.096	−22.5	15	—	—	—	—	—
		−0.066	5.096	−23.4	20	—	—	—	—	—
		−0.282	5.097	−24.2	25	—	—	—	—	—
		−0.549	5.099	−25.1	30	—	—	—	—	—
		−0.819	5.104	−26.0	35	—	—	—	—	—
		−1.094	5.117	−26.9	40	—	—	—	—	—
		−1.374	5.133	−27.8	45	—	—	—	—	—
		−1.659	5.149	−28.7	50	—	—	—	—	—
—, rac-2,3-di-tert-butyl-		−0.94	3.58 (pK$_1$)	−19	25	C	$\mu = 0.1$	59	—	—
—, imide-		4.0	10.2 (pK$_2$)	−33	25	C	$\mu = 0.1$	59	—	—
		6.510	9.623	−22.0	25	T	$\mu = 0$	300	—	—
Succinimide	Succinimide									
Sulfamic acid	See: Succinic acid, imide									
Sulfanilic acid	See: Sulfuric acid, monamide									
	See: Benzenesulfonic acid, 4-amino-									

HEATS OF PROTON IONIZATION, pK, AND RELATED THERMODYNAMIC QUANTITIES (continued)

Compound	Formula, Synonyms	ΔH (kcal/mole)	pK	ΔS cal/deg-mole	T°C	Method	Conditions	Reference	Other References	Remarks
Sulfide										
—, 2-aminoethyl-2'-hydroxyethyl-	$H_3NCH_2CH_2SCH_2CH_2OH$ 2-Aminoethyl-2'-hydroxyethyl sulfide 2-Amino-2'-hydroxydiethyl sulfide	11.7	9.12 (30°)	−3	10–40	T	$\mu = 0$	62	—	3.b.
Sulfone										
—, 4-amino-4'-chloro-diphenyl-	NH_3^+ — SO_2 — Cl	5.01	1.38 (25°)	10.5	—	T	$\mu = 0$	301	—	3.b.
Sulfuric acid	H_2SO_4	−4.86	2.00	−25.4	25	C	$\mu = 0$	408	258, 389, 409	—
—, 2-aminoethyl-	2-aminoethylsulfate	−5.54	1.08	−23.5	25	C	$\mu = 2.0$	407	—	3.b.
		11.10	9.18 (25°)	−5.10	5–45	T	$\mu = 0$	181	113, 302	—
—, monamide-	$^+H_3NSO_3H$ Sulfamic acid	0.47	0.988	−3.0	25	C	$\mu = 0$	330	—	—
Sulfurous acid	H_2SO_3	−3.860	1.997 (pK_1)	−22.1	25	T	$\mu = 0$	303	—	3.d.
		−2.91	7.29 (pK_2)	−42.2	25	—	—	329	—	—
5-Sulfosalicylic acid	See: Benzoic acid, 2-hydroxy-5-sulfo-									
Tartaric acid	HOOCCHOHCHOHCOOH d-2,3-Dihydroxybutanedioic acid d,2,3-Dihydroxysuccinic acid	0.741	3.036 (pK_1)	−11.4	25	T	$\mu = 0$	304	—	—
		1.01	4.39 (pK_2)	−16.5	10	T	$\mu = 0$	304	—	—
		0.753	4.381	−17.4	15	—	—	—	—	—
		0.497	4.372	−18.3	20	—	—	—	—	—
		0.237	4.366	−19.2	25	—	—	—	—	—
		−0.029	4.365	−20.1	30	—	—	—	—	—
		−0.299	4.367	−21.0	35	—	—	—	—	—
		−0.571	4.372	−21.8	40	—	—	—	—	—
Taurine	$H_3NCH_2CH_2SO_3H$ 2 Aminoethanesulfonic acid	9.99	9.06	−7.95	25	C	$\mu = 0$	113	305	—
TETA	See: Acetic acid (tetramethylenedinitrilo)-tetra-									
Tetracycline hydrochloride	[structure]	2.05	3.34	−8.4	25	C	$\mu = 0.01$	418	—	—
—, chloro-		2.35	3.27	−7.1	25	C	$\mu = 0.01$	418	—	—
—, dimethylchloro-		2.35	3.30	−7.2	25	C	$\mu = 0.01$	418	—	—
—, 4-epi-anhydro-		1.95	3.48	−9.4	25	C	$\mu = 0.01$	418	—	—
—, 4-epi-chloro-		2.25	3.59	−8.9	25	C	$\mu = 0.01$	418	—	—

HEATS OF PROTON IONIZATION, pK, AND RELATED THERMODYNAMIC QUANTITIES (continued)

Compound	Formula. Synonyms	ΔH (kcal/mole)	pK	ΔS cal/deg·mole	T°C	Method	Conditions	Reference	Other References	Remarks
Tetraethylenepentamine	$H_2^+N(CH_2CH_2\overset{H}{\underset{+}{N}}HCH_2CH_2NH_3^+)_2$	6.83	2.98 (pK$_1$)	9.3	25	C	0.1 M KCl	306	—	—
		7.89	4.72 (pK$_2$)	4.9	—	—	—	—	—	—
		10.71	8.08 (pK$_3$)	−1.0	—	—	—	—	—	—
		11.32	9.10 (pK$_4$)	−3.7	—	—	—	—	—	—
		10.76	9.67 (pK$_5$)	−8.2	—	—	—	—	—	—
Tetraglycine	See: Glycine,-N-glycylglycylglycyl-									
Tetrahydro-1,4-isoxazine	See: Morpholine									
Tetrahydropyrrole	See: Pyrrolidine									
N,N,N',N'-Tetrakis(2-aminoethyl)ethylenediamine	See: Ethane, 1,2-diamino-N,N,N',N'-tetrakis(2-aminoethyl)-									
2,3,5,6-Tetramethyl-4-aminopyridine	See: Pyridine, 4-amino-2,3,5,6-tetramethyl-									
Tetramethylenediamine	See: Butane, 1,4-diamino-									
Tetramethylenedinitrilotetraacetic acid	See: Acetic acid, tetramethylenedinitrilo-tetra-									
2,3,5,6-Tetramethyl-4-methylaminopyridine	See: Pyridine, 4(N-methylamino)-2,3,5,6-tetramethyl-									
2,4,5,7-Tetramethyl-1,8-phenanthroline	See: 1,8-phenanthroline, 2,4,5,7-tetramethyl-									
2,3,5,6-Tetramethylpyridine	See: Pyridine, 2,3,5,6-tetramethyl-									
Tetramethyleneimine	See: Pyrrollidine									
THAM	See: 1,3-Propanediol, 2-amino-2(hydroxymethyl)-									
2-Theonic acid	See: 2-Thiophenecarboxylic acid									
2-Thiabutane										
—, 4-amino-	$CH_3SCH_2CH_2\overset{+}{N}H_3$	12.3	9.18 (30°)	1	10-40	T	$\mu = 0$	169	—	3.b.
Thiazole	(4-3 NH ring structure with S)	2.02	2.55	−4.9	25	C	$\mu = 0.1$	307	—	—
—, 2,4-dimethyl-	—	4.21	3.98	−4.1	25	C	$\mu = 0.1$	307	—	—
—, 2,5-dimethyl-	—	3.99	3.91	−4.5	25	C	$\mu = 0.1$	307	—	—
—, 4,5-dimethyl-	—	3.97	3.73	−4.0	25	C	$\mu = 0.1$	307	—	—
—, 2-methyl-	—	3.21	3.40	−4.8	25	C	$\mu = 0.1$	307	—	—
—, 4-methyl-	—	2.99	3.16	−4.4	25	C	$\mu = 0.1$	307	—	—
—, 5-methyl-	—	2.88	3.03	−4.2	25	C	$\mu = 0.1$	307	—	—

HEATS OF PROTON IONIZATION, pK, AND RELATED THERMODYNAMIC QUANTITIES (continued)

Compound	Formula, Synonyms	ΔH (kcal/mole)	pK	ΔS cal/deg-mole	T°C	Method	Conditions	Reference	Other References	Remarks
Thiazole—(Continued)										
—, 2-*tert*-butyl-	[structure: $\overset{+}{N}H$ ring, S, C(CH$_3$)$_3$]	3.71	3.00	−1.3	25	C	$\mu = 0.1$	307	—	—
—, 4-*tert*-butyl-	—	3.86	3.04	−1.0	25	C	$\mu = 0.1$	307	—	—
—, 2,4,5-trimethyl-	—	4.95	4.55	−4.3	25	C	$\mu = 0.1$	307	—	—
Thioacetic acid	See: Acetic acid, thiolo-									
Thioglycolic acid	See: Acetic acid, mercapto-									
Thioglycolic acid, *S*-phenyl-	See: Acetic acid, mercaptophenyl-									
Thiolacetic acid	See: Acetic acid, thiolo-									
Thiooxine	See: Quinoline, 8-mercapto-									
2-Thiophenecarboxylic acid	[structure: ring $\overset{4}{5}\,\overset{3}{2}$, S, CO$_2$H], 2-Thenoic acid, α-Thiophenic acid	−1.3	3.529	−20.6	30	T	$\mu = 0$	70	—	—
		−1.4	3.546	−20.7	35	—	—	—	—	—
		−1.4	3.561	−20.9	40	—	—	—	—	—
α-Thiophenic acid	See: 2-Thiophenecarboxylic acid									
Thiosalicylic acid	See: Benzoic acid, 2-mercapto-									
Thiosemicarbazide	See: Semicarbazide, 3-thio-									
Thiosemicarbazide, 1,1-diacetic acid	See: Semicarbazide, 3-thio-1,1-diacetic acid									
Threonine	CH$_3$CHOHCH($\overset{+}{N}H_3$)COOH D-2-Amino-3-hydroxybutyric acid	1.36	2.09 (pK$_1$)	−5	25	C	$\mu = 0$	201	150	—
		1.823	2.200 (pK$_1$)	−3.4	1	T	$\mu = 0$	266	—	—
		1.549	2.132	−4.4	12.5	—	—	—	—	—
		1.180	2.088	−5.6	25	—	—	—	—	—
		0.728	2.070	−7.1	37.5	—	—	—	—	—
		0.191	2.055	−8.8	50	—	—	—	—	—
		10.04	9.10 (pK$_2$)	−8	25	C	$\mu = 0$	201	—	—
		10.075	9.748 (pK$_2$)	−8.0	1	T	$\mu = 0$	266	—	—
		10.085	9.420	−7.8	12.5	—	—	—	—	—
		9.960	9.100	−8.2	25	—	—	—	—	—
		9.712	8.812	−9.0	37.5	—	—	—	—	—
		9.320	8.548	−10.3	50	—	—	—	—	—
—, allo-	H$_3$CCH(OH)CH($\overset{+}{N}H_3$)COOH	1.478	2.178 (pK$_1$)	−4.6	1	T	$\mu = 0$	266	—	—
		1.176	2.138	−5.7	12.5	—	—	—	—	—
		0.772	2.108	−7.0	25	—	—	—	—	—

HEATS OF PROTON IONIZATION, pK, AND RELATED THERMODYNAMIC QUANTITIES (continued)

Compound	Formula, Synonyms	ΔH (kcal/mole)	pK	ΔS cal/deg-mole	T°C	Method	Conditions	Reference	Other References	Remarks
Threonine—(Continued)										
		0.287	2.090	-8.6	37.5	—	—	—	—	—
		-0.287	2.086	-10.4	50	—	—	—	—	—
		10.430	9.774 (pK$_1$)	-6.6	1	—	—	—	—	—
		10.465	9.432	-7.2	12.5	T	$\mu = 0$	**266**	—	—
		10.380	9.096	-6.8	25	—	—	—	—	—
		10.150	8.796	-7.5	37.5	—	—	—	—	—
		9.800	8.520	-8.7	50	—	—	—	—	—
Thymidine		7.75	9.79 (pK$_1$)	-18.8	25	C	$\mu = 0$	**163**	—	—
		12.4	12.85 (pK$_2$)	-17.4	25	C	$\mu = 0$	**163**	—	—
Thymine-2-desoxyriboside	Thymine-2-desoxyriboside									
Thymine	See: Thymidine Uracil, 5-methyl-									
TMTA	See: Acetic acid, propylenedinitrilotetra-									
Toluene		8.83	9.90	-15.7	25	C	$\mu = 0$	163	43	—
—, α-amino-		12.98	9.35	0.7	25	C	$\mu = 0$	330	—	—
		13.160	—	—	10	C	$C = 0.06\ M$	86	—	—
		13.180	9.37	—	20	—	—	—	—	—
		13.087	—	1.0	25	—	—	—	—	—
		12.995	—	—	30	—	—	—	—	—
—, α-amino-N,N-diethyl-		9.92	9.44	-9.8	25	C	$\mu = 0$	330	—	—
—, 2-amino-		7.367	4.447 (25°)	4.35	10–50	T	$\mu = 0$	340	88c	3.b.

HEATS OF PROTON IONIZATION, pK, AND RELATED THERMODYNAMIC QUANTITIES (continued)

Compound	Formula, Synonyms	ΔH (kcal/mole)	pK	ΔS cal/deg-mole	T°C	Method	Conditions	Reference	Other References	Remarks
Toluene—(Continued)										
—, 3-amino-	3-Methylaniline	7.467	4.712 (25°)	3.48	10–50	T	$\mu = 0$	340	88[c]	3.b.
—, 4-amino-	m-Toluidine	8.057	5.083 (25°)	3.75	10–50	T	$\mu = 0$	340	88[c]	3.b.
	p-Toluidine									
—, 2-hydroxy-	4-Methylaniline	5.73	10.333	−28.1	25	T	$\mu = 0$	106	107[d]	—
	o-Cresol									
—, 3-hydroxy-	2-Methylphenol	5.52	10.098	−27.7	25	T	$\mu = 0$	106	107[d]	—
	m-Cresol									
—, 3-hydroxy-4-nitro-	3-Methylphenol	4.58	7.41	−18.6	25	T	$\mu = 0.1$	398	—	—
		5.50	10.276	−28.6	25	T	$\mu = 0$	106	107[d]	—
—, 4-hydroxy-	p-Cresol									
	4-Methylphenol									
4-Toluenesulfonic acid	SO_3H — NH_3^+ — CH_3 (ring positions 1–6)									
—, 3-amino-		5.053	3.6327	0.33	25	T	$\mu = 0$	111	—	—
3-Toluenesulfonic acid	SO_3H — CH_3 — $^+NH_3$ (ring positions 1–6)									
—, 4-amino-		4.494	3.1239	0.78	25	T	$\mu = 0$	111	—	—
Toluhydroquinone	See: Benzene, 1,4-dihydroxy-2-methyl-									
α-Toluic acid	See: Acetic acid, phenyl-									
m-Toluic acid	See: Benzoic acid, 3-methyl-									
o-Toluic acid	See: Benzoic acid, 2-methyl-									
p-Toluic acid	See: Benzoic acid, 4-methyl-									
m-Toluidine	See: Toluene, 3-amino-									
o-Toluidine	See: Toluene, 2-amino-									
p-Toluidine	See: Toluene, 4-amino-									
2,2',2''-Triaminotriethylamine	$C_6H_{18}N_4$	10.3	7.71 (30°, pK_1)	−2	10–40	T	$\mu = 0$	145	—	3.b.
		12.3	9.00 (30°, pK_2)	0	10–40	T	$\mu = 0$	145	—	3.b.
		11.0	9.91 (30°, pK_3)	−9	10–40	T	$\mu = 0$	145	—	3.b.

HEATS OF PROTON IONIZATION, pK, AND RELATED THERMODYNAMIC QUANTITIES (continued)

Compound	Formula, Synonyms	ΔH (kcal/mole)	pK	ΔS cal/deg-mole	T°C	Method	Conditions	Reference	Other References	Remarks
1,2,3-Triazole	$NHN=CHCH=N$	8.88	9.26	−12.6	25	C	$\mu = 0$	352	—	—
—, 4-carboxylic acid-	—	0.84	3.22 (pK₁)	−11.5	25	C	$\mu = 0$	352	—	—
		5.89	8.73 (pK₂)	−20.2	25	C	$\mu = 0$	—		
—, 4,5-dibromo-	—	4.24	5.37	−10.4	25	C	$\mu = 0$	352		
—, 4,5-dicarboxylic acid-	—	0.11	1.86 (pK₁)	−8.2	25	C	$\mu = 0$	352		
		0.01	5.90 (pK₂)	−27.0	25	C	$\mu = 0$	—		
		2.26	9.30 (pK₃)	−35.0	25	C	$\mu = 0$	352		
—, 1-phenyl-4-carboxylic acid-	—	0.76	2.88	−10.6	25	C	$\mu = 0$	352		
—, 1-phenyl-4,5-dicarboxylic acid-	—	−0.67	2.13 (pK₁)	−11.1	25	C	$\mu = 0$	—		
		0.06	4.93 (pK₂)	−22.3	25	C	$\mu = 0$	352		
—, 1-phenyl-5-methyl-4-carboxylic acid-	—	0.17	3.73	−16.5	25	C	$\mu = 0$	352		
bis-(Tricyanovinyl)-amine	—	−7.5	~5.8	1	25	T	$\mu = 0$	225		
p-(Tricyanovinyl)-phenyl-dicyanomethane	See: 3-Pentene, 1,2,3,4,4-pentanitrile-2-phenyl-									
Triethanolamine	See: Amine, triethyl, 2,2',2''-trihydroxy-									
Triethylamine	See: Amine, triethyl-									
Triethylenediamine	See: Octane, 1,4-diazabicyclo[2,2,2]-									
Triethylenetetramine	See: Ethane, 1,2-diamino-N,N'-di(2-amino-ethyl)-									
Triglycine	See: Glycine, N-glycylglycyl									
Triglycolamic Acid	See: Acetic acid, nitrilo, tri-									
Tri(hydroxymethyl)-methylamine	See: 1,3-Propanediol, 2-amino-2(hydroxy-methyl)-									
2,4,6-Trihydroxymethyl-phenol	See: Phenol, 2,4,6-trihydroxymethyl-									
Triisopropanolamine	See: Amine, tripropyl-2,2',2''-trihydroxy-									
Trimethylacetic acid	See: Acetic acid, trimethyl-									
Trimethylamine	See: Amine, trimethyl-									
Trimethylenediamine	See: Propane, 1,3-diamino-									
2,3,5-Trimethylphenol	See: Benzene, 1-hydroxy-2,3,5-trimethyl-									
2,4,5-Trimethylphenol	See: Benzene, 1-hydroxy-2,4,5-trimethyl-									
2,4,6-Trimethylphenol	See: Benzene, 1-hydroxy-2,4,6-trimethyl-									
3,4,5-Trimethylphenol	See: Benzene, 1-hydroxy-3,4,5-trimethyl-									
Tripolyphosphoric acid	See: Phosphoric acid, tripoly-									
Tris(hydroxymethyl)amino-methane	See: 1,3-Propanediol, 2-amino-2(hydroxy-methyl)-									

HEATS OF PROTON IONIZATION, pK, AND RELATED THERMODYNAMIC QUANTITIES (continued)

Compound	Formula, Synonyms	ΔH (kcal/mole)	pK	ΔS cal/deg-mole	T°C	Method	Conditions	Reference	Other References	Remarks
TRIS	See: 1,3-Propanediol, 2-amino-2(hydroxymethyl)-									
Tryptophan	$CH_2CH(\overset{+}{N}H_3)CO_2H$	0	2.46 (pK$_1$)	−11.3	25	T	$\mu = 0.1$	308	—	—
		10.5	9.41 (pK$_2$)	−7.8	25	T	$\mu = 0.1$	308	—	—
TSDA	See: Thiosemicarbazide, 1,1-diacetic acid-									
Tyrosine	$CH_2CH(\overset{+}{N}H_3)COOH$ (HO-)	6.0	10.05 (pK$_2$)	−26	25	T	$\mu = 0.15$	60	—	—
—, aspartyl-	2-Amino-3(4-hydroxyphenyl)-propanoic acid									
	(carboxyl)	0.750	2.13 (25°, pK$_1$)	−7.23	0–25	T	$\mu = 0.01$	94	—	3.a.b.
	(carboxyl)	0	3.57 (25°, pK$_3$)	−16.3	0–25	T	$\mu = 0.01$	94	—	3.a.b.
	(amino)	10.16	8.92 (25°, pK$_3$)	−6.8	0–25	T	$\mu = 0.01$	94	—	3.a.b.
	(oxyphenol)	6.20	10.23 (25°, pK$_4$)	−26.0	0–25	T	$\mu = 0.01$	94	—	3.a.b.
—, 3,5-dibromo(L)-		1.700	2.17 (25°, pK$_1$)	−4.2	25–40	T	—	309	—	3.b.d.
		0.860	6.45 (25°, pK$_2$)	−26.6	25–40	T	—	309	—	3.b.d.
		9.120	7.60 (25°, pK$_3$)	−4.2	25–40	T	—	309	—	3.b.d.
—, 3,5-dichloro-		1.140	2.12 (25°, pK$_1$)	−5.8	25–40	T	—	309	—	3.b.d.
		1.420	6.47 (25°, pK$_2$)	−24.8	25–40	T	—	309	—	3.b.d.
		8.830	7.62 (25°, pK$_3$)	−5.3	25–40	T	—	309	—	3.b.d.
—, 3,5-diiodo-	Iodogorgoic acid	0.980	2.12 (25°, pK$_1$)	−6.4	0–40	T	—	309	54, 310	3.b.d.
		0.810	6.48 (25°, pK$_2$)	−26.9	0–40	T	—	309	—	3.b.d.
		8.790	7.82 (25°, pK$_3$)	−6.3	0–40	T	—	309	—	?.b.d.
Tyrosylarginine	See: Arginine, tyrosyl-									
UDP	See: Uridine, 5'-diphosphoric acid									
UMP	See: Uridine, 5'-monophosphoric acid									
Uracil	(uracil ring)	0.34	0.6 (pK$_1$)	−1.5	25	C	$\mu = 0$	163	—	—
		8.23	9.46 (pK$_2$)	−15.7	25	C	$\mu = 0$	163	—	—
—, 5-methyl-	See: Thymine									
Urea		0.0	0.58	−1.6	25	C	$\mu = 0.01$	405	—	—
—, 2-seleno-										
Ureidoacetic acid	See: Glycine, carbamoyl-									

HEATS OF PROTON IONIZATION, pK, AND RELATED THERMODYNAMIC QUANTITIES (continued)

Compound	Formula, Synonyms	ΔH (kcal/mole)	pK	ΔS cal/deg-mole	T°C	Method	Conditions	Reference	Other References	Remarks
Uridine										
—, 5'-diphosphoric acid	UDP	7.6	9.30 (pK$_1$)	−17.1	25	C	$\mu = 0$	163	—	—
—, 5'-monophosphoric acid	UMP	8.0	9.51 (pK$_1$)	−16.8	25	C	$\mu = 0.1$	39	—	—
		10.9	12.59 (pK$_2$)	−21.0	25	C	$\mu = 0$	163	—	—
	1(D)-Ribofuranosidouracil									
—, 5'-triphosphoric acid	UTP	−1.08	7.16 (pK$_4$)	−36.4	25	T	$\mu = 0$	46	—	—
		−1.12	6.63 (pK$_3$)	−34.0	25	T	$\mu = 0$	46	—	—
		6.6	9.71 (20°, pK$_4$)	−21.30	20–50	T	$\mu = 0.015$	411	—	3.b.
		−2.02	7.58 (pK$_5$)	−41.4	25	T	$\mu = 0$	46	—	—
Valerianic acid										
Valeric acid	See: Pentanoic acid									
—, α-amino-n-	See: Pentanoic acid, 2-amino-									
—, δ-amino-	See: Pentanoic acid, 5-amino-									
—, α-amino-β-methyl-	See: Pentanoic acid, 2-amino-3-methyl-									
Valine(DL)	(CH$_3$)$_2$CHCH(NH$_3^+$)COOH	0.890	2.320 (pK$_1$)	−7.3	1	T	$\mu = 0$	50	—	—
	α-Aminoisovaleric acid	0.540	2.297	−8.6	12.5				—	—
	l-2-Amino-3-methylbutanoic acid	0.080	2.286	−10.2	25				—	—
		−0.460	2.292	−12.0	37.5				—	—
		−1.100	2.310	−13.9	50				—	—
		0.1	3.12	−10.7	25	T		386	—	—
		0.5	3.72	−15.4	25	T		—	—	Solvent = 44.6% dioxane in water
								—	—	Solvent = 59.7% dioxane in water
		0.5	4.04	−16.8	25	T		—	—	Solvent = 69% dioxane in water
		10.73	9.719 (pK$_2$)	−8.5	25	C	$\mu = 0$	330	—	—

HEATS OF PROTON IONIZATION, pK, AND RELATED THERMODYNAMIC QUANTITIES (continued)

Compound	Formula, Synonyms	ΔH (kcal/mole)	pK	ΔS cal/deg-mole	T°C	Method	Conditions	Reference	Other References	Remarks
Valine(DL)—(Continued)		10.370	10.413 (pK$_2$)	−8.5	1	T	μ = 0	50	—	—
		10.800	10.064	−8.2	12.5	—	—	—	—	—
		10.740	9.719	−8.4	25	—	—	—	—	—
		10.560	9.405	−9.0	37.5	—	—	—	—	—
		10.230	9.124	−10.1	50	—	—	—	—	—
		8.9	10.18	−16.7	25	T		**386**	—	Solvent = 44.6% dioxane in water
		11.25	10.57	−10.6	25	T	—	386	—	Solvent = 59.7% dioxane in water
		10.00	11.18	−17.6	25	T	—	386	—	Solvent = 69% dioxane in water
Vanillic acid	See: Benzoic acid, 4-hydroxy-3-methoxy-									
Vanillin	See: Benzaldehyde, 4-hydroxy-3-methoxy-									
Vernine	See: Guanosine									
***o*-Vanillin**	See: Benzaldehyde, 2-hydroxy-3-methoxy-									
Veronal	See: Barbituric acid, 5,5-diethyl-									
Vitamin B$_2$	See: Riboflavin									
Vitamin G	See: Riboflavin									
Water	H$_2$O	13.335			25	C	μ = 0	311	5 11 26 81 316 317	For a complete list of references to earlier ΔH values for water see ref 311 and 312
		13.336	—	—	25	C	μ = 0	312	327	—
		14.513	14.939	−15.23	0	T	μ = 0	315	389, 413	—
		14.312	14.730	−15.95	5	—	—	—	415	—
		14.109	14.533	−16.67	10	—	—	—	416	—
		13.901	14.345	−17.40	15	—	—	—	—	—
		13.692	14.167	−18.12	20	—	—	—	—	—
		13.481	13.996	−18.83	25	—	—	—	—	—

HEATS OF PROTON IONIZATION, pK, AND RELATED THERMODYNAMICS QUANTITIES (continued)

Compound	Formula. Synonyms	ΔH (kcal/mole)	pK	ΔS cal/deg-mole	T °C	Method	Conditions	Reference	Other References	Remarks
Water—(Continued)										
		13.267	13.832	−19.53	30	—	—	—	—	—
		13.051	13.680	−20.24	35	—	—	—	—	—
		12.833	13.535	−20.96	40	—	—	—	—	—
		12.612	13.396	−21.66	45	—	—	—	—	—
		12.390	13.262	−22.35	50	—	—	—	—	—
		12.164	13.137	−23.05	55	—	—	—	—	—
		11.936	13.017	−23.74	60	—	—	—	—	—
		13.617	—	—	20	C	$\mu = 0.1$	412	—	—
		13.05	14.22	−21.3	25	C	$3M$ NaClO$_4$	414	—	—
		13.3	—	—	25	C	0	313	—	Conditions refer to wt % ethanol
		13.6	—	—	25	C	8.16	—	—	
		13.6	—	—	25	C	16.70	—	—	
		13.6	—	—	25	C	21.12	—	—	
		13.3	—	—	25	C	30.25	—	—	
		12.8	—	—	25	C	35.01	—	—	
		12.0	—	—	25	C	44.68	—	—	
		10.3	—	—	25	C	60.24	—	—	
		8.0	—	—	25	C	76.93	—	—	
		5.3	—	—	25	C	88.85	—	—	
		5.2	—	—	25	C	94.98	—	—	
		5.1	—	—	25	C	97.48	—	—	
		13.57	—	—	25	C	0	314	—	Conditions refer to wt % p-dioxane
		13.65	—	—	25	C	25.68	—	—	
		13.70	—	—	25	C	50.75	—	—	
		13.89	—	—	25	C	75.61	—	—	
		13.519	13.998	−18.7	25	T	0	185	—	Conditions refer to wt % dioxane
		13.515	14.622	−21.5	25	T	20	—	—	
		13.190	15.744	−27.8	25	T	45	—	—	
		12.662	17.859	−39.2	25	T	70	—	—	
Water, deuterated	D$_2$O Deuterium oxide	14.488	14.869	−19.45	25	C	$\mu = 0$	413	—	—
		15.371	15.740	−16.76	5	T	$\mu = 0$	318	319	—
		14.311	14.955	−20.43	25					
		12.884	14.182	−25.03	50					

2,3-Xylenol See: Benzene, 1,2-dimethyl-3-hydroxy-
2,4-Xylenol See: Benzene, 1,3-dimethyl-4-hydroxy-
2,5-Xylenol See: Benzene, 1,4-dimethyl-2-hydroxy-
2,6-Xylenol See: Benzene, 1,3-dimethyl-2-hydroxy-

HEATS OF PROTON IONIZATION, pK, AND RELATED THERMODYNAMIC QUANTITIES (continued)

Compound	Formula, Synonyms	ΔH (kcal/mole)	pK	ΔS cal/deg-mole	T°C	Method	Conditions	Reference	Other References	Remarks
3,4-Xylenol	See: Benzene, 1,2-dimethyl-4-hydroxy-									
3,5-Xylenol	See: Benzene, 1,5-dimethyl-3-hydroxy-									
2,3-Xylidine	See: Benzene, 1-amino-2,3-dimethyl-									
2,4-Xylidine	See: Benzene, 1-amino-2,4-dimethyl-									
2,5-Xylidine	See: Benzene, 2-amino-1,4-dimethyl-									
2,6-Xylidine	See: Benzene, 2-amino-1,3-dimethyl-									
3,4-Xylidine	See: Benzene, 4-amino-1,2-dimethyl-									
3,5-Xylidine	See: Benzene, 1-amino-3,5-dimethyl-									
9-β-D-Xylofuranosyl-adenine	See: Adenine, 9-β-D-xylofuranosyl-									
Xylose	HOCH$_2$—C—C—C—CHO (H OH H / OH OH OH)	8.2	12.29	−28.7	25	C	μ = 0	186	—	—

Prepared by Reed M. Izatt and James J. Christensen. Supported in part by National Institutes of Health, Grant No. RG-9430-08.

REFERENCES

1. Christensen, Izatt, and Hansen, *J. Am. Chem. Soc.,* 89, 213 (1967).
2. Harned and Ehlers, *J. Am. Chem. Soc.,* 55, 652 (1933).
3. Nelander, *Acta Chem. Scand.,* 18, 973 (1964).
4. Wadsö, *Acta Chem. Scand.,* 16, 479 (1962).
5. Aksnes, *Acta Chem. Scand.,* 16, 1967 (1962).
6. King, *J. Am. Chem. Soc.,* 79, 6151 (1957).
7. Paabo, Bates, and Robinson, *J. Phys. Chem.,* 70, 540 (1966).
8. Everett and Wynne-Jones, *Trans. Faraday Soc.,* 35, 1380 (1939).
9. Canady, Papée, and Laidler, *Trans. Faraday Soc.,* 54, 502 (1958).
10. Harned and Embree, *J. Am. Chem. Soc.,* 56, 1050 (1934).
11. Jenkins, *Trans. Faraday Soc.,* 41, 138 (1945).
12. Walde, *J. Phys. Chem.,* 43, 431 (1939).
13. Everett, Landsman, and Pinsent, *Proc. Roy. Soc. (A),* 215, 403 (1952).
14. Gary, Bates, and Robinson, *J. Phys. Chem.,* 69, 2750 (1965).
15. Paabo, Bates, and Robinson, *J. Phys. Chem.,* 70, 2073 (1966).
16. Ives and Pryor, *J. Chem. Soc.,* 2104 (1955).
17. Anderegg, *Helv. Chim. Acta,* 48, 1722 (1965).
18. Moeller and Thompson, *J. Inorg. Nucl. Chem.,* 24, 499 (1962).
19. Wright, *J. Am. Chem. Soc.,* 56, 314 (1934).
20. Anderegg, *Helv. Chim. Acta,* 47, 1801 (1964).
21. Boyd, Bryson, Nancollas, and Torrance, *J. Chem. Soc.,* 7353 (1965).
22. Nims, *J. Am. Chem. Soc.,* 58, 987 (1936).
23. Hughes and Martell, *J. Am. Chem. Soc.,* 78, 1319 (1956).
24. Hull, Davies, and Staveley, *J. Chem. Soc.,* 5422 (1964).
25. Moeller and Ferrus, *Inorg. Chem.,* 1, 49 (1962).
26. Edelin de la Praudiere and Staveley, *J. Inorg. Nucl. Chem.,* 26, 1713 (1964).
27. Milyukov and Polenova, *Izv. Vysshikh. Uchebn. Zavadenii, Khim. Khim. Technol.,* 8, 42 (1965).
28. Wrathall, Izatt, and Christensen, *J. Am. Chem. Soc.,* 86, 4779 (1964).
29. Irving, Nelander, and Wadsö, *Acta Chem. Scand.,* 18, 769 (1964).
30. Crockford and Douglas, *J. Am. Chem. Soc.,* 56, 1472 (1934).
31. King, *J. Am. Chem. Soc.,* 82, 3575 (1960).
32. Öckerbloom and Martell, *J. Am. Chem. Soc.,* 78, 267 (1956).
33. Anderegg, *Helv. Chim. Acta,* 48, 1718 (1965).
34. Ojelund and Wadsö, *Acta Chem. Scand.,* 21, 1408 (1967).
35. Sillen and Martell, *Stability Constants,* Publication No. 17, The Chemical Society, London, 1964.
36. Perkampus and Rössel, *Z. Elektrochem.,* 62, 94 (1958).
37. Sacconi, Paoletti, and Ciampolini, *J. Am. Chem. Soc.,* 82, 3831 (1960).
38. Christensen and Izatt, *J. Phys. Chem.,* 66, 1030 (1962).
39. Sukhorukov, Poltev, and Blyumenfel'd, *Abhandl. Deut. Akad. Wiss. Berlin Kl. Med.,* 381 (1964).
40a. Rawitscher and Sturtevant, *J. Am. Chem. Soc.,* 82, 3739 (1960).
41. Lewin and Tann, *J. Chem. Soc.,* 1466 (1962).
42. Harkins and Freiser, *J. Am. Chem. Soc.,* 80, 1132 (1958).
43. Lewin and Barnes, *J. Chem. Soc. B,* 478 (1966).
44. Christensen, Rytting, and Izatt, *J. Am. Chem. Soc.,* 88, 5105 (1966).
45. Izatt, Hansen, Rytting, and Christensen, *J. Am. Chem. Soc.,* 87, 2760 (1965).
46. Phillips, Eisenberg, George, and Rutman, *J. Biol. Chem.,* 240, 4393 (1965).
47. Phillips, Eisenberg, George, and Rutman, *J. Am. Chem. Soc.,* 88, 2631 (1966).
48. Taqui Khan and Martell, *J. Am. Chem. Soc.,* 88, 668 (1966).
49. Sturtevant, *J. Am. Chem. Soc.,* 64, 762 (1942).
50. Smith, Taylor, and Smith, *J. Biol. Chem.,* 122, 109 (1937).
51. Anderson, Newell, and Izatt, *Inorg. Chem.,* 5, 62 (1966).
52. Nims and Smith, *J. Biol. Chem.,* 101, 401 (1933).
53. Branch and Miyamoto, *J. Am. Chem. Soc.,* 52, 863 (1930).
54b. Miyamoto and Schmidt, *J. Biol. Chem.,* 90, 165 (1930).
55. King and King, *J. Am. Chem. Soc.,* 78, 1089 (1956).
56. Davies and Dunning, *J. Chem. Soc.,* 4168 (1965).
57. Ellenbogen, *J. Am. Chem. Soc.,* 78, 369 (1956).
58. King, *J. Am. Chem. Soc.,* 78, 6020 (1956).
59. Eberson and Wadsö, *Acta Chem. Scand.,* 17, 1552 (1963).
60. Tanford and Roberts, *J. Am. Chem. Soc.,* 74, 2509 (1952).

61.   Randall and Staveley, *J. Chem. Soc.*, 472 (1959).
62.   Lotz, Block, and Fernelius, *J. Phys. Chem.*, 63, 541 (1959).
63.   Wang, Bauman, and Murmann, *J. Phys. Chem.*, 68, 2296 (1964).
64.   Evans and Hamann, *Trans. Faraday Soc.*, 47, 25 (1951).
65.   Fyfe, *J. Chem. Soc.*, 1347 (1955).
66.   Evans and Hamann, *Trans. Faraday Soc.*, 47, 34 (1951).
67.   Ciampolini and Paoletti, *J. Phys. Chem.*, 65, 1224 (1961).
68.   Paoletti, Nuzzi, and Vacca, *J. Chem. Soc. A*, 1385 (1966).
69.   Bower, Robinson, and Bates, *J. Res. Natl. Bur. Standards*, 66A, 71 (1962).
70.   Lumme, Lahermo, and Tummavuori, *Acta Chem. Scand.*, 19, 2175 (1965).
71.   Datta and Grzybowski, *Trans. Faraday Soc.*, 54, 1188 (1958).
72.   Everett and Wynne-Jones, *Trans. Faraday Soc.*, 48, 531 (1952).
73.   Everett and Wynne-Jones, *Proc. R. Soc. Lond. Ser. A*, 177, 499 (1941).
74.   Schwabe, Graichen, and Spiethoff, *Z. Phys. Chem.*, 20, 68 (1959).
75.   Goldberg and Fernelius, *J. Phys. Chem.*, 63, 1328 (1959).
76.   Paoletti and Ciampolini, *Ric. Sci. Rend Sex A*, 3 (4), 405 (1963).
77.   Bates and Hetzer, *J. Phys. Chem.*, 65, 667 (1961).
78.   Paoletti, Stern, and Vacca, *J. Phys. Chem.*, 69, 3759 (1965).
79.   Ablard, McKinney, and Warner, *J. Am. Chem. Soc.*, 62, 2181 (1940).
80.   Bates and Allen, *J. Res. Natl. Bur. Standards*, 64A, 343 (1960).
81.   Pitzer, *J. Am. Chem. Soc.*, 59, 2365 (1937).
82.   Coulter, Sinclair, and Roper, *J. Am. Chem. Soc.*, 81, 2986 (1959).
83.   Bates and Pinching, *J. Res. Natl. Bur. Standards*, 42, 419 (1949).
84.   Vilallonga and Pouchan, *Biochim. Biophys. Acta*, 75, 449 (1963).
85.   Elliott and Mason, *J. Chem. Soc.*, 2352 (1959).
86.   Levi, McEwan, and Wolfenden, *J. Chem. Soc.*, 760 (1949).
87.   Biggs, *J. Chem. Soc.*, 2572 (1961).
88c. Zawidzki, Papée, and Canady, and Laidler, *Trans. Faraday Soc.*, 55, 1738 (1959).
89.   Bell, *The Proton in Chemistry*, Cornell University Press, Ithaca, New York, 1959, 64.
90.   Ko, O'Hara, Hu, and Hepler, *J. Am. Chem. Soc.*, 86, 1003 (1964).
91.   Bernhard, *J. Biol. Chem.*, 218, 961 (1956).
92.   Hanania, Irvine and Irvine, *J. Chem. Soc. A*, 296 (1966).
93.   Datta and Grzybowski, *Biochem. J.*, 78, 289 (1961).
94.   Greenstein, *J. Biol. Chem.*, 101, 603 (1933).
95.   Pelletier, *J. Chim. Phys.*, 57, 301 (1960).
96.   Sellers, Sunner, and Wadsö, *Acta Chem. Scand.*, 18, 202 (1964).
97.   Smith and Smith, *J. Biol. Chem.*, 146, 187 (1942).
98.   Kitzinger and Hems, *Biochem. J.*, 71, 395 (1959).
99.   Wynne-Jones and Saloman, *Trans. Faraday Soc.*, 34, 1321 (1938).
100. Jackson and Edwards, *J. Am. Chem. Soc.*, 83, 355 (1961).
101. Millero, Ahluwalia, and Hepler, *J. Chem. Eng. Data*, 10, 199 (1965).
102. Manov, Schuette, and Kirk, *J. Res. Natl. Bur. Standards*, 48, 84 (1952).
103. Rubaszewska and Grabowski, *Roczniki Chem.*, 33, 781 (1959).
104. Millero, Ahluwalia, and Hepler, *J. Chem. Eng. Data*, 9, 319 (1964).
105. Baxendale and Hardy, *Trans. Faraday Soc.*, 49, 1140 (1953).
106. Chen and Laidler, *Trans. Faraday Soc.*, 58, 480 (1962).
107d. Papée, Canady, Zawidzki, and Laidler, *Trans. Faraday Soc.*, 55, 1734 (1959).
108. Edwards and Sederstrom, *J. Phys. Chem.*, 65, 862 (1961).
109. Hamer, Pinching, and Acree, *J. Res. Natl. Bur. Standards*, 35, 539 (1945).
110. Hamer and Acree, *J. Res. Natl. Bur. Standards*, 35, 381 (1945).
111. Conn and Swinehart, *J. Phys. Chem.*, 69, 2653 (1965).
112. Diebel and Swinehart, *J. Phys. Chem.*, 61, 333 (1957).
113. Hopkins, Wu, and Hepler, *J. Phys. Chem.*, 69, 2244 (1965).
114. MacLoren and Swinehart, *J. Am. Chem. Soc.*, 73, 1822 (1951).
115. Morrett and Swinehart, *J. Phys. Chem.*, 67, 717 (1963).
116. Bates, Siegel, and Acree, *J. Res. Natl. Bur. Standards*, 31, 205 (1943).
117. Lane and Quinlan, *J. Am. Chem. Soc.*, 82, 2994 (1960).
118. Harkins and Freiser, *J. Am. Chem. Soc.*, 77, 1374 (1955).
119. Fernandez and Hepler, *J. Phys. Chem.*, 65, 110 (1959).
120. Hermans, Jr., Leach, and Scherega, *J. Am. Chem. Soc.*, 85, 1390 (1963).
121. Jones and Parton, *Trans. Faraday Soc.*, 48, 8 (1952).
122. Cottrell, Drake, Levi, Tully, and Wolfenden, *J. Chem. Soc.*, 1016 (1948).

123. Zawidzki, Papée, and Laidler, *Trans, Faraday Soc.*, 55, 1743 (1959).
124. Christensen, Wrathall, Izatt, and Tolman, *J. Phys. Chem.*, 71, 300 (1967).
125. Saraswat and Tripathi, *Bull. Chem. Soc. Jap.*, 38, 1555 (1965).
126. Davis and Hetzer, *J. Res. Natl. Bur. Standards*, 65A, 209 (1961).
127. Gupta and Soni, *J. Ind. Chem. Soc.*, 42, (6), 377 (1965).
128. Ernst, Irving, and Menashi, *Trans. Faraday Soc.*, 60, 56 (1964).
129. Banewicz, Reed, and Levitch, *J. Am. Chem. Soc.*, 79, 2693 (1957).
130. Bradley and Lewis, *J. Phys. Chem.*, 29, 782 (1925).
131. Pal'chevskii, Zakhar'evskii, and Malinina, *Vestnik Leningrad University*, 15 No 16, Ser. Fiz. i Khim, No. 3, 95 (1960).
132. Sturtevant, *J. Am. Chem. Soc.*, 64, 77 (1942).
133. Chaturvedi, Dinkar and Biswas, *Proc. Natl. Acad. Sci. India*, 34, 22 (1964).
134. Chaturfvedi and Katiyar, *Bull. Chem. Soc. Jap.*, 35, 1416 (1962).
135. Sager and Siewers, *J. Res. Natl. Bur. Standards*, 45, 489 (1950).
136. Perkampus and Köhler, *Z. Elektrochem.*, 64, 365 (1960).
137. Ahluwalia, Millero, Goldberg, and Hepler, *J. Phys. Chem.*, 70, 319 (1966).
138. Anderegg, *Helv. Chim. Acta*, 46, 2813 (1963).
139. Lahiri and Aditya, *Z. Phys. Chem.*, 41, 173 (1964).
140. Krumholz, *J. Am. Chem. Soc.*, 71, 3654 (1959).
141. Näsänen, *Suomen Kemistilehti*, 28B, 161 (1955).
142. Owen, *J. Am. Chem. Soc.*, 56, 1695 (1934).
143. Loeb and Scheraga, *J. Phys. Chem.*, 60, 1633 (1956).
144. Powell and Curtis, *J. Chem. Soc. B*, 1205 (1966).
145. Bertsch, Fernelius, and Block, *J. Phys. Chem.*, 62, 444 (1958).
146. Basolo, Murman, and Chen, *J. Am. Chem. Soc.*, 75, 1478 (1953).
147. Slovetskii, Shevelov, Fainzil'berg, and Novikov, *Zh. Vses. Khim. Okshchestva im D I Mendeleeva*, 6, 707 (1961).
148. Harned and Sutherland, *J. Am. Chem. Soc.*, 56, 2039 (1934).
149. Coates and Rigg, *Trans. Faraday Soc.*, 57, 1088 (1961).
150. Edsall and Wyman, *Biochemical Chemistry*, Vol I, Academic Press, New York, 1958, 452, 464.
151. King, *J. Am. Chem. Soc.*, 76, 1006 (1954).
152. Duculot, *Compt. Rend.*, 241, 1925 (1955).
153. Wissbrun, French, and Patterson, *J. Phys. Chem.*, 58, 693 (1954).
154. Shedlovsky and MacInnes, *J. Am. Chem. Soc.*, 57, 1705 (1935).
155. Stadie and Hawes, *J. Biol. Chem.*, 77, 241 (1928).
156. Scheurer, Brownell, and LuValle, *J. Phys. Chem.*, 62, 809 (1958).
157. Roughton, *J. Am. Chem. Soc.*, 63, 2930 (1941).
158. Gattow and Krebs, *Z. Anorg. Allgem. Chem.*, 323, 13 (1963).
159. Bates and Pinching, *J. Am. Chem.. Soc.*, 71, 1274 (1949).
160. Anderegg, *Helv. Chim. Acta*, 46, 1833 (1963).
161. Moeller and Hseu, *J. Inorg. Nucl. Chem.*, 24, 1635 (1962).
162. Cecil and McPhee, *Biochem. J.*, 60, 496 (1955).
163. Christensen, Rytting, and Izatt, *J. Phys. Chem.*, 71, 2700 (1967).
164. Lewin and Humphreys, *J. Chem. Soc. B*, 210 (1966).
165. Slovetskii, Belikov, Zavilovich, and Epishina, *Izv. Akad. Nauk. SSSR otd Khim. Nauk.*, 520 (1962).
166. Everett and Hyne, *J. Chem. Soc.*, 1636 (1958).
167. Ayräpäa, *Svensk, Kim. Tid.*, 62, 135 (1950).
168. Mickel and Andrews, *J. Am. Chem. Soc.*, 77, 5291 (1955).
169. McIntyre, Block, and Fernelius, *J. Am. Chem. Soc.*, 81, 529 (1959).
170. Clarke, Datta, and Rabin, *Biochem. J.*, 59, 209 (1955).
171. Partridge, Christensen, and Izatt, *J. Am. Chem. Soc.*, 88, 1649 (1966).
172. Davies, Singer, and Staveley, *J. Chem. Soc.*, 2304 (1954).
173. Everett and Pinsent, *Proc. Roy. Soc. Ser. A*, 215, 416 (1952).
174. Vacca and Arenare, *J. Phys. Chem.*, 71, 1495 (1967).
175. Paoletti, Ciampolini, and Vacca, *J. Phys. Chem.*, 67, 1065 (1963).
176. Jonassen, Bruce, LeBlanc. Meibohm, and Rogan, *J. Am. Chem. Soc.*, 72, 2430 (1950).
177. Moeller and Ferrus, *J. Inorg. Nucl. Chem.*, 20, 261 (1961).
178. Tillotson and Staveley, *J. Chem. Soc.*, 3613 (1958).
179. Moeller and Chu, *J. Inorg. Nucl. Chem.*, 28, 153 (1966).
180. Carini and Martell, *J. Am. Chem. Soc.*, 75, 4810 (1953).
181. Datta and Grzybowski, *J. Chem. Soc.*, 3068 (1962).
182. Bates and Pinching, *J. Res. Natl. Bur. Standards*, 46, 349 (1951).
183. Thorogood and Hanania, *Biochem. J.*, 87, 123 (1963).

184.   Harned and Embree, *J. Am. Chem. Soc.*, 56, 1042 (1934).
185.   Harned and Dedell, *J. Am. Chem. Soc.*, 63, 3308 (1941).
186.   Izatt, Rytting, Hansen, and Christensen, *J. Am. Chem. Soc.*, 88, 2641 (1966).
187.   Lumme, *Suomen Kem.*, 33B, 87 (1960).
188.   Miyamoto and Miohima, *Kyoritsu Yakka Daigaku Kankya Nempo*, 4, 12 (1958).
189.   Gruen, Laskowski, and Scheraga, *J. Am. Chem. Soc.*, 81, 3891 (1959).
190.   Ashby, Clarke, Crook, and Datta, *Biochem. J.*, 59, 203 (1955).
191.   Llopis and Ordonez, *J. Electroanal. Chem.*, 5, 129 (1963).
192.   Wilson and Cannon, *J. Biol. Chem.*, 119, 309 (1937).
193.   Datta and Grzybowski, *Biochem. J.*, 69, 218 (1958).
194.   Ashby, Crook, and Datta, *Biochem. J.*, 56, 198 (1954).
196.   King, *J. Am. Chem. Soc.*, 67, 2178 (1945).
197.   Anderson, Greenhalgh, and Izatt, *Inorg. Chem.*, 5, 2106 (1966).
198.   King, *J. Am. Chem. Soc.*, 73, 155 (1951).
199.   Murphy and Martell, *J. Biol. Chem.*, 226, 37 (1957).
200.   Harned and Owen, *The Physical Chemistry of Electrolytic Solutions*, 2nd ed., Reinhold, New York, 1950, 514.
201.   Izatt, Christensen, and Kothari, *Inorg. Chem.*, 3, 1565 (1964).
202.   Owen, *J. Am. Chem. Soc.*, 56, 24 (1934).
203.   D. Banerjea and S. D. Chaudhuri (communicated).
205.   Sturtevant, *J. Am. Chem. Soc.*, 63, 88 (1941).
206.   Datta, Grzybowski, and Bates, *J. Phys. Chem.*, 68, 275 (1964).
207.   Connor, Jones, and Tuleen, *Inorg. Chem.*, 4, 1129 (1965).
208.   Vaissermann, *Compt. Rend.*, 262C, 692 (1966).
209.   Bunville and Schwalbe, *Biochemistry*, 5, 3521 (1966).
210.   Iberson and Wadsö, *Acta Chem. Scand.*, 17, 1552 (1963).
211.   Laloi and Rumpf, *Bull. Soc. Chim. France*, 1645 (1961).
212.   Nicolas and Fernelius, *J. Phys. Chem.*, 65, 1047 (1961).
214.   Gibbs, *Arch. Biochem. Biophys.*, 51, 277 (1954).
215.   Sallavo and Lumme, *Suomen Kemistilehti*, 40, 155 (1967).
216.   Boughton and Keller, *J. Inorg. Nucl. Chem.*, 28, 2851 (1966).
217.   Izatt, Christensen, Pack, and Bench, *Inorg. Chem.*, 1, 828 (1962).
218.   Datta and Grzybowski, *J. Chem. Soc. B*, 136 (1966).
219.   Tanford and Wagner, *J. Am. Chem. Soc.*, 75, 434 (1953).
220.   Nozaki, Gurd, Chen, and Edsall, *J. Am. Chem. Soc.*, 79, 2123 (1957).
221.   Bauman and Wang, *Inorg. Chem.*, 3, 368 (1964).
222.   Hanania and Irvine, *J. Chem. Soc.*, 2750 (1962).
223.   Hanania, Irvine, and Shurayh, *J. Chem. Soc.*, 1149 (1965).
224.   Hamer, Burton, and Acree, *J. Res. Natl. Bur. Standards*, 24, 269 (1940).
225.   Boyd and Wang, *J. Am. Chem. Soc.*, 87, 430 (1965).
226.   Feates and Ives, *J. Chem. Soc.*, 2798 (1956).
227.   Beetlestone and Irvine, *J. Chem. Soc.*, 5090 (1964).
228.   Beetlestone and Irvine, *J. Chem. Soc.*, 3271 (1965).
229.   George and Hanania, *Biochem. J.*, 55, 236 (1953).
230.   Aston and Ziemer, *J. Am. Chem. Soc.*, 68, 1405 (1946).
231.   Slovetskii, Shevelov, Fainzil'berg, and Novikov, *Zh. Vses. Khim. Okshchestva im D I Mendeleeva*, 6, 599 (1961).
232.   Briere, Felici, and Piot, *Compt. Rend.*, 255, 107 (1962).
233.   Bjerrum, Unmack, and Zechmeisler, *Kgl. Danske. Videnskab. Selskab. Math. fys Medd.*, 5, No. 11 (1925).
234.   Pelletier and Quintin, *Compt. Rend.*, 244, 894 (1957).
235.   Pekkarinen, *Suomen Kemistilehti*, 38B (3), 63 (1965).
236.   Hetzer, Bates, and Robinson, *J. Phys. Chem.*, 70, 2869 (1966).
237.   Batzar, Chester, and Goldberg, *J. Chem. Eng. Data*, 8, 293 (1963).
238.   Hermans, Jr. and Rialdi, *Biochemistry*, 4, 1277 (1965).
239.   Blumberger, *Rec. Trav. Chim.*, 62, 753 (1943).
240.   Latimer and Zimmerman, *J. Am. Chem. Soc.*, 61, 1550 (1939).
241.   Larson, Bertrand, and Hepler, *J. Chem. Eng. Data*, 11, 595 (1966).
242.   Pinching and Bates, *J. Res. Natl. Bur. Standards*, 40, 405 (1948).
243.   Harned and Fallon, *J. Am. Chem. Soc.*, 61, 3111 (1939).
244.   Izatt, Fernelius, and Block, *J. Phys. Chem.*, 59, 235 (1955).
245.   Perlmann, *J. Biol. Chem.*, 239, 3762 (1964).
246.   Näsänen and Uusitalo, *Suomen Kemistilehti*, 29B No. 2, 11 (1956).
247.   Fernandez and Hepler, *J. Am. Chem. Soc.*, 81, 1783 (1959).
248.   Zavitsas, *J. Chem. Eng. Data*, 12, 94 (1967).

249. Millero, Ahluwalia, and Hepler, *J. Chem. Eng. Data,* 9, 192 (1964).
250. O'Hara and Hepler, *J. Phys. Chem.,* 65, 2107 (1961).
251. Crimmins, Dymek, Flood, and O'Hara, *J. Phys. Chem.,* 70, 931 (1966).
252. Robinson and Peiperyl, *J. Phys. Chem.,* 67, 2860 (1963).
253. Allen, Robinson, and Bower, *J. Phys, Chem.,* 66, 171 (1962).
254. Neale, *Trans. Faraday Soc.,* 17, 505 (1921).
255. Gary, Bates, and Robinson, *J. Phys. Chem.,* 68, 3806 (1964).
256. Bates, *J. Res. Natl. Bur, Standards,* 47, 127 (1951).
257. Bates and Acree, *J. Res. Natl. Bur, Standards,* 30, 129 (1943).
258. Christensen, Izatt, Hansen, and Partridge, *J. Phys. Chem.,* 70, 2003 (1966).
259. Irani and Taulli, *J. Inorg. Nucl. Chem.,* 28, 1011 (1966).
260. Nims, *J. Am. Chem. Soc.,* 56, 1110 (1934).
261. Grzybowski, *J. Phys. Chem.,* 62, 555 (1958).
262. Mitra, Malhotra, and Jain, *Trans. Faraday Soc.,* 62, 167 (1966).
263. Pagano, Goldberg, and Fernelius, *J. Phys. Chem.,* 65, 1062 (1961).
264. Bates and Bower, *J. Res. Natl. Bur. Standards,* 57, 153 (1956).
265. Held and Goldberg, *Inorg. Chem.,* 2, 585 (1963).
266. Smith, Gorham, and Smith, *J. Biol. Chem.,* 144, 737 (1942).
267. Hetzer, Robinson, and Bates, *J. Phys. Chem.,* 66, 2696 (1962).
268. Datta, Grzybowski, and Weston, *J. Chem. Soc.,* 792 (1963).
269. Woodhead, Paabo, Robinson, and Bates, *J. Res. Natl. Bur. Standards,* 69A, 263 (1965).
270. Irving and Wadsö, *Acta Chem. Scand.,* 18, 195 (1964).
271. Sturtevant, *J. Am. Chem. Soc.,* 77, 1495 (1955).
272. Hetzer and Bates, *J. Phys. Chem.,* 66, 308 (1962).
273. Harned and Ehlers, *J. Am. Chem. Soc.,* 55, 2379 (1933).
274. Moore and Felsing, *J. Am. Chem. Soc.,* 69, 2420 (1947).
275. Martin and Tartar, *J. Am. Chem. Soc.,* 59, 2672 (1937).
276. Nims and Smith, *J. Biol. Chem.,* 113, 145 (1936).
277. Cabani and Conti, *Gazz. Chim. Ital.,* 95, 533 (1965).
278. Mortimer and Laidler, *Trans. Faraday Soc.,* 55, 1731 (1959).
280. Andon, Cox, and Herington, *Trans. Faraday Soc.,* 50, 918 (1954).
281. Green and Freer, *J. Phys. Chem.,* 65, 2211 (1961).
282. Hanania and Irving, *J. Chem. Soc.,* 2745 (1962).
283. Bates and Hetzer, *J. Res. Natl. Bur. Standards,* 64A, 427 (1960).
284. Paabo, Robinson, and Bates, *Anal. Chem.,* 38, 1573 (1966).
285. Essery and Schofield, *J. Chem. Soc.,* 3939 (1961).
286. Goldberg and Fernelius, *J. Phys. Chem.,* 63, 1246 (1959).
287. Weimer and Fernelius, *J. Phys. Chem.,* 64, 1951 (1960).
289. Reichard and Fernelius, *J. Phys. Chem.,* 65, 380 (1961).
290. Millero, Ahluwalia, and Hepler, *J. Phys. Chem.,* 68, 3435 (1964).
291. Uusitalo, *Ann. Acad. Sci. Fennicae,* AII, 87 (1957).
292. Freasier, Oberg, and Wendlandt, *J. Phys. Chem.,* 62, 700 (1958).
293. Tanford, Havenstein, and Rands, *J. Am. Chem. Soc.,* 77, 6409 (1955).
294. Harkins and Freiser, *J. Phys. Chem.,* 63, 309 (1959).
295. Nair, *J. Inorg. Nucl. Chem.,* 26, 1911 (1964).
296. Pinching and Bates, *J. Res. Natl. Bur. Standards,* 45, 444 (1950).
297. Pinching and Bates, *J. Res. Natl. Bur. Standards,* 45, 322 (1950).
298. Cottrell and Wolfenden, *J. Chem. Soc.,* 1019 (1948).
299. Eden and Bates, *J. Res. Natl. Bur. Standards,* 62, 161 (1959).
300. Walton and Schilt, *J. Am. Chem. Soc.,* 74, 4995 (1952).
301. Sager and Byers, *J. Res. Natl. Bur. Standards,* 59, 245 (1957).
302. King and King, *J. Am. Chem. Soc.,* 74, 1212 (1952).
303. Johnstone and Leppla, *J. Am. Chem. Soc.,* 56, 2233 (1934).
304. Bates and Canham, *J. Res. Natl. Bur. Standards,* 47, 343 (1951).
305. King, *J. Am. Chem. Soc.,* 75, 2204 (1953).
306. Paoletti and Vacca, *J. Chem. Soc.,* 5051 (1964).
307. Goursot and Wadsö, *Acta Chem. Scand.,* 20, 1314 (1966).
308. Hermans, Jr., Donovan, and Scheraga, *J. Biol. Chem.,* 235, 91 (1960).
309. Winnek and Schmidt, *J. Gen. Physiol.,* 18, 889 (1935).
310. Dalton, Kirk, and Schmidt, *J. Biol. Chem.,* 88, 589 (1930).
311. Hale, Izatt, and Christensen, *J. Phys. Chem.,* 67, 2605 (1963).
312. Vanderzee and Swanson, *J. Phys. Chem.,* 67, 2608 (1963).

313. Bertrand, Millero, Wu, and Hepler, *J. Phys. Chem.,* 70, 699 (1966).
314. Kido and Fernelius, *J. Phys. Chem.,* 65, 574 (1961).
315. Harned and Hamer, *J. Am. Chem. Soc.,* 55, 2194 (1933).
316. Richards and Mair, *J. Am. Chem. Soc.,* 51, 737 (1929).
317. Rossini, *J. Res. Natl. Bur. Standards,* 6, 847 (1931).
318. Covington, Robinson, and Bates, *J. Phys. Chem.,* 70, 3820 (1966).
319. Wynne-Jones, *Trans. Faraday Soc.,* 32, 1397 (1936).
320. Das and Ives, *Proc. Chem. Soc.* (London), 373 (1961).
321. Morris, *J. Phys. Chem.,* 70, 3798 (1966).
322. Kern, *J. Chem. Educ.,* 37, 14 (1960).
323. Wrathall, Izatt, and Christensen, *J. Am. Chem. Soc.,* 87, 5809 (1965).
324. Christensen, Oscarson, and Izatt, *J. Am. Chem. Soc.,* 90, 5949 (1968).
325. Avedikian, *Bull. Soc. Chim. Franc,* 254 (1967).
326. Avedikian, *Bull. Soc. Chim. Franc,* 2570 (1966).
327. Avedikian, and Dollet, *Bull. Soc. Chim. France,* 4551 (1968).
328. Grenthe, *Acta Chem. Scand.,* 18, 283 (1964).
329. Larson and Hepler, in *Solute-Solvent Interactions,* Coetzee and Ritchie, Eds., Marcel Dekker, New York, 1969.
330. Christensen, Izatt, Wrathall, and Hansen, *J. Chem. Soc.,* A, 1212 (1969).
332. Agren, *Svensk. Kem. Tidskr.,* 86, 181 (1956).
333. Phillips, George, and Rutman, *Biochemistry,* 2, 501 (1963).
334. May and Felsing, *J. Am. Chem. Soc.,* 73, 406 (1951).
335. Timimi and Everett, *J. Chem. Soc.,* B, 1380 (1968).
336. Cox, Everett, Landsman, and Munn, *J. Chem. Soc.,* B, 1373 (1968).
337. Vasil'ev and Kockergina, *Russ. J. Phys. Chem.,* 42, 199 (1968).
338. O'Hara, Ko, Ackermann, and Hepler, *J. Phys. Chem.,* 71, 3107 (1967).
339. Aboul-Seoud and Doheim, *Can. J. Chem.,* 44, 521 (1966).
340. Bolton and Hall, *Aust. J. Chem.,* 20, 1797 (1967).
341. Bolton and Hall, *Aust. J. Chem.,* 21, 939 (1968).
342. Datta and Grzybowski, *J. Chem. Soc.,* 1091 (1954).
343. Searles, Tamres, Block, and Quarterman, *J. Am. Chem. Soc.,* 78, 4917 (1956).
344. Liotta, Leavell, and Smith, *J. Phys. Chem.,* 71, 3091 (1967).
345. Bolton, Hall, and Reece, *J. Chem. Soc.,* B, 709 (1967).
346. Desai and Milburn, *J. Am. Chem. Soc.,* 91, 1958 (1969).
347. Banerjee, Kundu, and Das, *J. Chem. Soc.,* A, 139 (1968).
348. Christensen, Wrathall, and Izatt, *Anal. Chem.,* 40, 175 (1968).
349. Vasil'ev and Kockergina, *Russ. J. Phys. Chem.,* 40, 1622 (1966).
350. Vasil'ev and Kockergina, *Russ. J. Phys. Chem.,* 41, 1149 (1967).
352. Hansen, West, Baca, and Blank, *J. Am. Chem. Soc.,* 90, 6588 (1968).
353. Holmes and Williams, *J. Chem. Soc.,* A, 1256 (1967).
354. Kul'ba and Makashev, *Russ. J. Inorg. Chem.,* 7, 661 (1962).
355. Jordan, *Rec. Chem. Prog.,* 19, 193 (1958).
356. Ojelund and Wadsö, *Acta Chem. Scand.,* 21, 2691 (1968).
357. Ojelund and Wadsö, *Acta Chem. Scand.,* 21, 1408 (1967).
358. Brunetti, Lim, and Nancollas, *J. Am. Chem. Soc.,* 90, 5120 (1968).
360. Coates, Marsden, and Rig, *Trans. Faraday Soc.,* 65, 863 (1969).
361. Hansen, Partridge, Izatt, and Christensen, *Inorg. Chem.,* 5, 569 (1966).
362. Raffa, Stern, and Malspeis, *Anal. Chem.,* 40, 70 (1968).
363. Popper, Roman, and Marcu, *Talanta,* 14, 1163 (1967).
364. Hetzer, Robinson, and Bates, *J. Phys. Chem.,* 72, 2081 (1968).
365. Fuger and Merciny, *Bull. Soc. Chim. Belges,* 77, 59 (1968).
366. Yatsimirskii and Prik, *Russ. J. Inorg. Chem.,* 7, 30 (1962).
367. Korobova and Prik, *Russ. J. Inorg. Chem.,* 10, 456 (1965).
368. Sacconi, Paoletti, and Ciampolini, *J. Chem. Soc.,* 5046 (1964).
369. Neuberger and Fletcher, *J. Chem. Soc.,* B, 178 (1969).
370. George and Tsou, *Biochem. J.,* 50, 440 (1952).
371. Beetlestone and Irving, *Proc. Roy. Soc.* (London), 277, 401 (1964).
372. Calmon, Cazaux-Maraval, and Maroni, *Bull. Soc. Chim. France,* 3779 (1968).
373. von Schalien, *Suomen. Kemistilehti,* B32, 148 (1959).
374. Schmidt, Kirk, and Appleman, *J. Biol. Chem.,* 88, 285 (1930).
375. Hearon, Burk, and Schade, *J. Natl. Cancer Inst.,* 9, 337 (1949).
376. Roberts, Meadows, and Jardetzky, *Biochemistry,* 8, 2053 (1969).
377. Ahrland, *Helv. Chim. Acta,* 50, 306 (1967).

378. Evans and Uri, *Trans. Faraday Soc.,* 45, 224 (1949).
379. Salvis, Mishchenko, and Flis, *Zh. Neorgan. Khim.,* 2, 1985 (1957).
380. Wagman, Evans, Halow, Parker, Baily, and Schumm, *Tech. Note 270-1,* U.S. Natl. Bur. Stand. (1965).
381. George, Hanania, Irvine, and Abu-Issa, *J. Chem. Soc.,* 5689 (1964).
382. Edsall, Felsenfeld, Goodman, and Gurd, *J. Am. Chem. Soc.,* 76, 3054 (1954).
383. Williams, *J. Chem. Soc.,* A, 2695 (1969).
384. Clarke and Glew, *Trans. Faraday Soc.,* 62, 539 (1966).
385. George and Hanania, *Biochem. J.,* 52, 517 (1952).
386. Pelletier, *J. Chim. Phys.,* 57, 311 (1960).
387. Tummavuoi and Lumme, *Acta Chem. Scand.,* 22, 2003 (1968).
388. Hughes and Stedman, *J. Chem. Soc.,* 1239 (1963).
389. Maksimova, *Russ. J. Phys. Chem.,* 41, 27 (1967).
390. Carson, Laye, and Smith, *J. Chem. Soc.,* A, 141 (1968).
391. Gentile and Dadger, *J. Chem. Eng. Data,* 13, 367 (1968).
392. Lahiri and Aditya, *J. Indian Chem. Soc.,* 41, 469 (1964).
393. Lahiri and Aditya, *Z. Phys. Chem. (Frankfurt),* 55, 6 (1967).
394. Lahiri and Aditya, *Z. Phys. Chem. (Frankfurt),* 43, 282 (1964).
395. Bolton, Hall, and Reese, *Spectrochim. Acta,* 22, 1149 (1966).
396. Bolton, Hall, and Reece, *J. Chem. Soc.,* B, 717 (1966).
397. Bolton, Hall, and Reece, *Spectrochim. Acta,* 22, 1825 (1966).
398. Robinson and Peiperl, *J. Phys. Chem.,* 67, 1723 (1963).
399. Levine and Wilson, *Inorg. Chem.* 7, 818 (1968).
400. Green, *Aust. J. Chem.,* 22, 721 (1969).
401. Green and Goodwin, *Aust. J. Chem.,* 21, 1165 (1968).
402. Hetzer, Bates, and Robinson, *J. Phys. Chem.,* 67, 1124 (1963).
403. Gutnikov and Freiser, *Anal. Chem.,* 40, 39 (1968).
404. Johnston and Freiser, *Anal. Chim. Acta,* 11, 201 (1954).
405. Goodard, Lodham, Ajayi, and Cambell, *J. Chem. Soc.,* A, 506 (1969).
406. Goddard, Nwankwo, and Stavley, *J. Chem. Soc.,* A, 1376 (1967).
407. Zielen, *J. Am. Chem. Soc.,* 81, 5022 (1959).
408. Izatt, Eatough, Christensen, and Bartholomew, *J. Chem. Soc.,* A, 45 (1969).
409. Davies, Jones, and Monk, *Trans. Faraday Soc.,* 48, 921 (1952).
410. Nancollas and Poulton, *Inorg. Chem.,* 8, 680 (1969).
411. Aylward, *J. Chem. Soc.,* B, 401 (1967).
412. Antikainen, *Suomen Kemistilehti,* B29, 14 (1956).
413. Goldberg and Hepler, *J. Phys. Chem.,* 72, 4654 (1968).
414. Arnek and Patel, *Acta Chem. Scand.,* 22, 1097 (1968).
415. Arnek and Wladyslaw, *Acta Chem. Scand.,* 21, 1449 (1967).
416. Vasil'ev and Lobanov, *Russ. J. Phys. Chem.,* 41, 434 (1967).
417. Hill, Ojelund, and Wadsö, *J. Chem. Thermodynamics,* 1, 111 (1969).
418. Benet and Goyan, *J. Pharm. Sci.,* 55, 1184 (1966).

This table originally appeared in Sober, Ed., *Handbook of Biochemistry and selected data for Molecular Biology,* 2nd ed., Chemical Rubber Co., Cleveland, 1970.

CALORIMETRIC ΔH VALUES ACCOMPANYING CONFORMATIONAL CHANGES OF MACROMOLECULES IN SOLUTION

Macromolecule	Mol wt	$S_{20,w}$	Solvent	pH[a]	Concentration[a]	Temperature[b] °C	Type of transition	Type of measurement	ΔH[e] kcal/mol	Ref.
Pepsin	3.5×10^5	—	0.05 M phosphate and about 0.15 M KCl	7.16	0.2–0.5%	15	Denaturation	Heat of mixing	22[c]	1, 2
				6.41		35			69[c]	
Trypsin	2.0×10^4	—	0.1 M NaCl	1.4–2.5	0.2–0.5%	25	Denaturation	Heat of mixing	8.0	3
Fibrin	3.3×10^5	—	1.0 M NaBr-acetate, phosphate	6.08	2.91%	25	Polymerization	Heat of mixing	−19	4
Fibrinogen	3.3×10^5	—	Phosphate	6.88	5 g/l	25	Clotting	Heat of mixing	−44.5	5
				6–8.5		25	Clotting	Heat of mixing	−44	
Mercaptalbumin	6.7×10^4	—	0.1 M NaCl	2.8–4.7	2–3.5%	25	Denaturation	Heat of mixing	1.5–3.4[d]	3, 6
						15			1.5[d]	6
Ferrihemoglobin	6.8×10^4	—	0.02 M sodium formate	3.2–3.8	0.76%	25	Denaturation	Heat of mixing	10 ± 0.3	7
Horse serum albumin	6.9×10^4	—	0.1 M glycine	7.0	0.61, 1.17%	15	Denaturation	Heat capacity	-76 ± 1.6	8
					2%	55			90 ± 15	
						68			75 ± 10	
						76			55 ± 7	
Myoglobin	1.78×10^4	—	0.15 M KCl	4.5	3 g/l	30	Denaturation	Heat of mixing	40	9
sperm whale	1.76×10^4	—	0.1 M glycine	9.5	3.0 g/l	85		Heat capacity	200	41, 42
				10.6		78			178	
				11.0		72			134	
				11.5		63			100	
				12.25		50			73	

[a] Final value in mixing experiments; m is used for molal concentration; M for molar.
[b] Transition temperature for heat capacity experiments.
[c] These values depend upon the choice of expressing pepsin.
[d] Value depended upon commercial source of protein.
[e] The manner of treating ionization changes and baseline shifts varies from worker to worker and may introduce differences between the results reported by different laboratories. The latter is discussed in some detail in Reference 48 for heat capacity measurements.

CALORIMETRIC ΔH VALUES ACCOMPANYING CONFORMATIONAL CHANGES OF MACROMOLECULES IN SOLUTION (continued)

Macromolecule	Mol wt	$S_{20,w}$	Solvent	pH[a]	Concentration[a]	Temperature[b] °C	Type of transition	Type of measurement	ΔH[e] kcal/mol	Ref.
Ribonuclease A	1.37×10^4	—	0.15 M KCl	2.8	1.385 and 2.69%	43	Denaturation	Heat capacity	70 ± 1	10
			0.1 M KCl	2.2	0.5–1.0 g/l	45		Heat of mixing	109 ± 5	11
			0.15 M KCl	2.8	1.5%	44		Heat capacity	86.5 ± 4.4	12
			Water	7.80	3.41–7.22%	60			99 ± 8	75
			1.5 M urea		7.14%	55			87 ± 4	
			2 M urea		4.86–7.43%	53.5			83 ± 5	76
			2.5 M urea		7.26%	52			81 ± 13	75
			3 M urea		5.73%	48.5			71 ± 4	76
			4 M urea		7.02%	46			68 ± 6	75
			1 M guanidine HCl		4.61%	50			79 ± 1	76
			2 M guanidine HCl		4.31%	37.5			55 ± 3	
			1 M hexa-methylene-tetramine		9.5	61			99 ± 2	75
			2 M hexa-methylene-tetramine		4.7–14.4%	60			105 ± 10	
			15 mM cacodylate	9.24	5.16–6.63%	61			99 ± 2	
			0.1 M glycine, acetate	2.4	0.1–1.0%	36			52	65
				3.3		47			66	
				3.7		50			73	
				4.44		54			77	
				6.0		59			89	
Ribonuclease — bovine pancreatic	1.37×10^4	—	0.04 M glycine	5.5	2.0 g/l	69.0	Denaturation	Heat capacity	115.0	43, 44
				4.0		57.0			108.0	
				3.3		49.0			97.0	
				2.75		42.0			91.5	
			HCl	0.36	0.5%	31.5			61	64
				1.05		29.9			59	
			0.2 M glycine	2.02		31.2			66	
				2.80	0.1–2.7%	40.6			88	
				3.28	0.5%	45.8			105	

CALORIMETRIC ΔH VALUES ACCOMPANYING CONFORMATIONAL CHANGES OF MACROMOLECULES IN SOLUTION (continued)

Macromolecule	Mol wt	$S_{20,w}$	Solvent	pH[a]	Concentration[a]	Temperature[b] °C	Type of transition	Type of measurement	ΔH[e] kcal/mol	Ref.
Ribonuclease – bovine pancreatic (continued)			0.2 M acetate	4.04		52.3			126	
				5.00		57.8			151	
				6.23		60.8			155	
				7.00		61.3			168	
				7.80		61.2			178	
			0.2 M NaCl	7.0	5.16–6.63%	47.1	Denaturation	Heat capacity	111	75
Ribonuclease S'	—		15 mM cacodylate	7.0	5.16–6.63%	37.6	Denaturation	Heat capacity	55	75
Ribonuclease S protein	—		15 mM cacodylate	7.0	5.16–6.63%	47.7	Denaturation	Heat capacity	107	75
Ribonuclease S	—		15 mM cacodylate 0.3 M NaCl		75 μM S-protein	5	S-protein + S-peptide = RNase S'	Heat of mixing	−23.6	55
						10			−25.2	
						15			−28.4	
						20			−33.3	
						25			−39.8	
						30			−47.9	
						35			−57.5	
						40			−68.8	
						0	S-protein + Met (O$_2$)-S-13-Peptide = Met (O$_2$)-13-RNase S'		−18.9	
						5			−19.8	
						10			−21.9	
						15			−25.0	
						20			−29.2	
						25			−34.4	
						30			−40.6	
						35			−47.8	
						40			−56.0	

CALORIMETRIC ΔH VALUES ACCOMPANYING CONFORMATIONAL CHANGES OF MACROMOLECULES IN SOLUTION (continued)

Macromolecule	Mol wt	$S_{20,w}$	Solvent	pH[a]	Concentration[a]	Temperature[b] °C	Type of transition	Type of measurement	ΔH[e] kcal/mol	Ref.
Tropocollagen										
Rat skin	3.6×10^5	—	Acetic acid, no salt	3.5	0.1–0.4 g/l	40.8	Denaturation	Heat capacity	1.53 residue	39, 40
Pike skin	3.6×10^5	—	Acetic acid, no salt	3.5	0.1–0.4 g/l	30.6	Denaturation	Heat capacity	1.24 residue	39, 40
Merlang skin	3.6×10^5	—	Acetic acid, no salt	3.5	0.1–0.4 g/l	21.5	Denaturation	Heat capacity	0.88 residue	39, 40
Cod skin	3.6×10^5	—	Acetic acid, no salt	3.5	0.1–0.4 g/l	20.0	Denaturation	Heat capacity	0.75 residue	39, 40
Lysozyme — egg white[h]	1.45×10^4	—	0.1 M phosphate	5.37	5.34–6.16%	76.5	Denaturation	Heat capacity	138 ± 7	62
			4 M urea		5.23–9.67%	65.5			103 ± 7	
			7 M urea		5.92%	55.0			80 ± 3	
			1 M guanidine HCl		4.40%	67.5			103 ± 6	
			2 M guanidine HCl		5.22%	58			85 ± 4	
			1 M hexa-methylene-tetramine		4.72–5.53%	—			121 ± 9	
			2 M hexa-methylene-tetramine		5.72%				121 ± 6	
	1.43×10^4		0.04 M glycine	4.5	1 0–5.0 g/l	78.5			141	44
				4.0		77.0			134	
				3.0		74.5			133	
				2.6		69.0			125	
				2.5		66.0			119	
				2.0		56.0			106	
				1.5		48.0			91	
			Water-HCl	1.0	2.5%	46			56 ± 8	13
			3 M guanidine HCl	1.25	4.47–22.4 g/l	25		Heat of mixing	30 ± 3	61

[h]Data are also found in Reference 74 for this protein, although complete experimental conditions were not given.

CALORIMETRIC ΔH VALUES ACCOMPANYING CONFORMATIONAL CHANGES OF MACROMOLECULES IN SOLUTION (continued)

Macromolecule	Mol wt	$S_{20,w}$	Solvent	pH[a]	Concentration[a]	Temperature[b] °C	Type of transition	Type of measurement	ΔH[e] kcal/mol	Ref.
Ovalbumin	4.5×10^4	—	0.1 M glycine	10.0	~2%	77.5	Denaturation	Heat capacity	210 ± 13	60
				9		73			172 ± 13	
				5		68			119 ± 13	
				4.5		62			95 ± 13	
				4		57			84 ± 13	
				3		52			45 ± 13	
Chymotrypsin	—	—	0.01 M KCl	2.0 ± 0.08	2.0–10.0 g/l	25	Denaturation	Heat of mixing	50 ± 10	50
						40			110	
Chymotrypsinogen	—	—	0.01 M KCl	2.0 ± 0.08	2.0–10.0 g/l	50	Denaturation	Heat of mixing	123	50
	2.57×10^4		Water HCl	1.95	0.21–0.26%	40.6			103	51
				2.03		42.0			102	
				2.08		42.0			99	
				2.59		48.0			126	
			0.05 M glycine HCl	2.99		53.9			145	
				3.02		54.2			135	
Me₂ SO-chymotrypsin	—	—	0.01 M KCl	2.0 ± 0.08	2.0–10.0 g/l	25	Denaturation	Heat of mixing	32	50
						40			73	
Chymotrypsin–bovine	—	—	0.04 M glycine	4.0	1.0–5.0 g/l	56.4	Denaturation	Heat capacity	162	44, 46
	2.52×10^4			3.4		55.4			155	
				3.1		52.0			149	
				2.8		48.6			142	
				2.6		44.8			132	
				2.2		38.2			108	
Chymotrypsinogen A	—	—	0.1 M NaCl, 0.1 M hydrocinnamate	7.4	0.4 mM	25	Activation to π chymotrypsin	Heat of mixing	0 ± 0.5	66
	2.51×10^4		HCl	3	5.92%	56	Denaturation	Heat capacity	154 ± 8	76
				2	6.77%	42.5			112 ± 3	
π Chymotrypsin	—	—	Various NaCl, CaCl₂, and buffers	7.4	0.06–0.7 mM	25	Conversion to δ-chymotrypsin	Heat of mixing	−2 ± 1	66

CALORIMETRIC ΔH VALUES ACCOMPANYING CONFORMATIONAL CHANGES OF MACROMOLECULES IN SOLUTION (continued)

Macromolecule	Mol wt	$S_{20,w}$	Solvent	pH^a	Concentration^a	Temperature^b °C	Type of transition	Type of measurement	ΔH^e kcal/mol	Ref.
α-Chymotrypsinogen	2.45×10^4	—	0.01 M glycine acetate	2.3	0.1–1.0%	43	Denaturation	Heat capacity	78	41
				2.6		49			102	
				2.8		51			110	
				3.4		58			130	
				4.0		61			140	
				5.0		62			148	
α-Chymotrypsin	—	—	0.05 M phosphate, 0.2 M KCl	7.8	0.1–0.4 mM	25	Dimerization	Heat of mixing	-17.1 ± 1.2	67
Cytochrome c bovine heart	1.24×10^4	—	0.04 M glycine	4.8	1.0–5.0 g/l	78.0	Denaturation	Heat capacity	107	44, 47
				4.5		77.0			103	
				3.9		72.0			96	
				3.7		70.0			93	
				3.4		66.0			84	
				3.2		62.0			77	
				3.0		59.0			70	
				2.8		52.5			60	
Poly (β-benzyl-L-aspartate)	—	—	5.2 mol % CHCl$_2$CO$_2$H, 94.8 mol % CHCl$_2$CHCl$_2$	—	0–1%	−0.8	Coil-helix	Heat of mixing	0.358 residue	13
			5.7 mol % CHCl$_2$CO$_2$H, 94.3 mol % CHCl$_2$CHCl$_2$			7.4			0.334 residue	
			6.0 mol % CHCl$_2$CO$_2$H, 94.0 mol % CHCl$_2$CHCl$_2$			17.8			0.298 residue	
			6.4 mol % CHCl$_2$CO$_2$H, 93.6 mol %			28.8			0.229 residue	

CALORIMETRIC ΔH VALUES ACCOMPANYING CONFORMATIONAL CHANGES OF MACROMOLECULES IN SOLUTION (continued)

Macromolecule	Mol wt	$S_{20,w}$	Solvent	pH[a]	Concentration[a]	Temperature[b] °C	Type of transition	Type of measurement	ΔH[e] kcal/mol	Ref.
Poly (β-benzyl-L-aspartate) (continued)			$CHCl_2 CHCl_2$ 93.3 mol %, $CHCl_2 CO_2 H$, 6.7 mol %			38.0			0.169 residue	
Poly (γ-benzyl-L-glutamate)	5×10^5	—	$CHCl_2 CHCl_2$ 47 mol%, $CHCl_2 CO_2 H$, 53 mol %	—	2–3 wt/vol%	0	Coil-helix	Heat capacity	0.75 residue	73
			$CHCl_3$ 51 mol %, $CHCl_2 CO_2 H$, 49 mol %			2			0.71 residue	
			$CHCl_3$ 56 mol %, $CHCl_2 CO_2 H$, 44 mol %			9			0.68 residue	
			$CHCl_3$ 62 mol %, $CHCl_2 CO_2 H$, 38 mol %			15			0.61 residue	
			$CHCl_3$ 78 mol %, $CHCl_2 CO_2 H$, 22 mol %			41			0.40 residue	
			$CHCl_3$ 46 mol %, $CHCl_2 CO_2 H$, 54 mol %			−21			0.84 residue	
			$CHCl_2 CHCl_2$ 52 mol %, $CHCl_2 CO_2 H$, 48 mol %			−15			0.80 residue	
			$CHCl_2 CHCl_2$ 65 mol %, $CHCl_2 CO_2 H$, 35 mol %			3			0.72 residue	

CALORIMETRIC ΔH VALUES ACCOMPANYING CONFORMATIONAL CHANGES OF MACROMOLECULES IN SOLUTION (continued)

Macromolecule	Mol wt	$S_{20,w}$	Solvent	pH[a]	Concentration[a]	Temperature[b] °C	Type of transition	Type of measurement	ΔH[e] kcal/mol	Ref.
Poly (γ-benzyl-L-glutamate) (continued)			$CHCl_2CHCl_2$ 74 mol %, $CHCl_2CO_2H$, 26 mol %			14			0.57 residue	
			$CHCl_2CHCl_2$ 79 mol %, $CHCl_2CO_2H$, 21 mol %			21			0.84 residue	
			$CHCl_2CHCl_2$ 81 mol %, $CHCl_2CO_2H$, 19 mol %			39			0.30 residue	
			$CHCl_2CHCl_2$ 37 mol %, $CHCl_2CO_2H$, 63 mol %			−24			0.86 residue	
			CH_2ClCH_2Cl 47 mol %, $CHCl_2CO_2H$, 53 mol %			−10			0.81 residue	
			CH_2ClCH_2Cl 56 mol %, $CHCl_2CO_2H$, 44 mol %			2			0.73 residue	
			CH_2ClCH_2Cl 63 mol %, $CHCl_2CO_2H$, 37 mol %			13			0.66 residue	
			CH_2ClCH_2Cl 69 mol %, $CHCl_2CO_2H$, 31 mol %			23			0.56 residue	
			CH_2ClCH_2Cl 74 mol %, $CHCl_2CO_2H$, 26 mol %			31			0.49 residue	

CALORIMETRIC ΔH VALUES ACCOMPANYING CONFORMATIONAL CHANGES OF MACROMOLECULES IN SOLUTION (continued)

Macromolecule	Mol wt	$S_{20,w}$	Solvent	pH[a]	Concentration[a]	Temperature[b] °C	Type of transition	Type of measurement	ΔH[e] kcal/mol	Ref.
Poly (γ-benzyl-L-glutamate) (continued)			CH_2ClCH_2Cl 77 mol % $CHCl_2CO_2H$, 23 mol %			40			0.36 residue	
	2.35×10^5		CH_2ClCH_2Cl 19 wt % CH_2ClCH_2Cl, 81 wt %		0.257 m residue	32			0.43 residue	14, 15
			$CHCl_2CO_2H$		0.132 m residue				0.68 residue	
					0.068 m residue				0.81 residue	
					$C \to 0$				0.95 ± 0.03 residue	
	2.7×10^5		25 vol % CH_2ClCH_2Cl, 75 vol % $CHCl_2CO_2H$ (CH_2ClCH_2Cl) → 100%		0.097 m residue	26			0.525 ± 0.08 residue	16
	1.6×10^5		$CHCl_2CO_2H$ CH_2ClCH_2Cl- $CHCl_2CO_2H$ 82 wt %		7 mM residue	30		Heat of solution	0.70 ± 0.05 residue	17
	3.5×10^5		$CHCl_2CO_2H$		0.25 m residue	37		Heat capacity	0.38 residue	18
					0.07 m residue	43			0.79 residue	
			83 wt % $CHCl_2CO_2H$		0.25 m residue	40			0.32 residue	
					0.13 m residue	44			0.62 residue	
					0.07 m residue	46			0.78 residue	
			85 wt % $CHCl_2CO_2H$		0.25 m residue	46			0.29 residue	

CALORIMETRIC ΔH VALUES ACCOMPANYING CONFORMATIONAL CHANGES OF MACROMOLECULES IN SOLUTION (continued)

Macromolecule	Mol wt	$S_{20,w}$	Solvent	pH[a]	Concentration[a]	Temperature[b] °C	Type of transition	Type of measurement	ΔH[e] kcal/mol	Ref.
Poly (γ-benzyl-L-glutamate) (continued)			88 wt % $CHCl_2CO_2H$		0.13 m residue	50			0.58 residue	
					0.07 m residue	53			0.76 residue	
					0.25 m residue	—			0.26 residue	
					0.13 m residue				0.55 residue	
					0.07 m residue				0.74 residue	
	2.9×10^5		25 vol % $CHCl_3$, 75 vol % $CHCl_2CO_2H$		0.139–0.082 m residue	25		Heat of mixing	1.0 ± 0.1 residue	31
	1.6×10^5		20 vol % CH_2ClCH_2Cl, 80 vol % $CHCl_2CO_2H$		2 g/l	30		Heat of solution	0.65 ± 0.3 residue	32
			25 vol % $CHCl_3$, 75 vol % $CHCl_2CO_2H$						0.65 ± 0.3 residue	
	2.0×10^5		25 vol % CH_2ClCH_2Cl, 75 vol % $CHCl_2CO_2H$		0.123–0.7 m residue	25		Heat of mixing	0.75 ± 0.2 residue	33
	3.5×10^4		75 vol % $CHCl_2CO_2H$, 25 vol % CH_2ClCH_2Cl		3%	—		Heat capacity	0.3 ± 0.8 residue	58
	4.5×10^4								0.38 ± 0.8 residue	
	9.9×10^4								0.49 ± 0.5 residue	
	29.0×10^4								0.615 ± 0.8 residue	

CALORIMETRIC ΔH VALUES ACCOMPANYING CONFORMATIONAL CHANGES OF MACROMOLECULES IN SOLUTION (continued)

Macromolecule	Mol wt	$S_{20,w}$	Solvent	pH[a]	Concentration[a]	Temperature[b] °C	Type of transition	Type of measurement	ΔH[e] kcal/mol	Ref.
Poly (γ-benzyl-L-glutamate) (continued)	33.5×10^4								0.59 ± 0.8 residue	
	43.5×10^4								0.60 ± 0.5 residue	
	55.0×10^4								0.515 ± 0.6 residue	
Poly-γ-benzyl-L-glutamate (deuterated)	2.7×10^5	—	34 vol % CH_2ClCH_2Cl, 66 vol % $CHCl_2CO_2H$	—	3%	8.5	Coil-helix	Heat capacity	0.67 ± 0.05 residue	19
			18 vol % CH_2ClCH_2Cl, 82 vol % $CHCl_2CO_2H$			40			0.38 ± 0.05 residue	
Poly (N-γ-carbobenzoxy-L-α, γ-diaminobutric acid)	—	—	CH_2ClCH_2Cl/ $CHCl_2CO_2H$	—	~2 g/l	30	Solvation	Heat of solution	-0.6 residue	68
							Order-disorder		0.255 ± .025 residue	
Poly(N-δ-carbobenzoxy-L-ornithine)	—	—	CH_2ClCH_2Cl/ $CHCl_2CO_2H$	—	~2 g/l	30	Order-disorder	Heat of solution	-0.65 residue	
Poly(L-lysine)	1.1×10^5	—	0.1 M KCl	6.0	0.25%	15, 25	Coil-helix α-β	Heat of mixing	1.2 residue 0 residue	63
Poly(ε-carbobenzoxy-L-lysine)	7.5×10^4 1.5×10^5	—	CH_2ClCH_2Cl/ $CHCl_2CO_2H$	—	0.1%	15, 25	Coil-helix	Heat of mixing	0.62 ± 0.04 residue	59
	2.75×10^5		37 vol % $CHCl_2CO_2H$, 63 vol % $CHCl_3$		3%	26		Heat capacity	0.21 ± 0.06 residue	20
Poly(L-glutamic acid)	$0.4–1.0 \times 10^5$	—	0.1 M KCl	4.5–5.5	0.5 g/l	30	Coil-helix	Heat of mixing	-1.1 ± 0.2 residue	21
Poly(γ-ethyl-L-glutamate)	1.3×10^5	—	40 vol % CH_2ClCH_2Cl, 60 vol % $CHCl_2CO_2H$	—	2 g/l	30	Coil-helix	Heat of solution	0.65 ± 0.3 residue	32

CALORIMETRIC ΔH VALUES ACCOMPANYING CONFORMATIONAL CHANGES OF MACROMOLECULES IN SOLUTION (continued)

Macromolecule	Mol wt	$S_{20,w}$	Solvent	pH[a]	Concentration[a]	Temperature[b] °C	Type of transition	Type of measurement	ΔH^e kcal/mol	Ref.
Poly(γ-ethyl-L-glutamate) (continued)	4.0×10^5	—	35 vol% CH_2ClCH_2Cl, 65 vol% $CHCl_2CO_2H$						0.65 ± 0.3 base pair	
Salmon DNA	—	21.7	0.1 M NaCl	6.0	0.15–0.6 g/l	25	Acid denaturation	Heat of mixing	8.31 base pair	23, 24
Herring sperm-atozoa DNA	—	—	0.015 M NaCl, 1.5 mM citrate	—	1%	75	Thermal denaturation	Heat capacity	5 base pair	15
Ps. fluorescens DNA	—	20.1	0.1 M NaCl	6.0	0.15–0.6 g/l	25	Acid denaturation	Heat of mixing	7.83 base pair	24
S. marcescens DNA	—	17.4	0.1 M NaCl	6.0	0.15–0.6 g/l	25	Acid denaturation	Heat of mixing	7.83 base pair	24
Sea urchin DNA	—	23.3	0.1 M NaCl	6.0	0.15–0.6 g/l	25	Acid denaturation	Heat of mixing	8.03 base pair	24
Calf thymus DNA	> 10^6	—	0.015 M NaCl, 1.5 mM citrate	6.0	10 g/l	72	Thermal denaturation	Heat capacity	7.0 ± 0.5 base pair	25
			0.15 M phosphate	11.3	~ 4 mM base pair	34	Denaturation		8.3 ± 0.5 base pair	52
				11.15		44.2			9.5 ± 0.5 base pair	
				11.00		47.2			9.5 ± 0.5 base pair	
				10.90		50.6			9.3 ± 0.5 base pair	
				10.70		56.6			10.0 ± 0.7 base pair	
				10.60		58.8			9.1 ± 0.5 base pair	
				10.45		63.5			9.2 ± 0.7 base pair	
				10.30		68.8			10.4 ± 0.7 base pair	
			1 mM phosphate, 1.5 mM Na$^+$	7.0		58.05			6.4 ± 0.3 base pair	
			1 mM			64.5			6.8 ± 0.3 base pair	

CALORIMETRIC ΔH VALUES ACCOMPANYING CONFORMATIONAL CHANGES OF MACROMOLECULES IN SOLUTION (continued)

Macromolecule	Mol wt	$S_{20,w}$	Solvent	pH[a]	Concentration[a]	Temperature[b] °C	Type of transition	Type of measurement	ΔH[e] kcal/mol	Ref.
Calf thymus DNA (continued)			phosphate, 6.5 mM Na+, 1 mM						base pair	
			phosphate, 11.2 mM Na+, 1 mM			68.8			6.9 ± 0.3 base pair	
			phosphate, 51 mM Na+, 1 mM			77.0			7.2 ± 0.3 base pair	
Cl. perfrigens DNA	—	—	1.0 mM KCl, 1.5 mM sodium citrate	7.0	5–6 g/l	55	Denaturation	Heat capacity	7.73 base pair	53
M. lysodeikticus DNA	—	—	1.0 mM KCl, 1.5 mM sodium citrate	7.0	5–6 g/l	79	Denaturation	Heat capacity	8.52 base pair	53
T₂ phage DNA	—	—	3 mM phosphate, 0.2 M NaCl	7.0	0.5 g/l	84.8	Denaturation	Heat capacity	9.65 base pair	37
			3 mM phosphate, 0.115 M NaCl			81.2			9.42 base pair	
			3 mM phosphate, 0.057 M NaCl			75.0			9.28 base pair	
			3 mM phosphate, 0.036 M NaCl			71.5			9.14 base pair	
			3 mM phosphate, 0.014 M NaCl			66.0			9.15 base pair	
			3 mM phosphate, 0.009 M NaCl			69.0			8.90 base pair	
			3 mM phosphate, 0.2 M NaCl	8.5		82.5			8.94 base pair	

CALORIMETRIC ΔH VALUES ACCOMPANYING CONFORMATIONAL CHANGES OF MACROMOLECULES IN SOLUTION (continued)

Macromolecule	Mol wt	$S_{20,w}$	Solvent	pH^a	Concentration^a	Temperature^b °C	Type of transition	Type of measurement	ΔH^e kcal/mol	Ref.
T_2 phage DNA (continued)			3 mM phosphate, glycine	8.9		76.5			8.03 base pair	
			3 mM glycine, 0.2 M NaCl	9.3		71.8			7.78 base pair	
				9.6		66.3			7.14 base pair	
			3 mM citrate, phosphate, 0.20 M NaCl	5.4		84.0			9.40 base pair	37, 38
			3 mM citrate, 0.20 M NaCl	4.8		82.3			8.57 base pair	
				4.3		76.0			6.60 base pair	
				4.0		71.5			5.43 base pair	
				3.8		68.0			5.00 base pair	
				3.5		64.0			4.70 base pair	
				3.2		55.0			7.00 base pair	
Salmon sperm DNA	—	—	1.0 mM KCl, 1.5 mM sodium citrate	7.0	5–6 g/l	60.6	Denaturation	Heat capacity	7.84 base pair	53
M_4 coliphage DNA	4×10^7	—	0.015 M HCl, 0.1 M KCl, 2.2 M urea, citrate	3.25	0.045–0.066 g/l	27	Denaturation	Heat of mixing	9.5 ± 1.5 base pair	69

CALORIMETRIC ΔH VALUES ACCOMPANYING CONFORMATIONAL CHANGES OF MACROMOLECULES IN SOLUTION (continued)

Macromolecule	Mol wt	$S_{20,w}$	Solvent	pH[a]	Concentration[a]	Temperature[b] °C	Type of transition	Type of measurement	ΔH[e] kcal/mol	Ref.
tRNA[Phe]-yeast	—	—	0.01 M tris, 50 μM Mg²⁺	7.0	~0.12%	66.5	Unfolding	Heat capacity	140	48
			0.01 M tris, 0.1 mM Mg²⁺			70			156	
			0.01 M tris, 0.25 mM Mg²⁺			73			216	
			0.01 M tris, 5 mM Mg²⁺			80.5			248	
			5 mM			68			175	
			phosphate, 0.1 M NaCl, 0.2 mM Mg²⁺	7.2	~10 μM	57		Heat of mixing	123 ± 25	49
			5 mM NaCl, 1 mM MgCl₂	6.5	0.06–0.08 mM	49		Heat capacity	200 ± 30	71
			5 mM citrate, 1 mM MgSO₄			70			250 ± 20	
			5 mM citrate, 5 mM MgSO₄			76.5			240 ± 20	
			5 mM citrate, 0.08 M MgSO₄, 0.5 M NaCl							
			5 mM citrate, 8 mM MgSO₄, 1.0 M NaCl			76.5			240 ± 20	
			5 mM citrate, 0.08 M MgSO₄, 0.5 M NaCl			76.5			240 ± 20	
			5 mM citrate, 0.02 M MgSO₄			76.5 79			220 ± 20 230 ± 20	
			5 mM citrate, 10 mM MgSO₄	6.5	1.25–2.5 g/l	60		Heat of mixing	310	71
Poly(A·U)	—	4.5–10.0 Poly(U), 8.0–12.0 Poly(A)	0.1 M KCl, 0.01 M cacodylate	6.6	20 mM–50 μM nucleotide	25	Poly (A) + Poly (U) = Poly (A·U)	Heat of mixing	−5.9 ± 0.2 base pair	26

CALORIMETRIC ΔH VALUES ACCOMPANYING CONFORMATIONAL CHANGES OF MACROMOLECULES IN SOLUTION (continued)

Macromolecule	Mol wt	$S_{20,w}$	Solvent	pH[a]	Concentration[a]	Temperature[b] °C	Type of transition	Type of measurement	ΔH[e] kcal/mol	Ref.
Poly(A·U) (continued)			0.5 M KCl, 0.01 M cacodylate						−5.9 ± 0.2 base pair	
			1.0 M KCl, 0.01 M cacodylate						−4.75 ± 0.3 base pair	
		2.1–12.2 Poly(A)	0.1 M KCl, 0.01 M cacodylate						−5.95 ± 0.1 base pair	
		6.1 Poly(A), 7.2 Poly(U)	0.1 M KCl, 0.01 M cacodylate	7.0	35 mM nucleotide	10			−6.29 ± 0.19 base pair[f]	2
						25			−6.97 ± 0.17 base pair[f]	
						40			−7.72 ± 0.29 base pair[f]	
			0.1 M NaCl, 0.01 M cacodylate	6.8	80 mM nucleotide	24			−5.95 ± 0.1 base pair	
			0.5 M NaCl, 0.01 M cacodylate			37			−6.50 ± 0.1 base pair	
						37			−6.69 ± 0.1 base pair	
	~10⁵	—	0.01 M citrate, 0.057 M NaCl		8.5 mM base pair	49		Heat capacity	−6.7 base pair	29
			0.01 M citrate, 0.10 M NaCl			54.8			−7.2 base pair	
			0.01 M citrate, 15 M NaCl			58.4			−7.7 base pair	
						85–90		Extrapolated	−8.5 ± 0.5 base pair	

[f] Heat change corrected for unfolding poly A before reaction.

CALORIMETRIC ΔH VALUES ACCOMPANYING CONFORMATIONAL CHANGES OF MACROMOLECULES IN SOLUTION (continued)

Macromolecule	Mol wt	$S_{20,w}$	Solvent	pH[a]	Concentration[a]	Temperature[b] °C	Type of transition	Type of measurement	ΔH[e] kcal/mol	Ref.
Poly(A·U) (continued)	—	9.53 Poly(A), 6.15 Poly(U)	0.018 M NaCl, 5 mM cacodylate	6.9–7.0	5.0 mM nucleotide	44.5		Heat capacity	−7.38 ± 0.08 base pair	34
			0.043 M NaCl, 5 mM cacodylate			51.3			−7.95 ± 0.07 base pair	
			0.103 M NaCl, 0.01 M cacodylate		6.04 mM nucleotide	58.3			−8.20 ± 0.2 base pair	
			0.104 M NaCl, 0.01 M cacodylate		5.0 mM nucleotide	58.2			−8.20 ± 0.24 base pair	
		7.56 Poly(A), 5.62 Poly(U)	0.011 M KCl, 5 mM cacodylate			35.9			−6.44 ± 0.22 base pair	
			0.012 M KCl, 5 mM cacodylate			36.2			−6.44 ± 0.22 base pair	
			0.040 M KCl, 5 mM cacodylate		2.28 mM nucleotide	47			−6.83 ± 0.33 base pair	
			0.054 M KCl, 5 mM cacodylate		5.0 mM nucleotide	48.7			−6.85 ± 0.11 base pair	
			0.055 M KCl, 5 mM cacodylate			48.8			−6.85 ± 0.11 base pair	
	~10⁵		0.06 M cations, 3.3 mM citrate	6.5	3.76 mM base pair	49.4			6.8 ± 0.4 base pair	3
			0.063 M cations, 3.3 mM citrate		1.88 mM base pair	51.2			−6.9 ± 0.4 base pair	

CALORIMETRIC ΔH VALUES ACCOMPANYING CONFORMATIONAL CHANGES OF MACROMOLECULES IN SOLUTION (continued)

Macromolecule	Mol wt	$S_{20,w}$	Solvent	pH^a	Concentration^a	Temperature^b °C	Type of transition	Type of measurement	ΔH^e kcal/mol	Ref.
Poly(A·U) (continued)			0.06 M cations, 3.3 mM citrate		0.984 mM base pair	51			−6.9 ± 0.4 base pair	3
			0.46 M cations, 0.01 M citrate	6.8	Not given	56.1			−8.2 base pair	36
						70.0			−8.8 base pair	
			0.50 M cations, 0.01 M citrate			54.1			−8.1 base pair	
						71.6			−8.7 base pair	
			0.57 M cations, 0.01 M citrate			53.5			−8.0 base pair	
						74.5			−8.9 base pair	
						95		Extrapolated	−9.5 ± 0.5 base pair	
			5 mM NaCl		8.5 mM nucleotide	45.8	Poly(A) + Poly(U) = Poly(A·U)	Heat capacity	−6.6 base pair	70
			5 mM NaCl, D₂O			47.7			−6.6 base pair	
			0.1 M NaCl, 0.01 M cacodylate			24	Poly(A·U) + Poly(U) = Poly(A·2U)	Heat of mixing	−3.82 ± 0.1 (A·2U) residue	28
						37			−3.5 ± 0.5 (A·2U) residue	

CALORIMETRIC ΔH VALUES ACCOMPANYING CONFORMATIONAL CHANGES OF MACROMOLECULES IN SOLUTION (continued)

Macromolecule	Mol wt	$S_{20,w}$	Solvent	pH[a]	Concentration[a]	Temperature[b] °C	Type of transition	Type of measurement	ΔH[e] kcal/mol	Ref.
Poly(A·U) (continued)	—	—	0.5 M NaCl, 0.01 M cacodylate			24			−3.80 ± 01 (A·2U) residue	
						37			−4.09 ± 0.1 (A·2U) residue	36
		7.56 Poly(A), 5.62 Poly(U)	0.015 M KCl, 5 mM cacodylate		7.72 mM nucleotide	28.4		Heat capacity	−1.24 ± 0.1 (A·2U) residue	
					7.50 mM nucleotide	28.6			−1.24 ± 0.15 (A·2U) residue	
		9.53 Poly(A), 6.15 Poly(U)	0.18 M NaCl, 5 mM cacodylate	6.9–7.0	7.0 mM nucleotide	32.6			−1.29 ± 0.16 (A·2U) residue	
			0.019 M NaCl, 5 mM cacodylate			31.5			−1.29 ± 0.16 (A·2U) residue	
	—	~10⁵	0.46 M cations, 0.01 M citrate	6.8	Not given	56.1			−4.1 ± 2 (A·2U) residue	
			0.50 M cations, 0.01 M cacodylate			54.1			−4.3 ± 2 (A·2U) residue	
			0.57 M cations, 0.01 M cacodylate			53.1			−4.2 ± 2 (A·2U) residue	
		~10⁵	0.01 M cacodylate, 0.01 M citrate, 0.5 M NaCl			54.3	2 Poly(A·U) = Poly(A·2U) + Poly(A)		3.2 (A·2U) residue	29
						85–90		Extrapolated	4.5 ± 0.5 (A·2U) residue	

CALORIMETRIC ΔH VALUES ACCOMPANYING CONFORMATIONAL CHANGES OF MACROMOLECULES IN SOLUTION (continued)

Macromolecule	Mol wt	$S_{20,w}$	Solvent	pH[a]	Concentration[a]	Temperature[b] °C	Type of transition	Type of measurement	ΔH[e] kcal/mol	Ref.
Poly(A·U) (continued)	—	9.53 Poly(A), 6.15 Poly(U)	0.263 M NaCl, 0.01 M cacodylate	6.9–7.0	5.0 mM nucleotide	57.5		Heat capacity	2.76 ± 0.1 (A·2U) residue	34
			0.46 M cations, .01 M citrate	6.8	Not given	56.1			4.1 (A·2U) residue	36
			0.50 M cations, .0 M citrate			54.1			3.8 (A·2U) residue	
			0.57 M cations, .01 M citrate			53.5			3.8 (A·2U) residue	
			0.01 M citrate, 0.50 M cations			95		Extrapolated	5.1 ± 0.5 (A·2U) residue	
						72.1	Poly(A·2U) = Poly(A) + 2 Poly(U)		11.9 (A·2U) residue	
						85–90			12.5 ± 0.5 (A·2U) residue	
		7.56 Poly A, 5.62 Poly U	0.268 M NaCl, 0.01 M cacodylate	6.9–7.0	7.0 mM nucleotide	67.9		Heat capacity	12.7 ± 0.13 (A·2U) residue	34
			5.5 mM KCl, 5 mM cacodylate		6.0 mM nucleotide	49.1			10.0 ± 0.25 (A·2U) residue	
			5.6 mM KCl, 5 mM cacodylate			49.3			10.0 ± 0.25 (A·2U) residue	

CALORIMETRIC ΔH VALUES ACCOMPANYING CONFORMATIONAL CHANGES OF MACROMOLECULES IN SOLUTION (continued)

Macromolecule	Mol wt	$S_{20,w}$	Solvent	pH[a]	Concentration[a]	Temperature[b] °C	Type of transition	Type of measurement	ΔH[e] kcal/mol	Ref.
Poly(A·U) (continued)	~10^5	—	0.46 M cations, 0.01 M citrate	6.8	Not given	70.0			12.9 (A·2U) residue	36
			0.50 M cations, 0.01 M citrate			71.6			13.0 (A·2U) residue	
			0.57 M cations, 0.01 M citrate			74.5			13.1 (A·2U) residue	
						95		Extrapolated	13.5 ± 0.5 (A·2U) residue	
	—	9.53 Poly(A), 6.15 Poly(U)	1.8 mM NaCl, 5 mM cacodylate	6.9–7.0	7.5 mM nucleotide	45.5	Poly(A) + 2 Poly(U) = Poly(A·U) + Poly(U)	Heat capacity	-8.38 ± 0.14 (A·U) residue	34
			1.9 mM NaCl, 5 mM cacodylate			45.0			-8.38 ± 0.14 (A·U) residue	
		7.56 Poly(A), 5.62 Poly(U)	1.5 mM KCl, 5 mM cacodylate		7.72 mM nucleotide	38.8			-7.49 ± 0.23 (A·U) residue	
					7.50 mM nucleotide	38.6			-7.49 ± 0.23 (A·U) residue	

CALORIMETRIC ΔH VALUES ACCOMPANYING CONFORMATIONAL CHANGES OF MACROMOLECULES IN SOLUTION (continued)

Macromolecule	Mol wt	$S_{20,w}$	Solvent	pH[a]	Concentration[a]	Temperature[b] °C	Type of transition	Type of measurement	ΔH[e] kcal/mol	Ref.
Poly A	∼10^5	4.23	0.1 M NaCl, 0.01 M tris	7.30	C → O from 0.37–1.34 g/l	35	Helix-coil	Heat capacity	9.4 residue	3
		—	Various salt concentrations	6.8	7.8 mM base pair	90–95		Extrapolated	4.5 ± 2 residue	
			0.20 M citrate, 0.15 M NaCl, HCl	5.50	0.0132 m nucleotide	31.5	Double helix-coil	Heat capacity	3.36 base pair	2
				5.30		39.2			4.09 base pair	
				5.06		47.1			4.67 base pair	
				4.89		56.6			5.13 base pair	
				4.70		65.5			5.57 base pair	
				4.20		85.5			5.90 base pair	
	—	6.1	0.1 M KCl, 0.01 M cacodylate	4.0		10		Heat of mixing	1.80 ± 0.25[g] residue	27
						25			2.74 ± 0.20[g] residue	
Poly(dA·dT)	—	—	5 mM NaCl$_4$, 1 mM cacodylate, 1 mM citrate	7.0	5.9–11.6 mM base pair	40	Helix-coil	Heat capacity	7.9 ± 0.14 base pair	56

[g]These calorimetric data were recalculated by Stevens and Felsenfeld[77] to yield values of 6.5 and 8.5 kcal mol nucleotide for the single helix-coil transition.

CALORIMETRIC ΔH VALUES ACCOMPANYING CONFORMATIONAL CHANGES OF MACROMOLECULES IN SOLUTION (continued)

Macromolecule	Mol wt	$S_{20,w}$	Solvent	pH^a	Concentration^a	Temperature^b °C	Type of transition	Type of measurement	ΔH^e kcal/mol	Ref.
Poly(I·C)	5×10^5	—	.01 M citrate, .063 m Na$^+$	6.9 ± 0.1	1.8 mm base pair	54.1	Poly(I) + Poly(C) = Poly(I·C)	Heat capacity	−6.5 ± 0.4 base pair	54
			.01 M citrate, 0.104 mNa$^+$		1.78 mm base pair	60.8			−6.8 ± 0.4 base pair	
			.01 M citrate, 0.303 m Na$^+$			67.6			−7.6 ± 0.4 base pair	
			.01 M citrate, 0.503 m Na$^+$			70.7			−7.9 ± 0.4 base pair	
			.01 M citrate, 1.003 m Na$^+$			73.9			−8.0 ± 0.4 base pair	
	—		0.02–0.2 M NaCl	8.0	70 mM base pair	20		Heat of mixing	−5.59 ± 0.02 base pair	57
			0.1–0.4 M NaCl			37			−5.59 ± 0.01 base pair	
Poly(I)	—	—	0.1 M citrate, 1.0 m Na$^+$	6.9 ± 0.1	1.79 mM base pair	43.6	Triple helix-coil	Heat capacity	−1.9 ± 0.4 residue	54

CALORIMETRIC ΔH VALUES ACCOMPANYING CONFORMATIONAL CHANGES OF MACROMOLECULES IN SOLUTION (continued)

Macromolecule	Mol wt	$S_{20,w}$	Solvent	pH^a	Concentration^a	Temperature^b °C	Type of transition	Type of measurement	ΔH^e kcal/mol	Ref.
Poly C	—	—	1 mM acetate, 0.01 M Na$^+$	3.68	~2.8 g/l	40	Double helix-coil	Heat capacity	4.06 base pair	72
				4.33		63			5.20 base pair	
				4.48		75			5.25 base pair	
				4.55		74			5.27 base pair	
				4.85		72			5.12 base pair	
				5.20		61			4.95 base pair	
				5.50		54			4.34 base pair	
				5.76		47			3.62 ɔase pair	

Compiled by Gordon C. Kresheck.

CALORIMETRIC ΔH VALUES ACCOMPANYING CONFORMATIONAL CHANGES OF MACROMOLECULES IN SOLUTION (continued)

REFERENCES

1. Buzzell and Sturtevant, *J. Am. Chem. Soc.*, 74, 1983 (1952).
2. Sturtevant, *J. Phys. Chem.*, 58, 97 (1954).
3. Gutfreund and Sturtevant, *J. Am. Chem. Soc.*, 75, 5447 (1953).
4. Sturtevant, Laskowski, Donnelly, and Scheraga, *J. Am. Chem. Soc.*, 77, 6163 (1955).
5. Laki and Kitzinger, *Nature* (Lond.), 178, 985 (1956).
6. Bro and Sturtevant, *J. Am. Chem. Soc.*, 80, 1789 (1958).
7. Forrest and Sturtevant, *J. Am. Chem. Soc.*, 82, 585 (1960).
8. Privalov and Monaselidze, *Biofizika*, 8, 420 (1963).
9. Hermans and Rialdi, *Biochemistry*, 4, 1277 (1965).
10. Beck, Gill, and Downing, *J. Am. Chem. Soc.*, 87, 901 (1965).
11. Kresheck and Scheraga, *J. Am. Chem. Soc.*, 88, 4588 (1966).
12. Danforth, Krakauer, and Sturtevant, *Rev. Sci. Instrum.*, 38, 484 (1967).
13. McKnight, Ph.D. thesis, University of Massachusetts, 1974.
14. Ackermann and Ruterjans, *Z. Phys. Chem.*, 41, 116 (1964).
15. Ackermann and Ruterjans, *Ber Bunsenges Phys. Chem.*, 68, 850 (1964).
16. Karasz, O'Reilly, and Bair, *Nature* (Lond.), 202, 693 (1964).
17. Giacometti and Turolla, *Z. Phys. Chem.*, 51, 108 (1966).
18. Ackermann and Neumann, *Biopolymers*, 5, 649 (1967).
19. Karasz and O'Reilly, *Biopolymers*, 4, 1015 (1966).
20. Karasz, O'Reilly, and Bair, *Biopolymers*, 3, 241 (1965).
21. Rialdi and Hermans, *J. Am. Chem. Soc.*, 88, 5719 (1966).
22. Klump, Neumann, and Ackermann, *Biopolymers*, 7, 423 (1969).
23. Sturtevant and Geiduschek, *J. Am. Chem. Soc.*, 80, 2911 (1958).
24. Bunville, Geiduschek, Rawitscher, and Sturtevant, *Biopolymers*, 3, 213 (1965).
25. Ruterjans, thesis, University of Munster, Germany, 1965.
26. Steiner and Kitzinger, *Nature* (Lond.), 194, 1172 (1962).
27. Rawitscher, Ross, and Sturtevant, *J. Am. Chem. Soc.*, 85, 1915 (1963).
28. Ross and Scruggs, *Biopolymers*, 3, 491 (1965).
29. Neumann and Ackermann, *J. Phys. Chem.*, 71, 2377 (1967).
30. Epand and Scheraga, *J. Am. Chem. Soc.*, 89, 3888 (1967).
31. Kagemoto and Fugishiro, *Makromol. Chem.*, 114, 139 (1968).
32. Giacometti, Turolla, and Boni, *Biopolymers*, 6, 441 (1968).
33. Kagemoto and Jujishiro, *Biopolymers*, 6, 1753 (1968).
34. Krakauer and Sturtevant, *Biopolymers*, 6, 491 (1968).
35. Hinz, Schmitz, and Ackermann, *Biopolymers*, 7, 611 (1969).
36. Neumann and Ackermann, *J. Phys. Chem.*, 73, 2170 (1969).
37. Privalov, *Mol. Biol.* (Mosc.), 3, 690 (1969).
38. Privalov, Ptitsyn, and Birstein, *Biopolymers*, 8, 559 (1969).
39. Privalov, *Biofizika*, 13, 955 (1968).
40. Privalov and Tiktopulo, *Biopolymers*, 9, 127 (1970).
41. Privalov, Khechinashvili, and Atanasov, *Biopolymers*, 10, 1865 (1971).
42. Atanasov, Khechinashvili, and Privalov, *Mol. Biol.* (Mosc.), 6, 33 (1972).
43. Privalov, Tiktopulo, and Khechinashvili, *Int. J. Protein Peptide Res.*, 5, 229 (1973).
44. Privalov and Khechinashvili, *J. Mol. Biol.*, 86, 665 (1974).
45. Khechinashvili, Privalov, and Tiktopulo, *FEBS Lett.*, 30, 57 (1973).
46. Tischenko, Tiltopulo, and Privalov, *Biofizika*, 19, 400 (1974).
47. Khechinashvili and Privalov, *Biofizika*, 19, 14 (1974).
48. Brandts, Jackson, and Ting, *Biochemistry*, 13, 3595 (1974).
49. Levy, Rialdi, and Biltonen, *Biochemistry*, 11, 4138 (1972).
50. Biltonen, Schwartz, and Wadsö, *Biochemistry*, 10, 3417 (1971).
51. Jackson and Brandts, *Biochemistry*, 9, 2294 (1970).
52. Shiao and Sturtevant, *Biopolymers*, 12, 1829 (1973).
53. Klump and Ackermann, *Biopolymers*, 10, 513 (1971).
54. Hinz, Haar, and Ackermann, *Biopolymers*, 9, 923 (1970).
55. Hearn, Richards, Sturtevant, and Watt, *Biochemistry*, 10, 806 (1971).
56. Scheffler and Sturtevant, *J. Mol. Biol.*, 42, 577 (1969).

CALORIMETRIC ΔH VALUES ACCOMPANYING CONFORMATIONAL CHANGES OF MACROMOLECULES IN SOLUTION (continued)

57. Ross and Scruggs, *J. Mol. Biol.,* 45, 567 (1969).
58. Kagemoto and Karasz, *Analytical Calorimetry,* Vol. 2, Plenum Press, New York, 1970, 147.
59. Giacometti, Turolla, and Boni, *Biopolymers,* 9, 979 (1970).
60. Privalov, *Biofizika,* 3, 308 (1963).
61. Atha and Ackers, *J. Biol. Chem.,* 246, 5845 (1971).
62. Delben and Crescenzi, *Biochim. Biophys. Acta,* 194, 615 (1969).
63. Chou and Scheraga, *Biopolymers,* 10, 657 (1971).
64. Tsong, Hearn, Warthall, and Sturtevant, *Biochemistry,* 9, 2666 (1970).
65. Gerassimov and Mikhailov, *Soobshch. Akad. Nauk Gruz. SSSR,* 64, 185 (1971).
66. Sturtevant and Beres, *Biochemistry,* 10, 2120 (1971).
67. Shiao and Sturtevant, *Biochemistry,* 8, 4910 (1969).
68. Giacometti, Turolla, and Verdini, *J. Am. Chem. Soc.,* 93, 3092 (1971).
69. Rialdi and Profumo, *Biopolymers,* 6, 899 (1968).
70. Klump, *Biopolymers,* 11, 2331 (1972).
71. Bode, Schernau, and Ackermann, *Biophys. Chem.,* 1, 214 (1974).
72. Klump, in press.
73. Simon and Karasz, *Thermochim. Acta,* 8, 97 (1974).
74. McKnight and Karasz, *Thermochim. Acta,* 5, 339 (1973).
75. Delben, Crescenzi, and Quadrifoglio, *Int. J. Protein Peptide Res.,* 3, 57 (1971).
76. Crescenzi and Delben, *Int. J. Protein Peptide Res.,* 3, 57 (1971).
77. Stevens and Felsenfeld, *Biopolymers* 2, 293 (1964).

FREE ENERGIES OF HYDROLYSIS AND DECARBOXYLATION

William P. Jencks

One of the reasons that there has been so much confusion and disagreement regarding the free energies of hydrolysis of "energy-rich" compounds of biochemical interest is that it is uncommon for any two workers to express their results according to the same nomenclature and conventions. The following summary may be helpful in making use of these tables of free energies of hydrolysis and decarboxylation.

The equilibrium constant, K_I, for the hydrolysis of an ester may be expressed according to Equation 1 and the free energy of hydrolysis according to Equation 2, using the convention that the concentration of water is expressed in the same units as the other reactants and pure water is 55.5 M. For glycine ethyl ester the values of K_I = 0.43 and $\Delta G°$ = + 500 cal/mol at 39° reflect

$$K_I = \frac{[RCOOH][HOR']}{[RCOOR'][HOH]} \tag{1}$$

$$\Delta G_I° = -RT \ln K_I \tag{2}$$

the fact that –OH and $-OC_2H_5$ have approximately the same affinity for the carbonyl group.

For biochemical reactions, which usually take place in dilute aqueous solution, it is generally more convenient to take the activity of pure water as 1.0, and this convention will be adopted here. For glycine ethyl ester the values of K_I = 24 and $\Delta G_I°$ = - 1,970 cal/mol according to this convention reflect the fact that the driving force toward hydrolysis which results from the high concentration of water compared to the other reactants is hidden in the equilibrium expression by the convention that the activity of liquid water is 1.0. This extra driving force amounts to $-RT$ in 55.5 $\cong -2,400$ cal/mol and is one reason that free energies of hydrolysis expressed according to this convention are unlikely to be equal to heats of hydrolysis. The standard states of the other reactants according to this convention are ideal 1 M solutions of the non-ionized species. This commonly leads to difficulty for a compound such as glycine which does not exist in a non-ionized form in appreciable concentration and the standard state is commonly modified, as in the case of the values for glycine ethyl ester given here, to refer to a species in which the *reacting* groups are non-ionized; i.e., to $H_3^+NCH_2COOH$ instead of H_2NCH_2COOH. This convention, which we shall call convention I, gives a single value of $\Delta G°$ which is true regardless of the pH. Its use requires that only the actual concentrations (or activities) of the *particular ionic species* which are given in the equilibrium expression be included in calculations. For example, in order to calculate the free energy of hydrolysis of glycine ethyl ester from the results of an experiment carried out at pH 3.0, it is necessary to insert the concentration of $H_3^+NCH_2COOH$ which is present at equilibrium at this pH, not the stoichiometric concentration of total glycine.

It is often convenient, especially when the ionization constants of the reactants are not accurately known, to use convention II in which the concentrations (or activities) of the reactants are given in terms of some convenient ionic species that is present under the conditions of the experiments and any hydrogen ions present in the equilibrium expression are included. For glycine ethyl ester this is shown in Equations 3 and 4 and the value of $\Delta G_{II}°$ is + 1,440 cal/mol. As in the case of convention I, this convention refers only to the concentrations of the particular

$$K_{II} = \frac{[RCOO^-][H^+][HOR']}{[RCOOR'][HOH]} \tag{3}$$

$$\Delta G_{II}^\circ = -RT \ln K_{II} \tag{4}$$

ionic species given in the equilibrium expression which may be present in a given solution. The value of ΔG_{II}° is independent of pH.

To interconvert these pH-independent free energies with free energies which hold for stoichiometric concentrations of reactants and products at a given pH, which would be found experimentally, it is only necessary to substitute in Equation 5 the actual concentrations present at the desired pH of the particular ionic species of the reactants which are present at that pH,

$$\Delta G' = \Delta G^\circ + RT \ln \frac{[\text{products}]}{[\text{reactants}]} \tag{5}$$

including any hydrogen ions given off or taken up in the reaction. Equation 5 is the basic equation which relates concentrations (or activities) to free energies. When a reaction is at equilibrium $\Delta G' = 0$ and the standard free energy is then a logarithmic function of the *equilibrium* concentrations of the reactants and products (Equation 6). When the reactants and products are all in the standard state of activity 1.0 the concentration term drops out and the free energy of the system is equal to the standard free energy of the reaction (Equation 7).

$$\Delta G^\circ = -RT \ln \frac{[C]_{eq}[D]_{eq}}{[A]_{eq}[B]_{eq}} = -RT \ln K \tag{6}$$

$$\Delta G' = \Delta G^\circ + RT \ln \frac{1 \times 1}{1 \times 1} = \Delta G^\circ \tag{7}$$

Thus, the standard free energy is the difference in free energy between a system in which all the reactants and products are in the standard states ($\Delta G' = \Delta G^\circ$) and the same system at equilibrium ($\Delta G' = 0$).

A useful special case of Equation 5 is the situation in which the total concentrations of all reactants and products, except hydrogen ion, are 1.0 M at a given pH. This gives a value of $\Delta G^{\circ\prime}$ which refers to 1.0 M *total concentrations of all the ionic species* of the reactants and products and is *true only at the specified pH*. This convention III is the convention which refers most directly to experimental results and is most useful in making comparisons of free energies of hydrolysis under physiological conditions. $\Delta G^{\circ\prime}$ is also sometimes referred to as $\Delta G'$, ΔG_{anal}°, ΔG_{exp}° and (unfortunately) as ΔG°.

For example, the value of $\Delta G^{\circ\prime}$ for glycine ethyl ester is calculated from the ΔG° of convention I by inserting the fraction of the 1 M total glycine that is present as the free acid at pH 7.0, as shown in Equation 8.

$$\Delta G^{\circ\prime} = \Delta G_I^\circ + RT \ln \frac{[\text{RCOOH}][\text{HOR}']}{[\text{RCOOR}'][\text{HOH}]} = +500 + 1{,}420 \log \frac{(1 \times 10^{-4.6})(1)}{(0.85)(1)} = -8{,}400 \text{ cal/mol} \tag{8}$$

The fact that glycine ethyl ester is only 85% in the protonated form at pH 7.0 introduces a further small correction. The same value of $\Delta G^{\circ\prime}$ may be calculated from the ΔG° of convention II by inserting the hydrogen ion activity at pH 7.0 into the equilibrium expression (Equation 9), because at pH 7 glycine is entirely in the form of the carboxylate anion, which is the form which is used in the equilibrium expression according to this convention.

$$\Delta G^{\circ\prime} = \Delta G_{II}^\circ + RT \ln \frac{[\text{RCOO}^-][\text{H}^+][\text{HOR}']}{[\text{RCOOR}'][\text{HOH}]} = 1{,}440 + 1{,}420 \log \frac{(1)(10^{-7})(1)}{(0.85)(1)} = -8{,}400 \text{ cal/mol} \tag{9}$$

These interconversions between pH-independent and pH-dependent free energies may generally be carried out without difficulty if the following two simple rules are followed:

1. The equilibrium expression for the pH-independent equilibrium constant and free energy of hydrolysis may include any desired ionic species of the reactants, but must be based on a *balanced equation* for the reaction which includes any *hydrogen ions* which are given off or taken up.

2. The actual concentrations (or activities) of the *particular ionic species* given in this expression and which are present at a given pH value must be substituted in the expression for the pH-independent equilibrium constant or free energy.

A final convention gives the molar free energy of hydrolysis under conditions in which the reactants are at concentrations other than 1.0 M. For glycine ethyl ester at pH 7.0 under conditions in which the reactants and products, except for water and hydrogen ion, are present at a concentration of 10^{-3} M, the free energy of hydrolysis is $-12,660$ cal/mol, as shown in Equation 10. The importance of specifying and understanding the particular convention that is being used is illustrated by the range of values from $+1,440$ to $-12,660$ calories/mole for the free energy of hydrolysis of glycine ethyl ester according to the different conventions. At pH 7 and under physiological conditions, glycine ethyl ester has a free energy of hydrolysis which clearly places it in the category of "high-energy" or "energy-rich" compounds. All of these conventions are correct and are useful for different purposes.

$$\Delta G' = \Delta G^{\circ\prime}_{pH\ 7} + RT \ln \frac{[gly]_{tot}[HOEt]_{tot}}{[glyOEt]_{tot}[HOH]} = -8,400 + 1,420 \log \frac{(10^{-3})(10^{-3})}{(10^{-3})(1)} = 12,660 \text{ cal/mol} \quad (10)$$

Complexation of the compounds which are involved in an equilibrium with other compounds which may be present in the solution is a common cause of difficulty in the determination of equilibria and free energies of hydrolysis. The most important example of this in biochemical reactions is the binding of magnesium and other ions to phosphate and polyphosphates. This problem may be dealt with in several ways:

1. The reaction may be carried out under conditions in which the complexing ions are present in negligible concentrations compared to the compounds involved in the equilibrium under study.

2. The concentrations of the free and complexed species of the reactants and ions may be calculated from equilibrium constants for complex formation, which must be determined in separate experiments or be obtained from the literature. There is still some disagreement in the literature as to the correct values for these complexing constants for many compounds and ions of biochemical importance.

3. The reaction may be carried out in the presence of an excess of ions under conditions in which most of the reactants exist in the form of the complex. The equilibrium constant and free energy are then obtained for this particular set of experimental conditions or may be expressed in terms of reactions of the complexed species. For example, the affinity of Mg^{2+} toward ATP^{4-} and toward ATP^{3-} is much larger than that toward $HPO_4{}^{2-}$, so that the equilibrium in the presence of excess Mg^{2+} may be expressed according to Equation 11. Most of the free energies of hydrolysis of polyphosphate compounds which are given in these tables refer to conditions in which Mg^{2+} is present in excess and most or all of the polyphosphates exist as the magnesium complexes. There is still no general agreement regarding these values.

$$K = \frac{[Mg \cdot ADP^-][HPO_4^=][H^+]}{[Mg \cdot ATP^=][H_2O]} \quad (11)$$

The free energies in these tables generally refer to concentrations rather than activities of the reactants. Thermodynamic values extrapolated to zero ionic strength are of theoretical interest, but have not often been obtained for reactions of biochemical importance.

It is worth noting that these equilibria refer only to aqueous solutions, and that a large fraction of a cell or cell particle is not aqueous. The perturbation of equilibria that may occur in nonaqueous systems is illustrated by the fact that esters of long chain fatty acids can be formed at equilibrium in a nonaqueous phase which is in contact with neutral buffer, although the equilibrium in the buffer solution is far toward hydrolysis.

It cannot be pointed out too often that thermodynamic measurements and conventions say nothing about the *pathway* by which a reaction takes place; i.e., the equilibrium state of a system under a given set of experimental conditions is the same regardless of the pathway by which equilibrium is attained. Thus, the common practice of calling a particular ionic species the "reactive" species from an observed change in the stoichiometric equilibrium position of a reaction with changing pH is incorrect. It is this independence of reaction pathway that makes it equally legitimate to specify the equilibrium constant of a reaction according to any of a number of equations which contain different ionic species of reactants and products; the only requirement is that the equations balance. Different equilibrium constants will be obtained, of course, from the different equations.

Further information regarding the methods for dealing with free energies of hydrolysis may be found in References 1–4. Carpenter[5] has prepared a useful summary of the dependence on pH of $\Delta G^{\circ\prime}$ for several classes of compounds of biochemical interest.

FREE ENERGIES OF HYDROLYSIS OF ESTERS OF ACETIC ACID
AND RELATED COMPOUNDS AT 25°

Compound	$-\Delta G^{\circ\,a}$	$-G^{\circ\prime}_{pH\,7}{}^{b}$	Reference
Acetic anhydride	15,700	21,800	6
p-Nitrophenyl acetate	9,430	13,010	3
m-Nitrophenyl acetate	8,550	11,610	3
p-Chlorophenyl acetate	7,590	10,650	3
Phenyl acetate	7,390	10,450	3
p-Methylphenyl acetate	6,890	9,950	3
p-Methoxyphenyl acetate	6,590	9,650	3
Acetyl hypochlorite	ca. 5,950	ca. 9,214	13, 14[f]
Acetyl phosphate	6,690[c]	10,300	7, 3
N,O-Diacetyl-N-methylhydroxylamine[d]	6,190	9,250	3
4-Pyridinealdoxime acetate[e] (37°)	5,670	8,730	8
Glycine ethyl ester (39°)	1,970	8,400	9
Valyl RNA (30°)	2,000[g]	8,400[g]	10
Trifluoroethyl acetate	4,970	8,030	4
Acetylcarnitine (35°)	4,150	7,210	4, 11
Acetylcholine	2,940	6,000	4
Chloroethyl acetate	2,840	5,900	4
Methoxyethyl acetate	2,180	5,240	4
Ethyl acetate	1,660	4,720	4

Compiled by **William P. Jencks.**

[a]Standard free energy of hydrolysis based on a standard state of 1 *M* concentrations of the *uncharged* reactants and products and an activity of pure water of 1.0 (convention I).

[b]Standard free energy of hydrolysis at pH 7.0 based on a standard state of 1 *M* total stoichiometric concentration of reactants and products, except hydrogen ion, and on an activity of pure water of 1.0 (convention III). Values for derivatives of acetic acid are based on a thermodynamic pK_a of 4.76 for acetic acid and a ΔG for ionization of acetic acid at pH 7.0 of 3,060 cal/mol. Values for $\Delta G^{\circ\prime}_{pH\,7}$ for acetate derivatives based on a pK'_a of 4.63 ± 0.02 at ionic strength 0.2 to 1.0[12] are 180 cal/mol more negative.

[c]For the dianions of acetyl phosphate and phosphate.

[d]For hydrolysis of the ester.

[e]Based on (closely similar) equilibrium constants with several thiol esters and the ΔG° for N,S-diacetyl-β-mercaptoethylamine.

[f]From the data of De la Mare in acetic acid containing traces of water[13] and an ionization constant of 4.1×10^{-8} for hypochlorous acid.[14]

[g]Based on Reference 10. $\Delta G^{\circ\prime}_{pH\,7.0}$ = −7,700 for ATP (→ PP and AMP), and pK_a valine = 2.32.

FREE ENERGIES OF HYDROLYSIS OF THIOL ESTERS

Compound	$-\Delta G^{\circ}$ [a]	$-G^{\circ\prime}_{pH\,7}$ [b]	Reference
N,S-Diacetyl-β-mercaptoethylamine	4,460	7,520	4
S-Acetylmercaptoacetate	4,140	7,200	4
S-Acetylmercaptopropanol	4,400	7,460	9
2-Diethylaminoethane thioacetate	—	7,470[c]	8
S-Acetylthiophenol	—	7,450[c]	8
2-Di*iso*propylaminoethane thioacetate	—	6,720[c]	8
S-Acetylglutathione	—	7,500[c]	8
	—	7,830[d]	7
Acetyl coenzyme A,	—	7,520[e]	4
pH 7.2	—	7,100[f]	15
	—	7,370[g]	7, 16

Compiled by William P. Jencks.

[a]Standard free energy of hydrolysis based on a standard state of 1 *M* concentrations of the *uncharged* reactants and products and an activity of pure water of 1.0 (convention I).

[b]Standard free energy of hydrolysis at pH 7.0 based on a standard state of 1 *M* total stoichiometric concentration of reactants and products, except hydrogen ion, and on an activity of pure water of 1.0 (convention III). Values for derivatives of acetic acid are based on a thermodynamic pK_a of 4.76 for acetic acid and a ΔG for ionization of acetic acid at pH 7.0 of 3,060 cal/mol. Values for $\Delta G^{\circ\prime}_{pH\,7}$ for acetate derivatives based on a pK'_a of 4.63 ± 0.02 at ionic strength 0.2 to 1.0[12] are 180 cal/mol more negative.

[c]Based on Reference 8 and 4-pyridinealdoxime acetate (see previous table).

[d]Based on Reference 7 and acetylimidazole (see following table).

[e]Based on *N,S*-diacetyl-β-mercaptoethylamine.

[f]Based on the equilibria for the condensation of acetate and acetyl coenzyme A with oxaloacetate to give citrate with the correction for citrate ionization recalculated as described in the text above.

[g]Based on equilibria with acetyl phosphate and acetylimidazol.[7,16]

FREE ENERGIES OF HYDROLYSIS OF AMIDES

Compound	$-\Delta G^{\circ\prime}_{pH\,7}$ [a]	Reference
Acetylimidazole	12,970	3
10-Formyltetrahydrofolic acid (pH 7.7, 37°)	5,830[b]	17
Asparagine	3,600	18
Glutamine	3,400	18
N-Dimethylpropionamide	2,100[c]	19
Propionamide	2,100[c]	19
Hippurylanilide (pH 5.0, 39°)	1,470	20
Benzoyltyrosyl-glycylanilide (pH 6.5, 23°)	1,360	21
Benzoyltyrosyl-glycinamide	420[d]	22
N-Acetyltyrosine hydroxamic acid	1,870	23
Acetohydroxamic acid	−200	23
N-Methylpropionamide	−300[c]	19

Compiled by William P. Jencks.

[a]Standard free energy of hydrolysis at pH 7.0 based on a standard state of 1 *M* total stoichiometric concentration of reactants and products, except hydrogen ion, and on an activity of pure water of 1.0 (convention III).

[b]Based on Reference 17 and $\Delta G^{\circ\prime}_{pH\,7.7} = -8,030$ cal/mol for ATP.

[c]Data for other simple amides at elevated temperatures are given in Reference 19.

[d]For cleavage of the peptide bond, to fully ionized products.

FREE ENERGIES OF HYDROLYSIS OF PHOSPHATES[a]

Compound	$-\Delta G^{\circ\prime}$ [b]	Reference
Phosphoenolpyruvate, pH 7.0	14,800	24
pH 7.4–8.4	12,800	2
β-Aspartyl phosphate, pH 8.0, 15°	13,000[c]	25
Carbamyl phosphate, pH 9.5	ca. 12,300[d]	26
3-Phosphoglyceroyl phosphate, pH 6.9	11,800	2
Acetyl phosphate, pH 7.0	10,300	7, 3
Creatine phosphate, pH 7.0, 7.5, 37°	10,300	27
Phosphoarginine, pH 8.0, excess Mg^{++}	7,700[e]	28
Uridine diphosphate glucose (glycoside cleavage), pH 7.6	7,300	29
Adenosine triphosphate ($\rightarrow$ AMP, PP), pH 7.0, excess Mg^{++}	7,700[f]	4, 9
pH 7.5, excess Mg^{++}	10,300[g]	30
Adenosine triphosphate ($\rightarrow$ ADP, Pi), 37°, pH 7.0, excess Mg^{++}	7,300	31–33
25°, pH 7.4, $10^{-3}\ M\ Mg^{++}$	8,800	44
25°, pH 7.4, 0 Mg^{++}	9,600	44
Pyrophosphate, pH 7.0	8,000	24
pH 7.0, 0.005 $M\ Mg^{++}$	4,500	24
Cytidine-2'-3'-phosphate (cyclic) $\rightarrow$ 3'-phosphate, pH 7.0	5,000	34
Glucose-1-phosphate, 25°, pH 7.0	5,000	35
N-Acetylethanolamine phosphate, pH 7.0	2,900	36
Glucose-6-phosphate, 25°, pH 7.0	3,300	35
α-Glycerophosphate, 38°, pH 8.5	2,200	37
pH 5.8	2,600	3
Hexose-6-phosphates, 38°, pH 8.5	2,800 ± 200	37
pH 5.8	3,200 ± 200	37

Compiled by William P. Jencks.

[a]For a more detailed compilation, see Reference 2.

[b]Standard free energy of hydrolysis based on a standard state of 1 M total stoichiometric concentration of reactants and products, except hydrogen ion, and on an activity of pure water of 1.0 (convention III).

[c]Based on Reference 25 and $\Delta G^{\circ\prime}_{pH\ 8.0} = -8,400$ cal/mol for ATP.

[d]Based on Reference 26 and $\Delta G^{\circ\prime}_{pH\ 9.5} = -10,440$ cal/mol for ATP.

[e]Based on Reference 28 and $\Delta G^{\circ\prime}_{pH\ 8.0} = -8,400$ cal/mol for ATP.

[f]Based on acetyl coenzyme A and the ATP-activated synthesis of acetyl coenzyme A.

[g]Based on a series of equilibria and the assumption of the similarity of the terminal phosphates of ATP and ADP. This value is not consistent with the preceding estimate.

FREE ENERGIES OF HYDROLYSIS OF GLYCOSIDES

Compound	$-\Delta G^{\circ\prime}_{pH\ 7}$ [a]	Reference
Uridine diphosphoglucose (pH 7.6)	7,300[b,c,d]	29
Sucrose	7,000[b]	38
Levan (fructofuranoside 2-6-fructose)	5,000[b]	39
Glucose-1-phosphate	5,000	35
Maltose	4,000	39
Glycogen	4,000	39
Amylose (α(1-4)glucosidic)	3,400	39

Compiled by William P. Jencks.

[a] Standard free energy of hydrolysis at pH 7.0 based on a standard state of 1 M total stoichiometric concentration of reactants and products, except hydrogen ion, and on an activity of pure water of 1.0 (convention III).

[b] Based on a corrected value for sucrose hydrolysis in Reference 38.

[c] Recalculated from K = 1.6, $\Delta G^{\circ\prime}$ = -300 from the data of Reference 29.

[d] The values for thymidine diphosphoglucose and adenosine diphosphoglucose are very similar.[45,46]

FREE ENERGIES OF DECARBOXYLATION

Compound	$-\Delta G^{\circ}$ [a]	Reference
Oxaloacetate$^{=}$ ⇄ Pyruvate^{-} + HCO_3^{-}	6,200	24
Methylmalonyl-CoA ⇄ Propionyl-CoA + HCO_3^{-}		
30°	6,200[b]	24, 40
28°	7,510[c]	41
Enzyme-biotin-CO_2 ⇄ Enzyme-biotin + HCO_3^{-} (pH 7.0, 0°)	4,700	43

Compiled by William P. Jencks.

[a] Based on the indicated ionic species of reactants and products. Units are cal/mole.

[b] Based on oxaloacetate decarboxylation and K for carbon dioxide transfer; K = [pyruvate] [D-methyl-malonyl-CoA] / [propionyl-CoA] = 1.0.[47]

[c] Based on ATP-coupled carboxylation and $\Delta G^{\circ\prime}_{pH\ 8.1}$ = -8.540 for ATP hydrolysis. The corresponding value at 37° from the data of Reference 42 is -11.500 cal/mol.[42]

REFERENCES

1. Johnson, in *The Enzymes,* Vol. 3, 2nd ed., Boyer, Lardy and Myrbäck, Eds., Academic, New York, 1960, chap. 21, 407.
2. Atkinson and Morton, in *Comparative Biochemistry,* Vol. 2, Florkin and Mason, Eds., Academic, New York, 1960, chap. 1, 1.
3. Gerstein and Jencks, *J. Am. Chem. Soc.,* 86, 4655 (1964).
4. Jencks and Gilchrist, *J. Am. Chem. Soc.,* 86, 4651 (1964).
5. Carpenter, *J. Am. Chem. Soc.,* 82, 1111 (1960).
6. Jencks, Barley, Barnett, and Gilchrist, *J. Am. Chem. Soc.,* 88, 4464 (1966).
7. Stadtman, in *The Mechanism of Enzyme Action,* McElroy and Glass, Eds., Johns Hopkins, Baltimore, 1954, 581.
8. O'Neill, Kohl, and Epstein, *Biochem. Pharm.,* 8, 399 (1961).
9. Jencks, Cordes, and Carriuolo, *J. Biol. Chem.,* 235, 3608 (1960).
10. Berg, Bergmann, Ofengand, and Dieckmann, *J. Biol. Chem.,* 236, 1726 (1961).

11.   Fritz, Schultz, and Srere, *J. Biol. Chem.*, 238, 2509 (1963).
12.   Bjerrum, Schwarzenbach, and Sillén, *Stability Constants.* Part I. Chemical Society, London, 1957.
13.   De la Mare, Hilton, and Vernon, *J. Chem. Soc.,* p. 4039 (1960).
14.   Mauger and Soper, *J. Chem. Soc.,* p. 71 (1946).
15.   Tate and Datta, *Biochem. J.,* 94, 470 (1965).
16.   Sly and Stadtman, *J. Biol. Chem.,* 238, 2639 (1963).
17.   Himes and Rabinowitz, *J. Biol. Chem.,* 237, 2903 (1962).
18.   Benzinger, Kitzinger, Hems, and Burton, *Biochem. J.,* 71, 400 (1959).
19.   Morawetz and Otaki, *J. Am. Chem. Soc.,* 85, 463 (1963).
20.   Carty and Kirschenbaum, *Biochim. Biophys. Acta,* 110, 399 (1965).
21.   Gawron, Glaid, Boyle, and Odstrchel, *Arch. Biochem. Biophys.,* 95, 203 (1961).
22.   Dobry, Fruton, and Sturtevant, *J. Biol. Chem.,* 195, 149 (1952).
23.   Jencks, Caplow, Gilchrist, and Kallen, *Biochemistry,* 2, 1313 (1963).
24.   Wood, Davis, and Lochmüller, *J. Biol. Chem.,* 241, 5692 (1966).
25.   Black and Wright, *J. Biol. Chem.,* 213, 27 (1955).
26.   Jones and Lipmann, *Proc. Natl. Acad. Sci. U.S.A.,* 46, 1194 (1960).
27.   Kuby and Noltmann, in *The Enzymes,* Vol. 4, 2nd ed., Boyer, Lardy, and Myrbäck, Eds., Academic, New York, 1962, chap. 31, 515.
28.   Uhr, Marcus, and Morrison, *J. Biol. Chem.,* 241, 5428 (1966).
29.   Avigad, *J. Biol. Chem.,* 239, 3613 (1964).
30.   Schuegraf, Ratner, and Warner, *J. Biol. Chem.,* 235, 3597 (1960).
31.   Atkinson, Johnson, and Morton, *Nature,* 184, 1925 (1959).
32.   Robbins and Boyer, *J. Biol. Chem.,* 224, 121 (1957).
33.   Benzinger, Kitzinger, Hems, and Burton, *Biochem. J.,* 71, 400 (1959).
34.   Bahr, Cathou, and Hammes, *J. Biol. Chem.,* 240, 3372 (1965).
35.   Atkinson, Johnson, and Morton, *Biochem. J.,* 79, 12 (1961).
36.   Dayan and Wilson, *Biochim. Biophys. Acta,* 77, 446 (1963).
37.   Meyerhof and Green, *J. Biol. Chem.,* 178, 655 (1949).
38.   Neufeld and Hassid, *Adv. Carbohyd. Chem.,* 18, 309 (1963) (footnote 166, p. 329).
39.   Dedonder, *Ann. Rev. Biochem.,* 30, 347 (1961).
40.   Wood and Stjernholm, *Proc. Natl. Acad. Sci. U.S.A.,* 47, 289 (1961).
41.   Kaziro, Grossman, and Ochoa, *J. Biol. Chem.,* 240, 64 (1965).
42.   Halenz, Feng, Hegre, and Lane, *J. Biol. Chem.,* 237, 2140 (1962).
43.   Wood, Lochmüller, Riepertinger, and Lynen, *Biochem. Z.,* 337, 247 (1963).
44.   Alberty, R. A., personal communication.
45.   Avigad and Milner, *Meth. Enzymol.,* 8, 341 (1966).
46.   Murata, Sugiyama, Minimikawa, and Akazawa, *Arch. Biochem. Biophys.,* 113, 34 (1966).
47.   Wood, H., personal communication.

This section originally appeared in Sober, Ed., *Handbook of Biochemistry and selected data for Molecular Biology,* 2nd ed., Chemical Rubber Co., Cleveland, 1970.

IONIZATION CONSTANTS OF ACIDS AND BASES

W. P. Jencks and J. Regenstein

These pK'_a values were taken from the original literature and from several extensive compilations of such data, of which the most important are

Albert, *Ionization Constants of Acids and Bases,* Methuen, London, 1962.

Bell, *The Proton in Chemistry,* 2nd ed., Cornell, Ithaca, New York, 1973.

Brown, McDaniel, and Häfliger, in Braude and Nachod, *Determination of Organic Structures by Physical Methods,* Academic Press, New York, 1955.

Kortum, Vogel, and Andrussow, *Dissociation Constants of Organic Acids in Aqueous Solution,* Butterworths, London, 1961.

Perrin, *Dissociation Constants of Organic Bases in Aqueous Solution,* Butterworths, London, 1965.

Yukawa, Ed., *Handbook of Organic Structural Analysis,* Benjamin, New York, 1965.

A particularly valuable source of dissociation constants obtained under a variety of experimental conditions is provided by Sillen L. G. and Martell, A. E., Eds., *Stability Constants,* Special Publications No. 17 and 25, Chemical Society, London, 1964 and 1971. This compilation also lists association constants of metals for a variety of inorganic and organic ligands.

The compounds selected were those which were thought most likely to be useful to biochemists and chemists and these compilations should be consulted for information on compounds which are not included here.

All values are reported as $pK'_a = -\log K'_a = 14 - pK'_b$. K'_a is the ionization constant

$$\frac{[H^+][A^-]}{[HA]} \quad \text{or} \quad \frac{[H^+][B]}{[HB^+]} \quad \text{or} \quad \frac{[A^{n-1}][H^+]}{[HA^n]}$$

Temperatures are not indicated because variations of pK'_a with temperature are generally smaller than the variations of the data from different sources for other reasons, but most of the data were obtained at or near 25°. Ionization constants which are reported as thermodynamic values at 25° are indicated with an asterisk, *, but some of these may only represent values measured at low ionic strength.

These pK'_a values and a measured pH should not be used to obtain an *exact* measure of the ratio of acid to base in a given solution. Ionic strength and specific salt effects, as well as possible errors in the reported pK'_a values, are likely to make such estimates inaccurate. It should be kept in mind that the effect of increasing ionic strength is generally to decrease the apparent pK'_a of neutral and anionic acids and to increase the pK'_a of cationic acids. These effects are particularly large for polyanions, such as phosphates.

There is some intentional redundancy in the tables to facilitate the location of listings for compounds that might be listed in several sections. The pK'_a values for amines refer to the ionization of the conjugate acids of the amines except for a few nitrogen acids, which undergo an acidic ionization.

The pH of a solution at a given ionic strength and temperature is given by

$$pH = pK'_a + \log \frac{[\text{base}]}{[\text{acid}]}$$

in which the pK_a' is measured under the same experimental conditions. The following relationships are useful to have readily available to estimate the ratio of acid to base at a given pH or to estimate the buffer ration of acid required to give a given pH; the compiler keeps a copy of these numbers on his desk.

Fraction base or acid	pH	Fraction base or acid	pH
5% or 95%	$pK_a' \pm 1.25$	30% or 70%	$pK_a' \pm 0.37$
10% or 90%	$pK_a' \pm 0.95$	35% or 65%	$pK_a' \pm 0.27$
15% or 85%	$pK_a' \pm 0.75$	40% or 60%	$pK_a' \pm 0.18$
20% or 80%	$pK_a' \pm 0.60$	45% or 55%	$pK_a' \pm 0.09$
25% or 75%	$pK_a' \pm 0.48$	50% or 50%	$pK_a' \pm 0$

INORGANIC ACIDS

Compound	pK'_a	Reference
AgOH	3.96	4
$Al(OH)_3$	11.2	28
$As(OH)_3$	9.22	28
H_3AsO_4	2.22, 7.0, 13.0	28
$H_2AsO_4^-$	6.98*	77
$HAsO_4^=$	11.53*	77
H_3AsO_3	9.22*	—
H_3AuO_3	13.3, 16.0	78
H_3BO_3	9.23	28
$H_2B_4O_7$	4.00	34
$HB_4O_7^-$	9.00	34
$Be(OH)_2$	3.7	4
HBr	−9.00	31
HOBr	8.7	28
HOCl	7.53, 7.46	28,33
$HClO_2$	2.0	28
$HClO_3$	−1.00	28
HCN	9.40	34
H_2CO_3	6.37, 6.35,* 3.77*	34, 23
HCO_3^-	10.33*	—
H_2CrO_4	−0.98, 0.74	30, 77
$HCrO_4^-$	6.50*	2, 30
HOCN	3.92	34
HF	3.17*	77
H_3GaO_3	10.32, 11.7	78
H_2GeO_3	8.59, 12.72	34, 78
$Ge(OH)_4$	8.68, 12.7	28
HI	−10.0	31
HOI	11.0	28
HIO_3	0.8	28
$H_4IO_6^-$	6.00	34
H_5IO_6	1.64, 1.55, 8.27	34, 28, 78
	3.29, 6.70, 15.0	—
$HMnO_4$	−2.25	30
NH_3OH^+	5.98*	12
NH_4^+	9.24*	77
HN_3	4.72*	77
H_3N	33	153
HNCS	~−2.0	143
HNO_2	3.29	28
HNO_3	−1.3	28
$N_2H_5^+$	7.99*	77
$H_2N_2O_2$	7.05	34
$H_2N_2O_2^-$	11.0	34
H_2NSO_3H	1.0	80
H_2OsO_5	12.1	34
H_2O	15.7	—
H_3O^+	−1.7	—
$Pb(OH)_2$	6.48 (10.92)	4, 78
PH_3	27	156
H_3PO_2	2.0, 2.23,* 1.07	28, 77
H_3PO_4	2.12*	77
$H_2PO_4^-$	7.21*	77
$HPO_4^=$	12.32*	77
H_3PO_5	1.12, 5.51, 12.80	102
H_3PO_3	2.0, 1.07	28, 77
$H_2PO_3^-$	6.58*	77
$H_4P_2O_7$	1.52*	77
$H_3P_2O_7^-$	2.36*	77

Compound	pK'_a	Reference
$H_2P_2O_7^=$	6.60*	77
$HP_2O_7^{\equiv}$	9.25*	77
$HReO_4$	−1.25	30
HSCN	0.85	77
H_3SiO_3	10.0	34
H_2S	7.00*	77
HS^-	12.92*	77
H_2SO_3	1.9, 7.0, 1.76*	28, 77
H_2SO_4	1.9	28
HSO_3^-	7.21*	77
HSO_4^-	1.99*	77
$H_2S_2O_3$	0.60,* 1.72*	77
$H_2S_2O_4$	1.9	29
H_2Se	3.89*	77
HSe^-	11.00*	77
H_2SeO_3	2.6, 8.3; 2.62*	28
$HSeO_3^-$	8.32	77
H_2SeO_4	Strong, 2.0	28
$HSeO_4^-$	2.00	34
$HSbO_2$	11.0	34
HTe	5.00	34
H_2Te	2.64, 11.0	34, 78
H_2TeO_3	2.7, 8.0	28
$Te(OH)_6$	6.2, 8.8	28
H_2VO_4	8.95	30
$HVO_4^=$	14.4	30
$H_4V_6O_{17}$	1.96	78
Cacodylic acid $(CH_3)_2As(O)OH$	1.57,* 6.27*	99

Substituted AsO_3H_2

CH_3-	3.61,* 8.18*	97
CH_3CH_2-	3.89,* 8.35*	—
$CH_3(CH_2)_4$-	4.14,* 10.07*	—
$CH_3(CH_2)_5$-	4.16,* 9.19*	—
$COOH(CH_2)$-	2.94,* 4.67,* 7.68*	—
$COOH(CH_2)_4$-	2.00,* 4.89,* 7.74*	—
o-$CH_3C_6H_4$-	3.82,* 8.85*	—
m-$CH_3C_6H_4$-	3.82,* 8.60*	—
p-$CH_3C_6H_4$-	3.70,* 8.68*	—
o-$NH_2C_6H_4$-	3.77,* 8.66*	—
m-$NH_2C_6H_4$-	4.05,* 8.62*	—
p-$NH_2C_6H_4$-	4.05,* 8.92*	—

Hydrated Metal Ions

Ti^{3+}	1.15	98
Bi^{+3}	1.58	—
Fe^{+3}	2.80	—
Hg^{+2}	2.60, 3.70	—
Sn^{+2}	4.00	—
Cr^{+3}	3.80	—
Al^{+3}	4.96	—
Sc^{+3}	4.96	—
Fe^{+2}	8.30	—
Cu^{+2}	8.30	—
Ni^{+2}	9.30	—
Zn^{+2}	9.60, 10.84	—

* Thermodynamic value.

PHOSPHATES AND PHOSPHONATES

Compound	pK_a'	Reference	Compound	pK_a'	Reference
Phosphates			**Phosphonates—(Continued)**		
Phosphate	212,* 7.21,* 12.32*	77	$CHCl_2$-	1.14, 5.61	57
			CH_2Cl-	1.40, 6.30	57
Glyceric acid 2-phosphate	3.6, 7.1	53	CH_2Br-	1.14, 6.52	57
Enolpyruvic acid	3.5, 6.4	53	$(^-OOCCH_2)_2NH^+$-$(CH_2)_2$-	6.54	57
Methyl-	1.54, 6.31	55			
Ethyl-	1.60, 6.62	55	CH_2I-	1.30, 6.72	57
n-Propyl-	1.88, 6.67	55	$NH_3^+CH_2CH_2$-	2.45, 7.00	57
n-Butyl-	1.80, 6.84	55	$C_6H_5CH{=}CH$-	2.00, 7.1	57
Dimethyl-	1.29	55	$HOCH_2$-	1.91, 7.15	57
Diethyl-	1.39	55	$C_6H_5NH_2^+(CH_2)_3$-	2.1	57
Di-n-propyl-	1.59	55	$C_6H_5NH(CH_2)_3$-	7.17	57
Di-n-butyl-	1.72	55	$Br(CH_2)_2$-	2.25, 7.3	57
Glucose-3-	0.84, 5.67	56	$CH_3(CH_2)_5CH(COO^-)$-	7.5	57
Glucose-4-	0.84, 5.67	56	$C_6H_5CH_2$-	2.3, 7.55	57
α-Glycero-	1.40, 6.44	54	$NH_3^+(CH_2)_4$-	2.55, 7.55	57
β-Glycero-	1.37, 6.34	54	$NH_3^+(CH_2)_5$-	2.6, 7.6	57
3-Phosphoglyceric acid	1.42, 3.42	54	$NH_3^+(CH_2)_{10}$-	8.00	57
2-Phosphoglyceric acid	1.42, 3.55, 7.1	–	$^-OOC(CH_2)_{10}$-	8.25	57
Peroxymonophosphoric acid	4.85	69	$(CH_3)_3SiCH_2$-	3.22, 8.70	57
			$C_6H_5CH_2$-	3.3†, 8.4†	57
Diphosphoglyceric acid	7.40, 7.99	54	$(C_6H_5)_3C$-	3.85†, 9.00†	57
Glyceraldehyde-	2.10, 6.75	54	**Arylphosphonic acids (57)**		
Dioxyacetone-	1.77, 6.45	54	$2X\text{-}RC_6H_3PO_3H_2$	–	–
Hexose di-	1.52, 6.31	54	X R		
Fructose-6-	0.97, 6.11	54	Cl $4\text{-}O_2N$	1.12, 6.14	–
Glucose-6-	0.94, 6.11	54	Br $5\text{-}O_2N$	6.14	–
Glucose-1-	1.10, 6.13	54	Cl 5-Cl	6.63	–
Pyrophosphoric acid	0.9, 2.0, 6.6, 9.4	54	Cl H	1.63, 6.98	–
Phosphopyruvic acid	3.5, 6.38	54	Br H	1.64, 7.00	–
DL-Phosphoserine	6.19	145	Br $5\text{-}CH_3$	1.81, 7.15	–
Creatine phosphate	2.7, 4.5	54	Cl $4\text{-}NH_2$	7.33	–
Arginine phosphate	2.8, 4.5, 9.6, 11.2	54	CH_3O $4\text{-}O_2N$	1.53, 6.96	–
Amino phosphate	(-0.9), 2.8, 8.2	54	CH_3O H	2.16, 7.77	–
Trimetaphosphate	2.05	77	CH_3O $4\text{-}NH_2$	8.22	–
Trimethyl phosphine	8.80*	99	HO $4\text{-}O_2N$	1.22, 5.39	–
Triphosphate	8.90, 6.26, 2.30	77	O_2N H	1.45, 6.74	–
Tetrametaphosphate	2.74	77	F H	1.64, 6.80	–
Fluorophosphate	0.55, 4.8	56	I H	1.74, 7.06	–
See also under *Nucleic Acid Derivatives*			NH_2 H	7.29	–
			CH_3 H	2.10, 7.68	–
Phosphonates			C_6H_5 H	8.13	–
$H_2O_3P(CH_2)_4PO_3H_2$	<2, 2.75, 7.54, 8.38	57	HOOC H	1.71, 9.17	–
$H_2O_3P(CH_2)_3PO_3H_2$	<2, 2.65, 7.34, 8.35	57			
$H_2O_3PCH_2CH(CH_3)$-PO_3H_2	<2, 2.6, 7.00, 9.27	57	**Substituted-PO_2H_2**		
			CH_3-	3.08*	97
$H_2O_3PCH_2PO_3H_2$	<2, 2.57, 6.87, 10.33	57	CH_3CH_2-	3.29*	97
Methyl-	2.35, 7.1*	57, 97	$CH_3(CH_2)_2$-	3.46*	97
Ethyl-	2.43, 7.85*	57, 97	$(CH_3)_2CH$-	3.56*	97
n-Propyl-	2.45, 8.18*	57, 97	$CH_3(CH_2)_3$-	3.41*	97
Isopropyl-	2.55, 7.75	57	$(CH_3)_3C$-	4.24*	97
n-Butyl-	2.59, 8.19	57	C_6H_5-	2.1*	97
Isobutyl-	2.70, 8.43	57	$p\text{-}BrC_6H_4$-	2.1*	97
s-Butyl-	2.74, 8.48	57	$p\text{-}CH_3OC_6H_4$-	2.35*	97
t-Butyl-	2.79, 8.88	57	$p\text{-}(CH_3)_2N$-	2.1*, 4.1*	97
Neopentyl-	2.84, 8.65	57			
1,1-Dimethylpropyl-	2.88, 8.96	57			
n-Hexyl-	2.6, 7.9	57			
n-Dodecyl-	8.25	57			
$CH_3(CH_2)_5CH(COOH)$-	1	57			
CF_3-	1.16, 3.93	57			
CCl_3-	1.63, 4.81	57			
$NH_3^+CH_2$-	2.35, 5.9	57			
$(^-OOCCH_2)_2NH^+CH_2$-	5.57	57			

	pK_a'				
X =	−H		−NH$_3^+$		Ref. 2
$X(CH_2)PO_3H_2$	2.35	7.1	1.85	5.35	–
$X(CH_2)_2PO_3H_2$	2.45	7.85	2.45	7.00	–
$X(CH_2)_4PO_3H_2$	–	–	2.55	7.55	–
$X(CH_2)_5PO_3H_2$	–	–	2.6	7.65	–
$X(CH_2)_6PO_3H_2$	2.6	7.9	–	–	–
$X(CH_2)_{10}PO_3H_2$	–	–	–	8.00	–

*Thermodynamic value.

†These values were obtained in 50 per cent ethanol.

For graphical plots of a large number of substituted phosphorus compounds see Ref. (83).

CARBOXYLIC ACIDS

ALIPHATIC

Acetic Acids, Substituted

Compound	pK_a'	Reference
H-	4.76*	2
O_2N-	1.68*	2
$(CH_3)_3N^+-$	1.83*	2
$(CH_3)_2NH^+-$	1.95	2
$CH_3NH_2^+-$	2.16*	2
NH_3^+-	2.31*	2
CH_3SO_2-	2.36*	2
NC-	2.43*	2
$C_6H_5SO_2-$	2.44	2
HO_2C-	2.83*	2
C_6H_5SO-	2.66	2
F-	2.66	2
Cl-	2.86*	2
Br-	2.86	2
Cl_2-	1.29	2
F_2-	1.24	2
Br_2-	1.48	142
Br_3-	0.66	2
Cl_3-	0.65	2
F_3-	0.23, (−0.26)	2
HON-	3.01	2
F_3C-	3.07*	2
ClF_2-	0.46	159
N_3-	3.03	2
I-	3.12	2
C_6H_5O-	3.12	2
$C_2H_5O_2C-$	3.35	2
C_6H_5S-	3.52*	2
CH_3O-	3.53	2
NCS-	3.58	2
CH_3CO-	3.58*	2
C_2H_5O-	3.60	2
$n\text{-}C_3H_7O-$	3.65	2
$n\text{-}C_4H_9O-$	3.66	2
$sec.\text{ -}C_4H_9O-$	3.67	2
HS-	3.67*	2
$i\text{-}C_3H_7O-$	3.69*	2
CH_3S-	3.72*	2
$i\text{-}C_3H_7S-$	3.72*	2
$C_6H_5CH_2S-$	3.73*	2
C_2H_5S-	3.74*	2
$n\text{-}C_3H_7S-$	3.77*	2
$n\text{-}C_4H_9S-$	3.81*	2
HO-	3.83*	2
$^-O_3S-$	4.05	2
$(C_6H_5)_3CS-$	4.30*	2
C_6H_5-	4.31*	2
$CH_2=CH-$	4.35*	2
CH_3-	4.88*	2
$^-O_2Se-$	5.43	2
$^-O_2C-$	5.69*	2
$(CH_3)_2-$	4.86	2
$(C_6H_5CH_2)_2-$	4.57	2
$(C_6H_5)_2$	3.96	2
$(CH_3)_3-$	5.01	2
CH_3CHOH-	3.9	2
$(CH_3CH_2)_2-$	4.74	
$(CH_3)_2(CN)-$	2.43	
$(CH_3)_2C(CN)-$	2.40	
$HC\equiv C-$	1.84	142
$CH_3C\equiv C-$	2.60	142
$C_6H_5C\equiv C-$	2.23	142
$CH_3CH=CH-$	4.69	142
$C_6H_5CH=CH-$	4.44	142
$3,5\text{-Di-}NO_2C_6H_5\text{-}$	2.82	142
OHC-	3.32	142
C_6H_5CO-	1.32	142

Straight-chain, Substituted

Substituent	Propionic		Butyric			Valeric	
	α	β	α	β	γ	α	δ
H-	4.88*	—	4.82*	—	—	4.86*	—
O_2N-	—	3.81	—	—	—	—	—
$(CH_3)_3N^+-$	—	—	—	—	—	—	—
H_3N^+-	2.34	3.60	—	—	4.23	—	4.27*
NC-	2.43	3.99*	—	—	4.44*	—	—
HO_2C-	—	4.19*	—	—	4.34*	—	4.42*
Cl-	2.80	4.08	2.84	4.06	4.52	—	4.70
Br-	2.98	4.02	2.99	—	4.58	—	4.72
HON=	3.32	4.01	3.15	—	—	3.19	4.64
F_3C-	—	4.18*	—	—	4.49	—	—
I-	3.12	4.06	—	—	4.64	—	4.77
C_6H_5O-	3.11	4.27	3.17	—	—	—	—
$C_2H_5O_2C-$	—	4.52	—	—	—	—	4.60
CH_3O-	3.52	4.46	—	—	4.68	—	4.72
CH_3CO-	—	4.60	—	—	4.67	—	4.72
C_2H_5O-	3.61	4.50	—	—	4.70	—	—
HS-	3.70*	4.34*	—	—	—	—	—
HO-	3.86*	4.51	4.22	4.52	4.72	3.89	—
$^-O_3S-$	4.22						
C_6H_5-	4.31	4.66*	—	—	4.76*	—	—
H_2CCH-	—	4.68*	—	—	4.72*	—	—
CH_3-	4.86*	4.82*	4.78	4.78*	4.86*	—	4.88*
$^-O_2Se-$	5.48	6.00	5.48	—	—	5.48	—
$^-O_2C-$	—	5.48*	—	—	5.42*	—	5.41*

General Aliphatic

Compound	pK_a'	Reference
Acetoacetic	3.58	6
Acetopyruvic	2.61, 7.85 (enol)	6
Aconitic, *trans*-	2.80, 4.46	6
Adipamic	4.37	101
Aminomalonic	3.32, 9.83	77
Betaine	1.84	6
α-Bromobutyric	2.97	77
N-Butylaminoacetic	2.29, 10.07	77
Caproic	4.88	101
Caprylic	4.89	101
N-(Carbamoylmethyl)-iminodiacetic	2.30, 6.60	77
β-Carboxymethylamino-propionic	3.61, 9.46	77
2-Carboxyethylimino-diacetic	2.06, 3.69, 9.66	77
Citric	3.09, 4.75, 5.41	6
Crotonic	4.69	6
Cyanomethyliminodiacetic	3.06, 4.34	77
Cyclohexane carboxylic	4.90	153
Cyclopentane carboxylic	4.99	153
Cyclopropane carboxylic	4.83	153
α-Diaminobutyric	1.85, 8.24, 10.44	77
α,β-Diaminopropionic	1.23, 6.69	77
Di-(carboxymethyl)-aminomethyl phosphonic	2.25, 5.57, 10.76	77
Diethylaminoacetic	2.04, 10.47	77
Dihydroxyfumaric	1.14	6
α,β-Dimercaptosuccinic	2.40, 3.46, 9.44, 11.82	77
Dimethylaminoacetic	2.08, 9.80	77
2,2-Dimethylbutanoic	4.93	131
2,2-Dimethylpropionic	5.03	131
2-Ethylbutanoic	4.75	131
α-Ethylbutyric	4.74	130
Ethyl hydrogen adipate	4.60	101
Ethyl hydrogen diethyl-malonate	3.64	101
Ethyl hydrogen dimethyl-malonate	3.52	101

* Thermodynamic value.

CARBOXYLIC ACIDS (continued)

Compound	pK'_a	Reference	Compound	pK'_a	Reference
General Aliphatic—(Continued)			**General Aliphatic—(Continued)**		
Ethyl hydrogen ethyl-malonate	3.40	101	Methyl hydrogen succinate	4.49	101
Ethyl hydrogen malonate	3.35	101	Methyliminodiacetic	2.81, 10.18	77
Ethyl hydrogen methyl-malonate	3.41	101	2-Methylthioethylimino-diacetic	2.1, 8.91	77
Ethyl hydrogen sebacate	4.84	101	Nitrilotriacetic	3.03, 3.07, 10.70	77
Ethyl hydrogen suberate	4.84	101	Octanoic	4.90	131
Ethyl hydrogen succinate	4.52	101	Oenanthylic	4.89	77
N-Ethylaminoacetic	2.30, 10.10	77	Oxalic	1.25, 4.14	77
Ethylenediaminetetraacetic	2.00, 2.67, 0.26, 6.16, 10.26, 0.96	6, 94	Oxalacetic (*trans*-enol)	2.56, 4.37	6, 97
Ethylenediamine-*N*,*N*-diacetic	5.58, 11.05	77	(*cis*-enol)	2.15, 4.06	6
Formic	3.77*	2	Pelargonic	4.95	101
Fumaric	3.03, 4.54	6	Pentanoic	4.84	131
Gluconic	3.86*	77	2-Methyl-	4.78	131
Glutaramic	4.40	101	3-Methyl-	4.77	131
Glyceric	3.55	6	4-Methyl-	4.85	131
Glycolic	3.82	6	2,2–Dimethyl-	4.97	131
Glyoxylic	3.32	6	2-Phosphonoethylimino-diacetic	1.95, 2.45, 6.54, 10.46	77
Heptanoic	4.89	131	Pivalic	5.05	153
Hexanoic	4.86	129	*N*-*n*-Propylaminoacetic	2.25, 10.03	77
Homogentisic	4.40	6	Protocatechuic	4.48	6
α-Hydroxybutyric	3.65	77	Pyruvic	2.50	6
β-Hydroxybutyric	4.39	77	Succinamic	4.54	101
N-2-Hydroxyethylimino-diacetic	2.2, 8.73	77	*N*-2-Sulfoethylimino-diacetic	1.92, 2.28, 8.16	77
β-Hydroxypropionic	3.73	77	Tartaric D or L	2.89, 4.16	6
3-Hydroxypropylimino-diacetic	2.06, 9.24	77	*meso*-	3.22, 4.85	6
Iminodiacetic	2.98*, 9.89*	77	Thiophen-2-carboxylic	3.53	129
Iminodipropionic	4.11, 9.61	77	Vinylacetic	4.42	6
β-Iodopropionic	4.04*	77	$CH_3CH_2OCH_2COOH$	3.65*	97
Isobutyric	4.86*	77	$o\text{-}CH_3C_6H_4OCH_2COOH$	3.23*	97
Isocaproic	4.85	130	$m\text{-}CH_3C_6H_4OCH_2COOH$	3.20*	97
Isohexanoic	4.85	129	$p\text{-}CH_3C_6H_4OCH_2COOH$	3.22*	97
Isovaleric	4.78	129	$2,6\text{-}(CH_3)_2C_6H_3OCH_2\text{-}COOH$	3.36*	97
N-Isopropylaminoacetic	2.36, 10.06	77	$o\text{-}CH_3OC_6H_4OCH_2COOH$	3.23*	97
α-Keto-β-methyl valeric	2.3	6	$m\text{-}CH_3OC_6H_4OCH_2\text{-}COOH$	3.14*	97
Lactic	3.86	6	$p\text{-}CH_3OC_6H_4OCH_2COOH$	3.21	97
Maleic	1.93, 6.58	6	$CH_3COCH_2COCOOH$	2.58*, 8.50*	97
Malic	3.40, 5.2	6	$CH_3CH(OH)COOCH(CH_3)COOH$	2.95*	97
Malonamic	3.64	101	$CH_2ClCH(OH)COOH$	3.12*	97
Mandelic	3.41	77	$CH_3CH(OH)CHClCOOH$	2.59*	97
α-Mercaptobutyric	3.53	77	$CH_3CHClCH(OH)COOH$	3.08*	97
2-Mercaptoethylimino-diacetic	−2.14, 8.17, 10.79	77	$CH_2ClC(CH_3)(OH)COOH$	3.20*	97
2-Methoxyethylimino-diacetic	2.2, 8.96	77	$COOHCHClCH(OH)COOH$	2.32*	97
N-Methylaminoacetic	2.24, 10.01	77	$CH_3COOC(CH_2COOH)_2COOH$	2.49*	97

* Thermodynamic value.

>

>

CARBOXYLIC ACIDS (continued)

Compound	pK'_a	Reference
Sulfur Containing Carboxylic Acids		
HOOCH$_2$SCH$_2$COOH	3.30*, 4.50*	97
HOOCCH$_2$SCH$_2$SCH$_2$-COOH	3.31*, 4.34*	97
HOOCCH$_2$S(CH$_2$)$_5$SCH$_2$-COOH	3.49*, 4.41*	97
CH$_3$SCH(CH$_3$)SCH$_2$-COOH	3.77*	97
(CH$_3$)$_2$CHSCH(CH$_3$)-COOH	3.78*	97

CH$_3$-CH-COOH   4.62*   97
|
S
\
HOOC-CH-CH$_3$

CH$_3$-CH-COOH   4.57*   97
|
S
|
CH$_3$-CH-COOH

CH$_3$-CH-COOH   3.14*   97
|
S-S
/
CH$_3$-CH-COOH

CH$_3$-CH-COOH   3.15*   97
|
S-S
/
COOH-CH-CH$_3$

HOOC(CH$_2$)$_9$S(CH$_2$)$_2$-NH$_2$	4.00*, 8.30*	97
HOOC(CH$_2$)$_{10}$S(CH$_2$)$_2$-NH$_2$	2.6*, 9.6*	97
CH$_3$SO$_2$CH(CH$_3$)COOH	2.44*	97

* Thermodynamic value.

† In 40% acetone.

Unsaturated Acids, *cis* and *trans*

$$R_1CH{=}CR_2COOH$$

R_1	R_2	*cis*-acid	*trans*-acid	Reference
H-	H-	4.25*	4.25*	2
CH$_3$-	H-	4.44*	4.69*	2
Cl-	H-	3.32	3.65	2
C$_6$H$_5$-	H-	3.88*	4.44*	2
o-ClC$_6$H$_4$-	H-	3.91	4.41	2
o-BrC$_6$H$_4$-	H-	4.02	4.41	2
CH$_3$-	CH$_3$-	4.30	5.02	2
C$_6$H$_5$-	H-	5.26†	5.58†	2
2,4,6-(CH$_3$)$_3$-C$_6$H$_2$-	H-	6.12†	5.70†	2
C$_6$H$_5$-	CH$_3$-	4.98†	5.98†	2

CARBOXYLIC ACIDS (continued)

Compound	pK'_a	Reference
Unsaturated Dicarboxylic Acids*		
Acetylenedicarboxylic	1.73, 4.40	2
Bromofumaric	1.46, 3.57	2
Bromomaleic	1.45, 4.62	2
Chlorofumaric	1.78, 3.81	2
Chloromaleic	1.72, 3.86	2
Citraconic (Dimethyl-maleic acid)	2.29, 6.15	2
Fumaric	3.02, 4.38	2
Itaconic (1-Propene-2,3-dicarboxylic acid)	3.85, 5.45	2
Maleic	1.92, 6.23	2
Mesaconic (Dimethyl-fumaric acid)	3.09, 4.75	2
Phthalic	2.95, 5.41	2
Δ^1-Tetrahydrophthalic	3.01, 5.34	2
Alicyclic Dicarboxylic Acids		
cis-Caronic (1,1-dimethyl-cyclopropane-2,3-dicarboxylic acid)	2.34*, 8.31*	2
trans-Caronic	3.83*, 5.32*	2
1,2-*trans*-Cyclopropane-dicarboxylic	3.65*, 5.13*	2
1,2-*cis*-Cyclopropane-dicarboxylic	3.33*, 6.47*	2
trans-Ethyleneoxide-dicarboxylic	1.93, 3.25	2
1,2-*trans*-Cyclobutane-dicarboxylic	3.94, 5.55	132
1,3-*trans*-Cyclobutane-dicarboxylic	3.81, 5.28	2
1,2-*trans*-Cyclopentane-dicarboxylic	3.89, 5.91	2
1,3-*trans*-Cyclopentane-dicarboxylic	4.40, 5.45	2
1,2-*trans*-Cyclohexane-dicarboxylic	4.18, 5.93	2
1,3-*trans*-Cyclohexane-dicarboxylic	4.31, 5.73	2
1,4-*trans*-Cyclohexane-dicarboxylic	4.18, 5.42	2
cis-Ethyleneoxide-dicarboxylic	1.94, 3.92	2
1,2-*cis*-Cyclobutane-dicarboxylic	4.16, 6.23	132
1,3-*cis*-Cyclobutane-dicarboxylic	4.03, 5.31	2
1,2-*cis*-Cyclopentane-dicarboxylic	4.37, 6.51	2
1,3-*cis*-Cyclopentane-dicarboxylic	4.23, 5.53	2
1,2-*cis*-Cyclohexane-dicarboxylic	4.34, 6.76	2
1,3-*cis*-Cyclohexane-dicarboxylic	4.10, 5.46	2
1,4-*cis*-Cyclohexane-dicarboxylic	4.44, 5.79	2

Compound	pK'_a	Reference
Hydroxycyclohexanecarboxylic Acids		
Cyclohexanecarboxylic	4.90	2
cis-1,2-	4.80	2
cis-1,3-	4.60	2
cis-1,4-	4.84	2
trans-1,2-	4.68	2
trans-1,3-	4.82	2
trans-1,4-	4.68	2
Bicyclo(2·2·2) octane-1-carboxylic Acids, 4-Substituted		
H-	6.75	2
$C_2H_5O_2C$-	6.31	2
NC-	5.90	2
HO-	6.33	2
Br-	6.08	2
Dicarboxylic Acids and Derivatives*		
Oxalic	1.23, 4.19	2
Malonic	2.83, 5.69	2
Methyl-	3.05, 5.76	2
Ethyl-	2.99, 5.83	2
Ethylisoamyl-	2.50, 7.31	129
n-Propyl-	3.00, 5.84	2
i-Propyl-	2.94, 5.88	2
Dimethyl-	3.17, 6.06	2
Methylethyl-	2.86, 6.41	2
Diethyl-	2.21, 7.29	2
Diisopropyl-	2.12, 8.85	136
Ethyl-*n*-propyl-	2.15, 7.43	2
Di-*n*-propyl-	2.07, 7.51	2
Glutaric	4.34, 5.42	2
β-Methyl-	4.25, 6.22	2
β-Ethyl-	4.29, 6.33	2
β-Isopropyl-	4.30, 5.51	129
β-*n*-Propyl-	4.31, 6.39	2
β,β-Dimethyl-	3.70, 6.29	2
β,β-Methylethyl-	3.62, 6.70	2
β,β-Diethyl-	3.62, 7.12	2
β,β-Di-*n*-propyl-	3.69, 7.31	2
β,β-Pentamethylene-	3.49, 6.96	129
Succinic	4.19, 5.48	2
Methyl-	4.07, 5.64	101
Ethyl-	4.07, 5.89	101
Tetramethyl-	3.50, 7.28	2
DL-1:2-Dichloro-	1.68, 3.18	20
meso-1:2-Dichloro-	1.74, 3.24	20
DL-1:2-Dibromo-	1.48	20
meso-1:2-Dibromo-	1.42, 2.97	20
DL-1:2-Dimethyl-	3.93, 6.00	20
meso-1:2-Dimethyl-	3.77, 5.36	20
D-Tartaric	3.03, 4.45	20
meso-Tartaric	3.29, 4.92	20
Adipic	4.42, 5.41	2
Pimelic	4.48, 5.42	2
Suberic	4.52, 5.40	2
Azelaic	4.55, 5.41	2

* Thermodynamic value.

CARBOXYLIC ACIDS (continued)

Compound	pK_a'	Reference
Lysergic Acid and Derivatives		
Ergometrine	6.8	2
Ergometrinine	7.3	2
Dihydroergometrine	7.4	2
α-Dihydrolysergol	8.3	2
β-Dihydrolysergol	8.2	2
6-Methylergoline	8.85	2
Lysergic acid	7.8, 3.3	2
Isolysergic acid	8.4, 3.4	2
α-Dihydrolysergic	8.3, 3.6	2
γ-Dihydrolysergic	8.6, 3.6	2

For complex chelating agents, see also Ref. (84).

AROMATIC

Compound	pK_a'	Reference
Anthracene-1-COOH	3.69	2
Anthracene-9-COOH	3.65	2
2-Furan-COOH	3.16	153
3-Furan-COOH	3.95	153
Naphthalene-2-COOH	4.17	2
Naphthalene-1-COOH	3.69	2
Naphthol-1-COOH	9.85	153
Naphthol-2-COOH	9.63	153
1-Phenyl-5-methyl-1,2,3-triazole-4-COOH	3.73	126
1-Phenyl-1,2,3-triazole-4-COOH	2.88	126
1-Phenyl-1,2,3-triazole-4,5-(COOH)$_2$	2.13, 4.93	126
2-Pyrrole-COOH	4.45	153
2-Thiophen-COOH	3.53	153
3-Thiophen-COOH	4.10	153
1,2,3-Triazole-4-COOH	3.22, 8.73	126
1,2,3-Triazole-4,5-(COOH)$_2$	1.86, 5.90, 9.30	126
Substituted Benzoic Acids		2, 97, 100, 101

$$X \text{—} \langle \text{benzene ring} \rangle \text{—COOH}$$

Benzoic acid	ortho	meta	para	Reference
Br-	2.85*	3.81*	4.00*	—
F-	3.27*	3.87*	4.14*	—
CH$_3$O-	4.09*	4.09*	4.47*	—
n-C$_3$H$_7$O-	4.24*	4.20*	4.46*	—
n-C$_4$H$_9$O-	—	4.25*	4.53*	—
C$_6$H$_5$O-	3.53*	3.95*	4.52*	—
CH$_3$-	3.91*	4.24*	4.34*	—
(CH$_3$)$_2$CH-	—	—	4.35*	—
(CH$_3$)$_3$N$^+$-	1.37	3.45	3.43	—
NC-	3.14*	3.60*	3.55*	—
HO$_2$C-	2.95*	3.54	3.51	—
F$_3$C-	—	3.79	—	—
HO-	2.98*	4.08*	4.58*	—
I-	2.86*	3.86*	3.93	—
Cl-	2.94*	3.83*	3.99*	—
(CH$_3$)$_3$Si-	—	4.24*	4.27*	—
C$_2$H$_5$O-	4.21*	4.17*	4.45*	—
i-C$_3$H$_7$O-	4.24*	4.15*	4.68*	—
n-C$_5$H$_{11}$O-	—	—	4.55*	—
C$_6$H$_5$-	3.46*	—	—	—
CH$_3$CH$_2$-	3.77	—	4.35*	—
(CH$_3$)$_3$C-	3.46	4.28	4.40*	—
NH$_2$-	2.05*	3.07*	2.38*	—
	4.95*	4.73*	4.89*	—
SO$_2$NH$_2$-	—	3.54	3.47	—
CH$_3$CO$_2$-	3.48	4.00	4.38	—
CH$_3$CONH-	3.63	4.07	4.28	—
$^-$HO$_3$P-	3.78	4.03	3.95	—
$^-$O$_3$S-	—	4.15	4.11	—
(CH$_3$)$_2$N-	8.42	5.10	5.03	—
$^-$HO$_3$As-	—	—	4.22	—
$^-$O$_2$C-	5.41	4.60	4.82	—
CH$_3$NH-	5.33	5.10	5.04	—

Compound	pK_a'	Reference
***Ortho*-Substituted Benzoic Acids**		
2-CH$_3$-	3.91*	2
2-t-C$_4$H$_9$-	3.46	2
2,6-(CH$_3$)$_2$-	3.21	2
2,3,4,6-(CH$_3$)$_4$-	4.00	2
2,3,5,6-(CH$_3$)$_4$-	3.52	2
2-C$_2$H$_5$-	3.77	2
2-C$_6$H$_5$-	3.46*	2
2,4,6-(CH$_3$)$_3$-	3.43	2
2,3,4,5-(CH$_3$)$_4$-	4.22	2
2,4-OH-	3.22*	97
2,6-OH-	1.22*	97

Benzoic acid	ortho	meta	para	Reference
H-	4.20*	—	—	—
O$_2$N-	2.17*	3.45*	3.44	—
CH$_3$CO-	4.14	3.83	3.70	—
CH$_3$SO$_2$-	—	3.64*	3.52*	—
CH$_3$S-	—	5.53	5.74	—
HS-	5.02	5.42	5.56	—

Acid	Position of carboxyl	pK^I	pK^{II}	pK^{III}	pK^{IV}	pK^V	pK^{VI}
Benzene Polycarboxylic Acids (2)							
Benzoic	1	4.17*	—	—	—	—	—
Phthalic	1, 2	2.98*	5.28*	—	—	—	—
Isophthalic	1, 3	3.46*	4.46*	—	—	—	—
Terephthalic	1, 4	3.51*	4.82*	—	—	—	—
Hemimellitic	1, 2, 3	2.80*	4.20*	5.87*	—	—	—
Trimellitic	1, 2, 4	2.52*	3.84*	5.20*	—	—	—
Trimesic	1, 3, 5	3.12*	3.89*	4.70*	—	—	—
Mellophanic	1, 2, 3, 4	2.06*	3.25*	4.73*	6.21*	—	—
Prehnitic	1, 2, 3, 5	2.38*	3.51*	4.44*	5.81*	—	—
Pyromellitic	1, 2, 4, 5	1.92*	2.87*	4.49*	5.63*	—	—
Benzenepentacarboxylic	1, 2, 3, 4, 5	1.80*	2.73*	3.97*	5.25*	6.46*	—
Mellitic	1, 2, 3, 4, 5, 6	1.40*	2.19*	3.31*	4.78*	5.89*	6.96*

* Thermodynamic value.

PHENOLS

Phenol	ortho	meta	para	Reference
H-	9.95*	—	—	52, 97, 100, 153
(CH₃)₃N⁺-	7.42	8	8	—
CH₃SO₂-	—	8.40	7.83	—
CH₃CO-	—	9.19	8.05	—
C₂H₅O₂C-	—	—	8.50*	—
C₃H₅CH₂O₂C-	—	—	8.41*	—
Br-	8.42*	9.11*	9.34	—
F-	8.81*	9.28*	9.95*	—
HO-	9.48	9.44	9.96	—
CH₃-	10.28*	10.08	10.19*	—
CH₃O-	9.93	9.65	10.20	—
⁻O₂C-	13.82	9.94*	9.39*	153
⁻⁻O₃P-	—	10.2	9.9	—
C₆H₅-	9.93	9.59	9.51	—
O₂N-	7.23*	8.35*	7.14*	—
OCH-	6.79	8.00	7.66	—
NC-	—	8.61	7.95	—
CH₃O₂C-	—	—	8.47*	—
n-C₄H₉O₂C-	—	—	8.47*	—
I-	8.51	9.17*	9.31	—
Cl-	8.48*	9.02*	9.38*	—
CH₃S-	—	9.53	9.53	—
HOCH₂-	9.92*	9.83*	9.82*	—
C₂H₅-	10.2	9.9	10.0	—
H₂N-	9.71	9.87	10.30	—
⁻O₃S-	—	9.29	9.03	—
⁻⁻O₃As-	—	—	8.37	—
NO-	—	—	6.35	—
H₂NCO-	8.37*	—	—	—

Name	pK′ₐ	Reference
POLYSUBSTITUTED PHENOLS		
2,3-Dimethyl-	10.54	101
2,4-Dimethyl-	10.60	101
2,5-Dimethyl-	10.41	101
2,6-Dimethyl-	10.63	101
2,4,5-Trimethyl-	10.88	101
2,3,5-Trimethyl-	10.69	101
3,4-Dimethyl-	10.36	101

* Thermodynamic value.

Name	pK′ₐ	Reference
Polysubstituted Phenols—(Continued)		
3,5-Dimethyl-	10.19	101
2,4-Dichloro-	7.85	101
3-Chloro-2-nitro-	6.75	101
3-Chloro-4-nitro-	6.80	101
3-Bromo-2-nitro-	6.78	101
3-Bromo-4-nitro-	6.84	101
3-Iodo-2-nitro-	6.89	101
3-Iodo-4-nitro-	6.94	101
2,4-Dinitro-	4.11	101
2,5-Dinitro-	5.22	101
2,6-Dinitro-	5.23	101
3,4-Dinitro-	5.42	101
2,4,6-Trinitro-	0.96	101
2-Chloro-4-nitro-	5.45	79
2-Nitro-4-chloro-	6.46	79
2-OCH₃-4-CH₂CH=CH₂-	10.00*	97
2-OCH₃-4-OHC-	7.40*	97
2-OCH₃-6-OHC-	7.91*	97
2-OCH₃-5-OHC-	8.89*	97
Chromotropic acid	5.36. 15.6	6
2-Amino-4,5-dimethylphenol hydrochloride	10.4, 5.28	51
4,5-Dihydroxybenzene-1,3-disulphonic acid	7.66, 12.6	77
Kojic acid	9.40	77
Resorcinol	9.15 (30°)	50
3-Hydroxyanthranilic acid	10.09, 5.20	51
2-Aminophenol hydrochloride	9.99, 4.86	51

SUBSTITUTED CATECHOLS

Name	pK′ₐ	Reference
3-Nitro-	6.66	101
4-Nitro-	6.89	101
3,4-Dinitro-	4.39	101
4-Formyl-	7.36	101
4-Hydroxyiminomethyl-	8.68	101
3-Methyl-	9.28	101
3-Methoxy-	9.28	101
4-Benzoyl-	7.74	101
4-Cyano-	7.72	101

ALCOHOLS AND OTHER OXYGEN ACIDS

Compound	pK_a'	Reference
ALCOHOLS, SIMPLE		
Choline	13.9	6
CF_3CH_2OH	12.43	63
$CF_3CH(OH)CH_3$	11.8	63
$C_3F_7CH_2OH$	11.4†	63
$(C_3F_7)_2CHOH$	10.6†	63
$CH \equiv CCH_2OH$	13.55	64
$C(CH_2OH)_4$	14.1	64
$CH_2OHCHOHCH_2OH$	14.4	64
CH_2OHCH_2OH	14.77	64
$CH_3OCH_2CH_2OH$	14.82	64
CH_3OH	15.54	64
$CH_2=CHCH_2OH$	~15.52	64
H_2O	15.74	64
CH_3CH_2OH	16	64
CCl_3CH_2OH	12.24	64
CHF_2CH_2OH	12.74	64
$CHCl_2CH_2OH$	12.89	64
CH_2ClCH_2OH	14.31	64
$CF_3C(CH_3)_2OH$	11.6	64
$HOCH_2CF_2CF_2CH_2OH$	11	64
$C_3F_7CH(C_2F_5)OH$	10.48	65
$(C_3F_7)_2CHOH$	10.52	65
$(CF_3)_2CHOH$	9.3	108
$(CF_3)_3COH$	5.4	108
$(CF_3)C(OH)CF_2NO$	3.9	108
$(CF_3)_2C(CH_3)OH$	9.6	122
$(CF_3)_2CHOH$	9.3	122
$(CF_3)_2C(CClF_2)OH$	5.3	122
$(CF_3)_2C(CCl_3)OH$	5.1	122
F_2CHCH_2OH	13.11	142
FCH_2CH_2OH	14.20	142
Br_3CCH_2OH	12.70	142
Br_2CHCH_2OH	13.29	142
$BrCH_2CH_2OH$	14.38	142
ICH_2CH_2OH	14.56	142
$NCCH_2CH_2OH$	14.03	142
$C_2H_5OCH_2CH_2OH$	14.98	142
$C_6H_5OCH_2CH_2OH$	14.60	142
$C_6H_5CH_2CH_2OH$	15.48	142
$HOCH_2CH_2OH$	15.11	142
$C_2H_5CH_2OH$	15.92	142
$C_3H_7CH_2OH$	15.87	142
$i\text{-}C_3H_7CH_2OH$	15.91	142
$(CH_3)_3CCH_2OH$	16.04	142
$CH_3C\equiv CCH_2OH$	14.16	142
$C_6H_5C\equiv CCH_2OH$	13.87	142
$CH_2=CHCH_2OH$	15.48	142
$CH_3CH=CHCH_2OH$	15.80	142
$C_6H_5CH=CHCH_2OH$	15.62	142
$C_6H_5CH_2OH$	15.44	142
3,5-Di-$NO_2C_6H_3CH_2OH$	14.43	142
$OHCCH_2OH$	14.80	142
CH_3COCH_2OH	14.19	142
$C_6H_5COCH_2OH$	13.33	142

Substituted Triphenylmethanols in	H_2SO_4	Acid $HClO_4$	Ref. HNO_3 66
CARBONIUM IONS			
4,4′,4″-Trimethoxy-	0.82	0.82	0.80
4,4′-Dimethoxy-	−1.24	−1.14	−1.11
4-Methoxy-	−3.40	−3.59	−3.41
4-Methyl-	−5.41	−5.67	—
4-Trideuteriomethyl-	−5.43	−5.67	—
3,3′,3″-Trimethyl-	−6.35	−5.95	—
Unsubstituted triphenyl-methanol	−6.63	−6.89	−6.60
4,4′,4″-Trichloro-	−7.74	−8.01	—
4-Nitro-	−9.15	−9.76	—

Compound	pK_a'	Reference
HYDROXAMIC ACIDS		
Aceto-	9.40	68
o-Aminobenzo-	9.17	93
p-Aminobenzo-	9.32	93
Benzo-	8.88	68
n-Butyro-	9.48	68
Chloroaceto-	8.40	93
p-Chlorobenzo-	9.59	68
p-Chlorophenoxyaceto-	8.75	93
Cyclohexano-	9.75	93
Formo-	8.65	93
Furo-	8.45	72
Glycine-	7.40	72
Hexano-	9.75	93
Hippuro-	8.80	72
p-Hydroxybenzo-	8.93	93
N-Hydroxyphthalimide	7.00, 6.10	71, 72
Indole-3-aceto-	9.58	93
L-Lacto-	9.35	93
D-Lysine-	7.93	93
L-Lysine-	7.93	93
p-Methylbenzo-	8.90	72
p-Methoxybenzo-	9.00	68
α-Naphtho-	~7.7	68
Nicotin-	8.30	72
isoNicotin-	7.85	72
Nicotin-methiodide	6.46	72
m-Nitrobenzo-	8.07	72
p-Nitrobenzo-	8.01	93
Phenylaceto-	9.19	68
N-Phenylbenzo-	9.15	93
N-Phenylnicotino-	8.00	93
Phthalo-	9.48	93
Picolin-	8.50	72
Propiono-	9.46	68
Pyrimidine-2-carbox-	7.88	72
Salicyl-	7.43	72
Tropo-	9.09	72
L-Tyrosine	9.20	93

* Thermodynamic value.

† 50% aqueous methanol.

ALCOHOLS AND OTHER OXYGEN ACIDS (continued)

Compound	pK_a'	Reference
OXIMES		
Acet-	12.42	18
Acetophenone	11.48	18
Benzophenone	11.3	18
Benzoquinoline mon-	6.25	93
1,2,3-Cyclohexanetrionetri-	8.0	76
Diethyl ket-	12.6	18
Isonitrosoacetone (INA)	8.3	76
Isonitrosoacetylacetone (INAA)	7.4	76
5-Methyl-1,2,3-cyclo-hexanetrione-1,3-di-	8.3	76
5-Methyl-1,2,3-cyclo-hexanetrionetri-	8.0	76
Phenylglyoxald-	8.30	93
Pyridine-2-ald-	3.56*, 10.17*	99
Pyridine-3-ald-	3.94*, 10.32*	99
Pyridine-4-ald-	4.58*, 9.91*	99
Pyridine-4-aldoxime dodeciodide	8.50	93
Pyridine-2-aldoxime heptiodide	8.00	93
Pyridine-2-aldoxime methiodide	8.00	93
Pyridine-3-aldoxime methiodide	9.20	93
Pyridine-4-aldoxime methiodide	8.50	93
Pyridine-4-aldoxime pentiodide	8.50	93
3-Pyridine-1,2-ethanedione-2-oxime methiodide	7.20	93
4-Pyridine-1,2-ethanedione-2-oxime methiodide	7.10	93

Compound	pK_a'	Reference
PEROXYACIDS		
Acetic	8.2	70
n-Butyric	8.2	70
Formic	7.1	70
Peroxydiphosphoric	5.18, 7.68	85
Peroxymonophosphoric	4.85	90
Peroxymonosulfuric	9.4	69
Propionic	8.1	70

$$X-\!\!\!\!\bigcirc\!\!\!\!-N^+\!-O^-$$

X 2—   3—   4—

PYRIDINE 1-OXIDES AND DERIVATIVES
(Ref. 99, 47, 67)

	2—	3—	4—
H-	0.79	—	—
CH_3CONH-	−0.42*	0.99*	1.59*
H_2N-	2.67*	1.47*	3.59*
$C_6H_5CH_2S-$	−0.23*	—	2.09*
HOOC-	—	0.09	−0.48
	—	2.73*	2.86*
$(CH_3)_2N-$	2.27*	—	3.88*
HO-	5.97*.	—	5.76*
CH_3O-	1.23*	—	2.05*
CH_3NH-	2.61*	—	3.85*
C_6H_5-	0.77*	0.74*	0.83*
CH_3-	2.61*	1.08*	1.29*
C_2H_5O-	1.18*	—	—
$C_6H_5CH_2O-$	—	—	1.99
NO_2-	—	—	−1.7
$COOC_4H_9-$	—	0.03	—

R =	H	CH_3	C_2H_5	*iso*-C_3H_7	*tert*-C_4H_9	*iso*-C_4H_9	Reference
PEROXIDES, ROOH							
	11.6	11.5	11.8	12.1	12.8	12.8	70

* Thermodynamic value.

ALCOHOLS AND OTHER OXYGEN ACIDS (continued)

Compound	pK'_a	Reference
Pyridine 1-oxides and Derivatives—(Continued)		
2-Amino-1-methoxy pyridinium perchlorate	12.4	67
4-Amino-1-methoxy-pyridinium perchlorate	>11	67
1-Benzyloxypyrid-2-one	−1.7	67
1-Benzyloxypyrid-4-one	2.58	67
4-Dimethylamino-1-methoxy-pyridinium perchlorate	>11	67
2-Methylamino-1-methoxy-pyridinium toluene-*p*-sulfonate	>11	67
1-Methoxypyrid-2-one	−1.3	67
1-Methoxypyrid-4-one	2.57	67
3-R-Pyrazine-1-Oxides		
R = CN-	−1.12	121
Cl-	−1.05	121
CH$_3$O-	−0.45	121
NH$_2$-	−1.92, 1.50	121
H-	0.05	121
CH$_3$-	0.46	121
(CH$_3$)$_2$N-	−1.77, 1.34	121
N ⌇	−1.80, 1.34	121

SULFINIC ACIDS

Compound	pK'_a	Reference
Benzene-	1.84, 2.16	73
p-Bromobenzene-	1.89	73
p-Chlorobenzene-	1.81	73
m-Nitrobenzene-	1.88	73
p-Nitrobenzene-	1.86	73
p-Toluene-	1.99	73

OTHER OXYGEN ACIDS

Compound	pK'_a	Reference
CF$_3$CH$_2$NHOH	11.3	108
(CF$_3$)$_2$CHNHOH	8.5	108
(CF$_3$)$_3$CNHOH	5.9	108
CF$_3$CHNOH	8.9	108
(CF$_3$)$_2$CNOH	6.0	108
Glutaconic dialdehyde	5.75	153
Hydroxylamine	13.7	133

OTHER OXYGEN ACIDS (continued)

Compound	pK'_a	Reference
Mannitol	13.5	100
Sucrose	12.7	100
Phenylboric acid	8.86	100
Pyridine-4-aldehyde	12.20	153
β-Phenylethylboric acid	10	100
Lyxose	12.11	25
Ribose	12.11	25
2-Deoxyribose	12.61	25
Xylose	12.15	25
Arabinose	12.34	25
Fructose	12.03	25
2-Deoxyglucose	12.52	25
Galactose	12.35	25
(CF$_3$)$_2$C(OH)$_2$	6.58	108
(CF$_2$Cl)$_2$C(OH)$_2$	6.67	108
(CF$_2$H)$_2$C(OH)$_2$	8.79	108
CF$_2$ClCF$_2$HC(OH)$_2$	7.90	108
Trimethylamine-*N*-oxide	4.6	18
Triethylamine-*N*-oxide	5.13*	99
Acetaldehyde hydrate	13.48	91
Formaldehyde hydrate	13.29	91
Glucose	12.43*	97
Mannose	13.50*	97
Sorbose	13.57	97
Acetamide-H$^+$	−0.025	149, 150
Chloroacetamide-H$^+$	−0.26	149, 150
Dichloroacetamide-H	−0.26	149, 150
N-Methylacetamide-H$^+$	0.26	149, 150
N,N-Dimethylacetamide-H$^+$	0.62	149, 150
Biotin-H$^+$	−1.13	149, 150
Desthiobiotin-H$^+$	−0.97	149, 150
Dimethylurea-H$^+$	−0.20	149, 150
Formamide-H$^+$	0.12	149, 150
N,N-Diethylformamide-H$^+$	0.36	149, 150
N,N-Dimethylformamide-H$^+$	0.18	149, 150
N-Methylformamide-H$^+$	0.52	149, 150
Imidazolidone-H$^+$	−1.05	149, 150
(CH$_3$)$_2$CHCH(OH)$_2$	13.77	160
CH$_3$CH(OH)$_2$	13.57	160
CH$_2$(OH)$_2$	13.27	160
CF$_3$CH(OH)$_2$	10.20	160
CCl$_3$CH(OH)$_2$	10.04	160
C$_6$H$_5$C(OH)$_2$CF$_3$	10.00	160

* Thermodynamic value.

AMINO ACIDS

Compound	pK'_a	Reference
Alanine	2.34, 9.69	6
N-Acetyl-	3.72	97
Amide	8.02*	99
3-(2-Aminoethyldithio)-	8.28, 9.30	99
Carbamyl-	3.89	99
N-Ethyl-	2.22, 10.22	99
N-Methyl-	2.22, 10.19	99
N-n-Propyl-	2.21, 10.19	99
β-(2-Pyridyl)-	1.37, 4.02, 9.22	99
β-(3-Pyridyl)-	1.77, 4.64, 9.10	99
β-(4-Pyridyl)-	4.85	99
β-Alanine	3.60, 10.19	6
N-acetyl-	**4.44**	129
Carbamyl-	**4.49**	129
Allothreonine	2.11, 9.01	99
O-Methyl-	1.92, 8.90	99
γ-Aminoacetoacetic acid	2.9, 8.3	99
α-Aminoadipic acid	2.14, 4.21	101
2-Aminobenzoic acid	2.19, 4.95	99
N,N-Dimethyl-	1.4, 8.49	99
3-Hydroxy-	5.19, 10.12	99
N-Methyl-	1.97, 5.34	99
3-Aminobenzoic acid	3.29, 5.10	99
4-Aminobenzoic acid	2.50, 4.87	99
4-Aminobutylphosphonic acid	2.55, 7.55, 10.9	99
4-Aminobutylsulphonic acid	10.65	99
α-Aminobutyric acid	2.55, 9.60	6
Carbamoyl-α-amino-n-butyric	3.89	129
γ-Aminobutyric acid	**4.23, 10.43**	6
Carbamyl-	**4.68**	129
2-Aminobutyric acid	**2.27, 9.68**	99
α-Amino-n-caproic acid	**2.33**	129
ε-Aminocaproic acid	**4.37**	129
10-Aminodecylphosphonic acid	8.0, 11.25	99
10-Aminodecylsulphonic acid	11.35	99
10-Amino-n-dodecanoic acid	4.648	99
Aminoethylphosphoric acid	2.45, 7.0, 10.8	99
2-Aminoethylsulphonic acid	8.95	99
ω-Aminoheptanoic acid	4.50	136
6-Aminohexanoic acid	4.37, 10.81	99
α-Aminoisobutyric acid	2.36, 10.21	6
Carbamyl-	**4.46**	129
α-Aminoisocaproic acid	**2.33**	129
α-Aminoisovaleric acid	**2.29**	129
δ-Aminolaevulinic acid	4.05, 8.90	99
Aminomethylphosphonic acid	2.35, 5.9	99
Aminomethylsulphonic acid	5.75	99
α-Amino-β-methyl-n-valeric	**2.32**	129
acid		
1-Aminonaphthalene-2-	1.71	99
sulphonic acid		
2-Aminonaphthalene-1-	2.35	99
sulphonic acid		
3-Amino-1-naphthoic acid	2.61, 4.39	99
4-Aminopentanoic acid	3.97, 10.46	99
5-Aminopentylsulphonic acid	10.95	99
4-Aminophenylacetic acid	3.60, 5.26	99
2-Aminophenylarsonic acid	3.77, 8.66	99
2-Aminophenylboric acid	4.53, 9.31	99
β-Aminopropionic acid	3.55*, 10.23*	97
4-Aminosalicylic acid	1.78, 3.63	99

Compound	pK'_a	Reference
α-Aminotricarballylic acid	2.10, 3.60, 4.60, 9.82	99
α-Aminovaleric acid	**4.20**	130
2-Anilinoethylsulphonic acid	3.80	99
Arginine	12.48, 2.17, 9.04	6
Argininosuccinic acid	>12, 1.62, 9.58, 2.70, 4.26	–
Asparagine	2.02, 8.8	6
α-Hydroxy-	2.28, 7.20	99
β-Hydroxy-	2.09, 8.29	99
Aspartic acid	2.09, 3.86, 9.82	99
Diamide	7.00	99
Hydroxy-	1.91, 3.51, 9.11	99
Azaserine	8.55	101
		99
		99
		99
γ-Butyrobetaine	3.94	99
Canaline	2.40-, 3.70, 9.20	99
Canavanine	2.50, 6.60, 9.25	99
L-Citrulline	2.43, 9.41	99
Creatine	2.67, 11.02	6
Creatinine	4.84, 9.2	6
Cycloserine	4.4, 7.4	101
Cysteine	10.78, 1.71, 8.33	6
Ethyl ester	6.69, 9.17	99
Methyl ester	6.56, 8.99	99
S-Ethyl-	1.94, 8.69	99
S-Methyl-	8.75	99
Cystine	1.65, 7.85	6
L-Cystine diamide	5.93, 6.90	99
2,4-Diaminobutyric acid	1.85, 8.28, 10.50	99
2,3-Diaminopropionic acid	1.23, 6.73, 9.56	99
2,7-Diaminosuberic acid	1.84, 2.64, 9.23, 9.89	99
3-Dimethylaminopropionic acid	9.85	99
Formamidinoglutaric acid	2.7, 4.4, 11.3	99
Formamidinoacetic acid	2.6, 11.5	99
Glutamic acid	2.19, 4.25, 9.67	6
Diethyl ester	7.04	99
γ-Monobenzyl ester	2.17, 9.00	99
α-Monoethyl ester	3.85, 7.84	99
γ-Monoethyl ester	2.15, 9.19	99
Glutamine	2.17, 9.13	6
Glycine	2.34, 9.6	6
N-Acetyl-	3.67	99
N,N-bis(2-Hydroxyethyl)-	2.50, 8.11	99
N-n-Butyl-	2.35, 10.25	99
Carbamyl-	3.88*	97
Chloroacetyl-	3.38*	97
N,N-Diethyl-	2.04, 10.47	99
Dihydroxyethyl-	8.08*	97
N,N-Dimethyl-	2.08-, 9.80	99
N-Ethyl-	2.34*, 10.23	99
Ethyl ester	7.83	99
Formyl-	3.43*	97
N-Isobutyl-	2.35, 10.12	99
Methyl ester	7.73	99

* Thermodynamic value.

AMINO ACIDS (continued)

Compound	pK_a'	Reference
Histamine	5.0, 9.7	6
Histidine	6.0, 1.82, 9.17	6
Amide	5.78, 7.64	99
2-Mercapto-	1.84, 8.47, 11.4	99
1-Methyl-	6.58, 8.60	99
2-Methyl-	1.7, 7.2, 9.5	99
Methyl ester	7.33, 5.38	99
Homocysteine	2.22, 8.87	101
β-Hydroxyglutamate	2.27, 4.29, 9.66	99
N-Hydroxyethylethylene-diamine-triacetic acid		
Hydroxylysine	2.13, 8.62, 9.67	6
Hydroxyproline	1.92, 9.73	6
Imidazolelactic acid	2.96, 7.35	99
Isoasparagine	2.97, 8.02	99
N-Acetyl	3.99	151
N-Carbobenzoxy-	4.05	151
Isocreatine	2.84	99
Isoglutamine	3.81, 7.88	99
N-Acetyl-	4.34	151
N-Carbobenzoxy-	4.39	151
Isoleucine	2.36, 9.68	6
Isoserine	2.72, 9.33	99
Leucine	2.36, 9.60	6
Amide	7.80	99
Ethyl ester	7.57	99
Lombricine	8.9	99
Lysine	2.18, 8.95	6
Hydroxy-	2.13, 8.62, 9.67	99
Methionine	2.28, 9.21	6
Amide	7.53	99
N-Methylaminodiacetic acid	2.15	129
Nitrilotriacetic acid	1.88, 2.48, 4.28	129
Norleucine	2.39, 9.76	6
Norvaline	2.30, 9.78	99
Octopine	1.40, 2.30, 8.72, 11.34	99
Ornithine	1.71, 8.69	6
Phenylalanine	1.83, 9.13	6
Amide	7.22	99
o-Chloro-	2.23, 8.94	99
m-Chloro-	2.17, 8.91	99
p-Chloro-	2.08, 8.96	99
3,4-Dihydroxy-	2.32*, 8.68*, 9.88*	97
2,4-Diiodo-3-hydroxy-	2.12*, 6.48*, 7.82*	97
o-Fluoro-	2.12*, 9.01*	97
m-Fluoro-	2.10*, 8.98*	97
p-Fluoro-	2.13*, 9.05*	97
p-HOSO$_2$NH-	1.99*, 8.64*, 10.26*	97
α-Methyl-	9.57	99
Methyl ester	7.00	99
Phenylglycine	1.83, 4.39	99
m-Chloro-	1.05, 3.93	99
p-Chloro-	1.46, 4.04	99
m-Cyano-	0.28, 3.78	99
m-Methyl-	1.89, 4.60	99
p-Methyl-	1.97, 4.85	99
Proline	1.99, 10.60	6
2-Pyrrolidone-5-carboxylic acid (glutamic acid)	3.32	6
Sarcosine	2.23, 10.01	6
Amide	8.31*	99
N-Dimethylamide	8.82*	99
N-Methylamide	8.24*	99
Serine	2.21, 9.15	6
Amide	7.30	99
Methyl ester	7.10	99
Taurine	1.5, 8.74	6
Thiolhistidine	<1.5, 11.4, 1.84, 8.47	6
Threonine	2.63, 10.43	6
O-Methyl-	2.02, 9.00	101
5,5,5-Trifluoroleucine	2.05, 8.92	111
6,6,6-Trifluoronorleucine	2.164, 9.46	111
4,4,4-Trifluorothreonine	1.55, 7.82	111
4,4,4-Trifluorovaline	1.54, 8.10	111
Tryptophan	2.38, 9.39	6
Amide	7.5	99
Tyrosine	10.07, 2.20, 9.11	6
Amide	7.48, 9.89	99
3,5-Dibromo-	2.17, 6.45, 7.60	99
3,5-Dichloro-	2.22, 6.47, 7.62	99
Diiodo-	6.48, 2.12, 7.82	6
Ethyl ester	7.33, 9.80	22
O-Methyl-	9.27	21
O-Methyl, ethyl ester	7.31	22
N-Trimethyl-	9.75	21
Urocanic acid	5.8, 3.5	—
Valine	2.32, 9.62	6
Amide	8.00	99
Hydroxy-	2.55, 9.77	99
β-Mercapto-	2.0, 8.0, 10.5	99

* Thermodynamic value.

AMINO ACIDS (continued)

Compound	pK_a''	Reference
$CH_3CH_2OCONHCH_2COOH$	3.66*	97
$CH_3CONH(CH_2)_2COOH$	4.45	97
$NH_2CONH(CH_2)_2COOH$	4.49	97
DL-$CH_3CH_2CH(NH_2)COOH$	2.29, 9.83	97
$CH_3CONHCH(CH_2CH_3)COOH$	3.72	97
$NH_2CONHCH(CH_2CH_3)COOH$	3.89	97
$NH_2CONH(CH_2)_3COOH$	4.68	97
$(CH_3)_2C(NH_2)COOH$	2.36*, 10.205*	97
$NH_2CONHC(CH_3)_2COOH$	4.46	97
DL-$CH_3(CH_2)_2CH(NH_2)COOH$	4.36*, 9.72*	97
$CH_3CH_2CH(NH_2)CH_2COOH$	4.02*, 10.40*	97
$NH_2(CH_2)_4COOH$	4.20*, 10.69*	97
DL-$(CH_3)_2CHCH(NH_2)COOH$	2.29, 9.74	97
$NH_2(CH_2)_5COOH$	4.43*, 10.75*	97
$NH_2(CH_2)_{11}COOH$	4.65*	97
$NH_2(CH_2)_3CH(NH_2)COOH$	1.94*, 8.65*, 10.76*	97
$NH_2CONH(CH_2)_3CH(NH_2)COOH$	2.43*, 9.41*	97
$COOH(CH_2)_2CH(NH_2)COOCH_2CH_3$	3.85*, 7.84*	97
$CH_3CH_2OCO(CH_2)_2CH(NH_2)COOH$	2.15*, 9.19*	97
$COOHCH_2NHCH_2COOH$	2.54*, 9.12*	97
$COOHCH_2N(CH_3)CH_2COOH$	2.15*, 10.09*	97
$N(CH_2COOH)_3$	2.96*, 10.23	97
$COOH(CH_2)_2NHCH_2COOH$	3.61*, 9.46*	97
$COOH(CH_2)_2NH(CH_2)_2COOH$	4.11*, 9.61*	97
$COOHCH_2NH(CH_2)_2NHCH_2COOH$	6.42*, 9.46*	97
$(CH_3)_2N(CH_2)_2N(CH_2COOH)_2$	6.05*, 10.07*	97
$(COOHCH_2)_2N(CH_2)_2N(CH_2COOH)_2$	6.27*, 10.95*	97
$COOH(CH_2)_2N(CH_2COOH)(CH_2)_2N(CH_2COOH-$ $(CH_2)_2COOH$	3.00*, 3.79*, 5.98*, 9.83*	97
$COOH(CH_2)_2NH(CH_2)_2NH(CH_2)_2COOH$	6.87*, 9.60*	97
$(COOHCH_2CH_2)_2N(CH_2)_2N(CH_2CH_2COOH)$	3.00*, 3.43*, 6.77*, 9.60*	97
NH_2COCH_2COOH	3.64*	97
$NH_2CO(CH_2)_2COOH$	4.54*	97
$NH_2CO(CH_2)_3COOH$	4.60*	97
$NH_2CO(CH_2)_4COOH$	4.63*	97

Compound	pK_a''	Reference
Oxyproline	1.92*, 9.73*	97
$COOHCH_2CH(OH)CH(NH_2)COOH$	2.32*, 4.23*, 9.56*	97
$CH_3CH_2SCH_2CH(NH_2)COOH$	2.03*, 8.60*	97
$CF_3(CH_2)_3CH(NH_2)COOH$	2.16*, 9.46*	97
$CF_3CH(CH_3)CH_2CH(NH_2)COOH$	2.05*, 8.94*	97
$CF_3(CH_2)_2CH(NH_2)COOH$	2.04*, 8.92*	97
$CF_3CH(CH_3)CH(NH_2)COOH$	1.54*, 8.10*	97
$CF_3CH_2CH(NH_2)COOH$	1.60*, 8.17*	97
$CF_3CH(OH)CH(NH_2)COOH$	1.55*, 7.82*	97
$CF_3CH(NH_2)CH_2COOH$	2.76*, 5.82*	97

* Thermodynamic value.

PEPTIDES

Compound	pK_a'	Reference
Ala-Ala-(LD)	3.12, 8.30	27
Ala-Ala-(LL)	3.30, 8.14	27
Ala-Ala-Ala-(3D)	3.39, 8.06	27
Ala-Ala-Ala-(DLL)	3.37, 8.06	27
Ala-Ala-Ala-Ala-(DLLL)	3.42, 7.99	27
Ala-Ala-Ala-(3L)	3.39, 8.03	27
Ala-Ala-Ala-Ala-(4L)	3.42, 7.94	27
Ala-Ala-Ala-(LDL)	3.31, 8.13	27
Ala-Ala-Ala-Ala-(LDLL)	3.22, 7.99	27
Ala-Ala-Ala-(LLD)	3.37, 8.05	27
Ala-Ala-Ala-Ala-(LLDL)	3.24, 7.93	27
Ala-Gly	3.16, 8.24	27
Ala-Gly-Gly	3.19, 8.15	99
Ala-Lys-Ala-(3L)	3.15, 7.65, 10.30	27
Ala-Lys-Ala-(LDL)	3.33, 7.97, 10.36	27
Ala-Lys-Ala-(LDLL)	3.32, 8.01, 10.37	27
Ala-Lys-Ala-(LLD)	3.29, 7.84, 10.49	27
Ala-Lys-Ala-Ala-(4L)	3.58, 8.01, 10.58	27
Ala-Lys-Ala-Ala-Ala-(5L)	3.53, 7.75, 10.35	27
Ala-Lys-Ala-Ala-Ala-(LDLLL)	3.30, 7.85, 10.29	27
β-Ala-1-methylhistidine	2.64, 7.04, 9.49	99
Ala-Pro	3.04, 8.38	99
β-Ala-Bis	2.73, 6.87, 9.73	99
Anserine	7.0, 2.65, 9.5	6
Asparaginyl-Gly	2.90, 7.25	99
Asp-Asp	2.70, 3.40, 4.70 8.26	99
α-Aspartyl-histidine	2.45, 3.02, 6.82, 7.98	99
β-Aspartyl-histidine	1.93, 2.95, 6.93, 8.72	99
Asp-Gly	2.10, 4.53, 9.07	99
Asp-Tyr	2.13, 3.57, 8.92, 10.23	99
Carnosine	6.83, 9.51	6
Cys-Cys	2.65, 7.27, 9.35, 10.85	99
Cys-Gly-Gly	3.13, 6.36, 6.95	99
Cys-Gly-Gly-Gly-Gly	3.21, 6.01, 6.87	99
L-Cystinylcystine	1.87, 2.94, 6.53, 7.66	99
N,N-Dimethylglycyl-glycine	3.11, 8.09	99
N,N-Dimethyl-leucyl-glycine	7.78	99
Glutaminyl-glutamic acid	3.14, 4.38, 7.62	99
Glutaminyl-glycine	3.15, 7.52	99
Glutathione	3.59, 8.75, 9.65	77
Glutathione, oxidized	3.15, 4.03, 8.57, 9.54	77
Gly-Ala (L), (D)	3.17, 8.23	27

Compound	pK_a'	Reference
Gly-Ala-Ala (LD)	3.30, 8.17	27
Gly-Ala-Ala (LL)	3.38, 8.10	27
Gly-Ala-Ala-Gly	3.30, 7.93	99
Gly-Asp	2.81, 4.45, 8.60	99
Gly-asparagine	2.82, 7.20	99
Gly-Gly	3.06, 8.13	6
Gly-Gly-cystine	2.71, 7.94	99
Gly-Gly-Gly	3.26, 7.91	23
Gly-His	6.79, 8.20	99
Gly-Leu	3.10, 8.41	99
Gly-Pro	2.81, 8.65	99
Gly-sarcosine	2.98, 8.57	99
Gly-Ser	2.92, 8.10	99
Gly-Ser-Gly	3.23, 7.99	99
Gly-Trp	8.06	99
Gly-Tyr	2.93, 8.45, 10.49	99
Gly-Val	3.15, 8.18	99
His-Gly	2.36, 6.27, 8.57	99
His-His	5.54, 6.80, 7.82	99
Leu-asparagine	2.83, 8.23	99
Leu-Tyr	2.87, 8.36, 10.28	99
Lys-Ala-(LD)	3.00, 7.74, 10.63	27
Lys-Ala-(LL)	3.22, 7.62, 10.70	27
Lys-Glu	2.98, 4.47, 8.45, 11.30	99
Lys-Lys-(LD)	2.85, 7.53, 9.92 10.98	27
Lys-Lys-(LL)	3.01, 7.53, 10.05 11.01	27
Lys-Lys-Lys-(3L)	3.08, 7.34, 9.80, 10.54, 11.32	27
Lys-Lys-Lys-(LDD)	2.94, 7.14, 9.60, 10.38, 11.09	27
Lys-Lys-Lys-(LDL)	2.91, 7.29, 9.79, 10.54, 11.42	27
Met-Met	2.22, 9.27	99
Methyl-Leu-Gly	3.29, 7.82	99
Phe-Ala-Arg	2.60, 7.54, 12.43	99
Phe-Gly	3.13, 7.62	99
Phenylalanylglycine amide	6.72	99
Pro-Gly	3.19, 8.97	99
Sarcosyl-Gly	3.14, 8.66	99
Sarcosyl-Leu	3.15, 8.67	99
Sarcosylsarcosine	2.89, 9.18	99
Ser-Gly	3.10, 7.33	99
Ser-Leu	3.08, 7.45	99
Tyr-Tyr	3.52, 7.68, 9.80, 10.26	99
Val-Gly	3.23, 8.00	99

NITROGEN COMPOUNDS

X	XNH_3^+	$XCH_2NH_3^+$	$X(CH_2)_2NH_3^+$	$X(CH_2)_3NH_3^+$	$X(CH_2)_4NH_3^+$	$X(CH_2)_5NH_3^+$	Reference
ALIPHATIC AMINES, SIMPLE							
Primary Amines							
H-	9.25*	10.64*	10.67*	10.58*	10.61*	10.63	2
HF$_2$C-	—	7.52	9.13	9.71	10.15*	10.37	2
RO$_2$C-	—	7.75	9.50*				—
HO-	5.96*	—	9.83*	10.20*	10.39*	10.49*	2
C$_6$H$_5$-	4.58*	9.37*	9.98	10.65*	10.84*	11.05*	2
H$_2$N-	8.12*	—					2
H$_2$C=CH-	—	9.69					—
CH$_3$-	10.64*	10.67*	10.58*	10.61*	10.63*	10.64*	2

X	Me$_2$N	-H	-NH$_3^+$	-CO$_2^-$	-SO$_3^-$	-PO$_3^-$	Reference
X-NH$_3^+$		9.25*	−0.88	—	1	10.25	2
X(CH$_2$)NH$_3^+$		10.64	—	9.77	5.75	10.8	2
X(CH$_2$)$_2$NH$_3^+$	5.98, 9.30	10.67	—	10.19	9.20	10.8	2, 118
X(CH$_2$)$_3$NH$_3^+$	9.91, 7.67	10.58	8.59	10.43	10.05	—	2, 118
X(CH$_2$)$_4$NH$_3^+$	8.44, 10.17	10.61	9.31	10.77	10.65	10.9	2, 118
X(CH$_2$)$_5$NH$_3^+$	9.07, 10.44	10.63	9.74	10.75	10.95	11.0	2, 118
X(CH$_2$)$_8$NH$_3^+$		10.65	10.10	—	—	—	2
X(CH$_2$)$_{10}$NH$_3^+$		10.64	—	—	11.35	11.25	2

For complex chelating agents of aliphatic amines, see also Reference 77.

Compound	pK_a'	Reference
Primary Amines		
1-Acetamido-2-aminoethane	9.05*	99
1-Acetoxy-2-aminoethane	9.1	99
Acetylhydrazine	3.24*	99
β-Alanine ester	9.13	1
Allyl-amine	9.49	1
1-Amino-2-benzamidoethane	9.13*	99
1-Amino-2-benzylamino-ethane	6.48*, 9.41*	99
1-Amino-2-bromoethane	8.49*	99
1-Amino-3-bromopropane	8.93*	99
1-Amino-2-butylamino-ethane	7.53*, 10.30*	99
(3-Amino)butylbenzene	9.79*	99
(4-Amino)butylbenzene	10.36*	99
γ-Amino-*n*-butyric acid ester	9.71	1
1-Amino-2-diethylamino-ethane	7.07*, 10.02*	99
1-Amino-2-dimethylamino-ethane	6.63*, 9.53*	99
1-Amino-2,2-dimethyl-propane	10.24*	99
1-Amino-2-ethylamino-ethane	7.63*, 10.56*	99
1-(2-Aminoethyl)-piperidine	6.38*, 9.89*	99
2-Aminoethylsulphonic acid	9.08	77
(2-Aminoethyl)trimethyl-ammonium chloride	7.1	99
1-Aminofluorene	3.87*	99
2-Aminofluorene	4.64*	99
3-Aminofluorene	4.82*	99
4-Aminofluorene	3.39*	99
1-Amino-2-furfurylamino-ethane	6.20*, 9.72*	99

Compound	pK_a'	Reference
Primary Amines—(Continued)		
1-Aminoheptane	10.66*	99
2-Aminoheptane	10.67*	99
1-Aminohexane	10.56*	99
1-Amino-4-hydroxybutane	10.35*	99
2-Amino-1-hydroxybutane	9.52*	99
1-Amino-2-hydroxycyclo-heptane	9.25*	99
1-Amino-2-hydroxycyclo-pentane	9.70* *cis* 9.28 *trans*	99
2-Amino-2'-hydroxydiethyl sulfide	9.04, 9.41	77, 99
1-Amino-2-(2-hydroxyethyl)-aminoethane	6.83*, 9.82*	99
2-Amino-3-hydroxyindan	8.13*	99
1-Amino-2-hydroxy-2-methylpropane	9.25*	99
2-Amino-1-hydroxy-2-methylpropane	9.71*	99
1-Amino-5-hydroxypentane	10.46*	99
1-Amino-3-hydroxypropane	9.96*	99
2-Amino-1-hydroxypropane	9.43*	99
1-Amino-2-(2-hydroxy-propyl)aminoethane	6.94*, 9.86*	99
1-Amino-2-(3-hydroxy-propyl)aminoethane	6.78*, 9.67*	99
2-Aminoindan	9.57*	99
5-Aminoindan	5.31*	99
1-Amino-2-isopropylamino-ethane	7.70*, 10.62*	99
Aminomalonic acid	3.32, 9.83	77
1-Amino-2-mercaptoethane	8.27*, 10.53*	99
1-Amino-2-methylbutane	10.64	99
1-Amino-3-methylbutane	10.60*	99
2-Amino-2-methylbutane	10.72*	99

* Thermodynamic value.

Compound	pK_a'	Reference
Primary Amines—(Continued)		
1-Amino-3-methylcyclo-hexane	10.56* *cis*, 10.61* *trans*	99
1-Amino-2-methylpropane	10.72*	99
2-Amino-2-methylpropane	10.68*	99
1-Amino-2-methylamino-ethane	6.86*, 10.15*	99
1-Amino-2-methylthioethane	9.49*	99
1-Aminononane	10.64*	99
1-Aminooctane	10.65*	99
2-Aminooctane	10.49*	99
1-Aminopentane	10.63*	99
3-Aminopentane	10.42*	99
1-Amino-2-propylamino-ethane	8.24*, 11.04*	99
1-Aminoprop-2-ene	9.49*	99
1-Aminoprop-2-yne	8.15*	99
1-Aminotetradecane	10.62*	99
1-Amino-3,3,3-trichloro-propane	9.65*	99
Benzamide	−1.85	99
Benzoylhydrazine	2.97	114
Benzyl-	9.34	1
bis(2-Aminoethyl)disulphide	8.82*, 9.16*	99
1,2-bis(2-Aminoethyl)thio-ethane	8.69*, 9.62*	99
1,2-bis Glycylamidoethane	7.63*, 8.35*	99
n-Butyl-	10.59	1
t-Butyl-	10.55	1
N-*n*-Butylethylenedi-	7.53, 10.30	77
Carbamylmethyl-	7.93	153
N-(Carbamoylmethyl)-iminodiacetic acid	2.30, 6.60	77
Cyanoethyl-	7.7	153
Cyanomethyl-	153	5.34
Ethoxycarbonylethyl-	9.13	153
Cyclohexyl-	10.64	1
Cyclohexylmethyl	10.49	1
1,4-Diaminobutane-	9.24, 10.72	146
2,3-Diaminobutane, *meso*	6.92, 9.97	77
2,3-Diaminobutane, *racemic*	6.91, 10.00	77
2,2′-Diaminodiethyl-	3.58, 8.86, 9.65	77
2,2′-Diaminodiethyl sulfide	8.84, 9.64	77
2,3-Diamino-2,3-dimethyl-butane	6.56, 10.13	77
1,3-Diamino-2,2-dimethyl-propane	8.18, 10.22	77
3,3′-Diaminodi-*n*-propyl-	8.02, 9.70, 10.70	77
N,*N*′-Di-(2-aminoethyl)-ethylenedi-	3.32, 6.67, 9.20, 9.92	77
1,2-Di-(2-aminoethylthio)-ethane	8.42, 9.32	77
1,3-Diamino-2-hydroxy-propane	7.93*, 9.69*	99
1,2-Diamino-2-methyl-propane	6.79, 10.00	77
1,3-Diaminopropan-2-ol	8.23, 9.68	77
1,2-Diaminopropane	7.13*, 10.00*	99
1,3-Diaminopropane	8.64*, 10.62*	99
β-Difluorethyl-	7.52	1
N,*N*′-Diglycylethylenedi-	7.63, 8.35	77
2,3-Dimethoxybenzyl-	9.41*	99
3,4-Dimethoxybenzyl-	9.39*	99
N,*N*-Diethylethylenedi-	7.07, 10.02	77
N,*N*-Dimethylethylenedi-	6.63, 9.53	77
4-4′-Diaminostilbene	3.9*, 5.2*	99
1-Dimethylamino-3-hydrazinobutane	5.90*, 9.23*	99
Ethanol-	9.50	1
Ethyl-	10.63	1
Ethylenedi-	9.98, 7.52	1, 77

Compound	pK_a'	Reference
Primary Amines—(Continued)		
Ethylenediamine-*N*,*N*-diacetic acid	7.63, 8.35	77
N-Ethylethylenedi-	7.63, 10.56	77
Furfuryl-	8.89	77
D-Glucos-	7.75*	99
Glycine ester	7.75	1
Hexadecyl-	10.61	153
Hydrazine	8.10	1
N-(2-Hydroxyethyl)-ethylenedi-	6.83, 9.82	77
Hydroxyl-	5.97	1
2-Hydroxy-3-methoxy-benzylamine	8.70, 11.06*	99
3-Hydroxy-2-methoxy-benzylamine	8.89*, 10.54*	99
4-Hydroxy-3-methoxy-benzylamine	8.94*, 10.52*	99
2-(2-Hydroxypropylamino)-ethyl-	6.94, 9.86	77
2-(3-Hydroxypropylamino)-ethyl-	6.78, 9.76	77
N-Isopropylethylenedi-	7.70, 10.62	77
Isopropyl-	10.63	1
Mercaptoethyl-	8.27, 10.53	77
Methoxy-	4.60*	12
Methoxycarbonylmethyl-	7.66	153
2-Methoxyethyl-	9.20	77
Methyl-	10.62	1
N-Methylaminoacetic acid	2.24, 10.01	77
Methyl-α-amino-β-mercapto-propionate	6.56, 8.99	77
Methyl benzimidate	5.8*	99
N-Methylethylenedi-	7.56, 10.40	77
2-Methylthioethyl-	9.18	77
Octyl-	10.65	153
neo-Pentyl-	10.21	1
Phenylamyl-	10.49	2
δ-Phenylbutyl-	10.40	2
β-Phenylethyl-	9.83	1
Phenylmethyl-	9.34	153
γ-Phenylpropyl-	10.20	1
Propionamide	−0.49*	99
n-Propyl-	10.53	1
N-*n*-Propylethylenedi-	7.54, 10.34	77
sec-Butyl-	10.56	1
Semicarbazide	3.65*	99
1,2,3-Triaminopropane	3.72, 7.95, 9.59	77
Triaminotriethyl-	8.56, 9.59, 10.29	77
β,β′,β″-Triaminotriethyl-	8.42, 9.44, 10.13	87
2,2,2-Trichloroethyl-	5.47*	99
2-Trienylmethyl-	8.92	77
Triethylenedi-	8.8*	—
Trifluoroethyl-	5.7	10
Trimethylsilylmethyl-	10.96	1
Thioacetamide	−1.76*	99
Tris-(hydroxymethyl)-aminomethane	8.10	77
Undecyl-	10.63	153
Vinylmethyl-	9.69	153
$CF_3SO_2^-$	5.8	128
Secondary Amines		
N-(2-Acetamido)-2-amino-ethane-sulfonic acid	6.9	108
Acetamidoglycine	7.7	108
Acetanilide	0.61	4
1-Acetylpiperazine	7.94*	99
Allylmethyl-	10.11	1
N-2-Aminoethylpiperazine	8.51*, 9.63*	99

* Thermodynamic value.

NITROGEN COMPOUNDS—(continued)

Compound	pK'_a	Reference	Compound	pK'_a	Reference
Secondary Amines—(Continued)			**Tertiary Amines**		
Aminomethylcyclohexane	10.59*	99	N-(2-Acetamido)imino-diacetic acid	6.6	99
1-4-Benzoquinoneimine	3.9*	99	1-Acetyl-2-diethylamino-ethane	9.04*	99
N-Benzoylpiperazine	7.78	1	Allyldimethyl-	8.73	1
Benzylethyl-	9.68	1	N-Allylmorpholine	7.05	1
Benzylmethyl-	9.58	1	N-Allylpiperidine	9.68	1
α-Benzylpyrrolidine	10.36	2	4-(2-Aminoethyl)morpholine	4.84, 9.45	77
α-Benzylpyrroline-	7.08	2	1-Benzoyl-4-methyl-piperazine	6.78*	99
1,2-Bisethylaminoethane	7.70*, 10.46*	99	1-Benzylcarbonyl-2-diethyl-aminoethane	9.40*	99
1,2-Bisfurfurylaminoethane	5.74*, 8.61*	99	1-Acetyl-2-dimethylamino-ethane	8.37*	99
1,2-Bisisopropylaminoethane	7.59*, 10.40*	99	1-Benzylcarbonyl-2-dimethyl-aminoethane	8.30*	99
1,2-Bismethylaminoethane	7.40*, 10.16*	99			
1,2-Bispropylaminoethane	7.53*, 10.27*	99	Benzyldiethyl-	9.48	1
N-Butylaminoacetic acid	2.29, 10.07	77	Benzyldimethyl-	8.93	1
t-Butylcyclohexyl-	11.23	1	Bis(2-chloroethyl)amino-ethane	6.55*	99
N-Carbethoxypiperazine	8.28	1	1-Bis(2-chloroethyl)-amino-2-methoxyethane	5.45*	99
β-Carboxymethylamino-propionic acid	3.61, 9.46	77	1,2-Bisdimethylamino-propane	5.40*, 9.49*	99
cis-2,6-Dimethylpiperidine	10.92	3	1,3-Bisdimethylamino-propane	7.7*, 9.8*	99
α-Cyclohexylpyrrolidine	10.80	2	N,N-Bis(2-hydroxyethyl)-2-aminoethanesulfonic acid	7.15	99
α-Cyclohexylpyrroline	7.95	2	N,N-Bis(2-hydroxyethyl)-glycine	8.35	99
Diallyl-	9.29	1	1-n-Butylpiperidine	10.47*	99
Di-n-butyl-	11.25	1	1-n-Butyl-2-methyl-Δ²-pyrroline	11.90	2
Diethyl-	10.98	1	1-Chloro-2-diethylamino-ethane	8.80*	99
N,N'-Diethylethylenedi-	7.70, 10.46	77	N-Chloro-N-ethylamino-ethane	1.02*	99
Di-(trimethylsilylmethyl)-	11.40	1	N-Chloro-N-methylamino-methane	0.46*	99
Di(hydroxyethyl)amine	8.88*	99			
Diisobutyl-	10.50	1	1-Cyanomethylpiperidine	4.55*	99
Diisopropyl-	11.05	1	Diallylmethyl-	8.79	1
Dimethyl-	10.64	1	Diethylaminoacetic acid	2.04, 10.47	77
N,N'-Dimethylethylenedi-	7.40, 10.16	77	1-Diethylaminobutan-(4)-	10.1	5
N,O-Dimethylhydroxyl-	4.75*	12	1-Diethylaminohexan-(6)-	10.1	5
1-Diphenylmethoxy-2-methylaminoethane	9.12*	99	1-Diethylaminohexane-thiol-(6)-	10.1	5
Di-n-propyl-	11.00	1	1-Diethylaminopropan-(3)-	8.0, 10.5	5
N,N'-Di-n-propylethylenedi-	8.14, 10.97	77	N-Diethylcysteamine	7.8, 10.75	5
Di-sec-butyl-	11.01	1	Di-(2-hydroxyethyl)amino-acetic acid	8.08	77
N-Ethylaminoacetic acid	2.30, 10.10	77			
Ethylenediamine-N,N'-diacetic acid	6.42, 9.46	77	Dimethylaminoacetic acid	2.08, 9.80	77
α-Ethylpyrrolidine	10.43	2	1-Dimethylamino-2-hydroxy-ethane	9.31*	99
α-Ethylpyrroline	7.43	2	1-Dimethylaminoprop-2-ene	8.64*	99
Iminodiacetic acid	2.98, 9.89	77	1-Dimethylaminoprop-2-yne	6.97*	99
Iminodipropionic acid	4.11, 9.61	77	Dimethyl-n-butyl-	10.02	1
N-Isopropylaminoacetic acid	2.36, 10.06	77	Dimethyl-t-butyl-	10.52	1
Methylaminocyclopentane	10.85*	99	N-Dimethylcysteamine	7.95, 10.7	5
1-Methylamino-prop-2-ene	10.11*	99	Dimethylethyl-	9.99	1
N-Methylglucamine	9.62*	99	N-Dimethylhydroxylamine	5.20*	12
N-Methylhydroxyl-	5.96*	12	Dimethylisobutyl-	9.91	1
N-Methylmethoxy-	4.75	1	Dimethylisopropyl-	10.30	1
2-Methylpiperidine	11.08*	99	N,N-Dimethylmethoxy-	3.65	1
3-Methylpiperidine	11.07*	99	1,2-Dimethylpiperidine	10.26	2
Methyltrifluoroethyl-	6.05	10			
Morpholine	8.36	1			
Piperazine	5.68, 9.82	77			
Piperidine	11.22	1			
N-n-Propylaminoacetic acid	2.28, 10.03	77			
1-Propylaminopropane	11.00	11			
Pyrrolidine	11.27	1			
α-(p-Tolyl)-pyrrolidine	10.01	2			
α-(p-Tolyl)pyrroline	7.59	2			
1-Tosylpiperazine	7.39	3			
Trimethyleneimine	11.29	1			
N-Tris(hydroxymethyl)-methylglycine	8.15	108			

* Thermodynamic value.

NITROGEN COMPOUNDS—(continued)

Compound	pK_a'	Reference
Tertiary Amines—(Continued)		
Dimethyl-n-propyl	9.99	1
1,2-Dimethylpyrrolidine	10.26	2
1,2-Dimethyl-Δ^2-pyrroline	11.94	2
Dimethyl-sec-butyl-	10.40	1
1,2-Dimethyl-Δ^2-tetrahydro-pyridine	11.57	2
Dimethyltrifluoroethyl-	4.75	10
1-Dipropylaminopropane	10.26*	99
N-Dipropylcysteamine	8.00, 10.8	5
1-Ethoxycarbonyl-4-methyl-piperazine	7.31*	99
N-Ethyl-cis-2,3-iminobutane	8.56	7
N-Ethyl-1,2-iminobutane	8.18	7
1-Ethyl-2-methylpiperidine	10.70	2
1-Ethyl-2-methyl-Δ^2-pyrroline	11.92	2
1-Ethyl-2-methylpyrrolidine	10.64	2
1-Ethyl-2-methyl-Δ^2-tetra-hydropyridine	11.57	2
N-Ethylmorpholine	7.70	1
N-Ethylpiperidine	10.40	1
N-Ethyl-trans-2,3-imino-butane	9.47	7
Hexamethylenetetra-	5.13	77
N-2-Hydroxyethylimino-diacetic acid	2.2, 8.73	77
N-2-Hydroxyethylpiperazine-N'-2-ethanesulfonic acid	7.55	99
1-Hydroxy-2-(2-hydroxy-ethylmethyl)-aminoethane	8.52*	99
1,2-Iminoethane	7.93	7
1-Methyl-2-n-butyl-pyrrolidine	10.24	2
1-Methyl-2-n-butyl-Δ^2-pyrroline	11.90	2
N-β-Mercaptoethyl-morpholine	6.65, 9.8	5
N-β-Mercaptoethyl-piperidine	7.95, 11.05	5
Methyl-β-diethylaminoethyl-sulfide	9.8	5
Methyldiethyl-	10.29	1
Methyliminodiacetic acid	2.81, 10.18	77
1-Methyl-4-nitrosopiperazine	5.93*	99
N-Methylmorpholine	7.41	1
N-Methylpiperidine	10.08	1
N-Methylpyrrolidine	10.46	1
N-Methyltrimethyleneimine	10.40	1
2-(N-Morpholino)ethane-sulfonic acid	6.15	99
Piperazine-N,N'-bis(2-ethanesulfonic acid	6.8	99
Propargyldimethyl-	7.05	1
Propargylethyldimethyl-	8.88	1
Propargylmethyldimethyl-	8.33	1
1-n-Propylpiperidine	10.48	2
Triallyl-	8.31	1
Tri-n-butyl-	10.89	1
Triethanol-	7.77	1
Triethyl-	10.65	1
Triethylenedi-	4.18, 8.19, 2.97, 8.82	77, 116
Trimethyl-	9.76	1
Trimethylhydroxylamine	3.65	12
Tri-n-propyl-	10.65	1

* Thermodynamic value.

Compound	pK_a'	Reference
Tertiary Amines—(Continued)		
N-Tris(hydroxymethyl)-methyl-2-aminoethane-sulfonic acid	7.5	99
$(CH_3)_2NCH_3$	9.76	119
$(CH_3CH_2)_2NCH_3$	10.29	119
$\underbrace{(CH_2)_2}NCH_3$	7.86	119
$\underbrace{(CH_2)_3}NCH_3$	10.40	119
$\underbrace{(CH_2)_4}NCH_3$	10.46	119
$\underbrace{(CH_2)_5}NCH_3$	10.08	119

Benzylamines, Monosubstituted	2	3	4	99
Chloro-	5.20*	—	—	—
Methoxy-	9.70*	9.15*	9.47*	—
Methyl-	9.19*	9.33*	9.36*	—
Sulphamoyl-	8.53*	8.55*	8.52*	—
	10.11*	10.14*	10.08*	—

Compound	pK_a	Reference
Cyanoamines		
N-Piperidine-CH_2CN		8
Et_2NCN	−2.0	8
$Et_2N(CH_2)_2CN$	7.65	8
$Et_2N(CH_2)_4CN$	10.08	8
$Et_2NC(CH_3)_2CN$	9.13	8
$EtN(CH_2CN)_2$	−0.6	8
$EtN(CH_2CH_2CN)_2$	4.55	8
H_2NCH_2CN	5.34	8
N-Amphetamine-$(CH_2)_2$-CN	7.23	8
N-Norcodeine-$(CH_2)_2CN$	5.68	8
Dimethylcyanimide	1.2	9
Diethylcyanimide	1.2	9
Aminoacetonitrile	5.3	9
Diethylaminoacetonitrile	4.5	9
2-Amino-2-cyanopropane	5.3	9
β-Isopropylamino-propionitrile	8.0	9
β-Diethylaminopropionitrile	7.6	9
1-Amino-2-cyanoethane	7.7*	99
Et_2NCH_2CN	4.55	8
$Et_2N(CH_2)_3CN$	9.29	8
$Et_2N(CH_2)_5CN$	10.46	8
$HN(CH_2CN)_2$	0.2	8
$HN(CH_2CH_2CN)_2$	5.26	8
$N(CH_2CH_2CN)_3$	1.1	8
N-Piperidine-$C(CH_3)_2CN$	9.22	8
N-Methamphetamine-$(CH_2)_2CN$	6.95	8
Methyl cyanamide	1.2	9
Ethyl cyanamide	1.2	9
Cyanamide	1.1	9
Dimethylaminoacetonitrile	4.2	9
β-Aminopropionitrile	7.7	9
β-Dimethylamino-propionitrile	7.0	9
β,β''-Dicyanodiethylamine	5.2	9
Cyclic Amines		
1,2-Iminoethane	7.98	7
cis-2,3-Iminobutane	8.72	7
1,2-Imino-2-methylpropane	8.61	7
1,2-Iminobutane	8.29	7
trans-2,3-Iminobutane	8.69	7

NITROGEN COMPOUNDS—(continued)

Compound	pK_a'	Reference
Cyclic Amines—(Continued)		
In 80 per cent methyl cellosolve:		
Pentamethylene-	9.99	2
Hexamethylene-	10.00	2
Heptamethylene-	9.77	2
Octamethylene	9.39	2
Nonamethylene-	9.14	2
Decamethylene-	9.04	2
Undecamethylene-	9.31	2
Dodecamethylene-	9.31	2
Tridecamethylene-	9.35	2
Tetradecamethylene-	9.35	2
Hexadecamethylene-	9.29	2
Heptadecamethylene-	9.27	2
Cyclohexyl-	9.82	2
Cycloheptyl-	9.99	2
Cycloöctyl-	10.01	2
Cyclononyl-	9.95	2
Cyclodecyl-	9.85	2
Cycloundecyl-	9.71	2
Cyclododecyl-	9.62	2
Cyclotridecyl-	9.63	2
Cyclotetradecyl-	9.54	2
Cyclopentadecyl-	9.54	2
Cycloheptadecyl-	9.57	2
Cyclooctadecyl-	9.54	2

Phenylethylamines

2-Phenylethylamine	9.78	11
N-Methyl-2-(3,4 dihydroxy-phenyl)-ethylamine	8.78	11
N-Methyl-2-phenyl-	10.31	11
Epinephrine	8.55	11
Arterenol	8.55	11

$$R_2 \underset{R_3}{\overset{R_1}{\diagup}} CHCH_2NHR_4 \qquad 11$$

R_1	R_2	R_3	R_4	pK_a'
H	H	H	H	9.78
H	H	OH	H	8.90
H	OH	OH	H	8.81
OH	H	OH	H	8.67
H	OH	H	H	9.22
OH	OH	H	H	8.93
OH	OH	OH	H	8.58
H	H	H	CH$_3$	10.31
H	H	OH	CH$_3$	9.31
H	OH	OH	CH$_3$	8.62
OH	H	OH	CH$_3$	8.89
H	OH	H	CH$_3$	9.36
OH	OH	H	CH$_3$	8.78
OH	OH	OH	CH$_3$	8.55

Compound	pK_a'	Reference
ALKALOIDS AND DERIVATIVES		
Acetylscopolamine	7.35*	99
Aconitine	8.35*	99
Alypine	3.8*, 9.5*	99

Compound	pK_a'	Reference
Alkaloids and Derivatives—(Continued)		
Anhydroplatynecine	9.40*	99
Apomorphine	7.20*, 8.92*	99
Aposcopolamine	7.72*	99
Arecaidine	9.07*	99
Arecaidine methyl ester	7.64*	99
Arecoline	7.41*	99
Aspidospermine	7.63*	99
Atropine	9.85*	99
Benzoylecgonine	11.80*	99
Benzoylecgonine methyl ester	8.74*	99
N-Benzyltriacanthine	5.94*	99
Berberine	11.73*	99
Brucine	2.50*, 8.16*	99
N-Butylveratramine	7.20*	99
Cevadine	9.05*	99
Cinchonidine	4.17*, 8.40*	99
Cinchonine	4.28*, 8.35*	99
Cocaine	8.39*	99
Codeine	8.21*	99
Colchicine	1.85*	99
Cupreine	7.63*	99
Cytisine	1.20*, 8.12*	99
Deacetylaspidospermine	2.70*, 8.45*	99
Desoxyretronecine	9.51*	99
Dicodide	7.95*	99
Dihydroarecaidine	9.70*	99
Dihydroarecaidine, methyl ester	8.39*	99
Dihydrocodeine	8.75	99
α-Dihydrolysergic acid	3.57*, 8.45*	99
γ-Dihydrolysergic acid	3.60*, 8.71*	99
Dihydromorphine	9.35*	99
Dihydroergonovine	7.38*	99
α-Dihydrolysergol	8.30*	99
β-Dihydrolysergol	8.23*	99
Dihydronicotyrine	7.07*	99
Dilaudide	7.8*	99
Ecgonine	10.91	99
Ecgonine methyl ester	9.16*	99
Emetine	7.56*, 8.43*	99
Ergobasine	6.79*	99
Ergobasinine	7.43*	99
Ergometrinine	7.32*	99
Ergonovine	6.73*	99
Ethylmorphine	8.08*	99
N-Ethylveratramine	7.40*	99
β-Eucaine	9.35*	99
Eucodal	8.6*	99
Gelsemine	9.79*	99
Harmine	7.61*	99
Harmol	7.86*, 9.51*	99
Heliotridane	11.40*	99
Heliotridene	10.55*	99
Heroin	7.6*	99
Homatropine	9.7*	99
Hydrastine	6.63*	99
Hydrastinine	11.58*	99
Hydroquinine	8.87*	99
10-Hydroxycodeine	7.12*	99
Hyoscyamine	9.68*	99
Isolysergic acid	3.33*, 8.46*	99
Isopilocarpine	7.18*	99
Isoretronecanol	10.83*	99
Lysergic acid	3.32*, 7.82*	99

* Thermodynamic value.

NITROGEN COMPOUNDS—(continued)

Compound	pK_a'	Reference
Alkaloids and Derivatives—(Continued)		
N-Methyl-1-benzoyl-lecgonine	8.65*	99
6-Methyl ergoline	8.87*	99
Morphine	8.07*, 9.85*	99
Myosmine	5.26*	99
Narceine	3.5*, 9.3*	99
Narcotine	5.86*	99
Nicotine	3.13*, 8.02*	99
Nicotine dimethohydroxide	7.88*, 10.23*	99
Nicotine isomethohydroxide	5.35*, 11.72*	99
Nicotine methohydroxide	8.54*, 12.04*	99
Nicotine monomethobromide	3.09*	99
Nicotine oxide	5.00*	99
Nicotyrine	4.76*	99
Norcodeine	5.68*	99
Norcurarine	8.5*	99
Norhyoscyamine	10.28*	99
Nornicotyrine	4.35*	99
Optochine	4.05*, 8.5*	99
Papaverine	6.40*	99
Pelletierine	9.45*	99
N-Pentylveratramine	7.28*	99
Perlolidine	4.01*, 11.39*	99
Perloline	8.54*	99
Physostigmine	1.96*, 8.08*	99
Pilocarpate	7.47*	99
Pilocarpine	1.63*, 7.05*	99
Piperine	1.98*	99
Platynecine	10.20*	99
N-Propylveratramine	7.20*	99
Pseudoecgonine	9.70*	99
Pseudoecgonine methyl ester	8.15*	99
Pseudotropine	9.86*	99
Quinidine	4.2*, 8.77*	99
Quinine	4.32*, 8.4*	99
Retronecanol	10.88*	99
Retronecine	8.88*	99
Scopolamine	7.55*	99
Scopoline	8.20*	99
Sempervirine	10.6*	99
Solanine	7.54*	99
Sparteine	4.80*, 11.96*	99
Stovaine	7.9*	99
Strychnine	8.26*	99
Tetradehydroyohimbine	10.69*	99
Tetrahydro-α-morphimethine	8.65*	99
Tetrahydroserpentine	10.55*	99
Thebaine	8.15*	99
Theobromine	10.00*	99
Theophylline	8.6*	99
Triacanthine	6.0*	99
Tropacocaine	9.88*	99
Tropine	10.33*	99
Veratramine	7.49*	99
Yohimbine	3*, 7.45*	99

Monosubstituted substituent	ortho	meta	para
ANILINES (2, 99)			
H-	4.62*	4.64*	4.58*
(CH$_3$)$_3$N$^+$-	—	2.26	2.51
CH$_3$O$_2$C-	2.16	3.56	2.30
CH$_3$SO$_2$-	—	2.68*	1.48

* Thermodynamic value.

Monosubstituted substituent	ortho	meta	para
Anilines—(Continued)			
CH$_3$S-	—	4.05	4.40
Br-	2.60*	3.51*	3.91*
F-	2.96*	3.38*	4.52*
CH$_3$O-	4.49*	4.20*	5.29*
C$_6$H$_5$-	3.78*	4.18	4.27*
(CH$_3$)$_3$C-	3.78		
$^-$O$_3$S-	—	3.80	3.32
H$_3$N$^+$-	1.3	2.65	3.29
O$_2$N-	−0.28*	2.45*	0.98*, 1.11*
HO$_2$C-	2.04	3.05	2.32
C$_2$H$_5$O$_2$C-	2.10		2.38
F$_3$C-	—	3.49*	2.57*
HO-	4.72	4.17	5.50
Cl-	2.62*	3.32*	3.81*
(CH$_3$)$_3$Si-		4.64*	4.36*
C$_2$H$_5$O-	4.47*	4.17*	5.25*
CH$_3$-	4.38*	4.67*	5.07*
$^-$HO$_3$As-	3.77	4.05	4.05
H$_2$N-	4.47	4.88	6.08
CH$_3$CO-	2.22*	3.59*	2.19*
CN-	0.95*	2.75	1.74
C$_2$H$_5$-	4.37*	4.70*	—
C$_6$H$_5$CO-	—	—	2.17*
n-Butyl	4.26*	—	—
t-Butyl	5.03*	4.66*	4.95*
HCO-	—	—	1.76*
I-	2.60*	3.61*	3.78*
Isopropyl-	4.42*	4.67*	—
HS-	3.00*	—	—
	6.59*	—	—
Sulphamoyl	1.0*	2.90*	2.02*

Compound	pK_a'	Reference
3-Amino-2,6-dihydroxy-aniline	2.9*, 5.6*, 9.3*, 11.5*	99
3-Amino-4,6-dihydroxy-aniline	3.8*, 6.0*, 9.8*, 12.0*	99
3-Amino-2-hydroxyaniline	2.7*, 5.5*, 10.5*	99
3-Amino-4-hydroxyaniline	3.1*, 5.7*, 10.5*	99
4-Bromo-2,6-dimethylaniline	3.54*	99
3-Bromo-4-methylaniline	3.98*	99
4-Bromo-2-methylaniline	3.58*	99
3,5-di-t-Butylaniline	4.97	88
3-Chloro-5-methoxyaniline	3.10	88
3-Chloro-4-methylaniline	4.05*	88
2,4-Diaminoaniline	3.7*, 6.1*	99
2,4-Dibromoaniline	2.3*	99
2,6-Dibromoaniline	0.38*	99
3,5-Dibromoaniline	2.34*	99
2,4-Dichloroaniline	2.05*	99
2,5-Dichloroaniline	1.57*	99
3,5-Dichloroaniline	**2.37**	138
2,4-Dihydroxyaniline	5.7*, 9.3* 11.3*	99
2,6-Dihydroxyaniline	5.1*, 9.3*, 11.6*	99
3,5-Diiodoaniline	**2.37**	138
3,5-Dimethoxyaniline	3.82	88
3-Methoxy-5-nitroaniline	2.11	88
2,3-Dimethylaniline	4.70*	99
2,4-Dimethylaniline	4.84*	99
2,5-Dimethylaniline	4.57*	99

NITROGEN COMPOUNDS—(continued)

Compound	pK_a'	Reference	Compound	pK_a'	Reference
Anilines—(Continued)			**Anilines—(Continued)**		
2,6-Dimethylaniline	3.89*	99	3,5-Dinitroaniline	0.23	138
3,4-Dimethylaniline	5.17*	99	2,3,5,6-Tetramethylaniline	4.30*	99
3,5-Dimethylaniline	4.91*	99	2,4,6-Trimethylaniline	4.38*	99
2,4-Dinitroaniline	−4.27	120	3,4,5-Trimethylaniline	5.12*	99

	pK_1	pK_2	Ref.
Anilines (continued)			
3-Amino-5-nitrobenzoic acid	1.55	3.55	147
methyl ester	1.47	–	147
ethyl ester	1.52	–	147
Ethyl *N*-(*m*-carboxyphenyl)glycinate	1.15	4.30	147
Ethyl *N*-(*m*-methoxycarbonylphenyl)-glycinate	1.06	–	147
Ethyl *N*(*m*-ethoxycarbonylphenyl)gly-cinate	1.11	–	147
Ethyl *N*-phenylglycinate	2.08	–	147
Methyl *N*-phenylglycinate	2.07	–	147
Ethyl *N*-(*p*-carboxyphenyl)glycinate	–	4.88	147

R	C_6H_5NHR	$C_6H_5N(CH_3)R$	$C_6H_5NR_2$	$2\text{-}CH_3C_6H_4NHR$	$2\text{-}CH_3C_6H_4NR_2$
N-Substituted Anilines[2]					
H-	4.58	4.85	4.58	4.39	4.39
CH$_3$-	4.85	5.06	5.06	4.59	5.86
C$_2$H$_5$-	5.11	5.98	6.56	4.92	7.18
n-C$_3$H$_7$-	5.02	—	5.59	—	—
n-C$_4$H$_9$-	4.95	—	~5.7	—	—
i-C$_4$H$_9$-	—	5.20	—	—	—
sec-C$_4$H$_9$-	—	6.03	—	—	—
t-C$_6$H$_{12}$-	6.30	—	—	—	—
Cyclopentyl-	5.30	6.71	—	5.07	—
Cyclohexyl-	5.60	6.35	—	5.34	—
t-C$_4$H$_9$	6.95	7.52	—	6.49	—

	$4\text{-}NO_2C_6H_4R$	$4\text{-}HOOCC_6H_4R$	C_6H_5R
N-Substituted Anilines[119]			
R=			
(CH$_3$)$_2$N-	0.65	1.40	4.22
			4.39
CH$_3$CH$_2$N-	1.75	2.45	5.71
			5.85
			5.59
(CH$_2$)$_3$-N-	0.34		4.08
(CH$_2$)$_4$-N-	−0.42	0.39	3.71
			3.45
			3.24
(CH$_2$)$_5$-N-	2.46	2.67	4.60
			5.22
			4.93
(CH$_2$)$_6$-N-	−0.15		

* Thermodynamic value.

NITROGEN COMPOUNDS—(continued)

Compound	pK_a'	Reference
N,N-Diethyl-		
2,4-Dinitro-	0.18*	99
2-Methyl-	7.23*	99
3-Methyl-	7.12*	99
4-Methyl-	7.13*	99
4-Nitroso-	4.11*	99
N,N-Dimethyl-		
H-	5.07	52
m-NO$_2$-	2.63	52
m-CN-	2.97	52
p-NO$_2$-	0.61	52
p-CN-	1.78	52
p-NO-	4.54	52
N-Dimethyl-, in 50% Ethanol		
H-	4.21, 4.09	2
m-CH$_3$-	4.66	2
p-C$_2$H$_5$-	4.69	2
o-(CH$_3$)$_2$CH-	5.05	2
p-CH$_3$CH$_2$CH$_2$CH$_2$-	4.62	2
o-(CH$_3$)$_3$C-	4.26	2
p-I-	3.43, 2.73	2
p-Br-	3.52, 2.82	2
p-Cl-	3.33	2
m-(CH$_3$)$_3$Si-	4.41	2
o-CH$_3$O-	5.49	2
o-CH$_3$-	5.15, 5.07	2
p-CH$_3$-	4.94	2
p-CH$_3$CH$_2$CH$_2$-	4.43	2
p-(CH$_3$)$_2$CH-	4.77	2
p-(CH$_3$)$_2$CHCH$_2$-	4.19	2
p-(CH$_3$)$_3$C-	4.65	2
m-Br-	3.08	2
m-Cl-	3.09	2
p-F-	4.01	2
p-(CH$_3$)$_3$Si-	3.99	2
p-CH$_3$O-	5.14, 5.16	2
N-Methyl		
4-Chloro-	3.9*	99
4-Chloro, 2-nitro-	−1.49*	99
2-Methyl-	4.62*	99
3-Methyl-	5.00*	99
4-Methyl-	5.36*	99
2-Methoxycarbonyl-	3.53*	99
4-Methoxycarbonyl-	2.32*	99
Ortho-substituted, in 50% Ethanol		
H-	4.25	2
2-CH$_3$-	3.98, 4.09	2
2,3-(CH$_3$)$_2$-	4.42	2
2,4-(CH$_3$)$_2$-	4.61	2
2,5-(CH$_3$)$_2$-	4.17, 4.23	2
2,6-(CH$_3$)$_2$-	3.42, 3.49	2
3,5-(CH$_3$)$_2$-	4.48	2
2-CH$_3$-	4.09	2

Compound	pK_a'	Reference
Ortho-substituted, in 50% Ethanol (continued)		
2-(CH$_3$)$_2$CH-	4.06	2
2-(CH$_3$)$_3$C-	3.38	2
2,6-(CH$_3$)$_2$-4-(CH$_3$)$_3$C-	3.88	2
2,4-(CH$_3$)$_2$-6-(CH$_3$)$_3$C-	3.43	2
2-CH$_3$-4,6-(CH$_3$)$_3$C-	3.31	2
2,4,6-[(CH$_3$)$_3$C]$_3$-	< 2	2
2-Nitroaniline		
R=H-	−0.29	120
4-CH$_3$O-	0.77	120
4-CH$_3$-	0.43	120
4-F-	−0.44	120
4-Cl-	−1.03	120
4-Br-	−1.05	120
4-CF$_3$-	−2.25	120
4-CH$_3$OCO-	−2.61	120
4-NO$_2$-	−4.27	120
4-CH$_3$CO-	−2.85	120
4-HO-	1.20	120
3-CH$_3$-	−0.09	120
3-CH$_3$O-	−0.72	120
3-Cl-	−1.48	120
3-Br-	−1.48	120
3-NO$_2$-	−2.49	120
3-HO-	−0.55	120
6-Nitroaniline		
R=2-Cl-	−2.41	120
2-NO$_2$-	−5.56	120
2,4-Cl$_2$-	−3.16	120
4-CH$_3$-2-NO$_2$-	−4.45	120
2,4-(NO$_2$)$_2$-	−10.23	120
Other aniline derivatives, in 50% Ethanol		
Unhindered aniline	4.19	40
p-Aminodiphenyl	3.81	40
2-Naphthylamine	3.77	40
3-Phenanthrylamine	3.59	40
Hindered o-aminodiphenyl	3.03	40
peri		
1-Naphthylamine	3.40	40
9-Phenanthrylamine	3.19	40
3-Aminopyrene	2.91	40
meso		
9-Anthrylamine	2.7	40
m-Aminodiphenyl	3.82	40
2-Aminofluorene	4.21	40
2-Phenanthrylamine	3.60	40
2-Anthrylamine	3.40	40
1-Phenanthrylamine	3.23	40
1-Anthrylamine	3.22	40

* Thermodynamic value.

NITROGEN COMPOUNDS—(continued)

Compound		pK'_a	Reference
o-Aminophenols			
3-Hydroxyanthranilic acid		10.09, 5.20	51
2-Aminophenol hydrochloride		9.99, 4.86	51
2-Amino-4,5-dimethylphenolhydrochloride		10.40, 5.28	51
Indicators			
p-Aminoazobenzene		2.82, 2.76	
4-Chloro-2-nitroaniline		−1.02, −1.03	60
4:6 Dichloro-2-nitroaniline		−3.61, −3.32	60
6-Bromo-2,4-dinitroaniline		−6.64, −6.71	60
N,N-Dimethyl-2,4-dinitroaniline		−1.00	60
p-Nitrodiphenylamine		−2.4 to −2.9, −2.50	60
4-Methyl-2,6-dinitroaniline		−3.96, −4.44	60
Substituted Naphthylamines			
1-NH$_2$-		3.92*	2
1-NH$_2$-2-NO$_2$-		−1.6	2
1-NH$_2$-3-NO$_2$-		2.22	2
1-NH$_2$-4-NO$_2$-		0.54	2
1-NH$_2$-5-NO$_2$-		2.80	2
1-NH$_2$-6-NO$_2$-		3.15	2
1-NH$_2$-7-NO$_2$-		2.83	2
1-NH$_2$-8-NO$_2$-		2.79	2
1-NH$_2$-2-SO$_3^-$-		1.71	2
1-NH$_2$-3-SO$_3^-$-		3.20*	2
1-NH$_2$-4-SO$_3^-$-		2.81*	2
1-NH$_2$-5-SO$_3^-$-		3.69*	2
1-NH$_2$-6-SO$_3^-$-		3.80*	2
1-NH$_2$-7-SO$_3^-$-		3.66	2
1-NH$_2$-8-SO$_3^-$-		5.03*	2
2-NH$_2$-		4.11*	2
2-NH$_2$-1-NO$_2$-		−1.0	2
2-NH$_2$-3-NO$_2$-		2.93	2
2-NH$_2$-4-NO$_2$-		2.63	2
2-NH$_2$-5-NO$_2$-		3.16	2
2-NH$_2$-6-NO$_2$-		2.75	2
2-NH$_2$-7-NO$_2$-		3.13	2
2-NH$_2$-8-NO$_2$-		2.86	2
2-NH$_2$-1-SO$_3^-$-		2.35	2
2-NH$_2$-4-SO$_3^-$-		3.70	2
2-NH$_2$-5-SO$_3^-$-		3.96*	2
2-NH$_2$-6-SO$_3^-$-		3.74*	2
2-NH$_2$-7-SO$_3^-$-		3.95*	2
2-NH$_2$-8-SO$_3^-$-		3.89*	2
2-Naphthylamine X		4.16	88
1-NH$_2$,3-X	NO$_2$	2.07	88
	CN	2.26	88
	Cl	2.66	88
	Br	2.67	88
	I	2.82	88
	COOCH$_3$	3.12	88
	OCH$_3$	3.26	88
	OH	3.30	88
	CH$_3$	3.96	88
2-NH$_2$,5-X	Cl	2.71	88
	NO$_2$	3.01	88
	OH	4.07	88
1-NH$_2$,5-X	NO$_2$	2.73	88
	OH	3.96	88
	Cl	3.34	88
	NH$_2$	4.21	88

Compound		pK'_a	Reference
Substituted Naphthylamines—(Continued)			
1-NH$_2$,7-X	NO$_2$	2.55	88
	Cl	3.48	88
	OCH$_3$	4.07	88
	OH	4.20	88
1-NH$_2$,2-X	NO$_2$	−1.74	88
1-X,2-NH$_2$	NO$_2$	−0.85	88
1-NH$_2$,8-X	NO$_2$	2.79	88
2-NH$_2$,4-X	NO$_2$	2.43	88
	CN	2.66	88
	Cl	3.38	88
	Br	3.40	88
	I	3.41	88
	COOCH$_3$	3.38	88
	OCH$_3$	4.05	88
1-NH$_2$,6-X	NO$_2$	2.89	88
	Cl	3.48	88
	OCH$_3$	3.90	88
	OH	3.97	88
2-NH$_2$,7-X	NO$_2$	3.10	88
	Cl	3.71	88
	OCH$_3$	4.19	88
	OH	4.25	88
	NH$_2$	4.66	88
2-NH$_2$,6-X	NO$_2$	2.62	88
	OCH$_3$	4.64	88
2-NH$_2$,8-X	NO$_2$	2.73	88
1-NH$_2$,4-X	NO$_2$	0.54	88
	Br	3.21	88
2-NH$_2$,3-X	NO$_2$	1.48	88

HETEROCYCLIC COMPOUNDS

Nucleosides, Nucleotides, and Related Compounds

Compound	pK'_a	Reference
Adenine	4.15, 9.80	6
Adenine deoxyriboside-5'-phosphoric acid	4.4, 6.4	99
Adenosine	3.63, 12.5	6, 99
1-Oxide	2.25, 12.86	99
ADP	3.95, 6.3, (7.20*)	36, 113
5-Amino-1(β-D-ribosyluronic acid)-uracil	3.06	99
2'-AMP	3.81, 6.17	6
3'-AMP	3.74, 5.92	6
5'-AMP	3.74, 6.2–6.4	6
5-Aminouridine	3.11	99
1-D-Arabinosyl-5-methyl-cytosine	4.1	99
ATP	4.00 (4.1), 6.5 (7.68*)	36, 113
Barbital	7.85, 12.7	37
Barbituric acid	3.9, 12.5	37
N-n-Butyl-5-fluoro-2'-deoxycytidine	2.21	99
CDP	4.44, (7.18*)	6, 113
CDP (deoxy)	4.8, 6.6	6
2'-CMP	4.3–4.4, 6.19*	6
3'-CMP	4.16–4.31, 6.04	6
5'-CMP	4.5, 6.3	6
CTP	4.6, 6.4, (7.65*)	6, 113
Cytidine	4.22, 12.5	35
Cytosine	4.45, 12.2	6

* Thermodynamic value.

Compound	pK_a'	Reference
Nucleosides, Nucleotides and Related Compounds— (Continued)		
Cytosine (deoxy)	4.25, ~13	6, 101
Deoxycytidine-5'-phosphoric acid	4.6, 6.6	99
2,6-Diaminopurine	5.09, 10.77	6
N,N-Dimethylcytidine	3.58	99
N,N-Dimethyl-2'-deoxycytidine	3.75	99
N-Ethyl-5-fluoro-2'-deoxycytidine	2.21	99
5-Fluorocytidine	2.22	99
5-Fluoro-2'-deoxycytidine	2.39	99
5-Fluoro-N,N-dimethyl-2'-deoxycytidine	1.89	99
5-Fluoro-N-methyl-2'-deoxycytidine	2.14	99
GDP	2.9, 9.6, 6.3, (7.19*)	6, 113
1-D-Glucopyranosylcytosine	3.78	99
GMP (2' + 3')	2.3, 9.36, 0.7, 5.9	6
5'-GMP	2.4, 9.4, 6.1	6
5'-GMP (deoxy)	2.9, 9.7, 6.4	6
GTP	3.3, 9.3, 6.5, (7.65*)	6, 113
Guanine	3.3, 9.2, 12.3	6
Guanine deoxyriboside-3'-phosphoric acid	2.9, 6.4, 9.7	99
Guanosine	1.6, 9.16, 12.5	35
Guanosine-3'-phosphoric acid	0.7, 2.3, 5.92, 9.38	99
Guanosine (deoxy)	1.6–2.2, 9.16–9.5	6
Hypoxanthine	1.98, 8.94, 12.10	6
5'-IMP	8.9, 1.54, 6.04	6
Inosine	1.2, 8.75, 12.5	6, 35
ITP	7.68*	113
IDP	7.18*	113

Heterocyclic Bases

Compound	pK_a'	Reference
Nucleosides, Nucleotides and Related Compounds— (Continued)		
N-Methylcytidine	3.88	99
1-Methylcytidine	8.7	99
5-Methylcytidine	4.21	99
5-Methylcytidylic acid	4.4	99
5-Methylcytosine	4.6, 12.4	6
5-Methylcytosine deoxyriboside	4.5, 13.0	6
5-Methylcytosine deoxyriboside 5'-phosphate	4.4, 13	6, 100
N-Methyl-2'-deoxycytidine	3.97	99
5-Methyl-2'-deoxycytidine	4.33	99
1-Methyluracil	9.95	37
3-Methyluracil	9.75	37
1-Methylxanthine	7.7, 12.05	38
3-Methylxanthine	8.5, (8.1), 11.3	38
7-Methylxanthine	8.5, (8.3)	38
9-Methylxanthine	6.3	38
Orotic acid	2.8, 9.45, 13	6
Purine	2.52, 8.90	37
Pyrimidine	1.30	37
Thymidine	9.8	6
Thymine	0, 9.9, > 13.0	6
5'-TMP	10.0, 1.6, 6.5	6
UDP	9.4, 6.5, (7.16*)	6, 113
UMP (2' + 3')	9.43, 1.02, 5.88	6
5'-UMP	9.5, 6.4	6
Uracil	0.5, 9.5, 13.0	6
Uracil deoxyriboside	9.3, >13	6, 101
Uric acid	5.4, 10.3	6
Uridine	9.17, 12.5	35
UTP	9.5, 6.6, (7.58*)	6, 113
Xanthine	0.8, 7.44, 11.12	6
Xanthosine	0, 5.5, 13.0	6

Pyridine 5.14*

Quinoline 4.85*

Isoquinoline 5.14*

Benzoquinoline 5.05*

Acridine 5.60

5,6-Benzoquinoline 5.15*

7,8-Benzoquinoline 4.25*

Phenanthridine 3.30[a]

2,3-Benzacridine 4.52[a]

3,4-Benzacridine 4.70*

1,2-Benzacridine 3.45[a]

Pyridazine 2.10*

* Thermodynamic value.
[a] In 50% EtOH.

NITROGEN COMPOUNDS–(continued)

Heterocyclic Bases—(Continued)

Pyrimidine 1.10*

Pyrazine 0.37*

Cinnoline 2.64*

Phthalazine 3.39*

Quinazoline 3.31*

Quinoxaline 0.6*

Acridine	1-	2-	3-	4-	5-	9-
H-	5.60*	4.11[a]	–	–	–	–
H₂N-	4.40*	8.04*	5.88*	6.04*	9.99*	9.95*
	3.59[a]	7.61[a]	5.03[a]	5.50[a]	9.45[a]	–
HO-	4.18[a]	4.86[a]	5.52	4.45[a]	-0.32	–
	10.7[a]	9.9[a]	8.81	9.4*	>12	–
CH₃-	3.95[a]	–	4.60[a]	–	4.70[a]	–
H₂N-(1-CH₃-)-	–	–	–	4.79[a]	9.73[a]	3.22[a]
1,9-(CH₃)₂-	2.88[a]	–	–	–	–	–
COOH-	–	5.26*	–	7.76*	–	5.0*

Reference 3, 39, 99

[a]In 50% ethanol.

Compound	pK'_a	Reference
Heterocyclics—(Continued)		
Acridine	5.62	39
3-Amino-7-carboxy-	2.3*, 9.0*	99
3-Amino-6-chloro-	7.22*	99
3-Amino-7-chloro-	6.91*	99
9-Amino-1-hydroxy-	5.57*	99
9-Amino-2-hydroxy-	7.67*	99
9-Amino-3-hydroxy-	6.59*	99
9-Amino-4-hydroxy-	7.01*	99
3-Amino-7-sulpho-	7.6*	99
2,7-Diamino-	6.14*	99
3,6-Diamino-	9.65*	99
3,7-Diamino-	8.11	99
3,9-Diamino-	8.11	99
5-Methoxy-	7	39
9-Methoxy-	-0.32*	99
2-Sulpho-	4.78*	99
Aureomycin	3.30, 7.44, 9.27	77
Azacycloundecane		
1-Methyl-6-hydroxy-7-oxo-	9.1*	99
Azaindole		
4-	6.9	154
5-	8.3	154
6-	8.0	154
7-	4.6	154

Compound	pK'_a	Reference
Heterocyclics—(Continued)		
3,4-Di-	4.0, 11.1	154
3,5-Di-	6.1, 10.9	154
2,5,7-Tri-	2.8, 9.5	154
4-Amino-2,5,7-tri-	4.6, 10.8	154
Azepine		
Hexahydro-	11.07*	99
Azetidine	11.29*	99
Aziridine	8.01*	99
1:2-Benzacridine	3.45[a]	19
5-Amino-	8.13[a]	19
7-Amino-	4.05[a]	19
8-Amino-	6.72, 5.97[a]	19
4':5-Diamino-	8.44[a]	19
2:3-Benzacridine	4.52[a]	19
5-Acetamido-	4.56[a]	19
5-Amino-	9.72[a]	19
7-Amino-	5.38[a]	19
5-Amino-6,7,8,9-tetrahydro-	9.66[a]	19
3:4-Benzacridine	4.70, 4.16[a]	19
8-Acetamido-	4.48[a]	19
5-Amino-	8.41[a]	19
7-Amino-	5.03[a]	19
8-Amino-	7.42, (651)[a]	19
8-Dimethylamino-	7.31, 6.99	19
Benzimidazole	5.4, 12.78	43, 86, 99
		107
2-Amino-	7.51*	—
5-Amino-	3.04*, 6.07*	—
6-Amino-	3.0, 6.0	—
2-Aminomethyl-	7.69*, 3.46*	—
1-α-D-Arabopyranosyl-	4.19*	—
1-α-L-Arabopyranosyl-	4.06*	—
1-α-L-Arabopyranosyl-5,6-dimethyl-	4.56*	—
1-α-L-Arabopyranosyl-5-methyl-	4.30	—

*Thermodynamic value.

[a]In 50% alcohol.

NITROGEN COMPOUNDS—(continued)

Heterocyclics—(Continued)

Compound	pK'_a	Reference
1-Ethyl-	5.59*	—
2-Ethyl-	6.18*	—
1-β-D-Glucopyranosyl-	3.97*	—
1-β-D-Glucopyranosyl-5,6-dimethyl-	4.60*	—
1-β-D-Glucopyranosyl-5-methyl-	4.29*	—
4-Hydroxy-	5.3, 9.5	—
4-Hydroxy-6-amino-	5.9	—
1-Hydroxymethyl-	5.41*	—
2-Hydroxymethyl-	5.40*, 11.55*	—
4-Hydroxy-6-nitro-	3.05	—
4-Methoxy-	5.1	—
1-Methyl-	5.54*	—
2-Methyl-	5.58	—
4-Methyl-	5.65*	—
5-Methyl-	5.78*	—
2-Methyl-4-hydroxy-6-amino-	6.65	—
1-Methyl-2-hydroxymethyl-	5.55, 11.45	—
2-Methyl-4-hydroxy-6-nitro-	3.9	—
4-Nitro-	3.33*	—
5-Nitro-	3.48*	—
6-Nitro-	3.05, 10.6	—
5-F-	1.67*, 4.92*	123, 124
5-Br-	1.98*, 4.66*	123, 124
5-CF$_3$-	2.28*, 4.22*	123, 124
5-Cl-	1.94*, 4.70*	123, 124
Benzo(c)cinnoline	2.20*	99
6,7-Benzoquinazoline	5.2*	99
5:6-Benzoquinoline	5.15, 3.90[a]	19
4-Amino-	7.99[a]	19
2-Amino-4-methyl-	7.14, 6.51[a]	19
4-Amino-2-methyl-	8.45[a]	19
1′-Amino-	5.03	19
3′-Amino-	4.02[a]	19
4′-Amino-	5.20, 4.10[a]	19
2′:4′-Diamino-	4.91[a]	19
2-Methyl-	4.44[a]	19
6:7-Benzoquinoline	5.05, 3.84[a]	19
3-Amino-	4.78, 3.73[a]	19
4-Amino-	8.75[a]	19
4-Amino-2-methyl-	9.45[a]	19
4-Amino-2-methyl-8-chloro-	5.95[a]	19
8-Chloro-	2.5[a]	19
3:4-Diamino-	8.15[a]	19
7:8-Benzoquinoline	4.25, 3.15[a]	19
4-Amino-	7.68[a]	19
2-Amino-4-methyl-	6.74, 6.02[a]	19
4-Amino-2-methyl-	7.96[a]	19
6-Amino-2-methyl-	5.23[a]	19
1′-Amino-2-methyl-	4.75[a]	19
Benzoxazole	(decomp.)	19
2-Amino-	3.73	19
Benzthiazole	1.2, 0.1[a]	19
2-Amino-	4.51	19
Benztriazole	1.6, 8.64*	19
Bispidine	10.25*	99

Heterocyclics—(Continued)

Compound	pK'_a	Reference
Caffeine	0.61	4
Cinchonine	7.2	4
Cinnoline	2.70, 2.29	19, 39
4-Amino-	6.84	19
3-Hydroxy-	8.64, 0.21	39
4-Hydroxy-	9.27, 0.35	39
5-Hydroxy-	7.40, 1.92	39
6-Hydroxy-	7.52, 3.65	39
7-Hydroxy-	7.56, 3.31	39
8-Hydroxy-	8.20, 2.74	39
4-Methoxy-	3.21	39
4-Methylthio-	3.09*	99
5,6,7,8-Tetrahydro-	4.30*	99
α,α′-Dipyridyl	4.43	6
4.5-Diazaindan	4.12*	99
1,4-Diazaindene	6.92*	99
1,5-Diazaindene	8.24*	99
1,6-Diazaindene	7.93*	99
1,7-Diazaindene	4.57*	99
Flavone	−1.2	154
Δ^2-Dihydro-2-methyl-	11.1	154
Furan,2-(2-aminoethyl-aminomethyl)-	6.54*, 9.87*	99
Furan, 3-(2-aminoethyl-aminomethyl)-	6.70*, 9.86*	99
Furan, 2-aminomethyl-	8.89	99
Gramine	16.00	152
5-Benzyloxy-	16.90	152
Histamine	6.0	43
Histidine methyl ester	5.2, 7.1	43
4-Hydroxymethyl-	6.39	99
2-(2-Imidazolyl)-	4.53*	99
4-(3-Methoxycarbonyl-propyl)-	7.3	99
1-Methyl-	6.95*	99
2-Methyl-	7.85	99
4-Methyl-	7.51	99
1-Methyl-4-chloro-	3.10	106
1-Methyl-5-chloro-	4.75	106
1-Methyl-4-nitro-	−0.53*	99
1-Methyl-5-nitro-	2.13*	99
1-Methyl-4-phenyl-	5.78*	99
2-Nitro-	7.15	106
4-Nitro-	−0.05*	99
5-Nitro-	9.20	106
4-Nitro-5-chloro-	5.85	106
5-Nitro-4-chloro-	5.85	106
2-Phenyl-	6.48*, 13.32*	99
4-Phenyl-	6.10*, 13.42*	99
5-Phenyl-	6.10, 13.42	107
4-(2-Pyridyl)-	5.42*	99
2,4,5-Trimethyl-	8.92*	99
Hydantoin	9.16	42
5:5-Dimethyl-2-thio-	8.71	42
5:5-Diphenyl-2-thio-	7.69	42
5-Isopropyl-2-thio-	8.70	42
1-Methyl-5:5-pentamethyl-ene-2-thio-	9.25	42
3-Methyl-5:5-pentamethyl-ene-2-thio-	11.23	42
5:5-Pentamethylene-2-thio-	8.79	42
3:5:5-Trimethyl-2-thio-	10.80	42

*Thermodynamic value.
[a]In 50% alcohol.

NITROGEN COMPOUNDS—(continued)

Compound	pK_a'	Reference
Heterocyclics—(Continued)		
Imidazoles	7.05, 14.52	107
4-(2-Acetoxyethyl)-	6.97*	99
N-Acetyl-	3.6	99
N-Acetylhistidine	7.05	43
4-Aminomethyl-	9.37*, 4.71*	99
5-Amino-4-(N-methyl-carboxamidino)-	9.5*	99
4-Bromo-	3.7	43
4-Carbamoyl-	3.7*, 11.8*	99
4-(3-Carbamoylpropyl)-	6.52*	99
Carbobenzoxy-L-histidyl-L-tyrosine ethyl ester	6.25	43
4-(2',4'-Dihydroxyphenyl)-	6.45	43
4-Chloro-1-methyl-	6.23*	99
2,4-Dimethyl-	8.36*	99
2,4-Dinitro-	2.85	106
2,5-Dinitro-	2.85	106
2,4-Diphenyl-	5.64*, 12.53*	99
2,5-Diphenyl-	5.64, 12.53	107
4,5-Diphenyl-	5.90*, 12.80*	99
1-Ethyl-	7.30*	99
2-Ethyl-	8.00	99
1H-Imidazo(4,5-b)pyridine	3.92*, 11.11*	99
1H-Imidazo(4,5-c)pyridine	6.10*, 10.88*	99
Imidazolidines		
2-Imino-1-methyl-4-oxo-	4.80*	99
2-Imino-4-oxo-	4.76*	99
n-Nitrimino-	-1.36*	99
2-Imidazoline		
2-(N-Benzylanilinomethyl)-	2.45*, 10.13*	99
2-(3-Diethylamino-1-phenyl)-propyl-	8.41*, 10.09*	99
2-(3-Dimethylamino-1-phenyl)-propyl-	7.98*, 9.99*	99
2-Diphenylmethyl-	9.78*	99
Imidazoline		
2-(3-(2-Hydroxynaphthyl))-	7.01*, 10.85*	99
2-(2-Hydroxyphenyl)-	6.63*, 12.58*	99
4-Methyl-5-carboxylic acid-	**2.49, 7.02**	144
1-Methyl-2-carboxylic acid-	**1.26, 7.25**	144
1H-Indazole	1.22*, 14*	99
3-Amino-	3.12*	99
4-Amino-	3.26*	99
5-Amino-	5.12*	99
6-Amino-	3.99*	99
7-Amino-	3.02*	99
Indole	-2.4*	99

Compound	pK_a'	Reference
Heterocyclics—(Continued)		
Indole	-3.6, 16.97	154, 152
2-Amino-	8.15*	99
2-Amino-1-methyl-	9.60*	99
1,2-Dimethyl-	0.34*	99
2,3-Dimethyl-	-1.10*	99
1-Methyl-	-1.80	99
2-Methyl-	-0.10*	99
3-Methyl-	-3.35*	99
3-Formyl-	12.36	152
3-Acetyl-	12.99	152
5-Nitro-	14.75	152
5-Cyano-	15.24	152
5-Bromo-	16.13	152
5-Fluoro-	16.30	152
4-Fluoro-	16.30	152
2-Carboxylate-	17.13	152
3-Carboxylate-	15.59	152
5-Carboxylate-	16.92	152
3-Acetic acid-	16.90	152
3-Carbinol-	16.50	152
L-Tryptophanol	16.91	152
L-Tryptophan	16.82	152
4-Methyltryptophan	16.90	152
5-Hydroxytryptophan	19.20	152
6-Methoxytryptophan	16.70	152
Tryptamine	16.60	152
Serotonine	19.50	152
Gramine	16.00	152
5-Benzyloxydramine	16.90	152
Skatole	16.60	152
5-NO$_2$-2-Carboxylate	14.91	152
5-Br-2-Carboxylate	16.10	152
5-MeO-2-Carboxylate	17.03	152
Isoalloxazine, 7,8 dichloro-		
10-(3-Dibutylaminopropyl)-	8.0*	99
10-(4-Diethylaminobutyl)-	9.7*	99
10-(2-Diethylaminoethyl)-	7.7*	99
10-(5-Diethylaminopropyl)-	9.1*	99
10-(3-Piperidinopropyl)-	9.0*	99
Isoxazole	1.3*	27
3,5-Dimethyl-	-2*	27
4,5-Dimethyl-	0*	27
3-Methyl-	-1*	27
5-Methyl-	2.3*	27
3,4,5-Trimethyl-	-1*	27

	1	3	4	5	6	7	8	Reference
								19, 44, 99
Isoquinoline	5.46	5.14						
OH-	-1.2	2.18*	4.78*	5.40	5.85	5.68*	5.66	—
	—	9.62*	8.70*	8.45	9.15	8.90*	8.40	—
NH$_2$-	7.59*	5.05	6.26*	5.59	7.16*	6.20	6.04*	—
CH$_3$-	—	5.64*	—	—	—	—	—	—
Br-	—	—	3.31*	—	—	—	—	—
SH-	-1.9*	0.39*	—	—	—	—	—	—
	10.86*	8.62*	—	—	—	—	—	—
CH$_3$O-	3.01*	—	—	—	—	—	—	—
CH$_3$S-	3.89*	3.37*	—	—	—	—	—	—
NO$_2$-	—		1.35*	3.49*	3.43*	3.57*	3.55*	—

*Thermodynamic value.

aIn 50% alcohol.

NITROGEN COMPOUNDS—(continued)

Compound	pK_a'	Reference
Heterocyclics—(Continued)		
Isoquinoline		
1,2,3,4-Tetrahydro-	9.4	154
Isoquinoline-N-oxide	1.01	47
Morpholine	8.39*	99
N-(3-Acetyl-3,3-diphenyl)-propyl-	6.83*	99
N-(2-Acetyl-2-phenyl)ethyl-	6.23*	99
N-Allyl-	7.02*	99
N-(2-Amino)ethyl-	4.84*, 9.45*	99
N-(2-Benzylcarbonyl-2-phenyl)ethyl-	6.17*	99
N-(2-Bis-2-hydroxypropyl)-aminoethyl-	7.9*	99
N-(3-Cyano-3,3-diphenyl)-propyl-	6.04*	99
N-(3-Cyano-1-methyl-3,3-diphenyl)propyl-	5.5*, 8.4*	99
N-(2-Diphenylmethyl-carbonyl)ethyl-	6.39*	99
N-(3,3-Diphenyl)propyl-	7.20*	99
N-(3,3-Diphenyl-3-propyl-carbonyl)propyl-	7.17*	99
N-Ethyl-	7.67*	99
N-(3-Ethylcarbonyl-3,3-diphenyl)propyl-	6.95*	99
N-(3-Ethylcarbonyl-1-methyl-3,3-diphenyl)-propyl-	6.68*	99
N-(3-Ethylcarbonyl-2-methyl-3,3-diphenyl)-propyl-	7.12*	99
N-(2-Hydroxy-3-morpholino)propyl-	5.00*, 6.98*	99
N-Methyl-	7.38*	99
N-(1-Methyl-3,3-diphenyl)-propyl-	6.85*	99
N-(2-Morpholino)ethyl-	3.63*, 6.65*	99
N-(3-Morpholino)propyl-	6.25*, 7.25*	99
1(3)H-Naphth(1,2-d)-imidazole	5.27*	99
1H-Naphth(2,3-d)imidazole	5.21*, 12.58*	99
2-Amino-	6.99*	99
2-Carboxymethylthio-	1.9*, 4.72*	99
2-Ethyl-	6.13*, 12.9*	99
2-Hydroxy-	11.66*	99
2-Methyl-	6.10*, 12.9*	99
1,5-Naphthyridine	2.84*	99
1,6-Naphthyridine	3.76*	99

Compound	pK_a'	Reference
Heterocyclics—(Continued)		
1,7-Naphthyridine	3.61*	99
1,8-Naphthyridine	3.36*	99
8-Hydroxy-6-methyl-1:6-naphthyridinium chloride	4.34	44
8-Hydroxy-1:6-naphthyridine	4.08	44
Oxazoline		
4-Carbamoyl-2-phenyl-	2.9	96
2-Methyl-Δ^2-	5.5	96
2-Phenyl-Δ^2-	4.4	96
4-Methyl-	1.07	144
Ethyl-4-Methyl-5-carboxylate-	0.83	144
4-Methyl-5-carboxylic acid-	0.95, 2.88	144
Perimidine	6.35*	99
3,4-Pentamethylene,5,6,7,8-tetrahydrocinnoline	6.03*	99
1,5-Phenanthroline	0.75*, 4.10	99
1,7-Phenanthroline	1, 4	99
1,10-Phenanthroline	4.84*	99
4-Bromo-	4.01*	99
3-Chloro-	3.97*	99
4-Chloro-	4.30*	99
5-Chloro-	4.24*	99
4-Cyano-	3.56*	99
3-Ethyl-	4.96*	99
4-Ethyl-	5.42*	99
4-Hydroxy-	2.17*	99
2-Methyl-	4.98*	99
3-Methyl-	4.98*	99
5-Methyl-	5.26*	99
5-Nitro-	3.55*	99
2-Phenyl-	4.88*	99
3-Phenyl-	4.80*	99
4-Phenyl-	4.88*	99
5-Phenyl-	4.70*	99
o-Phenanthroline	4.27[a], 5.2	19
p-Phenanthroline	3.12[a]	19
1:10-Diamino-3:8-dimethyl-	8.76[a], 6.31[a]	19
6 m-Phenanthroline	3.11[a]	19
1-Amino-	ca. 7.3, 7.29[a]	19
Phenazine	1.23	39
1-Amino-	2.6[a]	19
2-Amino-	4.75, 3.46[a]	19
1:3-Diamino-	5.64[a]	
2:3-Diamino-	4.74	19
2:7-Diamino-	4.63, 3.9[a]	19

* Thermodynamic value.

NITROGEN COMPOUNDS—(continued)

Compound	pK'_a	Reference
Heterocyclics (continued)		
1-Hydroxy-	**1.61*, 8.33***	
2-Hydroxy-	**7.5, 2.6**	99
Phenanthridine	**4.65**	39
2-Amino-9-methyl-	**5.66[a]**	44
6-Amino-	**6.88**	19
7-Amino-9-methyl-	**5.23[a]**	40
9-Amino-	**7.31, 6.75[a]**	19
2:7-Diamino-9-methyl-	**6.26[a]**	19
2-Hydroxy-	**8.79, 4.82**	19
6-Hydroxy-	**8.43, 5.35**	44
7-Hydroxy-	**4.38, 8.68**	44
9-Hydroxy (phenanthridone)	**<−1.5**	44
9-Methoxy-	**2.38**	44
2:7:9-Triamino-	**8.06[a]**	44
Phenothiazine		19
10-(2-Diethylaminoethyl)-	9.06*	99
10-(2-Dimethylaminobutyl)-	9.02*	99
10-(2-Dimethylaminoethyl)-	8.66*	99
3,7-Diamino-	**5.3, 4.4**	154
Phthalazine	3.47	19
1-Amino-	6.60	19
1-Hydroxy-	11.99, −2	39
1-Mercapto-	−3.43*, 9.99*	99
1-Methoxy-	3.73*	99
1-Methyl-	4.37*	99
1-Methylthio-	3.44*	99
1-Phenyl-	3.51*	99
Picolinic acid	5.52	4
Trimethyl-(2,6-di-tert-butyl-4-picolyl)ammonium	**3.51**	125
Pteridines	4.12	101
6-Cl-	3.68*	99
2-Me-	4.87	101
4-Me-	2.94	101
7-Me-	3.49	101
6,7-di-Me-	2.93	101
2,6,7-tri-Me-	3.76	101
2-OH-	<2, 11.13	101
4-OH-	<1.5, 7.98	101
6-OH-	3.67, 6.7	101
7-OH-	1.2, 6.41	101
2,4-di-OH-	<1.0, 7.91	101
2,6-di-OH-	6.7, 11.6	101
2,7-di-OH-	5.83, 10.07	101
4,6-di-OH-	6.08, 9.73	101
4,7-di-OH-	6.82, 10.02	101
6,7-di-OH-	6.87, 10.0	101
2,4,6-tri-OH-	5.73, 9.41	101
3,4,7-tri-OH-	3.61	101
4-OH-6-Me-	8.19	101
4-OH-7-Me-	8.09	101
6-OH-5-Me-	3.73, 10.6	101
7-OH-8-Me-	1.1	101
2-OH-1-Me-	<1, 11.43	101
2-OH-3-Me-	1, 11.01	101
2-OH-3,6,7-tri-Me-	<2, 11.36	101
2-OH-6,7,8-tri-Me-	<2, 10.26	101
4-OH-1-Me-	1.25	101
4-OH-3-Me-	−0.47	101
4-OH-6,7-di-Me-	8.39	101
4-OH-6,7,8-tri-Me-	4.70, 9.46	101

Compound	pK'_a	Reference
Heterocyclics (continued)		
7,8-Dihydro-6-OH-	4.78, 10.54	101
5,6-Dihydro-7-OH-	3.36, 9.94	101
5,6-Dihydro-6,7-di-OH-5-Me-	2.91, 9.33	101
5,6-Dihydro-4,7-di-OH-6-Me-	8.43, 11.40	101
4,7-di-OH-6-CHO-	5.93, 9.31	101
4,7-di-OH-6-COOH-	ca. 3, 6.69, 10.15	101
2-MeO-	2.13	101
4-MeO-	1.04	101
6-MeO-	3.60	101
7-MeO-	1.64	101
2-NH$_2$-	4.29	101
4-NH$_2$-	3.51	101
6-NH$_2$-	4.15	101
7-NH$_2$-	2.96	101
2,4-di-NH$_2$-	5.32	101
4,6-di-NH$_2$-	4.37	101
4,7-di-NH$_2$-	4.97	101
2,4,7-tri-NH$_2$-	6.03	101
4,6,7-tri-NH$_2$-	5.57	101
2,4,6,7-tetra-NH$_2$-	6.86	101
2-MeNH-	3.64	101
4-MeNH-	4.33	101
2-Me$_2$N-	3.03	101
4-Me$_2$N-	4.33	101
6-Me$_2$N-	4.31	101
7-Me$_2$N-	2.56	101
2-NH$_2$-7-OH-	1.5*, 7.50*	99
2-NH$_2$-4-CH$_3$O-	3.44*	99
2-NH$_2$-4-CH$_3$-	2.81*	99
2-NH$_2$-6-CH$_3$-	4.03*	99
2-NH$_2$-7-CH$_3$-	3.73*	99
4-NH$_2$-2-CH$_3$-	4.28*	99
2-NH$_2$-4-OH-	2.31, 7.92	101
4-NH$_2$-2-OH-	3.21, 9.97	101
2-NH$_2$-4,6-di-OH-	1.6, 6.3, 9.23	101
2-MeCONH-	2.67	101
4-MeCONH-	1.21	101
4-H$_2$NNH-	4.00	101
2-SH-	9.98	101
4-SH-	6.81	101
7-SH-	5.5	101
2-MeS-	2.2	101
4-MeS-	2.59	101
7-MeS-	2.49	101
4-MeS-7-Me-	<2	101
4-SH-7-Me-	7.02	101
3,4-Dihydro-2-OH-	0*, 12.6*	99
3,4-Dihydro-4-OH-	4.75*, 11.25*	99
5,6-Dihydro-4-OH-	2.94*, 10.24*	99
7,8-Dihydro-2-OH-	3.46*	99
7,8-Dihydro-4-OH-	0.32*, 12.13*	99
6-OH-2-CH$_3$-	4.65*, 6.33*	99
6-OH-4-CH$_3$-	4.08*, 6.41*	99
6-OH-7-CH$_3$-	3.69*, 7.20*	
7,8-Dihydro-4,6-dimethyl-	6.0	154
5,6,7,8-Tetrahydro-	6.6	154
5,6,7,8-Tetrahydro-5-formyl-	5.0	154
5,6,7,8-Tetrahydro-4-methyl-	6.7	154
5,6,7,8-Tetrahydro-4-hydroxy-	3.9, 10.1	154

*Thermodynamic value.

[a]In 50% alcohol.

NITROGEN COMPOUNDS—(continued)

Compound	pK_a'	Reference
Heterocyclics—(Continued)		
Pteridines		
7-OH-2-CH$_3$-	1.68*, 6.70*	99
3-Methyl-4-pteridone	−0.47	44
1-Methyl-4-pteridone	1.25	44
Pteroylglutamic acid	8.26	77
Purine	2.39, 8.93	101
2-Amino-6,8-bistrifluoro-methyl-	0.3*, 5.02*	99
6-Amino-9-cyclohexyl-amino-	4.19*	99
2-Amino-8-phenyl-	3.97*, 9.21*	99
2-Amino-6-trifluoromethyl-	1.85*, 8.87*	99
8-Carboxy-	0*, 2.93*, 9.41*	99
6-Chloro-	7.85*	99
6-Cyclohexylamino-	4.2*, 10.2*	99
9-Cyclohexyl-6-cyclohexyl-amino-	4.4*	99
2,6-Diamino-8-trifluoro-methyl-	3.68*, 7.55*	99
1,6-Dihydro-1,7-dimethyl-6-oxo-	2.13*	99
7,8-Dihydro-7,9-dimethyl-8-oxo-	2.8*	99
1,2-Dihydro-8-hydroxy-1-methyl-2-oxo-	−0.5*, 7.0*, 13.0*	99
1,6-Dihydro-8-hydroxy-1-methyl-6-oxo-	8.54*, 11.87*	99
2,3-Dihydro-8-hydroxy-3-methyl-2-oxo-	1.25*, 8.0*, 13.0*	99
1,6-Dihydro-6-imino-1-methyl-	11.0*, 7.0*	99
1,2-Dihydro-1-methyl-2-oxo-	1.8*, 8.80*	99
1,2-Dihydro-2-oxo-1-β-D-ribofuranosyl-	1.5*, 8.55*	99
6-Dimethylcarbamoyl-	0*, 7.9*	99
6-(Ethoxycarbonyl)amino-	2.4*, 9.63*, 12.2*	99
6-Ethylamino-2-methyl-	5.01*	99
6-Hydrazonomethyl-	2.8*, 9.2*	99
8-Hydroxymethyl-	2.58*, 8.83*	99
6-Hydroxy-2-trifluoro-methyl-	1.1*, 5.1*, 11.2*	99
7-Methyl-	2.25*	99
8-Methylthio-	2.92*, 7.70*	99
Purine 1-oxide, 6-amino-	2.69*, 8.845*, 15.4*	99
2-SH-	7.15*, 10.4	101
6-SH-	7.77, 10.84	101
8-SH-	6.64, 11.64	101
2-MeS-	8.91	101
6-MeS-	8.74	101
2-NH$_2$-	−0.28, 3.80, 9.93	101
6-NH$_2$- (adenine)	<1, 4.22, 9.8	101
8-NH$_2$-	4.68, 9.36	101
2,6-di-NH$_2$-	<1, 5.09, 10.7	101
2,6,8-tri-NH$_2$-	2.41, 6.23, 10.96	101
6-MeNH-	1, 4.18, 9.99	101
8-MeNH-	4.78, 9.56	101
2-Me$_2$N-	4.02, 10.22	101
6-Me$_2$N-	<1, 3.84, 10.5	101

Compound	pK_a'	Reference
Heterocyclics—(Continued)		
8-Me$_2$N-	1, 4.80, 9.73	101
2-C$_6$H$_5$NH-	4.2, 10.1	101
2-Me-6-MeNH-	5.08	101
2-NH$_2$-6-OH- (Guanine)	3.3, 9.2, 12.3	101
6-NH$_2$-2-OH-	4.51, 8.99	101
2-NH$_2$-6-SH-	8.2, 11.6	101
6-CHO-	2.4, 8.8	101
6-HONH-	3.88, 9.88, >12	101
6-NH$_2$CONH-	2.35, 9.95	101
6-CN-	ca. 0.3, 6.88	101
6-CF$_3$-	0, 7.35	101
8-CF$_3$-	1.0, 5.12	101
6-Me-	2.6, 9.02	101
8-Me-	2.85, 9.37	101
9-Me-	2.36	101
8-Ph-	2.68, 8.09	101
2-OH-	8.43, 11.90	101
6-OH- (Hypoxanthine)	8.94, 12.10	101
8-OH-	8.24, >12	101
2,8-di-OH-	7.65, 9.7	101
2,6-di-OH- (Xanthine)	7.44, 11.12	101
1-Me-2,6-di-OH-	7.7, 12.5	101
3-Me-2,6-di-OH-	8.33, 11.9	101
7-Me-2,6-di-OH-	8.7, 10.7	191
1,3-di-Me-2,6-di-OH-	8.81	101
1,7-di Me-2,6-di-OH-	8.71	101
3,7-di Me-2,6-di-OH-	9.97	101
2,6,8-tri-OH- (Uric acid)	5.75, 10.3	101
2,6,8-tri-OH-3-Me-	5.75, >12	101
2,6,8-tri-OH-1-Me-	5.75, 10.6	101
2,6,8-tri-OH-1,3-di-Me-	5.75	101
2,6,8-tri-OH-3,7-di-Me-	5.5, 12	101
2,6,8-tri-OH-1,3,7-tri-Me-	6.0	101
2,6,8-tri-OH-3,7,9-tri-Me-	8.35	101
6-OH-9-Me-	1.86, 9.32	101
8-OH-7-Me-	2.69, 8.20	101
8-OH-9-Me-	2.80, 9.05	101
2-MeO-	2.44, 9.2	101
6-MeO-	1.98, 8.94	101
8-MeO-	3.14, 7.73	101
Xanthine	7.53, 11.63	101
1-Me-	7.7, 12.05	101
3-Me-	8.10, 11.3	101
7-Me-	8.30	101
9-Me-	6.25	101
Hypoxanthine	8.8, 12.0	101
Uric acid	5.78, 5.85	101
Pyrazine	1.1 (0.6)	49, 39
3-Acetyl-2-aminomethylene-amino-	5.49*	99
2-Amidino-3-methylamino-	8.96*	99
2-Amino-	3.14	19
2-Amino-3-carboxy-	3.70*	99
2-Carbomoyl-	−0.5*	99
2-Carbamoyl-3-methyl-amino-	2.09*	99
2,3-Dicarboxy-	0.9*, 3.57*	99
2,5-Dimethyl-	2.1	49
2,6-Dimethyl-	2.5*	99
2-Dimethylamino-	3.24*	99
2-Hydroxy-	−0.1*, 8.25*	99
2-Mercapto-	−0.73*, 6.34*	99

* Thermodynamic value.

NITROGEN COMPOUNDS—(continued)

Compound	pK_a'	Reference
Heterocyclics—(Continued)		
2-Methoxy-	0.75	39
2-Methyl-	−5.25*, 1.45*	99
1-Methyl-2-pyrazone	−0.04	39
2-Methylamino-	3.39*	99
2-Methylthio-	0.48*	99
2-Sulphanilamido-	6.04*	99
2,3,5,6-Tetramethyl-	2.8	49
Trimethyl-	−0.35*, 2.8*	99
Pyrazole	2.48	99
3-(2-aminoethyl)-	2.02*, 9.61*	99
1,3-Dimethyl-	3.11*	99
3,5-Dimethyl-	4.38*	99
1-Methyl-	2.04*	99
3-Methyl-	3.56*	99
Pyrazolo(4;,5;-4,5)pyrimidine		
6-Amino-	4.96*, 10.19*	99
Pyrazolo(5′,4′-4,5)pyrimidine	2.80*, 9.58*	99
6-Amino-	4.55*, 10.88*	99
6-Amino-1′-methyl-	4.28*	99
6-Amino-2′-methyl-	5.37*, 11.34*	99
6-Amino-3′-methyl-	4.57*, 11.15*	99
6-Anilino-	3.88*	99
6-Benzylamino-	4.12*, 10.97*	99
2,6-Diamino-	4.63*, 11.25*	99
6-Diethylamino-	4.67*	99
6-Dimethylamino-	4.49*, 11	99
6-Ethylamino-	4.56*, 10.94*	99
6-Furfurylamino-	3.97*	99
6-Isopropylamino-	4.58*, 11.03*	99
6-Methylamino-	4.49*, 10.59*	99
1-Methyl-	2.46*	99
6-Methylthio-	1.0*, 9.69*	99
6-Phenethylamino-	4.34*	99
Pyridazine	2.33	19
3-Amino-	5.19	19
4-Amino-	6.65*	99
3-Amino-6-methyl-	5.29*	99
3-*n*-Butyl-	3.49*	99
3-Carbamoyl-	1.0*	99
4-Carbamoyl-	1.0*	99
3,5-Dihydroxy-	−2.2*	99
3:6-Dihydroxy-	5.67, −2.2, 13	39
3:6-Dimethoxy-	1.61	39
3,4-Dimethyl-	4.10*	99
3,5-Dimethyl-	4.11*	99
3,6-Dimethyl-	3.99*	99
4,5-Dimethyl-	4.13*	99
3-Hydroxy-	10.46, −1.8	39
4-Hydroxy-	8.68, 1.07	39
4-Hydroxy-2-methyl- pyridazinium chloride	1.74	44
3-Mercapto-	−2.68*, 8.32*	99
4-Mercapto-	−0.75*, 6.55*	99
3-Methoxy-	2.52	39
4-Methoxy-	3.70	39
3-Methyl-	3.46*	99
4-Methyl-	3.53*	99
3-Methylthio-	2.26*	99
4-Methylthio-	3.24*	99
3-Sulphanilamido-	7.06*	99

Compound	pK_a'		
Heterocyclics—(Continued)			
Pyridine (2, 46–48, 88, 99, 105, 117, 140)	2–	3–	4–
H-	5.17*	—	—
Cl-	0.72*	2.84*	3.88
I-	1.82*	3.25*	4.02*
CH₃CH₂-	5.97*	5.70*	6.02*
(CH₃)₃C-	5.76*	5.82*	5.99*
HO-	0.75	4.86	3.27
	11.62	8.72	11.09
NO₂-	−2.06	0.81	1.23
SO₃-	—	2.9	
CH₃O-	3.28	4.88	6.62
C₂H₅O-			6.67
F-	−0.44*	2.97*	
Br-	0.90*	2.84*	3.82
CH₃-	5.97*	5.68*	6.02*
(CH₃)₂CH-	5.83*	5.72*	6.02*
CH₃CO-	—	3.18*	—
CONH₂-	2.10	3.40	3.61
NC-	−0.26*	1.45	1.90*
CH₃CONH-	4.09*	4.46	5.87
EtOOC-	—	3.35	3.45
NH₂-	6.71*	6.03*	9.11
	−7.6	−1.5	−6.3
C₆H₅CONH-	3.33*	3.80*	5.32*
COOH-	0.99*	2.00*	1.77*
	5.39*	4.83*	4.84*
HCO-	3.80*	3.80*	4.77*
	12.80*	13.10*	12.20*
H₂NNHCO-	—	1.86*	1.82*
	3.86*	3.29*	3.52*
	12.27*	11.47*	10.79*
HS-	−1.07*	2.26*	1.43*
	10.00*	7.03*	8.86*
CH₃OCO-	2.21*	3.13*	3.26*
CH₃NH-	—	—	9.66
CH₃S-	3.59*	4.42*	5.94*
C₆H₅-	4.48*	4.80*	5.55*
CH₂=CH-	4.98*	—	5.62*
Benzyl-	5.13*	—	—
Benzylthio-	3.23*	—	5.41*
t-Butyl-	5.76*	5.82*	5.99*
Dimethylaminoethyl-	3.46*	4.30*	4.66*
	8.75*	8.86*	8.70*
Dimethylaminomethyl-	2.58*	3.17*	3.39*
	8.12*	8.00*	7.66*
Hexyl-	5.95*	—	—
Methanesulphonamido-	1.10*	3.43*	3.64*
	8.02*	7.02*	9.07*
N-Methylacetamido-	2.01*	3.52*	4.62*
N-Methylbenzamido-	1.44*	3.66*	4.68*
N-Methylmethanesulphon- amido-	1.73*	3.94*	5.14*
Piperidineoethyl-	3.59*	4.25*	4.68*
	9.29*	8.81*	9.06*
Piperidinomethyl-	2.61*	3.16*	3.90*
	8.51*	8.30*	7.88*
2-Pyridyl-	4.44*	—	—
3-Pyridyl-	4.42*	4.60*	—
	1.52*	3.0*	
4-Pyridyl-	4.77*	4.85*	4.82*
	1.19*	3.0*	3.17*

* Thermodynamic value.

NITROGEN COMPOUNDS—(continued)

Heterocyclics—(Continued)

Pyridine-N-oxides: see oxygen acids

Pyridines

Compound	pK'$_a$	Reference
2,3-Me$_2$-	6.60	48
2,4-Me$_2$-	6.72	48
2,5-Me$_2$-	6.47	48
2,6-Me$_2$-	6.77	48
3,4-Me$_2$-	6.52	48
3,5-Me$_2$-	6.14	48
2,4,6-Me$_3$-	7.48	48
2-Me,5-Et	6.51	48
2-Amino-3-nitro-	2.38	105
2-Amino-3-nitro-	2.42, −12.4	141
2-Amino-5-nitro-	2.80, −12.1	141
3-Amino-2-nitro-	0.02, −9.07	141
4-Amino-3-nitro-	5.04	105
3-Amino-4-methylamino-	0.38, 9.57	105
4-Amino-3-methylamino-	0.12, 9.37	105
2-Amyl-	6.00*	45
2-Benzamido-	3.33	46
2-Benzyl-	5.13	45
2-Chloro-3-nitro-	−2.6	105
2,3-Diamino-	7.00, −0.01	105
3,4-Diamino-	9.14, 0.49	105
4,5-Diamino-2-chloro-	4.79, 0.08	105
2,3-Diamino-6-chloro-	3.02, −0.91	105
2,4-Dichloro-5-amino-	0.73	105
2:4-Dihydroxy-	13, 1.37, 6.50	39
4-Ethoxy-3-nitro-	2.67	105
2-Hydroxy-3-nitro-	−4.00, 8.52	105
4-Hydroxy-3-nitro-	−0.70, 7.65	105
1-Methyl-2-pyridone	0.32	39
1-Methyl-4-pyridone	3.33	39
4-Methylamino-3-nitro-	5.19	105
1-Methylpyrid-2-one acetylimine	7.12	46
1-Methylpyrid-4-one acetylimine	11.03	46
1-Methylpyrid-4-one benzylimine	9.89	46
2:4:6-Trihydroxy-	4.6, 9.0, 13.00	39
Trimethyl(2,6-di-tert-butyl-4-pyridyl)ammonium ion	1.65	125
Trimethyl(2-tert-butyl-6-pyridyl)ammonium ion	<−1	125
Dimethyl(2,6-di-tert-butyl-4-pyridyl)ammonium ion	1.6	125
Δ'-Tetrahydro-2-Methyl-	9.6	154
Δ'-Tetrahydro-1,2-Dimethyl-	11.4	154
1,4-Dihydro-1,4,4-Trimethyl-	7.4	154

Pyrimidines

	pK$_a$	3,4-(NH$_2$)$_2$	3-NH$_2$	4-NH$_2$	5-NH$_2$	6-NH$_2$	2-NH$_2$
2-OH[140,141]		−0.87	2.78	−5.14	−0.61	−6.12	
		4.16		2.65	3.77	2.32	
		13.43		13.54	11.65	11.38	
2-OMe		5.68, 1.06	3.35	−5.86			
				7.05	4.28	4.64	
1-Me-2=O	0.32		2.94				
1-Me-4=O	3.33		3.88		0.05	2.09	
4-OH			0.04				−6.58
			3.84				5.04
			11.38				10.69
4-OMe			7.30				
2-NH$_2$			−12.1		−10.7	−3.73	
			0.38		1.97	6.00	7.62
			6.73		6.46		
3-NH$_2$				−10.7			
				0.80			
				9.19			

Monosubstituted[99]	2	4	5
Amino-	3.45*	5.69*	2.51*
n-Butylamino-	4.14*	—	—
Carboxy-	−1.13*	—	—
	2.85*	—	—
Dimethylamino-	3.93*	6.32*	—
Ethylamino-	3.89*	—	—
Hydroxy-	2.15*	1.66*	1.85*
	9.20*	8.63*	6.80*
Mercapto-	1.35*	0.68*	—
	7.10*	6.94*	—
Methoxy-	<1	2.5*	—
Methyl-	—	1.91*	—
Methylamino-	3.79*	6.09*	—
Methylthio-	0.59*	—	—
Sulphanilamido-	6.34*	6.17*	6.22*
	2.0*	—	—

Compound	pK'$_a$	Reference
Cytidine	4.08, 12.5	134
Cytosine	4.60, 12.16	101
5-Me-cytosine	4.6, 12.4	101
Isocytosine	4.01, 9.42	101
Thymine	9.90	134
Thymidine	9.79	134
Uracil	9.45	101
Uridine	9.30, 12.59	134
1-Me-	9.99	101
3-Me-	9.71	101
5-Me-	9.94	101
1,3-di-Me-	none	101
Orotic acid	2.40, 9.45	101

*Thermodynamic value.

NITROGEN COMPOUNDS—(continued)

Compound	pK_a'	Reference	Compound	pK_a'	Reference
Heterocyclics (continued)			**Heterocyclics (continued)**		
Pyrimidines, substituted			2,4,5-Trihydroxy-	8.11, 11.48	39
4,6-Bisdimethylamino-	6.34*	99	(Bobarbituric acid)		
4,5-Bismethylamino-	6.77*	99	Pyrimidines, 2-amino		
4,6-Bismethylamino-	6.32*	99	substituted		
5-Bromo-2,4-dihydroxy-	−7.25*, 7.83*	99	5-Bromo-	1.95*	99
5-Bromo-2-methylamino-	2.09*	99	4-Chloro-1,6-dihydro-6-	9.87*	99
4-Butylamino-2-hydroxy-	4.67*	99	imino-1-methyl-		
4-Butylamino-2-mercapto-	3.2*, 11.15*	99	4-Diethylaminoethylamino-	7.9*, 9.7*	99
4-Carboxy-6-hydroxy-	2.8*, 8.4*	99	5,6-dimethyl-		
2-Chloro-4-methylamino-	2.90*	99	4-Diethylaminoethylamino-	7.5*, 9.55*	99
4-Chloro-2-methylamino-	2.59*	99	6-methyl-		
4-Chloro-6-dimethylamino-	2.42*	99	1,4-Dihydro-4-imino-1-	12.9*	99
4-Chloro-6-methylamino-	2.24*	99	methyl-		
1,2-Dihydro-2-imino-1-	10.71*	99	4,5-Dimethyl-	5.0*	99
methyl-			4,6-Dimethyl-	4.85	39
1,4-Dihydro-4-imino-1-	12.18	99	4-Dimethylamino-	7.94*	99
methyl-			4,6-Diphenyl-	3.74*	99
1,2-Dihydro-1-methyl-2-	2.45*	99	4-Hydroxy-	3.91*, 9.54*	99
oxo-			4-Methyl-	4.11*	99
1,4-Dihydro-1-methyl-4-	2.02*	99	5-Nitro-	0.35	99
oxo-			1,4,5,6-Tetrahydro-	14.11*	99
1,6-Dihydro-1-methyl-6-	1.79*	99	4-Amino-	7.23*	−
oxo-			5-Amino-	0.99*, 3.93*	99
1,2-Dihydro-1-methyl-2-	1.66*	99	4-Amino-6-chloro-	3.55*	99
thio-			5-Amino-4,6-dihydroxy-	3.6*, 8.9*	99
1,4-Dihydro-1-methyl-6-	0.56*	99	6-Amino-5-formylamino-4-	2.5*, 9.9*	99
thio-			hydroxy-		
2,4-Dihydroxy-	−3.38*, 9.45*	99	4-Amino-6-hydroxy-	3.30*, 10.81*	99
2,4-Dihydroxy-5-amino-	**3.20, 8.52**	**140**	4-Amino-6-hydroxy-5-	3.58*, 11.10*	99
2,6-Dihydroxy-4-amino-	**0.00, 8.69, 15.32**	**140**	methyl-		
2,6-Dihydroxy-4,5-diamino-	**4.56**	**140**	4-Amino-6-methyl-	7.7*	99
2,6-(=0)-1,3-Methyl-4,5-diamino-	**4.44**	**140**	5-Amino-6-methylthio-	5.44*	99
4,5-Dihydroxy-	1.99*, 7.52*, 11.69*	99	4,5-Diamino-	2.50*, 7.60*	99
			4,6-Diamino-	6.81*	99
4,6-Dihydroxy-	5.4	39	Pyrimidines, 4-amino		
4,6-Dimethyl-	2.7*	99	substituted		
4-Dimethylamino-2-	4.21*	99	5-Aminomethyl-2-methyl-	4.0*, 7.1*	99
hydroxy-			5-Carboxymethylamino-	2.99*, 6.70*	99
4-Dimethylamino-6-	1.22*, 10.49*	99	6-Chloro-	2.10*	−
hydroxy-			6-Chloro-2-methylamino-	3.79*	99
4,6-Dimethyl-2-methyl-	5.23*	99	1,2-Dihydro-1,5-dimethyl-2-	4.69*	99
amino-			oxo-		
4-Dimethylamino-2-	6.13*	99	1,6-Dihydro-6-imino-1-	11.94*	99
methoxy-			methyl-		
4-Dimethylamino-6-	4.27*	99	1,2-Dihydro-1-methyl-2-oxo-	4.57*	99
methoxy-			1,6-Dihydro-1-methyl-6-oxo-	0.98*	99
4-Ethoxy-2-hydroxy-	1.00*, 10.7*	99	2,3-Dihydro-3-methyl-2-oxo-	7.4*	99
4-Ethylamino-2-mercapto-	3.08*, 11.15*	99	5-Fluoro-2-hydroxy-	2.90*	99
4-Ethylamino-2-hydroxy-	4.56*	99	2-Hydroxy-	4.60*, 12.16*	99
5-Ethyl-2-methylamino-	4.29*	99	6-Hydroxy-	1.36*, 10.08*	99
4-Hydroxy-5-methoxy-	1.75*, 8.64*	99	2-Hydroxy-5-methyl-	4.6*, 12.4*	99
2-Hydroxy-4-methyl-	3.06*, 9.9*	99	6-Hydroxy-2-methylamino-	3.20*, 11.06*	99
4-Hydroxy-6-methyl-	2.06*, 9.1*	99	2-Hydroxy-5-nitro-	7.40*	99
2-Mercapto-4-methyl-	2.1*, 8.1*	99	2-Mercapto-	3.29*, 10.66*	99
4-Mercapto-6-methyl-	1.8*, 7.3*	99	2-Methoxy-	5.3*	99
2-Mercapto-4-methyl-	3.07*, 11.12*	99	6-Methoxy-	4.00*	99
amino-			6-Methyl-	6.16*	99
4-Methyl-2-methylthio-	1.86*	99	2-Methylamino-	7.53*	99
4-Methyl-6-methylthio-	3.16*	99	6-Methylamino-	6.30*	99
4-Methyl-2-sulphanilamido-	7.06*	99	6-Methylamino-5-nitro-	2.73*	99
1,4,5,6-Tetrahydro-	13.0	99	5-Trifluoroacetamido-	3.91*	99
1,4,5,6-Tetrahydro-2-Amino-	**14.1**	**154**	5-Amino-	6.00*	99
2,4,6-Trihydroxy-	3.9, 12.5	39	6-Amino-	5.99*	99
(Barbituric acid)			6-Amino-5-bromo-	4.20*	99

* Thermodynamic value.

NITROGEN COMPOUNDS—(continued)

Compound	pK_a'	Reference	Compound	pK_a'	Reference
Heterocyclics—(Continued)			**Heterocyclics—(Continued)**		
5-Amino-2-chloro-	2.63*	99	Pyrrole	17.51	152
5-Amino-2-chloro-6-ethoxycarbonyl-	1.27*	99	2,4-Dimethyl-	2.55*	99
			2,5-Dimethyl-	−0.71*	99
5-Amino-2-dimethylamino-6-ethoxycarbonyl-	−2.03*, 6.47*	99	3,4-Dimethyl-	0.66*	99
5-Amino-6-ethoxycarbonyl-2-hydroxy-	−3.21*, 3.22*	99	3-Ethyl-2,4-dimethyl-	3.54*	99
			1-Methyl-	−2.90*	99
5-Amino-6-ethoxycarbonyl-2-mercapto-	2.11*	99	2-Methyl-	−0.21	99
			3-Methyl-	−1.00	99
5-Amino-6-hydroxy-	1.28*, 3.54*,	99	2,3,4,5-Tetramethyl-	3.77	99
	9.89*	99	2,3,4-Trimethyl-	3.94	99
5-Amino-2-hydroxy-	4.34*, 11.48*	99	2,3,5-Trimethyl-	2.00	99
6-Amino-2-hydroxy-	6.47*, 11.98*	99	Δ'-Dihydro-1,2-dimethyl-	11.9	154
5-Amino-2-mercapto-	2.93*, 10.42*	99	Δ'-Dihydro-2-ethyl-	7.9	154
5-Amino-2-methylthio-	5.03*	99	Δ³-Dihydro-1-methyl-	9.9	154
5,6-Diamino-	1.41*, 5.75*	99	Pyrrolidine	11.27*	99
Pyrimidines, 5-amino substituted			3-Amino-2,5-dioxo-	5.9*, 9.0*	99
			1-(2-Aminoethyl)-	6.56*, 9.74*	99
4-Carboxy-6-carboxy-methylamino-2-chloro-	3.05*, 4.48*	99	2-Benzyl-	10.31*	99
			1-n-Butyl-2-methyl-	10.61*	99
4-Carboxy-6-carboxy-methylamino-2-dimethyl-amino-	9.85*	99	2-n-Butyl-1-methyl-	10.20*	99
			2-Carbamoyl-	8.82*	99
			1-Cyanomethyl-	4.8*	99
4-(1-Carboxyethylidene)-imino-	3.04*, 7.10*	99	2-Cyclohexyl-	10.76*	99
			1,2-Dimethyl-	10.20*	99
4-Carboxymethylamino-	2.9*, 6.59*	99	1-Ethyl-2-methyl-	10.56*	99
2-Ethoxy-4-ethoxycarbonyl-6-ethoxycarbonylmethyl-amino-	4.58*	99	1-Methyl-	10.32*	99
			2-(p-Tolyl)-	9.95	99
			2-Pyrrolines		
4-Methyl-	3.06*	99	2-Benzyl-	7.06*	99
1-Methyl-2-pyrimidone	2.50	39	2-n-Butyl-1-methyl-	11.84*	99
1-Methyl-4-pyrimidone	1.8	39	2-Cyclohexyl-	7.91*	99
3-Methyl-4-pyrimidone	1.84	39	1,2-Dimethyl-	11.90*	99
4-Pyrone	0.1	154	1-Ethyl-2-methyl-	11.84*	99
2,6-Dimethyl-	0.4	154	2-Phenyl-	6.7	99
3-Hydroxy-	7.9	154			

Quinazoline[19,39,99]	2	4	5	6	7	8
NH_2-	4.43	5.73	3.56*	3.2[a]	4.59*	2.4[a]
OH-	1.30	2.12	3.62*	3.12	3.20*	3.41
CH_3O-	1.31	3.13	3.39*	2.83*	2.87*	3.49*
Cl-	−1.6	—	3.73*	3.53*	3.25*	3.28*
SH-	0.26*	1.51*	—	—	—	—
	8.18*	8.47*	—	—	—	—
CH_3-	4.50*	2.44*	3.61*	3.39*	3.15*	3.18*
CH_3S-	1.60*	2.97*	—	—	—	—
NO_2-	—	—	3.73*	4.16*	4.03*	3.98*

Compound	pK_a'	Reference
3,4-Dihydro-	1.47, 9.2	99, 154
3,4-Dihydro-2-methyl-	10.2	154
3,4-Dihydro-3-methyl-	9.2	154
3,4-Dihydro-3-methyl-4-hydroxy	7.6	154
2:4-Dihydroxy-	9.78, 2.5	39
2,4-Dimethyl-	3.58*	99
3-Methiodide	7.26	39
3-Oxide	1.47*	99
1,2,3,4-Tetrahydro-	10.2	154

*Thermodynamic value.
[a] In 50% ethanol or methanol.

NITROGEN COMPOUNDS—(continued)

Quinoline[2],[44],[99]		3	4	5	6	7	8
H-	4.85*	4.80	4.69*	—	—	—	—
H₂N-	7.25*	4.86*	9.08*	5.37*	5.54*	6.56*	3.90*
HO-	−0.36	4.30	2.27	5.20	5.17	5.48	5.13
	11.74	8.06	11.25	8.54	8.88	8.85	9.89
CH₃-	5.42	5.14	5.20	4.62	4.92	5.08	4.60
	5.8	—	5.6	—	—	—	—
F-	—	2.36*	—	3.68*	4.00*	4.04*	3.08*
HO₂C-	4.96*	4.62*	4.53*	4.81*	4.98*	4.97*	7.20*
Br-	1.05*	2.75*	—	3.62*	3.91*	3.87*	3.33*
Cl-	—	2.63*	3.72*	3.65*	3.99*	3.85*	3.12*
HS-	1.44*	2.29*	0.77*	—	—	—	—
	10.25*	6.17*	8.87*	—	—	—	—
CH₃O-	3.16*	—	6.45*	—	5.03*	—	—
CH₃S-	3.67*	3.84*	5.85*	4.46*	4.71*	—	3.46*
O₂N-	—	1.03	—	2.69*	2.72*	2.40*	2.55*

Compound	pK_a'	Reference	Compound	pK_a'	Reference
Heterocyclics—(Continued)			**Heterocyclics—(Continued)**		
Quinoline			2-Mercapto-	−1.24*, 7.20*	99
4-Amino-7-chloro-	8.23*	99	1-Methiodide-	5.74	39
4-Amino-8-hydroxy-	6.91*, 10.71*	99	2-Methoxy-	0.28*	99
5-Amino-8-hydroxy-	5.67*, 11.24*	99	2-Methyl-	0.95*	99
4-Amino-6-methoxy-	8.93*	99	2-Methylamino-	4.07*	99
8-Amino-6-methoxy-	3.38*	99	2-Methylthio-	0.29*	99
6-Bromo-4-chloro-	2.83*	99	1:5-Naphthyridine	2.91	39
4-Chloro-6-ethoxy-	3.82*	99	Quinuclidine	10.95*	99
4-Chloro-6-fluoro-	2.95*	99	3-Carbamoyl-	9.67*	99
4-Chloro-6-methoxy-	3.93	99	3-Cyano-	7.81*	99
4-Chloro-6-methyl-	3.96	99	3-Phenyl-	10.23*	99
4,6-Dichloro-	2.81	99	4-CH₃-	10.88	158
2:4-Diamino-	9.45	19	4-CH₂CH₃-	10.95	158
1,4-Dihydro-4-imino-1-methyl-	12.4*	99	4-CH(CH₃)₂-	11.02	158
			4-C(CH₃)₃-	11.07	158
1,2-Dihydro-1-methyl-2-oxo-	−0.71*	99	4-CH=CH₂-	10.60	158
			4-C₆H₅-	10.20	158
1,2-Dihydro-1-methyl-2-thio-	−1.6	99	4-CH₂OH-	10.45	158
			4-CH₂OCH₃-	10.50	158
2:4-Dihydroxy-	5.86, 0.76	39	4-CH₂OCOCH₃-	10.27	158
2,3-Dimethyl-	4.94*	99	4-CH₂OTs-	9.87	158
2,4-Dimethyl-	5.12*	99	4-CH₂Cl-	10.19	158
2,6-Dimethyl-	6.1*	99	4-CH₂Br-	10.13	158
2,7-Dimethyl-	5.02*	99	4-CH₂I-	10.12	158
2,8-Dimethyl-	4.11*	99	4-CH(OH)₂-	9.90	158
1-Methyl-2-quinolone	−0.71	39	4-COO⁻	10.55	158
1-Methyl-4-quinolone	2.46	39	4-COOCH₃-	9.46	158
1,2,3,4-Tetrahydro-	5.0	154	4-COOC₂H₅-	9.44	158
Quinoxaline	0.8, 0.56	19, 39	4-COCH₃-	9.45	158
2-Amino-	3.96	19	4-CONH₂-	9.38	158
5-Amino-	2.62	19	4-CN-	8.07	158
6-Amino-	2.95	19	4-OH-	9.44	158
2-Carbamoyl-	−0.4*	99	4-OCH₃-	9.31	158
2:3-Diamino-	4.70	19	4-OCOCH₃-	8.99	158
2:3-Dihydroxy-	9.52	39	4-Cl-	8.62	158
2-Hydroxy-	9.08, −1.37	39	4-Br-	8.49	158
4-Hydroxy-	10.01, 2.85	39	4-I-	8.70	158
5-Hydroxy-	8.65, 0.9	39	4-NH₂-	10.10	158
6-Hydroxy-	7.92, 1.40	39	4-NHCH₃-	10.28	158
5-Hydroxy-1-methyl-quinoxalinium chloride	5.74	44	4-N(CH₃)₂-	10.11	158
			4-NHCOCH₃-	9.54	158
1,2,3,4-Tetrahydro-	2.1, 4.8	154	4-NHCOOC₂H₅-	9.57	158

* Thermodynamic value.

NITROGEN COMPOUNDS—(continued)

Compound	pK'_a	Reference	Compound	pK'_a	Reference
Heterocyclics (continued)			**Heterocyclics (continued)**		
4-NO$_2$-	7.65	158	1,3,5-Triazanaphthalene		
Riboflavin	9.93	77	4-Hydroxy-	8.98*	99
Serotonine	19.50	152	1,3,8-Triazanaphthalene		
Sulphadiazine	6.48	ε	2-Hydroxy-	1.81*, 10.06*	99
Sulphapyridine	8.43	6	1,4,5-Triazanaphthalene	1.20	39
Sulphaguanidine	11.25	6	8-Hydroxy-	8.76, 0.60	39
Sulphathiazole	7.12	6	1,4,6-Triazanaphthalene	3.05*	99
Terramycin	3.10, 7.26, 9.11	77	5-Hydroxy-	11.05, 0.78	39
1,4,5,8-Tetra-aza-	2.47*	99	1,2,4-Triazine		
naphthalene			3-Amino-	3.09*	99
Tetramethylenediamine	10.7	4	1,3,5-Triazine		
1,2,4,5-Tetrazine			2-Amino-4,6-bisethyl-	6.18*	99
3,6-Diethyl-1,4-dihydro-	4.23*	99	amino-		
1,4-Dihydro-	2.25*	99	2-Amino-4,6-dimethyl-	3.56*	99
Tetrazole	4.9	51	2,4-Diamino-	5.88*	99
5-Chloro-	2.1	51	2,4-Diamino-6-guanidino-	9.4*	99
5-Amino-	1.8, 6.0	51	2:4-Dihydroxy-	6.5	39
1,2,4-Thiadiazole			2,4,6-Triamino-	5.1*	99
3-Amino-5-phenyl-	0.1*	99	2,4,6-Trisdi(2-hydroxy)-	4.70*	99
5-Amino-3-phenyl-	1.4*	99	ethylamino-		
1,3,4-Thiadiazole			2,4,6-Trisguanidino-	4.6*, 7.6*,	99
2-Amino-5-phenyl-	2.9*	99		10.3*	
2-Benzylamino-5-phenyl-	2.5*	99	2,4,6-Trishydroxymethyl-	4.0*	99
2-Ethylamino-5-phenyl-	3.05*	99	amino-		
2-Methylamino-5-phenyl-	2.8*	99	**Triazole**		
Thiazine			Benzo-	8.38	126
Δ^2-Dihydro-2-methyl-	7.6	154	4,5 Dibromo-	5.37	126
1,4-Thiazine			1-Phenyl-5-methyl-	3.73	126
Tetrahydro-	8.40*	99	4-carboxylic acid-		
Thiazole	2.44*	99	1-Phenyl-4-carboxylic acid-	2.88	126
2-Amino-	5.36, 5.39	99, 41	1-Phenyl-4,5-dicarboxylic	2.13, 4.93	126
5-Carbamoyl-	0.6*	99	acid-		
4-Methyl-5-carboxylic	3.51	144	4-Carboxylic acid-	3.22, 8.73	126
acid-			4,5-Dicarboxylic acid-	1.86, 5.90, 9.30	126
4-Methyl-2-carboxylic	1.20, 3.18	144	**1,2,3-Triazole**	1.17*, 9.51*	99
acid-			1-Methyl-	1.25*	99
Δ^2-Dihydro-2-methyl-	5.2	154	**1,2,4-Triazole**	10.3, 2.3	154
1,3,4-Thiazole			3,5-Dimethyl-	3.8	154
2-Nitramino-	−2.5*	99	3,5-Diamino-	12.1, 4.4	154
Thiazolidine	6.22*, 6.31	99, 95	3-Methyl-	10.7, 3.3	154
4-Carboxy-	1.42*, 6.30*	99	3-Amino-	11.1, 4.0	154
4-Carboxylate methyl ester	4.00	95	3-Chloro-	8.1	154
4-Methoxycarbonyl-	3.91*	99	3,5-Dichloro-	5.2	154
Thiazolo(5,4-d)pyrimidine			**1,2,3-Triazolo(5′,4′-4,5)**		
7-Amino-	2.74*	99	Pyrimidine	2.03*, 4.96*	99
7-Methylamino-	2.81*	99	Tryptamine	16.60	152
Thiophene			Tryptophan	16.82	152
2-(2-Aminoethylamino-	6.29*, 9.77	99	4-Methyl-	16.90	152
methyl)-			5-Hydroxy-	19.20	152
2-Aminomethyl-	8.92	99			
1,2,4-Triazanaphthalene	−0.82*	99			

* Thermodynamic value.

SPECIAL NITROGEN COMPOUNDS

Compound	pK_a'	Reference
AMIDINES AND GUANIDINES		
Acetamidine	12.52	19
C-substituted Amidinium Ions		
BuO-	10.15	114
CH_2CHCH_2O-	9.70	114
EtO-	10.02	114
Me-	12.41	114
Me_2N-	13.4	114
MeO-	9.72	114
MeS-	9.83	114
NH_2-	13.86	114
Ph-	11.6	114
PrO-	10.16	114
1-Chloro-4-N^1-methyl-guanidino benzene	12.6*	99
1-Chloro-4-N^3-methyl-guanidino benzene	10.85*	99
Diguanide	3.07, 13.25	99
Ethyl-	3.08, 11.47	99
Ethylene-	11.34, 1.74, 2.88, 11.76	77
Methyl-	3.00, 11.42	99
Phenyl-	2.16, 10.71	77
Guanidine	13.6	99
Acetyl-	8.26	99
2-Anthryl-	11.0	99
Carbamoyl-	3.76	99
N,N-Dimethyl-	13.4	99
N,N'-Dimethyl-	13.6	99
N,N'-Diphenyl-	10.12	99
N-Methyl-	13.4	99
N-Methyl-N'-nitro-	12.40	99
2-Naphthyl-	10.7	99
Nitro-	−0.55, 12.20	99
Nitroamino-	−1, 10.60	99
Nitroso-	2.13, 11.70	99
Pentamethyl-	13.8	99
N-Phenyl-	10.77	99
N,N,N'-Trimethyl-	13.6	99
N,N',N''-Trimethyl-	13.9	99
Triphenyl-	9.10	99
1-substituted Guanidinium Ions		
CN-	−0.4	114
H_2NCO-	7.85	114
MeO-	7.46	114
$^-O_2CCH_2O-$	7.51	114
O-Methylisourea	9.80	20
N-Phenyl-	7.3	20
S-Methylisothiourea	9.83	20
N-Phenyl-	7.14	20
1-substituted 3-Nitroguanidines		
Bz-	8.10	114
EtO_2CCH_2-	11.20	114
H_2NCO-	7.50	114
$NCCH_2-$	9.30	114
NH_2-	10.60	114
Ph-	10.50	114
C-substituted-N-Phenylamidinium Ions		
EtO-	7.71	114

Compound	pK_a'	Reference
Amidines and Guanidines—(Continued)		
H_2N-	10.77	114
MeO-	7.41	114
MeS-	7.14	114
PhNH-	10.42	144
AMIDOXIMES		
Benz-	4.99	17
Malon-	~4.77	17
Ox-	3.02	17
α-Phenylacet-	5.24	17
Succin-	3.11, 5.97	17
o-Tolu-	4.03	17
p-Tolu-	5.14	17
HYDRAZINES (30°)		
Hydrazine	8.07	13
Acet-	3.24	15
N,N-Diethyl-	7.71	13
N,N'-Diethyl-	7.78	13
N,N-Dimethyl-	7.21	13
N,N'-Dimethyl-	7.52	13
Ethyl-	7.99	13
Glycylhydrazide	2.38, 7.69	15
Isonicotinhydrazide	1.85, 3.54, 10.77	77
Methyl-	7.87	13
Phenyl-	5.21 (15°)	14
Tetramethyl-	6.30	13
Trimethyl-	6.56	13
$HOCH_2CH_2-$	7.12	148
$C_6H_5-CH_2-$	6.83	148
$C_6H_5O(CH_2)_2$	6.80	148
$C_2H_5OOC-CH_2-$	5.97	148
$NC-(CH_2)_2-$	5.91	148
$CHF_2-CF_2-CH(CH_3)-$	5.59	148
$HC\equiv C-CH_2-$	5.46	148
CF_3-CH_2-	5.38	148
$CHF_2-(CF_2)_3-CH_2-$	5.34	148
$C_6H_5-CH(CF_3)-$	4.88	148
HYDRAZONES Hydrazone of		
Benzophenone	3.85	
p-Chloro-	3.53	16
p,p'-Dimethoxy-	4.38	16
p,p'-Dichloro-	3.13	16
p-Methoxyacetophenone	4.94	16
Phenyl-2-thienyl ketone	3.80	16
SEMICARBAZONES of		16
Acetone	1.33	14
Acetaldehyde	1.10	14
Benzaldehyde	0.96	14
Furfural	1.44	14
Pyruvic acid	0.59	14
Semicarbazide	3.66	14
NITROGEN ACIDS		
Dimedone	5.23	18
Diphenylthiocarbazone	4.5	6
Nitrourea	4.57	18
Nitrourethane	3.28	18
Phthalimide	8.30	18

* Thermodynamic value.

SPECIAL NITROGEN COMPOUNDS (continued)

Compound	pK_a'	Reference	Compound	pK_a'	Reference
OTHER			**Other—(Continued)**		
Acetamide	−0.51	4	Thiourea	−0.96	4
Azobenzene	−2.48*	99	Urea	0.18	4, 99
4-Amino-3'-methyl-	−2.88*	99	O-Allyl-	9.70	99
4-Amino-4'-methyl-	3.04	99	O-n-Butyl-	10.15	99
4-Dimethylamino-	−1.3*, 3.226*	99	O-Cyclohexyl-	10.19	99
4-Hydroxy-	−0.93*, 8.2*	99	O-Isobutyl-	10.30	99
Benzamide			O-Isopentyl-	10.11	99
3,5-Dinitro-4-methyl-	−2.77	110	N-Methyl	0.9	99
4-Methoxy-	−1.46	110	O-Methyl-	9.72	99
3-Nitro-	−2.25	110	O-Phenethyl-	10.03	99
2,3,6-Trichloro-	−3.10	110	Phenyl-	−0.3	99
3,4,5-Trimethoxy-	−1.86	110	N-Phenyl-O-methyl-	7.3	20
·Diphenylthiocarbazone	4.5	6	N-Propyl-	10.16	99

* Thermodynamic value.

THIOLS

Compound	pK_a'	Reference	Compound	pK_a'	Reference
N-Acetylcysteine	9.52	112	o-Mercaptophenylacetic acid	4.28, 7.67	59
N-Acetyl-β-mercaptoiso-leucine	10.30	112	2-Mercaptopropionic acid	4.32, 10.30	153
N-Acetylpenicillamine	9.90	112	Methyl cysteine	6.5, (7.5)	81
O-Aminothiophenol	6.59	81	Methyl-[β-diethylaminoethyl]-sulfide	9.8	5
p-Chlorothiophenol	7.50	81	Methyl thioglycolate	7.8	23
Cysteine	1.8, 8.3, 10.8	23	p-Nitrobenzenethiol	5.1	58
Cysteine ethyl ester	6.53, 9.05	112	Penicillamine	7.90, 10.42	112
Cysteinylcysteine	2.65, 7.27, 9.35, 10.85	23	Thiocyanic acid	−1.84	104
			Thioglycolic acid	3.67, 10.31	23
1-Diethylamino-butane-(4)	10.1	5	Thiophenol	7.8, 6.52	59, 81, 82
1-Diethylamino-hexane-(6)	10.1	5	Pentafluoro-	2.68	155
1-Diethylamino-propane-(3)	8.0, 10.5	5	p-Me-	6.82	157
N-Diethyl-cysteamine	7.8, 10.75	5	p-OMe-	6.77	157
N-Dimethyl-cysteamine	7.95, 10.7	5	m-Me-	6.66	157
N-Dipropyl-cysteamine	8.00, 10.8	5	m-OMe-	6.38	157
Ethyl mercaptan	10.50	81	p-Cl-	6.13	157
Glutathione	2.12, 3.59, 8.75, 9.65	23	p-Br-	6.02	157
			m-Cl-	5.78	157
DL-Homocysteine	8.70, 10.46	112	p-COMe-	5.33	157
2-Mercaptoethanesulfonate	7.53 (9.1)	81	m-NO$_2$-	5.24	157
Mercaptoethanol	9.5	23	p-NO$_2$-	4.71, 4.50	157
Mercaptoethylamine	8.6, 10.75	23	l-Thio-D-sorbitol	9.35	81
N-β-Mercaptoethylmorpholine	6.65, 9.8	5	N-Trimethyl cysteine	8.6	23
N-β-Mercaptoethylpiperidine	7.95, 11.05	5			
β-Mercaptoisoleucine	8.10, 10.6	112			

X =	−H	−S$^-$	−SH	X =	−H	−S$^-$	−SH
X(CH$_2$)$_2$SH	12.0	13.96	10.75	X(CH$_2$)$_3$SH	—	13.24	11.14
X(CH$_2$)$_4$SH	12.4	13.25	11.50	X(CH$_2$)$_5$SH	—	13.27	11.82

Compound	pK_a'	Reference	Compound	pK_a'	Reference
Mercaptans, RSH			t-C$_4$H$_9$-	11.05	82
R			(CH$_3$)$_2$CH-	10.86*	103
C$_6$H$_5$CH$_2$-	9.43	82	(CH$_3$)$_3$C-	11.22*	103
HOCH$_2$CH(OH)CH$_2$-	9.51	82	HOCH$_2$CH$_2$-	9.72	103
CH$_2$=CHCH$_2$-	9.96	82	CH$_3$CONHCH$_2$CH$_2$-	9.92	103
n-C$_4$H$_9$-	10.66	82	$^-$OCOCH$_2$-	10.68*	103
t-C$_5$H$_{11}$-	11.21	82	$^-$OCOCH$_2$CH$_2$-	10.84*	103
C$_2$H$_5$OCOCH$_2$-	7.95	82	o-$^-$OCOC$_6$H$_4$-	8.88*	103
C$_2$H$_5$OCH$_2$CH$_2$-	9.38	82	p-$^-$OCOC$_6$H$_4$-	5.80*	103
HOCH$_2$CH(OH)CH$_2$-	9.66	82	CH$_3$CO-	3.62*	103
n-C$_3$H$_7$-	10.65	82			

* Thermodynamic value.

CARBON ACIDS

Compound	pK'_a	Reference	Compound	pK'_a	Reference
Acetone	c. 20	24	$C_6H_5COCH_2NC_5H_5$	10.51	74
Acetonitrile	c. 25	24	$CH(COCH_3)_3$	5.85	74
Acetylacetone	8.95	24	$CH_3SO_2CH_3$	c. 23	74
Benzoylacetone (enol)	8.23	24	$CH(SO_2CH_3)_3$	Strong	74
Diacetylacetone	7.42	153	$C_2H_5O_2CCH_2CN$	9	74
Dihydroresorcinol	5.26	153	$CH_3CO_2C_2H_5$	c. 24.5	74
Ethyl acetoacetate	10.68	153	$CHC_2H_5(CO_2C_2H_5)_2$	15	74
Dimethylsulfone	14	24	CH_3CONH_2	c. 25	74
Hydrocyanic acid	9.21	25	$CH_2(CO_2C_2H_5)_2$	13.3	74
Nitroethane	8.44	153	CH_3CO_2H	c. 24	74
Nitromethane	10.21	153			
2-Nitropropane	7.74	18			
Saccharin	1.6	18			
Triacetylmethane	5.81	153			
CH_4	40	127			
$CH(NO_2)_3$	0	127			
CH_3CN	25	127			
$CH_2(CN)_2$	12	127			
$CH(CN)_3$	0	127			
$CH_3CHClNO_2$	7	74			
$CH_3COCH_2NO_2$	5.1	74			
$CH(NO_2)_3$	Strong	74			
$CH_3COCHCl_2$	15	74			
$CH_3COCHC_2H_5CO_2C_2H_5$	12.7	74			
$CH_3COCHCH_3COCH_3$	11	74			
$CH_3COCH_2COC_6H_5$	9.4	74			
$C_6H_5COCH_2COCF_3$	6.82	74			
CH_3COCH_2CHO	5.92	74			
$CH_3COCH_2CO_2CH_3$	10	74			
$CH_3SO_2CH_2SO_2CH_3$	14	74			
$CH_3SO_2CH(COCH_3)_2$	4.3	74			
$C_2H_5O_2CCH_2NO_2$	5.82	74			
$CH_2(NO_2)_2$	3.57	74			
CH_3COCH_2Cl	c. 16.5	74			
$CH_3COCH_2CO_2C_2H_5$	10.68	74			
$CH_3COCH_2COCH_3$	9	74			
$CH_3COCHBrCOCH_3$	7	74			
$CH_3COCH_2COCF_3$	4.7	74			

Compound	pK'_a	Reference

thiophene-CCH_2CCF_3 (with two C=O groups) — 6.10 — 74

cyclohexanone with H and $CO_2C_2H_5$ — 10.96 — 74

cyclopentanone with H and $COCH_3$ — 7.82 — 74

cyclopentanone with H and $CO_2C_2H_5$ — 10.5 — 74

cyclohexanone with H and CCH_3 (C=O) — 10.1 — 74

$CH_2(CHO)_2$ — 5 — 74

RC(NO_2)_2H R=	pK'_a	References
CH_3-	5.13	127
CH_3CH_2-	5.49	127
$CH_3CH_2CH_2-$	5.35	127
$(CH_3)_2CHCH_2-$	5.36	127
$CH_3(CH_2)_2CH_2-$	5.34	127
$CH_3(CH_2)_3CH_2-$	5.37	127
$CH_3(CH_2)_4CH_2-$	5.46	127
$CH_3(CH_2)_5CH_2-$	5.46	127
$CH_3(CH_2)_6CH_2-$	5.46	127
$CH_3(CH_2)_7CH_2-$	5.45	127
$CH_2=CHCH_2-$	4.95	127
$HOCH_2CH_2-$	4.44	127
$H-$	3.57	127
C_6H_5-	3.71	127
$CH(NO_2)_2CH_2-$	1.09	127
$CH_3C(NO_2)_2CH_2-$	1.35	127
$CH_3OOCCH_2CH_2-$	4.34	127
CH_3OCH_2-	3.48	127
$N\equiv CCH_2CH_2-$	3.45	127
$O_2NCH_2CH_2-$	3.24	127
CH_3OOCCH_2-	3.08	127
$(CH_3)_3\overset{+}{N}CH_2-$	-1.87	127
$N\equiv CCH_2-$	2.27	127
$(CH_3)_2CH-$	6.71	127

MISCELLANEOUS

Compound	pK'_a	Reference	Compound	pK'_a	Reference
ANTIBIOTICS AND VITAMINS			**INDICATORS AND DYES** (continued)		
Chlorotetracycline	3.30, 7.44, 9.27	99	Methyl red (2)	5.0	28
			Methyl yellow	3.25	28
5-Desoxypyridoxal	4.17, 8.14	99	Neutral red	7.4	28
Dimethyloxytetracycline	7.5, 9.4	99	Nile blue A	2.4	99
Isochlorotetracycline	3.1, 6.7, 8.3	99	Phenol blue	−6.5, 4.8	99
O-Methylpyridoxal	4.75	99	Phenolindophenol	−5.3, 0.95, 8.1	99
Oxytetracycline	3.27, 7.32, 9.11	99	Phenol red	8.03	97
Pyridoxal	4.20, 8.66, 13	99	Pinachrom (M)	7.31	99
Pyridoxal 5-phosphate	4.14, 6.20 8.69	99	N-Propylanilinesulpho-nephthalein	1.57, 13.11	99
Pyridoxamine	3.31, 7.90, 10.4	99	Propylhelianthin	3.95	99
			Pyronine B	7.7	99
Pyridoxamine 5-phosphate	2.5, 3.69, 5.76, 8.61, 10.92	99	Quinaldine red	2.63	99
			Rhodamine B	3.2	99
Pyridoxine	5.00, 8.96	99	Safranin O	6.4	99
Riboflavin	10.02	99	Thioflavine T	2.7	99
Tetracycline	3.30, 7.68, 9.69	99	Thionine	6.9	99
			Thymol blue (1)	1.65	28
			Thymol blue (2)	9.2	28
INDICATORS AND DYES			Toluidine blue 0	7.5	99
Acridine red	3.1	99	Tropeoline 00	2.0	28
Anilinesulphonephthalein	1.59, 12.26	99	**PORPHYRINS, BILE PIGMENTS AND STEROIDS**		
N-Benzylanilinesulpho-nephthalein	0.30, 12.76	99	Biliverdin	3	99
Bindschedler's green	−2.5	99	Chlorin e_6	1.9	99
Bismark brown Y	5.0	99	Chlorin p_6 trimethyl ester	1.4	99
Brilliant cresyl blue	3.2	99	Coproporphyrin I	7.13, 4.2	99
Bromocresol green	4.9	28	Deuteroporphyrin IX dimethyl		
Bromocresol purple	6.46	97	ester disulphonic acid	0.3, 4.7	99
Bromophenol blue	4.1	28	Dipyrrylmethene	8.50	99
Bromothymol blue	7.3	28	Mesobiliviolin	4.0	99
Butylhelianthin	4.01	99	N-Methyl coproporphyrin I	0.7, 11.3	99
Chlorophenol blue	4.43	97	N-Methyl coproporphyrin I		
Chlorophenol red	7.96	97	methyl ester	0.7, 8.3	99
Chrysoidin Y	5.3	99	Methylphaeophorbid a	0.2	99
Congo red	4.19	97	Methylphaeophorbid b	−0.1, 1.9	99
m-Cresol purple	1.70	97	Phaeopurpurin 18 methyl	−0.2, 2.1	99
Cyanine	8.62	99	ester		
N-Ethylanilinesulpho-nephthalein	1.73, 13.20	99	Phyllochlorin	2.1, 4.6	99
			Phylloporphyrin	2.5, 5	99
Ethylhelianthin	4.34	99	Pyrrochlorin	2.0, 4.5	99
Hexylhelianthin	3.71	99	Pyrroporphyrin	2.0, 4.5	99
Iodophenol blue	2.19	97	Rhodin g_7	1.6	99
N-Methylanilinesulphone-phthalein	1.36, 12.94	99	Rhodochlorin dimethyl ester	0.9	99
			b-Rhodochlorin dimethyl	0.2, 2.8	99
Methylene blue	3.8	99	ester		
Methylene green	3.2	99	Rhodoporphyrin	1.2, 3.7	99
Methylhelianthin	3.76	99	Stercobilin	7.60	99
Methyl orange	3.45	28	d-Urobilin	7.20	99
Methyl red (1)	2.3	28	i-Urobilin	7.40	99

REFERENCES

1. **Hall,** *J. Am. Chem. Soc.,* 79, 5441 (1957).
2. **Brown, McDaniel, and Häfliger,** in *Determination of Organic Structures by Physical Methods,* Braude and Nachod, Eds., Academic Press, New York, 1955, p. 567.
3. **Hall,** *J. Am. Chem. Soc.,* 79, 5439 (1957).
4. **Hodgman, Ed.,** *Handbook of Chemistry and Physics,* The Chemical Rubber Co., Cleveland, 1951, p. 1636.
5. **Franzen,** *Chem. Ber.,* 90, 623 (1957).
6. **Dawson, Elliott, Elliott, and Jones,** *Data for Biochemical Research,* Clarendon, Oxford, 1959.

7. Buist and Lucas, *J. Am. Chem. Soc.*, 79, 6157 (1957).
8. Stevenson and Williamson, *J. Am. Chem. Soc.*, 80, 5943 (1958).
9. Soloway and Lipschitz, *J. Org. Chem.*, 23, 613 (1958).
10. Bissell and Finger, *J. Org. Chem.*, 24, 1256 (1959).
11. Tuckerman, Mayer, and Nachod, *J. Am. Chem. Soc.*, 81, 92 (1959).
12. Bissot, Parry, and Campbell, *J. Am. Chem. Soc.*, 79, 796 (1957).
13. Hinman, *J. Org. Chem.*, 23, 1587 (1958).
14. Conant and Bartlett, *J. Am. Chem. Soc.*, 54, 2881 (1932).
15. Lindegreen and Niemann, *J. Am. Chem. Soc.*, 71, 1504 (1949).
16. Harnsberger, Cochran, and Szmant, *J. Am. Chem. Soc.*, 77, 5048 (1955).
17. Pearse and Pflaum, *J. Am. Chem. Soc.*, 81, 6505 (1959).
18. Bell and Higginson, *Proc. Roy. Soc.*, 197A, 141 (1949).
19. Albert, Goldacre, and Phillips, *J. Chem. Soc.*, 2240 (1948).
20. Dippy, Hughes, and Rozanski, *J. Chem. Soc.*, 2492 (1959).
21. Edsall, Martin, Bruce, and Hollingworth, *Proc. Natl. Acad. Sci. U.S.*, 44, 505 (1958).
22. Martin, Edsall, Wetlaufer, and Hollingworth, *J. Biol. Chem.*, 233, 1429 (1958).
23. Edsall and Wyman, *Biophysical Chemistry*, Academic Press, New York, 1958.
24. Pearson and Dillon, *J. Am. Chem. Soc.*, 75, 2439 (1953).
25. Ang, *J. Chem. Soc.*, 3822 (1959).
26. Martin and Fernelius, *J. Am. Chem. Soc.*, 81, 1509 (1959).
27. Ellenbogen, *J. Am. Chem. Soc.*, 74, 5198 (1952).
28. Kolthoff and Elving, *Treatise on Analytical Chemistry*, Interscience Encyclopedia, New York, 1959, 1.
29. Edwards, *J. Am. Chem. Soc.*, 76, 1540 (1954).
30. Bailey, Carrington, Lott, and Symons, *J. Chem. Soc.*, 290 (1960).
31. Brownstein and Stillman, *J. Phys. Chem.*, 63, 2061 (1959).
32. Meier and Schwarzenbach, *Helv. Chim. Acta*, 40, 907 (1957).
33. Ingham and Morrison, *J. Chem. Soc.*, 1200 (1933).
34. Hildebrand, *Principles of Chemistry*, Macmillan, New York, 1940.
35. Baddiley, in *The Nucleic Acids*, Chargaff and Davidson, Eds., Academic Press, New York, I, 1955, 137.
36. Circular OR-18, Pabst Laboratories, Milwaukee, Wisc., April 1961.
37. Bendich, in *The Nucleic Acids*, Chargaff and Davidson, Eds., Academic Press, New York, I, 1955, 81.
38. Jordan, in *The Nucleic Acids*, Chargaff and Davidson, Eds., Academic Press, New York, I, 1955, 447.
39. Albert and Phillips, *J. Chem. Soc.*, 1294 (1956).
40. Elliott and Mason, *J. Chem. Soc.*, 2352 (1959).
41. Angyal and Angyal, *J. Chem. Soc.*, 1461 (1952).
42. Edward and Nielsen, *J. Chem. Soc.*, 5075 (1957).
43. Bruice and Schmir, *J. Am. Chem. Soc.*, 80, 148 (1958).
44. Mason, *J. Chem. Soc.*, 674 (1958).
45. Linnell, *J. Org. Chem.*, 25, 290 (1960).
46. Jones and Katrizky, *J. Chem. Soc.*, 1317 (1959).
47. Jaffee and Doak, *J. Am. Chem. Soc.*, 77, 4441 (1955).
48. Clarke and Rothwell, *J. Chem. Soc.*, 1885 (1960).
49. Keyworth, *J. Org. Chem.*, 24, 1355 (1959).
50. Gawron, Duggan, and Grelecki, *Anal. Chem.*, 24, 969 (1952).
51. Sims, *J. Chem. Soc.*, 3648 (1959).
52. Fickling, Fischer, Mann, Packer, and Vaughan, *J. Am. Chem. Soc.*, 81, 4226 (1959).
53. Wold and Ballou, *J. Biol. Chem.*, 227, 301 (1957).
54. McElroy and Glass, *Phosphorus Metabolism*, Johns Hopkins, Baltimore, I, 1951.
55. Kumler and Eiler *J. Am. Chem. Soc.*, 65, 2355 (1943).
56. Van Wazer, *Phosphorus and its Compounds*, Inter-Science, New York, I, 1958.
57. Freedman and Doak, *Chem. Rev.*, 57, 479 (1957).
58. Ellman, *Arch. Biochem. Biophys.*, 74, 443 (1958).
59. Pascal and Tarbell, *J. Am. Chem. Soc.*, 79, 6015 (1957).
60. Bascombe and Bell, *J. Chem. Soc.*, 1096 (1959).
61. Gawron and Draus, *J. Am. Chem. Soc.*, 80, 5392 (1958).
62. Mukherjee and Grunwald, *J. Phys. Chem.*, 62, 1311 (1958).
63. Ballinger and Long, *J. Am. Chem. Soc.*, 81, 1050 (1959).
64. Ballinger and Long, *J. Am. Chem. Soc.*, 82, 795 (1960).
65. Haszeldine, *J. Chem. Soc.*, 1757 (1953).
66. Deno, Berkheimer, Evans, and Peterson, *J. Am. Chem. Soc.*, 81, 2344 (1959).
67. Gardner and Katritzky, *J. Chem. Soc.*, 4375 (1957).
68. Wise and Brandt, *J. Am. Chem. Soc.*, 77, 1058 (1955).

69. Fortnum, Battaglia, Cohen, and Edwards, *J. Am. Chem. Soc.,* 82, 778 (1960).
70. Everett and Minkoff, *Trans. Faraday. Soc.,* 49, 410 (1953).
71. Bauer and Miarka, *J. Am. Chem. Soc.,* 79, 1983 (1957).
72. Green, Sainsbury, Saville, and Stansfield, *J. Chem. Soc.,* 1583 (1958).
73. Burkhard, Sellers, DeCou, and Lambert, *J. Org. Chem.,* 24, 767 (1959).
74. Bell, *The Proton in Chemistry,* Cornell, Ithaca, 1959.
75. Stewart and Maeser, *J. Am. Chem. Soc.,* 46, 2583 (1924).
76. Jencks and Carriuolo, *J. Am. Chem. Soc.,* 82, 1778 (1960).
77. Bjerrum, Schwarzenbach, and Sillen, *Stability Constants of Metal-Ion Complexes, Part II, Inorganic Ligands,* Chemical Society, London, 1957.
78. Parsons, *Handbook of Electrochemical Constants,* Butterworths, London, 1959.
79. Bower and Robinson, *J. Phys. Chem.,* 64, 1078 (1960).
80. Candlin and Wilkins, *J. Chem. Soc.,* 4236 (1960).
81. Danehy and Noel, *J. Am. Chem. Soc.,* 82, 2511 (1960).
82. Kreevoy, Harper, Duvall, Wilgus, and Ditsch, *J. Am. Chem. Soc.,* 82, 4899 (1960).
83. Kabachnik, Mastrukova, Shipov, and Melentyeva, *Tetrahedron,* 9, 10 (1960).
84. Bjerrum, Schwarzenbach, and Sillen, *Stability Constants of Metal-Ion Complexes, Part I, Organic Ligands,* Chemical Society, London, 1957.
85. Crutchfield and Edwards, *J. Am. Chem. Soc.,* 82, 3533 (1960).
86. Lane and Quinlan, *J. Am. Chem. Soc.,* 82, 2994, 2997 (1960).
87. Moeller and Ferrús, *J. Phys. Chem.,* 64, 1083 (1960).
88. Bryson, *J. Am. Chem. Soc.,* 82, 4858, 4862, 4871 (1960).
89. Henderson and Streuli, *J. Am. Chem. Soc.,* 82, 5791 (1960).
90. Fortnum, Battaglia, Cohen, and Edwards, *J. Am. Chem. Soc.,* 82, 778 (1960).
91. Bell and McTigue, *J. Chem. Soc.,* 2983 (1960).
92. Li, Miller, Solony, and Gillis, *J. Am. Chem. Soc.,* 82, 3737 (1960).
93. Cohen and Erlanger, *J. Am. Chem. Soc.,* 82, 3928 (1960).
94. Olson and Margerum, *J. Am. Chem. Soc.,* 82, 5602 (1960).
95. Ratner and Clarke, *J. Am. Chem. Soc.,* 59, 200 (1937).
96. Porter, Rydon, and Schofield, *Nature,* 182, 927 (1958).
97. Kortum, Vogel, and Andrussow, *Dissociation Constants of Organic Acids and Aqueous Solution,* Butterworths, London, 1961.
98. King, *Qualitative Analysis and Electrolytic Solutions,* Harcourt Brace, New York, 1959.
99. Perrin, *Dissociation Constants of Organic Bases in Aqueous Solution,* Butterworths, London, 1965.
100. Albert and Serjeant, *Ionization Constants of Acids and Bases,* John Wiley & Sons, New York, 1962.
101. Yukawa, Ed., *Handbook of Organic Structural Analysis,* Benjamin, New York, 1965, 584.
102. Battaglia and Edwards, *Inorg. Chem.,* 4, 552 (1965).
103. Irving, Nelander, and Wadsö, *Acta Chem. Scand.,* 18, 769 (1964).
104. Morgan, Stedman, and Whincup, *J. Chem. Soc.,* 4813 (1965).
105. Barlin, *J. Chem. Soc.,* 2150 (1964).
106. Gallo, Pasqualucci, Radaelli, and Lancini, *J. Org. Chem.,* 29, 862 (1964).
107. Walba and Isensee, *J. Org. Chem.,* 26, 2789 (1961).
108. DYatkin, Mochalina, and Knunyants, *Tetrahedron,* 21, 2991 (1965).
109. Bell and McTigue, *J. Chem. Soc.,* 2985 (1960).
110. Yates and Riordan, *Can. J. Chem.,* 43, 2328 (1965).
111. *Tables for Identification of Organic Compounds,* Chemical Rubber Co., 3rd ed., 1967.
112. Friedman, Cavins, and Wall, *J. Am. Chem. Soc.,* 87, 3672 (1965).
113. Phillips, Eisenberg, George, and Rutman, *J. Biol. Chem.,* 240, 4393 (1965).
114. Charton, *J. Org. Chem.,* 30, 969 (1965).
115. Good, Winget, Winter, Connolly, Izawa, and Singh, *Biochemistry,* 5, 467 (1966).
116. Paelotti, Stern, and Vacca, *J. Phys. Chem.,* 69, 3759 (1965).
117. Spinner, *J. Chem. Soc.,* 3855 (1963).
118. Hine, Via, and Jensen, *J. Org. Chem.,* 36, 2926 (1971).
119. Eastes, Aldridge, Minesinger, and Kamlet, *J. Org. Chem.,* 36, 3847 (1971).
120. Kamlet and Minesinger, *J. Org. Chem.,* 36, 610 (1971).
121. Paulder and Humphrey, *J. Org. Chem.,* 35, 3467 (1970).
122. Filler and Schure, *J. Org. Chem.,* 32, 1217 (1967).
123. Walba, Stiggall, and Coutts, *J. Org. Chem.,* 32, 1954 (1967).
124. Walba and Ruiz-Velasco, *J. Org. Chem.,* 34, 3315 (1969).
125. Deutsch and Cheung, *J. Org. Chem.,* 38, 1124 (1973).
126. Hansen, West, Baca, and Blank, *J. Am. Chem. Soc.,* 90, 6588 (1971).
127. Sitzman, Adolph, and Kamlet, *J. Am. Chem. Soc.,* 90, 2815 (1968).

128. Hendrickson, Bergeron, Giga, and Sternbach, *J. Am. Chem. Soc.,* 95, 3412 (1973).
129. Christensen, Izatt, and Hansen, *J. Am. Chem. Soc.,* 89, 213 (1967).
130. Christensen, Oscarson, and Izatt, *J. Am. Chem. Soc.,* 90, 5949 (1968).
131. Christensen, Slade, Smith, Izatt, and Tsang, *J. Am. Chem. Soc.,* 92, 4164 (1970).
132. Bloomfield and Fuchs, *J. Chem. Soc., B,* 363 (1970).
133. Hughes, Nicklin, and Shrimanker, *J. Chem. Soc., B,* 3485 (1971).
134. Christensen, Rytting, and Izatt, *J. Chem. Soc., B,* 1643 (1970).
135. Christensen, Rytting, and Izatt, *J. Chem. Soc., B,* 1646 (1970).
136. Ives and Prasad, *J. Chem. Soc., B,* 1652 (1970).
137. Ives and Mosely, *J. Chem. Soc., B,* 1655 (1970).
138. Bolton and Hall, *J. Chem. Soc., B,* 1247 (1970).
139. Chuchani and Frohlich, *J. Chem. Soc., B,* 1417 (1970).
140. Barlin and Pfleinderer, *J. Chem. Soc., B,* 1425 (1971).
141. Bellobono and Favini, *J. Chem. Soc., B,* 2034 (1971).
142. Takahashi, Cohen, Miller, and Peake, *J. Org. Chem.,* 36, 1205 (1971).
143. Crowell and Hankins, *J. Phys. Chem.,* 73 1380 (1969).
144. Haake and Bausher, *J. Phys. Chem.,* 72, 2213 (1968).
145. Mäkitie and Mirttinen, *Suomen Kemistilehti, B,* 44, 155 (1971).
146. Koskinen and Nikkilä, *Suomen Kemistilehti, B,* 45, 89 (1971).
147. Serjeant, *Aust. J. Chem.,* 22, 1189 (1969).
148. Pollet and VandenEynde, *Bull. Soc. Chim. Belges,* 77, 341 (1968).
149. Wada and Takenaka, *Bull. Chem. Soc. Jap.,* 44, 2877 (1971).
150. Caplow, *Biochemistry,* 8, 2656 (1969).
151. Nozaki and Tanford, *J. Biol. Chem.,* 242, 4731 (1967).
152. Yagil, *Tetrahedron,* 23, 2855 (1967).
153. Albert, *Ionization Constants of Acids and Bases,* Methuen and Co. Ltd, London, 1962.
154. Albert, *Heterocyclic Chemistry,* Athalone Press, London, 1968.
155. Jencks and Salvesen, *J. Am. Chem. Soc.,* 93, 4433 (1971).
156. Bell, *The Proton in Chemistry,* 2nd ed., Cornell, Ithaca, N.Y. (1973).
157. DeMaria, Fini, and Hall, *J. Chem. Soc. Perkin II,* 1969 (1973).
158. Ceppi, Eckhardt, and Grob, *Tetrahedron Lett.,* 37, 3627 (1973).
159. Kurz and Farrar, *J. Am. Chem. Soc.,* 91, 6057 (1969).
160. Hine and Koser, *J. Org. Chem.,* 36, 1348 (1971).

DECI-NORMAL SOLUTIONS OF OXIDATION AND REDUCTION REAGENTS

Atomic and molecular weights in the following table are based upon the 1965 atomic weight scale and the isotope C-12. The weight in grams of the compound in 1 cc of the following deci-normal solutions is found by dividing the H equivalent in the last column by 1,000.

Name	Formula	Atomic or molecular weight	Hydrogen equivalent	0.1 Hydrogen equivalent in g
Antimony	Sb	121.75	$\frac{1}{3}$Sb	6.0875
Arsenic	As	74.9216	$\frac{1}{5}$As	3.7461
Arsenic trisulfide	As_2S_3	246.0352	$\frac{1}{4}As_2S_3$	6.1509
Arsenous oxide	As_2O_3	197.8414	$\frac{1}{4}As_2O_3$	4.9460
Barium peroxide	BaO_2	169.3388	$\frac{1}{2}BaO_2$	8.4669
Barium peroxide hydrate	$BaO_2 \cdot 8H_2O$	313.4615	$\frac{1}{2}BaO_2 \cdot 8H_2O$	15.6730
Calcium	Ca	40.08	$\frac{1}{2}$Ca	2.004
Calcium carbonate	$CaCO_3$	100.0894	$\frac{1}{2}CaCO_3$	5.0045
Calcium hypochlorite	$Ca(OCl)_2$	142.9848	$\frac{1}{4}Ca(OCl)_2$	3.5746
Calcium oxide	CaO	56.0794	$\frac{1}{2}$CaO	2.8040
Chlorine	Cl	35.453	Cl	3.5453
Chromium trioxide	CrO_3	99.9942	$\frac{1}{3}CrO_3$	3.3331
Ferrous ammonium sulfate	$FeSO_4(NH_4)SO_4 \cdot 6H_2O$	392.0764	$FeSO_4(NH_4)_2SO_4 \cdot 6H_2O$	39.2076
Hydroferrocyanic acid	$H_4Fe(CN)_6$	215.9860	$H_4Fe(CN)_6$	21.5986
Hydrogen peroxide	H_2O_2	34.0147	$\frac{1}{2}H_2O_2$	1.7007
Hydrogen sulfide	H_2S	34.0799	$\frac{1}{2}H_2S$	1.7040
Iodine	I	126.9044	I	12.6904
Iron	Fe	55.847	Fe	5.5847
Iron oxide (ferrous)	FeO	71.8464	FeO	7.1846
Iron oxide (ferric)	Fe_2O_3	159.6922	$\frac{1}{2}Fe_2O_3$	7.9846
Lead peroxide	PbO_2	239.1888	$\frac{1}{2}PbO_2$	11.9594
Manganese dioxide	MnO_2	86.9368	$\frac{1}{2}MnO_2$	4.3468
Nitric acid	HNO_3	63.0129	$\frac{1}{3}HNO_3$	2.1004
Nitrogen trioxide	N_2O_3	76.0116	$\frac{1}{4}N_2O_3$	1.9002
Nitrogen pentoxide	N_2O_5	108.0104	$\frac{1}{5}N_2O_5$	1.8001
Oxalic acid	$C_2H_2O_4$	90.0358	$\frac{1}{2}C_2H_2O_4$	4.5018
Oxalic acid hydrate	$C_2H_2O_4 \cdot 2H_2O$	126.0665	$\frac{1}{2}C_2H_2O_4 \cdot 2H_2O$	6.3033
Oxygen	O	15.9994	$\frac{1}{2}$O	0.8000
Potassium dichromate	$K_2Cr_2O_7$	294.1918	$\frac{1}{6}K_2Cr_2O_7$	4.9032
Potassium chlorate	$KClO_3$	122.5532	$\frac{1}{6}KClO_3$	2.0425
Potassium chromate	K_2CrO_4	194.1076	$\frac{1}{3}K_2CrO_4$	6.4733
Potassium ferrocyanide	$K_4Fe(CN)_6$	368.3621	$K_4Fe(CN)_6$	36.8362
Potassium ferrocyanide	$K_4Fe(CN)_6 \cdot 3H_2O$	422.4081	$K_4Fe(CN)_6 \cdot 3H_2O$	42.2408
Potassium iodide	KI	166.0064	KI	16.6006
Potassium nitrate	KNO_3	101.1069	$\frac{1}{3}KNO_3$	3.3702
Potassium perchlorate	$KClO_4$	138.5526	$\frac{1}{8}KClO_4$	1.7319
Potassium permanganate	$KMnO_4$	158.0376	$\frac{1}{5}KMnO_4$	3.1608
Sodium chlorate	$NaClO_3$	106.4410	$\frac{1}{6}NaClO_3$	1.7740
Sodium nitrate	$NaNO_3$	84.9947	$\frac{1}{3}NaNO_3$	2.8332
Sodium thiosulfate	$Na_2S_2O_3 \cdot 5H_2O$	248.1825	$Na_2S_2O_3 \cdot 5H_2O$	24.8183
Stannous chloride	$SnCl_2$	189.5960	$\frac{1}{2}SnCl_2$	9.4798
Stannous oxide	SnO	134.6894	$\frac{1}{2}$SnO	6.7345
Sulfur dioxide	SO_2	64.0628	$\frac{1}{2}SO_2$	3.2031
Tin	Sn	118.69	$\frac{1}{2}$Sn	5.935

This table originally appeared in Sober, Ed., *Handbook of Biochemistry and selected data for Molecular Biology*, 2nd ed., Chemical Rubber Co., Cleveland, 1970.

MEASUREMENT OF pH

Roger G. Bates and Maya Paabo

I. Definition of pH

The following definition of pH has received the endorsement of the International Union of Pure and Applied Chemistry.

1. *Operational definition.* In all existing national standards the definition of pH is an operational one. The electromotive force E_x of the cell:

Pt, H$_2$ |solution X|concentrated KCl solution| reference electrode

is measured and likewise the electromotive force E_s of the cell:

Pt, H$_2$ |solution S|concentrated KCl solution| reference electrode

both cells being at the same temperature throughout and the reference electrodes and bridge solutions being identical in the two cells. The pH of the solution X, denoted by pH(X), is then related to the pH of the solution S, denoted by pH(S), by the definition:

$$pH(X) = pH(S) + \frac{E_x - E_s}{(RT \ln 10)/F}$$

where R denotes the gas constant, T the thermodynamic temperature, and F the faraday constant. Thus defined the quantity pH is dimensionless.

To a good approximation, the hydrogen electrodes in both cells may be replaced by other hydrogen ionresponsive electrodes, e.g., glass or quinhydrone. The two bridge solutions may be of any molality not less than 3.5 mol kg^{-1}, provided they are the same (see *Pure. Appl. Chem.,* 1, 163 1960).

2. *Standards.* The difference between the pH of two solutions having been defined as above, the definition of pH can be completed by assigning a value of pH at each temperature to one or more chosen solutions designated as standards. A series of pH(S) values for seven suitable standard reference solutions is given in Table 1. The constants for calculating pH(S) values over the temperature range for 0 to 95°C are given in Table 2.

If the definition of pH given above is adhered to strictly, then the pH of a solution may be slightly dependent on which standard solution is used. These unavoidable deviations are caused not only by imperfections in the response of the hydrogen ion electrodes but also by variations in the liquid-junction potentials resulting from the different ionic compositions and mobilities of the several standards and from differences in the structure of the liquid-liquid boundary. In fact such variations in measured pH are usually too small to be of practical significance. Moreover, the acceptance of several standards allows the use of the following alternative definition of pH.

Table 1
VALUES OF pH(S) FOR SEVEN PRIMARY STANDARD SOLUTIONS

$t/°C$	A	B	C	D	E	F	G	$t/°C$	A	B	C	D	E	F	G
0	—	3.863	4.003	6.984	7.534	9.464	10.317	40	3.547	3.753	4.035	6.838	7.380	9.068	9.889
5	—	3.840	3.999	6.951	7.500	9.395	10.245	45	3.547	3.750	4.047	6.834	7.373	9.038	9.856
10	—	3.820	3.998	6.923	7.472	9.332	10.179	50	3.549	3.749	4.060	6.833	7.367	9.011	9.828
15	—	3.802	3.999	6.900	7.448	9.276	10.118	55	3.554	—	4.075	6.834	—	8.985	—
20	—	3.788	4.002	6.881	7.429	9.225	10.062	60	3.560	—	4.091	6.836	—	8.962	—
25	3.557	3.776	4.008	6.865	7.413	9.180	10.012	70	3.580	—	4.126	6.845	—	8.921	—
30	3.552	3.766	4.015	6.853	7.400	9.139	9.966	80	3.609	—	4.164	6.859	—	8.885	—
35	3.549	3.759	4.024	6.844	7.389	9.102	9.925	90	3.650	—	4.205	6.877	—	8.850	—
38	3.548	3.755	4.030	6.840	7.384	9.081	9.903	95	3.674	—	4.227	6.886	—	8.833	—

The compositions of the standard solutions are:
A: KH tartrate (saturated at 25°C)
B: KH_2 citrate, m = 0.05 mol kg^{-1}
C: KH phthalate, m = 0.05 mol kg^{-1}
D: $KH_2 PO_4$, m = 0.025 mol kg^{-1}; $Na_2 HPO_4$, m = 0.025 mol kg^{-1}
E: $KH_2 PO_4$, m = 0.008695 mol kg^{-1}; $Na_2 HPO_4$, m = 0.03043 mol kg^{-1}
F: $Na_2 B_4 O_7$, m = 0.01 mol kg^{-1}
G: $NaHCO_3$, m = 0.025 mol kg^{-1}; $Na_2 CO_3$, m = 0.025 mol kg^{-1} where m denotes molality.

Table 2
VALUES OF THE CONSTANTS OF THE EQUATION: $pH(S) = \frac{A}{T} + B + CT + DT^2$,
FOR SEVEN PRIMARY STANDARD BUFFER SOLUTIONS
FROM 0 TO 95°C

Solution	Temperature range °C	A	B	C	$10^5 D$	Standard deviation of the fitted curves
A. Tartrate	25 to 95	−1727.96	23.7406	−0.075947	9.2873	0.0016
B. Citrate	0 to 50	1280.4	−4.1650	0.012230	0	0.0010
C. Phthalate	0 to 95	1678.30	−9.8357	0.034946	−2.4804	0.0027
D. Phosphate	0 to 95	3459.39	−21.0574	0.073301	−6.2266	0.0017
E. Phosphate	0 to 50	5706.61	−43.9428	0.154785	−15.6745	0.0011
F. Borax	0 to 95	5259.02	−33.1064	0.114826	−10.7860	0.0025
G. Carbonate	0 to 50	2557.1	−4.2846	0.019185	0	0.0026

The electromotive force E_x is measured, and likewise the electromotive forces E_1 and E_2, of two similar cells with the solution X replaced by the standard solutions S_1 and S_2 such that E_1 and E_2 values are on either side of, and as near as possible to, E_x. The pH of solution X is then obtained by assuming linearity between pH and E, that is to say

$$\frac{pH(X) - pH(S_1)}{pH(S_2) - pH(S_1)} = \frac{E_x - E_1}{E_2 - E_1}$$

This procedure is especially recommended when the hydrogen-ion-responsive electrode is a glass electrode.

II. Standard Solutions

The pH meter or other electrometric pH assembly does not, strictly speaking, measure the pH but rather indicates a difference between the pH of an unknown solution (X) and a standard solution (S), both of which are at the same temperature. The pH meter should always be standardized routinely with two reference solutions of assigned pH, chosen if possible to bracket the pH of the test solution. These standards are prepared as indicated in Table 3. For convenience, air weights of the buffer salts are given. A good grade of distilled or de-ionized water should be used; for the four solutions of highest pH, the water should be freed of dissolved carbon dioxide by boiling or purging. For a detailed discussion of the properties of the primary standard buffer solutions, the reader is referred to chapter 4 of R. G. Bates, *Determination of pH*, 2nd ed., (John Wiley and Sons, Inc., New York, 1973).

Highly pure buffer materials should be used. These materials are obtainable commercially; they are also distributed as certified standard reference materials by the National Bureau of Standards. It should be noted that individual lots show slight variations; hence, the values certified for a particular lot may differ slightly from those given in Table 1.

Table 3
PREPARATION OF PRIMARY STANDARD BUFFER SOLUTIONS

Standard solution	NBS SRM No.[a]	Buffer substance	Weight in air[b] (g)	Standard solution	NBS SRM No.[a]	Buffer substance	Weight in air[b] (g)
A. Tartrate	188	$KHC_4H_4O_6$	(Satd. at 25°C)	E. Phosphate	186Ic	KH_2PO_4	1.179
					186IIb	Na_2HPO_4	4.302
B. Citrate	190	$KH_2C_6H_5O_7$	11.41	F. Borax	187a	$Na_2B_4O_7 \cdot$	3.80
C. Phthalate	185d	$KHC_8H_4O_4$	10.12			$10 H_2O$	
D. Phosphate	186Ic	KH_2PO_4	3.388	G. Carbonate	191	$NaHCO_3$	2.092
	186IIb	Na_2HPO_4	3.533		192	Na_2CO_3	2.640

[a]These materials may be ordered from the Office of Standard Reference Materials, National Bureau of Standards, Washington, D.C. 20234.
[b]This weight of salt to be dissolved in water and diluted to 1 liter at 25°C to provide concentrations indicated in Table 1.

The use of two or more standard reference solutions may disclose small inconsistencies in the standardization of the pH meter, depending on which standards are chosen. When this is the case, the best results are often obtained by assuming linearity between E and pH between the two calibrating points bracketing the pH of the unknown.

III. Electrodes

Although the hydrogen electrode is the ultimate standard on which the pH scale is based, in practice the convenient and versatile glass electrode is favored for the vast majority of pH measurements. New glass electrodes, or those that have been allowed to dry out, should be conditioned by soaking in water for several hours before use and after exposure to nonaqueous or dehydrating media. Some glass electrodes are designed especially for use at high temperatures, while others are best suited to low-temperature use. Special "high pH" electrodes are also available. For optimum results, careful attention should be paid to selection of the proper electrode for the problem at hand.

Glass electrodes of small dimensions are of great utility when sample volumes are limited. The pH-sensitive glasses are, however, moderately soluble, and small amounts of alkali are dissolved from the glass surface by the solutions in which the electrode is immersed. For this reason, the most accurate results are obtained when the ratio of the electrode area to sample volume is small.

The concentrated solution of potassium chloride that joins the reference electrode with the unknown or standard solution is reasonably effective in reducing the liquid-junction potential to small, fairly constant, values. It is important to assure that the flow of bridge solution into the test solution is neither excessive nor completely interrupted by crystallization of salt in the aperture where liquid-liquid contact is established.

Temperature gradients within the pH cell are a common source of difficulty, marked by variability and inaccuracy in the reading. Both of the electrodes, and the standard and test solutions as well, should be within a few degrees Celsius of the same temperature. For results of the highest reliability, temperature control should be provided. It is the function of the temperature compensator of the pH meter to adjust the pH-e.m.f. slope in such a manner that a difference of e.m.f. (in volts) is correctly converted to a difference of pH. This adjustment cannot compensate for inequalities of temperature through the cell or for differences between the temperature of the standard and test solutions.

IV. Techniques

Electrodes and sample cups should be washed carefully with distilled or de-ionized water and gently dried with clean absorbent tissue. The electrodes are immersed in the first standard solution and the temperature compensator of the measuring instrument is set at the temperature of the solutions whose pH is to be measured. The standardization control of the instrument is adjusted until the meter is balanced at the known pH of the standard, as given in Table 1. This procedure is repeated with successive portions of the same standard until replacement causes no change in the position of balance. The electrodes are then washed once more and dried.

A second standard solution is selected and the measurement repeated without altering the position of the standardization control. The pH reading of this second solution is noted and the sample replaced with a second portion of the same solution. This replacement is continued until successive readings agree within 0.02 pH unit, when the electrodes and meter may be judged to be functioning properly. It is advisable to make a final check with one of the buffers at the conclusion of a series of measurements.

After the instrument is properly standardized, a portion of the test solution is placed in the sample cup and the pH reading noted. Successive portions are again used until two measurements agree within the limits imposed by the reproducibility of the measuring instrument and the temperature control. With the best meters, measurements on buffered solutions should be reproducible to 0.01 unit or even better. With water or poorly buffered solutions, values agreeing to 0.1 unit may have to be accepted. Some improvement will result if poorly buffered solutions are protected from carbon dioxide of the atmosphere during the period of the measurements.

V. Interpretation of pH Numbers

The standard values of pH given in the table of an earlier section are based on hydrogen electrode potentials as measured in cells without a liquid junction. The uncertainty of the standard values is estimated at 0.005 unit. The accuracy of the results furnished by a given pH assembly adjusted with these primary standards is, however, further limited by inconsistencies which have their origin in defects of the glass electrode response and variations in the liquid-junction potential. For these reasons, the accuracy of experimental pH numbers can be considered to be better than 0.01 unit only under unusually favorable conditions.

The operational definition of pH fulfills adequately the need for an experimental scale capable of furnishing reproducible pH numbers. The interpretation of these numbers may be of secondary importance and should only be attempted when the standard and unknown solutions are matched so closely in composition that there is good reason to believe that the liquid-junction potential remains fairly constant when the standard is replaced by the unknown. In general, this will be the case when the unknowns are aqueous solutions of simple salts of total concentration not in excess of 0.2 M with pH values between 2.5 and 11.5.

When these "ideal" conditions prevail, the experimental pH can be considered to approach $-\log a_H$, where a_H is the conventional hydrogen ion activity defined in a manner consistent with the convention on which the standard values of pH(S) were based Bates, R. G., *J. Res. Natl. Bur. Standards,* 66A, 179 (1962)). All quantitative applications of pH measurements, when justifiable, should therefore be based on the approximation $pH(X) \approx -\log a_H = -\log m_H \gamma_H$, where m is molality and γ is the activity coefficient.

VI. Indicator Methods

Acid-base indicators have the property of altering the color of a solution in the region 1 to 2 pH units as the pH changes. They are therefore useful for pH measurements, although in general the accuracy is inferior to that obtainable by electrometric procedures. A list of suitable indicators, their pH ranges and color changes, is given in Table 4.

Table 4
ACID-BASE INDICATORS

Indicator	pH Range	Color change	Indicator	pH Range	Color change
Acid cresol red	0.2–1.8	Red–yellow	Metacresol purple	7.6–9.2	Yellow–purple
Acid metacresol purple	1.2–2.8	Red–yellow	Thymol blue	8.0–9.6	Yellow–blue
Acid thymol blue	1.2–2.8	Red–yellow	Phthalein red	8.6–10.2	Yellow–red
Bromophenol blue	3.0–4.6	Yellow–blue	Tolyl red	10.0–11.6	Red–yellow
Bromocresol green	3.8–5.4	Yellow–blue	Acyl red	10.0–11.6	Red–yellow
Methyl red	4.4–6.0	Red–yellow	Parazo orange	11.0–12.6	Yellow–orange
Chlorophenol red	5.2–6.8	Yellow–red	Acyl blue	12.0–13.6	Red–blue
Bromocresol purple	5.2–6.8	Yellow–purple	Benzo yellow	2.4–4.0	Red–yellow
Bromothymol blue	6.0–7.6	Yellow–blue	Benzo red	4.4–7.6	Red–blue
Phenol red	6.8–8.4	Yellow–red	Thymol red	8.0–11.2	Yellow–red
Cresol red	7.2–8.8	Yellow–red			

Courtesy of W. A. Taylor and Co.

Equal concentrations of the same indicator are added to the test solution and to each of a series of buffer solutions of known pH selected to bracket the pH of the test solution. Color comparisons are made with a colorimeter or spectrophotometer, and solutions of equal color are assumed to have the same pH. The pH of a series of suitable reference solutions can be determined in advance by electrometric methods. Alternatively, tables of pH as a function of composition can be utilized. The compositions and pH values of a set of useful solutions covering the range pH 1 to 13 are summarized in Table 5.

Table 5
BUFFER SOLUTIONS FOR INDICATOR MEASUREMENTS AND pH CONTROL

25 ml 0.2 *M* KCl, *x* ml 0.2 *M* HCl, diluted to 100 ml

pH	x	pH	x
1.00	67.0	1.50	20.7
1.10	52.8	1.60	16.2
1.20	42.5	1.70	13.0
1.30	33.6	1.80	10.2
1.40	26.6	1.90	8.1
—	—	2.00	6.5
—	—	2.10	5.1
—	—	2.20	3.9

50 ml 0.1 *M* KH Phthalate, *x* ml 0.1 *M* HCl, diluted to 100 ml

pH	x	pH	x
2.20	49.5	3.20	15.7
2.30	45.8	3.30	12.9
2.40	42.2	3.40	10.4
2.50	38.8	3.50	8.2
2.60	35.4	3.60	6.3
2.70	32.1	3.70	4.5
2.80	28.9	3.80	2.9
2.90	25.7	3.90	1.4
3.00	22.3	4.00	0.1
3.10	18.8	—	—

50 ml 0.1 *M* KH Phthalate, *x* ml 0.1 *M* NaOH, diluted to 100 ml

pH	x	pH	x
4.10	1.3	5.10	25.5
4.20	3.0	5.20	28.8
4.30	4.7	5.30	31.6
4.40	6.6	5.40	34.1
4.50	8.7	5.50	36.6
4.60	11.1	5.60	38.8
4.70	13.6	5.70	40.6
4.80	16.5	5.80	42.3
4.90	19.4	5.90	43.7
5.00	22.6	—	—

50 ml 0.1 *M* KH_2PO_4, *x* ml 0.1 *M* NaOH, diluted to 100 ml

pH	x	pH	x
5.80	3.6	6.80	22.4
5.90	4.6	6.90	25.9
6.00	5.6	7.00	29.1
6.10	6.8	7.10	32.1
6.20	8.1	7.20	34.7
6.30	9.7	7.30	37.0
6.40	11.6	7.40	39.1
6.50	13.9	7.50	41.1
6.60	16.4	7.60	42.8
6.70	19.3	7.70	44.2
—	—	7.80	45.3
—	—	7.90	46.1
—	—	8.00	46.7

50 ml of a mixture 0.1 *M* with respect to both KCl and H_3BO_3, *x* ml 0.1 *M* NaOH, diluted to 100 ml

pH	x	pH	x
8.00	3.9	9.00	20.8
8.10	4.9	9.10	23.6
8.20	6.0	9.20	26.4
8.30	7.2	9.30	29.3
8.40	8.6	9.40	32.1
8.50	10.1	9.50	34.6
8.60	11.8	9.60	36.9
8.70	13.7	9.70	38.9
8.80	15.8	9.80	40.6
8.90	18.1	9.90	42.2
—	—	10.00	43.7
—	—	10.10	45.0
—	—	10.20	46.2

50 ml 0.1 *M* Tris(hydroxmethyl)-aminomethane, *x* ml 0.1 *M* HCl, diluted to 100 ml

pH	x	pH	x
7.00	46.6	8.00	29.2
7.10	45.7	8.10	26.2
7.20	44.7	8.20	22.9
7.30	43.4	8.30	19.9
7.40	42.0	8.40	17.2
7.50	40.3	8.50	14.7
7.60	38.5	8.60	12.4
7.70	36.6	8.70	10.3
7.80	34.5	8.80	8.5
7.90	32.0	8.90	7.0
—	—	9.00	5.7

Table 5 (continued)
BUFFER SOLUTIONS FOR INDICATOR MEASUREMENTS AND pH CONTROL

50 ml 0.025 M Borax, x ml 0.1 M HCl, diluted to 100 ml

pH	x	pH	x
8.00	20.5	8.50	15.2
8.10	19.7	8.60	13.5
8.20	18.8	8.70	11.6
8.30	17.7	8.80	9.4
8.40	16.6	8.90	7.1
—	—	9.00	4.6
—	—	9.10	2.0

50 ml 0.05 M NaHCO$_3$, x ml 0.1 M NaOH, diluted to 100 ml

pH	x	pH	x
9.60	5.0	10.60	19.1
9.70	6.2	10.70	20.2
9.80	7.6	10.80	21.2
9.90	9.1	10.90	22.0
10.00	10.7	11.00	22.7
10.10	12.2	—	—
10.20	13.8	—	—
10.30	15.2	—	—
10.40	16.5	—	—
10.50	17.8	—	—

50 ml 0.025 M Borax, x ml 0.1 M NaOH, diluted to 100 ml

pH	x	pH	x
9.20	0.9	10.20	20.5
9.30	3.6	10.30	21.3
9.40	6.2	10.40	22.1
9.50	8.8	10.50	22.7
9.60	11.1	10.60	23.3
9.70	13.1	10.70	23.80
9.80	15.0	10.80	24.25
9.90	16.7	—	—
10.00	18.3	—	—
10.10	19.5		

50 ml 0.05 M Na$_2$HPO$_4$, x ml 0.1 M NaOH, diluted to 100 ml

pH	x	pH	x
10.90	3.3	11.40	9.1
11.00	4.1	11.50	11.1
11.10	5.1	11.60	13.5
11.20	6.3	11.70	16.2
11.30	7.6	11.80	19.4
—	—	11.90	23.0
—	—	12.00	26.9

25 ml 0.2 M KCl, x ml 0.2 M NaOH, diluted to 100 ml

pH	x	pH	x
12.00	6.0	12.50	20.4
12.10	8.0	12.60	25.6
12.20	10.2	12.70	32.2
12.30	12.8	12.80	41.2
12.40	16.2	12.90	53.0
—	—	13.00	66.0

Source: Bower and Bates, *J. Res. Natl. Bur. Standards,* 55, 197 (1955); Bates and Bower, *Anal. Chem.,* 28, 1322 (1956).

Table 6
pH VALUES FOR MISCELLANEOUS BUFFER SOLUTIONS OVER A RANGE OF TEMPERATURE

Composition of the buffer solution	m^a	0	5	10	15	20	25	30	35	40	45	50
Potassium dihydrogen phosphate (*m*) (1)	0.005	—	—	—	—	—	6.251	—	—	—	—	—
Sodium succinate (*m*)	0.015	—	—	—	—	—	6.162	—	—	—	—	—
	0.025	—	—	—	—	—	6.109	—	—	—	—	—
Piperazine phosphate (*m*) (2)	0.02	6.580	6.515	6.453	6.394	6.338	6.284	6.234	6.185	6.140	6.097	6.058
	0.05	6.589	6.525	6.463	6.404	6.348	6.294	6.243	6.195	6.149	6.106	6.066
2,2-Bis(hydroxymethyl)-2,2''-nitrilotriethanol (2*m*)	0.02	7.000	6.905	6.812	6.722	6.635	6.551	6.469	6.390	6.312	6.237	6.165
	0.04	7.029	6.932	6.839	6.748	6.662	6.577	6.495	6.415	6.336	6.262	6.190
	0.06	7.050	6.953	6.859	6.767	6.681	6.595	6.513	6.434	6.353	6.280	6.208
	0.08	7.067	6.969	6.876	6.783	6.696	6.610	6.528	6.448	6.367	6.294	6.222
Hydrochloric acid (*m*) (3)	0.10	7.082	6.983	6.889	6.796	6.710	6.623	6.540	6.460	6.378	6.306	6.235
Morpholine (1.5*m*) Hydrochloric acid (*m*) (4)	0.10	8.963	8.828	8.702	8.579	8.458	8.343	8.231	8.120	8.013	7.908	7.806
Tris(hydroxymethyl)amino-methane ("Tris") (*m*), Tris.HCl (*m*) (5)	0.05	8.946	8.774	8.614	8.461	8.313	8.173	8.036	7.904[b]	7.777	7.654	7.537
Tris (*m*), Tris.HCl (3*m*) (6)	0.01667	8.471	8.303	8.142	7.988	7.840	7.698	7.563	7.433	7.307	7.186	7.070

t/°C

[a] mol kg^{-1}
[b] 7.851 at 37°C

Table 6 (continued)
pH VALUES FOR MISCELLANEOUS BUFFER SOLUTIONS OVER A RANGE OF TEMPERATURE

Composition of the buffer solution	m^a	0	5	10	15	20	25	30	35	40	45	50
						$t/°C$						
N-Tris(hydroxymethyl)methyl-glycine ("Tricine") (m), Na Tricinate (m) (7)	0.05	—	8.485	8.375	8.271	8.175	8.079	7.988	7.902	7.817	7.740	7.663
Tricine ($3m$), Na Tri-cinate (m) (7)	0.02	—	8.023	7.916	7.813	7.713	7.621	7.527	7.437[c]	7.355	7.275	7.197

[c] 7.407 at 37°C

Compiled by Roger G. Bates and Maya Paabo.

Contribution from the National Bureau of Standards, not subject to copyright.

REFERENCES

1. Paabo, Bates, and Robinson, *J. Res. Natl. Bur. Standards*, 67A, 573 (1963).
2. Hetzer, Robinson, and Bates, *Anal. Chem.*, 40, 634 (1968).
3. Paabo and Bates, *J. Phys. Chem.*, 74, 702 (1970).
4. Hetzer, Bates, and Robinson, *J. Phys. Chem.*, 70, 2869 (1966).
5. Bates and Robinson, *Anal. Chem.*, 45, 420 (1973).
6. Durst and Staples, *Clin. Chem.*, 18, 206 (1972).
7. Bates, Roy and Robinson, *Anal. Chem.*, 45, 1663 (1973).

BUFFER SOLUTIONS

Unless otherwise stated, stock solutions and buffers should be prepared and made up with distilled water free from carbon dioxide. Standard reagents should be used. The strength of solutions made up with reagents of doubtful purity or degree of hydration must be checked by titration. The amounts x and y of the stock solutions required to yield a desired pH value are given in the table on pages 364 and 365. In the table below, the buffers are arranged in the alphabetical order of their chemical names.

No.	Buffer	pH range	Stock solutions A	B	Composition of the buffer
1	Walpole's acetate[1,2]	3.6–5.6	0.2 molar acetic acid (12.0 g/l)	0.2 molar sodium acetate (16.4 g $C_2H_3O_2Na$ or 27.2 g $C_2H_3O_2Na \cdot 3H_2O$ per liter)	x ml A + y ml B made up to 100 ml
2	Gomori's aconitate[2]	2.5–5.7	0.5 molar aconitic acid (87.1 g/l)	0.2-N NaOH	10 ml A + x ml B made up to 100 ml
3	Michaelis's barbital sodium[3]	6.8–9.6	0.1 molar barbital sodium (20.6 g/l)	0.1-N HCl	x ml A + (100−x) ml B
4	Michaelis's barbital sodium-acetate[4]	2.6–9.4	$\frac{1}{7}$ molar sodium acetate in $\frac{1}{7}$ molar barbital sodium (19.43 g $NaC_2H_3O_2 \cdot 3H_2O$ + 29.45 g barbital sodium in 1 liter)	0.1-N HCl	50 ml A + x ml B + 20 ml 8.5% NaCl solution made up to 250 ml
5	Clark and Lubs' borate[5]	7.8–10.0	0.1 molar boric acid in 0.1 molar KCl (6.2 g H_3BO_3 + 7.46 g KCl per liter)	0.1-N NaOH	50 ml A + x ml B made up to 100 ml
6	Kolthoff's borax-phosphate[6,7]	5.8–9.2	0.05 molar borax (19.1 g/l)	0.1 molar monopotassium phosphate (13.6 g KH_2PO_4 per liter)	x ml A + (100−x) ml B
7	Kolthoff's borax-succinic acid[7]	3.0–5.8	0.05 molar borax (19.1 g/l)	0.05 molar succinic acid (5.9 g/l)	x ml A + (100−x) ml B
8	Plumel's cacodylate[2,8]	5.0–7.4	0.2 molar sodium cacodylate (42.8 g $Na[CH_3]_2AsO_2 \cdot 3H_2O$ per liter)	0.2-N HCl	25 ml A + x ml B made up to 100 ml
9	Delory and King's carbonate-bicarbonate[2,9]	9.2–10.7	0.2 molar anhydrous sodium carbonate (21.2 g/l)	0.2 molar sodium bicarbonate (16.8 g/l)	x ml A + y ml B made up to 100 ml
10	Sørensen's citrate I[10,11]	2.2–4.8	0.1 molar disodium citrate (21.0 g citric acid [1H_2O] diss. in 200 ml 1-N NaOH and made up to 1 liter)	0.1-N HCl	x ml A + (100−x) ml B
11	Sørensen's citrate II[10,11]	5.0–6.8	As No. 10	0.1-N NaOH	x ml A + (100−x) ml B
12	Teorell and Stenhagen's citrate-phosphate-borate[12]	2.0–12.0	To citric acid and phosphoric acid solutions (ca. 100 ml), each corr. to 100 ml 1-N NaOH, add 3.54 g cryst. orthoboric acid and 343 ml 1-N NaOH, and make up the mixture to 1 liter	0.1-N HCl	20 ml A + x ml B made up to 100 ml
13	McIlvaine's citric acid-phosphate[13]	2.2–8.0	0.1 molar citric acid (21.0 g $C_6H_8O_7 \cdot 1H_2O$ per liter)	0.2 molar disodium phosphate (35.6 g $Na_2HPO_4 \cdot 2H_2O$ per liter)	x ml A + (100−x) ml B
14	Stafford, Watson and Rand's dimethyl-glutarate[14]	3.2–7.6	0.1 molar $\beta\beta$-dimethylglutaric acid (16.02 g/l)	0.2-N NaOH	(*a*) 100 ml A + x ml B made up to 1000 ml (*b*) 100 ml A + x ml B + 5.845 g NaCl made up to 1000 ml (Δ 0.1 molar NaCl)
15	Sørensen's glycine I[10,11]	1.2–3.6	0.1 molar glycine in 0.1-N NaCl (7.5 g glycine + 5.85 g NaCl in 1 liter)	0.1-N HCl	x ml A + (100−x) ml B
16	Sørensen's glycine II[10,11]	8.4–13.0	As No. 15	0.1-N NaOH	x ml A + (100−x) ml B
17	Sørensen's phosphate[11,15]	5.0–8.2	$\frac{1}{15}$ molar monopotassium phosphate (9.08 g KH_2PO_4 per liter)	$\frac{1}{15}$ molar disodium phosphate (11.88 g $Na_2HPO_4 \cdot 2H_2O$ per liter)	x ml A + (100−x) ml B

BUFFER SOLUTIONS (continued)

No.	Buffer	pH range	Stock solutions		Composition of the buffer
			A	B	
18	Clark and Lubs' phthalate I[5]	2.2–3.8	0.1 molar potassium biphthalate (20.4 g KHC$_8$H$_4$O$_4$ per liter)	0.1-N HCl	50 ml A + x ml B made up to 100 ml
19	Clark and Lubs' phthalate II[5]	4.0–6.2	As No. 18	0.1-N NaOH	50 ml A + x ml B made up to 100 ml
20	Smith and Smith's piperazine[16]	4.8–7.0 8.8–11.0	Molar piperazine dihydrochloride (159.1 g/l)	1-N NaOH	1000 ml A + x ml B
21	Smith and Smith's piperazine (sea-water)[16]	5.4–9.8	0.01 molar piperazine dihydrochloride in filtered sea water (pH 8.0) (1.591 g/l)	1-N NaOH	1000 ml A + x ml B
22	Smith and Smith's piperazine-glycylglycine[16]	4.4–10.8	0.01 molar piperazine dihydrochloride in 0.01 molar glycylglycine (1.591 g piperazine·2HCl + 1.321 g glycylglycine in 1 liter)	1-N NaOH	1000 ml A + x ml B
23	Clark and Lubs' potassium chloride-hydrochloric acid[5]	1.0–2.2	0.2-N KCl (14.9 g/l)	0.2-N HCl	25 ml A + x ml B made up to 100 ml
24	Gomori's succinate[2]	3.8–6.0	0.2 molar succinic acid (23.62 g/l)	0.2-N NaOH	25 ml A + x ml B made up to 100 ml
25	Gomori's tris[2,17]	7.2–9.0	0.2 molar tris (24.2 g tris[hydroxy-methyl]aminomethane per liter)	0.2-N HCl	25 ml A + x ml B made up to 100 ml
26	Gomori's tris-maleate[2,17]	5.2–8.6	0.2 molar tris acid maleate (24.2 g tris-[hydroxymethyl]aminomethane + 23.2 g maleic acid or 19.6 g maleic anhydride in 1 liter)	0.2-N NaOH	25 ml A + x ml B made up to 100 ml

BUFFER SOLUTIONS (continued)

pH	1 x (23°C)	1 y (23°C)	2 (23°C)	3 (23°C)	4 (23°C)	5 (20°C)	6 (18°C)	7 (18°C)	8 (23°C)	10 (*)	11 (18°C)	12 (20°C)	13 (21°C)	14a (21°C)	14b (21°C)	15 (18°C)	17 (18°C)	18 (20°C)	19 (20°C)	20 (25°C)	21 (25°C)	22 (25°C)	23 (20°C)	25 (23°C)	26 (23°C)
1.0																									
1.2																									
1.4																									
1.6																									
1.8																									
2.0												73.30	—												
2.2										33.0		67.85	98.00			15.0		46.70							
2.4										34.6		63.85	93.80			28.7		39.60							
2.6			9.0		160.5					36.4		60.80	89.10			38.2		32.95							
2.8			12.3		157.5					38.4		58.45	84.15			45.7		26.42							
3.0			16.0		154.0			1.4		40.5		56.50	79.45			52.3		20.32							
3.2			20.0		150.0			3.5		42.8		54.95	75.30	8.3	15.7	58.3		14.70							
3.4			24.0		146.0			6.0		46.0		53.70	71.50	14.7	22.1	64.5		9.90							
3.6	46.3	3.7	28.0		140.0			9.5		48.5		52.65	67.80	22.0	27.9	70.2		5.97							
3.8	44.0	6.0	32.0		133.5			13.7		52.0		51.55	64.50	27.4	33.3	75.6		2.63					7.5		
4.0	41.0	9.0	36.0		125.5			17.8		55.8	96.3	50.50	61.45	33.3	37.4	80.8			0.40				10.0		
4.2	36.8	13.2	39.8		117.5			22.2		61.2	85.0	49.45	58.60	36.8	40.9	85.6			3.70				13.3		
4.4	30.5	19.5	43.3		109.5			26.2		67.8	76.5	48.35	55.90	39.8	43.6	90.3			7.50				16.7		
4.6	25.5	24.5	46.8		101.5			30.0		76.6	69.3	47.26	53.25	41.7	46.2	94.5			12.15				20.0		
4.8	20.0	30.0	50.0		94.5			33.5		88.2	63.3	46.22	50.70	43.9	48.9				17.70	70.0			23.5		
5.0	14.8	35.2	52.8		88.0			36.8	23.5		59.5	45.18	48.50	46.2	52.0		98.8		23.85	111.0			26.7		3.5
5.2	10.5	39.5	55.3		82.5			39.5	22.5		56.7	44.05	46.40	49.0	54.3		98.0		29.95	174.0			30.3		5.4
5.4	8.8	41.2	58.0		78.0			42.1	21.5		54.5	42.94	44.25	52.0	59.2		96.7		35.45	258.0	0.88	0.07	34.2		7.8
5.6	4.8	45.2	61.3		75.0			44.3	19.6		53.0	41.80	42.00	55.8	64.6		94.8		39.85	355.0	2.00	0.54	37.5		10.3
5.8					72.5		7.9	46.0	17.4		51.8	40.61	39.55	59.7	69.7		91.9		43.00	466.0	3.53	1.20	40.7		
6.0					71.0		12.3		14.8			39.42	36.85	65.0	74.6		87.7		45.45	581.0	4.93	2.20	43.5		13.0
6.2					69.5		17.0		11.9			38.09	33.90	70.5	79.5		81.5		47.00	690.0	6.15	3.20			15.8
6.4					68.5		23.0		9.2			36.74	30.75	75.5	84.2		73.2			780.0	7.18	4.19			18.5
6.6					66.5		28.8		6.7			35.36	27.25	84.7	88.1		62.7			850.0	7.97	5.38			21.3
6.8				52.2	64.0		34.2		4.7			33.92	22.75	88.0	91.5		50.8			900.0	8.54	6.51			22.5
7.0				53.6	60.5		39.0		3.2			32.65	18.15	90.4	93.6		39.2			936.0	8.95	7.45		22.1	24.0
7.2				55.4	56.5		43.4		2.1			31.45	13.05	92.1	95.1		28.5			936.0	9.23	8.27		20.7	25.5
7.4				58.1	50.5		46.4		1.4			30.35	9.15	93.2	96.0		19.6			936.0	9.47	8.99		19.2	27.0
7.6				61.5	43.0		49.2					29.43	6.35	94.0	97.0		13.2			936.0	9.65	9.53		16.3	29.0
7.8				66.2	34.5	2.61	52.0					28.68	4.30				8.6			936.0	9.83	9.99			31.8

Inline buffer labels appearing within the table: **3** (23°C), **16** (18°C), **24** (23°C).

Note: This table gives the quantities (x or x, y) of stock solutions required to make up any of the numbered buffers listed on pages 362 and 363 to a desired pH value at the temperature given.

BUFFER SOLUTIONS (continued)

pH	1 (23°C)	2 (23°C)	4 (23°C)	5 (20°C)	6 (18°C)	7 (18°C)	8 (23°C)	9 (23°C) x	9 (23°C) y	10 (*)	11 (18°C)	12 (20°C)	13 (21°C)	14a (21°C)	14b (21°C)	15 (18°C)	17 (18°C)	18 (20°C)	19 (20°C)	20 (25°C)	21 (25°C)	22 (25°C)	23 (20°C)	25 (23°C)	26 (23°C)	pH (23°C)
8.0	—	71.6	26.0	3.97	55.0	—	—	—	—	—	—	28.02	2.75	—	—	—	5.5	—	—	936.0	10.01	14.45	—	13.4	34.5	8.0
2	—	76.9	18.5	5.90	57.6	—	—	—	—	—	—	27.45	—	—	—	—	3.3	—	—	936.0	10.27	15.71	—	11.0	37.5	2
4	—	82.3	13.0	8.50	62.0	—	—	—	—	—	—	26.90	—	—	—	96.5	—	—	—	936.0	10.65	16.97	—	8.3	40.5	4
6	—	87.1	9.0	12.00	68.0	—	—	—	—	—	—	26.10	—	—	—	94.8	—	—	—	936.0	11.13	18.25	—	6.1	43.3	6
8	—	90.8	6.0	16.30	75.2	—	—	—	—	—	—	24.90	—	—	—	92.1	—	—	—	1079.0	11.76	19.50	—	4.1	—	8
9.0	—	93.6	4.0	21.30	86.8	—	—	—	—	—	—	23.75	—	—	—	88.5	—	—	—	1121.0	12.62	20.64	—	2.5	—	9.0
2	—	95.2	2.0	26.70	100.0	—	—	2.0	23.0	—	—	22.38	—	—	—	84.2	—	—	—	1179.0	13.62	21.85	—	—	—	2
4	—	97.4	1.0	32.00	—	—	—	4.8	20.3	—	—	21.12	—	—	—	79.0	—	—	—	1252.0	14.88	23.13	—	—	—	4
6	—	98.5	—	36.85	—	—	—	8.0	17.0	—	—	19.94	—	—	—	73.2	—	—	—	1346.0	16.42	24.51	—	—	—	6
8	—	—	—	40.80	—	—	—	11.0	14.0	—	—	18.81	—	—	—	67.7	—	—	—	1456.0	1865	25.87	—	—	—	8
10.0	—	—	—	43.90	—	—	—	13.8	11.3	—	—	17.92	—	—	—	63.0	—	—	—	1566.0	—	27.17	—	—	—	10.0
2	—	—	—	—	—	—	—	16.5	8.5	—	—	16.97	—	—	—	59.0	—	—	—	1666.0	—	28.32	—	—	—	2
4	—	—	—	—	—	—	—	19.3	5.8	—	—	16.36	—	—	—	56.1	—	—	—	1762.0	—	29.37	—	—	—	4
6	—	—	—	—	—	—	—	21.3	3.8	—	—	15.95	—	—	—	53.7	—	—	—	1844.0	—	30.42	—	—	—	6
8	—	—	—	—	—	—	—	—	—	—	—	15.40	—	—	—	52.2	—	—	—	1905.0	—	31.53	—	—	—	8
11.0	—	—	—	—	—	—	—	—	—	—	—	14.52	—	—	—	51.0	—	—	—	1952.0	—	—	—	—	—	11.0
2	—	—	—	—	—	—	—	—	—	—	—	13.20	—	—	—	50.3	—	—	—	—	—	—	—	—	—	2
4	—	—	—	—	—	—	—	—	—	—	—	11.23	—	—	—	49.7	—	—	—	—	—	—	—	—	—	4
6	—	—	—	—	—	—	—	—	—	—	—	8.40	—	—	—	48.8	—	—	—	—	—	—	—	—	—	6
8	—	—	—	—	—	—	—	—	—	—	—	4.70	—	—	—	47.8	—	—	—	—	—	—	—	—	—	8
12.0	—	—	—	—	—	—	—	—	—	—	—	0.40	—	—	—	46.2	—	—	—	—	—	—	—	—	—	12.0
2	—	—	—	—	—	—	—	—	—	—	—	—	—	—	—	43.6	—	—	—	—	—	—	—	—	—	2
4	—	—	—	—	—	—	—	—	—	—	—	—	—	—	—	40.0	—	—	—	—	—	—	—	—	—	4
6	—	—	—	—	—	—	—	—	—	—	—	—	—	—	—	33.4	—	—	—	—	—	—	—	—	—	6
8	—	—	—	—	—	—	—	—	—	—	—	—	—	—	—	24.2	—	—	—	—	—	—	—	—	—	8
13.0	—	—	—	—	—	—	—	—	—	—	—	—	—	—	—	7.5	—	—	—	—	—	—	—	—	—	13.0

* pH variation negligible over the normal working temperature range.

BUFFER SOLUTIONS (continued)

REFERENCES

1.   **Walpole,** *J. Chem. Soc.,* 105, 2501 (1914).
2.   **Gomori,** in *Methods in Enzymology,* Colowick and Kaplan, Eds., New York, Vol. 1, p. 138.
3.   **Michaelis,** *J. Biol. Chem.,* 87, 33 (1930).
4.   **Michaelis,** *Biochem Z.,* 234, 139 (1931).
5.   **Clark and Lubs,** *J. Bacteriol.,* 2, 1 (1917).
6.   **Kolthoff,** *Säure-Basen-Indicatoren,* Berlin (1932), p 257.
7.   **Kolthoff,** *J. Biol. Chem.,* 63, 135 (1925).
8.   **Plumel,** *Bull. Soc. Chim. Biol. (Paris),* 30, 129 (1948).
9.   **Delory and King,** *Biochem. J.,* 39, 245 (1945).
10. **Sørensen,** *Biochem. Z.,* 21, 131 (1909).
11. **Sørensen,** *Ergebn Physiol.,* 12, 393 (1912).
12. **Teorell and Stenhagen,** *Biochem. Z.,* 299, 416 (1938).
13. **McIlvaine,** *J. Biol. Chem.,* 49, 183 (1921).
14. **Stafford et al.,** *Biochim. Biophys. Acta,* 18, 319 (1955); Krebs and Hems, personal communication (1957).
15. **Sørensen,** *Biochem. Z.,* 22, 352 (1909).
16. **Smith and Smith,** *Biol. Bull.,* 96, 233 (1949).
17. **Gomori,** *Proc. Soc. Exp. Biol.,* 68, 354 (1948).

Reprinted from *Documenta Geigy, Scientific Tables,* Diem, Ed., Geigy Chemical Corp., Ardsley, New York, 1962, 314, 315. With permission of the copyright owners, Geigy Pharmaceuticals.

AMINE BUFFERS USEFUL FOR BIOLOGICAL RESEARCH

Norman Good

All of these amines are highly polar, water-soluble substances. Their advantages and disadvantages must be determined empirically for each biological reaction system. For best buffering performance they should be used at pH's close to the pKa, preferably within ±0.5 pH units of the pKa and never more than ±1.0 unit from the pKa. Note that the pKa's, and therefore the pH's of buffered solutions, change with temperature in the manner indicated.

AMINE BUFFERS USEFUL FOR BIOLOGICAL RESEARCH

Chemical name	Trivial name or acronym	Structure	pKa at 20°C	ΔpKa/°C
2-(N-Morpholino)ethanesulfonic acid	MES		6.15	-0.011
Bis(2-Hydroxyethyl)imino-tris-(hydroxymethyl)methane	Bistris	$(HOCH_2CH_2)_2 = N-C \equiv (CH_2OH)_3$	6.5	—
N-(2-Acetamido)iminodiacetic acid	ADA[a]		6.6	-0.011
Piperazine-N,N'-bis(2-ethanesulfonic acid)	PIPES		6.8	-0.0085
1,3-Bis[tris(hydroxymethyl)-methylamino]propane	Bistrispropane	$(HOCH_2)_3 = C-NH(CH_2)_3 NH-C \equiv (CH_2OH)_3$	6.8 (9.0)	—
N-(Acetamido)-2-aminoethanesulfonic acid	ACES	$H_2NCOCH_2 N^+H_2 CH_2 CH_2 SO_3^-$	6.9	-0.020
3-(N-Morpholino)propanesulfonic acid	MOPS		7.15	-0.013
N,N'-Bis(2-Hydroxyethyl)-2-amino-ethanesulfonic acid	BES	$(HOCH_2CH_2)_2 = N^+HCH_2 CH_2 SO_3^-$	7.15	-0.016
N-Tris(hydroxymethyl)methyl-2-amino-ethanesulfonic acid	TES	$(HOCH_2)_3 = C-N^+H_2 CH_2 CH_2 SO_3^-$	7.5	-0.020
N-2-Hydroxyethylpiperazine-N'-ethanesulfonic acid	HEPES[b]		7.55	-0.014
N-2-Hydroxyethylpiperazine-N'-propanesulfonic acid	HEPPS[b]		8.1	-0.015

[a]These substances may bind certain di- and polyvalent cations and therefore they may sometimes be useful for providing constant, low level concentrations of free heavy metal ions (heavy metal buffering).
[b]These substances interfere with and preclude the Folin protein assay.

AMINE BUFFERS USEFUL FOR BIOLOGICAL RESEARCH (continued)

Chemical name	Trivial name or acronym	Structure	pKa at 20°C	ΔpKa/°C
N-Tris(hydroxymethyl)methylglycine	Tricine[a]	$(HOCH_2)_3 \equiv C-N^+H_2 CH_2 COO^-$	8.15	−0.021
Tris(hydroxymethyl)aminomethane	Tris	$(HOCH_2)_3 \equiv CNH_2$	8.3	−0.031
N,N-Bis(2-hydroxyethyl)glycine	Bicine[a]	$(HOCH_2 CH_2)_2 = N^+HCH_2 COO^-$	8.35	−0.018
Glycylglycine	Glycylglycine[a]	$H_3 N^+CH_2 CONHCH_2 COO^-$	8.4	−0.028
N-Tris(hydroxymethyl)methyl-3-amino-propanesulfonic acid	TAPS	$(HOCH_2)_3 \equiv C- N^+H_2 (CH_2)_3 SO_3^-$	8.55	−0.027
1,3-Bis[tris(hydroxymethyl)-methylamino]propane	Bistrispropane	$(HOCH_2)_3 \equiv C-NH(CH_2)_3 NH-C\equiv(CH_2 OH)_3$	9.0 (6.8)	—
Glycine	Glycine[a]	$H_3 N^+CH_2 COO^-$	9.9	—

Compiled by Norman Good.

For further information on these and other buffers, see Good and Izawa, in *Methods in Enzymology, Part B*, Vol. 24, Pietro, Ed., Academic Press, New York, 1972, 53.

PREPARATION OF BUFFERS FOR USE IN ENZYME STUDIES*

G. Gomori

The buffers described in this section are suitable for use either in enzymatic or histochemical studies. The accuracy of the tables is within ±0.05 pH at 23°. In most cases the pH values will not be off by more than ±0.12 pH even at 37° and at molarities slightly different from those given (usually 0.05 M).

The methods of preparation described are not necessarily identical with those of the original authors. The titration curves of the majority of the buffers recommended have been redetermined by the writer. The buffers are arranged in the order of ascending pH range. For more complete data on phosphate and acetate buffers over a wide range of concentrations, see Vol. I [10].*

*From Gomori, in *Methods in Enzymology,* Vol. 1, Colowick and Kaplan, Eds., Academic Press, New York, 1955, 138. With permission.

Table 1
HYDROCHLORIC ACID-POTASSIUM CHLORIDE BUFFER*

x	pH
97.0	1.0
78.0	1.1
64.5	1.2
51.0	1.3
41.5	1.4
33.3	1.5
26.3	1.6
20.6	1.7
16.6	1.8
13.2	1.9
10.6	2.0
8.4	2.1
6.7	2.2

*Stock solutions

A:0.2 M solution of KCl (14.91 g in 1,000 ml)
B:0.2 M HCl 50 ml of A + x ml of B, diluted to a total of 200 ml

REFERENCE

1. Clark and Lubs, *J. Bacteriol.,* 2, 1 (1917).

Table 2
GLYCINE-HCl BUFFER*

x	pH	x	pH
5.0	3.6	16.8	2.8
6.4	3.4	24.2	2.6
8.2	3.2	32.4	2.4
11.4	3.0	44.0	2.2

*Stock solutions

A:0.2 M solution of glycine (15.01 g in 1,000 ml)
B:0.2 M HCl 50 ml of A + x ml of B, diluted to a total of 200 ml

REFERENCE

1. Sφrensen, *Biochem. Z.,* 21, 131 (1909); 22, 352 (1909).

Table 3
PHTHALATE-HYDROCHLORIC ACID BUFFER*

x	pH	x	pH
46.7	2.2	14.7	3.2
39.6	2.4	9.9	3.4
33.0	2.6	6.0	3.6
26.4	2.8	2.63	3.8
20.3	3.0		

*Stock solutions

A:0.2 M solution of potassium acid phthalate (40.84 g in 1,000 ml)
B:0.2 M HCl 50 ml of A + x ml of B, diluted to a total of 200 ml

REFERENCE

1. **Clark and Lubs,** *J. Bacteriol.,* 2, 1 (1917).

Table 4
ACONITATE BUFFER*

x	pH	x	pH
15.0	2.5	83.0	4.3
21.0	2.7	90.0	4.5
28.0	2.9	97.0	4.7
36.0	3.1	103.0	4.9
44.0	3.3	108.0	5.1
52.0	3.5	113.0	5.3
60.0	3.7	119.0	5.5
68.0	3.9	126.0	5.7
76.0	4.1		

*Stock solutions

A:0.5 M solution of aconitic acid (87.05 g in 1,000 ml)
B:0.2 M NaOH 20 ml of A + x ml of B, diluted to a total of 200 ml

REFERENCE

1. **Gomori,** unpublished data.

Table 5
CITRATE BUFFER*

x	y	pH
46.5	3.5	3.0
43.7	6.3	3.2
40.0	10.0	3.4
37.0	13.0	3.6
35.0	15.0	3.8
33.0	17.0	4.0
31.5	18.5	4.2
28.0	22.0	4.4
25.5	24.5	4.6
23.0	27.0	4.8
20.5	29.5	5.0
18.0	32.0	5.2
16.0	34.0	5.4
13.7	36.3	5.6
11.8	38.2	5.8
9.5	41.5	6.0
7.2	42.8	6.2

*Stock solutions

A:0.1 M solution of citric acid (21.01 g in 1,000 ml)
B:0.1 M solution of sodium citrate (29.41 g $C_6H_5O_7Na_3 \cdot 2H_2O$ in 1,000 ml; the use of the salt with 5½ H_2O is not recommended). x ml of A + y ml of B, diluted to a total of 100 ml

REFERENCE

1. **Lillie,** *Histopathologic Technique,* Blakiston, Philadelphia and Toronto, 1948.

Table 6
ACETATE BUFFER*

x	y	pH
46.3	3.7	3.6
44.0	6.0	3.8
41.0	9.0	4.0
36.8	13.2	4.2
30.5	19.5	4.4
25.5	24.5	4.6
20.0	30.0	4.8
14.8	35.2	5.0
10.5	39.5	5.2
8.8	41.2	5.4
4.8	45.2	5.6

*Stock solutions

A:0.2 *M* solution of acetic acid (11.55 ml in 1,000 ml)

B:0.2 *M* solution of sodium acetate (16.4 g of $C_2H_3O_2Na$ or 27.2 g of $C_2H_3O_2Na\cdot3H_2O$ in 1,000 ml) *x* ml of A + *y* ml of B, diluted to a total of 100 ml

REFERENCE

1. Walpole, *J. Chem. Soc.,* 105, 2501 (1914).

Table 7
CITRATE-PHOSPHATE BUFFER*

x	y	pH
44.6	5.4	2.6
42.2	7.8	2.8
39.8	10.2	3.0
37.7	12.3	3.2
35.9	14.1	3.4
33.9	16.1	3.6
32.3	17.7	3.8
30.7	19.3	4.0
29.4	20.6	4.2
27.8	22.2	4.4
26.7	23.3	4.6
25.2	24.8	4.8
24.3	25.7	5.0
23.3	26.7	5.2
22.2	27.8	5.4
21.0	29.0	5.6
19.7	30.3	5.8
17.9	32.1	6.0
16.9	33.1	6.2
15.4	34.6	6.4
13.6	36.4	6.6
9.1	40.9	6.8
6.5	43.6	7.0

*Stock solutions

A:0.1 *M* solution of citric acid (19.21 g in 1,000 ml)

B:0.2 *M* solution of dibasic sodium phosphate (53.65 g of Na_2-$HPO_4\cdot7H_2O$ or 71.7 g of $Na_2HPO_4\cdot12H_2O$ in 1,000 ml) *x* ml of A + *y* ml of B, diluted to a total of 100 ml

REFERENCE

1. McIlvaine, *J. Biol. Chem.,* 49, 183 (1921).

Table 8
SUCCINATE BUFFER*

x	pH	x	pH
7.5	3.8	26.7	5.0
10.0	4.0	30.3	5.2
13.3	4.2	34.2	5.4
16.7	4.4	37.5	5.6
20.0	4.6	40.7	5.8
23.5	4.8	43.5	6.0

*Stock solutions

A:0.2 M solution of succinic acid (23.6 g in 1,000 ml)

B:0.2 M NaOH 25 ml of A + x ml of B, diluted to a total of 100 ml

REFERENCE

1. **Gomori,** unpublished data.

Table 9
PHTHALATE-SODIUM HYDROXIDE BUFFER*

x	pH	x	pH
3.7	4.2	30.0	5.2
7.5	4.4	35.5	5.4
12.2	4.6	39.8	5.6
17.7	4.8	43.0	5.8
23.9	5.0	45.5	6.0

*Stock solutions

A:0.2 M solution of potassium acid phthalate (40.84 g in 100 ml)

B:0.2 M NaOH 50 ml of A + x ml of B, diluted to a total of 200 ml

REFERENCE

1. **Clark and Lubs,** *J. Bacteriol.,* 2, 1 (1917).

Table 10
MALEATE BUFFER*

x	pH	x	pH
7.2	5.2	33.0	6.2
10.5	5.4	38.0	6.4
15.3	5.6	41.6	6.6
20.8	5.8	44.4	6.8
26.9	6.0		

*Stock solutions

A:0.2 M solution of acid sodium maleate (8 g of NaOH + 23.2 g of maleic acid or 19.6 g of maleic anhydride in 1,000 ml)

B:0.2 M NaOH 50 ml of A + x ml of B, diluted to a total of 200 ml

REFERENCE

1. **Temple,** *J. Am. Chem. Soc.,* 51, 1754 (1929).

Table 11
CACODYLATE BUFFER*

x	pH	x	pH
2.7	7.4	29.6	6.0
4.2	7.2	34.8	5.8
6.3	7.0	39.2	5.6
9.3	6.8	43.0	5.4
13.3	6.6	45.0	5.2
18.3	6.4	47.0	5.0
23.8	6.2		

*Stock solutions

A:0.2 M solution of sodium cacodylate (42.8 g of $Na(CH_3)_2 AsO_2 \cdot 3H_2O$ in 1,000 ml)

B:0.2 M HCl 50 ml of A + x ml of B, diluted to a total of 200 ml

REFERENCE

1. **Plumel,** *Bull. Soc. Chim. Biol.,* 30, 129 (1949).

Table 12
PHOSPHATE BUFFER*

x	y	pH	x	y	pH
93.5	6.5	5.7	45.0	55.0	6.9
92.0	8.0	5.8	39.0	61.0	7.0
90.0	10.0	5.9	33.0	67.0	7.1
87.7	12.3	6.0	28.0	72.0	7.2
85.0	15.0	6.1	23.0	77.0	7.3
81.5	18.5	6.2	19.0	81.0	7.4
77.5	22.5	6.3	16.0	84.0	7.5
73.5	26.5	6.4	13.0	87.0	7.6
68.5	31.5	6.5	10.5	90.5	7.7
62.5	37.5	6.6	8.5	91.5	7.8
56.5	43.5	6.7	7.0	93.0	7.9
51.0	49.0	6.8	5.3	94.7	8.0

*Stock solutions

A:0.2 M solution of monobasic sodium phosphate (27.8 g in 1,000 ml)

B:0.2 M solution of dibasic sodium phosphate (53.65 g of $Na_2HPO_4 \cdot 7H_2O$ or 71.7 g of $Na_2HPO_4 \cdot 12H_2O$ in 1,000 ml) x ml of A + y ml of B, diluted to a total of 200 ml

REFERENCE

1. Sφrensen, *Biochem. Z.*, 21, 131 (1909); 22, 352 (1909).

Table 13
TRIS(HYDROXYMETHYL)-AMINO METHANE-MALEATE (TRIS-MALEATE) BUFFER*[†]

x	pH	x	pH
7.0	5.2	48.0	7.0
10.8	5.4	51.0	7.2
15.5	5.6	54.0	7.4
20.5	5.8	58.0	7.6
26.0	6.0	63.5	7.8
31.5	6.2	69.0	8.0
37.0	6.4	75.0	8.2
42.5	6.6	81.0	8.4
45.0	6.8	86.5	8.6

*Stock solutions

A:0.2 M solution of Tris acid maleate (24.2 g. of tris(hydroxymethyl)aminomethane + 23.2 g of maleic acid or 19.6 g of maleic anhydride in 1,000 ml)

B:0.2 M NaOH 50 ml of A + x ml of B, diluted to a total of 200 ml

[†]A buffer-grade Tris can be obtained from the Sigma Chemical Co., St. Louis, MO., or From Matheson Coleman & Bell, East Rutherford, NJ.

REFERENCE

1. **Gomori,** *Proc. Soc. Exp. Biol. Med.*, 68, 354 (1948).

Table 14
BARBITOL BUFFER*[†]

x	pH
1.5	9.2
2.5	9.0
4.0	8.8
6.0	8.6
9.0	8.4
12.7	8.2
17.5	8.0
22.5	7.8
27.5	7.6
32.5	7.4
39.0	7.2
43.0	7.0
45.0	6.8

*Stock solutions

A:0.2 *M* solution of sodium barbital (veronal) (41.2 g in 1,000 ml)
B:0.2 *M* HCl 50 ml of A + *x* ml of B, diluted to a total of 200 ml

[†]Solutions more concentrated than 0.05 *M* may crystallize on standing, especially in the cold.

REFERENCE

1. **Michaelis,** *J. Biol. Chem.,* 87, 33 (1930).

Table 15
TRIS(HYDROXYMETHYL)-AMINOMETHANE (TRIS) BUFFER*[†]

x	pH
5.0	9.0
8.1	8.8
12.2	8.6
16.5	8.4
21.9	8.2
26.8	8.0
32.5	7.8
38.4	7.6
41.4	7.4
44.2	7.2

*Stock solutions

A:0.2 *M* solution of tris(hydroxymethyl)aminomethane (24.2 g in 1,000 ml)
B:0.2 *M* HCl 50 ml of A + *x* ml of B, diluted to a total of 200 ml

[†]A buffer-grade Tris can be obtained from the Sigma Chemical Co., St. Louis, MO., or from Matheson Coleman & Bell, East Rutherford, NJ.

Table 16
BORIC ACID-BORAX BUFFER*

x	pH	x	pH
2.0	7.6	22.5	8.7
3.1	7.8	30.0	8.8
4.9	8.0	42.5	8.9
7.3	8.2	59.0	9.0
11.5	8.4	83.0	9.1
17.5	8.6	115.0	9.2

*Stock solutions

A:0.2 *M* solution of boric acid (12.4 g in 1,000 ml)
B:0.05 *M* solution of borax (19.05 g in 1,000 ml; 0.2 *M* in terms of sodium borate) 50 ml of A + x ml of B, diluted to a total of 200 ml

REFERENCE

1. Holmes, *Anat. Rec.,* 86, 163 (1943).

Table 17
2-AMINO-2-METHYL-1,3-PROPANEDIOL (AMMEDIOL) BUFFER*

x	pH	x	pH
2.0	10.0	22.0	8.8
3.7	9.8	29.5	8.6
5.7	9.6	34.0	8.4
8.5	9.4	37.7	8.2
12.5	9.2	41.0	8.0
16.7	9.0	43.5	7.8

*Stock solutions

A:0.2 *M* solution of 2-amino-2-methyl-1,3-propanediol (21.03 g in 1,000 ml)
B:0.2 *M* HCl 50 ml of A + x ml of B, diluted to a total of 200 ml

REFERENCE

1. Gomori, *Proc. Soc. Exp. Biol. Med.,* 62, 33 (1946).

Table 18
GLYCINE-NaOH BUFFER*

x	pH	x	pH
4.0	8.6	22.4	9.6
6.0	8.8	27.2	9.8
8.8	9.0	32.0	10.0
12.0	9.2	38.6	10.4
16.8	9.4	45.5	10.6

*Stock solutions

A:0.2 *M* solution of glycine (15.01 g in 1,000 ml)
B:0.2 *M* NaOH 50 ml of A + x ml of B, diluted to a total of 200 ml.

REFERENCE

1. Sφrensen, *Biochem. Z.,* 21, 131 (1909); 22, 352 (1909).

Table 19
BORAX-NaOH BUFFER*

x	pH
0.0	9.28
7.0	9.35
11.0	9.4
17.6	9.5
23.0	9.6
29.0	9.7
34.0	9.8
38.6	9.9
43.0	10.0
46.0	10.1

*Stock solutions

A:0.05 *M* solution of borax (19.05 g in 1,000 ml; 0.02 *M* in terms of sodium borate)
B:0.2 *M* NaOH 50 ml of A + x ml of B, diluted to a total of 200 ml

REFERENCE

1. Clark and Lubs, *J. Bacteriol.,* 2, 1 (1917).

Table 20
CARBONATE-BICARBONATE
BUFFER*

x	y	pH
4.0	46.0	9.2
7.5	42.5	9.3
9.5	40.5	9.4
13.0	37.0	9.5
16.0	34.0	9.6
19.5	30.5	9.7
22.0	28.0	9.8
25.0	25.0	9.9
27.5	22.5	10.0
30.0	20.0	10.1
33.0	17.0	10.2
35.5	14.5	10.3
38.5	11.5	10.4
40.5	9.5	10.5
42.5	7.5	10.6
45.0	5.0	10.7

*Stock solutions

A:0.2 M solution of anhydrous sodium carbonate (21.2 g in 1,000 ml)
B:0.2 M solution of sodium bicarbonate (16.8 g in 1,000 ml) x ml of A + y ml of B, diluted to a total of 200

REFERENCE

1. **Delory and King,** *Biochem. J.,* 39, 245 (1945).

pH RANGES OF INDICATORS

pH scale: 0 1 2 3 4 5 6 7 8 9 10 11 12 13 14

Indicator	Color transitions (low pH → high pH)
Methyl violet	yellow — blue — → violet
Crystal violet	yellow — blue
Ethyl violet	yellow — blue
Malachite green oxalate	yellow — blue-green — → colorless
Methyl green	yellow — blue
2-(p-Dimethylaminophenylazo)pyridine	yellow — red
o-Cresolsulfonephthalein (cresol red)	red — yellow — red — yellow
Quinaldine red	colorless — red
p-(p-Dimethylaminophenylazo)benzoic acid sodium salt (para methyl red)	red — yellow
m-(p-Anilinophenylazo)benzenesulfonic acid sodium salt (metanil yellow)	red — yellow
4-Phenylazodiphenylamine	red — yellow
Thymolsulfonephthalein (thymol blue)	red — yellow — yellow — blue
m-Cresolsulfonephthalein (meta cresol purple)	red — yellow — yellow — purple
p-(p-Anilinophenylazo)benzenesulfonic acid sodium salt (orange iv)	orange — yellow
4-o-Tolylazo-o-toluidine	orange — yellow
Erythrosin disodium salt	orange — red
Benzopurpurin 4B	violet — red
N,N-Dimethyl-p-(m-tolylazo)aniline	red — yellow
2,4-Dinitrophenol	colorless — yellow
N,N-Dimethyl-p-phenylazoaniline (p-dimethylaminoazobenzene)	red — yellow
4,4'-bis(2-Amino-1-naphthylazo)-2,2'-stilbenedisulfonic acid	purple — red
Tetrabromophenolphthalein ethyl ester potassium salt	yellow — blue
3",3',5',5"-Tetrabromophenolsulfonephthalein (bromophenol blue)	yellow — blue — red
Congo red	blue — red
Methyl orange-xylene cyanole solution	purple — green
Methyl orange	red — yellow
Ethyl orange	red — yellow
4-(4-Dimethylamino-1-naphthylazo)-3-methoxybenzenesulfonic acid	violet — yellow
3",3',5',5"-Tetrabromo-m-cresolsulfonephthalein (bromocresol green)	yellow — blue
Resazurin	orange — violet
4-Phenylazo-1-naphthylamine	red — yellow
Ethyl red	colorless — red
2-(p-Dimethylaminophenylazo)pyridine	yellow — red — yellow

Note: The pH ranges shown are approximations intended to aid in selecting the proper indicator.

pH RANGES OF INDICATORS (continued)

Indicator	Acid color	Transition pH	Base color
4-(p-Ethoxyphenylazo)-m-phenylenediamine monohydrochloride	orange	4–5	yellow
Lacmoid	red	4–6	blue
Alizarin red S	yellow	4–6	red
Methyl red	red	4–6	yellow
Propyl red	red	4–6	yellow
5,5''-Dibromo-o-cresolsulfonephthalein (bromocresol purple)	yellow	5–7	purple
3',3''-Dichlorophenolsulfonephthalein (chlorophenol red)	yellow	5–7	red
p-Nitrophenol	colorless	5–7	yellow
Alizarin	yellow	5–7	red
2-(2,4-Dinitrophenylazo)-1-naphthol-3,6-disulfonic acid disodium salt	yellow	6–7	blue
3',3''-Dibromothymolsulfonephthalein (bromothymol blue)	yellow	6–8	blue
6,8-Dinitro-2,4-(1H,3H)quinazolinedione (m-dinitrobenzoylene urea)	colorless	6–8	yellow
Brilliant yellow	yellow	6–8	orange
Phenolsulfonephthalein (phenol red)	yellow	6–8	red
Neutral red	red	6–8	amber
m-Nitrophenol	colorless	7–8	yellow
o-Cresolsulfonephthalein (cresol red)	yellow	7–9	red
Curcumin	yellow	7–9	red
m-Cresolsulfonephthalein (meta cresol purple)	yellow	7–9	purple
4,4'-Bis(4-amino-1-naphthylazo)-2,2'-stilbenedisulfonic acid	blue	8	red
Thymolsulfonephthalein (thymol blue)	yellow	8–10	blue
o-Cresolphthalein	colorless	8–10	red
p-Naphtholbenzein	orange	8–10	blue
Phenolphthalein	colorless	8–10	pink
Ethyl bis(2,4-dinitrophenyl) acetate	colorless	8–10	blue
Thymolphthalein	colorless	9–11	blue
5-(p-Nitrophenylazo)salicylic acid sodium salt (alizarin yellow R)	yellow	9–11	red
Curcumin	red	10–12	orange
Alizarin	red	11–13	purple
p-(2,4-Dihydroxyphenylazo)benzenesulfonic acid sodium salt	yellow	11–12	orange
5,5'-Indigodisulfonic acid disodium salt	blue	11–13	yellow
2,4,6-Trinitrotoluene	colorless	12–13	orange
1,3,5-Trinitrobenzene	colorless	12–14	orange
Clayton yellow	yellow	12–14	amber

Additional notations: o-Cresolsulfonephthalein (cresol red): red → yellow (pH 1–2); m-Cresolsulfonephthalein (meta cresol purple): red → yellow (pH 1–3); Thymolsulfonephthalein (thymol blue): red → yellow (pH 1–3); p-Nitrophenol: colorless → yellow; p-Naphtholbenzein: colorless → orange.

*Courtesy of Eastman Kodak Co., Rochester, N.Y. 14650.

INDICATORS FOR VOLUMETRIC WORK AND pH DETERMINATIONS

Indicator		Acid color	pH range	Basic color	Preparation
Methyl violet 6B	Tetra and pentamethylated *p*-rosaniline hydrochloride	Y	0.1–1.5	B	pH: 0.25% water
Metacresol purple (acid range)	*m*-Cresolsulfonphthalein	R	0.5–2.5	Y	pH: 0.10 g. in 13.6 ml. 0.02 N NaOH, diluted to 250 ml. with water
Metanil yellow	4-Phenylamino-azobenzene-3′-sulfonic acid	R	1.2–2.3	Y	pH: 0.25% in ethanol
p-Xylenol blue (acid range)	1,4-Dimethyl-5-hydroxybenzene-sulfonphthalein	R	1.2–2.8	Y	pH: 0.04% in ethanol
Thymol blue (acid range)	Thymolsulfonphthalein	R	1.2–2.8	Y	pH: 0.1 g. in 10.75 ml. 0.02 N NaOH, diluted to 250 ml. with water
Tropaeolin OO	Sodium *p*-diphenylaminoazo-benzenesulfonate	R	1.4–2.6	Y	pH: 0.1% in water Vol.: 1% in water
Quinaldine red	2-(*p*-Dimethylaminostyryl)-quinoline ethiodide	C	1.4–3.2	R	Vol.: 0.1% in ethanol
Benzopurpurine 4B	Ditolyl-diazo-bis-α-naphthyl-amine-4-sulfonic acid	B-V	1.3–4.0	R	pH, vol.: 0.1% in water
Methyl violet 6B	Tetra and pentamethylated *p*-rosaniline hydrochloride	B	1.5–3.2	V	pH, vol.: 0.25% in water
2,4-Dinitrophenol		C	2.6–4.0	Y	pH, vol.: 0.1 g. in 5 ml. ethanol, diluted to 100 ml. with water
Methyl yellow	*p*-Dimethylaminoazobenzene	R	2.9–4.0	Y	pH, vol.: 0.05% in ethanol
Bromphenol blue	Tetrabromophenolsulfon-phthalein	Y	3.0–4.6	B	pH: 0.1 g. in 7.45 ml. 0.02 N NaOH, diluted to 250 ml. with water
Tetrabromophenol blue	Tetrabromophenol-tetrabromo-sulfonphthalein	Y	3.0–4.6	B	pH: 0.1 g. in 5.00 ml. 0.02 N NaOH, diluted to 250 ml. with water
Direct purple	Disodium 4,4′-bis(2-amino-1-naphthylazo)-2,2′-stilbene-disulfonate	B-P	3.0–4.6	R	Vol.: 0.1 g. in 7.35 ml. 0.02 N NaOH, diluted to 100 ml. with water
Congo red	Diphenyl-diazo-bis-1-naphthyl-amine-4-sodium sulfonate	B	3.0–5.2	R	pH: 0.1% in water
Methyl orange	4′-Dimethylaminoazobenzene-4-sodium sulfonate	R	3.1–4.4	Y	Vol.: 0.1% in water
Brom-chlorphenol blue	Dibromodichlorophenolsulfon-phthalein	Y	3.2–4.8	B	pH: 0.1 g. in 8.6 ml. 0.02 N NaOH, diluted to 250 ml. with water Vol.: 0.04% in ethanol
p-Ethoxychrysoidine	4′-Ethoxy-2,4-diaminoazo-benzene	R	3.5–5.5	Y	Vol.: 0.1% in ethanol
α-Naphthyl red		R	3.7–5.0	Y	Vol.: 0.1% in ethanol
Sodium alizarinsulfonate	Dihydroxyanthraquinone sodium sulfonate	Y	3.7–5.2	V	pH, vol.: 1% in water
Bromcresol green	Tetrabromo-*m*-cresolsulfon-phthalein	Y	3.8–5.4	B	pH: 0.10 g. in 7.15 ml. 0.02 N NaOH, diluted to 250 ml. with water
2,5-Dinitrophenol		C	4.0–5.8	Y	pH, vol.: 0.10 g. in 20 ml. ethanol, then dilute to 100 ml. with water
Methyl red	4′-Dimethylaminoazobenzene-2-carboxylic acid	R	4.2–6.2	Y	pH: 0.10 g. in 18.6 ml. 0.02 N NaOH, diluted to 250 ml. with water Vol.: 0.1% in ethanol
Lacmoid		R	4.4–6.2	B	Vol.: 0.5% in ethanol
Azolitmin		R	4.5–8.3	B	Vol.: 0.5% in water
Litmus		R	4.5–8.3	B	Vol.: 0.5% in water

Note: The indicator colors are abbreviated as follows: B, blue; Br, brown; C, colorless; G, green; L, lilac; O, orange; P, pink; Pu, purple; R, red; V, violet; and Y, yellow.

INDICATORS FOR VOLUMETRIC WORK AND pH DETERMINATIONS (continued)

Indicator	Chemical name	Acid color	pH range	Basic color	Preparation
Cochineal	Complex hydroxyanthraquinone derivative	R	4.8–6.2	V	Vol.: Triturate 1 g. with 20 ml. ethanol and 60 ml. water, let stand 4 days, and filter
Hematoxylin		Y	5.0–6.0	V	Vol.: 0.5 % in ethanol.
Chlorphenol red	Dichlorophenolsulfonphthalein	Y	5.0–6.6	R	pH: 0.1 g. in 11.8 ml. 0.02 N NaOH, diluted to 250 ml. with water Vol.: 0.04 % in ethanol
Bromcresol purple	Dibromo-o-cresolsulfonphthalein	Y	5.2–6.8	Pu	pH: 0.1 g. in 9.25 ml. 0.02 N NaOH, diluted to 250 ml. with water Vol.: 0.02 % in ethanol
Bromphenol red	Dibromophenolsulfonphthalein	Y	5.2–7.0	R	pH: 0.1 g. in 9.75 ml. 0.02 N NaOH, diluted to 250 ml. with water Vol.: 0.04 % in ethanol
Alizarin	1,2-Dihydroxyanthraquinone	Y	5.5–6.8	R	Vol.: 0.1 % in ethanol
Dibromophenoltetrabromo- phenolsulfonphthalein		Y	5.6–7.2	Pu	pH: 0.1 g. in 1.21 ml. 0.1 N NaOH, diluted to 250 ml. with water
p-Nitrophenol		C	5.6–7.6	Y	pH, vol.: 0.25 % in water
Bromothymol blue	Dibromothymolsulfonphthalein	Y	6.0–7.6	B	pH: 0.1 g. in 8 ml. 0.02 N NaOH, diluted to 250 ml. with water Vol.: 0.1 % in 50 % ethanol
Indo-oxine	5,8-Quinolinequinone-8-hydroxy-5-quinolyl-5-imide	R	6.0–8.0	B	Vol.: 0.05 % in ethanol
Cucumin		Y	6.0–8.0	Br-R	Vol: saturated aq. soln.
Quinoline blue	Cyanine	C	6.6–8.6	B	Vol.: 1 % in ethanol
Phenol red	Phenolsulfonphthalein	Y	6.8–8.4	R	pH: 0.1 g. in 14.20 ml. 0.02 N NaOH, diluted to 250 ml. with water Vol.: 0.1 % in ethanol
Neutral red	2-Methyl-3-amino-6-dimethyl-aminophenazine	R	6.8–8.0	Y	pH, vol.: 0.1 g. in 70 ml. ethanol, diluted to 100 ml. with water
Rosolic acid aurin; corallin		Y	6.8–8.2	R	pH, vol.: 1 % in 50 % ethanol
Cresol red	o-Cresolsulfonphthalein	Y	7.2–8.8	R	pH: 0.1 g. in 13.1 ml. 0.02 N NaOH, diluted to 250 ml. with water Vol.: 0.1 % in ethanol
α-Naphtholphthalein		P	7.3–8.7	G	pH, vol.: 0.1 % in 50 % ethanol
Metacresol purple (alkaline range)	m-Cresolsulfonphthalein	Y	7.4–9.0	P	pH: 0.1 g. in 13.1 ml. 0.02 N NaOH, diluted to 250 ml. with water Vol.: 0.1 % in ethanol
Ethylbis-2,4-dinitrophenylacetate		C	7.5–9.1	B	Vol.: saturated soln. in equal volumes of acetone and ethanol
Tropaeolin OOO No. 1	Sodium α-naphtholazobenzene-sulfonate	Y	7.6–8.9	R	Vol.: 0.1 % in water
Thymol blue (alkaline range)	Thymolsulfonphthalein	Y	8.0–9.6	B	pH: 0.1 g. in 10.75 ml. 0.02 N NaOH, diluted to 250 ml. with water Vol.: 0.1 % in ethanol
p-Xylenol blue	1,4-Dimethyl-5-hydroxybenzene-sulfonphthalein	Y	8.0–9.6	B	pH, vol.: 0.04 % in ethanol
o-Cresolphthalein		C	8.2–9.8	R	pH, vol.: 0.04 % in ethanol
α-Naphtholbenzein		Y	8.5–9.8	G	pH, vol.: 1 % in ethanol

INDICATORS FOR VOLUMETRIC WORK AND pH DETERMINATIONS (continued)

Indicator	Chemical name	Acid color	pH range	Basic color	Preparation
Phenolphthalein	3,3-Bis(*p*-hydroxyphenyl)-phthalide	C	8.2–10	R	Vol.: 1% in ethanol
Thymolphthalein		C	9.3–10.5	B	pH, vol.: 0.1% in ethanol
Nile blue A	Aminonaphthodiethylamino-phenoxazine sulfate	B	10–11	P	Vol.: 0.1% in water
Alizarin yellow GG	3-Carboxy-4-hydroxy-3'-nitro-azobenzene	Y	10–12	L	pH, vol.: 0.1% in 50% ethanol
Alizarin yellow R	3-Carboxy-4-hydroxy-4'-nitro-azobenzene sodium salt	Y	10.2–12.0	R	pH, vol.: 0.1% in water
Poirrer's blue C4B		B	11–13	R	pH: 0.2% in water
Tropaeolin O	*p*-Benzenesulfonic acid-azo-resorcinol	Y	11–13	O	pH: 0.1% in water
Nitramine	Picrylnitromethylamine	C	10.8–13	Br	pH: 0.1% in 70% ethanol
1,3,5-Trinitrobenzene		C	11.5–14	O	pH: 0.1% in ethanol
Indigo carmine	Sodium indigodisulfonate	B	11.6–14	Y	pH: 0.25% in 50% ethanol

MIXED INDICATORS

Composition	Solvent	Transition pH	Acid color	Transition color	Basic color
Dimethyl yellow, 0.05% + Methylene blue, 0.05%	alc.	3.2	Blue–violet	—	Green
Methyl orange, 0.02% + Xylene cyanole FF, 0.28%	50% alc.	3.9	Red	Gray	Green
Methyl yellow, 0.08% + Methylene blue, 0.004%	alc.	3.9	Pink	Straw–pink	Yellow–green
Methyl orange, 0.1% + Indigocarmine, 0.25%	aq.	4.1	Violet	Gray	Yellow–green
Bromcresol green, 0.1% + Methyl orange, 0.02%	aq.	4.3	Orange	Light green	Dark green
Bromcresol green, 0.075% + Methyl red, 0.05%	alc.	5.1	Wine–red	—	Green
Methyl red, 0.1% + Methylene blue, 0.05%	alc.	5.4	Red–violet	Dirty blue	Green
Bromcresol green, 0.05% + Chlorphenol red, 0.05%	aq.	6.1	Yellow–green	—	Blue–violet
Bromcresol purple, 0.05% + Bromthymol blue, 0.05%	aq.	6.7	Yellow	Violet	Violet–blue
Neutral red, 0.05% + Methylene blue, 0.05%	alc.	7.0	Violet–blue	Violet–blue	Green
Bromthymol blue, 0.05% + Phenol red, 0.05%	aq.	7.5	Yellow	Violet	Dark violet
Cresol red, 0.025% + Thymol blue, 0.15%	aq.	8.3	Yellow	Rose	Violet
Phenolphthalein, 0.033% + Methyl green, 0.067%	alc.	8.9	Green	Gray–blue	Violet
Phenolphthalein, 0.075% + Thymol blue, 0.025%	50% alc.	9.0	Yellow	Green	Violet
Phenolphthalein, 0.067% + Naphtholphthalein, 0.033%	50% alc.	9.6	Pale rose	—	Violet
Phenolphthalein, 0.033% + Nile blue, 0.133%	alc.	10.0	Blue	Violet	Red
Alizarin yellow, 0.033% + Nile blue, 0.133%	alc.	10.8	Green	—	Red–brown

UNIVERSAL INDICATORS FOR APPROXIMATE pH DETERMINATIONS

No. 1. Dissolve 60 mg methyl yellow, 40 mg methyl red, 80 mg bromthymol blue, 100 mg thymol blue and 20 mg phenolphthalein in 100 ml of ethanol and add enough 0.1 N NaOH to produce a yellow color.

No. 2. Dissolve 18.5 mg methyl red, 60 mg bromthymol blue and 64 mg phenolphthalein in 100 ml of 50% ethanol and add enough 0.1 N NaOH to produce a green color.

pH	Color		pH	Color	
	No. 1	No. 2		No. 1	No. 2
1	Cherry–red	Red	7	Yellowish–green	Greenish–yellow
2	Rose	Red	8	Green	Green
3	Red–orange	Red	9	Bluish–green	Greenish–blue
4	Orange–red	Deeper red	10	Blue	Violet
5	Orange	Orange–red	11	—	Reddish–violet
6	Yellow	Orange–yellow			

ADSORPTION INDICATORS

Indicator	Color change	Indicator for
Chromotrope F4B	Red to gray–green (bromides), pink to green (iodides)	Bromides, iodides
Dichlorofluorescein	Yellow–green to pink	Bromides, chlorides
Diiodofluorescein	Yellow to pink	Iodides
Fluorescein	Green to pink	Chlorides
Phenosafranine	Pink to blue	Chlorides, bromides
Rhodamine 6G	Pink to violet	Silver with bromide
Rose bengal	Deep pink to bluish–pink	Iodides
p-Ethoxychrysoidine	Red to yellow	Chlorides, thiocyanates, silver with iodides or thiocyanates
Eosin	Changes to pink	Bromides, iodides, thiocyanates
Indo-oxine	Red to blue	Halides
Tartrazine	Colorless to yellowish–green	Silver with halides

OXIDATION-REDUCTION INDICATORS

Common name	Transition potential volts (N Hydrogen Electrode = 0.000)	Color	
		Reduced form	Oxidized form
p-Ethoxychrysoidin	0.76	Red	Yellow
Diphenylamine	0.776	Colorless	Purple
Diphenylbenzidine	0.776	Colorless	Purple
Diphenylamine-sulfonic acid or barium salt	0.84	Colorless	Purple
Naphthidine	—	Colorless	Red
Dimethylferroin	0.97	Red	Yellowish–green
Eriogreen B	0.99	Yellow	Orange
Erioglaucin A	1.0	Yellowish–green	Red
Xylene cyanole FF	1.0	—	—
2,2'-Dipyridyl ferrous ion	1.03	Red	Colorless
N-Phenylanthranilic acid	1.08	Colorless	Pink
Methylferroin	1.08	Red	Pale-blue
Ferroin (o-Phenanthrolineferrous ion)	1.12	Red	Pale-blue
Chloroferroin	1.17	Red	Pale-blue
Nitroferroin	1.31	Red	Pale greenish–blue
α-Naphtholflavone	—	Pale straw	Brownish–orange

CONCENTRATION OF ACIDS AND BASES

COMMON COMMERCIAL STRENGTHS

	Molecular weight	Moles per liter	Grams per liter	Per cent by weight	Specific gravity
Acetic acid, glacial	60.05	17.4	1,045	99.5	1.05
Acetic acid	60.05	6.27	376	36	1.045
Butyric acid	88.1	10.3	912	95	0.96
Formic acid	46.02	23.4	1,080	90	1.20
	—	5.75	264	25	1.06
Hydriodic acid	127.9	7.57	969	57	1.70
	—	5.51	705	47	1.50
	—	0.86	110	10	1.1
Hydrobromic acid	80.92	8.89	720	48	1.50
	—	6.82	552	40	1.38
Hydrochloric acid	36.5	11.6	424	36	1.18
	—	2.9	105	10	1.05
Hydrocyanic acid	27.03	25	676	97	0.697
	—	0.74	19.9	2	0.996
Hydrofluoric acid	20.01	32.1	642	55	1.167
	—	28.8	578	50	1.155
Hydrofluosilicic acid	144.1	2.65	382	30	1.27
Hypophosphorous acid	66.0	9.47	625	50	1.25
	—	5.14	339	30	1.13
	—	1.57	104	10	1.04
Lactic acid	90.1	11.3	1,020	85	1.2
Nitric acid	63.02	15.99	1,008	71	1.42
	—	14.9	938	67	1.40
	—	13.3	837	61	1.37
Perchloric acid	100.5	11.65	1,172	70	1.67
	—	9.2	923	60	1.54
Phosphoric acid	80	18.1	1,445	85	1.70
Sulfuric acid	98.1	18.0	1,766	96	1.84
Sulfurous acid	82.1	0.74	61.2	6	1.02
Ammonium hydroxide	35.0	14.8	251	28	0.898
Potassium hydroxide	56.1	13.5	757	50	1.52
	—	1.94	109	10	1.09
Sodium carbonate	106.0	1.04	110	10	1.10
Sodium hydroxide	40.0	19.1	763	50	1.53
	—	2.75	111	10	1.11

Reprinted from *The Merck Index,* 7th ed., Merck and Co., Rahway, N.J., 1960, 1572. With permission of the copyright owners.

REFRACTIVE INDEX TABLES

Table 1
OPTICAL ROTATORY DISPERSION CALCULATIONS
DISPERSION OF THE REFRACTIVE INDEX OF WATER $(n^{20})^a$

λ, nm	n	n^2	$\dfrac{n^2+2}{3}$	$\dfrac{3}{n^2+2}$	λ, nm	n	n^2	$\dfrac{n^2+2}{3}$	$\dfrac{3}{n^2+2}$
182.9	1.4640	2.14330	1.38110	0.72406	415	1.3419	1.80070	1.26690	0.789329
199.0	1.4257	2.03262	1.34421	0.74393	420	1.3415	1.79962	1.26654	0.789552
231.3	1.3888	1.92877	1.30959	0.76360	425	1.3411	1.79855	1.26618	0.789775
240	1.3828	1.91214	1.30405	0.766845	430	1.3408	1.79774	1.26591	0.789943
245	1.3800	1.90440	1.30147	0.768364	435	1.3404	1.79667	1.26556	0.790166
250	1.3772	1.89668	1.29889	0.769886	440	1.3400	1.79560	1.26520	0.790389
255	1.3748	1.89008	1.29669	0.771193	445	1.3396	1.79453	1.26484	0.790612
260	1.3725	1.88376	1.29459	0.772448	450	1.3393	1.79372	1.26457	0.790780
265	1.3701	1.87717	1.29239	0.773759	455	1.3389	1.79265	1.26422	0.791003
270	1.3682	1.87197	1.29066	0.774799	460	1.3386	1.79185	1.26395	0.791171
275	1.3664	1.86705	1.28902	0.775785	465	1.3383	1.79105	1.26368	0.791338
280	1.3647	1.86241	1.28747	0.776718	470	1.3380	1.79024	1.26341	0.791506
285	1.3630	1.85777	1.28592	0.777652	475	1.3377	1.78944	1.26315	0.791673
290	1.3615	1.85368	1.28456	0.778476	480	1.3374	1.78864	1.26288	0.791841
295	1.3600	1.84960	1.28320	0.779302	485	1.3371	1.78784	1.26261	0.792009
300	1.3587	1.84607	1.28202	0.780018	490	1.3369	1.78730	1.26243	0.792121
305	1.3574	1.84253	1.28084	0.780735	495	1.3366	1.78650	1.26217	0.792288
310	1.3563	1.83955	1.27985	0.781342	500	1.3364	1.78597	1.26199	0.792400
315	1.3551	1.83630	1.27877	0.782004	505	1.3361	1.78516	1.26172	0.792568
320	1.3541	1.83359	1.27786	0.782557	510	1.3359	1.78463	1.26154	0.792680
325	1.3531	1.83088	1.27696	0.783110	515	1.3357	1.78409	1.26136	0.792792
330	1.3521	1.82817	1.27606	0.783663	520	1.3355	1.78356	1.26119	0.792904
335	1.3512	1.82574	1.27525	0.784162	525	1.3353	1.78303	1.26101	0.793016
340	1.3504	1.82358	1.27453	0.784605	530	1.3351	1.78249	1.26083	0.793128
345	1.3496	1.82142	1.27381	0.785048	535	1.3349	1.78196	1.26065	0.793240
350	1.3489	1.81953	1.27318	0.785437	540	1.3347	1.78142	1.26047	0.793352
355	1.3482	1.81764	1.27255	0.785825	545	1.3345	1.78089	1.26030	0.793464
360	1.3475	1.81576	1.27192	0.786214	550	1.3343	1.78036	1.26012	0.793576
365	1.3469	1.81414	1.27138	0.786547	555	1.3341	1.77982	1.25994	0.793688
370	1.3463	1.81252	1.27084	0.786880	560	1.3340	1.77956	1.25985	0.793744
375	1.3457	1.81091	1.27030	0.787214	565	1.3338	1.77902	1.25967	0.793856
380	1.3451	1.80929	1.26976	0.787548	570	1.3336	1.77849	1.25950	0.793968
385	1.3446	1.80795	1.26932	0.787826	575	1.3335	1.77822	1.25941	0.794024
390	1.3441	1.80660	1.26887	0.788104	580	1.3333	1.77769	1.25923	0.794136
395	1.3436	1.80526	1.26842	0.788382	585	1.3332	1.77742	1.25914	0.794192
400	1.3432	1.80419	1.26806	0.788605	590	1.3330	1.77689	1.25896	0.794305
405	1.3428	1.80311	1.26770	0.788828	595	1.3329	1.77662	1.25887	0.794361
410	1.3423	1.80177	1.26726	0.789106	600	1.3327	1.77609	1.25870	0.794473

Compiled by Elemer Mihalyi with data from *Int. Crit. Tables,* 7, 13 (1930).

Taken from Fasman, in *Methods in Enzymology,* Vol. 6, Colowick and Kaplan, Eds., Academic Press, New York, 1963, chap. 126.

Table 2
VARIATION OF REFRACTIVE INDEX OF H_2O $(n)^T$, WITH TEMPERATURE[a]

Temperature	λ_{434}	λ_{486}	λ_{589}	Temperature	λ_{434}	λ_{486}	λ_{589}
0°	—	—	1.3340	20°	1.3404	1.3371	1.3330
5°	—	—	1.3339	60°	1.3346	1.3315	1.3272
10°	1.3411	1.3378	1.3337	70°	1.3325	1.3294	1.3252

[a]From *Int. Crit. Tables,* 7, 13, 1930.

Taken from Fasman, in *Methods in Enzymology,* Vol. 6, Colowick and Kaplan, Eds., Academic Press, New York, 1963, chap. 126.

Table 3
REFRACTIVE INDEX
CALCULATIONS OF VARIOUS SOLVENTS[a]

Solvent	n_D^{20}	$\dfrac{n^2 + 2}{3}$	$\dfrac{3}{n^2 + 2}$
Acetic acid	1.3718	1.2939	0.7729
Chloroethanol	1.4419	1.3597	0.7355
Chloroform	1.446	1.3636	0.7334
Dichloroacetic acid[b]	1.4659	1.3830	0.7231
Dimethyl formamide[c]	1.4280	1.3464	0.7427
Dimethyl sulfoxide[d]	1.4787	1.3955	0.7166
Dioxane	1.4232	1.3418	0.7453
Ethylene dichloride	1.4443	1.3620	0.7342
Ethylene glycol[c]	1.4306	1.3489	0.7413
Trifluoroacetic acid	1.285	1.2171	0.8216
Water	1.3330	1.2590	0.7943

[a]From "Merck Index."
[b]22°.
[c]25°.
[d]21°.

Taken from Fasman, in *Methods in Enzymology,* Vol. 6, Colowick and Kaplan, Eds, Academic Press, New York, 1963, chap. 126.

Table 4
REFRACTIVE INDEX OF UREA SOLUTIONS VERSUS CONCENTRATION[a]

M[b]	$n_{589(D)}^{20}$	n_{546}^{20}	n_{436}^{20}	n_{D}^{35}
9.00	1.40794	1.41013	1.41886	1.40482
8.00	1.39908	—	—	
6.00	1.38322	—	—	
5.00	1.37488	1.37684	1.38381	1.37211
3.50	1.36256	1.36432	1.37123	1.36005
2.00	1.34998	1.35188	—	1.34788
1.10	1.34222	1.34394	1.34958	1.34032
0.50	1.33728	1.33871	1.34473	1.33548

[a]From Warren and Gordon, *J. Phys. Chem.*, 70, 297 (1966).
[b]M = molarity in aqueous solution.

Taken from Adler and Fasman, ORD as a means of determining nucleic acid conformation, in *Methods in Enzymology*, 12B, 268 (1968).

Table 5
REFRACTIVE INDEX OF SODIUM CHLORIDE (n_{D}^{25}) VERSUS CONCENTRATION[a]

Grams per 100 g of mixture	n_{D}^{25}	$\dfrac{n^2 + 2}{3}$	$\dfrac{3}{n^2 + 2}$	Grams per 100 g of mixture	n_{D}^{25}	$\dfrac{n^2 + 2}{3}$	$\dfrac{3}{n^2 + 2}$
0.5280	1.3334	1.2593	0.7941	1.1068	1.3344	1.2602	0.7935
0.5493	1.3334	1.2593	0.7941	5.3562	1.3417	1.2667	0.7895
0.9980	1.3342	1.2600	0.7937	5.4131	1.3418	1.2668	0.7894
1.0618	1.3343	1.2601	0.7936	14.344	1.3575	1.2809	0.7807

[a]From *International Critical Tables*, 7, 73, 1930.

From Fasman, in *Methods in Enzymology*, Vol. 6, Colowick and Kaplan, Eds., Academic Press, New York, 1963, chap. 126.

Table 6
REFRACTIVE INDEX OF LiBr (n_{D}^{25}) VERSUS CONCENTRATION[a]

Grams per 100 g of mixture	n_{D}^{25}	$\dfrac{n^2 + 2}{3}$	$\dfrac{3}{n^2 + 2}$	Grams per 100 g of mixture	n_{D}^{25}	$\dfrac{n^2 + 2}{3}$	$\dfrac{3}{n^2 + 2}$
0.1980	1.3327	1.2587	0.7945	3.7527	1.3383	1.2637	0.7913
0.4313	1.3331	1.2591	0.7942	4.2994	1.3391	1.2644	0.7909
1.0244	1.3340	1.2599	0.7937	14.966	1.3566	1.2801	0.7812
1.8718	1.3353	1.2610	0.7930	32.55	1.3919	1.3125	0.7619

[a]From Landolt-Bornstein, EII, 5th ed., p 534.

From Fasman, in *Methods in Enzymology*, Vol. 6, Colowick and Kaplan, Eds., Academic Press, New York, 1963, chap. 126.

Table 7
DISPERSION OF REFRACTIVE INDEX OF VARIOUS SOLVENTS[a]

λ, nm	n	$\dfrac{n^2 + 2}{3}$	$\dfrac{3}{n^2 + 2}$	λ, nm	n	$\dfrac{n^2 + 2}{3}$	$\dfrac{3}{n^2 + 2}$
CCl$_4$ (Carbon Tetrachloride)				**Ethylene Dichloride—(Continued)**			
265.5	—	—	—	365.0	1.4648	1.3819	0.7236
289.4	—	—	—	435.8	1.4539	1.3713	0.7292
313.1	1.4985	1.4152	0.7066	546.1	1.4447	1.3694	0.7340
365.0	1.4831	1.3999	0.7143	**Ethylene Glycol[b]**			
435.8	1.4706	1.3876	0.7207	435.8	1.4400	1.3583	0.7363
546.1	1.4603	1.3775	0.7260	546.1	1.4330	1.3515	0.7400
2-Chloroethanol				**Ethylene Glycol-Water, 1:1 (Vol.)[b]**			
265.5	1.4940	1.4107	0.7089	435.8	1.3923	1.3125	0.7619
289.4	1.4823	1.3991	0.7148	546.1	1.3858	1.3065	0.7652
313.1	1.4731	1.3900	0.7194	**Formamide**			
365.0	1.4608	1.3780	0.7257	265.5	1.5379	1.4551	0.6872
435.8	1.4511	1.3686	0.7307	289.4	1.5139	1.4306	0.6990
546.1	1.4423	1.3601	0.7352	313.1	1.4980	1.4147	0.7069
CHCl$_3$ (Chloroform)				365.0	1.4772	1.3940	0.7174
265.5	1.5051	1.4218	0.7033	435.8	1.4619	1.3791	0.7251
289.4	1.4911	1.4078	0.7103	546.1	1.4313	1.3495	0.7410
313.1	1.4806	1.3974	0.7156	**Formic acid**			
365.0	1.4661	1.3832	0.7230	265.5	1.4178	1.3367	0.7481
435.8	1.4546	1.3720	0.7289	289.4	1.4063	1.3259	0.7542
546.1	1.4454	1.3631	0.7336	313.1	1.3982	1.3183	0.7586
C$_6$H$_{12}$ (Cyclohexane)				365.0	1.3874	1.3083	0.7644
265.5	1.4741	1.3910	0.7189	435.8	1.3785	1.3001	0.7692
289.4	1.4631	1.3802	0.7245	546.1	1.3709	1.2931	0.7733
313.1	1.4549	1.3723	0.7287	**Furan**			
365.0	1.4432	1.3609	0.7348	265.5	—	—	—
435.8	1.4335	1.3516	0.7399	289.4	—	—	—
546.1	1.4249	1.3435	0.7443	313.1	1.4766	1.3935	0.7176
Dichloroacetic acid				365.0	1.4537	1.3711	0.7293
265.5	—	—	—	435.8	1.4369	1.3549	0.7381
289.4	—	—	—	546.1	1.4241	1.3427	0.7448
313.1	1.5045	1.4212	0.7036	**Hydrazine**			
365.0	1.4893	1.4060	0.7112	265.5	—	—	—
435.8	1.4776	1.3944	0.7172	289.4	—	—	—
546.1	1.4495	1.3670	0.7315	313.1	—	—	—
Dimethylformamide				365.0	1.4980	1.4147	0.7069
265.5	—	—	—	435.8	1.4837	1.4005	0.7140
289.4	1.4913	1.4080	0.7102	546.1	1.4714	1.3883	0.7203
313.1	1.4761	1.3930	0.7179	**Methylene Chloride**			
365.0	1.4564	1.3737	0.7280	265.5	1.4786	1.3954	0.7167
435.8	1.4419	1.3597	0.7355	289.4	1.4661	1.3832	0.7230
546.1	1.4495	1.3670	0.7315	313.1	1.4561	1.3734	0.7281
p-Dioxane				365.0	1.4431	1.3609	0.7348
265.5	1.4699	1.3869	0.7210	435.8	1.4323	1.3505	0.7405
289.4	1.4583	1.3755	0.7270	546.1	1.4237	1.3423	0.7450
313.1	1.4500	1.3675	0.7313	**8 M Urea (Aqueous Solution)**			
365.0	1.4384	1.3563	0.7373	265.5	1.4572	1.3745	0.7276
435.8	1.4293	1.3476	0.7421	289.4	1.4433	1.3610	0.7347
546.1	1.4207	1.3395	0.7466	313.1	1.4340	1.3521	0.7396
Ethylene Dichloride				365.0	1.4208	1.3396	0.7465
265.5	1.5002	1.4169	0.7058	435.8	1.4105	1.3298	0.7520
289.4	1.4878	1.4045	0.7120	546.1	1.4022	1.3221	0.7564
313.1	1.4778	1.3946	0.7171				

[a]We are indebted to Drs. J. Foss, Y. Yang, and J. Schellman for permission to publish these tables.
[b]Data from Adler and Fasman, *Meth. Enzymol,* (1968).

From Fasman, in *Methods in Enzymology,* Colowick and Kaplan, Eds., Academic Press, New York, 1963, chap. 126.

Table 8
CONSTANTS FOR
REFRACTIVE INDEX CALCULATIONS[a]

Solvent Systems	A	λ_0, nm
50% 2-propanol/50% water (v/v)	0.8396	98.5
75% 2-propanol/25% water (v/v)	0.8616	99.4
50% methanol/50% water (v/v)	0.7780	97.0
75% methanol/25% water (v/v)	0.7770	96.2
50% 3-butanol/50% water (v/v)	0.8603	99.2
75% 3-butanol/25% water (v/v)	0.8616	99.4

[a]Values of refractive indices needed to calculate (m') at various wavelengths were obtained from the Sellmeier approximation:

$$(n^2 - 1) = A^2/(\lambda^2 - \lambda_0^2)$$

where n is the index of refraction, λ is the wavelength of interest and A and λ_0 are constants to be determined. To evaluate the constants, the refractive indices of the above solvent systems were measured at 589 nm, 546 nm and 436 nm using a Bausch and Lomb refractometer with sodium and mercury light sources.

From Tooney and Fasman, *J. Mol. Biol.*, 36, 355 (1968). With permission of author G. Fasman and copyright owners Academic Press, New York.

CALIBRATION OF OPTICAL INSTRUMENTS

SPECTROPHOTOMETERS

Alice Adler and Gerald D. Fasman

Wavelength Accuracy

A number of compounds with sharp absorbance or emission lines at known wavelengths are suitable for periodic calibration of wavelength.

Benzene vapor: ultraviolet region — A drop of benzene is allowed to vaporize in a stoppered cell. The spectrophotometer is set for ultraviolet measurements, and a minimum slit width is obtained by adjusting the photomultiplier dynode voltage and slit width controls. At 266.7 nm the absorbance should be about 0.2; if it is lower, then another drop of benzene is added. The spectrum is scanned at very slow speed. Sharp absorption bands occur at 260.0 and 252.9 nm.

Hydrogen lamp emission: visible region — The hydrogen lamp (ultraviolet source) has emission maxima at 486.1 and 656.1 nm. This source is used, and the slit width is adjusted to give an absorbance slightly over 1 at a wavelength longer than 660 nm. [In the Cary 14 the selector switch is turned to "reference," the reference (but not the sample) compartment is uncovered, and the slit width is set near 0.01 mm.] The wavelength is then slowly adjusted to shorter values. Well-defined pen deflections should be found at the wavelengths indicated.

Holmium oxide filters — The use of filters made of holmium oxide glass allows the calibration of wavelength for a spectrophotometer over a broad range. (These filters are available from Arthur Thomas Co., Philadelphia, Pa., and from Unicam Instruments, Cambridge, England.) Strong, sharp absorbance lines are found (among others) at 279.4, 287.5, 360.9, 418.4, 536.2, and 637.5 nm.[1]

Absorbance Accuracy

Neutral density filters — The accuracy of optical density for a spectrophotometer in the visible and near-UV ranges can be measured by means of a set of neutral density filters, for example of O.D. 0.3, 0.5, and 1.0. High-precision neutral density filters, with a quartz substrate, obtained from Baird-Atomic, Bedford, Mass., are accurate to ±0.03 absorbance units over the wavelength range 275 to 1,200 nm. These filters give flat absorbance curves at $\geqslant$ 340 nm. They are available in round or square shapes and can be mounted to fit the sample holder.

Potassium dichromate solution — Absorbance accuracy can also be calibrated through the utilization of a standard solution of $K_2Cr_2O_7$ prepared by dissolving 120.0 mg of dry $K_2Cr_2O_7$ in one liter of 0.01 N H_2SO_4. The following values should be obtained for absorbance maxima in matched 1-cm cells with 0.01 N H_2SO_4 as the blank:[1] OD_{350} = 1.288 ±0.02 and OD_{257} = 1.738 ±0.03. The solution obeys Beer's Law upon dilution.

SPECTROPOLARIMETERS

The absolute accuracy of optical rotatory dispersion instruments can be monitored with a standard sucrose solution. Specific rotation values of a 0.25% solution of National Bureau of Standards sucrose can be compared to the following values[2] and, if necessary, the polarimeter can be recalibrated: $[\alpha]_{589}$ = 66.5, $[\alpha]_{546.1}$ = 78.3, $[\alpha]_{500}$ = 94.6, $[\alpha]_{400}$ = 156.2, $[\alpha]_{349.6}$ = 213.8, $[\alpha]_{303}$ = 315.0, $[\alpha]_{250}$ = 525.3. The presence of artifacts due to light absorption can be detected by placing optically inactive absorbing solutions in tandem with sucrose. The ORD should not be distorted. Suitable solutions are potassium dichromate (λ_{max} = 350 nm) or N,N-dimethyl acetamide (λ_{max} = 195 nm) made up to give OD ~ 1 at their absorption maxima.

CIRCULAR DICHROMETERS

The compound commonly used for calibration of circular dichroism instruments is d-10-camphorsulfonic acid (CSA) in 0.1% aqueous solution, which displays a large ellipticity band at 290 nm. However, CSA is not commercially available in purity sufficient for use as a primary standard. CSA forms a hydrate containing ~ 7% water under normal laboratory conditions,[3,4] so that weight may not be an accurate measure of concentration. Furthermore, yellow impurities were found[5] in some batches of reagent grade (Eastman Kodak) CSA. The acid can be purified by recrystallization from acetic acid,[3] followed by vacuum sublimation, drying at 80°C under vacuum, and storage in a desiccator.[5] It is then suitable as a CD standard with which the signal gain adjustment controlling the magnitude of the observed CD signal on an instrument can be manipulated.

The exact value of the peak molecular ellipticity, $[\theta]_{290}$, of CSA is not known with certainty, partially because of impurity problems; it can be calculated[3,4,6] from accurate ORD data on the same CSA sample obtained on a well-calibrated polarimeter. For this calculation to be valid the sample must not contain optically active impurities, although small amounts of water are tolerable. A simple way to obtain an absolute CD value for a rotationally pure aqueous solution of CSA standard is to use calculated[6] ratios of peak molecular ellipticity to peak and trough molecular rotations: $[\theta]_{290}/[M]_{306}$ = 1.76 and $[\theta]_{290}/[M]_{270}$ = −1.37. For example, one dry purified CSA sample[5] yielded measured rotations of $[M]_{306}$ = +4,480 and $[M]_{270}$ = −5,700, from which $[\theta]_{290}$ equals the average of 4,480 × 1.76 = 7,880 and −5,700 × −1.37 = 7,800, or $[\theta]_{290}$ = 7,840 (corresponding to $\epsilon_L - \epsilon_R$ = 2.37.) The resulting $[\theta]_{290}$ magnitude can then be used to calibrate the circular dichrometer, even though the CSA sample may contain some water. Note that the calibration setting recommended in the Cary 60 operating manual corresponds to $[\theta]_{290}$ = 7,150 and should not be used with CSA standards of unknown purity.

PATH LENGTH OF CELLS

The path length of thin (1 mm or less) sample cells used for optical measurements is not always given accurately by the manufacturer's specifications. The exact path length of such cells may be measured[7] by counting interference fringes in the near IR region; this procedure is recommended for cells thinner than 1 mm. When an empty cell is placed in the light path of a spectrophotometer, interference fringes can be seen if the faces of the cell are parallel and if the path length of the cell, b, is not greater than about 10^2 to 10^3 times the wavelength of the incident light. The absorption spectrum of the cell is taken over a suitable wavelength range. The spectrophotometer should be purged with N_2 to remove water vapor, and a sensitive or expanded absorbance scale, if available, should be utilized. The number of fringes, n, is counted, and the path length, b, in mm, is calculated[7] from the equation

$$b = \frac{n}{2000} \frac{(\lambda_L \lambda_S)}{(\lambda_L - \lambda_S)}$$

where

λ_L = the longest wavelength of the range measured, in microns;
λ_S = the shortest wavelength of the range measured, in microns.

For example, a 0.1-mm quartz cell was calibrated using the wavelength range 1 to 2 μm. 72 fringes were observed between 1.055 and 1.690 μm; therefore, b = 0.1011 mm. Since quartz and glass are not transparent at $\lambda > \sim 2$ μm, 1 mm is the upper limit of path length measurable by this method. More flexibility is obtained with IR-transparent cells such as CaF_2.

REFERENCES

1. Unicam SP 800 Operating Instruction, Pye Unicam Ltd., Cambridge, England.
2. **Brand, Washburn, Erlanger, Ellenbogen, Daniel, Zippmann, and Scheu,** *J. Am. Chem. Soc.,* 76, 3037 (1954).
3. **De Tar,** *Anal. Chem.,* 41, 1406 (1969).
4. **Krueger and Pschigoda,** *Anal. Chem.,* 43, 675 (1971).
5. **Fasman and Lituri,** unpublished.
6. **Cassim and Yang,** *Biochemistry,* 8, 1947 (1969).
7. **Potts,** *Chemical Infrared Spectroscopy,* Vol. I, John Wiley & Sons, New York, 1963, 119.

CIRCULAR DICHROIC SPECTRA OF NUCLEOPROTEINS

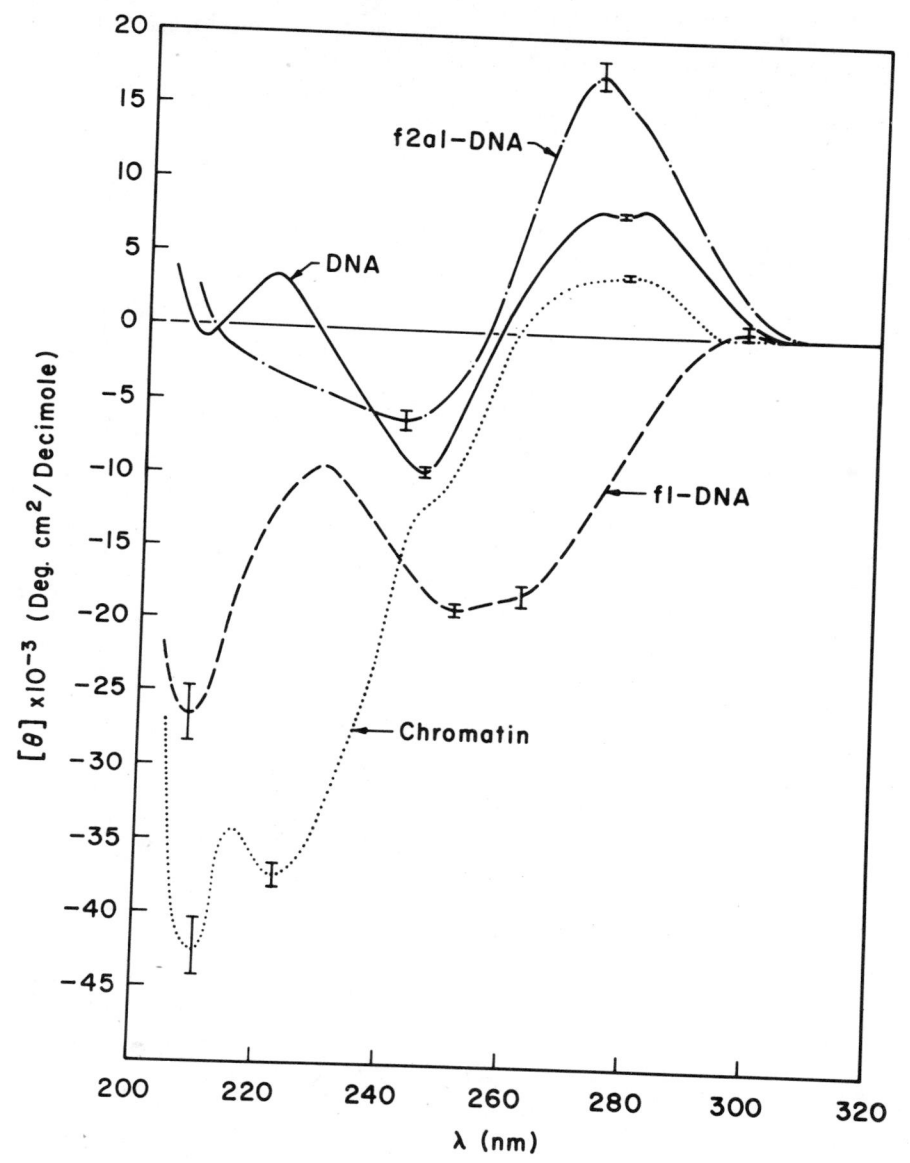

Circular dichroic spectra of calf thymus DNA and selected nucleoproteins derived from it. The solvent for all samples is $0.14M$ NaCl or NaF + $0.002M$ tris, pH 7. Concentrations 1 to 2 × $10^{-4}M$ nucleotide residues; path length 1 cm; temperature 23°. $[\theta]$ values are molar ellipticities per nucleotide residue. r for reconstituted complexes is the concentration ratio of calf thymus histone peptide residues to DNA nucleotide residues. The fl histone-DNA complex was formed by means of a stepwise NaF gradient;[1] all other histone-DAN complexes were reconstituted with a continuous linear guanidine · HCl gradient.[2] Curves: (——) native calf thymus DNA (1); (· · · ·) sheared calf thymus chromatin;[3] (- - -) fl-DNA, r = 1.0;[1] (— · — ·) f2al-DNA, r = 1.5.[4] Histone f2b and f2a2 yield DNA complexes with CD spectra similar to f2al-DNA; maximum ellipticity values at r = 1.5 are $[\theta]_{273}$ = 22,800 for f2b,[2] and $[\theta]_{272}$ = 17,800 for f2a2.[5] f3 histone causes smaller changes in the DNA CD pattern;[5] at r = 1.5 $[\theta]_{274}$ = 9,800 for f3 in the reduced (SH) state, and $[\theta]_{277}$ = 7,600 in the oxidized (SS) form.

Compiled by Alice Adler.

REFERENCES

1. Fasman, Valenzuela, and Adler, *Biochemistry,* 10, 3795 (1971).
2. Adler, Ross, Chen, Stafford, Woiszwillo, and Fasman, *Biochemistry,* 13, 616 (1974).
3. Slayter, Shih, Adler, and Fasman, *Biochemistry,* 11, 3044 (1972).
4. Adler, Fulmer, and Fasman, *Biochemistry,* 14, 1445 (1975).
5. Adler, Moran, and Fasman, *Biochemistry,* 14, 4179 (1975).

REFRACTIVE INDEX DEPENDENCE

Table 1
CONCENTRATION DEPENDENCE OF REFRACTIVE INDEX FOR METHANOL, 2-CHLOROETHANOL, AND 2-METHOXYETHANOL (WATER MIXTURE)[a]

Alcohol concentration % v/v	Methanol (27°)		2-Chloroethanol (25°)		2-Methoxyethanol (25°)	
	$n_{589.3}$	$n_{435.8}$	$n_{589.3}$	$n_{435.8}$	$n_{589.3}$	$n_{435.8}$
10	1.33373	1.34091	—	—	—	—
20	1.33582	1.34295	1.35631	1.36404	1.34907	1.35652
40	1.33947	1.34656	1.37886	1.38720	1.36604	1.37381
60	1.34034	1.34722	1.40001	1.40881	1.38103	1.38899
70	1.33941	1.34624				
80	1.33686	1.34372	1.42001	1.42938	1.39286	1.40106
90	1.33202	1.33870				
100	1.32557	1.33206	1.44064	1.45059	1.40000	1.40845

[a]Measurements performed with Bausch & Lomb Refractometer at indicated temperature.

Table 2
CONCENTRATION DEPENDENCE OF SELLMEIER CONSTANTS FOR METHANOL, 2-CHLOROETHANOL, AND 2-METHOXYETHANOL[a]

Alcohol concentration % v/v	Methanol (27°)		2-Chloroethanol (25°)		2-Methoxyethanol (25°)	
	a	λ_v^2	a	λ_v^2	a	λ_v^2
10	0.75682	9805.9	—	—	—	—
20	0.76250	9693.3	0.81547	9961.3	0.79686	9786.1
40	0.77231	9553.8	0.87485	10164	0.84164	9783.8
60	0.77523	9268.5	0.93174	10223	0.88194	9676
70	0.77291	9223				
80	0.76604	9322.6	0.98590	10419	0.91376	9705.2
90	0.75375	9197.3				
99	0.73726	9106.7	1.04258	10601	0.93277	9840.1

[a]Constants of the Sellmeier approximation for refractive index dispersion

$$n^2 = 1 + \frac{a\lambda^2}{\lambda^2 - \lambda_v^2}$$

were calculated from refractive index values, measured at 589.3 and 435.8 nm with a Bausch & Lomb Refractometer at indicated temperature.

Table 3
CONCENTRATION DEPENDENCE OF LORENTZ CORRECTION FACTORS
FOR METHANOL AT INDICATED WAVELENGTHS[a]

$$\frac{3}{n^2 + 2}$$

λ, nm	Methanol							
	10%	20%	40%	60%	70%	80%	90%	99%
600	0.7940	0.7928	0.7908	0.7903	0.7908	0.7923	0.7950	0.7986
590	0.7939	0.7927	0.7907	0.7902	0.7907	0.7921	0.7948	0.7984
580	0.7937	0.7925	0.7905	0.7900	0.7905	0.7920	0.7947	0.7983
570	0.7935	0.7924	0.7903	0.7898	0.7904	0.7918	0.7945	0.7981
560	9.7933	0.7922	0.7901	0.7897	0.7902	0.7916	0.7943	0.7980
550	0.7931	0.7920	0.7899	0.7895	0.7900	0.7914	0.7942	0.7978
540	0.7929	0.7918	0.7897	0.7893	0.7898	0.7912	0.7940	0.7976
530	0.7927	0.7916	0.7895	0.7891	0.7896	0.7910	0.7938	0.7974
520	0.7925	0.7913	0.7893	0.7889	0.7894	0.7908	0.7935	0.7972
510	0.7922	0.7911	0.7891	0.7886	0.7891	0.7906	0.7933	0.7970
500	0.7920	0.7908	0.7888	0.7884	0.7889	0.7903	0.7931	0.7967
490	0.7917	0.7906	0.7885	0.7881	0.7886	0.7901	0.7928	0.7965
480	0.7914	0.7903	0.7882	0.7878	0.7884	0.7898	0.7925	0.7962
470	0.7911	0.7899	0.7879	0.7875	0.7881	0.7895	0.7922	0.7959
460	0.7908	0.7896	0.7876	0.7872	0.7877	0.7891	0.7919	0.7956
450	0.7904	0.7893	0.7872	0.7869	0.7874	0.7888	0.7916	0.7953
440	0.7900	0.7889	0.7869	0.7865	0.7870	0.7884	0.7912	0.7949
430	0.7896	0.7885	0.7864	0.7861	0.7866	0.7880	0.7908	0.7946
420	0.7891	0.7880	0.7860	0.7857	0.7862	0.7876	0.7904	0.7941
410	0.7887	0.7875	0.7855	0.7852	0.7857	0.7871	0.7900	0.7937
400	0.7881	0.7870	0.7850	0.7847	0.7853	0.7866	0.7895	0.7932
395	0.7879	0.7867	0.7847	0.7844	0.7850	0.7864	0.7892	0.7930
390	0.7876	0.7864	0.7845	0.7842	0.7847	0.7861	0.7890	0.7927
385	0.7873	0.7861	0.7842	0.7839	0.7844	0.7858	0.7887	0.7924
380	0.7869	0.7858	0.7838	0.7836	0.7841	0.7855	0.7884	0.7922
375	0.7866	0.7855	0.7835	0.7833	0.7838	0.7852	0.7881	0.7919
370	0.7863	0.7852	0.7832	0.7829	0.7835	0.7849	0.7878	0.7915
365	0.7859	0.7848	0.7828	0.7826	0.7832	0.7845	0.7874	0.7912
360	0.7855	0.7844	0.7825	0.7822	0.7828	0.7842	0.7871	0.7909
355	0.7852	0.7841	0.7821	0.7819	0.7824	0.7838	0.7867	0.7905
350	0.7847	0.7836	0.7817	0.7815	0.7820	0.7834	0.7863	0.7902
345	0.7843	0.7832	0.7813	0.7811	0.7816	0.7830	0.7859	0.7898
340	0.7839	0.7828	0.7808	0.7806	0.7812	0.7826	0.7855	0.7894
335	0.7834	0.7823	0.7804	0.7802	0.7808	0.7821	0.7851	0.7889
330	0.7829	0.7818	0.7799	0.7797	0.7803	0.7817	0.7846	0.7885
325	0.7824	0.7813	0.7794	0.7792	0.7798	0.7812	0.7841	0.7880
320	0.7818	0.7807	0.7788	0.7787	0.7793	0.7806	0.7836	0.7875
319	0.7817	0.7806	0.7787	0.7786	0.7792	0.7805	0.7835	0.7874
318	0.7816	0.7805	0.7786	0.7785	0.7791	0.7804	0.7834	0.7873
317	0.7815	0.7804	0.7785	0.7784	0.7790	0.7803	0.7833	0.7872
316	0.7814	0.7803	0.7784	0.7783	0.7789	0.7802	0.7832	0.7871
315	0.7812	0.7802	0.7782	0.7781	0.7787	0.7801	0.7831	0.7870
314	0.7811	0.7801	0.7781	0.7780	0.7786	0.7800	0.7830	0.7869
313	0.7810	0.7799	0.7780	0.7779	0.7785	0.7799	0.7829	0.7868
312	0.7809	0.7798	0.7779	0.7778	0.7784	0.7797	0.7827	0.7867
311	0.7808	0.7797	0.7778	0.7777	0.7783	0.7796	0.7826	0.7865
310	0.7806	0.7796	0.7776	0.7776	0.7782	0.7795	0.7825	0.7864
309	0.7805	0.7794	0.7775	0.7774	0.7780	0.7794	0.7824	0.7863
308	0.7804	0.7793	0.7774	0.7773	0.7779	0.7793	0.7823	0.7862
307	0.7802	0.7792	0.7773	0.7772	0.7778	0.7791	0.7822	0.7861
306	0.7801	0.7791	0.7771	0.7771	0.7777	0.7790	0.7820	0.7860
305	0.7800	0.7789	0.7770	0.7769	0.7776	0.7789	0.7819	0.7859

[a]Lorentz factors calculated from Sellmeier approximation constants (Table 2).

Table 3 (continued)
CONCENTRATION DEPENDENCE OF LORENTZ CORRECTION FACTORS FOR METHANOL AT INDICATED WAVELENGTHS

$$\frac{3}{n^2 + 2}$$

λ, nm	Methanol							
	10%	20%	40%	60%	70%	80%	90%	99%
304	0.7798	0.7788	0.7769	0.7768	0.7774	0.7788	0.7818	0.7857
303	0.7797	0.7787	0.7767	0.7767	0.7773	0.7786	0.7817	0.7856
302	0.7796	0.7785	0.7766	0.7766	0.7772	0.7785	0.7815	0.7855
301	0.7794	0.7784	0.7765	0.7764	0.7770	0.7784	0.7814	0.7854
300	0.7793	0.7783	0.7763	0.7763	0.7769	0.7782	0.7813	0.7852
299	0.7792	0.7781	0.7762	0.7762	0.7768	0.7781	0.7812	0.7851
298	0.7790	0.7780	0.7761	0.7760	0.7767	0.7780	0.7810	0.7850
297	0.7789	0.7778	0.7759	0.7759	0.7765	0.7778	0.7809	0.7849
296	0.7787	0.7777	0.7758	0.7758	0.7764	0.7777	0.7808	0.7847
295	0.7786	0.7775	0.7756	0.7756	0.7762	0.7776	0.7806	0.7846
294	0.7784	0.7774	0.7755	0.7755	0.7761	0.7774	0.7805	0.7845
293	0.7783	0.7772	0.7753	0.7753	0.7760	0.7773	0.7803	0.7843
292	0.7781	0.7771	0.7752	0.7752	0.7758	0.7771	0.7802	0.7842
291	0.7780	0.7769	0.7750	0.7750	0.7757	0.7770	0.7801	0.7840
290	0.7778	0.7768	0.7749	0.7749	0.7755	0.7768	0.7799	0.7839
289	0.7777	0.7766	0.7747	0.7747	0.7754	0.7767	0.7798	0.7838
288	0.7775	0.7765	0.7746	0.7746	0.7752	0.7765	0.7796	0.7836
287	0.7773	0.7763	0.7744	0.7744	0.7751	0.7764	0.7795	0.7835
286	0.7772	0.7761	0.7743	0.7743	0.7749	0.7762	0.7793	0.7833
285	0.7770	0.7760	0.7741	0.7741	0.7748	0.7761	0.7792	0.7832
284	0.7768	0.7758	0.7739	0.7740	0.7746	0.7759	0.7790	0.7830
283	0.7767	0.7756	0.7738	0.7738	0.7744	0.7758	0.7789	0.7829
282	0.7765	0.7755	0.7736	0.7736	0.7743	0.7756	0.7787	0.7827
281	0.7763	0.7753	0.7734	0.7735	0.7741	0.7754	0.7785	0.7826
280	0.7761	0.7751	0.7732	0.7733	0.7740	0.7753	0.7784	0.7824
279	0.7760	0.7750	0.7731	0.7731	0.7738	0.7751	0.7782	0.7822
278	0.7758	0.7748	0.7729	0.7730	0.7736	0.7749	0.7780	0.7821
277	0.7756	0.7746	0.7727	0.7728	0.7734	0.7747	0.7779	0.7819
276	0.7754	0.7744	0.7725	0.7726	0.7733	0.7746	0.7777	0.7817
275	0.7752	0.7742	0.7724	0.7724	0.7731	0.7744	0.7775	0.7816
274	0.7750	0.7740	0.7722	0.7723	0.7729	0.7742	0.7774	0.7814
273	0.7748	0.7738	0.7720	0.7721	0.7727	0.7740	0.7772	0.7812
272	0.7747	0.7737	0.7718	0.7719	0.7726	0.7738	0.7770	0.7810
271	0.7745	0.7735	0.7716	0.7717	0.7724	0.7737	0.7768	0.7809
270	0.7743	0.7733	0.7714	0.7715	0.7722	0.7735	0.7766	0.7807
269	0.7741	0.7731	0.7712	0.7713	0.7720	0.7733	0.7764	0.7805
268	0.7738	0.7729	0.7710	0.7711	0.7718	0.7731	0.7763	0.7803
267	0.7736	0.7726	0.7708	0.7709	0.7716	0.7729	0.7761	0.7801
266	0.7734	0.7724	0.7706	0.7707	0.7714	0.7727	0.7759	0.7799
265	0.7732	0.7722	0.7704	0.7705	0.7712	0.7725	0.7757	0.7798
264	0.7730	0.7720	0.7702	0.7703	0.7710	0.7723	0.7755	0.7796
263	0.7728	0.7718	0.7700	0.7701	0.7708	0.7721	0.7753	0.7794
262	0.7726	0.7716	0.7697	0.7699	0.7706	0.7719	0.7751	0.7792
261	0.7723	0.7714	0.7695	0.7697	0.7704	0.7717	0.7749	0.7790
260	0.7721	0.7711	0.7693	0.7695	0.7702	0.7714	0.7747	0.7788
259	0.7719	0.7709	0.7691	0.7693	0.7699	0.7712	0.7744	0.7785
258	0.7716	0.7707	0.7688	0.7691	0.7697	0.7710	0.7742	0.7783
257	0.7714	0.7704	0.7686	0.7688	0.7695	0.7708	0.7740	0.7781
256	0.7712	0.7702	0.7684	0.7686	0.7693	0.7705	0.7738	0.7779
255	0.7709	0.7700	0.7681	0.7684	0.7691	0.7703	0.7736	0.7777
254	0.7707	0.7697	0.7679	0.7681	0.7688	0.7701	0.7733	0.7775
253	0.7704	0.7695	0.7676	0.7679	0.7686	0.7698	0.7731	0.7772
252	0.7702	0.7692	0.7674	0.7677	0.7683	0.7696	0.7729	0.7770

Table 3 (continued)
CONCENTRATION DEPENDENCE OF LORENTZ CORRECTION FACTORS FOR METHANOL AT INDICATED WAVELENGTHS

$$\frac{3}{n^2 + 2}$$

λ, nm	Methanol							
	10%	20%	40%	60%	70%	80%	90%	99%
251	0.7699	0.7690	0.7671	0.7674	0.7681	0.7694	0.7726	0.7768
250	0.7696	0.7687	0.7669	0.7672	0.7679	0.7691	0.7724	0.7765
249	0.7694	0.7684	0.7666	0.7669	0.7676	0.7689	0.7721	0.7763
248	0.7691	0.7682	0.7664	0.7667	0.7674	0.7686	0.7719	0.7761
247	0.7688	0.7679	0.7661	0.7664	0.7671	0.7684	0.7716	0.7758
246	0.7686	0.7676	0.7658	0.7661	0.7668	0.7681	0.7714	0.7756
245	0.7683	0.7673	0.7655	0.7659	0.7666	0.7678	0.7711	0.7753
244	0.7680	0.7671	0.7653	0.7656	0.7663	0.7676	0.7709	0.7751
243	0.7677	0.7668	0.7650	0.7653	0.7660	0.7673	0.7706	0.7748
242	0.7674	0.7665	0.7647	0.7651	0.7658	0.7670	0.7703	0.7745
241	0.7671	0.7662	0.7644	0.7648	0.7655	0.7667	0.7701	0.7743
240	0.7668	0.7659	0.7641	0.7645	0.7652	0.7664	0.7698	0.7740
239	0.7665	0.7656	0.7638	0.7642	0.7649	0.7661	0.7695	0.7737
238	0.7662	0.7653	0.7635	0.7639	0.7646	9.7659	0.7692	0.7734
237	0.7659	0.7650	0.7632	0.7636	0.7643	0.7656	0.7689	0.7732
236	0.7655	0.7646	0.7629	0.7633	0.7640	0.7652	0.7686	0.7729
235	0.7652	0.7643	0.7626	0.7630	0.7637	0.7649	0.7683	0.7726
234	0.7649	0.7640	0.7622	0.7627	0.7634	0.7646	0.7680	0.7723
233	0.7645	0.7637	0.7619	0.7624	0.7631	0.7643	0.7677	0.7720
232	0.7642	0.7633	0.7616	0.7620	0.7628	0.7640	0.7674	0.7717
231	0.7639	0.7630	0.7612	0.7617	0.7625	0.7637	0.7671	0.7714
230	0.7635	0.7626	0.7609	0.7614	0.7621	0.7633	0.7667	0.7710
229	0.7631	0.7623	0.7605	0.7611	0.7618	0.7630	0.7664	0.7707
228	0.7628	0.7619	0.7602	0.7607	0.7614	0.7626	0.7661	0.7704
227	0.7624	0.7615	0.7598	0.7604	0.7611	0.7623	0.7657	0.7701
226	0.7620	0.7612	0.7594	0.7600	0.7607	0.7619	0.7654	0.7697
225	0.7616	0.7608	0.7591	0.7596	0.7604	0.7616	0.7650	0.7694
224	0.7612	0.7604	0.7587	0.7593	0.7600	0.7612	0.7647	0.7690
223	0.7608	0.7600	0.7583	0.7589	0.7597	0.7608	0.7643	0.7687
222	0.7604	0.7596	0.7579	0.7585	0.7593	0.7605	0.7640	0.7683
221	0.7600	0.7592	0.7575	0.7581	0.7589	0.7601	0.7636	0.7680
220	0.7596	0.7588	0.7571	0.7577	0.7585	0.7597	0.7632	0.7676
219	0.7592	0.7584	0.7567	0.7573	0.7581	0.7593	0.7628	0.7672
218	0.7588	0.7579	0.7563	0.7569	0.7577	0.7589	0.7624	0.7668
217	0.7583	0.7575	0.7558	0.7565	0.7573	0.7585	0.7620	0.7664
216	0.7579	0.7571	0.7554	0.7561	0.7569	0.7580	0.7616	0.7660
215	0.7574	0.7566	0.7549	0.7557	0.7565	0.7576	0.7612	0.7656
214	0.7569	0.7561	0.7545	0.7552	0.7560	0.7572	0.7608	0.7652
213	0.7565	0.7557	0.7540	0.7548	0.7556	0.7567	0.7603	0.7648
212	0.7560	0.7552	0.7536	0.7544	0.7551	0.7563	0.7599	0.7644
211	0.7555	0.7547	0.7531	0.7539	0.7547	0.7558	0.7594	0.7639
210	0.7550	0.7542	0.7526	0.7534	0.7542	0.7554	0.7590	0.7635
209	0.7545	0.7537	0.7521	0.7530	0.7538	0.7549	0.7585	0.7630
208	0.7540	0.7532	0.7516	0.7525	0.7533	0.7544	0.7581	0.7626
207	0.7535	0.7527	0.7511	0.7520	0.7528	0.7539	0.7576	0.7621
206	0.7529	0.7522	0.7506	0.7515	0.7523	0.7534	0.7571	0.7616
205	0.7524	0.7516	0.7500	0.7510	0.7518	0.7529	0.7566	0.7611
204	0.7518	0.7511	0.7495	0.7504	0.7513	0.7524	0.7561	0.7606
203	0.7512	0.7505	0.7489	0.7499	0.7507	0.7518	0.7556	0.7601
202	0.7507	0.7499	0.7484	0.7494	0.7502	0.7513	0.7550	0.7596
201	0.7501	0.7494	0.7478	0.7488	0.7497	0.7507	0.7545	0.7591
200	0.7495	0.7488	0.7472	0.7483	0.7491	0.7502	0.7539	0.7586

Table 3 (continued)
CONCENTRATION DEPENDENCE OF LORENTZ CORRECTION FACTORS FOR METHANOL
AT INDICATED WAVELENGTHS

$$\frac{3}{n^2 + 2}$$

λ, nm	Methanol							
	10%	20%	40%	60%	70%	80%	90%	99%
199	0.7488	0.7481	0.7466	0.7477	0.7485	0.7496	0.7534	0.7580
198	0.7482	0.7475	0.7460	0.7471	0.7480	0.7490	0.7528	0.7575
197	0.7476	0.7469	0.7454	0.7465	0.7474	0.7484	0.7522	0.7569
196	0.7469	0.7463	0.7447	0.7459	0.7468	0.7478	0.7516	0.7563
195	0.7463	0.7456	0.7441	0.7453	0.7461	0.7472	0.7510	0.7557
194	0.7456	0.7449	0.7434	0.7446	0.7455	0.7465	0.7504	0.7551
193	0.7449	0.7442	0.7428	0.7440	0.7449	0.7459	0.7498	0.7545
192	0.7442	0.7435	0.7421	0.7433	0.7442	0.7452	0.7491	0.7539
191	0.7434	0.7428	0.7414	0.7427	0.7435	0.7446	0.7485	0.7532
190	0.7427	0.7421	0.7406	0.7420	0.7429	0.7439	0.7478	0.7526
189	0.7419	0.7413	0.7399	0.7413	0.7422	0.7432	0.7471	0.7519
188	0.7412	0.7406	0.7392	0.7405	0.7415	0.7424	0.7464	0.7512
187	0.7404	0.7398	0.7384	0.7398	0.7407	0.7417	0.7457	0.7505
186	0.7396	0.7390	0.7376	0.7391	0.7400	0.7409	0.7450	0.7498
185	0.7387	0.7382	0.7368	0.7383	0.7392	0.7402	0.7442	0.7491

Compiled by Murray Ettinger.

Table 4
CONCENTRATION DEPENDENCE OF LORENTZ CORRECTION FACTORS FOR 2-CHLOROETHANOL AT INDICATED WAVELENGTHS[a]

$$\frac{3}{n^2 + 2}$$

2-Chloroethanol

λ, nm	20%	40%	60%	80%	99%
600	0.7815	0.7691	0.7577	0.7471	0.7363
590	0.7813	0.7690	0.7575	0.7469	0.7361
580	0.7811	0.7688	0.7573	0.7467	0.7359
570	0.7809	0.7686	0.7571	0.7465	0.7357
560	0.7807	0.7684	0.7569	0.7463	0.7354
550	0.7805	0.7681	0.7567	0.7460	0.7352
540	0.7803	0.7679	0.7565	0.7458	0.7349
530	0.7801	0.7677	0.7562	0.7455	0.7346
520	0.7798	0.7674	0.7559	0.7452	0.7343
510	0.7796	0.7671	0.7556	0.7449	0.7340
500	0.7793	0.7668	0.7553	0.7446	0.7337
490	0.7790	0.7665	0.7550	0.7443	0.7333
480	0.7787	0.7662	0.7547	0.7439	0.7329
470	0.7784	0.7658	0.7543	0.7435	0.7325
460	0.7780	0.7655	0.7539	0.7431	0.7321
450	0.7776	0.7650	0.7535	0.7426	0.7316
440	0.7772	0.7646	0.7530	0.7422	0.7311
430	0.7768	0.7641	9.7525	0.7416	0.7306
420	0.7763	0.7636	0.7520	0.7411	0.7300
410	0.7758	0.7631	0.7514	0.7405	0.7294
400	0.7752	0.7625	0.7508	0.7399	0.7287
395	0.7749	0.7622	0.7505	0.7395	0.7284
390	0.7746	0.7619	0.7502	0.7392	0.7280
385	0.7743	0.7615	0.7498	0.7388	0.7276
380	0.7740	0.7612	0.7494	0.7384	0.7272
375	0.7736	0.7608	0.7491	0.7380	0.7268
370	0.7733	0.7604	0.7487	0.7376	0.7263
365	0.7729	0.7600	0.7482	0.7371	0.7259
360	0.7725	0.7596	0.7478	0.7367	0.7254
355	0.7721	0.7591	0.7473	0.7362	0.7249
350	0.7716	0.7587	0.7469	0.7357	0.7244
345	0.7712	0.7582	0.7464	0.7352	0.7238
340	0.7707	0.7577	0.7458	0.7346	0.7232
335	0.7702	0.7571	0.7453	0.7340	0.7226
330	0.7697	0.7566	0.7447	0.7334	0.7220
325	0.7691	0.7560	0.7441	0.7328	0.7213
320	0.7685	0.7754	0.7434	0.7321	0.7206
319	0.7684	0.7553	0.7433	0.7320	0.7204
318	0.7683	0.7551	0.7432	0.7318	0.7203
317	0.7682	0.7550	0.7430	0.7317	0.7201
316	0.7680	0.7549	0.7429	0.7315	0.7200
315	0.7679	0.7547	0.7428	0.7314	0.7198
314	0.7678	0.7546	0.7426	0.7312	0.7197
313	0.7677	0.7545	0.7425	0.7311	0.7195
312	0.7675	0.7543	0.7423	0.7309	0.7194
311	0.7674	0.7542	0.7422	0.7308	0.7192
310	0.7673	0.7540	0.7420	0.7306	0.7191
309	0.7671	0.7539	0.7419	0.7305	0.7189
308	0.7670	0.7538	0.7417	0.7303	0.7187
307	0.7669	0.7536	0.7416	0.7302	0.7186
306	0.7667	0.7535	0.7414	0.7300	0.7184
305	0.7666	0.7533	0.7413	0.7298	0.7182
304	0.7664	0.7532	0.7411	0.7297	0.7181

$$\frac{3}{n^2 + 2}$$

2-Chloroethanol

λ, nm	20%	40%	60%	80%	99%
303	0.7663	0.7530	0.7410	0.7295	0.7179
302	0.7661	0.7529	0.7408	0.7293	0.7177
301	0.7660	0.7527	0.7406	0.7292	0.7175
300	0.7659	0.7525	0.7405	0.7290	0.7173
299	0.7657	0.7524	0.7403	0.7288	0.7172
298	0.7655	0.7522	0.7402	0.7286	0.7170
297	0.7654	0.7521	0.7400	0.7285	0.7168
296	0.7652	0.7519	0.7398	0.7283	0.7166
295	0.7651	0.7517	0.7396	0.7281	0.7164
294	0.7649	0.7516	0.7395	0.7279	0.7162
293	0.7648	0.7514	0.7393	0.7277	0.7160
292	0.7646	0.7512	0.7391	0.7275	0.7158
291	0.7644	0.7510	0.7389	0.7274	0.7156
290	0.7643	0.7509	0.7387	0.7272	0.7154
289	0.7641	0.7507	0.7386	0.7270	0.7152
288	0.7639	0.7505	0.7384	0.7268	0.7150
287	0.7638	0.7503	0.7382	0.7266	0.7148
286	0.7636	0.7501	0.7380	0.7264	0.7146
285	0.7634	0.7500	0.7378	0.7262	0.7144
284	0.7632	0.7498	0.7376	0.7260	0.7142
283	0.7631	0.7496	0.7374	0.7258	0.7140
282	0.7629	0.7494	0.7372	0.7255	0.7137
281	0.7627	0.7492	0.7370	0.7253	0.7135
280	0.7625	0.7490	0.7368	0.7251	0.7133
279	0.7623	0.7488	0.7366	0.7249	0.7131
278	0.7621	0.7486	0.7364	0.7247	0.7128
277	0.7619	0.7484	0.7362	0.7244	0.7126
276	0.7617	0.7482	0.7359	0.7242	0.7123
275	0.7615	0.7480	0.7357	0.7240	0.7121
274	0.7613	0.7477	0.7355	0.7238	0.7119
273	0.7611	0.7475	0.7353	0.7235	0.7116
272	0.7609	0.7473	0.7350	0.7233	0.7114
271	0.7607	0.7471	0.7348	0.7230	0.7111
270	0.7605	0.7469	0.7346	0.7228	0.7109
269	0.7603	0.7466	0.7343	0.7225	0.7106
268	0.7601	0.7464	0.7341	0.7223	0.7103
267	0.7598	0.7462	0.7339	0.7220	0.7101
266	0.7596	0.7459	0.7336	0.7218	0.7098
265	0.7594	0.7457	0.7334	0.7215	0.7095
264	0.7592	0.7454	0.7331	0.7212	0.7092
263	0.7589	0.7452	0.7328	0.7210	0.7090
262	0.7587	0.7449	0.7326	0.7207	0.7087
261	0.7585	0.7447	0.7323	0.7204	0.7084
260	0.7582	0.7444	0.7321	0.7201	0.7081
259	0.7580	0.7442	0.7318	0.7199	0.7078
258	0.7577	0.7439	0.7315	0.7196	0.7075
257	0.7575	0.7436	0.7312	0.7193	0.7072
256	0.7572	0.7434	0.7310	0.7190	0.7069
255	0.7570	0.7431	0.7307	0.7187	0.7066
254	0.7567	0.7428	0.7304	0.7184	0.7062
253	0.7564	0.7425	0.7301	0.7181	0.7059
252	0.7562	0.7422	0.7298	0.7178	0.7056
251	0.7559	0.7419	0.7295	0.7174	0.7053

[a]Lorentz factors calculated from Sellmeier approximation constants (Table 2).

Table 4 (continued)
CONCENTRATION DEPENDENCE OF LORENTZ CORRECTION FACTORS FOR 2-CHLOROETHANOL AT INDICATED WAVELENGTHS

$$\frac{3}{n^2 + 2}$$

$$\frac{3}{n^2 + 2}$$

2-Chloroethanol

λ, nm	20%	40%	60%	80%	99%
250	0.7556	0.7417	0.7292	0.7171	0.7049
249	0.7553	0.7414	0.7289	0.7168	0.7046
248	0.7550	0.7410	0.7286	0.7165	0.7042
247	0.7547	0.7407	0.7282	0.7161	0.7039
246	0.7544	0.7404	0.7279	0.7158	0.7035
245	0.7541	0.7401	0.7276	0.7154	0.7032
244	0.7538	0.7398	0.7272	0.7151	0.7028
243	0.7535	0.7395	0.7269	0.7147	0.7024
242	0.7532	0.7391	0.7266	0.7144	0.7020
241	0.7529	0.7388	0.7262	0.7140	0.7016
240	0.7526	0.7384	0.7259	0.7136	0.7013
239	0.7523	0.7381	0.7255	0.7132	0.7009
238	0.7519	0.7377	0.7251	0.7129	0.7005
237	0.7516	0.7374	0.7248	0.7125	0.7000
236	0.7512	0.7370	0.7244	0.7121	0.6996
235	0.7509	0.7367	0.7240	0.7117	0.6992
234	0.7505	0.7363	0.7236	0.7113	0.6988
233	0.7502	0.7359	0.7232	0.7108	0.6983
232	0.7498	0.7355	0.7228	0.7104	0.6979
231	0.7495	0.7351	0.7224	0.7100	0.6975
230	0.7491	0.7347	0.7220	0.7096	0.6970
229	0.7487	0.7343	0.7216	0.7091	0.6965
228	0.7483	0.7339	0.7211	0.7087	0.6961
227	0.7479	0.7335	0.7207	0.7082	0.6956
226	0.7475	0.7331	0.7203	0.7077	0.6951
225	0.7471	0.7326	0.7198	0.7073	0.6946
224	0.7467	0.7322	0.7193	0.7068	0.6941
223	0.7463	0.7317	0.7189	0.7063	0.6936
222	0.7458	0.7313	0.7184	0.7058	0.6931
221	0.7454	0.7308	0.7179	0.7053	0.6925
220	0.7450	0.7303	0.7174	0.7048	0.6920
219	0.7445	0.7299	0.7169	0.7043	0.6914
218	0.7440	0.7294	0.7164	0.7037	0.6909

2-Chloroethanol

λ, nm	20%	40%	60%	80%	99%
217	0.7436	0.7289	0.7159	0.7032	0.6903
216	0.7431	0.7284	0.7154	0.7026	0.6897
215	0.7426	0.7279	0.7149	0.7021	0.6892
214	0.7421	0.7273	0.7143	0.7015	0.6886
213	0.7416	0.7268	0.7138	0.7009	0.6879
212	0.7411	0.7263	0.7132	0.7003	0.6873
211	0.7406	0.7257	0.7126	0.6997	0.6867
210	0.7401	0.7251	0.7120	0.6991	0.6861
209	0.7395	0.7246	0.7115	0.6985	0.6854
208	0.7390	0.7240	0.7108	0.6978	0.6847
207	0.7384	0.7234	0.7102	0.6972	0.6841
206	0.7378	0.7228	0.7096	0.6965	0.6834
205	0.7373	0.7222	0.7090	0.6959	0.6827
204	0.7367	0.7215	0.7083	0.6952	0.6819
203	0.7361	0.7209	0.7076	0.6945	0.6812
202	0.7355	0.7202	0.7070	0.6938	0.6805
201	0.7348	0.7196	0.7063	0.6930	0.6797
200	0.7342	0.7189	0.7056	0.6923	0.6789
199	0.7335	0.7182	0.7048	0.6915	0.6781
198	0.7329	0.7175	0.7041	0.6908	0.6773
197	0.7322	0.7167	0.7034	0.6900	0.6765
196	0.7315	0.7160	0.7026	0.6892	0.6756
195	0.7308	0.7153	0.7018	0.6883	0.6748
194	0.7301	0.7145	0.7010	0.6875	0.6739
193	0.7293	0.7137	0.7002	0.6866	0.6730
192	0.7286	0.7129	0.6994	0.6858	0.6721
191	0.7278	0.7121	0.6985	0.6849	0.6711
190	0.7270	0.7112	0.6977	0.6840	0.6702
189	0.7262	0.7104	0.6968	0.6830	0.6692
188	0.7254	0.7095	0.6959	0.6821	0.6682
187	0.7245	0.7086	0.6949	0.6811	0.6672
186	0.7237	0.7077	0.6940	0.6801	0.6661
185	0.7228	0.7068	0.6930	0.6791	0.6651

Compiled by Murray Ettinger.

Table 5
CONCENTRATION DEPENDENCE OF LORENTZ CORRECTION FACTORS FOR 2-METHOXYETHANOL AT INDICATED WAVELENGTHS[a]

λ, nm	$\dfrac{3}{n^2 + 2}$ 2-Methoxyethanol					λ, nm	$\dfrac{3}{n^2 + 2}$ 2-Methoxyethanol				
	20%	40%	60%	80%	99%		20%	40%	60%	80%	99%
600	0.7855	0.7761	0.7679	0.7615	0.7577	303	0.7708	0.7610	0.7526	0.7459	0.7417
590	0.7853	0.7759	0.7678	0.7614	0.7575	302	0.7706	0.7608	0.7525	0.7457	0.7415
580	0.7851	0.7758	0.7676	0.7612	0.7574	301	0.7705	0.7607	0.7523	0.7456	0.7413
570	0.7850	0.7756	0.7674	0.7610	0.7572	300	0.7704	0.7605	0.7522	0.7454	0.7412
560	0.7848	0.7754	0.7672	0.7608	0.7570	299	0.7702	0.7604	0.7520	0.7453	0.7410
550	0.7846	0.7752	0.7670	0.7606	0.7567	298	0.7701	0.7602	0.7519	0.7451	0.7409
540	0.7844	0.7750	0.7668	0.7604	0.7565	297	0.7699	0.7601	0.7517	0.7450	0.7407
530	0.7841	0.7747	0.7665	0.7601	0.7563	296	0.7698	0.7599	0.7516	0.7448	0.7405
520	0.7839	0.7745	0.7663	0.7599	0.7560	295	0.7696	0.7598	0.7514	0.7447	0.7404
510	0.7836	0.7742	0.7660	0.7596	0.7557	294	0.7695	0.7596	0.7512	0.7445	0.7402
500	0.7834	0.7740	0.7658	0.7593	0.7554	293	0.7693	0.7595	0.7511	0.7443	0.7400
490	0.7831	0.7737	0.7655	0.7590	0.7551	292	0.7691	0.7593	0.7509	0.7442	0.7399
480	0.7828	0.7734	0.7651	0.7587	0.7548	291	0.7690	0.7591	0.7508	0.7440	0.7397
470	0.7825	0.7730	0.7648	0.7583	0.7544	290	0.7688	0.7590	0.7506	0.7438	0.7395
460	0.7821	0.7727	0.7644	0.7580	0.7540	289	0.7687	0.7588	0.7504	0.7436	0.7394
450	0.7817	0.7723	0.7640	0.7576	0.7536	288	0.7685	0.7586	0.7502	0.7435	0.7392
440	0.7813	0.7719	0.7636	0.7571	0.7532	287	0.7683	0.7585	0.7501	0.7433	0.7390
430	0.7809	0.7714	0.7632	0.7567	0.7527	286	0.7682	0.7583	0.7499	0.7431	0.7388
420	0.7805	0.7709	0.7627	0.7562	0.7522	285	0.7680	0.7581	0.7497	0.7429	0.7386
410	0.7800	0.7704	0.7622	0.7557	0.7517	284	0.7678	0.7579	0.7495	0.7428	0.7384
400	0.7794	0.7699	0.7616	0.7551	0.7511	283	0.7676	0.7578	0.7494	0.7426	0.7382
395	0.7791	0.7696	0.7613	0.7548	0.7508	282	0.7675	0.7576	0.7492	0.7424	0.7381
390	0.7788	0.7693	0.7610	0.7545	0.7505	281	0.7673	0.7574	0.7490	0.7422	0.7379
385	0.7785	0.7690	0.7607	0.7541	0.7501	280	0.7671	0.7572	0.7488	0.7420	0.7377
380	0.7782	0.7686	0.7604	0.7538	0.7498	279	0.7669	0.7570	0.7486	0.7418	0.7375
375	0.7779	0.7683	0.7600	0.7534	0.7494	278	0.7667	0.7568	0.7484	0.7416	0.7373
370	0.7775	0.7679	0.7596	0.7531	0.7490	277	0.7666	0.7566	0.7482	0.7414	0.7371
365	0.7771	0.7675	0.7592	0.7527	0.7486	276	0.7664	0.7564	0.7480	0.7412	0.7368
360	0.7768	0.7671	0.7588	0.7523	0.7482	275	0.7662	0.7562	0.7478	0.7410	0.7366
355	0.7764	0.7667	0.7584	0.7518	0.7477	274	0.7660	0.7560	0.7476	0.7408	0.7364
350	0.7759	0.7663	0.7580	0.7514	0.7473	273	0.7658	0.7558	0.7474	0.7406	0.7362
345	0.7755 *	0.7658	0.7575	0.7509	0.7468	272	0.7656	0.7556	0.7472	0.7404	0.7360
340	0.7750	0.7654	0.7570	0.7504	0.7463	271	0.7654	0.7554	0.7470	0.7402	0.7358
335	0.7745	0.7649	0.7565	0.7499	0.7458	270	0.7652	0.7552	0.7468	0.7399	0.7355
330	0.7740	0.7643	0.7560	0.7494	0.7452	269	0.7650	0.7550	0.7466	0.7397	0.7353
325	0.7735	0.7638	0.7555	0.7488	0.7446	268	0.7648	0.7548	0.7464	0.7395	0.7351
320	0.7729	0.7632	0.7549	0.7482	0.7440	267	0.7645	0.7546	0.7461	0.7393	0.7349
319	0.7728	0.7631	0.7547	0.7481	0.7439	266	0.7643	0.7543	0.7459	0.7390	0.7346
318	0.7727	0.7630	0.7546	0.7479	0.7438	265	0.7641	0.7541	0.7457	0.7388	0.7344
317	0.7726	0.7628	0.7545	0.7478	0.7436	264	0.7639	0.7539	0.7455	0.7386	0.7341
316	0.7725	0.7627	0.7544	0.7477	0.7435	263	0.7637	0.7537	0.7452	0.7383	0.7339
315	0.7723	0.7626	0.7542	0.7476	0.7434	262	0.7634	0.7534	0.7450	0.7381	0.7337
314	0.7722	0.7625	0.7541	0.7474	0.7432	261	0.7632	0.7532	0.7447	0.7379	0.7334
313	0.7721	0.7623	0.7540	0.7473	0.7431	260	0.7630	0.7529	0.7445	0.7376	0.7331
312	0.7720	0.7622	0.7539	0.7472	0.7430	259	0.7627	0.7527	0.7443	0.7374	0.7329
311	0.7718	0.7621	0.7537	0.7470	0.7428	258	0.7625	0.7525	0.7440	0.7371	0.7326
310	0.7717	0.7619	0.7536	0.7469	0.7427	257	0.7622	0.7522	0.7438	0.7368	0.7324
309	0.7716	0.7618	0.7535	0.7467	0.7425	256	0.7620	0.7520	0.7435	0.7366	0.7321
308	0.7715	0.7617	0.7533	0.7466	0.7424	255	0.7618	0.7517	0.7432	0.7363	0.7318
307	0.7713	0.7615	0.7532	0.7465	0.7422	254	0.7615	0.7514	0.7430	0.7360	0.7315
306	0.7712	0.7614	0.7530	0.7463	0.7421	253	0.7612	0.7512	0.7427	0.7358	0.7313
305	0.7711	0.7613	0.7529	0.7462	0.7420	252	0.7610	0.7509	0.7424	0.7355	0.7310
304	0.7709	0.7611	0.7528	0.7460	0.7418	251	0.7607	0.7506	0.7422	0.7352	0.7307

[a] Lorentz factors calculated from Sellmeier approximation constants (Table 2).

Table 5 (continued)
CONCENTRATION DEPENDENCE OF LORENTZ CORRECTION FACTORS FOR 2-METHOXYETHANOL AT INDICATED WAVELENGTHS

$$\frac{3}{n^2 + 2}$$

λ, nm	2-Methoxyethanol				
	20%	40%	60%	80%	99%
250	0.7604	0.7504	0.7419	0.7349	0.7304
249	0.7602	0.7501	0.7416	0.7346	0.7301
248	0.7599	0.7498	0.7413	0.7344	0.7298
247	0.7596	0.7495	0.7410	0.7341	0.7295
246	0.7593	0.7492	0.7407	0.7338	0.7292
245	0.7590	0.7489	0.7404	0.7334	0.7289
244	0.7588	0.7486	0.7401	0.7331	0.7286
243	0.7585	0.7483	0.7398	0.7328	0.7282
242	0.7582	0.7480	0.7395	0.7325	0.7279
241	0.7578	0.7477	0.7392	0.7322	0.7276
240	0.7575	0.7474	0.7389	0.7319	0.7272
239	0.7572	0.7470	0.7385	0.7315	0.7269
238	0.7569	0.7467	0.7382	0.7312	0.7265
237	0.7566	0.7464	0.7379	0.7308	0.7262
236	0.7562	0.7460	0.7375	0.7305	0.7258
235	0.7559	0.7457	0.7372	0.7301	0.7255
234	0.7556	0.7453	0.7368	0.7298	0.7251
233	0.7552	0.7450	0.7365	0.7294	0.7247
232	0.7549	0.7446	0.7361	0.7290	0.7243
231	0.7545	0.7443	0.7357	0.7287	0.7239
230	0.7541	0.7439	0.7354	0.7283	0.7236
229	0.7538	0.7435	0.7350	0.7279	0.7232
228	0.7534	0.7431	0.7346	0.7275	0.7227
227	0.7530	0.7427	0.7342	0.7271	0.7223
226	0.7526	0.7423	0.7338	0.7267	0.7219
225	0.7522	0.7419	0.7334	0.7263	0.7215
224	0.7518	0.7415	0.7330	0.7258	0.7210
223	0.7514	0.7411	0.7326	0.7254	0.7206
222	0.7510	0.7407	0.7321	0.7250	0.7202
221	0.7506	0.7402	0.7317	0.7245	0.7197
220	0.7502	0.7398	0.7312	0.7241	0.7192
219	0.7497	0.7394	0.7308	0.7236	0.7188
218	0.7493	0.7389	0.7303	0.7231	0.7183

$$\frac{3}{n^2 + 2}$$

λ, nm	2-Methoxyethanol				
	20%	40%	60%	80%	99%
217	0.7488	0.7384	0.7299	0.7227	0.7178
216	0.7484	0.7380	0.7294	0.7222	0.7173
215	0.7479	0.7375	0.7289	0.7217	0.7168
214	0.7474	0.7370	0.7284	0.7212	0.7163
213	0.7470	0.7365	0.7279	0.7207	0.7158
212	0.7465	0.7360	0.7274	0.7202	0.7152
211	0.7460	0.7355	0.7269	0.7196	0.7147
210	0.7455	0.7350	0.7264	0.7191	0.7141
209	0.7449	0.7344	0.7258	0.7186	0.7136
208	0.7444	0.7339	0.7253	0.7180	0.7130
207	0.7439	0.7333	0.7247	0.7174	0.7124
206	0.7433	0.7328	0.7242	0.7168	0.7118
205	0.7428	0.7322	0.7236	0.7163	0.7112
204	0.7422	0.7316	0.7230	0.7157	0.7106
203	0.7416	0.7310	0.7224	0.7150	0.7099
202	0.7410	0.7304	0.7218	0.7144	0.7093
201	0.7404	0.7298	0.7212	0.7138	0.7086
200	0.7398	0.7291	0.7205	0.7131	0.7080
199	0.7392	0.7285	0.7199	0.7125	0.7073
198	0.7385	0.7278	0.7192	0.7118	0.7066
197	0.7379	0.7272	0.7185	0.7111	0.7059
196	0.7372	0.7265	0.7178	0.7104	0.7052
195	0.7365	0.7258	0.7171	0.7097	0.7044
194	0.7358	0.7251	0.7164	0.7090	0.7037
193	0.7351	0.7243	0.7157	0.7082	0.7029
192	0.7344	0.7236	0.7149	0.7074	0.7021
191	0.7336	0.7228	0.7142	0.7067	0.7013
190	0.7329	0.7221	0.7134	0.7059	0.7005
189	0.7321	0.7213	0.7126	0.7051	0.6997
188	0.7313	0.7204	0.7118	0.7042	0.6988
187	0.7305	0.7196	0.7110	0.7034	0.6979
186	0.7297	0.7188	0.7101	0.7025	0.6971
185	0.7288	0.7179	0.7092	0.7016	0.6961

Compiled by Murray Ettinger.

SUMMARY OF DATA ON DENSITY AND INDEX
OF REFRACTION OF GUANIDINE·HCl

Weight fraction (W)	Molar concn.	Density (d/d_0)	$\bar{V}$ (G·HCl)	$n_D^{22.5°}$
0.0	0.0	—	0.700	1.3327
0.09866	1.060	1.02690	0.731	1.3507
0.18661	2.055	1.05186	0.743	1.3672
0.25649	2.878	1.07184	0.749	1.3809
0.36223	4.177	1.10173	0.754	1.4020
0.44181	5.208	1.12617	0.758	1.4188
0.52271	6.296	1.15074	0.760	1.4362

$d/d_0 = 1 + 0.2720 \ W + 0.0315 \ W^2$
where d_0 is the absolute density of water at 20.15°C.

From Kielley and Harrington *Biochim. Biophys. Acta,* 41, 414 (1960). With permission of authors and copyright owners Elsevier Publishing Company.

$d/d_0 = 1 + 0.2710 \ W + 0.0330 W^2$
where d_0 is the absolute density of water at 25°C.
Using d_0 for each appropriate temperature will give density to within 1 part per thousand for any temperature from 15 to 35°C.

From Kawahara and Tanford, *J. Biol. Chem.,* 241, 3228 (1966). With permission of American Society of Biological Chemists, Inc.

VISCOSITY AND DENSITY OF AQUEOUS SOLUTIONS OF UREA AND GUANIDINE HYDROCHLORIDE[a]

Table 1
VISCOSITY OF AQUEOUS SOLUTIONS OF UREA, AT 25°

Concentration			Relative viscosity, η/η_0		
				Calculated	
Weight %	Moles/liter	Moles/kg H_2O	Observed	Equation 3	Equation 4
6.159	1.039	1.093	1.043	1.043	1.043
10.567	1.804	1.967	1.080	1.080	1.080
13.920	2.399	2.692	1.113	1.112	1.113
18.311	3.191	3.732	1.162	1.162	1.162
22.655	3.994	4.877	1.219	1.220	1.220
25.543	4.537	5.712	1.265	1.264	1.264
31.017	5.590	7.486	1.359	1.362	1.361
33.112	5.999	8.242	1.403	1.405	1.405
35.069	6.388	8.993	1.449	1.449	1.448
39.098	7.198	10.689	1.545	1.549	1.550
30.548	7.289	10.892	1.564	1.561	1.562
40.565	7.497	11.364	1.591	1.589	1.590
43.100	8.019	12.612	1.663	1.663	1.666

$$\eta/\eta_0 = 1 + 3.75 \times 10^{-2}C + 3.15 \times 10^{-3}C^2 + 3.10 \times 10^{-4}C^3 \qquad (3)$$

$$\eta/\eta_0 = 1 + 3.74 \times 10^{-2}m + 1.78 \times 10^{-3}m^2 - 4.4 \times 10^{-5}m^3 \qquad (4)$$

Table 2
VISCOSITY OF AQUEOUS SOLUTIONS OF GUANIDINE HYDROCHLORIDE, AT 25°

Concentration			Relative viscosity, η/η_0	
Weight %	Moles/liter	Moles/kg H_2O	Observed	Calculated, Equation 5
5.693	0.603	0.632	1.022	1.022
10.299	1.105	1.202	1.043	1.043
20.052	2.209	2.625	1.110	1.107
23.767	2.645	3.263	1.139	1.139
29.377	3.318	4.354	1.199	1.199
34.688	3.975	5.559	1.270	1.271
38.197	4.418	6.469	1.329	1.329
42.247	4.940	7.656	1.408	1.409
46.741	5.531	9.186	1.517	1.518
50.040	5.973	10.484	1.613	1.617
52.791	6.348	11.704	1.716	1.713
54.759	6.619	12.669	1.794	1.793
54.847	6.632	12.714	1.798	1.797
57.528	7.005	14.177	1.922	1.922
60.340	7.399	15.925	2.083	2.078

$$\eta/\eta_0 = 1 + 5.0 \times 10^{-3}m^{1/2} + 1.80 \times 10^{-2}m + 1.213 \times 10^{-2}m^{3/2} \qquad (5)$$

Table 3
DENSITIES OF AQUEOUS SOLUTIONS OF UREA WITH ADDED REAGENTS, AT 25°

| | Composition | | | | | Specific gravity | | |
| | Urea | | Other reagents | | | Calculated | | |
Solution	Weight %	Moles/liter	Reagent	Weight %	Moles/liter	Equation 8	Equation 9	Observed
1	33.116	6.013	Acetic acid	0.331	0.060	1.0939		1.0938
			Sodium acetate	0.302	0.040			
2	32.997	6.011	KH$_2$PO$_4$	0.249	0.020	1.0974		1.0973
			K$_2$HPO$_4$	0.479	0.030			
3	33.118	6.011	β-Mercaptoethanol	0.717	0.100	1.0935		1.0934
			NaCl	0.108	0.020			
4	11.273	2.003	NaCl	5.505	1.005	1.0716	1.0704	1.0700
5	21.927	4.001	NaCl	5.371	1.007	1.1022	1.0998	1.0991
6	21.226	4.001	NaCl	10.365	2.007	1.1405	1.1358	1.1354
7	32.089	6.003	NaCl	5.059	0.972	1.1304	1.1271	1.1269
8	31.102	6.000	NaCl	9.942	1.971	1.1681	1.1620	1.1621
9	31.074	6.000	NaCl	10.082	2.000	1.1699	1.1630	1.1631

Table 4
DENSITIES OF AQUEOUS SOLUTIONS OF GUANIDINE HYDROCHLORIDE WITH ADDED REAGENTS, AT 25°

| | Composition | | | | | Specific gravity | | |
| | Guanidine hydrochloride | | Other reagents | | | Calculated | | |
Solution	Weight %	Moles/liter	Reagent	Weight %	Moles/liter	Equation 8	Equation 9	Observed
1	50.240	6.017	Acetic acid	0.316	0.060	1.1467		1.1477
			Sodium acetate	0.289	0.040			
2	50.154	6.020	KH$_2$PO$_4$	0.238	0.020	1.1504		1.1501
			K$_2$HPO$_4$	0.457	0.030			
3	49.382	5.889	β-Mercaptoethanol	0.682	0.100	1.1429		1.1428
4	50.284	6.016	β-Mercaptoethanol	0.670	0.102	1.1465		1.1463
			NaCl	0.103	0.020			
5	49.785	5.948	NaCl	0.141	0.040	1.1449	1.1442	
6	17.745	2.018	NaCl	5.432	1.010	1.0964	1.0892	1.0895
7	32.936	4.024	NaCl	10.081	2.013	1.1778	1.1703	1.1706

$$\frac{d}{d_0} - 1 = \sum_i \left(\frac{d_i}{d_0} - 1\right) \quad (8)$$

d_i/d_0 is the relative density which each solute component would produce alone at the same molar concentration that it has in the final mixture, and the sum extends over all solute components.

$$d/d_0 = 1 + \alpha_1[W_1 + (\alpha_2/\alpha_1)W_2] + \beta_1(W_1 + \sqrt{\beta_2/\beta_1}\,W_2)^2 \quad (9)$$

Table 5
VISCOSITIES OF AQUEOUS SOLUTIONS OF UREA WITH ADDED REAGENTS, AT 25°

Solution (Table 3)	Added reagents alone (η/η_0)	Equivalent urea concentration	Effective total urea concentration	η/η_0 calculated Equation 10	η/η_0 calculated Empirical method[a]	η/η_0 observed
		moles/liter				
1	1.0206	0.525	6.538	1.427	1.466	1.456
2	1.0185	0.475	6.486	1.425	1.460	1.443
3	1.018	0.460	6.471	1.425	1.459	1.433
4	1.097	2.12	4.12	1.187	1.230	1.225
5	1.097	2.12	6.13	1.317	1.419	1.409
6	1.217	3.95	7.95	1.437	1.653	1.648
7	1.093	2.06	8.06	1.499	1.669	1.662
8	1.212	3.89	9.89	1.617	1.979	1.988
9	1.216	3.94	9.94	1.621	1.988	1.997

Table 6
VISCOSITIES OF AQUEOUS SOLUTIONS OF GUANIDINE HYDROCHLORIDE WITH ADDED REAGENTS, AT 25°

Solution (Table 4)	Added reagents alone (η/η_0)	Molality of guanidine hydrochloride — Equivalent to added reagents	Molality of guanidine hydrochloride — Guanidine hydrochloride added	Molality of guanidine hydrochloride — Effective total	η/η_0 calculated Equation 10	η/η_0 calculated Empirical method[a]	η/η_0 observed
1	1.0206	0.61	10.697	11.31	1.645	1.682	1.691
2	1.0185	0.55	10.680	11.23	1.643	1.675	1.675
3	1.012	0.35	10.350	10.70	1.604	1.633	1.627
4	1.018	0.54	10.753	11.29	1.641	1.680	1.665
5	1.012	0.35	10.419	10.77	1.611	1.639	1.627
6	1.097	2.42	2.417	4.84	1.192	1.227	1.233
7	1.217	4.66	6.048	10.71	1.493	1.634	1.804

$$\frac{\eta}{\eta_0} - 1 = \sum_i \left(\frac{\eta_i}{\eta_0} - 1 \right) \quad (10)$$

[a]C, concentration in moles per liter;
d_0, absolute density of water at designated temperature;
m, molality;
η, viscosity of the solution;
η_0, viscosity of water at designated temperature;
W, weight fraction of solute in solution.

Tables and equations were taken from Kawahara and Tanford, *J. Biol. Chem.*, 241, 3228 (1966). With permission of the authors and copyright owners.

REFRACTIVE INDEX (n_D) OF LIQUIDS

The indices given in the following table are for 20 °C except when otherwise indicated by a superscript.

(Items listed in the order of increasing indices)

Liquid	n_D	Liquid	n_D
Methanol	1.3276[25]	Butyl alcohol, secondary	1.3949
Acetaldehyde	1.3316	Butylamine, secondary	1.39501[16.7]
Water	1.3330	Butyl acetate	1.3951
Methylene chloride	1.3348[15]	Valeraldehyde	1.3952
Acetonitrile	1.34596[16.5]	Butyl chloride, secondary	1.3953[25]
Diethyl ether	1.34972[4.8]	Butyl methyl ketone	1.39694[17.4]
Dimethylamine	1.350[17]	Butyl alcohol, secondary	1.397
n-Pentane	1.3577	n-Propyl nitrate	1.3972
Acetone	1.35886[19.4]	n-Octane	1.3975
Methyl acetate	1.35935	Isobutyl alcohol	1.3976[15]
n-Propyl nitrite	1.3613	Isobutylamine	1.39878[17]
Ethyl alcohol, anhydrous	1.36242[18.35]	Ethyl propyl ketone	1.39899[22]
Dimethyl carbonate	1.3687	Butyric acid	1.39906
Propionitrile	1.36888[14.6]	n-Butyl alcohol	1.3993
Ethyl propyl ether	1.36948	Acrolein	1.39975
Formic acid	1.37137	Ethyl n-butyrate	1.400
Isobutyl nitrite	1.37151[22.1]	Methacrylonitrile	1.4002
Acetic acid	1.37182	Triethylamine	1.40032
Ethyl acetate	1.37216[18.9]	n-Butylamine	1.401
Isobutyraldehyde	1.37302	Isobutyl chloride	1.4010[15]
n-Hexane	1.37536	Amyl acetate	1.4012
Isopropylamine	1.37698[15.4]	n-Butyl nitrate	1.40130[23.2]
Isopropyl alcohol	1.37757	Isobutyl nitrate	1.40130[23.3]
Methyl propionate	1.37767[18.5]	n-Butyl chloride	1.4015
Butylamine, tertiary	1.37940[18]	1-Nitropropane	1.4015
Propyl ether	1.3807	Isoamyl acetate	1.40170[17.9]
Ethyl methyl ketone	1.38071[15.9]	Isovaleric acid	1.40178[22.4]
Ethyl borate	1.381	Methyl cellosolve	1.4028
Butyronitrile	1.3816[24]	Propionic anhydride	1.4038
Nitromethane	1.3818	Cyclopentane	1.4039
Acetal	1.3819	Tetrahydrofuran	1.4040[25]
Butyraldehyde	1.38433	Dipropylamine	1.40455[19.5]
n-Propyl acetate	1.38438	Paraldehyde	1.40486
Ethyl nitrate	1.38484[21.5]	n-Butyl n-butyrate	1.4049
Ethyl carbonate	1.38456	Propionyl chloride	1.40507
n-Amyl nitrite	1.38506	Amyl alcohol, tertiary	1.4052
n-Propyl alcohol	1.38543	Isoamyl alcohol	1.4075
n-Heptane	1.3855[25]	Isoamyl ether	1.408
Butyl chloride, tertiary	1.3869[18]	Valeric acid	1.4086
Isoamyl nitrite	1.38708[20.7]	Diisobutylamine	1.40934
Diethylamine	1.38730[18]	pri-n-Amyl alcohol	1.40994
Propionic acid	1.38736[19.9]	Diethyl oxalate	1.41011
Butyl alcohol, tertiary	1.38779	Isoamyl chloride	1.4103
Isopropyl methyl ketone	1.38788[16]	n-Amyl chloride	1.4119[18]
n-Propyl chloride	1.3886	n-Decane	1.41203
Methyl propyl ketone	1.38946[20.2]	Isoamyl nitrate	1.4122[22]
Acetyl chloride	1.38976	Tetramethylethylene	1.4128
n-Propylamine	1.39006[16.6]	Isovaleryl chloride	1.41361[24.3]
Nitroethane	1.39007[24.3]	Cyclopentanol	1.41530
Isovaleraldehyde	1.3902	Valeryl chloride	1.41555
Isobutyl acetate	1.3907[19]	n-Hexyl methyl ketone	1.41613
Valeronitrile	1.3909	n-Hexanol	1.4162[25]
Acrylonitrile	1.393	1,1-Dichloroethane	1.41655
Ethyl butyrate	1.39302[18]	Diethyl acetic acid	1.41788[10]
Diacetyl	1.3933[18]	Undecane	1.4184
n-Propyl propionate	1.3935	Allylamine	1.41943[22]
Diethyl ketone	1.3939[16.6]	Methyl urethan	1.4200[18.9]

REFRACTIVE INDEX (n_D) OF LIQUIDS (continued)

Liquid	n_D
Diethyl succinate	1.42007
Octyl acetate	1.4204
Ethyl acetoacetate	$1.42092^{16.6}$
Dibutyl ketone	1.421^{15}
Furan	1.42157
Acrylic acid	1.4224
Cyclopentene	**1.42246**
Ethyl chloracetate	**1.42274**
Diisoamylamine	**1.42289^{21}**
1,4-Dioxane	**1.4232**
***m* Heptylamine**	**1.424**
Ethyl bromide	**1.4241**
n-Heptyl alcohol	1.42410
Isopropyl bromide	1.42508
Triisobutylamine	$1.42519^{17.3}$
Succinaldehyde	1.4254
Vinyl acetic acid	1.4257^{15}
Cyclohexane	1.4264
Methyl carbitol	1.4264^{27}
Butyl bromide, tertiary	1.428
1,1-Dichloropropane	1.42887
Heptyl methyl carbinol	1.4290^{25}
n-Octylamine	1.430
Ethylene glycol	1.4311
Nonanoic acid	1.4330
Undecanal	1.4334
n-Propyl bromide	1.43414
Butyl bromide, secondary	1.4344^{25}
Glycerol tributyrate	1.4359
Diethyl malate	1.4362
Cyclopentanone	1.4366
Crotonaldehyde	$1.43838^{17.3}$
1,2-Dichloropropane	1.4388
Isobutyl bromide	1.4391^{15}
Tetranitromethane	$1.43976^{16.9}$
n-Butyl bromide	1.4398
Trimethylene glycol	1.4398
Isoamyl bromide	1.4412
Lactic acid (*dl*)	1.4414
Tridecane	$1.4419^{16.8}$
Epichlorohydrin	$1.44195^{11.5}$
Dimethyl malate	1.4425
Mesityl oxide	1.4425^{22}
Cycloheptane	1.4440
1,2-Dichloroethane	1.4443
Cyclohexene	1.44507^{22}
Formamide	$1.44530^{22.7}$
Tetradecane	1.4459
1,3-Dichloropropane	1.4469
Diethylene glycol	1.4475
Chloroform	1.4476
sym.-Dichloroethylene	1.4490^{15}
n-Octyl bromide	1.4503^{25}
Trichloroethyl acetate	1.45068
Cyclohexanone	1.4507
Piperidine	1.4534
Ethanolamine	1.4539
Ethylenediamine, anhyd.	1.4540^{26}
Trichloroethylene	1.4556^{25}
Chloral	1.45572
d-Citronellol	1.4566
sym.-Dichloroethyl ether	1.4570
Rose oil	$1.4570–1.4630^{30}$
Eucalyptus oil	1.4580–1.4700
Eucalyptol	1.4584^{15}
Lavender oil	1.4590–1.4700
Peppermint oil	1.4590–1.4650
Chloropicrin	1.46075^{23}
Coriander oil	1.4620–1.4720
dl-Bornyl acetate	1.4623^{22}
Peanut oil	$1.4625–1.4645^{40}$
Cardamom oil	1.4630–1.4660
Carbon tetrachloride	1.46305^{15}
d-Fenchone	1.4636^{18}
Phytol	1.46380
Rosemary oil	1.4640–1.4760
Bergamot oil	1.4650–1.4675
Allyl bromide	1.46545
dl-Pinene	1.4658
Dichloroacetic acid	**1.4659^{22}**
Turpentine oil	**1.4680–1.4780**
Myristica oil (West Indian)	**1.4690–1.4760**
Decalin (*trans.*)	**1.46994^{18}**
1,1,2-Trichloroethane	**1.4711**
Orange oil	1.4723–1.4737
Bitter orange oil	1.4725–1.4755
Glycerin	1.4729
Lemon oil	1.4738–1.4755
Chenopodium oil	1.4740–1.4790
Myristica oil (East Indian)	1.4740–1.4880
Crotonic anhydride	1.47446
Dwarf pine needle oil	1.4750–1.4800
Poppy seed oil	1.4766–1.4774
Pine oil	1.4780–1.4820
Geraniol	1.4798
Nitroglycerin	$1.482^{18.6}$
Decalin (*cis*)	1.4828
Caraway oil	1.4840–1.4880
Spearmint oil	1.4840–1.4910
Furfuryl alcohol	1.4850^{25}
Triethanolamine	1.4852
Geranial	1.48752
Toluene	1.4893^{24}
Furfuryl chloride	1.4941
sym.-Tetrachloroethane	1.4942
Cumene	1.4947^{15}
Thyme oil	1.4950–1.5050
m-Xylene	1.4973
α-Ionone	$1.49842^{22.3}$
Benzene	1.50142
Diethyl phthalate	1.5019
o-Cymene	$1.50206^{16.1}$
Isopropyl iodide	1.5026
2-Picoline	$1.50293^{16.7}$
Phenyl acetate	1.503
Pyrrole	1.5035
3-Picoline	1.50432^{24}
Butyl phenyl ether	1.5046
n-Propyl iodide	1.50508
o-Xylene	1.50545
Tetrachloroethylene	1.50547
Isoamyl salicylate	1.506
Bay oil	1.5070–1.5160
Phenetole	1.5076
Pyridine	1.50919^{21}
Ethyl isothiocyanate	1.5134

REFRACTIVE INDEX (n_D) OF LIQUIDS (continued)

Liquid	n_D	Liquid	n_D
Isovalerophenone	1.51385[15.8]	m-Dichlorobenzene	1.54570[20.9]
Dimethyl phthalate	1.51546[20.8]	Tetralin	1.54614
Anisole	1.51791	p-Tolualdehyde	1.54693[16.6]
Methyl benzoate	1.51810[16]	o-Chlorophenol	1.5473[40]
Isopropyl phenyl ketone	1.51919[16.6]	o-Nitrotoluene	1.54739[20.4]
β-Ionone	1.51977[18.9]	o-Tolualdehyde	1.54852[19]
p-Chlorotoluene	1.5199[19]	o-Dichlorobenzene	1.5518[22]
Phenyl propyl ketone	1.52016[18.3]	Nitrobenzene	1.55291
Propylene dibromide	1.5203	Anise oil	1.5530–1.5600
m-Chlorotoluene	1.5214[19]	N-Ethylaniline	1.55558[20.3]
Carvacrol	1.52295	N,N-Dimethylaniline	1.55819
Benzyl acetate	1.5232	Bromobenzene	1.5625[15]
Nicotine	1.52392[22.4]	Benzyl benzoate	1.5681[21]
Phenylethyl alcohol	1.5240	N-Methylaniline	1.57021[21.2]
Chlorobenzene	1.5248	Phthalyl chloride	1.57099[15.5]
Sassafras oil	1.5250–1.5350	Indene	1.57107[12.7]
Ethyl salicylate	1.52511[14.4]	m-Toluidine	1.5711[22]
Furfural	1.52608	o-Toluidine	1.57276
Clove oil	1.5270–1.5350	Salicylaldehyde	1.57358[19.7]
Pimenta oil	1.5270–1.5400	Isoeugenol	1.5739[19]
o-Tolunitrile	1.52720[23.1]	Dibenzylamine	1.57432[22]
Thiophene	1.5287	o-Anisidine	1.57536
o-Chlorotoluene	1.5288	Anisaldehyde	1.5764[13]
Benzonitrile	1.52892	Aniline	1.5863
Methyl iodide	1.5293[21]	o-Chloroaniline	1.5895
Acetophenone	1.53418[19]	m-Chloroaniline	1.59424[20.7]
Methyl salicylate	1.5369	Boromoform	1.5980[19]
Safrol	1.5383	Cinnamon oil	1.6020–1.6135
Indan	1.53877[16.4]	Phenylhydrazine	1.6081[20.3]
Benzyl alcohol	1.53956	m-Dibromobenzene	1.6083[17.5]
m-Cresol	1.5398	o-Dibromobenzene	1.6117[17.5]
Benzylamine	1.5401	Iodobenzene	1.62145[18.5]
m-Tolualdehyde	1.54068[21.4]	Isoquinoline	1.62233[25.1]
N,N-Diethylaniline	1.54105[22.3]	Quinoline	1.62450[24.9]
Benzyl chloride	1.5415[15]	Carbon disulfide	1.62950[18]
Eugenol	1.5416[19.4]	sym.-Tetrabromoethane	1.638
m-Nitrotoluene	1.5425[30]	Phenyl isothiocyanate	1.64918[23.1]
sym.-Dibromoethylene	1.5428	1-Naphthaldehyde	1.65464[19.3]
Carvacrylamine	1.543[19]	1-Bromonaphthalene	1.65876[19.4]
Styrene	1.54344[17]	Methylene iodide	1.7425[15]

SPECIFIC GRAVITY OF LIQUIDS

Specific gravity and density are not identical although the abbreviation "d" is frequently used to designate specific gravity. Specific gravity and density are numerically equal when water is the standard of reference for specific gravity and g/ml is the unit designation for density.

The numerical value for specific gravity is usually written with a superscript (indicating the temperature of the liquid) and a subscript (indicating the temperature of the liquid to which it is referred), thus d_4^{25} 1.724 or sp. gr. 1.724_4^{25}. When these are omitted in this table, the specific gravity at 20°C referred to water at 4°C is intended. When the standard of reference is not specified, for liquids and solids, it is understood to be water.

Water is most dense at 4°C, hence the sp. gr. of a liquid with reference to water will be higher at all other temperatures than it is at 4°C. To obtain the sp. gr. with reference to water at the same temperature as the liquid, multiply the sp. gr. of $_4^{15}$, $_4^{20}$, or $_4^{25}$ by 1.001, 1.002, or 1.003, respectively.

(Items listed in the order of increasing specific gravities)

Liquid	Specific gravity	Liquid	Specific gravity
n-Pentane	0.626	n-Heptylamine	0.777
n-Hexane	0.660	Cyclohexane	0.778
1-Butyne	0.668_4^0	n-Octylamine	0.779_{20}^{20}
Dimethylamine	0.680_4^0	Isoamyl ether	0.781_{15}^{15}
Isoprene	0.681	Propionitrile	0.783
n-Heptane	0.684	Acetonitrile	0.783_{25}^{25}
2-Butyne	0.688^{25}	n-Butyl ether	0.784_4^0
1,5-Hexadiene	0.688	Isopropyl alcohol	0.785
Isopropylamine	0.694_4^{15}	Isovaleronitrile	0.788
Butylamine, tertiary	0.696	Butyl alcohol, tertiary	0.789
Triethylboron	0.696^{23}	Methanol, anhydrous	0.791
Ethylamine	0.706_4^0	Acetone	0.792
Diethylamine	0.711_4^{18}	Isobutyraldehyde	0.794
2,4-Hexadiene	0.711	Acrylonitrile	0.797
Diethyl ether	0.713	Ethyl alcohol, anhydrous	$0.798_{15.56}^{15.56}$
n-Nonane	0.716	Valeronitrile	0.801
Triethylamine	0.723_4^{25}	Isovaleraldehyde	0.803_4^{17}
Butylamine, secondary	0.724	n-Propyl alcohol	0.804
Isopropyl ether	0.726	Allyl ether	0.805_6^{18}
Ethyl methyl ether	0.726_4^0	Ethyl methyl ketone	0.805
2,4-Heptadiene	$0.733_4^{21.5}$	Isobutyl alcohol	0.806_4^{15}
Isobutylamine	0.724_4^{25}	Propionaldehyde	0.807
Propyl ether	0.736	Butyl alcohol, secondary	0.808
Methyl propyl ether	0.738	Amyl alcohol, tertiary	0.809
Dipropylamine	0.738	Methyl propyl ketone	0.809
Ethyl n-propyl ether	0.739	n-Butyl alcohol	0.810
n-Butylamine	0.740	Cycloheptane	0.810
Undecane	0.741	Cyclohexene	0.810
N,N-Dimethylamylamine	0.743	Isoamyl alcohol	0.813_4^{15}
Ethyl isopropyl ether	0.745_4^0	Ethyl propyl ketone	$0.813_4^{21.8}$
Isoamylamine	0.751	pri-n-Amyl alcohol	0.814
Cyclopentane	0.751	Heptyl ether	0.815_4^0
Butyl ethyl ether	0.752	Diethyl ketone	0.816_4^{19}
Isohexylamine	0.758_4^{25}	Ethyl alcohol, 95 per cent	$0.816_{15.56}^{15.56}$
Isobutyl ether	0.761_4^{15}	Butyraldehyde	0.817
Allylamine	0.761	Dipropyl ketone	0.817
n-Amylamine	0.761	Ethyl butyl ketone	0.818
Butyl methyl ether	0.764_4^0	n-Hexyl methyl ketone	0.819
Allyl ether ether	0.765	1-Hexanol	0.819
n-Dodecane	0.766_4^0	3-Hexanol	0.819
Dibutylamine	0.767	Isoamyl alcohol, secondary	0.819
n-Butyl ether	0.769_{20}^{20}	Pinacolin	0.821_4^0
Cyclopentene	0.774	Amyl methyl ketone	0.822_4^{15}

SPECIFIC GRAVITY OF LIQUIDS (continued)

Liquid	Specific gravity	Liquid	Specific gravity
Cycloheptene	0.823	Ethyl nitrite	0.900^{15}_{15}
n-Octyl alcohol	0.825	Caraway oil	$0.900–0.910^{25}_{25}$
2-Undecanone	0.826	Ethyl acetate	0.902
Light Liquid Petrolatum	$0.828–0.880^{25}_{25}$	Linoleic acid	0.903
2-Hexanol	0.829^{0}_{4}	Cubeb oil	$0.905–0.925^{25}_{25}$
n-Decyl alcohol	0.829	Eucalyptus oil	$0.905–0.925^{25}_{25}$
n-Undecylaldehyde	0.830	Diethyl Carbitol	0.907
Butyl methyl ketone	0.830^{0}_{4}	Styrene	0.907
1-Undecanol	0.833^{23}_{4}	Undecylenic acid	0.908^{25}_{4}
Acrolein	0.841	Olive oil	$0.910–0.915^{25}_{25}$
Orange oil	$0.842–0.846^{25}_{25}$	Expressed almond oil	$0.910–0.915^{25}_{25}$
Bitter orange oil	$0.845–0.851^{25}_{25}$	Persic oil	$0.910–0.923^{25}_{25}$
Butyl chloride, tertiary	0.847^{15}	Thyme oil	$0.910–0.935^{25}_{25}$
Rose oil	$0.848–0.863^{30}_{15}$	n-Butyl nitrite	0.911^{0}_{4}
Lemon oil	$0.849–0.855^{25}_{25}$	Peanut oil	$0.912–0.920^{25}_{25}$
Amyl ether ketone	0.850^{0}_{4}	Mustard oil	$0.914–0.916^{15}_{15}$
n-Amyl nitrite	0.853	Corn oil	$0.914–0.921^{25}_{25}$
Rectified turpentine oil	$0.853–0.862^{25}_{25}$	Methyl propionate	0.915
Dwarf pine needle oil	$0.853–0.871^{25}_{25}$	Glycerin trioleate	0.915
Allyl alcohol	0.854	Cottonseed oil	$0.915–0.921^{25}_{25}$
Mesityl oxide	0.854	Sesame oil	$0.916–0.921^{25}_{25}$
Myristica oil	$0.854–0.910^{25}_{25}$	Spearmint oil	$0.917–0.934^{25}_{25}$
p-Cymene	0.857	Cardamom oil	$0.917–0.947^{25}_{25}$
dl-Pinene	0.858	Coconut oil	$0.918–0.923^{25}_{25}$
Isopropyl chloride	0.859	Cod liver oil	$0.918–0.927^{25}_{25}$
2-Diethylaminoethanol	0.860^{25}_{25}	Halibut liver oil	$0.920–0.930^{25}_{25}$
Liquid Petrolatum	$0.860–0.905^{25}_{25}$	Eucalyptol	$0.921–0.923^{25}_{25}$
Piperidine	0.861	Ethyl formate	0.924^{25}_{4}
Cumene	0.863	Soya oil	$0.924–0.927^{15}_{15}$
Coriander oil	$0.863–0.875^{25}_{25}$	Linseed oil	$0.925–0.935^{25}_{25}$
Orange flower oil	$0.863–0.880^{25}_{25}$	Pine oil	$0.927–0.940^{25}_{25}$
Phytol	0.864^{0}_{4}	Methyl acetate	0.928
m-Xylene	0.864	Cellosolve	0.930
Toluene	0.866	Ionone	$0.933–0.937^{25}_{25}$
Ethyl benzene	0.867	N,N-Diethylaniline	0.935
m-Cymene	0.870	Furan	0.937
Isoamyl acetate	0.870^{25}_{4}	Allyl chloride	0.938
Isopropyl acetate	0.870	Valeric acid	0.942
Isobutyl nitrite	0.870^{20}_{20}	Castor oil	$0.945–0.965^{25}_{25}$
Butyl chloride, secondary	0.871	Cyclohexanone	0.948
Octyl acetate	0.873^{20}_{20}	Pyrrole	0.948
Isobutyl acetate	0.875	Cyclopentanone	0.948
Isoamyl nitrite	0.875^{25}	Cyclopentanol	0.949
Bergamot oil	$0.875–0.880^{25}_{25}$	Isobutyric acid	0.949
Lavender oil	$0.875–0.888^{25}_{25}$	2-Picoline	0.950^{15}_{4}
o-Cymene	0.876	Chenopodium oil	$0.950–0.980^{25}_{25}$
Benzene	0.879^{15}_{4}	Myrcia oil	$0.950–0.990^{25}_{25}$
Amyl acetate	0.879^{20}_{20}	Cycloheptanone	0.951
Geraniol	0.881^{16}_{4}	Fennel oil	$0.953–0.973^{25}_{25}$
n-Amyl chloride	0.883	Dimethylaniline	0.956
Isobutyl chloride	0.883^{15}	4-Picoline	0.957^{15}_{4}
n-Butyl chloride	0.884	3-Picoline	0.961^{15}_{4}
Pine needle oil	$0.884–0.886^{15}_{15}$	Indan	0.965
Citronella oil	$0.885–0.912^{25}_{25}$	Methyl cellosolve	0.966
2-Dimethylaminoethanol	0.887	Phenetole	0.967
n-Propyl acetate	0.887	Vitamin K$_1$	0.967^{25}_{25}
1-Menthol	0.890^{15}_{15}	Tetralin	0.970
Propyl chloride	0.890^{20}_{20}	Carvacrol	0.976
Isoamyl chloride	0.893	Pyridine	0.978^{25}_{4}
Rosemary oil	$0.894–0.912^{25}_{25}$	Anise oil	$0.978–0.988^{25}_{25}$
Oleic acid	0.895^{18}	Ethyl urethan	0.981
Isodurene	0.896^{0}_{4}	Benzylamine	0.983^{19}_{4}
Peppermint oil	$0.896–0.908^{25}_{25}$	Benzyl acetone	0.989^{23}_{17}
o-Xylene	0.897	m-Toluidine	0.989

SPECIFIC GRAVITY OF LIQUIDS (continued)

Liquid	Specific gravity	Liquid	Specific gravity
Carbitol	0.990	Acetyl chloride	1.105
Dimethyl glyoxal	0.990^{15}_{15}	Ethyl nitrate	1.105
Isoamyl benzoate	0.993^{19}_4	Chlorobenzene	1.107
Paraldehyde	0.994	Polyethylene Glycol 400	$1.110-1.140^{25}_{25}$
Anisole	0.995	Cinnamaldehyde	1.112^{15}_4
Isoamyl nitrate	0.996^{22}_4	Benzyl benzoate	1.118^{25}_4
Morpholine	0.999	Diethylene glycol	1.118^{20}_{20}
Water	0.9970^0_4	Anisaldehyde	1.123
Water	0.9999^{20}_{20}	Diethyl phthalate	1.123^{25}_4
Water	$1.0000^{4.08}_{4.08}$	Triethanolamine	1.124
Isobutyl benzoate	1.002^{15}	Polyethylene Glycol 300	$1.124-1.130^{25}_{25}$
o-Toluidine	1.004	Furfuryl alcohol	1.130
Indene	1.006	Nitromethane	1.130
Nicotine	1.009	Formamide	1.134
Benzonitrile	1.010^{15}_{15}	Ethyl salicylate	1.136^{15}_4
Hydrazine	1.011^{15}_4	m-Nitrotoluene	1.157
Ethanolamine	1.018	Ethyl chloroacetate	1.159
Pimenta oil	$10.018-1.048^{25}_{25}$	Furfural	1.160
Aniline	1.022	Glycerol triacetate	1.161
Sparteine	1.023	o-Nitrotoluene	1.163
Phenylethyl alcohol	1.024^{15}_4	Salicylaldehyde	1.167
Dibenzylamine	1.026	Methyl salicylate	1.184^{25}_{25}
Chloroacetal	1.026^{16}_4	Dimethyl phthalate	1.189^{25}_{25}
Acetophenone	1.033^{15}_{15}	Nitrobenzene	1.205^{15}_4
1,4-Dioxane	1.034	Isoamyl bromide	1.210^{15}_4
m-Cresol	1.034	Benzoyl chloride	1.219^{15}_{15}
Glycerol tributyrate	1.035	sym.-Dichloroethyl ether	1.222
Propylene glycol	1.036^{25}_4	Butyl bromide, tertiary	1.222
Phlorol	1.037^{12}	Formic acid	1.226^{15}_4
Butyl nitrate, secondary	1.038^0_4	Methyl chloroacetate	1.238^{20}_{20}
Bitter almond oil	$1.038-1.060^{25}_{25}$	Amyl bromide	1.246^0_4
Clove oil	$1.038-1.060^{25}_{25}$	Lactic acid (dl)	1.249^{15}_4
Ethyl succinate	1.040	uns.-Ethylene dichloride	1.252
Benzyl ether	1.043	Butyl bromide, secondary	1.258
Benzyl alcohol	1.045^{25}_4	Glycerol	1.260
Cinnamon oil	$1.045-1.063^{25}_{25}$	Carbon disulfide	1.263
o-Cresol	1.047	n-Butyl bromide	1.269^{25}_4
n-Butyl phthalate	1.047	Isobutyl bromide	1.272^{15}_4
n-Butyl nitrate	1.048^0_4	sym.-Dichloroethylene	1.291^{15}_4
Acetic acid (glacial)	1.049^{25}_{25}	o-Dichlorobenzene	1.307^{20}_{20}
Benzaldehyde	1.050^{15}_4	Isopropyl bromide	1.310
Ethyl benzoate	1.051^{15}_4	Ethylsulfuric acid	1.316^{17}
Ethyl malonate	1.055	Methylene chloride	1.335^{15}_4
Benzyl acetate	1.057^{16}_4	n-Propyl bromide	1.353
Allyl benzoate	1.058^{15}_{15}	m-Xylyl bromide	1.371^{23}_4
n-Propyl nitrate	1.058	Benzotrichloride	1.380^{15}_4
Succinaldehyde	1.064	Ethyl trichloroacetate	1.383
Thiophene	1.064	Allyl bromide	1.398
Methyl carbonate	1.065^{17}_4	Ethyl bromide	1.430
Eugenol	1.066	Benzyl bromide	1.438^{22}_0
p-Chlorotoluene	1.070	Hydrogen peroxide, anhydrous	1.465^0_4
m-Chlorotoluene	1.072	Trichloroethylene	1.465
Diethyl maleate	1.074^{15}_{15}	Chloroform	1.498^{15}_4
Benzofuran	1.078^{15}_{15}	Bromobenzene	1.499^{15}_{15}
o-Chlorotoluene	1.082	Chloral	1.512
Acetic anhydride	1.087^{15}_4	Trichloroethanol	1.550^{20}_{20}
o-Anisidine	1.092	Dichloroacetic acid	1.563
Methyl benzoate	1.094^{15}_4	Benzoyl bromide	1.570^{15}_4
Quinoline	1.095	Glycerophosphoric acid	1.590^{19}_4
m-Anisidine	1.096	Nitroglycerin	1.592^{25}_4
Diethanolamine	1.097	Carbon tetrachloride	1.595
Benzyl chloride	1.103^{18}_4	Tetrachloroethane	1.600
Aldol	1.103	Tetrachloroethylene	1.631^{15}_4

SPECIFIC GRAVITY OF LIQUIDS (continued)

Liquid	Specific gravity	Liquid	Specific gravity
Chloropicrin	1.651	Methyl iodide	2.251
Diphosgene	1.653_4^{14}	Bromal	2.300_4^{15}
Thionyl chloride	$1.655^{10.4}$	Methylene bromide	2.495
Acetyl bromide	1.663_4^{16}	Bromoform	2.890
Isopropyl iodide	1.703	Tetrabromoethane	2.964
Ethyl iodide	1.933	Methylene iodide	3.325
Ethylene bromide	2.170_4^{25}	Mercury	13.546
Ethylene dibromide	2.172_{25}^{25}		

Reprinted from *The Merck Index* (1960), 7th ed., Merck and Co., Rahway, N.J., pp 1532–1535, with permission of the copyright owner.

VISCOSITY AND DENSITY TABLES

Sucrose in Water, 0.0°C

Sucrose %	Density[a] g/ml	Viscosity[b] cp	Sucrose %	Density[a] g/ml	Viscosity[b] cp	Sucrose %	Density[a] g/ml	Viscosity[b] cp
0	1.0004	1.780	24	1.1037	4.646	48	1.2269	34.57
1	1.0043	1.830	25	1.1085	4.912	49	1.2324	39.23
2	1.0082	1.884	26	1.1133	5.202	50	1.2380	44.74
3	1.0122	1.941	27	1.1181	5.519	51	1.2436	51.29
4	1.0162	2.002	28	1.1229	5.866	52	1.2493	59.11
5	1.0203	2.066	29	1.1278	6.246	53	1.2550	68.52
6	1.0244	2.135	30	1.1327	6.665	54	1.2607	79.92
7	1.0285	2.208	31	1.1376	7.126	55	1.2665	93.85
8	1.0326	2.286	32	1.1426	7.635	56	1.2723	111.0
9	1.0368	2.369	33	1.1476	8.201	57	1.2781	132.3
10	1.0411	2.458	34	1.1527	8.829	58	1.2840	158.9
11	1.0453	2.552	35	1.1578	9.530	59	1.2899	192.6
12	1.0496	2.653	36	1.1629	10.31	60	1.2958	235.7
13	1.0539	2.761	37	1.1680	11.20	61	1.3018	291.4
14	1.0583	2.877	38	1.1732	12.19	62	1.3078	364.2
15	1.0627	3.001	39	1.1784	13.31	63	1.3138	460.6
16	1.0671	3.134	40	1.1836	14.58	64	1.3199	589.9
17	1.0716	3.277	41	1.1889	16.03	65	1.3260	766.0
18	1.0760	3.430	42	1.1942	17.69	66	1.3321	1010.
19	1.0806	3.596	43	1.1996	19.59	67	1.3383	1352.
20	1.0852	3.774	44	1.2050	21.78	68	1.3445	1842.
21	1.0898	3.967	45	1.2104	24.31	69	1.3507	2556.
22	1.0944	4.175	46	1.2159	27.24	70	1.3570	3621.
23	1.0991	4.401	47	1.2213	30.66			

Sucrose in Water, 5.0°C

Sucrose %	Density[a] g/ml	Viscosity[b] cp	Sucrose %	Density[a] g/ml	Viscosity[b] cp	Sucrose %	Density[a] g/ml	Viscosity[b] cp
0	1.0004	1.516	24	1.1027	3.831	48	1.2250	25.97
1	1.0043	1.558	25	1.1074	4.042	49	1.2306	29.28
2	1.0082	1.603	26	1.1122	4.272	50	1.2361	33.16
3	1.0121	1.650	27	1.1169	4.523	51	1.2417	37.73
4	1.0161	1.700	28	1.1218	4.796	52	1.2474	43.16
5	1.0201	1.753	29	1.1266	5.094	53	1.2530	49.62
6	1.0241	1.809	30	1.1315	5.422	54	1.2587	57.39
7	1.0282	1.869	31	1.1364	5.781	55	1.2645	66.79
8	1.0323	1.933	32	1.1413	6.177	56	1.2702	78.24
9	1.0365	2.001	33	1.1463	6.614	57	1.2760	92.30
10	1.0406	2.073	34	1.1513	7.099	58	1.2819	109.7
11	1.0448	2.150	35	1.1563	7.637	59	1.2877	131.4
12	1.0491	2.232	36	1.1614	8.236	60	1.2936	158.9
13	1.0534	2.319	37	1.1665	8.905	61	1.2996	194.0
14	1.0577	2.413	38	1.1717	9.656	62	1.3056	239.1
15	1.0620	2.513	39	1.1768	10.50	63	1.3116	297.9
16	1.0664	2.621	40	1.1820	11.45	64	1.3176	375.6
17	1.0708	2.736	41	1.1873	12.53	65	1.3237	479.4
18	1.0753	2.859	42	1.1926	13.76	66	1.3298	620.3
19	1.0798	2.992	43	1.1979	15.16	67	1.3360	814.3
20	1.0843	3.135	44	1.2033	16.76	68	1.3422	1086.
21	1.0889	3.290	45	1.2087	18.60	69	1.3484	1473.
22	1.0934	3.456	46	1.2140	20.73	70	1.3546	2034.
23	1.0981	3.636	47	1.2195	23.18			

[a]Original data were stated to a precision of about 1 part in 10,000. Maximum deviation from original data is 7 parts in 10,000.

[b]Precision of original data was between 1 part in 1,000 and 1 part in 10,000. Maximum deviation from original data is 4 parts in 1,000 in the range covered in this set of tables.

VISCOSITY AND DENSITY TABLES (continued)

Sucrose in Water, 10.0°C

Sucrose %	Density[a] g/ml	Viscosity[b] cp	Sucrose %	Density[a] g/ml	Viscosity[b] cp	Sucrose %	Density[a] g/ml	Viscosity[b] cp
0	1.0002	1.308	24	1.1016	3.206	48	1.2231	19.96
1	1.0040	1.343	25	1.1062	3.377	49	1.2286	22.37
2	1.0079	1.380	26	1.1109	3.562	50	1.2341	25.17
3	1.0118	1.420	27	1.1157	3.763	51	1.2397	28.45
4	1.0157	1.462	28	1.1204	3.982	52	1.2453	32.32
5	1.0196	1.506	29	1.1252	4.220	53	1.2510	36.89
6	1.0236	1.553	30	1.1300	4.481	54	1.2566	42.34
7	1.0277	1.603	31	1.1349	4.767	55	1.2623	48.87
8	1.0317	1.655	32	1.1398	5.080	56	1.2681	56.75
9	1.0358	1.711	33	1.1448	5.424	57	1.2739	66.35
10	1.0400	1.771	34	1.1498	5.805	58	1.2797	78.11
11	1.0442	1.835	35	1.1548	6.225	59	1.2855	92.65
12	1.0484	1.902	36	1.1598	6.692	60	1.2914	110.8
13	1.0526	1.974	37	1.1649	7.211	61	1.2973	133.6
14	1.0569	2.051	38	1.1700	7.790	62	1.3033	162.7
15	1.0612	2.134	39	1.1752	8.438	63	1.3093	200.0
16	1.0655	2.222	40	1.1803	9.167	64	1.3153	248.6
17	1.0699	2.316	41	1.1856	9.988	65	1.3214	312.5
18	1.0743	2.417	42	1.1908	10.92	66	1.3275	397.7
19	1.0788	2.525	43	1.1961	11.97	67	1.3336	512.9
20	1.0833	2.642	44	1.2014	13.17	68	1.3398	671.1
21	1.0878	2.767	45	1.2068	14.54	69	1.3460	891.7
22	1.0924	2.903	46	1.2122	16.11	70	1.3522	1205.
23	1.0969	3.049	47	1.2176	17.92			

Sucrose in Water, 15.0°C

Sucrose %	Density[a] g/ml	Viscosity[b] cp	Sucrose %	Density[a] g/ml	Viscosity[b] cp	Sucrose %	Density[a] g/ml	Viscosity[b] cp
0	0.9996	1.140	24	1.1002	2.719	48	1.2211	15.65
1	1.0034	1.170	25	1.1048	2.859	49	1.2265	17.44
2	1.0073	1.202	26	1.1095	3.010	50	1.2320	19.52
3	1.0111	1.235	27	1.1142	3.174	51	1.2376	21.93
4	1.0150	1.271	28	1.1189	3.352	52	1.2432	24.75
5	1.0189	1.308	29	1.1237	3.546	53	1.2488	28.06
6	1.0229	1.348	30	1.1285	3.757	54	1.2544	31.98
7	1.0269	1.390	31	1.1334	3.987	55	1.2601	36.64
8	1.0309	1.434	32	1.1382	4.239	56	1.2658	42.22
9	1.0350	1.481	33	1.1432	4.515	57	1.2716	48.95
10	1.0391	1.531	34	1.1481	4.818	58	1.2774	57.12
11	1.0432	1.584	35	1.1531	5.153	59	1.2832	67.12
12	1.0474	1.640	36	1.1581	5.522	60	1.2891	79.48
13	1.0516	1.701	37	1.1632	5.932	61	1.2950	94.85
14	1.0558	1.765	38	1.1682	6.386	62	1.3009	114.2
15	1.0601	1.833	39	1.1734	6.894	63	1.3069	138.7
16	1.0644	1.906	40	1.1785	7.461	64	1.3129	170.2
17	1.0688	1.985	41	1.1837	8.097	65	1.3189	211.0
18	1.0732	2.068	42	1.1889	8.813	66	1.3250	264.6
19	1.0776	2.158	43	1.1942	9.621	67	1.3311	336.0
20	1.0820	2.255	44	1.1995	10.54	68	1.3373	432.2
21	1.0865	2.358	45	1.2048	11.58	69	1.3434	563.8
22	1.0910	2.470	46	1.2102	12.77	70	1.3497	746.7
23	1.0956	2.590	47	1.2156	14.13			

VISCOSITY AND DENSITY TABLES (continued)

Sucrose in Water, 20.0°C

Sucrose %	Density[a] g/ml	Viscosity[b] cp	Sucrose %	Density[a] g/ml	Viscosity[b] cp	Sucrose %	Density[a] g/ml	Viscosity[b] cp
0	0.9988	1.004	24	1.0987	2.333			
1	1.0026	1.030	25	1.1033	2.449	48	1.2189	12.50
2	1.0064	1.057	26	1.1079	2.575	49	1.2244	13.86
3	1.0102	1.086	27	1.1126	2.710	50	1.2299	15.42
4	1.0140	1.116	28	1.1173	2.857	51	1.2354	17.23
5	1.0179	1.148	29	1.1220	3.016	52	1.2409	19.33
6	1.0219	1.181	30	1.1268	3.189	53	1.2465	21.79
7	1.0258	1.217	31	1.1316	3.378	54	1.2522	24.67
8	1.0298	1.255	32	1.1365	3.583	55	1.2578	28.07
9	1.0339	1.295	33	1.1414	3.808	56	1.2635	32.11
10	1.0380	1.337	34	1.1463	4.053	57	1.2693	36.96
11	1.0421	1.382	35	1.1513	4.323	58	1.2750	42.78
12	1.0462	1.430	36	1.1563	4.621	59	1.2808	49.85
13	1.0504	1.480	37	1.1613	4.948	60	1.2867	58.50
14	1.0546	1.535	38	1.1663	5.311	61	1.2926	69.15
15	1.0588	1.592	39	1.1714	5.714	62	1.2985	82.39
16	1.0631	1.654	40	1.1766	6.163	63	1.3044	99.01
17	1.0674	1.720	41	1.1817	6.664	64	1.3104	120.1
18	1.0718	1.790	42	1.1870	7.226	65	1.3164	147.0
19	1.0762	1.865	43	1.1922	7.857	66	1.3225	182.0
20	1.0806	1.946	44	1.1975	8.570	67	1.3286	227.8
21	1.0851	2.032	45	1.2028	9.376	68	1.3347	288.5
22	1.0896	2.125	46	1.2081	10.29	69	1.3408	370.2
23	1.0941	2.225	47	1.2135	11.33	70	1.3470	481.8

Sucrose in Water, 25.0°C

Sucrose %	Density[a] g/ml	Viscosity[b] cp	Sucrose %	Density[a] g/ml	Viscosity[b] cp	Sucrose %	Density[a] g/ml	Viscosity[b] cp
0	0.9977	0.8913	24	1.0970	2.023			
1	1.0014	0.9139	25	1.1016	2.121	48	1.2167	10.14
2	1.0052	0.9376	26	1.1062	2.226	49	1.2221	11.19
3	1.0090	0.9625	27	1.1108	2.339	50	1.2276	12.39
4	1.0128	0.9886	28	1.1155	2.462	51	1.2331	13.77
5	1.0167	1.016	29	1.1202	2.595	52	1.2386	15.37
6	1.0206	1.045	30	1.1250	2.739	53	1.2442	17.22
7	1.0246	1.076	31	1.1298	2.895	54	1.2498	19.39
8	1.0285	1.108	32	1.1346	3.064	55	1.2554	21.93
9	1.0325	1.142	33	1.1395	3.249	56	1.2611	24.92
10	1.0366	1.179	34	1.1444	3.451	57	1.2668	28.48
11	1.0407	1.217	35	1.1493	3.672	58	1.2726	32.73
12	1.0448	1.258	36	1.1543	3.914	59	1.2784	37.85
13	1.0489	1.301	37	1.1593	4.181	60	1.2842	44.04
14	1.0531	1.347	38	1.1643	4.475	61	1.2901	51.61
15	1.0574	1.396	39	1.1694	4.799	62	1.2959	60.93
16	1.0616	1.449	40	1.1745	5.160	63	1.3019	72.50
17	1.0659	1.505	41	1.1797	5.560	64	1.3078	86.99
18	1.0702	1.564	42	1.1848	6.008	65	1.3138	105.3
19	1.0746	1.628	43	1.1901	6.509	66	1.3199	128.8
20	1.0790	1.696	44	1.1953	7.072	67	1.3259	159.1
21	1.0835	1.770	45	1.2006	7.706	68	1.3320	198.8
22	1.0879	1.848	46	1.2059	8.423	69	1.3382	251.4
23	1.0924	1.933	47	1.2113	9.236	70	1.3444	322.0

VISCOSITY AND DENSITY TABLES (continued)

Sucrose in Water, 30.0

Sucrose %	Density[a] g/ml	Viscosity[b] cp	Sucrose %	Density[a] g/ml	Viscosity[b] cp	Sucrose %	Density[a] g/ml	Viscosity[b] cp
0	0.9963	0.7978	24	1.0951	1.771	48	1.2144	8.344
1	1.0000	0.8176	25	1.0997	1.854	49	1.2198	9.168
2	1.0038	0.8384	26	1.1043	1.943	50	1.2252	10.10
3	1.0075	0.8601	27	1.1089	2.039	51	1.2307	11.18
4	1.0113	0.8830	28	1.1136	2.143	52	1.2362	12.42
5	1.0152	0.9069	29	1.1183	2.255	53	1.2418	13.84
6	1.0191	0.9322	30	1.1230	2.376	54	1.2474	15.50
7	1.0230	0.9588	31	1.1278	2.506	55	1.2530	17.43
8	1.0270	0.9868	32	1.1326	2.648	56	1.2586	19.69
9	1.0310	1.016	33	1.1374	2.802	57	1.2643	22.36
10	1.0350	1.048	34	1.1423	2.970	58	1.2701	25.52
11	1.0391	1.081	35	1.1472	3.153	59	1.2758	29.30
12	1.0432	1.116	36	1.1522	3.353	60	1.2816	33.84
13	1.0473	1.154	37	1.1572	3.572	61	1.2875	39.34
14	1.0515	1.193	38	1.1622	3.813	62	1.2933	46.05
15	1.0557	1.235	39	1.1672	4.079	63	1.2992	54.30
16	1.0599	1.280	40	1.1723	4.372	64	1.3052	64.53
17	1.0642	1.328	41	1.1775	4.697	65	1.3112	77.35
18	1.0685	1.380	42	1.1826	5.058	66	1.3172	93.54
19	1.0728	1.434	43	1.1878	5.461	67	1.3232	114.2
20	1.0772	1.493	44	1.1931	5.912	68	1.3293	140.9
21	1.0816	1.555	45	1.1983	6.418	69	1.3354	175.8
22	1.0861	1.622	46	1.2036	6.988	70	1.3416	222.0
23	1.0906	1.694	47	1.2090	7.632			

Compiled by Norman G. Anderson based on equations developed by E. J. Barber in *J. Nat. Cancer Inst. Monograph*, **21**, 219 (1966).

DENSITY AT 25°C OF CsCl SOLUTION AS A FUNCTION OF REFRACTIVE INDEX

Refractive Index (Sodium D line, 25°C)	Density (gm/cc)	Refractive Index (Sodium D line, 25°C)	Density (gm/cc)	Refractive Index (Sodium D line, 25°C)	Density (gm/cc)
1.34400	1.09857	1.35020	1.16591	1.35640	1.23324
1.34410	1.09966	1.35030	1.16699	1.35650	1.23433
1.34420	1.10075	1.35040	1.16808	1.35660	1.23541
1.34430	1.10183	1.35050	1.16917	1.35670	1.23650
1.34440	1.10292	1.35060	1.17025	1.35680	1.23758
1.34450	1.10400	1.35070	1.17134	1.35690	1.23867
1.34460	1.10509	1.35080	1.17242	1.35700	1.23976
1.34470	1.10618	1.35090	1.17351	1.35710	1.24084
1.34480	1.10726	1.35100	1.17460	1.35720	1.24193
1.34490	1.10835	1.35110	1.17568	1.35730	1.24301
1.34500	1.10943	1.35120	1.17677	1.35740	1.24410
1.34510	1.11052	1.35130	1.17785	1.35750	1.24519
1.34520	1.11161	1.35140	1.17894	1.35760	1.24627
1.34530	1.11269	1.35150	1.18003	1.35770	1.24736
1.34540	1.11378	1.35160	1.18111	1.35780	1.24844
1.34550	1.11486	1.35170	1.18220	1.35790	1.24953
1.34560	1.11595	1.35180	1.18328	1.35800	1.25062
1.34570	1.11704	1.35190	1.18437	1.35810	1.25170
1.34580	1.11812	1.35200	1.18546	1.35820	1.25279
1.34590	1.11921	1.35210	1.18654	1.35830	1.25387
1.34600	1.12029	1.35220	1.18763	1.35840	1.25496
1.34610	1.12138	1.35230	1.18871	1.35850	1.25605
1.34620	1.12247	1.35240	1.18980	1.35860	1.25713
1.34630	1.12355	1.35250	1.19089	1.35870	1.25822
1.34640	1.12464	1.35260	1.19197	1.35880	1.25930
1.34650	1.12572	1.35270	1.19306	1.35890	1.26039
1.34660	1.12681	1.35280	1.19414	1.35900	1.26148
1.34670	1.12790	1.35290	1.19523	1.35910	1.26256
1.34680	1.12898	1.35300	1.19632	1.35920	1.26365
1.34690	1.13007	1.35310	1.19740	1.35930	1.26473
1.34700	1.13115	1.35320	1.19849	1.35940	1.26582
1.34710	1.13224	1.35330	1.19957	1.35950	1.26691
1.34720	1.13333	1.35340	1.20066	1.35960	1.26799
1.34730	1.13441	1.35350	1.20175	1.35970	1.26908
1.34740	1.13550	1.35360	1.20283	1.35980	1.27016
1.34750	1.13658	1.35370	1.20392	1.35990	1.27125
1.34760	1.13767	1.35380	1.20500	1.36000	1.27234
1.34770	1.13876	1.35390	1.20609	1.36010	1.27342
1.34780	1.13984	1.35400	1.20718	1.36020	1.27451
1.34790	1.14093	1.35410	1.20826	1.36030	1.27559
1.34800	1.14201	1.35420	1.20935	1.36040	1.27668
1.34810	1.14310	1.35430	1.21043	1.36050	1.27777
1.34820	1.14419	1.35440	1.21152	1.36060	1.27885
1.34830	1.14527	1.35450	1.21261	1.36070	1.27994
1.34840	1.14636	1.35460	1.21369	1.36080	1.28102
1.34850	1.14744	1.35470	1.21478	1.36090	1.28211
1.34860	1.14853	1.35480	1.21586	1.36100	1.28320
1.34870	1.14962	1.35490	1.21695	1.36110	1.28428
1.34880	1.15070	1.35500	1.21804	1.36120	1.28537
1.34890	1.15179	1.35510	1.21912	1.36130	1.28645
1.34900	1.15287	1.35520	1.22021	1.36140	1.28754
1.34910	1.15396	1.35530	1.22129	1.36150	1.28863
1.34920	1.15505	1.35540	1.22238	1.36160	1.28971
1.34930	1.15613	1.35550	1.22347	1.36170	1.29080
1.34940	1.15722	1.35560	1.22455	1.36180	1.29188
1.34950	1.15830	1.35570	1.22564	1.36190	1.29297
1.34960	1.15939	1.35580	1.22672	1.36200	1.29406
1.34970	1.16048	1.35590	1.22781	1.36210	1.29514
1.34980	1.16156	1.35600	1.22890	1.36220	1.29623
1.34990	1.16265	1.35610	1.22998	1.36230	1.29731
1.35000	1.16373	1.35620	1.23107	1.36240	1.29840
1.35010	1.16482	1.35630	1.23215	1.36250	1.29949

DENSITY AT 25°C OF CsCl SOLUTION AS A FUNCTION OF REFRACTIVE INDEX (continued)

Refractive Index (Sodium D line, 25°C)	Density (gm/cc)	Refractive Index (Sodium D line, 25°C)	Density (gm/cc)	Refractive Index (Sodium D line, 25°C)	Density (gm/cc)
1.36260	1.30057	1.36900	1.37008	1.37550	1.44067
1.36270	1.30166	1.36910	1.37116	1.37560	1.44175
1.36280	1.30274	1.36920	1.37225	1.37570	1.44284
1.36290	1.30383	1.36930	1.37333	1.37580	1.44393
1.36300	1.30492	1.36940	1.37442	1.37590	1.44501
1.36310	1.30600	1.36950	1.37551	1.37600	1.44610
1.36320	1.30709	1.36960	1.37659	1.37610	1.44718
1.36330	1.30817	1.36970	1.37768	1.37620	1.44827
1.36340	1.30926	1.36980	1.37876	1.37630	1.44936
1.36350	1.31035	1.36990	1.37985	1.37640	1.45044
1.36360	1.31143	1.37000	1.38094	1.37650	1.45153
1.36370	1.31252	1.37010	1.38202	1.37660	1.45261
1.36380	1.31360	1.37020	1.38311	1.37670	1.45370
1.36390	1.31469	1.37030	1.38420	1.37680	1.45479
1.36400	1.31578	1.37040	1.38528	1.37690	1.45587
1.36410	1.31686	1.37050	1.38637	1.37700	1.45696
1.36420	1.31795	1.37060	1.38745	1.37710	1.45804
1.36430	1.31903	1.37070	1.38854	1.37720	1.45913
1.36440	1.32012	1.37080	1.38963	1.37730	1.46022
1.36450	1.32121	1.37090	1.39071	1.37740	1.46130
1.36460	1.32229	1.37100	1.39180	1.37750	1.46239
1.36470	1.32338	1.37110	1.39288	1.37760	1.46347
1.36480	1.32446	1.37120	1.39397	1.37770	1.46456
1.36490	1.32555	1.37130	1.39506	1.37780	1.46565
1.36500	1.32664	1.37140	1.39614	1.37790	1.46673
1.36510	1.32772	1.37150	1.39723	1.37800	1.46782
1.36520	1.32881	1.37160	1.39831	1.37810	1.46890
1.36530	1.32989	1.37170	1.39940	1.37820	1.46999
1.36540	1.33098	1.37180	1.40049	1.37830	1.47108
1.36550	1.33207	1.37190	1.40157	1.37840	1.47216
1.36560	1.33315	1.37200	1.40266	1.37850	1.47325
1.36570	1.33424	1.37210	1.40374	1.37860	1.47433
1.36580	1.33532	1.37220	1.40483	1.37870	1.47542
1.36590	1.33641	1.37230	1.40592	1.37880	1.47651
1.36600	1.33750	1.37240	1.40700	1.37890	1.47759
1.36610	1.33858	1.37250	1.40809	1.37900	1.47868
1.36620	1.33967	1.37260	1.40917	1.37910	1.47976
1.36630	1.34075	1.37270	1.41026	1.37920	1.48085
1.36640	1.34184	1.37280	1.41135	1.37930	1.48194
1.36650	1.34293	1.37290	1.41243	1.37940	1.48302
1.36660	1.34401	1.37300	1.41352	1.37950	1.48411
1.36670	1.34510	1.37310	1.41460	1.37960	1.48519
1.36680	1.34618	1.37320	1.41569	1.37970	1.48628
1.36690	1.34727	1.37330	1.41678	1.37980	1.48737
1.36700	1.34836	1.37340	1.41786	1.37990	1.48845
1.36710	1.34944	1.37350	1.41895	1.38000	1.48954
1.36720	1.35053	1.37360	1.42003	1.38010	1.49062
1.36730	1.35161	1.37370	1.42112	1.38020	1.49171
1.36740	1.35270	1.37380	1.42221	1.38030	1.49280
1.36750	1.35379	1.37390	1.42329	1.38040	1.49388
1.36760	1.35487	1.37400	1.42438	1.38050	1.49497
1.36770	1.35596	1.37410	1.42546	1.38060	1.49605
1.36780	1.35704	1.37420	1.42655	1.38070	1.49714
1.36790	1.35813	1.37430	1.42764	1.38080	1.49823
1.36800	1.35922	1.37440	1.42872	1.38090	1.49931
1.36810	1.36030	1.37450	1.42981	1.38100	1.50040
1.36820	1.36139	1.37460	1.43089	1.38110	1.50148
1.36830	1.36247	1.37470	1.43198	1.38120	1.50257
1.36840	1.36356	1.37480	1.43307	1.38130	1.50366
1.36850	1.36465	1.37490	1.43415	1.38140	1.50474
1.36860	1.36573	1.37500	1.43524	1.38150	1.50583
1.36870	1.36682	1.37510	1.43632	1.38160	1.50691
1.36880	1.36790	1.37520	1.43741	1.38170	1.50800
1.36890	1.36899	1.37530	1.43850	1.38180	1.50909
		1.37540	1.43958	1.38190	1.51017

DENSITY AT 25°C OF CsCl SOLUTION AS A FUNCTION OF REFRACTIVE INDEX (continued)

Refractive Index (Sodium D line, 25°C)	Density (gm/cc)	Refractive Index (Sodium D line, 25°C)	Density (gm/cc)	Refractive Index (Sodium D line, 25°C)	Density (gm/cc)
1.38200	1.51126	1.38850	1.58185	1.39500	1.65244
1.38210	1.51234	1.38860	1.58293	1.39510	1.65353
1.38220	1.51343	1.38870	1.58402	1.39520	1.65461
1.38230	1.51452	1.38880	1.58511	1.39530	1.65570
1.38240	1.51560	1.38890	1.58619	1.39540	1.65678
1.38250	1.51669	1.38900	1.58728	1.39550	1.65787
1.38260	1.51777	1.38910	1.58836	1.39560	1.65896
1.38270	1.51886	1.38920	1.58945	1.39570	1.66004
1.38280	1.51995	1.38930	1.59054	1.39580	1.66113
1.38290	1.52103	1.38940	1.59162	1.39590	1.66221
1.38300	1.52212	1.38950	1.59271	1.39600	1.66330
1.38310	1.52320	1.38960	1.59379	1.39610	1.66439
1.38320	1.52429	1.38970	1.59488	1.39620	1.66547
1.38330	1.52538	1.38980	1.59597	1.39630	1.66656
1.38340	1.52646	1.38990	1.59705	1.39640	1.66764
1.38350	1.52755	1.39000	1.59814	1.39650	1.66873
1.38360	1.52863	1.39010	1.59923	1.39660	1.66982
1.38370	1.52972	1.39020	1.60031	1.39670	1.67090
1.38380	1.53081	1.39030	1.60140	1.39680	1.67199
1.38390	1.53189	1.39040	1.60248	1.39690	1.67307
1.38400	1.53298	1.39050	1.60357	1.39700	1.67416
1.38410	1.53406	1.39060	1.60466	1.39710	1.67525
1.38420	1.53515	1.39070	1.60574	1.39720	1.67633
1.38430	1.53624	1.39080	1.60683	1.39730	1.67742
1.38440	1.53732	1.39090	1.60791	1.39740	1.67850
1.38450	1.53841	1.39100	1.60900	1.39750	1.67959
1.38460	1.53949	1.39110	1.61009	1.39760	1.68068
1.38470	1.54058	1.39120	1.61117	1.39770	1.68176
1.38480	1.54167	1.39130	1.61226	1.39780	1.68285
1.38490	1.54275	1.39140	1.61334	1.39790	1.68393
1.38500	1.54384	1.39150	1.61443	1.39800	1.68502
1.38510	1.54492	1.39160	1.61552	1.39810	1.68611
1.38520	1.54601	1.39170	1.61660	1.39820	1.68719
1.38530	1.54710	1.39180	1.61769	1.39830	1.68828
1.38540	1.54818	1.39190	1.61877	1.39840	1.68936
1.38550	1.54927	1.39200	1.61986	1.39850	1.69045
1.38560	1.55035	1.39210	1.62095	1.39860	1.69154
1.38570	1.55144	1.39220	1.62203	1.39870	1.69262
1.38580	1.55253	1.39230	1.62312	1.39880	1.69371
1.38590	1.55361	1.39240	1.62420	1.39890	1.69479
1.38600	1.55470	1.39250	1.62529	1.39900	1.69588
1.38610	1.55578	1.39260	1.62638	1.39910	1.69697
1.38620	1.55687	1.39270	1.62746	1.39920	1.69805
1.38630	1.55796	1.39280	1.62855	1.39930	1.69914
1.38640	1.55904	1.39290	1.62963	1.39940	1.70022
1.38650	1.56013	1.39300	1.63072	1.39950	1.70131
1.38660	1.56121	1.39310	1.63181	1.39960	1.70240
1.38670	1.56230	1.39320	1.63289	1.39970	1.70348
1.38680	1.56339	1.39330	1.63398	1.39980	1.70457
1.38690	1.56447	1.39340	1.63506	1.39990	1.70565
1.38700	1.56556	1.39350	1.63615	1.40000	1.70674
1.38710	1.56664	1.39360	1.63724	1.40010	1.70783
1.38720	1.56773	1.39370	1.63832	1.40020	1.70891
1.38730	1.56882	1.39380	1.63941	1.40030	1.71000
1.38740	1.56990	1.39390	1.64049	1.40040	1.71108
1.38750	1.57099	1.39400	1.64158	1.40050	1.71217
1.38760	1.57207	1.39410	1.64267	1.40060	1.71326
1.38770	1.57316	1.39420	1.64375	1.40070	1.71434
1.38780	1.57425	1.39430	1.64484	1.40080	1.71543
1.38790	1.57533	1.39440	1.64592	1.40090	1.71651
1.38800	1.57642	1.39450	1.64701	1.40100	1.71760
1.38810	1.57750	1.39460	1.64810	1.40110	1.71869
1.38820	1.57859	1.39470	1.64918	1.40120	1.71977
1.38830	1.57968	1.39480	1.65027	1.40130	1.72086
1.38840	1.58076	1.39490	1.65135	1.40140	1.72194

DENSITY AT 25°C OF CsCl SOLUTION AS A FUNCTION OF REFRACTIVE INDEX (continued)

Refractive Index (Sodium D line, 25°C)	Density (gm/cc)	Refractive Index (Sodium D line, 25°C)	Density (gm/cc)	Refractive Index (Sodium D line, 25°C)	Density (gm/cc)
1.40150	1.72303	1.40770	1.79036	1.41390	1.85770
1.40160	1.72412	1.40780	1.79145	1.41400	1.85878
1.40170	1.72520	1.40790	1.79253	1.41410	1.85987
1.40180	1.72629	1.40800	1.79362	1.41420	1.86095
1.40190	1.72737	1.40810	1.79471	1.41430	1.86204
1.40200	1.72846	1.40820	1.79579	1.41440	1.86313
1.40210	1.72955	1.40830	1.79688	1.41450	1.86421
1.40220	1.73063	1.40840	1.79796	1.41460	1.86530
1.40230	1.73172	1.40850	1.79905	1.41470	1.86638
1.40240	1.73280	1.40860	1.80014	1.41480	1.86747
1.40250	1.73389	1.40870	1.80122	1.41490	1.86856
1.40260	1.73498	1.40880	1.80231	1.41500	1.86964
1.40270	1.73606	1.40890	1.80339	1.41510	1.87073
1.40280	1.73715	1.40900	1.80448	1.41520	1.87181
1.40290	1.73823	1.40910	1.80557	1.41530	1.87290
1.40300	1.73932	1.40920	1.80665	1.41540	1.87399
1.40310	1.74041	1.40930	1.80774	1.41550	1.87507
1.40320	1.74149	1.40940	1.80882	1.41560	1.87616
1.40330	1.74258	1.40950	1.80991	1.41570	1.87724
1.40340	1.74366	1.40960	1.81100	1.41580	1.87833
1.40350	1.74475	1.40970	1.81208	1.41590	1.87942
1.40360	1.74584	1.40980	1.81317	1.41600	1.88050
1.40370	1.74692	1.40990	1.81425	1.41610	1.88159
1.40380	1.74801	1.41000	1.81534	1.41620	1.88267
1.40390	1.74909	1.41010	1.81643	1.41630	1.88376
1.40400	1.75018	1.41020	1.81751	1.41640	1.88485
1.40410	1.75127	1.41030	1.81860	1.41650	1.88593
1.40420	1.75235	1.41040	1.81969	1.41660	1.88702
1.40430	1.75344	1.41050	1.82077	1.41670	1.88810
1.40440	1.75452	1.41060	1.82186	1.41680	1.88919
1.40450	1.75561	1.41070	1.82294	1.41690	1.89028
1.40460	1.75670	1.41080	1.82403	1.41700	1.89136
1.40470	1.75778	1.41090	1.82512	1.41710	1.89245
1.40480	1.75887	1.41100	1.82620	1.41720	1.89353
1.40490	1.75995	1.41110	1.82729	1.41730	1.89462
1.40500	1.76104	1.41120	1.82837	1.41740	1.89571
1.40510	1.76213	1.41130	1.82946	1.41750	1.89679
1.40520	1.76321	1.41140	1.83055	1.41760	1.89788
1.40530	1.76430	1.41150	1.83163	1.41770	1.89896
1.40540	1.76538	1.41160	1.83272	1.41780	1.90005
1.40550	1.76647	1.41170	1.83380	1.41790	1.90114
1.40560	1.76756	1.41180	1.83489	1.41800	1.90222
1.40570	1.76864	1.41190	1.83598	1.41810	1.90331
1.40580	1.76973	1.41200	1.83706	1.41820	1.90439
1.40590	1.77081	1.41210	1.83815	1.41830	1.90548
1.40600	1.77190	1.41220	1.83923	1.41840	1.90657
1.40610	1.77299	1.41230	1.84032	1.41850	1.90765
1.40620	1.77407	1.41240	1.84141	1.41860	1.90874
1.40630	1.77516	1.41250	1.84249	1.41870	1.90982
1.40640	1.77624	1.41260	1.84358	1.41880	1.91091
1.40650	1.77733	1.41270	1.84466	1.41890	1.91200
1.40660	1.77842	1.41280	1.84575	1.41900	1.91308
1.40670	1.77950	1.41290	1.84684	1.41910	1.91417
1.40680	1.78059	1.41300	1.84792	1.41920	1.91525
1.40690	1.78167	1.41310	1.84901	1.41930	1.91634
1.40700	1.78276	1.41320	1.85009	1.41940	1.91743
1.40710	1.78385	1.41330	1.85118	1.41950	1.91851
1.40720	1.78493	1.41340	1.85227	1.41960	1.91960
1.40730	1.78602	1.41350	1.85335	1.41970	1.92068
1.40740	1.78710	1.41360	1.85444	1.41980	1.92177
1.40750	1.78819	1.41370	1.85552	1.41990	1.92286
1.40760	1.78928	1.41380	1.85661		

DENSITY AT 25°C OF CsCl SOLUTION AS A FUNCTION OF REFRACTIVE INDEX (continued)

Refractive Index (Sodium D line, 25°C)	Density (gm/cc)	Refractive Index (Sodium D line, 25°C)	Density (gm/cc)	Refractive Index (Sodium D line, 25°C)	Density (gm/cc)
1.42000	1.92394	1.42050	1.92937	1.42100	1.93480
1.42010	1.92503	1.42060	1.93046	1.42110	1.93589
1.42020	1.92611	1.42070	1.93154	1.42120	1.93697
1.42030	1.92720	1.42080	1.93263	1.42130	1.93806
1.42040	1.92829	1.42090	1.93372	1.42140	1.93915

Calculated from the equation $\rho^{25°C} = (10.8601 \times R.I.) - 13.4974$ of Ifft, Voet, and Vinograd *J. Phys. Chem.*, 65, 1138 (1961). R.I. is the refractive index (Sodium D line, 25°C).

Compiled by Norman G. Anderson and Norman L. Anderson.

DENSITY GRADIENT AT EQUILIBRIUM
FOR SOME SALT SOLUTIONS

Salt	Density of Solution	$10^{10} \dfrac{d\rho}{\omega^2 r \, dr}$[a]
Cs_2SO_4	1.3	13[b]
CsCl	1.30	6.47
RbCl	1.30	3.24
K acetate	1.41	0.71
LiBr	1.83	0.8
LiCl	1.33	0.4[c]
KBr	1.30	3.11
RbBr	1.30	6.12

[a]$d\rho/dr$ is the density gradient at distance r from the center of rotation at a speed of ω radians per second.
[b]From graph by Ludlum and Warner, *J. Biol. Chem.*, 240, 2961–2965 (1965).
[c]For solutions in D_2O.

From Myron K. Brakke for Iscotables. With permission of the author and copyright owner, Instrumentation Specialties Company, Inc., Lincoln, Nebraska.

COEFFICIENTS FOR CALCULATION OF DENSITY, $\rho_{25°}$; FROM REFRACTIVE INDEXES OF SOLUTIONS AT 25°, $\eta_D^{25°}$. BY FORMULA $\rho_{25°} = a\eta_D^{25°} - b$

Solute	Coefficients		Density Range
	a	b	
Cs_2SO_4	12.1200	15.1662	1.15–1.40
	13.6986	17.3233	1.40–1.70
CsBr	9.9667	12.2876	1.25–1.35
CsCl	10.8601	13.4974	1.25–1.90
Cs acetate	10.7527	13.4247	1.80–2.05
Cs formate	13.7363	17.4286	1.72–1.82
KBr	6.4786	7.6431	1.10–1.35
RbBr	9.1750	11.2410	1.15–1.65

[a]From Vinograd and Hearst, 1962.
[b]Ludlum and Warner (1965) give following formula for density, $\rho_{25°}$, of $CsSO_4$: $\rho_{25°} = 0.9954 + 11.1066(\eta - \eta_0) + 26.4460(\eta - \eta_0)^2$ where η is the refractive index of the solution and η_0 that of water.

From Myron K. Brakke for Iscotables. With permission of the author and copyright owner, Instrumentation Specialties Company, Inc., Lincoln, Nebraska.

DENSITIES AT 25° OF VARIOUS SALT SOLUTIONS USED FOR DENSITY GRADIENTS

Solute	Concentration, wt. %[a]					
	10	20	30	40	50	60
LiCl	1.054	1.113	1.178	1.250		
LiBr	1.073	1.160	1.261	1.381	1.53	1.716
KBr	1.072	1.158	1.257	1.371		
NaBr	1.078	1.172	1.281	1.410		
RbBr	1.079	1.174	1.285	1.419	1.582	
CsCl[b]	1.079	1.174	1.286	1.420	1.582	1.785
CsBr	1.081	1.180	1.297	1.440	1.616	
$CsSO_4$[b]	1.086	1.190				
K acetate	1.048	1.100	1.155	1.213	1.272	1.333
K citrate	1.066	1.140	1.221			
K tartrate	1.066	1.139	1.218	1.305	1.400	
glycerol	1.021	1.045	1.071	1.097	1.124	1.151
sucrose[c]	1.0381	1.081	1.127	1.176	1.230	1.289

[a]Values are from International Critical Tables. Highest values given are not necessarily for saturated solutions.

[b]The density of CsCl solutions may be calculated from the formula wt. % = $137.48 - 138.11(1\rho_4^{25})$ for 30–60% solutions (Vinograd and Hearst, 1962). The density of Cs_2SO_4 solutions may be calculated from the formula $P_{25} = 1.0047 + 0.28369m - 0.017428m^2$ $(0.5 \leqslant m \leqslant 3.5)$ where m is the molality (Ludlum and Warner, 1965).

[c]Specific gravity 20°/40°C.

From Myron K. Brakke for Iscotables. With permission of the author and copyright owner, Instrumentation Specialties Company, Inc., Lincoln, Nebraska.

VISCOSITIES OF VARIOUS SALT SOLUTIONS USED FOR DENSITY GRADIENTS[a]

Solute	Temperature	Relative viscosity for a molal concentration of							
		0.5	1	2	3	4	5	10	15
LiCl	0°	1.069	1.129		1.454				
	25°	1.069	1.142	1.302	1.479	1.673	1.895	3.73	8.23
KBr	0°		0.913	0.845	0.817				
	25°	0.984	0.969	0.967	1.007				
NaBr	25°	1.029	1.062	1.154					
RbBr	25°	0.979							
CsCl	25°	0.985	0.975						
Cs_2SO_4	25°	1.067	1.145						
K Acetate	18°	1.125	1.248	1.515	1.817	2.172			
K Tartrate[b]	18°	1.183							

Solute	Temperature	Viscosity in centipoise at, weight % concentration of					
		0%	5%	10%	20%	40%	60%
Sucrose	0°	1.79	2.042	2.436	3.72	14.79	
	5°	1.52	1.72	2.042	3.154	11.56	156
	25°	0.894			1.960	5.187	43.86
Glycerol	20°	1.002			1.769	3.75	10.96
	25°				1.542	3.18	8.82

[a]Values selected from International Critical Tables, and Landolt-Bornstein, Zehlenwerte und Funktionen aus Physik, Chemie, Geophysik und Technik, 7th ed.

[b]1.74 at 1.5 molal.

From Myron K. Brakke for Iscotables. With permission of the author and copyright owner, Instrumentation Specialties Company, Inc., Lincoln, Nebraska.

4,698 ISOKINETIC GLYCEROL AND SUCROSE GRADIENTS FOR DENSITY GRADIENT CENTRIFUGATION

Ben A. M. van der Zeijst and Henri P. J. Bloemers

Particles can be sedimented with constant velocities, proportional to their $S_{20,w}$ values, in properly constructed so-called *isokinetic* gradients. A total of 4,698 different sucrose and glycerol gradients has been calculated for 29 types of swinging bucket rotors. Data are given for particle densities at intervals of 0.10 from 1.20 to 2.00 g/ml, each at 5°C, 10°C, and 20°C and at 5%, 10%, and 15% (w/w) top concentration. Some practical instructions for the use of the gradients are given.

INTRODUCTION

Since its introduction twenty years ago,[1] density gradient centrifugation has been widely used in biochemistry and molecular biology. Initially, linear sucrose gradients have been used almost exclusively.

Generally, the sedimentation velocity of a particle in a concentration gradient is a complicated function of its position in the gradient: The sedimentation velocity increases with increasing distance from the rotor axis; on the other hand, it decreases as the particle moves to zones with higher viscosities and densities. In isokinetic gradients (see below) these effects are exactly matched. However, in linear sucrose gradients (and also in linear glycerol gradients) the latter effect usually prevails. This situation leads to two serious disadvantages: The resolution in the gradient is less than optimal and the calculation of $S_{20,w}$ values from sedimentation patterns is very complicated. It was demonstrated[2] that the sedimentation velocity of particles of a specified density is constant in a specified linear gradient run at a specified temperature in the SW 50L or SW 39 rotor (Beckman Instruments Inc., Spinco Division, Palo Alto, California). Martin and Ames give data for the calculation of sedimentation coefficients for this particular gradient.[2] The use of their method for other gradients, rotor types, and temperatures is generally not allowed and will result in wrong sedimentation coefficients. However, Fritsch has recently shown that the sedimentation velocity is approximately constant in linear 5 to 20% sucrose gradients or linear 10 to 30% glycerol gradients for some other rotor types, temperatures, and particle densities.[3]

Several methods have been described to calculate sedimentation coefficients from centrifugation in linear gradients.[4-12] These methods are tedious and time consuming without the use of a highspeed computer program. In one case, the method contains serious errors.[7] McEwen[13] has compiled tables which are rather easy to use for the calculation of sedimentation coefficients in linear sucrose gradients. A purely empirical method consists of the calibration of a gradient for a particular set of conditions with particles of known S-values and densities.[14-16]

In 1967, Noll introduced the so-called isokinetic gradients,[17] i.e., gradients that are so constructed that particles of a specified density sediment with a constant velocity proportional to their $S_{20,w}$ values. Thus $S_{20,w}$ values can be easily calculated. Moreover their resolving capacity is better. It was felt as a drawback[12] that for each particular situation, i.e., rotor dimension, particle density, temperature, and top concentration, a different isokinetic gradient has to be calculated. Indeed, gradients for only 6 out of the 29 rotor types now in use have been compiled and published.[15,18]

In this paper, we present data for the construction of a large number of gradients for all rotors available from Spinco-Beckman (Beckman Instruments, Inc., Spinco Division, Palo Alto, California, U.S.A.), International (International Equipment Co., Needham

Heights, Massachusetts, U.S.A.), MSE (Measuring and Scientific Equipment Ltd., London, England), and Christ (Heraeus-Christ, GMBH, Osterode am Harz, G. F. R). This compilation has been made with the aid of a computer program that we have used routinely for the calculation of sucrose and glycerol gradients.

Practical details about the construction and fractionation of gradients are given here only as far as necessary for a good understanding of our results. For a further discussion see References 15, 18, and 19.

For zonal rotors a similar set of tables for the construction of *equivolumetric* gradients was published elsewhere.[29]

NUMERICAL CALCULATIONS

Theory

The velocity (dx/dt) at which a particle of density ρ_p sediments in a medium of density ρ_m and viscosity η_m is

$$\frac{dx}{dt} = S_{20,w}\, \omega^2\, x \cdot \frac{\rho_p - \rho_m}{\rho_p - \rho_{20,w}} \cdot \frac{\eta_{20,w}}{\eta_m} \tag{1}$$

where $S_{20,w}$ is the sedimentation coefficient of the particle at $20°C$ in water, ω the angular velocity, $\rho_{20,w}$ the density and $\eta_{20,w}$ the viscosity of water at $20°C$, and x the distance from the axis of rotation to the particle. For an isokinetic gradient dx/dt is constant by definition. The value of this constant is defined by the conditions at the top of the gradient (η_t and ρ_t, the subscript t standing for top). The relation between these conditions and those (η_m and ρ_m) at other positions (x) in the gradient are given in Equation 2:

$$x = x_t \cdot \frac{\eta_m}{\eta_t} \cdot \frac{\rho_p - \rho_t}{\rho_p - \rho_m} \tag{2}$$

For the calculation of density and viscosity of the medium we used empirical relationships with constants given by Barber[20] for sucrose and by van der Zeijst and Bult for glycerol as explained in Reference 21.

The Dimensions of Rotors and Tubes

The radius at the top and the bottom of the centrifuge tube were obtained from data given by the manufacturers. In some cases, considerable discrepancies were found between the contents of tubes measured by us and the nominal values given by the manufacturer. In these cases we have used our values for the calculations. The values used for the calculation of the gradients are presented in Table 1.

No corrections have been made for the round bottom of the gradient tube. Thus, in this part of the tube S-values cannot be determined.

Construction of Gradients

Isokinetic gradients are convex and can be constructed with a constant-volume gradient mixer.[15,17,18] Three parameters determine the gradient: the initial concentration in the mixing vessel (= top concentration), the volume of the mixing vessel (V_m), and the concentration of the solution introduced into the vessel (C_h). The volume delivered to the tube (V_d) at the moment when the concentration of the gradient solution has reached the value C_x is given by

$$V_d = - V_m \cdot \ln \frac{C_h - C_x}{C_h - C_t} \tag{3}$$

Table 1

GEOMETRICAL PARAMETERS OF SWINGING-OUT ROTORS USED FOR THE CALCULATION OF ISOKINETIC GRADIENTS AND REFERENCE TO DATA TABLES FOR THE CONSTRUCTION OF GRADIENTS[a]

Manufacturer	Rotortype	Table number for gradients	Maximal speed (r/min)	Content of tubes (ml)	Material of tubes	Gradient volume (ml)	Radius at top of gradient (cm)
Beckman	SW 25.1	4	25,000	34.5	NC	30.0	6.23
	SW 25.2	5	25,000	62.4	NC	60.0	6.71
	SW 27	6	27,000	41.0	NC	37.5	7.90
	SW 27.1	7	27,000	18.7	NC	17.5	7.13
	SW 36	8	36,000	13.9	NC	13.0	6.16
	SW 40 Ti	9	40,000	13.7	NC	13.0	6.82
	SW 41 Ti	10	41,000	12.8	NC	12.0	6.88
	SW 50 L	11	50,000	5.8	NC	5.0	5.40
	SW 50.1	12	50,000	5.8	NC	5.0	6.33
	SW 56 Ti	13	56,000	4.7	NC	4.2	6.19
	SW 65 L Ti	14	65,000	5.8	NC	5.0	4.51
Christ	9740	15	30,000	30	NC	25.0	6.07
	9750	16	40,000	5.5	NC	5.0	5.07
	9792	17	52,000	5.5	NC	5.0	5.21
	9793	18	40,000	13	NC	12.0	7.08
	9795	19	20,000	70	PP	65.0	5.99
International	485 (SB-110)	20	25,000	38.4	PP	35.0	7.03
	488 (SB-283)	21	41,000	13.0	PP	12.0	6.20
	498 (SB-405)	22	60,000	3.6	PP	3.2	5.18
MSE	43127-101	23	50,000	3.0	PP	2.6	4.79
	43127-102	24	60,000	6.8	PP, PC	6.0	4.95
	43127-103	25	40,000	5.6	PP, PC	5.0	5.35
	43127-104	26	30,000	25.0	PP, PC	22.0	6.68
	43127-106	27	23,500	70.0	PP, PC	65.0	7.17
	43127-111	28	40,000	13.3	PP, PC	12.5	6.89
	43127-115	29	56,000	4.9	PP, PC	4.5	6.08
	43127-126	30	45,000	5.6	PP, PC	5.0	6.14
	43127-501	31	25,000	16.8	PP, PC	15.5	6.85
	43127-503	32	25,000	38.0	PP, PC	34.0	8.94

[a] NC = nitrocellulose, PP = polypropylene, and PC = polycarbonate.

in which all concentrations are expressed in percentages (w/v). For the construction of an isokinetic gradient with a given top concentration (C_t), V_m and C_h should be selected so that the difference between V_x (the volume at which the concentration should be C_x) and V_d is minimized.

Computer Program

The program performs the following steps.

a. Using Equation 2 and the relationships for the density and viscosity of the medium, beginning at the top, the sucrose (or glycerol) concentration is increased in steps of 0.5%, thus calculating the radii (x) corresponding to a number of concentrations. The volumes corresponding to these radii are computed using the dimensions of the rotor and the rotor tubes.

b. The concentrations are expressed in percentages (w/w). For the calculations under d, they are converted into percentages (w/v) by means of the relationship percentages (w/v) = percentages (w/w) density.

c. Using Equation 1 for the conditions at the top of the gradient, the velocity of a 100 S particle in ml/h is computed.

d. The sum of $|V_d - V_x|$ for all points considered is calculated for different combinations of V_m and C_h. By minimizing this sum the proper values for V_m and C_h are found.

e. In addition, some parameters for the preparation of sucrose solutions are calculated.

Density of Particles

For the particle density, the reciprocal of the effective or hydrodynamic partial specific volume as defined by Gagen[22] should be taken. This density is equal to the solvent density at which sedimentation no longer occurs. Its numerical value depends on the solute used for the preparation of the gradient, as has been found and discussed by several authors,[22-28] probably due to selective solvation effects. Plots of η_m S (the viscosity-corrected sedimentation coefficient) vs. the solution density give straight lines from which the particle density can be obtained by extrapolation. In Table 2 we give a summary of the results thus obtained for different types of macromolecules[22,23,25,27-31] together with some other data.[18,32]

Table 2
EFFECTIVE DENSITIES OF BIOLOGICAL PARTICLES
IN SUCROSE SOLUTIONS

Nature of particles	Effective density in sucrose solutions (g/ml)	References
Viruses		
Tobacco mosaic virus	1.27	23
Southern bean mosaic virus	1.26	25
Influenza A	1.19	24
T$_2$-Bacteriophage	1.27	30
Ribosomes		
Rat liver	1.41	18,31
DNA		
Calf thymus	1.42	28
RNA		
rRNA	1.89	18
Tobacco mosaic virus	1.49	32
Proteins		
Ovalbumin	1.27	22
Conalbumin	1.23	22
Lysozyme	1.32	22
Bovine serum albumin	1.32	22
Human hemoglobin	1.28	22

In the theory of isokinetic gradients it is implicitly assumed that during the centrifugation the density of sedimenting particles is constant and no osmotic or hydration effects exist. For the particles given in Table 2 this assumption appears to be justified.

It should be emphasized that the exact knowledge of the density of sedimenting particles is essential for every method of calculation of sedimentation coefficients. For isokinetic gradients, however, it is a problem a priori whereas for linear gradients it is a problem a postiori.

Integrated Centrifugation Time

The average ω^2, $\bar{\omega}^2$, value is required for the calculation of S-values. It is most easily determined by an electronic integrator. When an integrator is not available $\bar{\omega}^2$ has to be estimated during acceleration and deceleration. In most cases during these periods rotor velocities are approximately linear with time (results not shown). Thus, $\int \omega^2\, dt$ during acceleration and deceleration is approximately equal to one third of the maximal ω^2 multiplied by the sum of the acceleration time and the deceleration time.

RESULTS

Tables 4 to 32 summarize the results of the computer calculations. The values found for V_m and C_h give a good approximation of the desired gradient. The average value found for $|V_d - V_x|$ was always smaller than 1.0% of the gradient volume. In the tables data are given for the preparation of sucrose (or glycerol) solutions: They include the amount of solute to be used for the solution to be introduced in the mixing vessel as well as the amount of twofold concentrated buffer and the amount of water. To avoid unnecessary duplications of these values for the top concentrations, these are given in Table 3.

Table 3
PREPARATION OF LIGHT SUCROSE AND GLYCEROL SOLUTIONS

t (°C)	% (w/w)	ml buffer 2x	ml H$_2$O
		Glycerol	
5.0	5.0	49.39	45.61
5.0	10.0	48.77	41.23
5.0	15.0	48.16	36.84
10.0	5.0	49.42	45.58
10.0	10.0	48.81	41.19
10.0	15.0	48.21	36.79
20.0	5.0	49.51	45.49
20.0	10.0	48.92	41.08
20.0	15.0	48.33	36.67
		Sucrose	
5.0	5.0	49.01	45.99
5.0	10.0	48.05	41.95
5.0	15.0	47.08	37.92
10.0	5.0	49.04	45.96
10.0	10.0	48.08	41.92
10.0	15.0	47.12	37.88
20.0	5.0	49.12	45.88
20.0	10.0	48.17	41.83
20.0	15.0	47.22	37.78

Example of the Use of the Tables

Separation of particles of a density of 1.40% g/ml in a 15% (w/w) top concentration glycerol gradient at 5°C in the Beckman SW 50.1 rotor. Table 1 indicates that Table 12 contains the data for this rotor.

1. Preparation of 15% (w/w) glycerol (Table 3). Add 48.16 ml of double concentrated buffer solution and 36.84 ml of water at 15 g of glycerol to obtain 100 g of a 15% (w/w) glycerol solution.

2. Preparation of heavy glycerol solution. Table 12 indicates that a 41% (w/w) glycerol solution is required. This solution is prepared by adding together 41 g of glycerol, 45.12 ml of double concentrated buffer solution and 13.88 ml of water.

3. Construction of the gradient. The mixing vessel is filled with 8.0 ml of the 15% glycerol solution for each gradient to be constructed (Table 12). The outlet of the mixing vessel is put on the bottom of the centrifuge tube. The inlet is connected to a reservoir containing the 41% glycerol solution. The latter solution is introduced in the mixing vessel and the tubes are filled with the appropriate amount of gradient solution. In this case 5 ml gradients are to be used (Table 1).

4. Calculation of centrifugation time. If one wants to centrifuge 80 S particles 4 ml down into the gradient, a centrifugation time of $\frac{100}{80} \times \frac{4}{2.65} = 1.89$ h is required at 50,000 r/min (Tables 1 and 12).

Table 4
DATA FOR THE CONSTRUCTION OF ISOKINETIC SUCROSE AND GLYCEROL GRADIENTS FOR BECKMAN ROTOR SW 25.1

Particle density g/ml	Temperature °C	Top concentrations %(w/w)	Sucrose C_h %(w/w)	Sucrose Buffer 2x ml	Sucrose H_2O ml	Sucrose V_m ml	Sucrose Velocity of 100 S particle at top speed ml/h	Glycerol C_h %(w/w)	Glycerol Buffer 2x ml	Glycerol H_2O ml	Glycerol V_m ml	Glycerol Velocity of 100 S particle at top speed ml/h
1.20	5.0	5.0	22.00	45.73	32.27	31.00	3.82	30.50	46.32	23.18	37.50	3.90
1.20	5.0	10.0	25.50	45.05	29.45	33.00	2.86	34.00	45.92	20.08	36.00	3.10
1.20	5.0	15.0	28.50	44.48	27.02	33.50	2.04	38.00	45.46	16.54	37.00	2.45
1.20	10.0	5.0	22.50	45.68	31.82	31.50	4.46	31.00	46.33	22.67	37.00	4.56
1.20	10.0	10.0	25.50	45.10	29.40	32.00	3.36	35.00	45.88	19.12	36.50	3.65
1.20	10.0	15.0	28.50	44.53	26.97	32.00	2.42	38.50	45.48	16.02	36.50	2.90
1.20	20.0	5.0	23.50	45.60	30.90	32.00	5.88	33.00	46.26	20.74	37.50	6.04
1.20	20.0	10.0	26.50	45.03	28.47	33.00	4.49	36.50	45.87	17.63	37.00	4.90
1.20	20.0	15.0	29.50	44.47	26.03	33.50	3.28	39.00	45.60	15.40	34.50	3.95
1.30	5.0	5.0	25.00	45.15	29.85	31.50	3.96	33.50	45.97	20.53	37.00	3.99
1.30	5.0	10.0	28.00	44.57	27.43	32.00	3.10	38.00	45.46	16.54	37.50	3.25
1.30	5.0	15.0	31.50	43.90	24.60	33.50	2.35	40.50	45.18	14.32	35.00	2.64
1.30	10.0	5.0	25.50	45.10	29.40	31.50	4.62	34.50	45.93	19.57	37.50	4.66
1.30	10.0	10.0	29.00	44.44	26.56	33.50	3.64	38.50	45.48	16.02	37.00	3.82
1.30	10.0	15.0	32.00	43.87	24.13	34.00	2.77	41.00	45.20	13.80	34.50	3.13
1.30	20.0	5.0	26.50	45.03	28.47	32.00	6.08	36.50	45.87	17.63	37.50	6.16
1.30	20.0	10.0	29.50	44.47	26.03	32.50	4.85	40.50	45.43	14.07	37.50	5.12
1.30	20.0	15.0	32.50	43.90	23.60	33.00	3.74	42.50	45.21	12.29	34.50	4.24
1.40	5.0	5.0	27.00	44.76	28.24	32.50	4.03	35.50	45.74	18.76	37.50	4.03
1.40	5.0	10.0	29.50	44.29	26.21	32.00	3.22	39.50	45.29	15.21	37.00	3.33
1.40	5.0	15.0	33.00	43.62	23.38	33.50	2.50	41.50	45.07	13.43	34.00	2.74
1.40	10.0	5.0	27.50	44.72	27.78	32.50	4.70	36.50	45.71	17.79	37.50	4.71
1.40	10.0	10.0	30.00	44.25	25.75	32.00	3.78	40.00	45.32	14.68	36.50	3.91
1.40	10.0	15.0	33.50	43.58	22.92	33.50	2.95	43.00	44.98	12.02	35.00	3.24
1.40	20.0	5.0	28.50	44.66	26.84	32.50	6.18	38.50	45.65	15.85	37.50	6.23
1.40	20.0	10.0	31.00	44.18	24.82	32.00	5.03	42.50	45.21	12.29	37.50	5.23
1.40	20.0	15.0	34.00	43.62	22.38	32.50	3.98	44.00	45.05	10.95	34.00	4.39
1.50	5.0	5.0	28.00	44.57	27.43	32.50	4.07	36.50	45.63	17.87	37.50	4.06
1.50	5.0	10.0	30.50	44.10	25.40	32.00	3.29	40.50	45.18	14.32	37.00	3.37
1.50	5.0	15.0	33.50	43.52	22.98	32.50	2.59	42.50	44.95	12.55	34.00	2.80
1.50	10.0	5.0	27.50	44.72	27.78	30.00	4.75	37.50	45.59	16.91	37.50	4.74
1.50	10.0	10.0	31.00	44.06	24.94	32.00	3.86	41.50	45.15	13.35	37.00	3.96
1.50	10.0	15.0	34.00	43.49	22.51	32.50	3.06	43.50	44.93	11.57	34.00	3.31

Table 4 (continued)

DATA FOR THE CONSTRUCTION OF ISOKINETIC SUCROSE AND GLYCEROL GRADIENTS FOR BECKMAN ROTOR SW 25.1

Particle density g/ml	Temperature °C	Top concentrations %(w/w)	Sucrose					Glycerol				
			C_h %(w/w)	Buffer 2x ml	H_2O ml	V_m ml	Velocity of 100 S particle at top speed ml/h	C_h %(w/w)	Buffer 2x ml	H_2O ml	V_m ml	Velocity of 100 S particle at top speed ml/h
1.50	20.0	5.0	29.00	44.56	26.44	31.00	6.25	39.00	45.60	15.40	36.50	6.27
1.50	20.0	10.0	32.00	43.99	24.01	32.00	5.14	43.50	45.11	11.39	37.00	5.30
1.50	20.0	15.0	35.00	43.43	21.57	32.50	4.11	44.50	45.00	10.50	33.00	4.48
1.60	5.0	5.0	28.00	44.57	27.43	31.00	4.10	37.50	45.52	16.98	37.50	4.08
1.60	5.0	10.0	31.50	43.90	24.60	33.00	3.34	41.00	45.12	13.88	36.50	3.40
1.60	5.0	15.0	34.50	43.34	22.16	33.50	2.65	43.50	44.84	11.66	34.50	2.84
1.60	10.0	5.0	29.00	44.44	26.56	32.00	4.78	38.50	45.48	16.02	37.50	4.76
1.60	10.0	10.0	31.50	43.96	24.54	31.50	3.92	42.00	45.09	12.91	36.50	4.00
1.60	10.0	15.0	35.00	43.30	21.70	33.50	3.13	43.50	44.93	11.57	33.00	3.35
1.60	20.0	5.0	29.50	44.47	26.03	30.50	6.29	39.50	45.54	14.96	36.00	6.29
1.60	20.0	10.0	32.50	43.90	23.60	31.50	5.21	44.00	45.05	10.95	36.50	5.34
1.60	20.0	15.0	35.50	43.34	21.16	32.00	4.21	45.00	44.94	10.06	32.50	4.54
1.70	5.0	5.0	28.50	44.48	27.02	31.00	4.12	38.00	45.46	16.54	37.50	4.09
1.70	5.0	10.0	32.00	43.81	24.19	33.00	3.38	41.50	45.07	13.43	36.50	3.43
1.70	5.0	15.0	34.50	43.34	22.16	32.50	2.69	44.00	44.79	11.21	34.50	2.87
1.70	10.0	5.0	29.00	44.44	26.56	30.50	4.80	39.00	45.43	15.57	37.50	4.78
1.70	10.0	10.0	32.00	43.87	24.13	31.50	3.96	42.50	45.04	12.46	36.50	4.02
1.70	10.0	15.0	35.50	43.20	21.30	33.50	3.18	44.00	44.87	11.13	33.00	3.39
1.70	20.0	5.0	30.00	44.37	25.63	30.50	6.32	40.00	45.49	14.51	35.50	6.31
1.70	20.0	10.0	33.00	43.81	23.19	31.50	5.26	44.00	45.05	10.95	35.50	5.38
1.70	20.0	15.0	35.50	43.34	21.16	31.00	4.27	45.50	44.89	9.61	32.50	4.58
1.80	5.0	5.0	28.50	44.48	27.02	30.00	4.13	38.50	45.40	16.10	37.50	4.10
1.80	5.0	10.0	31.50	43.90	24.60	31.00	3.40	42.50	44.95	12.55	37.50	3.44
1.80	5.0	15.0	35.00	43.24	21.76	33.00	2.72	43.50	44.84	11.66	33.00	2.89
1.80	10.0	5.0	29.00	44.44	26.56	30.00	4.82	39.00	45.43	15.57	37.00	4.79
1.80	10.0	10.0	32.50	43.77	23.73	32.00	3.99	43.00	44.98	12.02	36.50	4.04
1.80	10.0	15.0	35.00	43.30	21.70	31.50	3.22	45.00	44.76	10.24	34.00	3.41
1.80	20.0	5.0	30.50	44.28	25.22	30.50	6.34	40.50	45.43	14.07	35.50	6.32
1.80	20.0	10.0	33.00	43.81	23.19	30.50	5.30	44.50	45.00	10.50	35.50	5.40
1.80	20.0	15.0	36.50	43.15	20.35	32.50	4.32	45.00	44.94	10.06	31.00	4.61
1.90	5.0	5.0	29.00	44.38	26.62	30.50	4.14	48.50	45.40	16.10	37.00	4.11
1.90	5.0	10.0	32.00	43.81	24.19	31.50	3.42	42.00	45.01	12.99	36.00	3.45
1.90	5.0	15.0	35.50	43.15	21.35	33.50	2.75	44.00	44.79	11.21	33.50	2.90

Table 4 (continued)

DATA FOR THE CONSTRUCTION OF ISOKINETIC SUCROSE AND GLYCEROL GRADIENTS FOR BECKMAN ROTOR SW 25.1

Particle density g/ml	Temperature °C	Top concentrations %(w/w)	Sucrose					Glycerol				
			C_h %(w/w)	Buffer 2x ml	H_2O ml	V_m ml	Velocity of 100 S particle at top speed ml/h	C_h %(w/w)	Buffer 2x ml	H_2O ml	V_m ml	Velocity of 100 S particle at top speed ml/h
1.90	10.0	5.0	29.50	44.34	26.16	30.50	4.83	39.50	45.37	15.13	37.00	4.80
1.90	10.0	10.0	32.50	43.77	23.73	31.50	4.01	43.00	44.98	12.02	36.00	4.06
1.90	10.0	15.0	35.00	43.30	21.70	31.00	3.25	45.00	44.76	10.24	33.50	3.43
1.90	20.0	5.0	30.50	44.28	25.22	30.00	6.36	40.50	45.43	14.07	35.00	6.34
1.90	20.0	10.0	33.50	43.71	22.79	31.00	5.33	44.50	45.00	10.50	35.00	5.42
1.90	20.0	15.0	36.50	43.15	20.35	32.00	4.36	45.50	44.89	9.61	31.50	4.64
2.00	5.0	5.0	29.00	44.38	26.62	30.00	4.15	38.00	45.35	15.65	37.50	4.11
2.00	5.0	10.0	32.00	43.81	24.19	31.00	3.44	42.50	44.95	12.55	36.50	3.46
2.00	5.0	15.0	35.50	43.15	21.35	33.00	2.77	44.00	44.79	11.21	33.00	2.92
2.00	10.0	5.0	29.50	44.34	26.16	30.00	4.84	39.50	45.37	15.13	36.50	4.80
2.00	10.0	10.0	33.00	43.68	23.32	32.00	4.03	43.00	44.98	12.02	35.50	4.07
2.00	10.0	15.0	35.50	43.20	21.30	31.50	3.27	45.00	44.76	10.24	33.00	3.45
2.00	20.0	5.0	30.50	44.28	25.22	29.50	6.37	41.00	45.38	13.62	35.50	6.34
2.00	20.0	10.0	33.50	43.71	22.79	30.50	5.35	45.00	44.94	10.06	35.50	5.44
2.00	20.0	15.0	37.00	43.06	19.94	32.50	4.39	45.50	44.89	9.61	31.00	4.66

Table 5

DATA FOR THE CONSTRUCTION OF ISOKINETIC SUCROSE AND GLYCEROL GRADIENTS FOR BECKMAN ROTOR SW 25.2

Particle density g/ml	Temperature °C	Top concentration %(w/w)	Sucrose					Glycerol				
			C_h %(w/w)	Buffer 2x ml	H_2O ml	V_m ml	Velocity of 100 S particle at top speed ml/h	C_h %(w/w)	Buffer 2x ml	H_2O ml	V_m ml	Velocity of 100 S particle at top speed ml/h
1.20	5.0	5.0	22.50	45.63	31.87	55.00	6.51	32.00	46.15	21.85	69.00	6.65
1.20	5.0	10.0	26.00	44.96	29.04	59.00	4.88	35.50	45.74	18.76	67.00	5.28
1.20	5.0	15.0	29.00	44.38	26.62	60.00	3.48	39.00	45.35	15.65	67.00	4.17
1.20	10.0	5.0	23.00	45.58	31.42	56.00	7.60	33.00	46.10	20.90	70.00	7.78
1.20	10.0	10.0	26.00	45.01	28.99	57.00	5.73	36.00	45.76	18.24	66.00	6.22
1.20	10.0	15.0	29.50	44.34	26.16	61.00	4.12	39.50	45.37	15.13	66.00	4.95
1.20	20.0	5.0	24.00	45.51	30.49	57.00	10.02	34.50	46.10	19.40	69.00	10.29
1.20	20.0	10.0	26.50	45.03	28.47	56.00	7.65	37.50	45.76	16.74	66.00	8.35
1.20	20.0	15.0	30.00	44.37	25.63	60.00	5.59	40.50	45.43	14.07	65.00	6.73
1.30	5.0	5.0	26.00	44.96	29.04	58.00	6.75	35.50	45.74	18.76	70.00	6.80
1.30	5.0	10.0	29.00	44.38	26.62	59.00	5.29	39.50	45.29	15.21	69.00	5.54
1.30	5.0	15.0	32.00	43.81	24.19	60.00	4.00	41.00	45.12	13.88	62.00	4.51
1.30	10.0	5.0	26.50	44.91	28.59	58.00	7.88	36.50	45.71	17.79	70.00	7.95
1.30	10.0	10.0	29.00	44.44	26.56	57.00	6.21	40.00	45.32	14.68	68.00	6.52
1.30	10.0	15.0	32.50	43.77	23.73	60.00	4.73	42.00	45.09	12.91	62.00	5.33
1.30	20.0	5.0	27.00	44.94	28.06	56.00	10.37	38.50	45.65	15.85	70.00	10.51
1.30	20.0	10.0	30.50	44.28	25.22	60.00	8.26	41.50	45.32	13.18	67.00	8.73
1.30	20.0	15.0	33.50	43.71	22.79	61.00	6.38	43.50	45.11	11.39	62.00	7.23
1.40	5.0	5.0	27.00	44.76	28.24	55.00	6.87	37.50	45.52	16.98	70.00	6.88
1.40	5.0	10.0	30.50	44.10	25.40	59.00	5.49	41.00	45.12	13.88	68.00	5.67
1.40	5.0	15.0	34.00	43.43	22.57	62.00	4.26	42.50	44.95	12.55	61.00	4.67
1.40	10.0	5.0	28.00	44.63	27.37	57.00	8.01	38.50	45.48	16.02	70.00	8.03
1.40	10.0	10.0	31.00	44.06	24.94	59.00	6.44	41.00	45.20	13.80	65.00	6.67
1.40	10.0	15.0	34.00	43.49	22.51	60.00	5.03	43.00	44.98	12.02	60.00	5.53
1.40	20.0	5.0	28.50	44.66	26.84	55.00	10.54	40.00	45.49	14.51	68.00	10.62
1.40	20.0	10.0	32.00	43.99	24.01	59.00	8.57	43.00	45.16	11.84	65.00	8.92
1.40	20.0	15.0	35.00	43.43	21.57	60.00	6.78	45.00	44.94	10.06	61.00	7.48
1.50	5.0	5.0	28.50	43.90	27.02	57.00	6.94	38.50	45.40	16.10	69.00	6.92
1.50	5.0	10.0	31.50	43.34	24.60	59.00	5.61	41.50	45.07	13.43	66.00	5.75
1.50	5.0	15.0	34.50	43.34	22.16	60.00	4.41	43.50	44.84	11.66	61.00	4.77
1.50	10.0	5.0	29.00	44.44	26.56	57.00	8.09	39.50	45.37	15.13	69.00	8.09
1.50	10.0	10.0	32.00	43.87	24.13	59.00	6.59	42.50	45.04	12.46	66.00	6.76

Table 5 (continued)

DATA FOR THE CONSTRUCTION OF ISOKINETIC SUCROSE AND GLYCEROL GRADIENTS FOR BECKMAN ROTOR SW 25.2

Particle density g/ml	Temperature °C	Top concentration %(w/w)	Sucrose					Glycerol				
			C_h %(w/w)	Buffer 2x ml	H_2O ml	V_m ml	Velocity of 100 S particle at top speed ml/h	C_h %(w/w)	Buffer 2x ml	H_2O ml	V_m ml	Velocity of 100 S particle at top speed ml/h
1.50	10.0	15.0	35.00	43.30	21.70	60.00	5.21	43.50	44.93	11.57	58.00	5.64
1.50	20.0	5.0	29.50	44.47	26.03	55.00	10.65	41.50	45.32	13.18	69.00	10.68
1.50	20.0	10.0	32.50	43.90	23.60	56.00	8.75	43.50	45.11	11.39	63.00	9.03
1.50	20.0	15.0	36.00	43.24	20.76	60.00	7.01	44.50	45.00	10.50	56.00	7.63
1.60	5.0	5.0	29.00	44.38	26.62	56.00	6.98	39.50	45.29	15.21	70.00	6.95
1.60	5.0	10.0	32.00	43.81	24.19	58.00	5.70	42.00	45.01	12.99	65.00	5.80
1.60	5.0	15.0	34.50	43.34	22.16	57.00	4.52	43.50	44.84	11.66	59.00	4.84
1.60	10.0	5.0	29.50	44.34	26.16	56.00	8.15	40.50	45.26	14.24	70.00	8.12
1.60	10.0	10.0	32.50	43.77	23.73	58.00	6.68	43.00	44.98	12.02	65.00	6.82
1.60	10.0	15.0	35.50	43.20	21.30	59.00	5.33	44.50	44.82	10.68	59.00	5.72
1.60	20.0	5.0	30.00	44.37	25.63	54.00	10.72	42.00	45.27	12.73	67.00	10.73
1.60	20.0	10.0	33.50	43.71	22.79	57.00	8.88	44.00	45.05	10.95	62.00	9.11
1.60	20.0	15.0	36.50	43.15	20.35	59.00	7.17	45.50	44.89	9.61	57.00	7.73
1.70	5.0	5.0	29.50	44.29	26.21	56.00	7.02	40.00	45.23	14.77	70.00	6.97
1.70	5.0	10.0	32.00	43.81	24.19	56.00	5.75	42.50	44.95	12.55	65.00	5.84
1.70	5.0	15.0	35.50	43.15	21.35	60.00	4.59	44.00	44.79	11.21	59.00	4.89
1.70	10.0	5.0	30.00	44.25	25.75	56.00	8.19	41.00	45.20	13.80	69.00	8.15
1.70	10.0	10.0	33.00	43.68	23.32	58.00	6.75	43.50	44.93	11.57	65.00	6.86
1.70	10.0	15.0	35.50	43.20	21.30	57.00	5.42	45.00	44.76	10.24	59.00	5.77
1.70	20.0	5.0	31.00	44.18	24.82	55.00	10.77	42.50	45.21	12.29	67.00	10.76
1.70	20.0	10.0	33.50	43.71	22.79	55.00	8.97	44.50	45.00	10.50	62.00	9.17
1.70	20.0	15.0	37.00	43.06	19.94	59.00	7.29	46.00	44.83	9.17	57.00	7.81
1.80	5.0	5.0	29.50	44.29	26.21	55.00	7.04	40.50	45.18	14.32	70.00	6.99
1.80	5.0	10.0	32.50	43.71	23.79	57.00	5.80	43.00	44.90	12.10	65.00	5.87
1.80	5.0	15.0	35.50	43.15	21.35	58.00	4.64	44.00	44.79	11.21	58.00	4.92
1.80	10.0	5.0	30.50	44.15	25.35	56.00	8.22	41.50	45.15	13.35	69.00	8.16
1.80	10.0	10.0	33.00	43.68	23.32	56.00	6.80	44.00	44.87	11.13	65.00	6.89
1.80	10.0	15.0	36.00	43.11	20.89	58.00	5.48	45.00	44.76	10.24	58.00	5.82
1.80	20.0	5.0	31.00	44.18	24.82	54.00	10.81	43.00	45.16	11.84	67.00	10.78
1.80	20.0	10.0	34.00	43.62	22.38	56.00	9.03	45.00	44.94	10.06	62.00	9.21
1.80	20.0	15.0	37.00	43.06	19.94	57.00	7.37	46.50	44.78	8.72	57.00	7.86
1.90	5.0	5.0	29.50	44.29	26.21	54.00	7.06	41.00	45.12	13.88	70.00	7.00

Table 5 (continued)

DATA FOR THE CONSTRUCTION OF ISOKINETIC SUCROSE AND GLYCEROL GRADIENTS FOR BECKMAN ROTOR SW 25.2

Particle density g/ml	Temperature °C	Top concentration %(w/w)	Sucrose					Glycerol				
			C_h %(w/w)	Buffer 2x ml	H_2O ml	V_m ml	Velocity of 100 S particle at top speed ml/h	C_h %(w/w)	Buffer 2x ml	H_2O ml	V_m ml	Velocity of 100 S particle at top speed ml/h
1.90	5.0	10.0	33.00	43.62	23.38	58.00	5.83	43.00	44.90	12.10	64.00	5.89
1.90	5.0	15.0	35.50	43.15	21.35	57.00	4.69	44.00	44.79	11.21	57.00	4.95
1.90	10.0	5.0	30.50	44.15	25.35	55.00	8.24	41.50	45.15	13.35	68.00	8.18
1.90	10.0	10.0	33.50	43.58	22.92	57.00	6.84	44.00	44.87	11.13	64.00	6.92
1.90	10.0	15.0	36.50	43.02	20.48	59.00	5.53	45.00	44.76	10.24	57.00	5.85
1.90	20.0	5.0	31.50	44.09	24.41	54.00	10.83	43.00	45.16	11.84	66.00	10.80
1.90	20.0	10.0	34.50	43.52	21.98	56.00	9.08	45.00	44.94	10.06	61.00	9.24
1.90	20.0	15.0	37.00	43.06	19.94	56.00	7.44	46.50	44.78	8.72	56.00	7.90
2.00	5.0	5.0	29.50	44.29	26.21	53.00	7.08	41.00	45.12	13.88	69.00	7.01
2.00	5.0	10.0	33.00	43.62	23.38	57.00	5.86	43.50	44.84	11.66	65.00	5.91
2.00	5.0	15.0	35.50	43.15	21.35	56.00	4.72	44.50	44.73	10.77	58.00	4.97
2.00	10.0	5.0	30.50	44.15	25.35	54.00	8.26	42.00	45.09	12.91	69.00	8.19
2.00	10.0	10.0	33.50	43.58	22.92	56.00	6.87	43.50	44.93	11.57	62.00	6.94
2.00	10.0	15.0	36.50	43.02	20.48	58.00	5.57	45.50	44.71	9.79	58.00	5.87
2.00	20.0	5.0	32.00	43.99	24.01	55.00	10.86	43.50	45.11	11.39	66.00	10.81
2.00	20.0	10.0	34.50	43.52	21.98	55.00	9.12	45.00	44.94	10.06	60.00	9.27
2.00	20.0	15.0	37.00	43.06	19.94	55.00	7.49	47.00	44.73	8.27	57.00	7.94

Table 6

DATA FOR THE CONSTRUCTION OF ISOKINETIC SUCROSE AND GLYCEROL GRADIENTS FOR BECKMAN ROTOR SW 27

Particle density g/ml	Temperature °C	Top concentration %(w/w)	Sucrose					Glycerol				
			C_h %(w/w)	Buffer 2x ml	H_2O ml	V_m ml	Velocity of 100 S particle at top speed ml/h	C_h %(w/w)	Buffer 2x ml	H_2O ml	V_m ml	Velocity of 100 S particle at top speed ml/h
1.20	5.0	5.0	22.50	45.63	31.87	41.00	5.65	30.00	46.38	23.62	46.50	5.77
1.20	5.0	10.0	25.50	45.05	29.45	42.00	4.23	34.00	45.92	20.08	46.00	4.59
1.20	5.0	15.0	28.00	44.57	27.43	40.00	3.02	37.00	45.57	17.43	44.00	3.62
1.20	10.0	5.0	22.00	45.77	32.23	38.00	6.60	31.00	46.33	22.67	46.50	6.75
1.20	10.0	10.0	25.50	45.10	29.40	40.50	4.97	35.00	45.88	19.12	46.50	5.40
1.20	10.0	15.0	28.50	44.53	26.97	41.00	3.58	37.50	45.59	16.91	43.50	4.29
1.20	20.0	5.0	23.00	45.70	31.30	39.00	8.69	32.50	46.32	21.18	46.50	8.93
1.20	20.0	10.0	26.00	45.13	28.87	40.00	6.64	36.50	45.87	17.63	46.50	7.24
1.20	20.0	15.0	29.00	44.56	26.44	40.50	4.85	38.50	45.65	15.85	42.50	5.84
1.30	5.0	5.0	25.00	45.15	29.85	40.00	5.85	33.00	46.03	20.97	46.00	5.90
1.30	5.0	10.0	28.00	44.57	27.43	40.50	4.59	37.50	45.52	16.98	46.50	4.81
1.30	5.0	15.0	31.50	43.90	24.60	42.50	3.47	40.00	45.23	14.77	43.50	3.91
1.30	10.0	5.0	25.50	45.10	29.40	40.00	6.83	34.00	45.99	20.01	46.50	6.90
1.30	10.0	10.0	28.50	44.53	26.97	41.00	5.38	38.00	45.54	16.46	46.00	5.65
1.30	10.0	15.0	31.50	43.96	24.54	41.00	4.10	41.00	45.20	13.80	44.00	4.63
1.30	20.0	5.0	26.00	45.13	28.87	39.00	8.99	36.00	45.93	18.07	46.50	9.12
1.30	20.0	10.0	29.50	44.47	26.03	41.50	7.17	40.00	45.49	14.51	46.50	7.57
1.30	20.0	15.0	32.50	43.90	23.60	41.50	5.54	42.00	45.27	12.73	42.50	6.28
1.40	5.0	5.0	26.50	44.86	28.64	39.50	5.96	35.00	45.80	19.20	46.50	5.97
1.40	5.0	10.0	29.50	44.29	26.21	40.50	4.76	39.00	45.35	15.65	46.00	4.92
1.40	5.0	15.0	33.00	43.62	23.38	42.50	3.69	41.50	45.07	13.43	43.00	4.05
1.40	10.0	5.0	26.50	44.91	28.59	38.50	6.95	36.00	45.76	18.24	46.50	6.97
1.40	10.0	10.0	30.00	44.25	25.75	40.50	5.59	40.00	45.32	14.68	46.00	5.78
1.40	10.0	15.0	33.00	43.68	23.32	41.00	4.36	43.00	44.98	12.02	44.50	4.79
1.40	20.0	5.0	27.50	44.85	27.65	38.50	9.15	38.00	45.71	16.29	46.50	9.21
1.40	20.0	10.0	31.00	44.18	24.82	40.50	7.43	42.00	45.27	12.73	46.50	7.74
1.40	20.0	15.0	34.00	43.62	22.38	41.00	5.88	43.50	45.11	11.39	42.00	6.49
1.50	5.0	5.0	27.50	44.67	27.83	39.50	6.02	36.00	45.69	18.31	46.50	6.01
1.50	5.0	10.0	30.50	44.10	25.40	40.50	4.87	40.00	45.23	14.77	45.50	4.99
1.50	5.0	15.0	33.50	43.52	22.98	41.00	3.83	42.50	44.95	12.55	43.00	4.14
1.50	10.0	5.0	28.00	44.63	27.37	39.50	7.02	37.00	45.65	17.35	46.50	7.02
1.50	10.0	10.0	31.00	44.06	24.94	40.50	5.71	41.00	45.20	13.80	46.00	5.86

Table 6 (continued)

DATA FOR THE CONSTRUCTION OF ISOKINETIC SUCROSE AND GLYCEROL GRADIENTS FOR BECKMAN ROTOR SW 27

Particle density g/ml	Temperature °C	Top concentration %(w/w)	Sucrose					Glycerol				
			C_h %(w/w)	Buffer 2x ml	H_2O ml	V_m ml	Velocity of 100 S particle at top speed ml/h	C_h %(w/w)	Buffer 2x ml	H_2O ml	V_m ml	Velocity of 100 S particle at top speed ml/h
1.50	10.0	15.0	34.00	43.49	22.51	41.00	4.52	43.00	44.98	12.02	42.00	4.89
1.50	20.0	5.0	29.00	44.56	26.44	39.50	9.24	39.00	45.60	15.40	46.00	9.27
1.50	20.0	10.0	32.00	43.99	24.01	40.50	7.60	43.00	45.16	11.84	46.00	7.84
1.50	20.0	15.0	34.50	43.52	21.98	39.50	6.09	44.00	45.05	10.95	40.50	6.62
1.60	5.0	5.0	28.00	44.57	27.43	39.00	6.06	37.00	45.57	17.43	46.50	6.03
1.60	5.0	10.0	31.00	44.00	25.00	40.00	4.94	41.00	45.12	13.88	46.50	5.03
1.60	5.0	15.0	34.00	43.43	22.57	41.00	3.92	42.50	44.95	12.55	41.50	4.20
1.60	10.0	5.0	28.50	44.53	26.97	39.00	7.07	38.00	45.54	16.46	46.50	7.05
1.60	10.0	10.0	31.50	43.96	24.54	40.00	5.80	42.00	45.09	12.91	46.50	5.91
1.60	10.0	15.0	35.00	43.30	21.70	42.50	4.63	44.50	44.82	10.68	44.00	4.96
1.60	20.0	5.0	29.00	44.56	26.44	37.50	9.30	39.50	45.54	14.96	45.50	9.31
1.60	20.0	10.0	32.50	43.90	23.60	40.00	7.70	43.50	45.11	11.39	45.00	7.91
1.60	20.0	15.0	35.50	43.34	21.16	40.50	6.22	45.00	44.94	10.06	41.00	6.71
1.70	5.0	5.0	28.50	44.48	27.02	39.00	6.09	37.50	45.52	16.98	46.50	6.05
1.70	5.0	10.0	31.00	44.00	25.00	39.00	4.99	41.50	45.07	13.43	46.00	5.07
1.70	5.0	15.0	34.00	43.43	22.57	39.50	3.98	43.00	44.90	12.10	41.50	4.24
1.70	10.0	5.0	29.00	44.44	26.56	39.00	7.10	38.50	45.48	16.02	46.50	7.07
1.70	10.0	10.0	32.00	43.87	24.13	40.00	5.85	42.50	45.04	12.46	46.00	5.95
1.70	10.0	15.0	35.00	43.30	21.70	41.00	4.70	44.50	44.82	10.68	43.00	5.01
1.70	20.0	5.0	30.00	44.37	25.63	38.50	9.34	39.50	45.54	14.96	44.50	9.33
1.70	20.0	10.0	33.00	43.81	23.19	40.00	7.78	44.00	45.05	10.95	45.00	7.95
1.70	20.0	15.0	36.00	43.24	20.76	41.00	6.32	45.00	44.94	10.06	40.00	6.77
1.80	5.0	5.0	29.00	44.38	26.62	39.50	6.11	38.00	45.46	16.54	46.50	6.06
1.80	5.0	10.0	31.50	43.90	24.60	39.50	5.03	42.00	45.01	12.99	46.50	5.09
1.80	5.0	15.0	34.50	43.34	22.16	40.00	4.03	43.50	44.84	11.66	42.00	4.27
1.80	10.0	5.0	29.00	44.44	26.56	38.00	7.13	38.50	45.48	16.02	46.00	7.08
1.80	10.0	10.0	32.50	43.77	23.73	40.50	5.90	43.00	44.98	12.02	46.50	5.98
1.80	10.0	15.0	35.50	43.20	21.30	41.50	4.76	44.50	44.82	10.68	42.00	5.05
1.80	20.0	5.0	30.00	44.37	25.63	37.50	9.38	40.00	45.49	14.51	44.50	9.35
1.80	20.0	10.0	33.00	43.81	23.19	39.00	7.84	44.00	45.05	10.95	44.00	7.99
1.80	20.0	15.0	36.00	43.24	20.76	39.50	6.40	45.50	44.89	9.61	40.50	6.82
1.90	5.0	5.0	28.50	44.48	27.02	37.50	6.13	38.00	45.46	16.54	46.00	6.08

Table 6 (continued)

DATA FOR THE CONSTRUCTION OF ISOKINETIC SUCROSE AND GLYCEROL GRADIENTS FOR BECKMAN ROTOR SW 27

Particle density g/ml	Temperature °C	Top concentration %(w/w)	Sucrose					Glycerol				
			C_h %(w/w)	Buffer 2x ml	H_2O ml	V_m ml	Velocity of 100 S particle at top speed ml/h	C_h %(w/w)	Buffer 2x ml	H_2O ml	V_m ml	Velocity of 100 S particle at top speed ml/h
1.90	5.0	10.0	32.00	43.81	24.19	40.00	5.06	42.00	45.01	12.99	45.50	5.11
1.90	5.0	15.0	35.00	43.24	21.76	41.00	4.07	44.00	44.79	11.21	42.50	4.30
1.90	10.0	5.0	29.50	44.34	26.16	38.50	7.15	39.00	45.43	15.57	46.00	7.10
1.90	10.0	10.0	32.00	43.87	24.13	38.50	5.93	43.00	44.98	12.02	45.50	6.00
1.90	10.0	15.0	35.00	43.30	21.70	39.00	4.80	44.50	44.82	10.68	41.50	5.07
1.90	20.0	5.0	30.50	44.28	25.22	38.00	9.40	40.00	45.49	14.51	43.50	9.37
1.90	20.0	10.0	33.00	43.81	23.19	38.00	7.88	44.50	45.00	10.50	44.50	8.02
1.90	20.0	15.0	36.00	43.24	20.76	39.00	6.45	45.50	44.89	9.61	40.00	6.86
2.00	5.0	5.0	29.00	44.38	26.62	38.00	6.14	38.50	45.40	16.10	46.50	6.08
2.00	5.0	10.0	32.00	43.81	24.19	39.50	5.08	42.50	44.95	12.55	46.00	5.12
2.00	5.0	15.0	35.00	43.24	21.76	40.50	4.10	44.00	44.79	11.21	42.00	4.31
2.00	10.0	5.0	30.00	44.25	25.75	39.00	7.16	39.00	45.43	15.57	45.50	7.11
2.00	10.0	10.0	32.50	43.77	23.73	39.00	5.96	43.00	44.98	12.02	45.00	6.02
2.00	10.0	15.0	35.50	43.20	21.30	40.00	4.83	44.50	44.82	10.68	41.00	5.10
2.00	20.0	5.0	30.50	44.28	25.22	37.50	9.42	40.50	45.43	14.07	44.00	9.38
2.00	20.0	10.0	34.00	43.62	22.38	40.00	7.92	44.50	45.00	10.50	44.00	8.04
2.00	20.0	15.0	36.50	43.15	20.35	40.00	6.50	45.50	44.89	9.61	39.50	6.89

Table 7

DATA FOR THE CONSTRUCTION OF ISOKINETIC SUCROSE AND GLYCEROL GRADIENTS FOR BECKMAN ROTOR SW 27.1

Particle density g/ml	Temperature °C	Top concentration %(w/w)	Sucrose					Glycerol				
			C_h %(w/w)	Buffer 2x ml	H_2O ml	V_m ml	Velocity of 100 S particle at top speed ml/h	C_h %(w/w)	Buffer 2x ml	H_2O ml	V_m ml	Velocity of 100 S particle at top speed ml/h
1.20	5.0	5.0	23.00	45.53	31.47	14.75	1.94	32.00	46.15	21.85	17.50	1.98
1.20	5.0	10.0	26.00	44.96	29.04	15.00	1.45	35.50	45.74	18.76	17.25	1.58
1.20	5.0	15.0	29.50	44.29	26.21	16.25	1.04	39.50	45.29	15.21	17.75	1.24
1.20	10.0	5.0	23.00	45.58	31.42	14.25	2.27	33.00	46.10	20.90	17.75	2.32
1.20	10.0	10.0	26.00	45.01	28.99	14.50	1.71	36.50	45.71	17.79	17.50	1.85
1.20	10.0	15.0	29.50	44.34	26.16	15.75	1.23	40.00	45.32	14.68	17.50	1.47
1.20	20.0	5.0	24.00	45.51	30.49	14.50	2.99	34.00	46.15	19.85	17.25	3.07
1.20	20.0	10.0	27.00	44.94	28.06	15.00	2.28	37.00	45.82	17.18	16.50	2.49
1.20	20.0	15.0	30.00	44.37	25.63	15.50	1.67	40.50	45.43	14.07	16.50	2.01
1.30	5.0	5.0	26.00	44.96	29.04	14.75	2.01	35.50	45.74	18.76	17.75	2.03
1.30	5.0	10.0	28.50	44.48	27.02	14.50	1.58	39.00	45.35	15.65	17.25	1.65
1.30	5.0	15.0	32.00	43.81	24.19	15.25	1.19	41.50	45.07	13.43	16.25	1.34
1.30	10.0	5.0	26.00	45.01	28.99	14.25	2.35	36.50	45.71	17.79	17.75	2.37
1.30	10.0	10.0	29.50	44.34	26.16	15.25	1.85	40.50	45.26	14.24	17.75	1.94
1.30	10.0	15.0	32.50	43.77	23.73	15.25	1.41	42.00	45.09	12.91	16.00	1.59
1.30	20.0	5.0	28.00	44.75	27.25	15.50	3.09	38.50	45.65	15.85	17.75	3.13
1.30	20.0	10.0	30.00	44.37	25.63	14.75	2.46	40.50	45.43	14.07	16.25	2.60
1.30	20.0	15.0	33.00	43.81	23.19	15.00	1.90	43.00	45.16	11.84	15.50	2.16
1.40	5.0	5.0	28.00	44.57	27.43	15.25	2.05	37.50	45.52	16.98	17.75	2.05
1.40	5.0	10.0	30.50	44.10	25.40	15.00	1.64	40.50	45.18	14.32	17.00	1.69
1.40	5.0	15.0	33.50	43.52	22.98	15.25	1.27	43.00	44.90	12.10	16.00	1.39
1.40	10.0	5.0	28.00	44.63	27.37	14.50	2.39	38.50	45.48	16.02	17.75	2.39
1.40	10.0	10.0	31.00	44.06	24.94	15.00	1.92	41.50	45.15	13.35	17.00	1.99
1.40	10.0	15.0	34.00	43.49	22.51	15.25	1.50	43.50	44.93	11.57	15.75	1.65
1.40	20.0	5.0	29.00	44.56	26.44	14.50	3.14	40.50	45.43	14.07	17.75	3.16
1.40	20.0	10.0	32.00	43.99	24.01	15.00	2.55	42.50	45.21	12.29	16.25	2.66
1.40	20.0	15.0	35.00	43.43	21.57	15.25	2.02	45.00	44.94	10.06	15.50	2.23
1.50	5.0	5.0	28.00	44.57	27.43	14.00	2.07	38.50	45.40	16.10	17.75	2.06
1.50	5.0	10.0	31.50	43.90	24.60	15.00	1.67	42.00	45.01	12.99	17.25	1.71
1.50	5.0	15.0	34.50	43.34	22.16	15.25	1.32	43.50	44.84	11.66	15.50	1.42
1.50	10.0	5.0	29.00	44.44	26.56	14.50	2.41	39.50	45.37	15.13	17.75	2.41

Table 7 (continued)

DATA FOR THE CONSTRUCTION OF ISOKINETIC SUCROSE AND GLYCEROL GRADIENTS FOR BECKMAN ROTOR SW 27.1

Particle density g/ml	Temperature °C	Top concentration %(w/w)	Sucrose					Glycerol				
			C_h %(w/w)	Buffer 2x ml	H_2O ml	V_m ml	Velocity of 100 S particle at top speed ml/h	C_h %(w/w)	Buffer 2x ml	H_2O ml	V_m ml	Velocity of 100 S particle at top speed ml/h
1.50	10.0	10.0	32.00	43.87	24.13	15.00	1.96	43.00	44.98	12.02	17.25	2.01
1.50	10.0	15.0	35.00	43.30	21.70	15.25	1.55	44.00	44.87	11.13	15.25	1.68
1.50	20.0	5.0	30.00	44.37	25.63	14.50	3.17	41.50	45.32	13.18	17.50	3.18
1.50	20.0	10.0	33.00	43.81	23.19	15.00	2.61	44.00	45.05	10.95	16.50	2.69
1.50	20.0	15.0	36.00	43.24	20.76	15.25	2.09	45.00	44.94	10.06	14.75	2.28
1.60	5.0	5.0	29.00	44.38	26.62	14.25	2.08	39.50	45.29	15.21	17.75	2.07
1.60	5.0	10.0	32.00	43.81	24.19	14.75	1.70	42.50	44.95	12.55	17.00	1.73
1.60	5.0	15.0	35.50	43.15	21.35	15.75	1.35	44.00	44.79	11.21	15.50	1.44
1.60	10.0	5.0	29.50	44.34	26.16	14.25	2.43	40.50	45.26	14.24	17.75	2.42
1.60	10.0	10.0	32.50	43.77	23.73	14.75	1.99	43.50	44.93	11.57	17.00	2.03
1.60	10.0	15.0	35.00	43.30	21.70	14.50	1.59	45.00	44.76	10.24	15.50	1.70
1.60	20.0	5.0	30.50	44.28	25.22	14.25	3.19	42.50	45.21	12.29	17.50	3.20
1.60	20.0	10.0	34.00	43.62	22.38	15.25	2.65	44.50	45.00	10.50	16.25	2.72
1.60	20.0	15.0	36.50	43.15	20.35	15.00	2.14	45.50	44.89	9.61	14.50	2.31
1.70	5.0	5.0	29.50	44.29	26.21	14.25	2.09	40.00	45.23	14.77	17.75	2.08
1.70	5.0	10.0	33.00	43.62	23.38	15.50	1.72	43.00	44.90	12.10	17.00	1.74
1.70	5.0	15.0	35.50	43.15	21.35	15.25	1.37	44.00	44.79	11.21	15.00	1.46
1.70	10.0	5.0	30.00	44.25	25.75	14.25	2.44	41.00	45.20	13.80	17.75	2.43
1.70	10.0	10.0	33.00	43.68	23.32	14.75	2.01	44.00	44.87	11.13	17.00	2.04
1.70	10.0	15.0	36.00	43.11	20.89	15.25	1.61	45.50	44.71	9.79	15.50	1.72
1.70	20.0	5.0	31.50	44.09	24.41	14.50	3.21	43.00	45.16	11.84	17.50	3.21
1.70	20.0	10.0	34.00	43.62	22.38	14.50	2.67	44.50	45.00	10.50	15.75	2.73
1.70	20.0	15.0	37.00	43.06	19.94	15.00	2.17	46.00	44.83	9.17	14.50	2.33
1.80	5.0	5.0	30.00	44.19	25.81	14.50	2.10	40.50	45.18	14.32	17.75	2.08
1.80	5.0	10.0	32.50	43.71	23.79	14.50	1.73	42.50	44.95	12.55	16.25	1.75
1.80	5.0	15.0	36.00	43.05	20.95	15.50	1.38	44.50	44.73	10.77	15.25	1.47
1.80	10.0	5.0	30.50	44.15	25.35	14.25	2.45	41.50	45.15	13.35	17.75	2.43
1.80	10.0	10.0	33.50	43.58	22.92	15.00	2.03	43.50	44.93	11.57	16.25	2.05
1.80	10.0	15.0	36.00	43.11	20.89	14.75	1.63	45.00	44.76	10.24	14.75	1.73
1.80	20.0	5.0	31.00	44.18	24.82	13.75	3.22	43.50	45.11	11.39	17.50	3.21
1.80	20.0	10.0	34.50	43.52	21.98	14.75	2.69	44.50	45.00	10.50	15.50	2.74

Table 7 (continued)

DATA FOR THE CONSTRUCTION OF ISOKINETIC SUCROSE AND GLYCEROL GRADIENTS FOR BECKMAN ROTOR SW 27.1

Particle density g/ml	Temperature °C	Top concentration %(w/w)	Sucrose					Glycerol				
			C_h %(w/w)	Buffer 2x ml	H_2O ml	V_m ml	Velocity of 100 S particle at top speed ml/h	C_h %(w/w)	Buffer 2x ml	H_2O ml	V_m ml	Velocity of 100 S particle at top speed ml/h
1.80	20.0	15.0	37.50	42.96	19.54	15.25	2.20	46.00	44.83	9.17	14.25	2.34
1.90	5.0	5.0	30.00	44.19	25.81	14.25	2.11	40.50	45.18	14.32	17.50	2.09
1.90	5.0	10.0	33.00	43.62	23.38	14.75	1.74	42.50	44.95	12.55	16.00	1.76
1.90	5.0	15.0	35.50	43.15	21.35	14.50	1.40	44.50	44.73	10.77	15.00	1.48
1.90	10.0	5.0	30.50	44.15	25.35	14.00	2.46	42.00	45.09	12.91	17.75	2.44
1.90	10.0	10.0	33.00	43.68	23.32	14.00	2.04	43.50	44.93	11.57	16.00	2.06
1.90	10.0	15.0	36.50	43.02	20.48	15.00	1.65	45.50	44.71	9.79	15.00	1.74
1.90	20.0	5.0	32.00	43.99	24.01	14.25	3.23	43.50	45.11	11.39	17.25	3.22
1.90	20.0	10.0	34.50	43.52	21.98	14.50	2.71	44.50	45.00	10.50	15.25	2.75
1.90	20.0	15.0	38.00	42.87	19.13	15.50	2.22	47.00	44.73	8.27	14.75	2.36
2.00	5.0	5.0	30.00	44.19	25.81	14.00	2.11	41.00	45.12	13.88	17.75	2.09
2.00	5.0	10.0	33.00	43.62	23.38	14.50	1.75	43.00	44.90	12.10	16.25	1.76
2.00	5.0	15.0	36.00	43.05	20.95	15.00	1.41	44.50	44.73	10.77	14.75	1.48
2.00	10.0	5.0	31.00	44.06	24.94	14.25	2.46	42.00	45.09	12.91	17.50	2.44
2.00	10.0	10.0	33.50	43.58	22.92	14.25	2.05	44.50	44.82	10.68	16.50	2.07
2.00	10.0	15.0	36.50	43.02	20.48	14.75	1.66	45.50	44.71	9.79	14.75	1.75
2.00	20.0	5.0	32.00	43.99	24.01	14.00	3.24	44.00	45.05	10.95	17.25	3.22
2.00	20.0	10.0	34.50	43.52	21.98	14.25	2.72	45.50	44.89	9.61	15.75	2.76
2.00	20.0	15.0	38.00	42.87	19.13	15.25	2.23	46.50	44.78	8.72	14.25	2.37

Table 8

DATA FOR THE CONSTRUCTION OF ISOKINETIC SUCROSE AND GLYCEROL GRADIENTS FOR BECKMAN ROTOR SW 36

Particle density g/ml	Temperature °C	Top concentration %(w/w)	Sucrose					Glycerol				
			C_h %(w/w)	Buffer 2x ml	H_2O ml	V_m ml	Velocity of 100 S particle at top speed ml/h	C_h %(w/w)	Buffer 2x ml	H_2O ml	V_m ml	Velocity of 100 S particle at top speed ml/h
1.20	5.0	5.0	23.00	45.53	31.47	12.75	2.98	32.00	46.15	21.85	15.25	3.04
1.20	5.0	10.0	25.00	45.15	29.85	11.75	2.23	34.50	45.86	19.64	14.00	2.42
1.20	5.0	15.0	28.50	44.48	27.02	12.50	1.59	38.00	45.46	16.54	14.00	1.91
1.20	10.0	5.0	23.00	45.58	31.42	12.25	3.48	33.00	46.10	20.90	15.50	3.56
1.20	10.0	10.0	25.50	45.10	29.40	12.00	2.62	35.50	45.82	18.68	14.25	2.85
1.20	10.0	15.0	29.00	44.44	26.56	12.75	1.89	38.50	45.48	16.02	13.75	2.26
1.20	20.0	5.0	23.50	45.60	30.90	12.00	4.59	34.50	46.10	19.40	15.25	4.71
1.20	20.0	10.0	26.00	45.13	28.87	11.75	3.50	37.00	45.82	17.18	14.25	3.82
1.20	20.0	15.0	29.50	44.47	26.03	12.50	2.56	40.00	45.49	14.51	13.75	3.08
1.30	5.0	5.0	25.50	45.05	29.45	12.25	3.09	35.00	45.80	19.20	15.00	3.11
1.30	5.0	10.0	28.50	44.48	27.02	12.50	2.42	38.50	45.40	16.10	14.50	2.54
1.30	5.0	15.0	32.00	43.81	24.19	13.25	1.83	40.50	45.18	14.32	13.25	2.06
1.30	10.0	5.0	26.00	45.01	28.99	12.25	3.61	36.50	45.71	17.79	15.50	3.64
1.30	10.0	10.0	28.50	44.53	26.97	12.00	2.84	40.00	45.32	14.68	15.00	2.98
1.30	10.0	15.0	32.00	43.87	24.13	12.75	2.16	42.00	45.09	12.91	13.75	2.44
1.30	20.0	5.0	26.50	45.03	28.47	12.00	4.75	37.50	45.76	16.74	14.75	4.81
1.30	20.0	10.0	29.50	44.47	26.03	12.25	3.78	40.00	45.49	14.51	13.75	4.00
1.30	20.0	15.0	33.00	43.81	23.19	13.00	2.92	42.50	45.21	12.29	13.00	3.31
1.40	5.0	5.0	27.00	44.76	28.24	12.25	3.14	37.50	45.52	16.98	15.50	3.15
1.40	5.0	10.0	30.00	44.19	25.81	12.50	2.51	40.00	45.23	14.77	14.25	2.60
1.40	5.0	15.0	34.00	43.43	22.57	13.75	1.95	42.50	44.95	12.55	13.50	2.14
1.40	10.0	5.0	27.50	44.72	27.78	12.25	3.67	38.00	45.54	16.46	15.25	3.68
1.40	10.0	10.0	30.50	44.15	25.35	12.50	2.95	40.50	45.26	14.24	14.00	3.05
1.40	10.0	15.0	34.00	43.49	22.51	13.25	2.30	43.00	44.98	12.02	13.25	2.53
1.40	20.0	5.0	28.50	44.66	26.84	12.25	4.83	39.50	45.54	14.96	14.75	4.86
1.40	20.0	10.0	31.50	44.09	24.41	12.50	3.92	42.00	45.27	12.73	13.75	4.08
1.40	20.0	15.0	34.50	43.52	21.98	12.75	3.10	44.00	45.05	10.95	12.75	3.43
1.50	5.0	5.0	27.50	44.67	27.83	11.75	3.18	38.50	45.40	16.10	15.25	3.17
1.50	5.0	10.0	31.00	44.00	25.00	12.50	2.57	41.00	45.12	13.88	14.25	2.63
1.50	5.0	15.0	34.00	43.43	22.57	12.75	2.02	42.50	44.95	12.55	12.75	2.18
1.50	10.0	5.0	28.00	44.63	27.37	11.75	3.71	39.00	45.43	15.57	15.00	3.70

Table 8 (continued)

DATA FOR THE CONSTRUCTION OF ISOKINETIC SUCROSE AND GLYCEROL GRADIENTS FOR BECKMAN ROTOR SW 36

Particle density g/ml	Temperature °C	Top concentration %(w/w)	Sucrose					Glycerol				
			C_h %(w/w)	Buffer 2x ml	H_2O ml	V_m ml	Velocity of 100 S particle at top speed ml/h	C_h %(w/w)	Buffer 2x ml	H_2O ml	V_m ml	Velocity of 100 S particle at top speed ml/h
1.50	10.0	10.0	31.50	43.96	24.54	12.50	3.01	42.00	45.09	12.91	14.25	3.09
1.50	10.0	15.0	34.50	43.39	22.11	12.75	2.39	43.00	44.98	12.02	12.50	2.58
1.50	20.0	5.0	29.00	44.56	26.44	11.75	4.87	40.50	45.43	14.07	14.50	4.89
1.50	20.0	10.0	32.00	43.99	24.01	12.00	4.01	44.00	45.05	10.95	14.25	4.14
1.50	20.0	15.0	35.50	43.34	21.16	12.75	3.21	45.00	44.94	10.06	12.75	3.49
1.60	5.0	5.0	28.50	44.48	27.02	12.00	3.20	39.00	45.35	15.65	15.25	3.18
1.60	5.0	10.0	32.00	43.81	24.19	12.75	2.61	42.50	44.95	12.55	14.75	2.66
1.60	5.0	15.0	34.50	43.34	22.16	12.50	2.07	43.50	44.84	11.66	13.00	2.22
1.60	10.0	5.0	29.00	44.44	26.56	12.00	3.73	39.50	45.37	15.13	14.75	3.72
1.60	10.0	10.0	32.00	43.87	24.13	12.25	3.06	42.50	45.04	12.46	14.00	3.12
1.60	10.0	15.0	35.00	43.30	21.70	12.50	2.44	44.50	44.82	10.68	13.00	2.62
1.60	20.0	5.0	29.50	44.47	26.03	11.50	4.91	41.00	45.38	13.62	14.25	4.91
1.60	20.0	10.0	33.00	43.81	23.19	12.25	4.06	44.00	45.05	10.95	13.75	4.17
1.60	20.0	15.0	36.00	43.24	20.76	12.50	3.28	45.00	44.94	10.06	12.25	3.54
1.70	5.0	5.0	29.00	44.38	26.62	12.00	3.21	39.00	45.35	15.65	14.75	3.19
1.70	5.0	10.0	31.50	43.90	24.60	12.00	2.63	42.00	45.01	12.99	14.00	2.67
1.70	5.0	15.0	35.00	43.24	21.76	12.75	2.10	44.00	44.79	11.21	13.00	2.24
1.70	10.0	5.0	29.50	44.34	26.16	12.00	3.75	40.00	45.32	14.68	14.75	3.73
1.70	10.0	10.0	32.50	43.77	23.73	12.25	3.09	43.00	44.98	12.02	14.00	3.14
1.70	10.0	15.0	35.50	43.20	21.30	12.50	2.48	44.50	44.82	10.68	12.75	2.64
1.70	20.0	5.0	31.00	44.18	24.82	12.25	4.93	41.00	45.38	13.62	14.00	4.92
1.70	20.0	10.0	33.50	43.71	22.79	12.25	4.10	45.00	44.94	10.06	14.00	4.20
1.70	20.0	15.0	36.50	43.15	20.35	12.50	3.34	45.50	44.89	9.61	12.25	3.57
1.80	5.0	5.0	29.00	44.38	26.62	11.75	3.22	39.50	45.29	15.21	14.75	3.20
1.80	5.0	10.0	32.50	43.71	23.79	12.50	2.65	42.00	45.01	12.99	13.75	2.69
1.80	5.0	15.0	35.00	43.24	21.76	12.25	2.13	44.00	44.79	11.21	12.75	2.25
1.80	10.0	5.0	30.00	44.25	25.75	12.00	3.76	40.00	45.32	14.68	14.50	3.74
1.80	10.0	10.0	32.50	43.77	23.73	12.00	3.11	43.00	44.98	12.02	13.75	3.15
1.80	10.0	15.0	35.50	43.20	21.30	12.25	2.51	44.50	44.82	10.68	13.75	2.66
1.80	20.0	5.0	30.50	44.28	25.22	11.50	4.95	41.50	45.32	13.18	12.50	4.94
1.80	20.0	10.0	33.00	43.81	23.19	11.50	4.13	45.50	44.89	9.61	14.00	4.22

Table 8 (continued)

DATA FOR THE CONSTRUCTION OF ISOKINETIC SUCROSE AND GLYCEROL GRADIENTS FOR BECKMAN ROTOR SW 36

Particle density g/ml	Temperature °C	Top concentration %(w/w)	Sucrose C_h %(w/w)	Buffer 2x ml	H_2O ml	V_m ml	Velocity of 100 S particle at top speed ml/h	Glycerol C_h %(w/w)	Buffer 2x ml	H_2O ml	V_m ml	Velocity of 100 S particle at top speed ml/h
1.80	20.0	15.0	36.50	43.15	20.35	12.25	3.37	45.50	44.89	9.61	12.00	3.60
1.90	5.0	5.0	30.00	44.19	25.81	12.25	3.23	39.50	45.29	15.21	14.50	3.21
1.90	5.0	10.0	32.50	43.71	23.79	12.25	2.67	43.50	44.84	11.66	14.50	2.70
1.90	5.0	15.0	35.50	43.15	21.35	12.50	2.15	43.50	44.84	11.66	12.25	2.27
1.90	10.0	5.0	30.00	44.25	25.75	11.75	3.77	40.50	45.26	14.24	14.50	3.74
1.90	10.0	10.0	33.00	43.68	23.32	12.25	3.13	43.00	44.98	12.02	13.50	3.17
1.90	10.0	15.0	36.00	43.11	20.89	12.50	2.53	44.50	44.82	10.68	12.25	2.68
1.90	20.0	5.0	31.50	44.09	24.41	12.00	4.96	42.00	45.27	12.73	14.00	4.94
1.90	20.0	10.0	34.00	43.62	22.38	12.00	4.16	45.00	44.94	10.06	13.50	4.23
1.90	20.0	15.0	36.50	43.15	20.35	12.00	3.40	47.00	44.73	8.27	12.75	3.62
2.00	5.0	5.0	29.00	44.38	26.62	11.25	3.24	40.00	45.23	14.77	14.75	3.21
2.00	5.0	10.0	33.00	43.62	23.38	12.50	2.68	43.00	44.90	12.10	14.00	2.70
2.00	5.0	15.0	35.00	43.24	21.76	12.00	2.16	44.50	44.73	10.77	12.75	2.28
2.00	10.0	5.0	30.50	44.15	25.35	12.00	3.78	40.50	45.26	14.24	14.25	3.75
2.00	10.0	10.0	33.00	43.68	23.32	12.00	3.14	44.00	44.87	11.13	14.00	3.17
2.00	10.0	15.0	36.00	43.11	20.89	12.25	2.55	45.00	44.76	10.24	12.50	2.69
2.00	20.0	5.0	31.50	44.09	24.41	11.75	4.97	42.00	45.27	12.73	13.75	4.95
2.00	20.0	10.0	34.50	43.52	21.98	12.25	4.18	44.50	45.00	10.50	13.00	4.24
2.00	20.0	15.0	37.00	43.06	19.94	12.25	3.43	46.50	44.78	8.72	12.25	3.63

Table 9

DATA FOR THE CONSTRUCTION OF ISOKINETIC SUCROSE AND GLYCEROL GRADIENTS FOR BECKMAN ROTOR SW 40 Ti

Particle density g/ml	Temperature °C	Top concentration %(w/w)	Sucrose					Glycerol				
			C_h %(w/w)	Buffer 2x ml	H_2O ml	V_m ml	Velocity of 100 S particle at top speed ml/h	C_h %(w/w)	Buffer 2x ml	H_2O ml	V_m ml	Velocity of 100 S particle at top speed ml/h
1.20	5.0	5.0	23.00	45.53	31.47	11.25	3.27	33.00	46.03	20.97	14.25	3.34
1.20	5.0	10.0	26.00	44.96	29.04	11.50	2.45	35.50	45.74	18.76	13.25	2.65
1.20	5.0	15.0	29.50	44.29	26.21	12.50	1.75	39.50	45.29	15.21	13.50	2.09
1.20	10.0	5.0	23.50	45.49	31.01	11.50	3.81	33.50	46.05	20.45	14.00	3.90
1.20	10.0	10.0	26.50	44.91	28.59	11.75	2.87	36.00	45.76	18.24	13.00	3.12
1.20	10.0	15.0	29.50	44.34	26.16	12.00	2.07	40.00	45.32	14.68	13.50	2.48
1.20	20.0	5.0	24.00	45.51	30.49	11.25	5.03	35.50	45.98	18.52	14.25	5.16
1.20	20.0	10.0	27.00	44.94	28.06	11.50	3.84	37.50	45.76	16.74	13.00	4.19
1.20	20.0	15.0	30.00	44.37	25.63	11.75	2.80	40.50	45.43	14.07	12.75	3.38
1.30	5.0	5.0	26.50	44.86	28.64	11.75	3.38	36.50	45.63	17.87	14.25	3.41
1.30	5.0	10.0	29.00	44.38	26.62	11.50	2.65	39.00	45.35	15.65	13.25	2.78
1.30	5.0	15.0	32.00	43.81	24.19	11.75	2.01	41.50	45.07	13.43	12.50	2.26
1.30	10.0	5.0	26.00	45.01	28.99	11.00	3.95	37.00	45.65	17.35	14.00	3.99
1.30	10.0	10.0	29.00	44.44	26.56	11.25	3.11	40.00	45.32	14.68	13.25	3.27
1.30	10.0	15.0	32.50	43.77	23.73	11.75	2.37	42.00	45.09	12.91	12.25	2.68
1.30	20.0	5.0	27.00	44.94	28.06	11.00	5.20	39.00	45.60	15.40	14.00	5.27
1.30	20.0	10.0	30.50	44.28	25.22	11.75	4.14	41.00	45.38	13.62	12.75	4.38
1.30	20.0	15.0	33.50	43.71	22.79	12.00	3.20	43.00	45.16	11.84	11.75	3.63
1.40	5.0	5.0	27.50	44.67	27.83	11.25	3.44	38.50	45.40	16.10	14.25	3.45
1.40	5.0	10.0	30.50	44.10	25.40	11.50	2.75	40.50	45.18	14.32	13.00	2.85
1.40	5.0	15.0	34.00	43.43	22.57	12.25	2.14	42.50	44.95	12.55	12.00	2.34
1.40	10.0	5.0	28.00	44.63	27.37	11.25	4.02	39.50	45.37	15.13	14.25	4.03
1.40	10.0	10.0	31.00	44.06	24.94	11.50	3.23	41.50	45.15	13.35	13.00	3.34
1.40	10.0	15.0	34.00	43.49	22.51	11.75	2.52	43.50	44.93	11.57	12.00	2.77
1.40	20.0	5.0	29.00	44.56	26.44	11.25	5.29	41.50	45.32	13.18	14.25	5.33
1.40	20.0	10.0	32.00	43.99	24.01	11.50	4.30	43.00	45.16	11.84	12.75	4.47
1.40	20.0	15.0	35.00	43.43	21.57	11.75	3.40	45.00	44.94	10.06	12.00	3.75
1.50	5.0	5.0	28.00	44.57	27.43	10.75	3.48	40.00	45.23	14.77	14.50	3.47
1.50	5.0	10.0	31.50	43.90	24.60	11.50	2.82	42.00	45.01	12.99	13.25	2.89
1.50	5.0	15.0	34.50	43.34	22.16	11.75	2.21	43.50	44.84	11.66	12.00	2.39
1.50	10.0	5.0	28.50	44.53	26.97	10.75	4.06	41.00	45.20	13.80	14.50	4.06

Table 9 (continued)

DATA FOR THE CONSTRUCTION OF ISOKINETIC SUCROSE AND GLYCEROL GRADIENTS FOR BECKMAN ROTOR SW 40 Ti

Particle density g/ml	Temperature °C	Top concentration %(w/w)	Sucrose					Glycerol				
			C_h %(w/w)	Buffer 2x ml	H_2O ml	V_m ml	Velocity of 100 S particle at top speed ml/h	C_h %(w/w)	Buffer 2x ml	H_2O ml	V_m ml	Velocity of 100 S particle at top speed ml/h
1.50	10.0	10.0	32.00	43.87	24.13	11.50	3.30	43.00	44.98	12.02	13.25	3.39
1.50	10.0	15.0	35.00	43.30	21.70	11.75	2.61	44.50	44.82	10.68	12.00	2.83
1.50	20.0	5.0	30.50	44.28	25.22	11.50	5.34	42.00	45.27	12.73	13.75	5.36
1.50	20.0	10.0	32.50	43.90	23.60	11.00	4.39	44.50	45.00	10.50	13.00	4.53
1.50	20.0	15.0	36.00	43.24	20.76	11.75	3.52	45.00	44.94	10.06	11.25	3.83
1.60	5.0	5.0	29.00	44.38	26.62	11.00	3.50	40.50	45.18	14.32	14.25	3.49
1.60	5.0	10.0	31.50	43.90	24.60	11.00	2.86	42.00	45.01	12.99	12.75	2.91
1.60	5.0	15.0	35.00	43.24	21.76	11.50	2.27	44.00	44.79	11.21	11.75	2.43
1.60	10.0	5.0	29.50	44.34	26.16	11.00	4.09	41.50	45.15	13.35	14.25	4.07
1.60	10.0	10.0	33.00	43.68	23.32	11.75	3.35	43.00	44.98	12.02	12.75	3.42
1.60	10.0	15.0	35.50	43.20	21.30	11.50	2.67	44.50	44.82	10.68	11.50	2.87
1.60	20.0	5.0	31.00	44.18	24.82	11.25	5.38	42.50	45.21	12.29	13.50	5.38
1.60	20.0	10.0	33.50	43.71	22.79	11.25	4.45	45.00	44.94	10.06	12.75	4.57
1.60	20.0	15.0	36.50	43.15	20.35	11.50	3.60	46.50	44.78	8.72	11.75	3.88
1.70	5.0	5.0	29.50	44.29	26.21	11.00	3.52	41.00	45.12	13.88	14.25	3.50
1.70	5.0	10.0	32.00	43.81	24.19	11.00	2.89	42.50	44.95	12.55	12.75	2.93
1.70	5.0	15.0	35.00	43.24	21.76	11.25	2.30	44.00	44.79	11.21	11.50	2.45
1.70	10.0	5.0	30.00	44.25	25.75	11.00	4.11	41.50	45.15	13.35	13.75	4.09
1.70	10.0	10.0	33.50	43.58	22.92	11.75	3.39	44.00	44.87	11.13	13.00	3.44
1.70	10.0	15.0	35.50	43.20	21.30	11.25	2.72	45.00	44.76	10.24	11.50	2.90
1.70	20.0	5.0	31.00	44.18	24.82	10.75	5.40	43.00	45.16	11.84	13.50	5.40
1.70	20.0	10.0	34.00	43.62	22.38	11.25	4.50	45.50	44.89	9.61	12.75	4.60
1.70	20.0	15.0	37.00	43.06	19.94	11.50	3.66	47.00	44.73	8.27	11.75	3.92
1.80	5.0	5.0	29.50	44.29	26.21	10.75	3.53	41.50	45.07	13.43	14.25	3.51
1.80	5.0	10.0	33.00	43.62	23.38	11.50	2.91	43.00	44.90	12.10	12.75	2.94
1.80	5.0	15.0	36.00	43.05	20.95	11.75	2.33	45.00	44.68	10.32	12.00	2.47
1.80	10.0	5.0	30.50	44.15	25.35	11.00	4.12	41.50	45.15	13.35	13.50	4.10
1.80	10.0	10.0	33.00	43.68	23.32	11.00	3.41	44.00	44.87	11.13	12.75	3.46
1.80	10.0	15.0	36.50	43.02	20.48	11.75	2.75	45.00	44.76	10.24	11.25	2.92
1.80	20.0	5.0	31.00	44.18	24.82	10.50	5.42	43.50	45.11	11.39	13.25	5.41
1.80	20.0	10.0	34.50	43.52	21.98	11.25	4.53	46.00	44.83	9.17	12.75	4.62

Table 9 (continued)

DATA FOR THE CONSTRUCTION OF ISOKINETIC SUCROSE AND GLYCEROL GRADIENTS FOR BECKMAN ROTOR SW 40 Ti

Particle density g/ml	Temperature °C	Top concentration %(w/w)	Sucrose					Glycerol				
			C_h %(w/w)	Buffer 2x ml	H_2O ml	V_m ml	Velocity of 100 S particle at top speed ml/h	C_h %(w/w)	Buffer 2x ml	H_2O ml	V_m ml	Velocity of 100 S particle at top speed ml/h
1.80	20.0	15.0	37.00	43.06	19.94	11.25	3.70	46.50	44.78	8.72	11.25	3.94
1.90	5.0	5.0	29.50	44.29	26.21	10.50	3.54	41.50	45.07	13.43	14.00	3.51
1.90	5.0	10.0	33.00	43.62	23.38	11.25	2.93	42.50	44.95	12.55	12.25	2.95
1.90	5.0	15.0	35.50	43.15	21.35	11.25	2.35	44.50	44.73	10.77	11.50	2.48
1.90	10.0	5.0	30.50	44.15	25.35	10.75	4.13	42.00	45.09	12.91	13.75	4.10
1.90	10.0	10.0	33.50	43.58	22.92	11.25	3.43	43.50	44.93	11.57	12.25	3.47
1.90	10.0	15.0	36.50	43.02	20.48	11.50	2.78	45.50	44.71	9.79	11.50	2.93
1.90	20.0	5.0	32.00	43.99	24.01	11.00	5.44	43.50	45.11	11.39	13.25	5.42
1.90	20.0	10.0	34.50	43.52	21.98	11.00	4.56	45.50	44.89	9.61	12.25	4.64
1.90	20.0	15.0	37.00	43.06	19.94	11.00	3.73	47.00	44.73	8.27	11.25	3.96
2.00	5.0	5.0	30.50	44.10	25.40	11.00	3.55	41.00	45.12	13.88	13.50	3.52
2.00	5.0	10.0	33.50	43.52	22.98	11.50	2.94	43.50	44.84	11.66	12.75	2.96
2.00	5.0	15.0	36.00	43.05	20.95	11.50	2.37	44.50	44.73	10.77	11.25	2.49
2.00	10.0	5.0	31.00	44.06	24.94	11.00	4.14	42.50	45.04	12.46	13.75	4.11
2.00	10.0	10.0	33.50	43.58	22.92	11.00	3.45	45.00	44.76	10.24	13.00	3.48
2.00	10.0	15.0	36.00	43.11	20.89	11.00	2.80	45.50	44.71	9.79	11.25	2.95
2.00	20.0	5.0	32.00	43.99	24.01	10.75	5.45	44.00	45.05	10.95	13.25	5.43
2.00	20.0	10.0	35.00	43.43	21.57	11.25	4.58	45.00	44.94	10.06	11.75	4.65
2.00	20.0	15.0	37.50	42.96	19.54	11.25	3.76	47.50	44.67	7.83	11.50	3.98

Table 10

DATA FOR THE CONSTRUCTION OF ISOKINETIC SUCROSE AND GLYCEROL GRADIENTS FOR BECKMAN ROTOR SW 41 Ti

Particle density g/ml	Temperature °C	Top concentration %(w/w)	Sucrose					Glycerol				
			Buffer 2x ml	H₂O ml	C_h %(w/w)	V_m ml	Velocity of 100 S particle at top speed ml/h	C_h %(w/w)	Buffer 2x ml	H₂O ml	V_m ml	Velocity of 100 S particle at top speed ml/h
1.20	5.0	5.0	45.63	31.87	22.50	10.75	3.46	32.50	46.09	21.41	14.00	3.54
1.20	5.0	10.0	45.05	29.45	25.50	11.00	2.59	34.50	45.86	19.64	12.50	2.81
1.20	5.0	15.0	44.38	26.62	29.00	12.00	1.85	38.50	45.40	16.10	12.75	2.22
1.20	10.0	5.0	45.58	31.42	23.00	11.00	4.04	32.50	46.16	21.34	13.50	4.13
1.20	10.0	10.0	45.01	28.99	26.00	11.25	3.05	35.50	45.82	18.68	12.75	3.31
1.20	10.0	15.0	44.44	26.56	29.00	11.50	2.19	39.00	45.43	15.57	12.75	2.63
1.20	20.0	5.0	45.60	30.90	23.50	10.75	5.33	35.00	46.04	18.96	14.00	5.47
1.20	20.0	10.0	45.03	28.47	26.50	11.00	4.07	37.00	45.82	17.18	12.75	4.44
1.20	20.0	15.0	44.47	26.03	29.50	11.25	2.97	40.50	45.43	14.07	12.75	3.58
1.30	5.0	5.0	45.05	29.45	25.50	11.00	3.59	35.50	45.74	18.76	13.75	3.62
1.30	5.0	10.0	44.48	27.02	28.50	11.25	2.81	38.50	45.40	16.10	13.00	2.95
1.30	5.0	15.0	43.90	24.60	31.50	11.25	2.13	41.00	45.12	13.88	12.25	2.40
1.30	10.0	5.0	44.91	28.59	26.50	11.50	4.19	36.50	45.71	17.79	13.75	4.23
1.30	10.0	10.0	44.34	26.16	29.50	11.75	3.30	39.00	45.43	15.57	12.75	3.47
1.30	10.0	15.0	43.77	23.73	32.50	12.00	2.51	41.50	45.15	13.35	12.00	2.84
1.30	20.0	5.0	45.03	28.47	26.50	10.75	5.51	38.00	45.71	16.29	13.50	5.59
1.30	20.0	10.0	44.47	26.03	29.50	11.00	4.39	41.50	45.32	13.18	13.25	4.64
1.30	20.0	15.0	43.81	23.19	33.00	11.50	3.39	43.00	45.16	11.84	12.00	3.85
1.40	5.0	5.0	44.86	28.64	26.50	10.50	3.65	37.50	45.52	16.98	13.75	3.66
1.40	5.0	10.0	44.29	26.21	29.50	10.75	2.92	40.00	45.23	14.77	12.75	3.02
1.40	5.0	15.0	43.52	22.98	33.50	11.75	2.26	42.00	45.01	12.99	11.75	2.48
1.40	10.0	5.0	44.82	28.18	27.00	10.50	4.26	38.50	45.48	16.02	13.75	4.27
1.40	10.0	10.0	44.15	25.35	30.50	11.25	3.43	42.00	45.09	12.91	13.50	3.54
1.40	10.0	15.0	43.49	22.51	34.00	11.75	2.67	42.50	45.04	12.46	11.50	2.94
1.40	20.0	5.0	44.75	27.25	28.00	10.50	5.61	40.00	45.49	14.51	13.50	5.64
1.40	20.0	10.0	44.09	24.41	31.50	11.25	4.56	42.00	45.27	12.73	12.25	4.74
1.40	20.0	15.0	43.62	22.38	34.00	11.00	3.60	44.50	45.00	10.50	11.75	3.98
1.50	5.0	5.0	44.48	27.02	28.50	11.25	3.69	39.00	45.35	15.65	14.00	3.68
1.50	5.0	10.0	44.00	25.00	31.00	11.25	2.99	41.00	45.12	13.88	12.75	3.06
1.50	5.0	15.0	43.52	22.98	33.50	11.00	2.35	43.00	44.90	12.10	11.75	2.54
1.50	10.0	5.0	44.44	26.56	29.00	11.25	4.30	39.50	45.37	15.13	13.75	4.30

Table 10 (continued)

DATA FOR THE CONSTRUCTION OF ISOKINETIC SUCROSE AND GLYCEROL GRADIENTS FOR BECKMAN ROTOR SW 41 Ti

Particle density g/ml	Temperature °C	Top concentration %(w/w)	Sucrose					Glycerol				
			C_h %(w/w)	Buffer 2x ml	H_2O ml	V_m ml	Velocity of 100 S particle at top speed ml/h	C_h %(w/w)	Buffer 2x ml	H_2O ml	V_m ml	Velocity of 100 S particle at top speed ml/h
1.50	10.0	10.0	31.00	44.06	24.94	10.75	3.50	42.00	45.09	12.91	12.75	3.59
1.50	10.0	15.0	35.00	43.30	21.70	11.75	2.77	43.50	44.93	11.57	11.50	3.00
1.50	20.0	5.0	29.50	44.47	26.03	10.75	5.66	41.00	45.38	13.62	13.25	5.68
1.50	20.0	10.0	33.00	43.81	23.19	11.50	4.65	44.00	45.05	10.95	12.75	4.80
1.50	20.0	15.0	36.00	43.24	20.76	11.75	3.73	44.50	45.00	10.50	11.00	4.06
1.60	5.0	5.0	28.50	44.48	27.02	10.75	3.71	39.50	45.29	15.21	13.75	3.70
1.60	5.0	10.0	32.00	43.81	24.19	11.50	3.03	43.00	44.90	12.10	13.50	3.09
1.60	5.0	15.0	34.50	43.34	22.16	11.25	2.40	44.00	44.79	11.21	12.00	2.57
1.60	10.0	5.0	29.00	44.44	26.56	10.75	4.33	40.00	45.32	14.68	13.50	4.32
1.60	10.0	10.0	32.00	43.87	24.13	11.00	3.55	42.00	45.09	12.91	12.25	3.62
1.60	10.0	15.0	35.00	43.30	21.70	11.25	2.83	45.00	44.76	10.24	12.00	3.04
1.60	20.0	5.0	30.50	44.28	25.22	11.00	5.70	41.50	45.32	13.18	13.00	5.70
1.60	20.0	10.0	33.00	43.81	23.19	11.00	4.72	44.00	45.05	10.95	12.25	4.84
1.60	20.0	15.0	36.00	43.24	20.76	11.25	3.81	45.50	44.89	9.61	11.25	4.11
1.70	5.0	5.0	29.00	44.38	26.62	10.75	3.73	40.00	45.23	14.77	13.75	3.71
1.70	5.0	10.0	32.50	43.71	23.79	11.50	3.06	42.00	45.01	12.99	12.50	3.11
1.70	5.0	15.0	35.50	43.15	21.35	11.75	2.44	44.50	44.73	10.77	12.00	2.60
1.70	10.0	5.0	29.50	44.34	26.16	10.75	4.35	40.50	45.26	14.24	13.50	4.33
1.70	10.0	10.0	32.50	43.77	23.73	11.00	3.59	43.00	44.98	12.02	12.50	3.65
1.70	10.0	15.0	36.00	43.11	20.89	11.75	2.88	45.50	44.71	9.79	12.00	3.07
1.70	20.0	5.0	30.50	44.28	25.22	10.50	5.73	42.00	45.27	12.73	13.00	5.72
1.70	20.0	10.0	33.50	43.71	22.79	11.00	4.77	44.50	45.00	10.50	12.25	4.87
1.70	20.0	15.0	36.50	43.15	20.35	11.25	3.87	46.00	44.83	9.17	11.25	4.15
1.80	5.0	5.0	29.00	44.38	26.62	10.50	3.75	40.00	45.23	14.77	13.50	3.72
1.80	5.0	10.0	32.50	43.71	23.79	11.25	3.08	43.50	44.84	11.66	13.25	3.12
1.80	5.0	15.0	35.50	43.15	21.35	11.50	2.47	44.50	44.73	10.77	11.75	2.62
1.80	10.0	5.0	30.00	44.25	25.75	10.75	4.37	41.00	45.20	13.80	13.50	4.34
1.80	10.0	10.0	32.50	43.77	23.73	10.75	3.62	43.00	44.98	12.02	12.25	3.66
1.80	10.0	15.0	35.50	43.20	21.30	11.00	2.91	45.00	44.76	10.24	11.50	3.09
1.80	20.0	5.0	31.50	44.09	24.41	11.00	5.75	42.00	45.27	12.73	12.75	5.73
1.80	20.0	10.0	34.00	43.62	22.38	11.00	4.80	45.00	44.94	10.06	12.25	4.90

Table 10 (continued)
DATA FOR THE CONSTRUCTION OF ISOKINETIC SUCROSE AND GLYCEROL GRADIENTS FOR BECKMAN ROTOR SW 41 Ti

Particle density g/ml	Temperature °C	Top concentration %(w/w)	Sucrose					Glycerol				
			C_h %(w/w)	Buffer 2x ml	H_2O ml	V_m ml	Velocity of 100 S particle at top speed ml/h	C_h %(w/w)	Buffer 2x ml	H_2O ml	V_m ml	Velocity of 100 S particle at top speed ml/h
1.80	20.0	15.0	37.50	42.96	19.54	11.75	3.92	46.00	44.83	9.17	11.00	4.18
1.90	5.0	5.0	29.00	44.38	26.62	10.25	3.76	40.50	45.18	14.32	13.50	3.72
1.90	5.0	10.0	32.50	43.71	23.79	11.00	3.10	43.00	44.90	12.10	12.75	3.13
1.90	5.0	15.0	35.50	43.15	21.35	11.25	2.49	44.00	44.79	11.21	11.25	2.63
1.90	10.0	5.0	30.00	44.25	25.75	10.50	4.38	41.00	45.20	13.80	13.25	4.35
1.90	10.0	10.0	32.50	43.77	23.73	10.50	3.64	44.50	44.82	10.68	13.00	3.68
1.90	10.0	15.0	36.00	43.11	20.89	11.25	2.94	44.50	44.82	10.68	11.00	3.11
1.90	20.0	5.0	31.50	44.09	24.41	10.75	5.76	42.50	45.21	12.29	12.75	5.74
1.90	20.0	10.0	34.00	43.62	22.38	10.75	4.83	45.00	44.94	10.06	12.00	4.91
1.90	20.0	15.0	36.50	43.15	20.35	10.75	3.95	45.50	44.89	9.61	10.50	4.20
2.00	5.0	5.0	30.00	44.19	25.81	10.75	3.76	40.50	45.18	14.32	13.50	3.73
2.00	5.0	10.0	33.00	43.62	23.38	11.25	3.12	42.50	44.95	12.55	12.25	3.14
2.00	5.0	15.0	36.00	43.05	20.95	11.50	2.51	45.00	44.68	10.32	11.75	2.64
2.00	10.0	5.0	30.50	44.15	25.35	10.75	4.39	41.00	45.20	13.80	13.00	4.35
2.00	10.0	10.0	33.00	43.68	23.32	10.75	3.65	44.00	44.87	11.13	12.50	3.69
2.00	10.0	15.0	36.50	43.02	20.48	11.50	2.96	45.50	44.71	9.79	11.50	3.12
2.00	20.0	5.0	31.00	44.18	24.82	10.25	5.77	42.50	45.21	12.29	12.75	5.75
2.00	20.0	10.0	34.50	43.52	21.98	11.00	4.85	46.00	44.83	9.17	12.50	4.93
2.00	20.0	15.0	37.00	43.06	19.94	11.00	3.98	46.50	44.78	8.72	11.00	4.22

Table 11

DATA FOR THE CONSTRUCTION OF ISOKINETIC SUCROSE AND GLYCEROL GRADIENTS FOR BECKMAN ROTOR SW 50 L

Particle density g/ml	Temperature °C	Top concentration %(w/w)	Sucrose					Glycerol				
			C_h %(w/w)	Buffer 2x ml	H_2O ml	V_m ml	Velocity of 100 S particle at top speed ml/h	C_h %(w/w)	Buffer 2x ml	H_2O ml	V_m ml	Velocity of 100 S particle at top speed ml/h
1.20	5.0	5.0	21.00	45.92	33.08	5.75	3.15	30.50	46.32	23.18	7.75	3.22
1.20	5.0	10.0	25.00	45.15	29.85	6.50	2.36	34.00	45.92	20.08	7.50	2.56
1.20	5.0	15.0	29.00	44.38	26.62	7.25	1.68	38.50	45.40	16.10	7.75	2.02
1.20	10.0	5.0	22.50	45.68	31.82	6.50	3.68	32.50	46.16	21.34	8.25	3.76
1.20	10.0	10.0	25.00	45.20	29.80	6.25	2.77	36.00	45.76	18.24	8.00	3.01
1.20	10.0	15.0	29.00	44.44	26.56	7.00	2.00	38.50	45.48	16.02	7.50	2.40
1.20	20.0	5.0	22.00	45.89	32.11	5.75	4.85	33.50	46.21	20.29	8.00	4.98
1.20	20.0	10.0	26.00	45.13	28.87	6.50	3.70	35.00	46.04	18.96	7.00	4.04
1.20	20.0	15.0	29.00	44.56	26.44	6.50	2.71	40.00	45.49	14.51	7.50	3.26
1.30	5.0	5.0	24.50	45.24	30.26	6.25	3.27	35.00	45.80	19.20	8.25	3.29
1.30	5.0	10.0	27.00	44.76	28.24	6.00	2.56	36.50	45.63	17.87	7.25	2.68
1.30	5.0	15.0	30.00	44.19	25.81	6.00	1.94	40.50	45.18	14.32	7.25	2.18
1.30	10.0	5.0	25.00	45.20	29.80	6.25	3.81	35.00	45.88	19.12	8.00	3.85
1.30	10.0	10.0	29.50	44.34	26.16	7.25	3.00	39.50	45.37	15.13	8.00	3.16
1.30	10.0	15.0	32.00	43.87	24.13	7.00	2.29	40.50	45.26	14.24	7.00	2.58
1.30	20.0	5.0	24.00	45.51	30.49	5.50	5.02	37.00	45.82	17.18	8.00	5.09
1.30	20.0	10.0	28.50	44.66	26.84	6.25	4.00	40.50	45.43	14.07	7.75	4.23
1.30	20.0	15.0	31.50	44.09	24.41	6.25	3.09	41.50	45.32	13.18	6.75	3.50
1.40	5.0	5.0	26.00	44.96	29.04	6.25	3.32	36.00	45.69	18.31	8.00	3.33
1.40	5.0	10.0	29.50	44.29	26.21	6.50	2.66	39.00	45.35	15.65	7.50	2.75
1.40	5.0	15.0	33.50	43.52	22.98	7.25	2.06	41.50	45.07	13.43	7.00	2.26
1.40	10.0	5.0	26.50	44.91	28.59	6.25	3.88	37.00	45.65	17.35	8.00	3.89
1.40	10.0	10.0	30.50	44.15	25.35	6.75	3.12	40.00	45.32	14.68	7.50	3.23
1.40	10.0	15.0	33.50	43.58	22.92	7.00	2.43	43.00	44.98	12.02	7.25	2.68
1.40	20.0	5.0	27.50	44.85	27.65	6.25	5.10	37.50	45.76	16.74	7.50	5.14
1.40	20.0	10.0	29.00	44.56	26.44	5.75	4.15	41.00	45.38	13.62	7.25	4.32
1.40	20.0	15.0	33.00	43.81	23.19	6.25	3.28	42.50	45.21	12.29	6.50	3.62
1.50	5.0	5.0	27.00	44.76	28.24	6.25	3.36	36.50	45.63	17.87	7.75	3.35
1.50	5.0	10.0	28.50	44.48	27.02	5.75	2.72	41.50	45.07	13.43	8.00	2.78
1.50	5.0	15.0	33.50	43.52	22.98	6.75	2.14	42.50	44.95	12.55	7.00	2.31
1.50	10.0	5.0	27.50	44.72	27.78	6.25	3.92	37.50	45.59	16.91	7.75	3.92

Table 11 (continued)

DATA FOR THE CONSTRUCTION OF ISOKINETIC SUCROSE AND GLYCEROL GRADIENTS FOR BECKMAN ROTOR SW 50 L

Particle density g/ml	Temperature °C	Top concentration %(w/w)	Sucrose C_h %(w/w)	Sucrose Buffer 2x ml	Sucrose H_2O ml	Sucrose V_m ml	Sucrose Velocity of 100 S particle at top speed ml/h	Glycerol C_h %(w/w)	Glycerol Buffer 2x ml	Glycerol H_2O ml	Glycerol V_m ml	Glycerol Velocity of 100 S particle at top speed ml/h
1.50	10.0	10.0	29.00	44.44	26.56	5.75	3.19	41.00	45.20	13.80	7.50	3.27
1.50	10.0	15.0	32.50	43.77	23.73	6.00	2.52	42.00	45.09	12.91	6.50	2.73
1.50	20.0	5.0	28.00	44.75	27.25	6.00	5.15	38.00	45.71	16.29	7.25	5.17
1.50	20.0	10.0	30.00	44.37	25.63	5.75	4.24	43.00	45.16	11.84	7.50	4.37
1.50	20.0	15.0	34.00	43.62	22.38	6.25	3.40	43.50	45.11	11.39	6.50	3.70
1.60	5.0	5.0	26.50	44.86	28.64	5.75	3.38	36.50	45.63	17.87	7.50	3.37
1.60	5.0	10.0	31.00	44.00	25.00	6.50	2.76	40.00	45.23	14.77	7.25	2.81
1.60	5.0	15.0	32.00	43.81	24.19	5.75	2.19	42.50	44.95	12.55	6.75	2.34
1.60	10.0	5.0	27.00	44.82	28.18	5.75	3.95	37.50	45.59	16.91	7.50	3.93
1.60	10.0	10.0	31.50	43.96	24.54	6.50	3.23	41.00	45.20	13.80	7.25	3.30
1.60	10.0	15.0	32.50	43.77	23.73	5.75	2.58	42.50	45.04	12.46	6.50	2.77
1.60	20.0	5.0	28.00	44.75	27.25	5.75	5.19	39.00	45.60	15.40	7.25	5.19
1.60	20.0	10.0	32.50	43.90	23.60	6.50	4.30	43.00	45.16	11.84	7.25	4.41
1.60	20.0	15.0	33.50	43.71	22.79	5.75	3.47	43.50	45.11	11.39	6.25	3.74
1.70	5.0	5.0	27.00	44.76	28.24	5.75	3.40	38.00	45.46	16.54	7.75	3.38
1.70	5.0	10.0	29.00	44.38	26.62	5.50	2.79	41.50	45.07	13.43	7.50	2.83
1.70	5.0	15.0	33.00	43.62	23.38	6.00	2.22	43.00	44.90	12.10	6.75	2.37
1.70	10.0	5.0	27.50	44.72	27.78	5.75	3.96	38.00	45.54	16.46	7.50	3.94
1.70	10.0	10.0	29.50	44.34	26.16	5.50	3.27	42.50	45.04	12.46	7.50	3.32
1.70	10.0	15.0	33.50	43.58	22.92	6.00	2.62	43.00	44.98	12.02	6.50	2.80
1.70	20.0	5.0	28.50	44.66	26.84	5.75	5.21	39.50	45.54	14.96	7.25	5.21
1.70	20.0	10.0	32.50	43.90	23.60	6.25	4.34	43.00	45.16	11.84	7.00	4.44
1.70	20.0	15.0	34.00	43.62	22.38	5.75	3.53	44.00	45.05	10.95	6.25	3.78
1.80	5.0	5.0	28.00	44.57	27.43	6.00	3.41	37.50	45.52	16.98	7.50	3.38
1.80	5.0	10.0	30.50	44.10	25.40	6.00	2.81	41.00	45.12	13.88	7.25	2.84
1.80	5.0	15.0	34.00	43.43	22.57	6.25	2.25	42.50	44.95	12.55	6.50	2.38
1.80	10.0	5.0	28.50	44.53	26.97	6.00	3.98	38.50	45.48	16.02	7.50	3.95
1.80	10.0	10.0	30.50	44.15	25.35	5.75	3.29	42.00	45.09	12.91	7.25	3.34
1.80	10.0	15.0	34.50	43.39	22.11	6.25	2.65	43.50	44.93	11.57	6.50	2.82
1.80	20.0	5.0	29.00	44.56	26.44	5.75	5.23	39.00	45.60	15.40	7.00	5.22
1.80	20.0	10.0	31.50	44.09	24.41	5.75	4.37	42.50	45.21	12.29	6.75	4.46

Table 11 (continued)

DATA FOR THE CONSTRUCTION OF ISOKINETIC SUCROSE AND GLYCEROL GRADIENTS FOR BECKMAN ROTOR SW 50 L

Particle density g/ml	Temperature °C	Top concentration %(w/w)	Sucrose C_h %(w/w)	Buffer 2x ml	H_2O ml	V_m ml	Velocity of 100 S particle at top speed ml/h	Glycerol C_h %(w/w)	Buffer 2x ml	H_2O ml	V_m ml	Velocity of 100 S particle at top speed ml/h
1.80	20.0	15.0	35.50	43.34	21.16	6.25	3.57	44.50	45.00	10.50	6.25	3.81
1.90	5.0	5.0	27.00	44.76	28.24	5.50	3.42	38.00	45.46	16.54	7.50	3.39
1.90	5.0	10.0	32.00	43.81	24.19	6.50	2.82	40.50	45.18	14.32	7.00	2.85
1.90	5.0	15.0	33.00	43.62	23.38	5.75	2.27	43.50	44.84	11.66	6.75	2.40
1.90	10.0	5.0	27.50	44.72	27.78	5.50	3.99	38.00	45.54	16.46	7.25	3.96
1.90	10.0	10.0	32.00	43.87	24.13	6.25	3.31	42.50	45.04	12.46	7.25	3.35
1.90	10.0	15.0	33.50	43.58	22.92	5.75	2.68	44.50	44.82	10.68	6.75	2.83
1.90	20.0	5.0	30.00	44.37	25.63	6.00	5.25	39.50	45.54	14.96	7.00	5.23
1.90	20.0	10.0	32.50	43.90	23.60	6.00	4.40	44.50	45.00	10.50	7.25	4.47
1.90	20.0	15.0	34.00	43.62	22.38	5.50	3.60	45.50	44.89	9.61	6.50	3.83
2.00	5.0	5.0	28.50	44.48	27.02	6.00	3.43	38.00	45.46	16.54	7.50	3.40
2.00	5.0	10.0	31.00	44.00	25.00	6.00	2.84	42.50	44.95	12.55	7.50	2.86
2.00	5.0	15.0	35.00	43.24	21.76	6.50	2.29	43.00	44.90	12.10	6.50	2.41
2.00	10.0	5.0	29.00	44.44	26.56	6.00	4.00	39.00	45.43	15.57	7.50	3.97
2.00	10.0	10.0	31.00	44.06	24.94	5.75	3.33	43.50	44.93	11.57	7.50	3.36
2.00	10.0	15.0	35.50	43.20	21.30	6.50	2.70	44.00	44.87	11.13	6.50	2.84
2.00	20.0	5.0	29.50	44.47	26.03	5.75	5.26	40.00	45.49	14.51	7.00	5.24
2.00	20.0	10.0	32.00	43.99	24.01	5.75	4.42	44.00	45.05	10.95	7.00	4.49
2.00	20.0	15.0	36.00	43.24	20.76	6.25	3.63	45.00	44.94	10.06	6.25	3.84

Table 12
DATA FOR THE CONSTRUCTION OF ISOKINETIC SUCROSE AND GLYCEROL GRADIENTS FOR BECKMAN ROTOR SW 50.1

Particle density g/ml	Temperature °C	Top concentration %(w/w)	Sucrose					Glycerol				
			C_h %(w/w)	Buffer 2x ml	H_2O ml	V_m ml	Velocity of 100 S particle at top speed ml/h	C_h %(w/w)	Buffer 2x ml	H_2O ml	V_m ml	Velocity of 100 S particle at top speed ml/h
1.20	5.0	5.0	22.00	45.73	32.27	7.50	3.69	28.50	46.55	24.95	8.25	3.77
1.20	5.0	10.0	24.50	45.24	30.26	7.25	2.77	34.00	45.92	20.08	8.75	3.00
1.20	5.0	15.0	27.50	44.67	27.83	7.25	1.97	37.00	45.57	17.43	8.25	2.37
1.20	10.0	5.0	21.50	45.87	32.63	7.00	4.31	30.50	46.39	23.11	8.75	4.41
1.20	10.0	10.0	24.50	45.29	30.21	7.00	3.25	33.00	46.10	20.90	8.00	3.53
1.20	10.0	15.0	28.50	44.53	26.97	7.75	2.34	37.50	45.59	16.91	8.25	2.81
1.20	20.0	5.0	21.00	46.08	32.92	6.25	5.69	32.00	46.37	21.63	8.75	5.84
1.20	20.0	10.0	25.50	45.22	29.28	7.25	4.34	35.00	46.04	18.96	8.25	4.74
1.20	20.0	15.0	28.50	44.66	26.84	7.25	3.17	39.50	45.54	14.96	8.50	3.82
1.30	5.0	5.0	23.00	45.53	31.47	6.50	3.83	32.00	46.15	21.85	8.50	3.86
1.30	5.0	10.0	28.00	44.57	27.43	7.75	3.00	36.50	45.63	17.87	8.50	3.15
1.30	5.0	15.0	29.50	44.29	26.21	6.75	2.27	40.50	45.18	14.32	8.50	2.56
1.30	10.0	5.0	26.00	45.01	28.99	8.00	4.47	33.50	46.05	20.45	8.75	4.51
1.30	10.0	10.0	28.00	44.63	27.37	7.50	3.52	38.00	45.54	16.46	8.75	3.70
1.30	10.0	15.0	31.00	44.06	24.94	7.50	2.68	39.50	45.37	15.13	7.75	3.03
1.30	20.0	5.0	24.00	45.51	30.49	6.50	5.88	34.00	46.15	19.85	8.25	5.96
1.30	20.0	10.0	27.50	44.85	27.65	6.75	4.69	38.50	45.65	15.85	8.25	4.95
1.30	20.0	15.0	31.00	44.18	24.82	7.00	3.62	41.00	45.38	13.62	7.75	4.10
1.40	5.0	5.0	25.50	45.05	29.45	7.00	3.90	34.50	45.86	19.64	8.75	3.90
1.40	5.0	10.0	29.50	44.29	26.21	7.75	3.12	39.00	45.35	15.65	8.75	3.22
1.40	5.0	15.0	32.00	43.81	24.19	7.50	2.42	41.00	45.12	13.88	8.00	2.65
1.40	10.0	5.0	26.50	44.91	28.59	7.25	4.55	34.50	45.93	19.57	8.50	4.56
1.40	10.0	10.0	29.50	44.34	26.16	7.50	3.66	38.50	45.48	16.02	8.25	3.78
1.40	10.0	15.0	32.00	43.87	24.13	7.25	2.85	42.50	45.04	12.46	8.25	3.14
1.40	20.0	5.0	25.00	45.32	29.68	6.25	5.98	36.00	45.93	18.07	8.25	6.02
1.40	20.0	10.0	29.50	44.47	26.03	7.00	4.86	41.00	45.38	13.62	8.50	5.06
1.40	20.0	15.0	32.50	43.90	23.60	7.00	3.85	43.50	45.11	11.39	8.00	4.25
1.50	5.0	5.0	27.00	44.76	28.24	7.25	3.94	35.50	45.74	18.76	8.75	3.93
1.50	5.0	10.0	29.50	44.29	26.21	7.25	3.19	40.00	45.23	14.77	8.75	3.26
1.50	5.0	15.0	32.00	43.81	24.19	7.00	2.50	42.00	45.01	12.99	8.00	2.71
1.50	10.0	5.0	27.50	44.72	27.78	7.25	4.59	36.00	45.76	18.24	8.50	4.59

Table 12 (continued)

DATA FOR THE CONSTRUCTION OF ISOKINETIC SUCROSE AND GLYCEROL GRADIENTS FOR BECKMAN ROTOR SW 50.1

Particle density g/ml	Temperature °C	Top concentration %(w/w)	Sucrose					Glycerol				
			C_h %(w/w)	Buffer 2x ml	H_2O ml	V_m ml	Velocity of 100 S particle at top speed ml/h	C_h %(w/w)	Buffer 2x ml	H_2O ml	V_m ml	Velocity of 100 S particle at top speed ml/h
1.50	10.0	10.0	29.50	44.34	26.16	7.00	3.74	39.50	45.37	15.13	8.25	3.83
1.50	10.0	15.0	32.50	43.77	23.73	7.00	2.96	43.50	44.93	11.57	8.25	3.20
1.50	20.0	5.0	26.00	45.13	28.87	6.25	6.04	37.00	45.82	17.18	8.25	6.06
1.50	20.0	10.0	30.50	44.28	25.22	7.00	4.97	41.50	45.32	13.18	8.25	5.13
1.50	20.0	15.0	33.00	43.81	23.19	6.75	3.98	44.00	45.05	10.95	7.75	4.33
1.60	5.0	5.0	26.50	44.86	28.64	6.75	3.96	35.50	45.74	18.76	8.50	3.95
1.60	5.0	10.0	31.50	43.90	24.60	8.00	3.23	40.00	45.23	14.77	8.50	3.29
1.60	5.0	15.0	33.00	43.62	23.38	7.25	2.56	42.00	45.01	12.99	7.75	2.75
1.60	10.0	5.0	27.00	44.82	28.18	6.75	4.62	36.00	45.76	18.24	8.25	4.61
1.60	10.0	10.0	28.50	44.53	26.97	6.25	3.79	41.00	45.20	13.80	8.50	3.87
1.60	10.0	15.0	33.50	43.58	22.92	7.25	3.03	43.00	44.98	12.02	7.75	3.25
1.60	20.0	5.0	28.00	44.75	27.25	6.75	6.08	37.00	45.82	17.18	8.00	6.09
1.60	20.0	10.0	29.50	44.47	26.03	6.25	5.04	41.50	45.32	13.18	8.00	5.17
1.60	20.0	15.0	34.50	43.52	21.98	7.25	4.07	44.00	45.05	10.95	7.50	4.39
1.70	5.0	5.0	27.00	44.76	28.24	6.75	3.98	36.00	45.69	18.31	8.50	3.96
1.70	5.0	10.0	29.00	44.38	26.62	6.50	3.27	40.50	45.18	14.32	8.50	3.31
1.70	5.0	15.0	33.00	43.62	23.38	7.00	2.60	42.50	44.95	12.55	7.75	2.77
1.70	10.0	5.0	27.50	44.72	27.78	6.75	4.65	36.50	45.71	17.79	8.25	4.62
1.70	10.0	10.0	29.50	44.34	26.16	6.50	3.83	41.00	45.20	13.80	8.25	3.89
1.70	10.0	15.0	33.50	43.58	22.92	7.00	3.07	43.50	44.93	11.57	7.75	3.28
1.70	20.0	5.0	28.50	44.66	26.84	6.75	6.11	37.00	45.82	17.18	7.75	6.11
1.70	20.0	10.0	30.00	44.37	25.63	6.25	5.09	41.50	45.32	13.18	7.75	5.20
1.70	20.0	15.0	34.50	43.52	21.98	7.00	4.14	44.50	45.00	10.50	7.50	4.43
1.80	5.0	5.0	25.50	45.05	29.45	6.00	4.00	36.00	45.69	18.31	8.25	3.97
1.80	5.0	10.0	30.00	44.19	25.81	6.75	3.29	41.00	45.12	13.88	8.50	3.33
1.80	5.0	15.0	35.00	43.24	21.76	8.00	2.64	43.50	44.84	11.66	8.00	2.79
1.80	10.0	5.0	28.50	44.53	26.97	7.00	4.66	36.00	45.76	18.24	8.00	4.63
1.80	10.0	10.0	30.50	44.15	25.35	6.75	3.86	41.50	45.15	13.35	8.25	3.91
1.80	10.0	15.0	33.00	43.68	23.32	6.50	3.11	44.50	44.82	10.68	8.00	3.30
1.80	20.0	5.0	29.00	44.56	26.44	6.75	6.13	37.50	45.76	16.74	7.75	6.12
1.80	20.0	10.0	31.00	44.18	24.82	6.50	5.13	42.00	45.27	12.73	7.75	5.23

Table 12 (continued)

DATA FOR THE CONSTRUCTION OF ISOKINETIC SUCROSE AND GLYCEROL GRADIENTS FOR BECKMAN ROTOR SW 50.1

Particle density g/ml	Temperature °C	Top concentration %(w/w)	Sucrose					Glycerol				
			C_h %(w/w)	Buffer 2x ml	H_2O ml	V_m ml	Velocity of 100 S particle at top speed ml/h	C_h %(w/w)	Buffer 2x ml	H_2O ml	V_m ml	Velocity of 100 S particle at top speed ml/h
1.80	20.0	15.0	34.50	43.52	21.98	6.75	4.18	45.00	44.94	10.06	7.50	4.46
1.90	5.0	5.0	27.00	44.76	28.24	6.50	4.01	36.00	45.69	18.31	8.25	3.97
1.90	5.0	10.0	29.00	44.38	26.62	6.25	3.31	40.00	45.23	14.77	8.00	3.34
1.90	5.0	15.0	33.50	43.52	22.98	7.00	2.66	42.50	44.95	12.55	7.50	2.81
1.90	10.0	5.0	27.50	44.72	27.78	6.50	4.68	37.00	45.65	17.35	8.25	4.64
1.90	10.0	10.0	29.00	44.44	26.56	6.00	3.88	41.00	45.20	13.80	8.00	3.92
1.90	10.0	15.0	34.50	43.39	22.11	7.25	3.14	44.00	44.87	11.13	7.75	3.32
1.90	20.0	5.0	28.00	44.75	27.25	6.25	6.15	38.00	45.71	16.29	7.75	6.13
1.90	20.0	10.0	32.50	43.90	23.60	7.00	5.15	41.50	45.32	13.18	7.50	5.24
1.90	20.0	15.0	34.00	43.62	22.38	6.50	4.22	44.50	45.00	10.50	7.25	4.49
2.00	5.0	5.0	26.00	44.96	29.04	6.00	4.02	36.50	45.63	17.87	8.25	3.98
2.00	5.0	10.0	30.50	44.10	25.40	6.75	3.33	41.00	45.12	13.88	8.25	3.35
2.00	5.0	15.0	35.00	43.24	21.76	7.75	2.68	44.00	44.79	11.21	8.00	2.82
2.00	10.0	5.0	26.50	44.91	28.59	6.00	4.69	36.50	45.71	17.79	8.00	4.65
2.00	10.0	10.0	31.00	44.06	24.94	6.75	3.90	40.50	45.26	14.24	7.75	3.94
2.00	10.0	15.0	36.00	43.11	20.89	8.00	3.16	45.00	44.76	10.24	8.00	3.33
2.00	20.0	5.0	29.00	44.56	26.44	6.50	6.16	38.00	45.71	16.29	7.75	6.14
2.00	20.0	10.0	31.50	44.09	24.41	6.50	5.18	42.50	45.21	12.29	7.75	5.26
2.00	20.0	15.0	35.50	43.34	21.16	7.00	4.25	45.50	44.89	9.61	7.50	4.50

Table 13

DATA FOR THE CONSTRUCTION OF ISOKINETIC SUCROSE AND GLYCEROL GRADIENTS FOR BECKMAN ROTOR SW 56 Ti

Particle density g/ml	Temperature °C	Top concentration %(w/w)	Sucrose					Glycerol				
			C_h %(w/w)	Buffer 2x ml	H_2O ml	V_m ml	Velocity of 100 S particle at top speed ml/h	C_h %(w/w)	Buffer 2x ml	H_2O ml	V_m ml	Velocity of 100 S particle at top speed ml/h
1.20	5.0	5.0	22.00	45.73	32.27	5.50	3.41	32.00	46.15	21.85	7.25	3.49
1.20	5.0	10.0	24.00	45.34	30.66	5.00	2.56	33.50	45.97	20.53	6.25	2.77
1.20	5.0	15.0	27.50	44.67	27.83	5.25	1.82	38.50	45.40	16.10	6.75	2.19
1.20	10.0	5.0	23.00	45.58	31.42	5.75	3.99	32.00	46.22	21.78	7.00	4.08
1.20	10.0	10.0	26.00	45.01	28.99	6.00	3.00	35.50	45.82	18.68	6.75	3.26
1.20	10.0	15.0	28.50	44.53	26.97	5.75	2.16	38.50	45.48	16.02	6.50	2.59
1.20	20.0	5.0	22.50	45.79	31.71	5.25	5.25	33.00	46.26	20.74	6.75	5.39
1.20	20.0	10.0	25.50	45.22	29.28	5.25	4.01	37.00	45.82	17.18	6.75	4.38
1.20	20.0	15.0	29.50	44.47	26.03	6.00	2.93	40.00	45.49	14.51	6.50	3.53
1.30	5.0	5.0	23.50	45.44	31.06	5.00	3.54	34.50	45.86	19.64	7.00	3.57
1.30	5.0	10.0	27.50	44.67	27.83	5.50	2.77	38.00	45.46	16.54	6.75	2.91
1.30	5.0	15.0	31.00	44.00	25.00	5.75	2.10	40.50	45.18	14.32	6.25	2.36
1.30	10.0	5.0	24.50	45.29	30.21	5.25	4.13	35.50	45.82	18.68	7.00	4.17
1.30	10.0	10.0	28.00	44.63	27.37	5.50	3.25	39.00	45.43	15.57	6.75	3.42
1.30	10.0	15.0	31.00	44.06	24.94	5.50	2.48	40.50	45.26	14.24	6.00	2.80
1.30	20.0	5.0	26.00	45.13	28.87	5.50	5.43	36.50	45.87	17.63	6.75	5.51
1.30	20.0	10.0	29.50	44.47	26.03	5.75	4.33	40.00	45.49	14.51	6.50	4.58
1.30	20.0	15.0	30.50	44.28	25.22	5.00	3.34	42.00	45.27	12.73	6.00	3.79
1.40	5.0	5.0	25.00	45.15	29.85	5.00	3.60	36.50	45.63	17.87	7.00	3.60
1.40	5.0	10.0	29.00	44.38	26.62	5.50	2.88	40.00	45.23	14.77	6.75	2.97
1.40	5.0	15.0	32.50	43.71	23.79	5.75	2.23	40.50	45.18	14.32	5.75	2.45
1.40	10.0	5.0	25.50	45.10	29.40	5.00	4.20	36.50	45.71	17.79	6.75	4.21
1.40	10.0	10.0	29.50	44.34	26.16	5.50	3.38	40.00	45.32	14.68	6.50	3.49
1.40	10.0	15.0	33.00	43.68	23.32	5.75	2.64	43.00	44.98	12.02	6.25	2.90
1.40	20.0	5.0	27.00	44.94	28.06	5.25	5.53	37.50	45.76	16.74	6.50	5.57
1.40	20.0	10.0	30.50	44.28	25.22	5.50	4.49	41.00	45.38	13.62	6.25	4.68
1.40	20.0	15.0	34.50	43.52	21.98	6.00	3.55	43.00	45.16	11.84	5.75	3.92
1.50	5.0	5.0	26.00	44.96	29.04	5.00	3.64	36.50	45.63	17.87	6.75	3.63
1.50	5.0	10.0	30.00	44.19	25.81	5.50	2.94	41.00	45.12	13.88	6.75	3.01
1.50	5.0	15.0	33.50	43.52	22.98	5.75	2.31	41.50	45.07	13.43	5.75	2.50
1.50	10.0	5.0	26.50	44.91	28.59	5.00	4.24	37.50	45.59	16.91	6.75	4.24

Table 13 (continued)

DATA FOR THE CONSTRUCTION OF ISOKINETIC SUCROSE AND GLYCEROL GRADIENTS FOR BECKMAN ROTOR SW 56 Ti

Particle density g/ml	Temperature °C	Top concentration %(w/w)	Sucrose C_h %(w/w)	Buffer 2x ml	H_2O ml	V_m ml	Velocity of 100 S particle at top speed ml/h	Glycerol C_h %(w/w)	Buffer 2x ml	H_2O ml	V_m ml	Velocity of 100 S particle at top speed ml/h
1.50	10.0	10.0	30.50	44.15	25.35	5.50	3.45	42.00	45.09	12.91	6.75	3.54
1.50	10.0	15.0	34.00	43.49	22.51	5.75	2.73	42.50	45.04	12.46	5.75	2.96
1.50	20.0	5.0	27.50	44.85	27.65	5.00	5.58	38.00	45.71	16.29	6.25	5.60
1.50	20.0	10.0	31.50	44.09	24.41	5.50	4.59	43.00	45.16	11.84	6.50	4.74
1.50	20.0	15.0	35.50	43.34	21.16	6.00	3.68	45.00	44.94	10.06	6.00	4.00
1.50	5.0	5.0	28.00	44.57	27.43	5.50	3.66	37.50	45.52	16.98	6.75	3.64
1.60	5.0	10.0	30.00	44.19	25.81	5.25	2.99	41.00	45.12	13.88	6.50	3.04
1.60	5.0	15.0	34.00	43.43	22.57	5.75	2.37	43.00	44.90	12.10	6.00	2.54
1.60	5.0	5.0	28.50	44.53	26.97	5.50	4.27	37.50	45.59	16.91	6.50	4.26
1.60	10.0	10.0	30.50	44.15	25.35	5.25	3.50	42.00	45.09	12.91	6.50	3.57
1.60	10.0	15.0	34.50	43.39	22.11	5.75	2.79	44.00	44.87	11.13	6.00	3.00
1.60	10.0	5.0	29.00	44.56	26.44	5.25	5.62	39.00	45.60	15.40	6.25	5.62
1.60	20.0	10.0	31.50	44.09	24.41	5.25	4.65	43.00	45.16	11.84	6.25	4.78
1.60	20.0	15.0	35.00	43.43	21.57	5.50	3.76	44.00	45.05	10.95	5.50	4.06
1.60	20.0	5.0	27.00	44.76	28.24	5.00	3.68	37.00	45.57	17.43	6.50	3.66
1.70	5.0	10.0	31.00	44.00	25.00	5.50	3.02	41.50	45.07	13.43	6.50	3.06
1.70	5.0	15.0	34.50	43.34	22.16	5.75	2.41	43.50	44.84	11.66	6.00	2.56
1.70	5.0	5.0	27.50	44.72	27.78	5.00	4.29	38.00	45.54	16.46	6.50	4.27
1.70	10.0	10.0	31.00	44.06	24.94	5.25	3.54	42.50	45.04	12.46	6.50	3.60
1.70	10.0	15.0	35.00	43.30	21.70	5.75	2.84	44.50	44.82	10.68	6.00	3.03
1.70	20.0	5.0	29.50	44.47	26.03	5.25	5.64	39.50	45.54	14.96	6.25	5.64
1.70	20.0	10.0	32.00	43.99	24.01	5.25	4.70	43.50	45.11	11.39	6.25	4.81
1.70	20.0	15.0	36.00	43.24	20.76	5.75	3.82	44.50	45.00	10.50	5.50	4.09
1.80	5.0	5.0	27.50	44.67	27.83	5.00	3.69	37.50	45.52	16.98	6.50	3.66
1.80	5.0	10.0	32.00	43.81	24.19	5.75	3.04	42.00	45.01	12.99	6.50	3.08
1.80	5.0	15.0	33.50	43.52	22.98	5.25	2.44	43.00	44.90	12.10	5.75	2.58
1.80	5.0	5.0	28.00	44.63	27.37	5.00	4.31	38.50	45.48	16.02	6.50	4.28
1.80	10.0	10.0	32.00	43.87	24.13	5.50	3.56	43.00	44.98	12.02	6.50	3.61
1.80	10.0	15.0	34.00	43.49	22.51	5.25	2.87	44.00	44.87	11.13	5.75	3.05
1.80	20.0	5.0	29.00	44.56	26.44	5.00	5.66	39.00	45.60	15.40	6.00	5.65
1.80	20.0	10.0	32.50	43.90	23.60	5.25	4.73	44.00	45.05	10.95	6.25	4.83
1.80	20.0	15.0	34.50	43.52	21.98	5.00	3.86	44.00	45.05	10.95	5.25	4.12

Table 13 (continued)

DATA FOR THE CONSTRUCTION OF ISOKINETIC SUCROSE AND GLYCEROL GRADIENTS FOR BECKMAN ROTOR SW 56 Ti

Particle density g/ml	Temperature °C	Top concentration %(w/w)	Sucrose					Glycerol				
			C_h %(w/w)	Buffer 2x ml	H_2O ml	V_m ml	Velocity of 100 S particle at top speed ml/h	C_h %(w/w)	Buffer 2x ml	H_2O ml	V_m ml	Velocity of 100 S particle at top speed ml/h
1.90	5.0	5.0	28.50	44.48	27.02	5.25	3.70	38.00	45.46	16.54	6.50	3.67
1.90	5.0	10.0	31.00	44.00	25.00	5.25	3.06	41.50	45.07	13.43	6.25	3.09
1.90	5.0	15.0	35.00	43.24	21.76	5.75	2.46	42.50	44.95	12.55	5.50	2.60
1.90	10.0	5.0	29.00	44.44	26.56	5.25	4.32	38.00	45.54	16.46	6.25	4.29
1.90	10.0	10.0	31.50	43.96	24.54	5.25	3.58	42.50	45.04	12.46	6.25	3.63
1.90	10.0	15.0	35.00	43.30	21.70	5.50	2.90	43.50	44.93	11.57	5.50	3.07
1.90	20.0	5.0	29.50	44.47	26.03	5.00	5.68	39.50	45.54	14.96	6.00	5.66
1.90	20.0	10.0	32.00	43.99	24.01	5.00	4.76	43.50	45.11	11.39	6.00	4.84
1.90	20.0	15.0	36.00	43.24	20.76	5.50	3.90	46.00	44.83	9.17	5.75	4.14
2.00	5.0	5.0	28.00	44.57	27.43	5.00	3.71	38.00	45.46	16.54	6.50	3.68
2.00	5.0	10.0	30.50	44.10	25.40	5.00	3.07	42.50	44.95	12.55	6.50	3.10
2.00	5.0	15.0	34.00	43.43	22.57	5.25	2.48	43.50	44.84	11.66	5.75	2.61
2.00	10.0	5.0	28.50	44.53	26.97	5.00	4.33	39.00	45.43	15.57	6.50	4.29
2.00	10.0	10.0	32.50	43.77	23.73	5.50	3.60	43.50	44.93	11.57	6.50	3.64
2.00	10.0	15.0	34.50	43.39	22.11	5.25	2.92	44.50	44.82	10.68	5.75	3.08
2.00	20.0	5.0	29.00	44.56	26.44	4.75	5.69	39.50	45.54	14.96	6.00	5.67
2.00	20.0	10.0	33.00	43.81	23.19	5.25	4.78	44.00	45.05	10.95	6.00	4.86
2.00	20.0	15.0	35.00	43.43	21.57	5.00	3.93	45.50	44.89	9.61	5.50	4.16

Table 14

DATA FOR THE CONSTRUCTION OF ISOKINETIC SUCROSE AND GLYCEROL GRADIENTS FOR BECKMAN ROTOR SW 65 L Ti

Particle density g/ml	Temperature °C	Top concentration %(w/w)	Sucrose					Glycerol				
			C_h %(w/w)	Buffer 2x ml	H_2O ml	V_m ml	Velocity of 100 S particle at top speed ml/h	C_h %(w/w)	Buffer 2x ml	H_2O ml	V_m ml	Velocity of 100 S particle at top speed ml/h
1.20	5.0	5.0	22.00	45.73	32.27	5.25	4.45	32.00	46.15	21.85	7.00	4.54
1.20	5.0	10.0	26.00	44.96	29.04	6.00	3.33	36.00	45.69	18.31	7.00	3.61
1.20	5.0	15.0	30.00	44.19	25.81	6.75	2.38	38.50	45.40	16.10	6.50	2.85
1.20	10.0	5.0	22.00	45.77	32.23	5.25	5.19	32.00	46.22	21.78	6.75	5.31
1.20	10.0	10.0	26.00	45.01	28.99	5.75	3.92	35.50	45.82	18.68	6.50	4.25
1.20	10.0	15.0	29.50	44.34	26.16	6.25	2.82	39.00	45.43	15.57	6.50	3.38
1.20	20.0	5.0	23.50	45.60	30.90	5.50	6.85	34.50	46.10	19.40	7.00	7.03
1.20	20.0	10.0	27.00	44.94	28.06	6.00	5.23	35.50	45.98	18.52	6.00	5.70
1.20	20.0	15.0	30.00	44.37	25.63	6.00	3.82	40.50	45.43	14.07	6.50	4.60
1.30	5.0	5.0	24.50	45.24	30.26	5.25	4.61	34.50	45.86	19.64	6.75	4.65
1.30	5.0	10.0	28.00	44.57	27.43	5.50	3.61	38.00	45.46	16.54	6.50	3.79
1.30	5.0	15.0	31.00	44.00	25.00	5.50	2.73	41.00	45.12	13.88	6.25	3.08
1.30	10.0	5.0	25.00	45.20	29.80	5.25	5.38	37.00	45.65	17.35	7.25	5.43
1.30	10.0	10.0	29.00	44.44	26.56	5.75	4.24	38.00	45.54	16.46	6.25	4.45
1.30	10.0	15.0	31.50	43.96	24.54	5.50	3.23	41.00	45.20	13.80	6.00	3.64
1.30	20.0	5.0	26.50	45.03	28.47	5.50	7.08	37.50	45.76	16.74	6.75	7.18
1.30	20.0	10.0	28.00	44.75	27.25	5.00	5.65	39.00	45.60	15.40	6.00	5.96
1.30	20.0	15.0	32.00	43.99	24.01	5.50	4.36	42.00	45.27	12.73	5.75	4.94
1.40	5.0	5.0	25.50	45.05	29.45	5.00	4.69	36.50	45.63	17.87	6.75	4.70
1.40	5.0	10.0	29.50	44.29	26.21	5.50	3.75	39.00	45.35	15.65	6.25	3.88
1.40	5.0	15.0	32.50	43.71	23.79	5.50	2.91	42.00	45.01	12.99	6.00	3.19
1.40	10.0	5.0	26.50	44.91	28.59	5.25	5.48	38.00	45.54	16.46	7.00	5.49
1.40	10.0	10.0	30.00	44.25	25.75	5.50	4.40	40.00	45.32	14.68	6.25	4.56
1.40	10.0	15.0	33.50	43.58	22.92	5.75	3.44	42.00	45.09	12.91	5.75	3.78
1.40	20.0	5.0	27.50	44.85	27.65	5.25	7.20	38.50	45.65	15.85	6.50	7.25
1.40	20.0	10.0	31.50	44.09	24.41	5.75	5.86	43.50	45.11	11.39	6.75	6.10
1.40	20.0	15.0	35.00	43.43	21.57	6.00	4.63	44.50	45.00	10.50	6.00	5.11
1.50	5.0	5.0	26.50	44.86	28.64	5.00	4.74	37.50	45.52	16.98	6.75	4.73
1.50	5.0	10.0	30.50	44.10	25.40	5.50	3.84	41.00	45.12	13.88	6.50	3.93
1.50	5.0	15.0	34.50	43.34	22.16	6.00	3.02	43.00	44.90	12.10	6.00	3.26
1.50	10.0	5.0	27.00	44.82	28.18	5.00	5.53	38.50	45.48	16.02	6.75	5.53

Table 14 (continued)

DATA FOR THE CONSTRUCTION OF ISOKINETIC SUCROSE AND GLYCEROL GRADIENTS FOR BECKMAN ROTOR SW 65 L Ti

Particle density g/ml	Temperature °C	Top concentration %(w/w)	Sucrose					Glycerol				
			C_h %(w/w)	Buffer 2x ml	H_2O ml	V_m ml	Velocity of 100 S particle at top speed ml/h	C_h %(w/w)	Buffer 2x ml	H_2O ml	V_m ml	Velocity of 100 S particle at top speed ml/h
1.50	10.0	10.0	31.00	44.06	24.94	5.50	4.50	41.00	45.20	13.80	6.25	4.62
1.50	10.0	15.0	35.00	43.30	21.70	6.00	3.56	43.00	44.98	12.02	5.75	3.85
1.50	20.0	5.0	28.00	44.75	27.25	5.00	7.28	40.00	45.49	14.51	6.50	7.30
1.50	20.0	10.0	32.00	43.99	24.01	5.50	5.98	43.00	45.16	11.84	6.25	6.17
1.50	20.0	15.0	33.50	43.71	22.79	5.00	4.79	44.00	45.05	10.95	5.50	5.22
1.60	5.0	5.0	28.50	44.48	27.02	5.50	4.77	38.50	45.40	16.10	6.75	4.75
1.60	5.0	10.0	30.50	44.10	25.40	5.25	3.89	42.50	44.95	12.55	6.75	3.97
1.60	5.0	15.0	34.50	43.34	22.16	5.75	3.09	43.00	44.90	12.10	5.75	3.31
1.60	10.0	5.0	29.00	44.44	26.56	5.50	5.57	38.50	45.48	16.02	6.50	5.55
1.60	10.0	10.0	31.00	44.06	24.94	5.25	4.57	43.50	44.93	11.57	6.75	4.66
1.60	10.0	15.0	35.00	43.30	21.70	5.75	3.64	43.00	44.98	12.02	5.50	3.91
1.60	20.0	5.0	29.50	44.47	26.03	5.25	7.32	40.50	45.43	14.07	6.50	7.33
1.60	20.0	10.0	32.00	43.99	24.01	5.25	6.07	43.00	45.16	11.84	6.00	6.23
1.60	20.0	15.0	36.00	43.24	20.76	5.75	4.90	45.50	44.89	9.61	5.75	5.29
1.70	5.0	5.0	27.50	44.67	27.83	5.00	4.80	39.00	45.35	15.65	6.75	4.77
1.70	5.0	10.0	31.50	43.90	24.60	5.50	3.93	41.50	45.07	13.43	6.25	3.99
1.70	5.0	15.0	33.00	43.62	23.38	5.00	3.14	43.50	44.84	11.66	5.75	3.34
1.70	10.0	5.0	29.50	44.34	26.16	5.50	5.60	39.00	45.43	15.57	6.50	5.57
1.70	10.0	10.0	31.50	43.96	24.54	5.25	4.61	42.50	45.04	12.46	6.25	4.69
1.70	10.0	15.0	35.50	43.20	21.30	5.75	3.70	43.50	44.93	11.57	5.50	3.95
1.70	20.0	5.0	30.00	44.37	25.63	5.25	7.36	40.50	45.43	14.07	6.25	7.35
1.70	20.0	10.0	32.50	43.90	23.60	5.25	6.13	44.50	45.00	10.50	6.25	6.26
1.70	20.0	15.0	36.50	43.15	20.35	5.75	4.98	46.00	44.83	9.17	5.75	5.34
1.80	5.0	5.0	28.00	44.57	27.43	5.00	4.81	38.50	45.40	16.10	6.50	4.78
1.80	5.0	10.0	30.50	44.10	25.40	5.00	3.96	43.50	44.84	11.66	6.75	4.01
1.80	5.0	15.0	34.00	43.43	22.57	5.25	3.17	44.50	44.73	10.77	6.00	3.36
1.80	10.0	5.0	28.50	44.53	26.97	5.00	5.62	39.50	45.37	15.13	6.50	5.58
1.80	10.0	10.0	32.50	43.77	23.73	5.50	4.65	42.00	45.09	12.91	6.00	4.71
1.80	10.0	15.0	34.50	43.39	22.11	5.25	3.75	45.50	44.71	9.79	6.00	3.97
1.80	20.0	5.0	30.50	44.28	25.22	5.25	7.38	41.00	45.38	13.62	6.25	7.37
1.80	20.0	10.0	33.00	43.81	23.19	5.25	6.17	44.00	45.05	10.95	6.00	6.29
1.80	20.0	15.0	35.00	43.43	21.57	5.00	5.04	45.50	44.89	9.61	5.50	5.37

Table 14 (continued)
DATA FOR THE CONSTRUCTION OF ISOKINETIC SUCROSE AND GLYCEROL GRADIENTS FOR BECKMAN ROTOR SW 65 L Ti

Particle density g/ml	Temperature °C	Top concentration %(w/w)	Sucrose					Glycerol				
			C_h %(w/w)	Buffer 2x ml	H_2O ml	V_m ml	Velocity of 100 S particle at top speed ml/h	C_h %(w/w)	Buffer 2x ml	H_2O ml	V_m ml	Velocity of 100 S particle at top speed ml/h
1.90	5.0	5.0	29.00	44.38	26.62	5.25	4.83	39.50	45.29	15.21	6.75	4.79
1.90	5.0	10.0	31.50	43.90	24.60	5.25	3.99	43.00	44.90	12.10	6.50	4.02
1.90	5.0	15.0	35.50	43.15	21.35	5.75	3.20	44.00	44.79	11.21	5.75	3.38
1.90	10.0	5.0	29.50	44.34	26.16	5.25	5.63	40.00	45.32	14.68	6.50	5.59
1.90	10.0	10.0	32.00	43.87	24.13	5.25	4.67	44.00	44.87	11.13	6.50	4.73
1.90	10.0	15.0	36.00	43.11	20.89	5.75	3.78	44.00	44.87	11.13	5.50	4.00
1.90	20.0	5.0	30.00	44.37	25.63	5.00	7.40	41.50	45.32	13.18	6.25	7.38
1.90	20.0	10.0	32.50	43.90	23.60	5.00	6.21	44.50	45.00	10.50	6.00	6.31
1.90	20.0	15.0	36.50	43.15	20.35	5.50	5.08	45.00	44.94	10.06	5.25	5.40
2.00	5.0	5.0	28.50	44.48	27.02	5.00	4.84	39.00	45.35	15.65	6.50	4.79
2.00	5.0	10.0	31.00	44.00	25.00	5.00	4.00	41.50	45.07	13.43	6.00	4.04
2.00	5.0	15.0	35.00	43.24	21.76	5.50	3.23	45.00	44.68	10.32	6.90	3.40
2.00	10.0	5.0	29.00	44.44	26.56	5.00	5.64	40.00	45.32	14.68	6.50	5.60
2.00	10.0	10.0	33.00	43.68	23.32	5.50	4.69	43.50	44.93	11.57	6.25	4.74
2.00	10.0	15.0	35.00	43.30	21.70	5.25	3.81	46.00	44.65	9.35	6.00	4.01
2.00	20.0	5.0	31.00	44.18	24.82	5.25	7.42	41.50	45.32	13.18	6.25	7.39
2.00	20.0	10.0	33.50	43.71	22.79	5.25	6.24	45.50	44.89	9.61	6.25	6.33
2.00	20.0	15.0	36.00	43.24	20.76	5.25	5.12	46.00	44.83	9.17	5.50	5.42

Table 15

DATA FOR THE CONSTRUCTION OF ISOKINETIC SUCROSE AND GLYCEROL GRADIENTS FOR CHRIST ROTOR 9740

Particle density g/ml	Temperature °C	Top concentration %(w/w)	Sucrose					Glycerol				
			C_h %(w/w)	Buffer 2x ml	H_2O ml	V_m ml	Velocity of 100 S particle at top speed ml/h	C_h %(w/w)	Buffer 2x ml	H_2O ml	V_m ml	Velocity of 100 S particle at top speed ml/h
1.20	5.0	5.0	22.50	45.63	31.87	31.50	5.36	31.00	46.26	22.74	37.50	5.47
1.20	5.0	10.0	25.50	45.05	29.45	32.00	4.01	35.00	45.80	19.20	37.50	4.35
1.20	5.0	15.0	28.50	44.48	27.02	32.50	2.86	38.00	45.46	16.54	36.00	3.43
1.20	10.0	5.0	22.50	45.68	31.82	30.50	6.26	31.50	46.27	22.23	37.00	6.40
1.20	10.0	10.0	25.50	45.10	29.40	31.00	4.72	35.50	45.82	18.68	37.00	5.12
1.20	10.0	15.0	28.50	44.53	26.97	31.50	3.39	38.00	45.54	16.46	34.50	4.07
1.20	20.0	5.0	23.00	45.70	31.30	30.00	8.24	33.50	46.21	20.29	37.50	8.47
1.20	20.0	10.0	26.00	45.13	28.87	30.50	6.29	36.00	45.93	18.07	35.00	6.87
1.20	20.0	15.0	29.00	44.56	26.44	31.00	4.60	39.00	45.60	15.40	33.50	5.54
1.30	5.0	5.0	25.50	45.05	29.45	32.00	5.55	34.00	45.92	20.08	37.00	5.60
1.30	5.0	10.0	28.50	44.48	27.02	32.50	4.35	38.50	45.40	16.10	37.50	4.56
1.30	5.0	15.0	31.50	43.90	24.60	32.50	3.29	40.50	45.18	14.32	34.00	3.71
1.30	10.0	5.0	26.00	45.01	28.99	32.00	6.48	35.00	45.88	19.12	37.50	6.54
1.30	10.0	10.0	28.50	44.53	26.97	31.50	5.11	39.00	45.43	15.57	37.00	5.36
1.30	10.0	15.0	32.00	43.87	24.13	33.00	3.89	41.50	45.15	13.35	34.50	4.39
1.30	20.0	5.0	26.50	45.03	28.47	31.00	8.53	37.00	45.82	17.18	37.50	8.65
1.30	20.0	10.0	29.00	44.56	26.44	30.50	6.80	40.50	45.43	14.07	36.50	7.18
1.30	20.0	15.0	33.00	43.81	23.19	33.50	5.25	42.50	45.21	12.29	33.50	5.95
1.40	5.0	5.0	26.50	44.86	28.64	30.50	5.65	36.00	45.69	18.31	37.50	5.66
1.40	5.0	10.0	29.50	44.29	26.21	31.00	4.52	40.00	45.23	14.77	37.00	4.67
1.40	5.0	15.0	33.00	43.62	23.38	32.50	3.50	42.00	45.01	12.99	34.00	3.85
1.40	10.0	5.0	27.00	44.82	28.18	30.50	6.59	37.00	45.65	17.35	37.50	6.61
1.40	10.0	10.0	30.50	44.15	25.35	32.50	5.30	41.00	45.20	13.80	37.00	5.49
1.40	10.0	15.0	34.00	43.49	22.51	34.00	4.14	42.00	45.09	12.91	32.50	4.55
1.40	20.0	5.0	28.00	44.75	27.25	30.50	8.68	39.00	45.60	15.40	37.50	8.74
1.40	20.0	10.0	31.50	44.09	24.41	32.50	7.05	42.50	45.21	12.29	36.50	7.34
1.40	20.0	15.0	34.50	43.52	21.98	33.00	5.58	44.00	45.05	10.95	33.00	6.16
1.50	5.0	5.0	27.50	44.67	27.83	30.50	5.71	37.00	45.57	17.43	37.00	5.70
1.50	5.0	10.0	31.00	44.00	25.00	32.50	4.62	40.50	45.18	14.32	36.00	4.73
1.50	5.0	15.0	34.00	43.43	22.57	33.00	3.63	43.00	44.90	12.10	34.00	3.93
1.50	10.0	5.0	28.50	44.53	26.97	31.50	6.66	38.00	45.54	16.46	37.00	6.66

Table 15 (continued)

DATA FOR THE CONSTRUCTION OF ISOKINETIC SUCROSE AND GLYCEROL GRADIENTS FOR CHRIST ROTOR 9740

Particle density g/ml	Temperature °C	Top concentration %(w/w)	Sucrose					Glycerol				
			C_h %(w/w)	Buffer 2x ml	H_2O ml	V_m ml	Velocity of 100 S particle at top speed ml/h	C_h %(w/w)	Buffer 2x ml	H_2O ml	V_m ml	Velocity of 100 S particle at top speed ml/h
1.50	10.0	10.0	31.50	43.96	24.54	32.50	5.42	41.50	45.15	13.35	36.00	5.56
1.50	10.0	15.0	34.50	43.39	22.11	33.00	4.29	44.00	44.87	11.13	34.00	4.64
1.50	20.0	5.0	29.50	44.47	26.03	31.50	8.76	40.00	45.49	14.51	37.00	8.79
1.50	20.0	10.0	32.00	43.99	24.01	31.00	7.20	43.50	45.11	11.39	36.00	7.44
1.50	20.0	15.0	35.50	43.34	21.16	33.00	5.77	44.50	45.00	10.50	32.00	6.28
1.60	5.0	5.0	28.00	44.57	27.43	30.00	5.75	38.00	45.46	16.54	37.50	5.72
1.60	5.0	10.0	31.50	43.90	24.60	32.00	4.69	41.50	45.07	13.43	36.50	4.78
1.60	5.0	15.0	34.50	43.34	22.16	32.50	3.72	43.50	44.84	11.66	33.50	3.98
1.60	10.0	5.0	29.00	44.44	26.56	31.00	6.71	39.00	45.43	15.57	37.50	6.68
1.60	10.0	10.0	32.00	43.87	24.13	32.00	5.50	42.00	45.09	12.91	35.50	5.61
1.60	10.0	15.0	35.00	43.30	21.70	32.50	4.39	43.50	44.93	11.57	32.00	4.71
1.60	20.0	5.0	30.00	44.37	25.63	30.50	8.82	40.50	45.43	14.07	36.50	8.83
1.60	20.0	10.0	32.50	43.90	23.60	30.50	7.31	44.00	45.05	10.95	35.50	7.50
1.60	20.0	15.0	36.00	43.24	20.76	32.50	5.90	45.50	44.89	9.61	32.50	6.37
1.70	5.0	5.0	28.50	44.48	27.02	30.00	5.78	38.50	45.40	16.10	37.50	5.74
1.70	5.0	10.0	31.50	43.90	24.60	31.00	4.74	41.50	45.07	13.43	35.50	4.81
1.70	5.0	15.0	34.50	43.34	22.16	31.50	3.78	43.00	44.90	12.10	32.00	4.02
1.70	10.0	5.0	29.50	44.34	26.16	31.00	6.74	39.50	45.37	15.13	37.50	6.70
1.70	10.0	10.0	32.50	43.77	23.73	32.00	5.55	42.50	45.04	12.46	35.50	5.64
1.70	10.0	15.0	35.00	43.30	21.70	31.50	4.46	44.00	44.87	11.13	32.00	4.75
1.70	20.0	5.0	30.00	44.37	25.63	29.50	8.86	40.50	45.43	14.07	35.50	8.85
1.70	20.0	10.0	33.00	43.81	23.19	30.50	7.38	44.00	45.05	10.95	34.50	7.54
1.70	20.0	15.0	36.50	43.15	20.35	32.50	6.00	46.00	44.83	9.17	32.50	6.43
1.80	5.0	5.0	29.00	44.38	26.62	30.50	5.80	39.00	45.35	15.65	37.50	5.75
1.80	5.0	10.0	32.50	43.71	23.79	32.50	4.77	42.50	44.95	12.55	36.50	4.83
1.80	5.0	15.0	35.00	43.24	21.76	32.00	3.82	44.00	44.79	11.21	33.00	4.05
1.80	10.0	5.0	29.50	44.34	26.16	30.00	6.76	39.50	45.37	15.13	36.50	6.72
1.80	10.0	10.0	32.50	43.77	23.73	31.00	5.60	43.00	44.98	12.02	35.50	5.67
1.80	10.0	15.0	35.50	43.20	21.30	32.00	4.51	44.00	44.87	11.13	31.50	4.79
1.80	20.0	5.0	30.00	44.37	25.63	29.00	8.89	41.00	45.38	13.62	35.50	8.87
1.80	20.0	10.0	33.50	43.71	22.79	31.00	7.43	45.00	44.94	10.06	35.50	7.58

Table 15 (continued)

DATA FOR THE CONSTRUCTION OF ISOKINETIC SUCROSE AND GLYCEROL GRADIENTS FOR CHRIST ROTOR 9740

Particle density g/ml	Temperature °C	Top concentration %(w/w)	Sucrose					Glycerol				
			C_h %(w/w)	Buffer 2x ml	H_2O ml	V_m ml	Velocity of 100 S particle at top speed ml/h	C_h %(w/w)	Buffer 2x ml	H_2O ml	V_m ml	Velocity of 100 S particle at top speed ml/h
1.80	20.0	15.0	36.00	43.24	20.76	30.50	6.07	45.50	44.89	9.61	31.00	6.47
1.90	5.0	5.0	29.00	44.38	26.62	29.50	5.81	39.00	45.35	15.65	37.00	5.76
1.90	5.0	10.0	31.50	43.90	24.60	29.50	4.80	42.50	44.95	12.55	36.00	4.85
1.90	5.0	15.0	35.00	43.24	21.76	31.50	3.86	44.00	44.79	11.21	32.50	4.07
1.90	10.0	5.0	30.00	44.25	25.75	30.50	6.78	40.00	45.32	14.68	37.00	6.73
1.90	10.0	10.0	32.50	43.77	23.73	30.50	5.63	43.00	44.98	12.02	35.00	5.69
1.90	10.0	15.0	36.00	43.11	20.89	32.50	4.55	45.00	44.76	10.24	32.50	4.81
1.90	20.0	5.0	31.00	44.18	24.82	30.00	8.92	41.00	45.38	13.62	35.00	8.89
1.90	20.0	10.0	33.50	43.71	22.79	30.00	7.48	45.00	44.94	10.06	35.00	7.60
1.90	20.0	15.0	36.50	43.15	20.35	31.00	6.12	45.50	44.89	9.61	30.50	6.50
2.00	5.0	5.0	29.00	44.38	26.62	29.00	5.83	39.50	45.29	15.21	37.50	5.77
2.00	5.0	10.0	32.50	43.71	23.79	31.50	4.82	42.50	44.95	12.55	35.50	4.86
2.00	5.0	15.0	35.00	43.24	21.76	31.00	3.89	44.50	44.73	10.77	33.00	4.09
2.00	10.0	5.0	30.00	44.25	25.75	30.00	6.80	40.00	45.32	14.68	36.50	6.74
2.00	10.0	10.0	33.00	43.68	23.32	31.00	5.65	43.50	44.93	11.57	35.50	5.71
2.00	10.0	15.0	36.00	43.11	20.89	32.00	4.59	44.50	44.82	10.68	31.50	4.83
2.00	20.0	5.0	31.00	44.18	24.82	29.50	8.94	41.50	45.32	13.18	35.00	8.90
2.00	20.0	10.0	33.50	43.71	22.79	29.50	7.51	45.00	44.94	10.06	34.50	7.63
2.00	20.0	15.0	36.50	43.15	20.35	30.50	6.17	46.00	44.83	9.17	31.00	6.53

Table 16

DATA FOR THE CONSTRUCTION OF ISOKINETIC SUCROSE AND GLYCEROL GRADIENTS FOR CHRIST ROTOR 9750

Particle density g/ml	Temperature °C	Top concentration %(w/w)	Sucrose					Glycerol				
			Buffer 2x ml	C_h %(w/w)	H_2O ml	V_m ml	Velocity of 100 S particle at top speed ml/h	C_h %(w/w)	Buffer 2x ml	H_2O ml	V_m ml	Velocity of 100 S particle at top speed ml/h
1.20	5.0	5.0	45.53	23.00	31.47	7.00	2.02	31.00	46.26	22.74	8.00	2.06
1.20	5.0	10.0	45.15	25.00	29.85	6.50	1.51	34.00	45.92	20.08	7.50	1.64
1.20	5.0	15.0	44.57	28.00	27.43	6.50	1.08	38.50	45.40	16.10	7.75	1.30
1.20	10.0	5.0	45.68	22.50	31.82	6.50	2.36	32.50	46.16	21.34	8.25	2.41
1.20	10.0	10.0	45.20	25.00	29.80	6.25	1.78	36.00	45.76	18.24	8.00	1.93
1.20	10.0	15.0	44.63	28.00	27.37	6.25	1.28	38.50	45.48	16.02	7.50	1.54
1.20	20.0	5.0	45.79	22.50	31.71	6.00	3.11	33.50	46.21	20.29	8.00	3.19
1.20	20.0	10.0	45.13	26.00	28.87	6.50	2.37	35.00	46.04	18.96	7.00	2.59
1.20	20.0	15.0	44.66	28.50	26.84	6.25	1.74	40.50	45.43	14.07	7.75	2.09
1.30	5.0	5.0	45.24	24.50	30.26	6.25	2.09	35.00	45.80	19.20	8.25	2.11
1.30	5.0	10.0	44.76	27.00	28.24	6.00	1.64	38.50	45.40	16.10	8.00	1.72
1.30	5.0	15.0	43.90	31.50	24.60	7.00	1.24	41.00	45.12	13.88	7.50	1.40
1.30	10.0	5.0	45.20	25.00	29.80	6.25	2.45	35.00	45.88	19.12	8.00	2.47
1.30	10.0	10.0	44.34	29.50	26.16	7.25	1.93	38.00	45.54	16.46	7.50	2.02
1.30	10.0	15.0	43.87	32.00	24.13	7.00	1.47	40.50	45.26	14.24	7.00	1.66
1.30	20.0	5.0	45.41	24.50	30.09	5.75	3.22	37.00	45.82	17.18	8.00	3.26
1.30	20.0	10.0	44.56	29.00	26.44	6.50	2.57	40.50	45.43	14.07	7.75	2.71
1.30	20.0	15.0	43.99	32.00	24.01	6.50	1.98	41.50	45.32	13.18	6.75	2.25
1.40	5.0	5.0	44.96	26.00	29.04	6.25	2.13	36.00	45.69	18.31	8.00	2.13
1.40	5.0	10.0	44.38	29.00	26.62	6.25	1.70	39.00	45.35	15.65	7.50	1.76
1.40	5.0	15.0	43.52	33.50	22.98	7.25	1.32	41.50	45.07	13.43	7.00	1.45
1.40	10.0	5.0	44.91	26.50	28.59	6.25	2.49	37.00	45.65	17.35	8.00	2.49
1.40	10.0	10.0	44.25	30.00	25.75	6.50	2.00	40.00	45.32	14.68	7.50	2.07
1.40	10.0	15.0	43.58	33.50	22.92	7.00	1.56	43.00	44.98	12.02	7.25	1.72
1.40	20.0	5.0	44.85	27.50	27.65	6.25	3.27	38.50	45.65	15.85	7.75	3.30
1.40	20.0	10.0	44.47	29.50	26.03	6.00	2.66	42.50	45.21	12.29	7.75	2.77
1.40	20.0	15.0	43.62	34.00	22.38	6.75	2.10	44.00	45.05	10.95	7.00	2.32
1.50	5.0	5.0	44.76	27.00	28.24	6.25	2.15	36.50	45.63	17.87	7.75	2.15
1.50	5.0	10.0	44.48	28.50	27.02	5.75	1.74	41.50	45.07	13.43	8.00	1.79
1.50	5.0	15.0	43.52	33.50	22.98	6.75	1.37	42.50	44.95	12.55	7.00	1.48
1.50	10.0	5.0	44.72	27.50	27.78	6.25	2.51	38.00	45.54	16.46	8.00	2.51

Table 16 (continued)
DATA FOR THE CONSTRUCTION OF ISOKINETIC SUCROSE AND GLYCEROL GRADIENTS FOR CHRIST ROTOR 9750

Particle density g/ml	Temperature °C	Top concentration %(w/w)	Sucrose					Glycerol				
			C_h %(w/w)	Buffer 2x ml	H_2O ml	V_m ml	Velocity of 100 S particle at top speed ml/h	C_h %(w/w)	Buffer 2x ml	H_2O ml	V_m ml	Velocity of 100 S particle at top speed ml/h
1.50	10.0	10.0	29.00	44.44	26.56	5.75	2.04	41.00	45.20	13.80	7.50	2.10
1.50	10.0	15.0	34.00	43.49	22.51	6.75	1.62	44.00	44.87	11.13	7.25	1.75
1.50	20.0	5.0	28.50	44.66	26.84	6.25	3.31	39.00	45.60	15.40	7.50	3.32
1.50	20.0	10.0	30.00	44.37	25.63	5.75	2.72	43.00	45.16	11.84	7.50	2.80
1.50	20.0	15.0	34.50	43.52	21.98	6.50	2.18	43.50	45.11	11.39	6.50	2.37
1.60	5.0	5.0	27.00	44.76	28.24	6.00	2.17	38.00	45.46	16.54	8.00	2.16
1.60	5.0	10.0	31.00	44.00	25.00	6.50	1.77	41.50	45.07	13.43	7.75	1.80
1.60	5.0	15.0	35.00	43.24	21.76	7.25	1.40	42.50	44.95	12.55	6.75	1.50
1.60	10.0	5.0	27.50	44.72	27.78	6.00	2.53	38.50	45.48	16.02	7.75	2.52
1.60	10.0	10.0	31.50	43.96	24.54	6.50	2.07	42.50	45.04	12.46	7.75	2.12
1.60	10.0	15.0	35.50	43.20	21.30	7.25	1.66	44.00	44.87	11.13	7.00	1.78
1.60	20.0	5.0	28.00	44.75	27.25	5.75	3.33	39.50	45.54	14.96	7.50	3.33
1.60	20.0	10.0	32.50	43.90	23.60	6.50	2.76	43.00	45.16	11.84	7.25	2.83
1.60	20.0	15.0	34.00	43.62	22.38	6.00	2.23	45.00	44.94	10.06	6.75	2.40
1.70	5.0	5.0	27.50	44.67	27.83	6.00	2.18	38.50	45.40	16.10	8.00	2.17
1.70	5.0	10.0	32.00	43.81	24.19	6.75	1.79	42.00	45.01	12.99	7.75	1.81
1.70	5.0	15.0	34.50	43.34	22.16	6.75	1.42	43.00	44.90	12.10	6.75	1.52
1.70	10.0	5.0	27.50	44.72	27.78	5.75	2.54	39.00	45.43	15.57	7.75	2.53
1.70	10.0	10.0	32.00	43.87	24.13	6.50	2.10	42.50	45.04	12.46	7.50	2.13
1.70	10.0	15.0	33.50	43.58	22.92	6.00	1.68	44.50	44.82	10.68	7.00	1.79
1.70	20.0	5.0	28.50	44.66	26.84	5.75	3.34	39.50	45.54	14.96	7.25	3.34
1.70	20.0	10.0	33.00	43.81	23.19	6.50	2.78	44.50	45.00	10.50	7.50	2.85
1.70	20.0	15.0	34.50	43.52	21.98	6.00	2.26	44.00	45.05	10.95	6.25	2.42
1.80	5.0	5.0	28.00	44.57	27.43	6.00	2.19	38.50	45.40	16.10	7.75	2.17
1.80	5.0	10.0	30.50	44.10	25.40	6.00	1.80	42.50	44.95	12.55	7.75	1.82
1.80	5.0	15.0	34.00	43.43	22.57	6.25	1.44	42.50	44.95	12.55	6.50	1.53
1.80	10.0	5.0	28.50	44.53	26.97	6.00	2.55	38.50	45.48	16.02	7.50	2.53
1.80	10.0	10.0	31.00	44.06	24.94	6.00	2.11	42.00	45.09	12.91	7.25	2.14
1.80	10.0	15.0	35.00	43.30	21.70	6.50	1.70	43.50	44.93	11.57	6.50	1.81
1.80	20.0	5.0	29.00	44.56	26.44	5.75	3.35	40.00	45.49	14.51	7.25	3.35
1.80	20.0	10.0	31.50	44.09	24.41	5.75	2.80	44.00	45.05	10.95	7.25	2.86

Table 16 (continued)
DATA FOR THE CONSTRUCTION OF ISOKINETIC SUCROSE AND GLYCEROL GRADIENTS FOR CHRIST ROTOR 9750

Particle density g/ml	Temperature °C	Top concentration %(w/w)	Sucrose					Glycerol				
			C_h %(w/w)	Buffer 2x ml	H₂O ml	V_m ml	Velocity of 100 S particle at top speed ml/h	C_h %(w/w)	Buffer 2x ml	H₂O ml	V_m ml	Velocity of 100 S particle at top speed ml/h
1.80	20.0	15.0	35.50	43.34	21.16	6.25	2.29	46.00	44.83	9.17	6.75	2.44
1.90	5.0	5.0	27.50	44.67	27.83	5.75	2.19	38.50	45.40	16.10	7.75	2.17
1.90	5.0	10.0	32.00	43.81	24.19	6.50	1.81	42.00	45.01	12.99	7.50	1.83
1.90	5.0	15.0	33.50	43.52	22.98	6.00	1.46	43.50	44.84	11.66	6.75	1.54
1.90	10.0	5.0	29.50	44.34	26.16	6.25	2.56	39.50	45.37	15.13	7.75	2.54
1.90	10.0	10.0	32.00	43.87	24.13	6.25	2.12	42.50	45.04	12.46	7.25	2.15
1.90	10.0	15.0	33.50	43.58	22.92	5.75	1.72	44.50	44.82	10.68	6.75	1.82
1.90	20.0	5.0	30.00	44.37	25.63	6.00	3.36	40.50	45.43	14.07	7.25	3.35
1.90	20.0	10.0	32.50	43.90	23.60	6.00	2.82	44.50	45.00	10.50	7.25	2.87
1.90	20.0	15.0	34.50	43.52	21.98	5.75	2.31	45.50	44.89	9.61	6.50	2.45
2.00	5.0	5.0	28.50	44.48	27.02	6.00	2.20	39.00	45.35	15.65	7.75	2.18
2.00	5.0	10.0	31.00	44.00	25.00	6.00	1.82	41.50	45.07	13.43	7.25	1.83
2.00	5.0	15.0	34.50	43.34	22.16	6.25	1.47	43.00	44.90	12.10	6.50	1.54
2.00	10.0	5.0	29.00	44.44	26.56	6.00	2.56	39.00	45.43	15.57	7.50	2.54
2.00	10.0	10.0	31.50	43.96	24.54	6.00	2.13	43.50	44.93	11.57	7.50	2.15
2.00	10.0	15.0	35.00	43.30	21.70	6.25	1.73	44.00	44.87	11.13	6.50	1.82
2.00	20.0	5.0	29.50	44.47	26.03	5.75	3.37	40.50	45.43	14.07	7.25	3.36
2.00	20.0	10.0	34.00	43.62	22.38	6.50	2.83	44.00	45.05	10.95	7.00	2.88
2.00	20.0	15.0	36.00	43.24	20.76	6.25	2.33	45.00	44.94	10.06	6.25	2.46

Table 17

DATA FOR THE CONSTRUCTION OF ISOKINETIC SUCROSE AND GLYCEROL GRADIENTS FOR CHRIST ROTOR 9792

Particle density g/ml	Temperature °C	Top concentration %(w/w)	Sucrose					Glycerol				
			C_h %(w/w)	Buffer 2x ml	H_2O ml	V_m ml	Velocity of 100 S particle at top speed ml/h	C_h %(w/w)	Buffer 2x ml	H_2O ml	V_m ml	Velocity of 100 S particle at top speed ml/h
1.20	5.0	5.0	21.50	45.82	32.68	6.25	3.51	30.50	46.32	23.18	8.00	3.59
1.20	5.0	10.0	25.50	45.05	29.45	7.00	2.63	33.50	45.97	20.53	7.50	2.85
1.20	5.0	15.0	28.00	44.57	27.43	6.75	1.88	38.00	45.46	16.54	7.75	2.25
1.20	10.0	5.0	23.00	45.58	31.42	7.00	4.10	32.50	46.16	21.34	8.50	4.19
1.20	10.0	10.0	25.00	45.20	29.80	6.50	3.09	36.00	45.76	18.24	8.25	3.35
1.20	10.0	15.0	28.00	44.63	27.37	6.50	2.22	38.50	45.48	16.02	7.75	2.67
1.20	20.0	5.0	22.50	45.79	31.71	6.25	5.40	33.50	46.21	20.29	8.25	5.55
1.20	20.0	10.0	26.50	45.03	28.47	7.00	4.12	37.00	45.82	17.18	8.00	4.50
1.20	20.0	15.0	29.00	44.56	26.44	6.75	3.01	40.00	45.49	14.51	7.75	3.63
1.30	5.0	5.0	24.00	45.34	30.66	6.25	3.64	35.00	45.80	19.20	8.50	3.67
1.30	5.0	10.0	27.50	44.67	27.83	6.50	2.85	38.00	45.46	16.54	8.00	2.99
1.30	5.0	15.0	30.50	44.10	25.40	6.50	2.16	40.50	45.18	14.32	7.50	2.43
1.30	10.0	5.0	25.00	45.20	29.80	6.50	4.25	36.00	45.76	18.24	8.50	4.29
1.30	10.0	10.0	27.50	44.72	27.78	6.25	3.35	39.50	45.37	15.13	8.25	3.51
1.30	10.0	15.0	30.50	44.15	25.35	6.25	2.55	40.00	45.32	14.68	7.00	2.88
1.30	20.0	5.0	26.00	45.13	28.87	6.50	5.59	36.50	45.87	17.63	8.00	5.66
1.30	20.0	10.0	30.00	44.37	25.63	7.25	4.45	40.00	45.49	14.51	7.75	4.71
1.30	20.0	15.0	32.50	43.90	23.60	7.00	3.44	41.00	45.38	13.62	6.75	3.90
1.40	5.0	5.0	25.00	45.15	29.85	6.00	3.70	36.00	45.69	18.31	8.25	3.71
1.40	5.0	10.0	29.00	44.38	26.62	6.50	2.96	40.50	45.18	14.32	8.25	3.06
1.40	5.0	15.0	31.50	43.90	24.60	6.25	2.30	41.00	45.12	13.88	7.00	2.52
1.40	10.0	5.0	26.00	45.01	28.99	6.25	4.32	37.00	45.65	17.35	8.25	4.33
1.40	10.0	10.0	29.50	44.34	26.16	6.50	3.47	40.00	45.32	14.68	7.75	3.59
1.40	10.0	15.0	32.00	43.87	24.13	6.25	2.71	42.50	45.04	12.46	7.25	2.98
1.40	20.0	5.0	27.00	44.94	28.06	6.25	5.68	37.50	45.76	16.74	7.75	5.72
1.40	20.0	10.0	30.50	44.28	25.22	6.50	4.62	42.00	45.27	12.73	7.75	4.81
1.40	20.0	15.0	34.50	43.52	21.98	7.25	3.65	43.50	45.11	11.39	7.00	4.04
1.50	5.0	5.0	26.00	44.96	29.04	6.00	3.74	36.50	45.63	17.87	8.00	3.73
1.50	5.0	10.0	30.00	44.19	25.81	6.50	3.03	39.50	45.29	15.21	7.50	3.10
1.50	5.0	15.0	34.50	43.34	22.16	7.50	2.38	42.00	45.01	12.99	7.00	2.57
1.50	10.0	5.0	26.50	44.91	28.59	6.00	4.36	37.50	45.59	16.91	8.00	4.36

Table 17 (continued)
DATA FOR THE CONSTRUCTION OF ISOKINETIC SUCROSE AND GLYCEROL GRADIENTS FOR CHRIST ROTOR 9792

Particle density g/ml	Temperature °C	Top concentration %(w/w)	Sucrose					Glycerol				
			C_h %(w/w)	Buffer 2x ml	H_2O ml	V_m ml	Velocity of 100 S particle at top speed ml/h	C_h %(w/w)	Buffer 2x ml	H_2O ml	V_m ml	Velocity of 100 S particle at top speed ml/h
1.50	10.0	10.0	30.50	44.15	25.35	6.50	3.55	40.50	45.26	14.24	7.50	3.64
1.50	10.0	15.0	35.00	43.30	21.70	7.50	2.81	43.50	44.93	11.57	7.25	3.04
1.50	20.0	5.0	27.50	44.85	27.65	6.00	5.74	39.00	45.60	15.40	7.75	5.76
1.50	20.0	10.0	32.00	43.99	24.01	6.75	4.72	42.50	45.21	12.29	7.50	4.87
1.50	20.0	15.0	36.00	43.24	20.76	7.50	3.78	44.50	45.00	10.50	7.00	4.12
1.60	5.0	5.0	26.00	44.96	29.04	5.75	3.77	37.50	45.52	16.10	8.00	3.75
1.60	5.0	10.0	30.50	44.10	25.40	6.50	3.07	41.00	45.12	13.88	7.75	3.13
1.60	5.0	15.0	33.00	43.62	23.38	6.25	2.43	42.50	44.95	12.55	7.00	2.61
1.60	10.0	5.0	28.50	44.53	26.97	6.50	4.39	37.50	45.59	16.91	7.75	4.38
1.60	10.0	10.0	31.00	44.06	24.94	6.50	3.60	42.00	45.09	12.91	7.75	3.68
1.60	10.0	15.0	33.50	43.58	22.92	6.25	2.87	43.50	44.93	11.57	7.00	3.08
1.60	20.0	5.0	29.00	44.56	26.44	6.25	5.78	39.00	45.60	15.40	7.50	5.78
1.60	20.0	10.0	31.50	44.09	24.41	6.25	4.79	42.50	45.21	12.29	7.25	4.91
1.60	20.0	15.0	35.00	43.43	21.57	6.50	3.87	44.50	45.00	10.50	6.75	4.17
1.70	5.0	5.0	26.50	44.86	28.64	5.75	3.78	38.00	45.46	16.54	8.00	3.76
1.70	5.0	10.0	31.00	44.00	25.00	6.50	3.10	41.50	45.07	13.43	7.75	3.15
1.70	5.0	15.0	35.50	43.15	21.35	7.50	2.47	43.00	44.90	12.10	7.00	2.64
1.70	10.0	5.0	29.00	44.44	26.56	6.50	4.41	38.00	45.54	16.46	7.75	4.39
1.70	10.0	10.0	31.50	43.96	24.54	6.50	3.64	42.50	45.04	12.46	7.75	3.70
1.70	10.0	15.0	34.00	43.49	22.51	6.25	2.92	44.00	44.87	11.13	7.00	3.11
1.70	20.0	5.0	29.50	44.47	26.03	6.25	5.81	39.50	45.54	14.96	7.50	5.80
1.70	20.0	10.0	32.00	43.99	24.01	6.25	4.83	43.00	45.16	11.84	7.25	4.94
1.70	20.0	15.0	36.00	43.24	20.76	6.75	3.93	45.00	44.94	10.06	6.75	4.21
1.80	5.0	5.0	27.50	44.67	27.83	6.00	3.80	37.50	45.52	16.98	7.75	3.77
1.80	5.0	10.0	29.50	44.29	26.21	5.75	3.13	42.00	45.01	12.99	7.75	3.16
1.80	5.0	15.0	34.50	43.34	22.16	6.75	2.50	43.50	44.84	11.66	7.00	2.65
1.80	10.0	5.0	28.00	44.63	27.37	6.00	4.43	38.50	45.48	16.02	7.75	4.40
1.80	10.0	10.0	32.50	43.77	23.73	6.75	3.67	43.00	44.98	12.02	7.75	3.72
1.80	10.0	15.0	34.50	43.39	22.11	6.50	2.96	44.50	44.82	10.68	7.00	3.14
1.80	20.0	5.0	28.50	44.66	26.84	5.75	5.83	40.00	45.49	14.51	7.50	5.81
1.80	20.0	10.0	33.00	43.81	23.19	6.50	4.87	43.50	45.11	11.39	7.25	4.96
1.80	20.0	15.0	34.50	43.52	21.98	6.00	3.97	45.50	44.89	9.61	6.75	4.24

Table 17 (continued)
DATA FOR THE CONSTRUCTION OF ISOKINETIC SUCROSE AND GLYCEROL GRADIENTS FOR CHRIST ROTOR 9792

Particle density g/ml	Temperature °C	Sucrose						Glycerol				
		Top concentration %(w/w)	C_h %(w/w)	Buffer 2x ml	H_2O ml	V_m ml	Velocity of 100 S particle at top speed ml/h	C_h %(w/w)	Buffer 2x ml	H_2O ml	V_m ml	Velocity of 100 S particle at top speed ml/h
1.90	5.0	5.0	28.50	44.48	27.02	6.25	3.81	38.50	45.40	16.10	8.00	3.78
1.90	5.0	10.0	31.00	44.00	25.00	6.25	3.14	41.50	45.07	13.43	7.50	3.18
1.90	5.0	15.0	34.00	43.43	22.57	6.25	2.53	43.00	44.90	12.10	6.75	2.67
1.90	10.0	5.0	29.00	44.44	26.56	6.25	4.44	39.00	45.43	15.57	7.75	4.41
1.90	10.0	10.0	31.50	43.96	24.54	6.25	3.69	42.50	45.04	12.46	7.50	3.73
1.90	10.0	15.0	35.00	43.30	21.70	6.50	2.98	44.00	44.87	11.13	6.75	3.15
1.90	20.0	5.0	29.50	44.47	26.03	6.00	5.84	39.50	45.54	14.96	7.25	5.82
1.90	20.0	10.0	32.00	43.99	24.01	6.00	4.90	43.00	45.16	11.84	7.00	4.98
1.90	20.0	15.0	36.00	43.24	20.76	6.50	4.01	45.00	44.94	10.06	6.50	4.26
2.00	5.0	5.0	28.00	44.57	27.43	6.00	3.82	38.00	45.46	16.54	7.75	3.78
2.00	5.0	10.0	30.00	44.19	25.81	5.75	3.16	42.50	44.95	12.55	7.75	3.18
2.00	5.0	15.0	34.50	43.34	22.16	6.50	2.55	44.00	44.79	11.21	7.00	2.68
2.00	10.0	5.0	28.00	44.63	27.37	5.75	4.45	38.50	45.48	16.02	7.50	4.42
2.00	10.0	10.0	33.00	43.68	23.32	6.75	3.70	43.50	44.93	11.57	7.75	3.74
2.00	10.0	15.0	35.00	43.30	21.70	6.50	3.00	45.00	44.76	10.24	7.00	3.17
2.00	20.0	5.0	29.00	44.56	26.44	5.75	5.85	40.00	45.49	14.51	7.25	5.83
2.00	20.0	10.0	33.50	43.71	22.79	6.50	4.92	44.00	45.05	10.95	7.25	5.00
2.00	20.0	15.0	35.00	43.43	21.57	6.00	4.04	46.00	44.83	9.17	6.75	4.28

Table 18

DATA FOR THE CONSTRUCTION OF ISOKINETIC SUCROSE AND GLYCEROL GRADIENTS FOR CHRIST ROTOR 9793

Particle density g/ml	Temperature °C	Top concentration %(w/w)	Sucrose					Glycerol				
			C_h %(w/w)	Buffer 2x ml	H_2O ml	V_m ml	Velocity of 100 S particle at top speed ml/h	C_h %(w/w)	Buffer 2x ml	H_2O ml	V_m ml	Velocity of 100 S particle at top speed ml/h
1.20	5.0	5.0	22.50	45.63	31.87	11.50	3.49	33.00	46.03	20.97	15.25	3.57
1.20	5.0	10.0	25.50	45.05	29.45	11.75	2.61	36.00	45.69	18.31	14.50	2.83
1.20	5.0	15.0	29.00	44.38	26.62	12.50	1.87	39.00	45.35	15.65	14.00	2.24
1.20	10.0	5.0	23.50	45.49	31.01	12.25	4.08	33.50	46.05	20.45	15.00	4.17
1.20	10.0	10.0	26.00	45.01	28.99	12.00	3.07	35.50	45.82	18.68	13.50	3.33
1.20	10.0	15.0	29.50	44.34	26.16	12.75	2.21	39.50	45.37	15.13	13.75	2.65
1.20	20.0	5.0	24.00	45.51	30.49	12.00	5.37	34.50	46.10	19.40	14.50	5.52
1.20	20.0	10.0	26.50	45.03	28.47	11.75	4.10	37.00	45.82	17.18	13.50	4.47
1.20	20.0	15.0	30.00	44.37	25.63	12.50	3.00	40.50	45.43	14.07	13.50	3.61
1.30	5.0	5.0	26.00	44.96	29.04	12.00	3.62	36.50	45.63	17.87	15.25	3.65
1.30	5.0	10.0	29.50	44.29	26.21	13.00	2.83	38.50	45.40	16.10	13.75	2.97
1.30	5.0	15.0	32.00	43.81	24.19	12.50	2.14	41.50	45.07	13.43	13.25	2.42
1.30	10.0	5.0	26.00	45.01	28.99	11.75	4.22	37.00	45.65	17.35	15.00	4.26
1.30	10.0	10.0	29.50	44.34	26.16	12.50	3.33	40.00	45.32	14.68	14.25	3.49
1.30	10.0	15.0	32.00	43.87	24.13	12.00	2.53	42.00	45.09	12.91	13.00	2.86
1.30	20.0	5.0	27.00	44.94	28.06	11.75	5.56	39.00	45.60	15.40	15.00	5.63
1.30	20.0	10.0	30.00	44.37	25.63	12.00	4.43	41.50	45.32	13.18	14.00	4.68
1.30	20.0	15.0	33.00	43.81	23.19	12.25	3.42	42.50	45.21	12.29	12.25	3.88
1.40	5.0	5.0	27.50	44.67	27.83	12.00	3.68	38.00	45.46	16.54	15.00	3.69
1.40	5.0	10.0	30.50	44.10	25.40	12.25	2.94	41.00	45.12	13.88	14.25	3.04
1.40	5.0	15.0	33.50	43.52	22.98	12.50	2.28	42.50	44.95	12.55	12.75	2.51
1.40	10.0	5.0	28.00	44.63	27.37	12.00	4.30	39.50	45.37	15.13	15.25	4.31
1.40	10.0	10.0	31.00	44.06	24.94	12.25	3.45	42.00	45.09	12.91	14.25	3.57
1.40	10.0	15.0	34.00	43.49	22.51	12.50	2.70	43.00	44.98	12.02	12.50	2.96
1.40	20.0	5.0	29.00	44.56	26.44	12.00	5.65	40.50	45.43	14.07	14.50	5.69
1.40	20.0	10.0	32.00	43.99	24.01	12.25	4.59	43.50	45.11	11.39	14.00	4.78
1.40	20.0	15.0	35.00	43.43	21.57	12.50	3.63	45.00	44.94	10.06	12.75	4.01
1.50	5.0	5.0	28.50	44.48	27.02	12.00	3.72	39.50	45.29	15.21	15.25	3.71
1.50	5.0	10.0	31.50	43.90	24.60	12.25	3.01	41.00	45.12	13.88	13.50	3.08
1.50	5.0	15.0	34.50	43.34	22.16	12.50	2.37	43.50	44.84	11.66	12.75	2.56
1.50	10.0	5.0	28.50	44.53	26.97	11.50	4.34	40.00	45.32	14.68	14.75	4.34
1.50	10.0	10.0	32.00	43.87	24.13	12.25	3.53	42.00	45.09	12.91	13.50	3.62

Table 18 (continued)

DATA FOR THE CONSTRUCTION OF ISOKINETIC SUCROSE AND GLYCEROL GRADIENTS FOR CHRIST ROTOR 9793

Particle density g/ml	Temperature °C	Top concentration %(w/w)	Sucrose					Glycerol				
			C_h %(w/w)	Buffer 2x ml	H_2O ml	V_m ml	Velocity of 100 S particle at top speed ml/h	C_h %(w/w)	Buffer 2x ml	H_2O ml	V_m ml	Velocity of 100 S particle at top speed ml/h
1.50	10.0	15.0	35.00	43.30	21.70	12.50	2.79	44.00	44.87	11.13	12.50	3.02
1.50	20.0	5.0	29.50	44.47	26.03	11.50	5.71	41.50	45.32	13.18	14.25	5.73
1.50	20.0	10.0	32.50	43.90	23.60	11.75	4.69	44.00	45.05	10.95	13.50	4.84
1.50	20.0	15.0	35.50	43.34	21.16	12.00	3.76	45.00	44.94	10.06	12.00	4.09
1.60	5.0	5.0	29.00	44.38	26.62	11.75	3.74	40.00	45.23	14.77	15.00	3.73
1.60	5.0	10.0	31.50	43.90	24.60	11.75	3.05	42.50	44.95	12.55	14.00	3.11
1.60	5.0	15.0	35.00	43.24	21.76	12.50	2.42	43.50	44.84	11.66	12.25	2.59
1.60	10.0	5.0	29.50	44.34	26.16	11.75	4.37	40.50	45.26	14.24	14.50	4.35
1.60	10.0	10.0	32.00	43.87	24.13	11.75	3.58	42.50	45.04	12.46	13.25	3.65
1.60	10.0	15.0	35.00	43.30	21.70	12.00	2.86	44.00	44.87	11.13	12.00	3.07
1.60	20.0	5.0	30.00	44.37	25.63	11.25	5.75	42.00	45.27	12.73	14.00	5.75
1.60	20.0	10.0	33.50	43.71	22.79	12.00	4.76	44.00	45.05	10.95	13.00	4.88
1.60	20.0	15.0	36.50	43.15	20.35	12.25	3.85	46.00	44.83	9.17	12.25	4.15
1.70	5.0	5.0	29.50	44.29	26.21	11.75	3.76	40.00	45.23	14.77	14.50	3.74
1.70	5.0	10.0	32.00	43.81	24.19	11.75	3.09	42.00	45.01	12.99	13.25	3.13
1.70	5.0	15.0	35.50	43.15	21.35	12.50	2.46	43.50	44.84	11.66	12.00	2.62
1.70	10.0	5.0	30.00	44.25	25.75	11.75	4.39	41.00	45.20	13.80	14.50	4.37
1.70	10.0	10.0	32.50	43.77	23.73	11.75	3.62	43.00	44.98	12.02	13.25	3.68
1.70	10.0	15.0	35.50	43.20	21.30	12.00	2.90	44.50	44.82	10.68	12.00	3.10
1.70	20.0	5.0	31.00	44.18	24.82	11.50	5.77	42.50	45.21	12.29	14.00	5.77
1.70	20.0	10.0	34.00	43.62	22.38	12.00	4.81	45.00	44.94	10.06	13.25	4.91
1.70	20.0	15.0	37.00	43.06	19.94	12.25	3.91	46.50	44.78	8.72	12.25	4.19
1.80	5.0	5.0	29.50	44.29	26.21	11.50	3.78	40.50	45.18	14.32	14.50	3.75
1.80	5.0	10.0	32.00	43.81	24.19	11.50	3.11	43.50	44.84	11.66	14.00	3.15
1.80	5.0	15.0	35.50	43.15	21.35	12.25	2.49	45.00	44.68	10.32	12.75	2.64
1.80	10.0	5.0	30.50	44.15	25.35	11.75	4.41	41.50	45.15	13.35	14.50	4.38
1.80	10.0	10.0	33.00	43.68	23.32	11.75	3.65	43.00	44.98	12.02	13.00	3.69
1.80	10.0	15.0	36.00	43.11	20.89	12.00	2.94	44.50	44.82	10.68	11.75	3.12
1.80	20.0	5.0	31.00	44.18	24.82	11.25	5.79	43.00	45.16	11.84	14.00	5.78
1.80	20.0	10.0	33.50	43.71	22.79	11.25	4.84	45.00	44.94	10.06	13.00	4.94
1.80	20.0	15.0	37.00	43.06	19.94	12.00	3.95	46.50	44.78	8.72	12.00	4.21

Table 18 (continued)

DATA FOR THE CONSTRUCTION OF ISOKINETIC SUCROSE AND GLYCEROL GRADIENTS FOR CHRIST ROTOR 9793

Particle density g/ml	Temperature °C	Top concentration %(w/w)	Sucrose					Glycerol				
			C_h %(w/w)	Buffer 2x ml	H_2O ml	V_m ml	Velocity of 100 S particle at top speed ml/h	C_h %(w/w)	Buffer 2x ml	H_2O ml	V_m ml	Velocity of 100 S particle at top speed ml/h
1.90	5.0	5.0	29.50	44.29	26.21	11.25	3.79	41.00	45.12	13.88	14.75	3.75
1.90	5.0	10.0	33.00	43.62	23.38	12.00	3.13	43.00	44.90	12.10	13.50	3.16
1.90	5.0	15.0	35.50	43.15	21.35	12.00	2.51	44.50	44.73	10.77	12.25	2.65
1.90	10.0	5.0	30.50	44.15	25.35	11.50	4.42	41.50	45.15	13.35	14.25	4.38
1.90	10.0	10.0	33.50	43.58	22.92	12.00	3.67	44.50	44.82	10.68	13.75	3.71
1.90	10.0	15.0	36.50	43.02	20.48	12.25	2.97	45.50	44.71	9.79	12.25	3.13
1.90	20.0	5.0	32.00	43.99	24.01	11.75	5.81	43.00	45.16	11.84	13.75	5.79
1.90	20.0	10.0	34.50	43.52	21.98	11.75	4.87	45.00	44.94	10.06	12.75	4.95
1.90	20.0	15.0	37.00	43.06	19.94	11.75	3.99	46.50	44.78	8.72	11.75	4.24
2.00	5.0	5.0	29.50	44.29	26.21	11.00	3.80	41.00	45.12	13.88	14.50	3.76
2.00	5.0	10.0	32.50	43.71	23.79	11.50	3.14	42.50	44.95	12.55	13.00	3.17
2.00	5.0	15.0	36.00	43.05	20.95	12.25	2.53	44.00	44.79	11.21	11.75	2.67
2.00	10.0	5.0	30.00	44.25	25.75	11.00	4.43	41.50	45.15	13.35	14.25	4.39
2.00	10.0	10.0	33.50	43.58	22.92	11.75	3.68	44.00	44.87	11.13	13.25	3.72
2.00	10.0	15.0	36.00	43.11	20.89	11.75	2.99	45.00	44.76	10.24	11.75	3.15
2.00	20.0	5.0	32.00	43.99	24.01	11.50	5.82	43.00	45.16	11.84	13.75	5.80
2.00	20.0	10.0	35.00	43.43	21.57	12.00	4.89	46.00	44.83	9.17	13.25	4.97
2.00	20.0	15.0	37.50	42.96	19.54	12.00	4.02	47.00	44.73	8.27	12.00	4.26

Table 19

DATA FOR THE CONSTRUCTION OF ISOKINETIC SUCROSE AND GLYCEROL GRADIENTS FOR CHRIST ROTOR 9795

Particle density g/ml	Temperature °C	Top concentration %(w/w)	Sucrose					Glycerol				
			C_h %(w/w)	Buffer 2x ml	H₂O ml	V_m ml	Velocity of 100 S particle at top speed ml/h	C_h %(w/w)	Buffer 2x ml	H₂O ml	V_m ml	Velocity of 100 S particle at top speed ml/h
1.20	5.0	5.0	23.50	45.44	31.06	58.00	4.01	33.50	45.97	20.53	72.00	4.10
1.20	5.0	10.0	27.00	44.76	28.24	63.00	3.00	36.50	45.63	17.87	69.00	3.26
1.20	5.0	15.0	29.50	44.29	26.21	61.00	2.14	39.50	45.29	15.21	67.00	2.57
1.20	10.0	5.0	24.50	45.29	30.21	61.00	4.68	34.00	45.99	20.01	71.00	4.79
1.20	10.0	10.0	27.00	44.82	28.18	60.00	3.53	37.00	45.65	17.35	67.00	3.83
1.20	10.0	15.0	30.00	44.25	25.75	62.00	2.54	40.00	45.32	14.68	66.00	3.05
1.20	20.0	5.0	24.50	45.41	30.09	57.00	6.17	35.50	45.98	18.52	70.00	6.34
1.20	20.0	10.0	27.50	44.85	27.65	59.00	4.71	38.50	45.65	15.85	67.00	5.14
1.20	20.0	15.0	30.50	44.28	25.22	61.00	3.44	40.50	45.43	14.07	62.00	4.15
1.30	5.0	5.0	27.00	44.76	28.24	60.00	4.16	37.00	45.57	17.43	72.00	4.19
1.30	5.0	10.0	30.00	44.19	25.81	62.00	3.26	40.00	45.23	14.77	68.00	3.42
1.30	5.0	15.0	33.00	43.62	23.38	63.00	2.46	42.00	45.01	12.99	63.00	2.78
1.30	10.0	5.0	27.50	44.72	27.78	60.00	4.85	37.50	45.59	16.91	70.00	4.90
1.30	10.0	10.0	30.50	44.15	25.35	62.00	3.82	40.50	45.26	14.24	67.00	4.02
1.30	10.0	15.0	33.50	43.58	22.92	63.00	2.91	43.00	44.98	12.02	63.00	3.29
1.30	20.0	5.0	28.00	44.75	27.25	58.00	6.39	39.00	45.60	15.40	68.00	6.47
1.30	20.0	10.0	31.00	44.18	24.82	60.00	5.09	42.00	45.27	12.73	66.00	5.38
1.30	20.0	15.0	34.00	43.62	22.38	61.00	3.93	44.00	45.05	10.95	61.00	4.46
1.40	5.0	5.0	28.50	44.48	27.02	59.00	4.23	38.00	45.46	16.54	69.00	4.24
1.40	5.0	10.0	31.50	43.90	24.60	61.00	3.38	41.50	45.07	13.43	67.00	3.50
1.40	5.0	15.0	34.50	43.34	22.16	62.00	2.62	43.50	44.84	11.66	62.00	2.88
1.40	10.0	5.0	29.50	44.34	26.16	61.00	4.94	39.00	45.43	15.57	69.00	4.95
1.40	10.0	10.0	32.00	43.87	24.13	61.00	3.97	42.50	45.04	12.46	67.00	4.11
1.40	10.0	15.0	35.00	43.30	21.70	62.00	3.10	44.00	44.87	11.13	61.00	3.40
1.40	20.0	5.0	30.00	44.37	25.63	59.00	6.50	40.50	45.43	14.07	67.00	6.54
1.40	20.0	10.0	33.50	43.71	22.79	63.00	5.28	43.50	45.11	11.39	64.00	5.50
1.40	20.0	15.0	36.00	43.24	20.76	62.00	4.17	45.50	44.89	9.61	60.00	4.61
1.50	5.0	5.0	30.00	44.19	25.81	61.00	4.27	39.00	45.35	15.65	68.00	4.26
1.50	5.0	10.0	32.50	43.71	23.79	61.00	3.46	42.00	45.01	12.99	65.00	3.54
1.50	5.0	15.0	35.00	43.24	21.76	60.00	2.72	44.50	44.73	10.77	62.00	2.94
1.50	10.0	5.0	30.50	44.15	25.35	61.00	4.99	39.50	45.37	15.13	66.00	4.98

Table 19 (continued)

DATA FOR THE CONSTRUCTION OF ISOKINETIC SUCROSE AND GLYCEROL GRADIENTS FOR CHRIST ROTOR 9795

Particle density g/ml	Temperature °C	Top concentration %(w/w)	Sucrose					Glycerol				
			C_h %(w/w)	Buffer 2x ml	H_2O ml	V_m ml	Velocity of 100 S particle at top speed ml/h	C_h %(w/w)	Buffer 2x ml	H_2O ml	V_m ml	Velocity of 100 S particle at top speed ml/h
1.50	10.0	10.0	33.50	43.58	22.92	63.00	4.06	43.00	44.98	12.02	65.00	4.16
1.50	10.0	15.0	36.00	43.11	20.89	62.00	3.21	44.50	44.82	10.68	59.00	3.48
1.50	20.0	5.0	31.00	44.18	24.82	58.00	6.56	41.00	45.38	13.62	64.00	6.58
1.50	20.0	10.0	34.00	43.62	22.38	60.00	5.39	45.00	44.94	10.06	65.00	5.57
1.50	20.0	15.0	37.00	43.06	19.94	62.00	4.32	46.50	44.78	8.72	60.00	4.70
1.60	5.0	5.0	30.50	44.10	25.40	60.00	4.30	39.50	45.29	15.21	67.00	4.28
1.60	5.0	10.0	33.00	43.62	23.38	60.00	3.51	42.50	44.95	12.55	64.00	3.58
1.60	5.0	15.0	36.00	43.05	20.95	62.00	2.78	44.50	44.73	10.77	60.00	2.98
1.60	10.0	5.0	31.00	44.06	24.94	59.00	5.02	40.50	45.26	14.24	67.00	5.00
1.60	10.0	10.0	34.00	43.49	22.51	62.00	4.12	43.50	44.93	11.57	64.00	4.20
1.60	10.0	15.0	37.00	42.92	20.08	64.00	3.28	45.50	44.71	9.79	60.00	3.52
1.60	20.0	5.0	32.00	43.99	24.01	59.00	6.60	42.00	45.27	12.73	65.00	6.61
1.60	20.0	10.0	34.50	43.52	21.98	59.00	5.47	45.50	44.89	9.61	64.00	5.61
1.60	20.0	15.0	37.50	42.96	19.54	61.00	4.42	47.00	44.73	8.27	59.00	4.77
1.70	5.0	5.0	30.50	44.10	25.40	58.00	4.32	40.00	45.23	14.77	67.00	4.30
1.70	5.0	10.0	33.50	43.52	22.98	60.00	3.55	43.00	44.90	12.10	64.00	3.60
1.70	5.0	15.0	36.50	42.96	20.54	62.00	2.83	45.00	44.68	10.32	60.00	3.01
1.70	10.0	5.0	31.00	44.06	24.94	57.00	5.04	41.00	45.20	13.80	66.00	5.02
1.70	10.0	10.0	34.00	43.49	22.51	60.00	4.16	44.00	44.87	11.13	64.00	4.23
1.70	10.0	15.0	37.00	42.92	20.08	62.00	3.34	46.00	44.65	9.35	60.00	3.56
1.70	20.0	5.0	32.50	43.90	23.60	58.00	6.63	42.50	45.21	12.29	64.00	6.63
1.70	20.0	10.0	35.00	43.43	21.57	59.00	5.52	45.50	44.89	9.61	62.00	5.65
1.70	20.0	15.0	38.00	42.87	19.13	61.00	4.49	47.50	44.67	7.83	59.00	4.81
1.80	5.0	5.0	31.00	44.00	25.00	58.00	4.34	40.50	45.18	14.32	67.00	4.31
1.80	5.0	10.0	34.00	43.43	22.57	61.00	3.57	43.50	44.84	11.66	64.00	3.62
1.80	5.0	15.0	37.00	42.86	20.14	63.00	2.86	45.50	44.62	9.88	60.00	3.03
1.80	10.0	5.0	31.50	43.96	24.54	58.00	5.06	41.00	45.20	13.80	65.00	5.03
1.80	10.0	10.0	34.50	43.39	22.11	60.00	4.19	44.50	44.82	10.68	64.00	4.25
1.80	10.0	15.0	37.00	42.92	20.08	60.00	3.38	46.00	44.65	9.35	59.00	3.58
1.80	20.0	5.0	32.50	43.90	23.60	57.00	6.66	42.50	45.21	12.29	63.00	6.64
1.80	20.0	10.0	35.50	43.34	21.16	59.00	5.56	45.50	44.89	9.61	61.00	5.67

Table 19 (continued)

DATA FOR THE CONSTRUCTION OF ISOKINETIC SUCROSE AND GLYCEROL GRADIENTS FOR CHRIST ROTOR 9795

Particle density g/ml	Temperature °C	Top concentration %(w/w)	Sucrose					Glycerol				
			C_h %(w/w)	Buffer 2x ml	H_2O ml	V_m ml	Velocity of 100 S particle at top speed ml/h	C_h %(w/w)	Buffer 2x ml	H_2O ml	V_m ml	Velocity of 100 S particle at top speed ml/h
1.80	20.0	15.0	38.00	42.87	19.13	59.00	4.54	48.00	44.62	7.38	60.00	4.84
1.90	5.0	5.0	31.50	43.90	24.60	59.00	4.35	40.50	45.18	14.32	66.00	4.31
1.90	5.0	10.0	34.00	43.43	22.57	59.00	3.59	44.00	44.79	11.21	65.00	3.63
1.90	5.0	15.0	36.50	42.96	20.54	59.00	2.89	45.50	44.62	9.88	59.00	3.05
1.90	10.0	5.0	32.00	43.87	24.13	58.00	5.08	41.50	45.15	13.35	65.00	5.04
1.90	10.0	10.0	34.50	43.39	22.11	59.00	4.21	44.50	44.82	10.68	63.00	4.26
1.90	10.0	15.0	37.50	42.83	19.67	61.00	3.41	46.50	44.60	8.90	60.00	3.60
1.90	20.0	5.0	33.00	43.81	23.19	57.00	6.68	43.00	45.16	11.84	63.00	6.65
1.90	20.0	10.0	35.50	43.34	21.16	58.00	5.60	45.50	44.89	9.61	60.00	5.69
1.90	20.0	15.0	38.00	42.87	19.13	58.00	4.58	48.00	44.62	7.38	59.00	4.87
2.00	5.0	5.0	31.50	43.90	24.60	58.00	4.36	40.50	45.18	14.32	65.00	4.32
2.00	5.0	10.0	34.50	43.34	22.16	61.00	3.61	44.00	44.79	11.21	64.00	3.64
2.00	5.0	15.0	37.00	42.86	20.14	61.00	2.91	45.50	44.62	9.88	59.00	3.06
2.00	10.0	5.0	32.00	43.87	24.13	57.00	5.09	41.50	45.15	13.35	65.00	5.05
2.00	10.0	10.0	35.00	43.30	21.70	60.00	4.23	44.50	44.82	10.68	65.00	4.27
2.00	10.0	15.0	37.50	42.83	19.67	60.00	3.43	46.50	44.60	8.90	62.00	3.62
2.00	20.0	5.0	33.50	43.71	22.79	58.00	6.69	43.00	45.16	11.84	59.00	6.66
2.00	20.0	10.0	36.00	43.24	20.76	59.00	5.62	45.50	44.89	9.61	62.00	5.71
2.00	20.0	15.0	38.50	42.78	18.72	59.00	4.62	48.50	44.57	6.93	60.00	4.89

Table 20

DATA FOR THE CONSTRUCTION OF ISOKINETIC SUCROSE AND GLYCEROL GRADIENTS FOR INTERNATIONAL ROTOR 485

Particle density g/ml	Temperature °C	Top concentration %(w/w)	Sucrose					Glycerol				
			C_h %(w/w)	Buffer 2x ml	H_2O ml	V_m ml	Velocity of 100 S particle at top speed ml/h	C_h %(w/w)	Buffer 2x ml	H_2O ml	V_m ml	Velocity of 100 S particle at top speed ml/h
1.20	5.0	5.0	23.00	45.53	31.47	37.00	4.17	31.00	46.26	22.74	42.00	4.26
1.20	5.0	10.0	25.50	45.05	29.45	36.00	3.12	35.00	45.80	19.20	42.00	3.39
1.20	5.0	15.0	28.50	44.48	27.02	36.50	2.23	38.00	45.46	16.54	40.50	2.67
1.20	10.0	5.0	22.50	45.68	31.82	34.00	4.87	32.00	46.22	21.78	42.50	4.98
1.20	10.0	10.0	26.00	45.01	28.99	36.50	3.67	36.00	45.76	18.24	42.50	3.99
1.20	10.0	15.0	29.00	44.44	26.56	37.00	2.64	38.50	45.48	16.02	40.00	3.17
1.20	20.0	5.0	23.50	45.60	30.90	35.00	6.42	33.50	46.21	20.29	42.00	6.59
1.20	20.0	10.0	26.50	45.03	28.47	36.00	4.90	37.50	45.76	16.74	42.50	5.35
1.20	20.0	15.0	29.50	44.47	26.03	36.50	3.58	39.50	45.54	14.96	39.00	4.32
1.30	5.0	5.0	25.50	45.05	29.45	35.50	4.32	34.50	45.86	19.64	42.50	4.36
1.30	5.0	10.0	28.50	44.48	27.02	36.50	3.39	38.50	45.40	16.10	42.00	3.55
1.30	5.0	15.0	32.00	43.81	24.19	38.50	2.56	41.00	45.12	13.88	39.50	2.89
1.30	10.0	5.0	26.00	45.01	28.99	36.00	5.05	35.50	45.82	18.68	42.50	5.09
1.30	10.0	10.0	29.00	44.44	26.56	36.50	3.98	39.50	45.37	15.13	42.50	4.18
1.30	10.0	15.0	32.00	43.87	24.13	37.00	3.03	42.00	45.09	12.91	40.00	3.42
1.30	20.0	5.0	27.00	44.94	28.06	36.00	6.64	37.00	45.82	17.18	42.00	6.73
1.30	20.0	10.0	29.50	44.47	26.03	35.50	5.29	40.50	45.43	14.07	41.00	5.59
1.30	20.0	15.0	33.00	43.81	23.19	37.50	4.09	42.50	45.21	12.29	37.50	4.63
1.40	5.0	5.0	27.00	44.76	28.24	35.50	4.40	36.00	45.69	18.31	42.00	4.41
1.40	5.0	10.0	30.00	44.19	25.81	36.00	3.52	40.50	45.18	14.32	42.50	3.64
1.40	5.0	15.0	33.00	43.62	23.38	36.50	2.73	42.00	45.01	12.99	38.00	2.99
1.40	10.0	5.0	27.50	44.72	27.78	35.50	5.13	37.00	45.65	17.35	42.00	5.15
1.40	10.0	10.0	30.00	44.25	25.75	35.00	4.13	41.00	45.20	13.80	41.50	4.27
1.40	10.0	15.0	33.50	43.58	22.92	36.50	3.22	43.50	44.93	11.57	39.50	3.54
1.40	20.0	5.0	28.50	44.66	26.84	35.50	6.75	39.50	45.54	14.96	42.50	6.80
1.40	20.0	10.0	31.00	44.18	24.82	35.00	5.49	42.00	45.27	12.73	40.00	5.72
1.40	20.0	15.0	34.50	43.52	21.98	37.00	4.34	44.00	45.05	10.95	37.00	4.80
1.50	5.0	5.0	28.00	44.57	27.43	35.50	4.44	37.50	45.52	16.98	42.50	4.44
1.50	5.0	10.0	31.00	44.00	25.00	36.50	3.60	40.50	45.18	14.32	40.50	3.68
1.50	5.0	15.0	34.00	43.43	22.57	37.00	2.83	43.00	44.90	12.10	38.00	3.06
1.50	10.0	5.0	28.00	44.63	27.37	34.00	5.19	38.50	45.48	16.02	42.50	5.18

Table 20 (continued)

DATA FOR THE CONSTRUCTION OF ISOKINETIC SUCROSE AND GLYCEROL GRADIENTS FOR INTERNATIONAL ROTOR 485

Particle density g/ml	Temperature °C	Top concentration %(w/w)	Sucrose					Glycerol				
			C_h %(w/w)	Buffer 2x ml	H_2O ml	V_m ml	Velocity of 100 S particle at top speed ml/h	C_h %(w/w)	Buffer 2x ml	H_2O ml	V_m ml	Velocity of 100 S particle at top speed ml/h
1.50	10.0	10.0	31.00	44.06	24.94	35.00	4.22	41.50	45.15	13.35	40.50	4.33
1.50	10.0	15.0	34.50	43.39	22.11	37.00	3.34	43.50	44.93	11.57	37.00	3.61
1.50	20.0	5.0	29.50	44.47	26.03	35.00	6.82	40.50	45.43	14.07	42.00	6.84
1.50	20.0	10.0	32.50	43.90	23.60	36.00	5.61	43.50	45.11	11.39	40.50	5.79
1.50	20.0	15.0	35.50	43.34	21.16	37.00	4.49	44.50	45.00	10.50	36.00	4.89
1.60	5.0	5.0	28.50	44.48	27.02	35.00	4.48	38.00	45.46	16.54	42.00	4.45
1.60	5.0	10.0	31.50	43.90	24.60	36.00	3.65	41.00	45.12	13.88	40.00	3.72
1.60	5.0	15.0	34.50	43.34	22.16	36.50	2.89	43.50	44.84	11.66	37.50	3.10
1.60	10.0	5.0	29.50	44.34	26.16	36.00	5.22	39.50	45.37	15.13	42.50	5.20
1.60	10.0	10.0	31.50	43.96	24.54	34.50	4.28	42.00	45.09	12.91	40.00	4.37
1.60	10.0	15.0	35.00	43.30	21.70	36.50	3.42	44.00	44.87	11.13	37.00	3.66
1.60	20.0	5.0	30.00	44.37	25.63	34.50	6.87	41.00	45.38	13.62	41.50	6.87
1.60	20.0	10.0	33.00	43.81	23.19	35.50	5.69	43.50	45.11	11.39	39.00	5.84
1.60	20.0	15.0	36.00	43.24	20.76	36.50	4.60	45.00	44.94	10.06	35.50	4.96
1.70	5.0	5.0	29.00	44.38	26.62	35.00	4.50	39.00	45.35	15.65	42.50	4.47
1.70	5.0	10.0	32.00	43.81	24.19	36.00	3.69	42.50	44.95	12.55	41.50	3.74
1.70	5.0	15.0	34.50	43.34	22.16	35.50	2.94	43.50	44.84	11.66	36.50	3.13
1.70	10.0	5.0	29.00	44.44	26.56	33.50	5.25	40.00	45.32	14.68	42.50	5.22
1.70	10.0	10.0	32.00	43.87	24.13	34.50	4.32	43.50	44.93	11.57	41.50	4.39
1.70	10.0	15.0	35.50	43.20	21.30	36.50	3.47	44.00	44.87	11.13	36.00	3.70
1.70	20.0	5.0	31.00	44.18	24.82	35.50	6.90	41.50	45.32	13.18	41.50	6.89
1.70	20.0	10.0	33.50	43.71	22.79	35.50	5.74	45.00	44.94	10.06	40.50	5.87
1.70	20.0	15.0	36.50	43.15	20.35	36.50	4.67	45.50	44.89	9.61	35.50	5.00
1.80	5.0	5.0	29.00	44.38	26.62	34.00	4.51	39.00	45.35	15.65	42.00	4.48
1.80	5.0	10.0	32.00	43.81	24.19	35.00	3.72	42.00	45.01	12.99	40.00	3.76
1.80	5.0	15.0	35.00	43.24	21.76	36.00	2.98	44.00	44.79	11.21	37.00	3.15
1.80	10.0	5.0	30.00	44.25	25.75	35.00	5.27	40.00	45.32	14.68	42.00	5.23
1.80	10.0	10.0	32.50	43.77	23.73	35.00	4.36	43.00	44.98	12.02	40.00	4.42
1.80	10.0	15.0	35.00	43.30	21.70	34.50	3.51	44.50	44.82	10.68	36.00	3.73
1.80	20.0	5.0	31.00	44.18	24.82	34.50	6.92	41.50	45.32	13.18	40.50	6.91
1.80	20.0	10.0	33.50	43.71	22.79	34.50	5.79	44.00	45.05	10.95	38.00	5.90

Table 20 (continued)

DATA FOR THE CONSTRUCTION OF ISOKINETIC SUCROSE AND GLYCEROL GRADIENTS FOR INTERNATIONAL ROTOR 485

Particle density g/ml	Temperature °C	Top concentration %(w/w)	Sucrose					Glycerol				
			C_h %(w/w)	Buffer 2x ml	H_2O ml	V_m ml	Velocity of 100 S particle at top speed ml/h	C_h %(w/w)	Buffer 2x ml	H_2O ml	V_m ml	Velocity of 100 S particle at top speed ml/h
1.80	20.0	15.0	36.50	43.15	20.35	35.50	4.72	46.00	44.83	9.17	35.50	5.04
1.90	5.0	5.0	29.50	44.29	26.21	34.50	4.53	39.50	45.29	15.21	42.50	4.49
1.90	5.0	10.0	32.00	43.81	24.19	34.50	3.74	43.00	44.90	12.10	41.00	3.77
1.90	5.0	15.0	35.50	43.15	21.35	36.50	3.00	44.00	44.79	11.21	36.50	3.17
1.90	10.0	5.0	29.50	44.34	26.16	33.00	5.28	40.50	45.26	14.24	42.00	5.24
1.90	10.0	10.0	33.00	43.68	23.32	35.50	4.38	44.00	44.87	11.13	41.00	4.43
1.90	10.0	15.0	36.00	43.11	20.89	36.50	3.54	45.00	44.76	10.24	36.50	3.75
1.90	20.0	5.0	30.50	44.28	25.22	32.50	6.94	42.00	45.27	12.73	40.50	6.92
1.90	20.0	10.0	34.00	43.62	22.38	35.00	5.82	45.00	44.94	10.06	39.00	5.92
1.90	20.0	15.0	37.00	43.06	19.94	36.00	4.77	46.50	44.78	8.72	36.00	5.06
2.00	5.0	5.0	30.00	44.19	25.81	35.00	4.54	40.00	45.23	14.77	42.50	4.49
2.00	5.0	10.0	32.50	43.71	23.79	35.00	3.75	43.50	44.84	11.66	41.50	3.78
2.00	5.0	15.0	35.50	43.15	21.35	36.00	3.03	44.00	44.79	11.21	36.00	3.19
2.00	10.0	5.0	30.00	44.25	25.75	33.50	5.29	40.50	45.26	14.24	41.50	5.25
2.00	10.0	10.0	33.00	43.68	23.32	35.00	4.40	44.00	44.87	11.13	40.50	4.44
2.00	10.0	15.0	36.00	43.11	20.89	36.00	3.57	45.00	44.76	10.24	36.00	3.76
2.00	20.0	5.0	31.00	44.18	24.82	33.00	6.96	42.00	45.27	12.73	40.00	6.93
2.00	20.0	10.0	34.00	43.62	22.38	34.50	5.85	45.50	44.89	9.61	39.50	5.94
2.00	20.0	15.0	37.00	43.06	19.94	35.50	4.80	46.50	44.78	8.72	35.50	5.09

Table 21

DATA FOR THE CONSTRUCTION OF ISOKINETIC SUCROSE AND GLYCEROL GRADIENTS FOR INTERNATIONAL ROTOR 488

Particle density g/ml	Temperature °C	Top concentration %(w/w)	Sucrose					Glycerol				
			C_h %(w/w)	Buffer 2x ml	H_2O ml	V_m ml	Velocity of 100 S particle at top speed ml/h	C_h %(w/w)	Buffer 2x ml	H_2O ml	V_m ml	Velocity of 100 S particle at top speed ml/h
1.20	5.0	5.0	23.00	45.53	31.47	10.00	3.03	32.50	46.09	21.41	12.25	3.09
1.20	5.0	10.0	26.50	44.86	28.64	10.75	2.27	36.00	45.69	18.31	12.00	2.46
1.20	5.0	15.0	30.00	44.19	25.81	11.75	1.62	40.00	45.23	14.77	12.50	1.94
1.20	10.0	5.0	23.50	45.49	31.01	10.00	3.54	34.00	45.99	20.01	12.75	3.62
1.20	10.0	10.0	26.50	44.91	28.59	10.25	2.67	36.50	45.71	17.79	11.75	2.89
1.20	10.0	15.0	30.00	44.25	25.75	11.25	1.92	40.50	45.26	14.24	12.25	2.30
1.20	20.0	5.0	24.50	45.41	30.09	10.25	4.66	35.00	46.04	18.96	12.25	4.79
1.20	20.0	10.0	27.00	44.94	28.06	10.00	3.56	38.00	45.71	16.29	11.75	3.88
1.20	20.0	15.0	30.50	44.28	25.22	11.00	2.60	41.50	45.32	13.18	12.00	3.13
1.30	5.0	5.0	26.00	44.96	29.04	10.00	3.14	37.50	45.52	16.98	13.25	3.16
1.30	5.0	10.0	29.00	44.38	26.62	10.25	2.46	39.50	45.29	15.21	12.00	2.58
1.30	5.0	15.0	32.50	43.71	23.79	10.75	1.86	42.50	44.95	12.55	11.75	2.10
1.30	10.0	5.0	26.50	44.91	28.59	10.00	3.66	38.50	45.48	16.02	13.25	3.70
1.30	10.0	10.0	29.50	44.34	26.16	10.25	2.89	40.00	45.32	14.68	11.75	3.03
1.30	10.0	15.0	33.00	43.68	23.32	11.00	2.20	43.00	44.98	12.02	11.50	2.48
1.30	20.0	5.0	28.00	44.75	27.25	10.50	4.82	40.50	45.43	14.07	13.25	4.89
1.30	20.0	10.0	31.00	44.18	24.82	10.75	3.84	41.00	45.38	13.62	11.25	4.06
1.30	20.0	15.0	33.50	43.71	22.79	10.50	2.97	44.00	45.05	10.95	11.00	3.37
1.40	5.0	5.0	28.00	44.57	27.43	10.25	3.19	40.00	45.23	14.77	13.50	3.20
1.40	5.0	10.0	31.00	44.00	25.00	10.50	2.55	41.00	45.12	13.88	11.75	2.64
1.40	5.0	15.0	34.00	43.43	22.57	10.75	1.98	44.00	44.79	11.21	11.50	2.17
1.40	10.0	5.0	28.50	44.53	26.97	10.25	3.73	41.00	45.20	13.80	13.50	3.74
1.40	10.0	10.0	31.50	43.96	24.54	10.50	3.00	41.50	45.15	13.35	11.50	3.10
1.40	10.0	15.0	34.50	43.39	22.11	10.75	2.34	44.50	44.82	10.68	11.25	2.57
1.40	20.0	5.0	29.50	44.47	26.03	10.25	4.90	41.50	45.32	13.18	12.50	4.94
1.40	20.0	10.0	32.50	43.90	23.60	10.50	3.99	43.00	45.16	11.84	11.25	4.15
1.40	20.0	15.0	35.50	43.34	21.16	10.75	3.15	45.50	44.89	9.61	10.75	3.48
1.50	5.0	5.0	29.00	44.38	26.62	10.25	3.23	40.00	45.23	14.77	12.75	3.22
1.50	5.0	10.0	32.00	43.81	24.19	10.50	2.61	42.50	44.95	12.55	12.00	2.68
1.50	5.0	15.0	35.00	43.24	21.76	10.75	2.05	45.00	44.68	10.32	11.50	2.22
1.50	10.0	5.0	30.00	44.25	25.75	10.50	3.77	41.50	45.15	13.35	13.00	3.76

Table 21 (continued)

DATA FOR THE CONSTRUCTION OF ISOKINETIC SUCROSE AND GLYCEROL GRADIENTS FOR INTERNATIONAL ROTOR 488

Particle density g/ml	Temperature °C	Top concentration %(w/w)	Sucrose C_h %(w/w)	Buffer 2x ml	H_2O ml	V_m ml	Velocity of 100 S particle at top speed ml/h	Glycerol C_h %(w/w)	Buffer 2x ml	H_2O ml	V_m ml	Velocity of 100 S particle at top speed ml/h
1.50	10.0	10.0	32.50	43.77	23.73	10.50	3.06	43.50	44.93	11.57	12.00	3.14
1.50	10.0	15.0	35.50	43.20	21.30	10.75	2.42	45.50	44.71	9.79	11.25	2.62
1.50	20.0	5.0	30.00	44.37	25.63	9.75	4.95	42.50	45.21	12.29	12.25	4.97
1.50	20.0	10.0	33.50	43.71	22.79	10.50	4.07	45.00	44.94	10.06	11.75	4.20
1.50	20.0	15.0	36.50	43.15	20.35	10.75	3.26	46.50	44.78	8.72	10.75	3.55
1.60	5.0	5.0	29.50	44.29	26.21	10.00	3.25	40.50	45.18	14.32	12.50	3.23
1.60	5.0	10.0	32.00	43.81	24.19	10.00	2.65	42.50	44.95	12.55	11.50	2.70
1.60	5.0	15.0	35.00	43.24	21.76	10.25	2.10	44.50	44.73	10.77	10.75	2.25
1.60	10.0	5.0	30.00	44.25	25.75	10.00	3.79	42.00	45.09	12.91	12.75	3.78
1.60	10.0	10.0	32.50	43.77	23.73	10.00	3.11	43.50	44.93	11.57	11.50	3.17
1.60	10.0	15.0	35.50	43.20	21.30	10.25	2.48	45.50	44.71	9.79	10.75	2.66
1.60	20.0	5.0	31.50	44.09	24.41	10.25	4.99	44.00	45.05	12.50	12.50	4.99
1.60	20.0	10.0	34.00	43.62	22.38	10.25	4.13	44.50	45.00	10.95	11.00	4.24
1.60	20.0	15.0	37.00	43.06	19.94	10.50	3.34	47.50	44.67	7.83	11.00	3.60
1.70	5.0	5.0	30.00	44.19	25.81	10.00	3.27	41.50	45.07	13.43	12.75	3.24
1.70	5.0	10.0	32.50	43.71	23.79	10.00	2.68	43.00	44.90	12.10	11.50	2.72
1.70	5.0	15.0	36.00	43.05	20.95	10.75	2.14	45.00	44.68	10.32	10.75	2.27
1.70	10.0	5.0	30.50	44.15	25.35	10.00	3.81	42.50	45.04	12.46	12.75	3.79
1.70	10.0	10.0	33.00	43.68	23.32	10.00	3.14	44.00	44.87	11.13	11.50	3.19
1.70	10.0	15.0	36.00	43.11	20.89	10.25	2.52	46.00	44.65	9.35	10.75	2.69
1.70	20.0	5.0	31.00	44.18	24.82	9.50	5.01	44.00	45.05	10.95	12.25	5.01
1.70	20.0	10.0	34.50	43.52	21.98	10.25	4.17	45.50	44.89	9.61	11.25	4.26
1.70	20.0	15.0	37.50	42.96	19.54	10.50	3.39	47.50	44.67	7.83	10.75	3.63
1.80	5.0	5.0	30.00	44.19	25.81	9.75	3.28	41.50	45.07	13.43	12.50	3.25
1.80	5.0	10.0	33.50	43.52	22.98	10.50	2.70	43.00	44.90	12.10	11.25	2.73
1.80	5.0	15.0	36.00	43.05	20.95	10.50	2.16	45.50	44.62	9.88	10.75	2.29
1.80	5.0	5.0	30.50	44.15	25.35	9.75	3.82	43.00	44.98	12.02	12.75	3.80
1.80	10.0	10.0	34.00	43.49	22.51	10.50	3.16	44.00	44.87	11.13	11.25	3.21
1.80	10.0	15.0	37.00	42.92	20.08	10.75	2.55	46.50	44.60	8.90	10.75	2.71
1.80	20.0	5.0	32.50	43.90	23.60	10.25	5.03	44.50	45.00	10.50	12.25	5.02
1.80	20.0	10.0	34.50	43.52	21.98	10.00	4.20	45.50	44.89	9.61	11.00	4.28

Table 21 (continued)

DATA FOR THE CONSTRUCTION OF ISOKINETIC SUCROSE AND GLYCEROL GRADIENTS FOR INTERNATIONAL ROTOR 488

Particle density g/ml	Temperature °C	Top concentration %(w/w)	Sucrose					Glycerol				
			C_h %(w/w)	Buffer 2x ml	H₂O ml	V_m ml	Velocity of 100 S particle at top speed ml/h	C_h %(w/w)	Buffer 2x ml	H₂O ml	V_m ml	Velocity of 100 S particle at top speed ml/h
1.80	20.0	15.0	37.50	42.96	19.54	10.25	3.43	48.00	44.62	7.38	10.75	3.66
1.90	5.0	5.0	31.00	44.00	25.00	10.25	3.29	42.50	44.95	12.55	13.00	3.26
1.90	5.0	10.0	33.00	43.62	23.38	10.00	2.71	44.50	44.73	10.77	12.00	2.74
1.90	5.0	15.0	36.00	43.05	20.95	10.25	2.18	45.50	44.62	9.88	10.75	2.30
1.90	10.0	5.0	31.00	44.06	24.94	9.75	3.83	43.00	44.98	12.02	12.50	3.81
1.90	10.0	10.0	34.00	43.49	22.51	10.25	3.18	44.00	44.87	11.13	11.00	3.22
1.90	10.0	15.0	36.50	43.02	20.48	10.25	2.57	46.50	44.60	8.90	10.75	2.72
1.90	20.0	5.0	32.50	43.90	23.60	10.00	5.04	45.00	44.94	10.06	12.25	5.02
1.90	20.0	10.0	34.50	43.52	21.98	9.75	4.23	45.50	44.89	9.61	10.75	4.30
1.90	20.0	15.0	38.00	42.87	19.13	10.50	3.46	48.50	44.57	6.93	10.75	3.68
2.00	5.0	5.0	30.50	44.10	25.40	9.75	3.29	42.50	44.95	12.55	12.75	3.26
2.00	5.0	10.0	33.00	43.62	23.38	9.75	2.73	43.50	44.84	11.66	11.25	2.75
2.00	5.0	15.0	36.50	42.96	20.54	10.50	2.20	46.00	44.57	9.43	10.75	2.31
2.00	10.0	5.0	31.50	43.96	24.54	10.00	3.84	43.50	44.93	11.57	12.75	3.81
2.00	10.0	10.0	33.50	43.58	22.92	9.75	3.20	45.00	44.76	10.24	11.50	3.23
2.00	10.0	15.0	36.50	43.02	20.48	10.00	2.59	47.00	44.54	8.46	10.75	2.73
2.00	20.0	5.0	32.00	43.99	24.01	9.50	5.05	44.00	45.05	10.95	11.75	5.03
2.00	20.0	10.0	35.50	43.34	21.16	10.25	4.25	45.50	44.89	9.61	10.75	4.31
2.00	20.0	15.0	38.00	42.87	19.13	10.25	3.49	48.50	44.57	6.93	10.75	3.69

Table 22

DATA FOR THE CONSTRUCTION OF ISOKINETIC SUCROSE AND GLYCEROL GRADIENTS FOR INTERNATIONAL ROTOR 498

Particle density g/ml	Temperature °C	Top concentration %(w/w)	Sucrose					Glycerol				
			C_h %(w/w)	Buffer 2x ml	H_2O ml	V_m ml	Velocity of 100 S particle at top speed ml/h	C_h %(w/w)	Buffer 2x ml	H_2O ml	V_m ml	Velocity of 100 S particle at top speed ml/h
1.20	5.0	5.0	22.00	45.73	32.27	3.70	2.63	30.00	46.38	23.62	4.40	2.68
1.20	5.0	10.0	24.50	45.24	30.26	3.60	1.97	33.50	45.97	20.53	4.20	2.13
1.20	5.0	15.0	28.50	44.48	27.02	4.00	1.40	36.50	45.63	17.87	4.00	1.68
1.20	10.0	5.0	22.00	45.77	32.23	3.60	3.07	31.00	46.33	22.67	4.40	3.14
1.20	10.0	10.0	25.50	45.10	29.40	3.80	2.31	35.00	45.88	19.12	4.40	2.51
1.20	10.0	15.0	28.00	44.63	27.37	3.60	1.66	37.50	45.59	16.91	4.10	2.00
1.20	20.0	5.0	22.50	45.79	31.71	3.50	4.04	32.50	46.32	21.18	4.40	4.15
1.20	20.0	10.0	25.50	45.22	29.28	3.60	3.09	36.50	45.87	17.63	4.40	3.37
1.20	20.0	15.0	29.00	44.56	26.44	3.80	2.26	38.50	45.65	15.85	4.00	2.72
1.30	5.0	5.0	24.50	45.24	30.26	3.60	2.72	33.50	45.97	20.53	4.40	2.74
1.30	5.0	10.0	28.50	44.48	27.02	4.00	2.13	37.50	45.52	16.98	4.40	2.24
1.30	5.0	15.0	31.50	43.90	24.60	4.00	1.61	40.00	45.23	14.77	4.10	1.82
1.30	10.0	5.0	25.50	45.10	29.40	3.80	3.18	34.00	45.99	20.01	4.40	3.21
1.30	10.0	10.0	28.00	44.63	27.37	3.70	2.50	38.00	45.54	16.46	4.30	2.63
1.30	10.0	15.0	31.00	44.06	24.94	3.70	1.91	40.50	45.26	14.24	4.00	2.15
1.30	20.0	5.0	26.00	45.13	28.87	3.70	4.18	35.50	45.98	18.52	4.30	4.24
1.30	20.0	10.0	28.50	44.66	26.84	3.60	3.33	40.00	45.49	14.51	4.40	3.52
1.30	20.0	15.0	31.50	44.09	24.41	3.60	2.58	41.50	45.32	13.18	3.90	2.92
1.40	5.0	5.0	26.00	44.96	29.04	3.60	2.77	35.00	45.80	19.20	4.40	2.78
1.40	5.0	10.0	29.50	44.29	26.21	3.80	2.22	39.00	45.35	15.65	4.30	2.29
1.40	5.0	15.0	33.00	43.62	23.38	4.00	1.72	42.50	44.95	12.55	4.30	1.89
1.40	10.0	5.0	26.50	44.91	28.59	3.60	3.23	35.50	45.82	18.68	4.30	3.24
1.40	10.0	10.0	29.50	44.34	26.16	3.70	2.60	39.50	45.37	15.13	4.20	2.69
1.40	10.0	15.0	32.50	43.77	23.73	3.70	2.03	43.00	44.98	12.02	4.20	2.23
1.40	20.0	5.0	27.00	44.94	28.06	3.50	4.25	37.00	45.82	17.18	4.20	4.28
1.40	20.0	10.0	30.50	44.28	25.22	3.70	3.46	41.00	45.38	13.62	4.20	3.60
1.40	20.0	15.0	34.00	43.62	22.38	3.90	2.73	45.00	44.94	10.06	4.30	3.02
1.50	5.0	5.0	26.50	44.86	28.64	3.50	2.80	36.00	45.69	18.31	4.40	2.79
1.50	5.0	10.0	30.00	44.19	25.81	3.70	2.27	40.50	45.18	14.32	4.40	2.32
1.50	5.0	15.0	33.50	43.52	22.98	3.90	1.78	43.50	44.84	11.66	4.30	1.93
1.50	10.0	5.0	27.50	44.72	27.78	3.60	3.27	36.50	45.71	17.79	4.30	3.26
1.50	10.0	10.0	30.50	44.15	25.35	3.70	2.66	41.00	45.20	13.80	4.30	2.73

Table 22 (continued)

DATA FOR THE CONSTRUCTION OF ISOKINETIC SUCROSE AND GLYCEROL GRADIENTS FOR INTERNATIONAL ROTOR 498

Particle density g/ml	Temperature °C	Top concentration %(w/w)	Sucrose					Glycerol				
			C_h %(w/w)	Buffer 2x ml	H_2O ml	V_m ml	Velocity of 100 S particle at top speed ml/h	C_h %(w/w)	Buffer 2x ml	H_2O ml	V_m ml	Velocity of 100 S particle at top speed ml/h
1.50	10.0	15.0	34.00	43.49	22.51	3.90	2.10	44.00	44.87	11.13	4.20	2.28
1.50	20.0	5.0	28.50	44.66	26.84	3.60	4.30	37.50	45.76	16.74	4.10	4.31
1.50	20.0	10.0	31.50	44.09	24.41	3.70	3.53	42.00	45.27	12.73	4.10	3.65
1.50	20.0	15.0	35.00	43.43	21.57	3.90	2.83	44.50	45.00	10.50	3.90	3.08
1.60	5.0	5.0	28.00	44.57	27.43	3.70	2.82	36.50	45.63	17.87	4.30	2.81
1.60	5.0	10.0	31.00	44.00	25.00	3.80	2.30	40.00	45.23	14.77	4.20	2.34
1.60	5.0	15.0	33.50	43.52	22.98	3.70	1.82	42.50	44.95	12.55	3.90	1.95
1.60	10.0	5.0	28.50	44.53	26.97	3.70	3.29	37.00	45.65	17.35	4.20	3.28
1.60	10.0	10.0	31.50	43.96	24.54	3.80	2.70	41.00	45.20	13.80	4.20	2.75
1.60	10.0	15.0	34.00	43.49	22.51	3.70	2.15	43.50	44.93	11.57	3.90	2.31
1.60	20.0	5.0	28.50	44.66	26.84	3.40	4.33	38.00	45.71	16.29	4.00	4.33
1.60	20.0	10.0	31.50	44.09	24.41	3.50	3.58	42.50	45.21	12.29	4.10	3.68
1.60	20.0	15.0	35.00	43.43	21.57	3.70	2.90	44.50	45.00	10.50	3.80	3.12
1.70	5.0	5.0	28.50	44.48	27.02	3.70	2.83	36.50	45.63	17.87	4.20	2.81
1.70	5.0	10.0	31.50	43.90	24.60	3.80	2.32	40.50	45.18	14.32	4.20	2.36
1.70	5.0	15.0	33.50	43.52	22.98	3.60	1.85	43.00	44.90	12.10	3.90	1.97
1.70	10.0	5.0	29.50	44.34	26.16	3.80	3.30	37.00	45.65	17.35	4.10	3.29
1.70	10.0	10.0	32.00	43.87	24.13	3.80	2.72	41.00	45.20	13.80	4.10	2.77
1.70	10.0	15.0	34.50	43.39	22.11	3.70	2.19	44.00	44.87	11.13	3.90	2.33
1.70	20.0	5.0	29.00	44.56	26.44	3.40	4.35	38.50	45.65	15.85	4.00	4.34
1.70	20.0	10.0	32.00	43.99	24.01	3.50	3.62	42.50	45.21	12.29	4.00	3.70
1.70	20.0	15.0	35.50	43.34	21.16	3.70	2.94	44.50	45.00	10.50	3.70	3.15
1.80	5.0	5.0	28.50	44.48	27.02	3.60	2.84	37.00	45.57	17.43	4.20	2.82
1.80	5.0	10.0	31.50	43.90	24.60	3.70	2.34	41.00	45.12	13.88	4.20	2.37
1.80	5.0	15.0	34.50	43.34	22.16	3.80	1.87	44.00	44.79	11.21	4.00	1.99
1.80	10.0	5.0	29.00	44.44	26.56	3.60	3.32	37.50	45.59	16.91	4.10	3.30
1.80	10.0	10.0	32.00	43.87	24.13	3.70	2.74	42.00	45.09	12.91	4.20	2.78
1.80	10.0	15.0	34.00	43.49	22.51	3.50	2.21	43.00	44.98	12.02	3.70	2.35
1.80	20.0	5.0	29.00	44.56	26.44	3.30	4.36	39.00	45.60	15.40	4.00	4.35
1.80	20.0	10.0	33.50	43.71	22.79	3.80	3.65	43.00	45.16	11.84	4.00	3.72
1.80	20.0	15.0	35.50	43.34	21.16	3.60	2.98	45.50	44.89	9.61	3.80	3.17
1.90	5.0	5.0	28.00	44.57	27.43	3.40	2.85	37.00	45.57	17.43	4.20	2.83
1.90	5.0	10.0	31.00	44.00	25.00	3.50	2.35	41.50	45.07	13.43	4.20	2.38

Table 22 (continued)

DATA FOR THE CONSTRUCTION OF ISOKINETIC SUCROSE AND GLYCEROL GRADIENTS FOR INTERNATIONAL ROTOR 498

Particle density g/ml	Temperature °C	Top concentration %(w/w)	Sucrose					Glycerol				
			C_h %(w/w)	Buffer 2x ml	H_2O ml	V_m ml	Velocity of 100 S particle at top speed ml/h	C_h %(w/w)	Buffer 2x ml	H_2O ml	V_m ml	Velocity of 100 S particle at top speed ml/h
1.90	5.0	15.0	34.50	43.34	22.16	3.70	1.89	43.50	44.84	11.66	3.90	2.00
1.90	10.0	5.0	29.00	44.44	26.56	3.50	3.33	37.50	45.59	16.91	4.10	3.30
1.90	10.0	10.0	31.50	43.96	24.54	3.50	2.76	42.00	45.09	12.91	4.10	2.79
1.90	10.0	15.0	35.00	43.30	21.70	3.70	2.23	44.50	44.82	10.68	3.90	2.36
1.90	20.0	5.0	30.50	44.28	25.22	3.60	4.37	39.00	45.60	15.40	4.00	4.36
1.90	20.0	10.0	33.00	43.81	23.19	3.60	3.67	43.00	45.16	11.84	3.90	3.73
1.90	20.0	15.0	36.50	43.15	20.35	3.80	3.00	44.00	45.05	10.95	3.50	3.19
2.00	5.0	5.0	29.00	44.38	26.62	3.60	2.86	37.00	45.57	17.43	4.10	2.83
2.00	5.0	10.0	31.50	43.90	24.60	3.60	2.37	41.50	45.07	13.43	4.20	2.38
2.00	5.0	15.0	35.00	43.24	21.76	3.80	1.91	44.50	44.73	10.77	4.00	2.01
2.00	10.0	5.0	30.00	44.25	25.75	3.70	3.33	38.00	45.54	16.46	4.10	3.31
2.00	10.0	10.0	32.50	43.77	23.73	3.70	2.77	42.00	45.09	12.91	4.10	2.80
2.00	10.0	15.0	35.50	43.20	21.30	3.80	2.25	43.00	44.98	12.02	3.60	2.37
2.00	20.0	5.0	30.00	44.37	25.63	3.40	4.38	39.00	45.60	15.40	3.90	4.36
2.00	20.0	10.0	32.50	43.90	23.60	3.40	3.68	43.00	45.16	11.84	3.90	3.74
2.00	20.0	15.0	36.00	43.24	20.76	3.60	3.02	45.50	44.89	9.61	3.70	3.20

Table 23

DATA FOR THE CONSTRUCTION OF ISOKINETIC SUCROSE AND GLYCEROL GRADIENTS FOR MSE ROTOR 43127-101

Particle density g/ml	Temperature °C	Top concentration %(w/w)	Sucrose					Glycerol				
			C_h %(w/w)	Buffer 2x ml	H_2O ml	V_m ml	Velocity of 100 S particle at top speed ml/h	C_h %(w/w)	Buffer 2x ml	H_2O ml	V_m ml	Velocity of 100 S particle at top speed ml/h
1.20	5.0	5.0	21.50	45.82	32.68	2.80	1.46	30.50	46.32	23.18	3.60	1.49
1.20	5.0	10.0	25.00	45.15	29.85	3.00	1.09	35.50	45.74	18.76	3.80	1.18
1.20	5.0	15.0	28.00	44.57	27.43	3.00	.78	38.50	45.40	16.10	3.60	.93
1.20	10.0	5.0	22.50	45.68	31.82	3.00	1.70	32.50	46.16	21.34	3.80	1.74
1.20	10.0	10.0	25.00	45.20	29.80	2.90	1.28	35.00	45.88	19.12	3.50	1.39
1.20	10.0	15.0	28.00	44.63	27.37	2.90	.92	39.00	45.43	15.57	3.60	1.11
1.20	20.0	5.0	22.50	45.79	31:71	2.80	2.24	33.50	46.21	20.29	3.70	2.30
1.20	20.0	10.0	26.00	45.13	28.87	3.00	1.71	36.50	45.87	17.63	3.50	1.87
1.20	20.0	15.0	29.50	44.47	26.03	3.20	1.25	39.50	45.54	14.96	3.40	1.51
1.30	5.0	5.0	24.50	45.24	30.26	2.90	1.51	35.00	45.80	19.20	3.80	1.52
1.30	5.0	10.0	27.50	44.67	27.83	2.90	1.18	37.50	45.52	16.98	3.50	1.24
1.30	5.0	15.0	31.50	43.90	24.60	3.20	.89	41.50	45.07	13.43	3.50	1.01
1.30	10.0	5.0	25.00	45.20	29.80	2.90	1.76	36.00	45.76	18.24	3.80	1.78
1.30	10.0	10.0	29.00	44.44	26.56	3.20	1.39	39.50	45.37	15.13	3.70	1.46
1.30	10.0	15.0	31.50	43.96	24.54	3.10	1.06	41.00	45.20	13.80	3.30	1.19
1.30	20.0	5.0	25.50	45.22	29.28	2.80	2.32	36.50	45.87	17.63	3.60	2.35
1.30	20.0	10.0	29.00	44.56	26.44	3.00	1.85	39.50	45.54	14.96	3.40	1.95
1.30	20.0	15.0	32.00	43.99	24.01	3.00	1.43	42.00	45.27	12.73	3.20	1.62
1.40	5.0	5.0	26.50	44.86	28.64	3.00	1.53	36.00	45.69	18.31	3.70	1.54
1.40	5.0	10.0	28.50	44.48	27.02	2.80	1.23	40.50	45.18	14.32	3.70	1.27
1.40	5.0	15.0	33.00	43.62	23.38	3.20	.95	42.50	44.95	12.55	3.40	1.04
1.40	10.0	5.0	27.00	44.82	28.18	3.00	1.79	37.00	45.65	17.35	3.70	1.80
1.40	10.0	10.0	29.00	44.44	26.56	2.80	1.44	39.50	45.37	15.13	3.40	1.49
1.40	10.0	15.0	33.50	43.58	22.92	3.20	1.12	42.00	45.09	12.91	3.20	1.24
1.40	20.0	5.0	27.50	44.85	27.65	2.90	2.36	38.50	45.65	15.85	3.60	2.37
1.40	20.0	10.0	32.00	43.99	24.01	3.30	1.92	41.50	45.32	13.18	3.40	1.99
1.40	20.0	15.0	34.00	43.62	22.38	3.10	1.51	44.50	45.00	10.50	3.30	1.67
1.50	5.0	5.0	27.50	44.67	27.83	3.00	1.55	38.00	45.46	16.54	3.80	1.55
1.50	5.0	10.0	31.50	43.90	24.60	3.30	1.26	41.00	45.12	13.88	3.60	1.29
1.50	5.0	15.0	33.50	43.52	22.98	3.10	.99	43.50	44.84	11.66	3.40	1.07
1.50	10.0	5.0	27.00	44.82	28.18	2.80	1.81	38.50	45.48	16.02	3.70	1.81
1.50	10.0	10.0	30.00	44.25	25.75	2.80	1.47	42.00	45.09	12.91	3.60	1.51
1.50	10.0	15.0	34.00	43.49	22.51	3.10	1.16	42.50	45.04	12.46	3.10	1.26

Table 23 (continued)

DATA FOR THE CONSTRUCTION OF ISOKINETIC SUCROSE AND GLYCEROL GRADIENTS FOR MSE ROTOR 43127-101

Particle density g/ml	Temperature °C	Top concentration %(w/w)	Sucrose					Glycerol				
			C_h %(w/w)	Buffer 2x ml	H_2O ml	V_m ml	Velocity of 100 S particle at top speed ml/h	C_h %(w/w)	Buffer 2x ml	H_2O ml	V_m ml	Velocity of 100 S particle at top speed ml/h
1.50	20.0	5.0	27.50	44.85	27.65	2.70	2.38	39.00	45.60	15.40	3.50	2.39
1.50	20.0	10.0	31.00	44.18	24.82	2.80	1.96	44.00	45.05	10.95	3.60	2.02
1.50	20.0	15.0	35.00	43.43	21.57	3.10	1.57	45.00	44.94	10.06	3.20	1.71
1.60	5.0	5.0	27.50	44.67	27.83	2.80	1.56	38.00	45.46	16.54	3.70	1.55
1.60	5.0	10.0	30.00	44.19	25.81	2.80	1.27	40.50	45.18	14.32	3.40	1.30
1.60	5.0	15.0	34.50	43.34	22.16	3.20	1.01	43.00	44.90	12.10	3.20	1.08
1.60	10.0	5.0	28.50	44.53	26.97	2.90	1.82	39.00	45.43	15.57	3.70	1.82
1.60	10.0	10.0	31.00	44.06	24.94	2.90	1.49	41.50	45.15	13.35	3.40	1.52
1.60	10.0	15.0	35.00	43.30	21.70	3.20	1.19	44.50	44.82	10.68	3.30	1.28
1.60	20.0	5.0	29.50	44.47	26.03	2.90	2.40	40.00	45.49	14.51	3.50	2.40
1.60	20.0	10.0	32.00	43.99	24.01	2.90	1.98	43.50	45.11	11.39	3.40	2.04
1.60	20.0	15.0	36.00	43.24	20.76	3.20	1.60	45.00	44.94	10.06	3.10	1.73
1.70	5.0	5.0	28.50	44.48	27.02	2.90	1.57	38.50	45.40	16.10	3.70	1.56
1.70	5.0	10.0	30.50	44.10	25.40	2.80	1.29	41.00	45.12	13.88	3.40	1.31
1.70	5.0	15.0	34.50	43.34	22.16	3.10	1.03	43.50	44.84	11.66	3.20	1.09
1.70	10.0	5.0	29.00	44.44	26.56	2.90	1.83	39.00	45.43	15.57	3.60	1.82
1.70	10.0	10.0	31.50	43.96	24.54	2.90	1.51	42.00	45.09	12.91	3.40	1.53
1.70	10.0	15.0	35.00	43.30	21.70	3.10	1.21	45.00	44.76	10.24	3.30	1.29
1.70	20.0	5.0	30.00	44.37	25.63	2.90	2.41	40.50	45.43	14.07	3.50	2.40
1.70	20.0	10.0	32.50	43.90	23.60	2.90	2.00	44.00	45.05	10.95	3.40	2.05
1.70	20.0	15.0	36.50	43.15	20.35	3.20	1.63	45.50	44.89	9.61	3.10	1.75
1.80	5.0	5.0	27.50	44.67	27.83	2.70	1.57	39.00	45.35	15.65	3.70	1.56
1.80	5.0	10.0	32.50	43.71	23.79	3.20	1.30	42.00	45.01	12.99	3.50	1.31
1.80	5.0	15.0	34.00	43.43	22.57	2.90	1.04	44.50	44.73	10.77	3.30	1.10
1.80	10.0	5.0	27.50	44.72	27.78	2.60	1.84	39.50	45.37	15.13	3.60	1.83
1.80	10.0	10.0	31.50	43.96	24.54	2.80	1.52	42.50	45.04	12.46	3.40	1.54
1.80	10.0	15.0	34.50	43.39	22.11	2.90	1.23	43.50	44.93	11.57	3.00	1.30
1.80	20.0	5.0	30.50	44.28	25.22	2.90	2.42	41.00	45.38	13.62	3.50	2.41
1.80	20.0	10.0	33.00	43.81	23.19	3.00	2.02	44.50	45.00	10.50	3.40	2.06
1.80	20.0	15.0	36.00	43.24	20.76	2.80	1.65	44.50	45.00	10.50	2.90	1.76
1.90	5.0	5.0	28.50	44.48	27.02	2.90	1.58	39.00	45.35	15.65	3.60	1.57
1.90	5.0	10.0	31.50	43.90	24.60	2.90	1.30	43.00	44.90	12.10	3.60	1.32
1.90	5.0	15.0	35.50	43.15	21.35	3.20	1.05	43.50	44.84	11.66	3.10	1.11

Table 23 (continued)

DATA FOR THE CONSTRUCTION OF ISOKINETIC SUCROSE AND GLYCEROL GRADIENTS FOR MSE ROTOR 43127-101

Particle density g/ml	Temperature °C	Top concentration %(w/w)	Sucrose					Glycerol				
			C_h %(w/w)	Buffer 2x ml	H_2O ml	V_m ml	Velocity of 100 S particle at top speed ml/h	C_h %(w/w)	Buffer 2x ml	H_2O ml	V_m ml	Velocity of 100 S particle at top speed ml/h
1.90	10.0	5.0	29.00	44.44	26.56	2.80	1.84	39.00	45.43	15.57	3.50	1.83
1.90	10.0	10.0	32.00	43.87	24.13	2.90	1.53	43.50	44.93	11.57	3.50	1.55
1.90	10.0	15.0	36.00	43.11	20.89	3.20	1.24	45.00	44.76	10.24	3.20	1.31
1.90	20.0	5.0	29.50	44.47	26.03	2.70	2.42	41.00	45.38	13.62	3.40	2.41
1.90	20.0	10.0	32.00	43.99	24.01	2.70	2.03	45.00	44.94	10.06	3.40	2.07
1.90	20.0	15.0	35.50	43.34	21.16	2.80	1.66	45.50	44.89	9.61	3.00	1.77
2.00	5.0	5.0	27.50	44.67	27.83	2.60	1.58	39.00	45.35	15.65	3.60	1.57
2.00	5.0	10.0	33.00	43.62	23.38	3.20	1.31	42.00	45.01	12.99	3.40	1.32
2.00	5.0	15.0	34.00	43.43	22.57	2.80	1.06	43.00	44.90	12.10	3.00	1.11
2.00	10.0	5.0	28.00	44.63	27.37	2.60	1.85	40.00	45.32	14.68	3.60	1.83
2.00	10.0	10.0	32.00	43.87	24.13	2.80	1.54	43.00	44.98	12.02	3.40	1.55
2.00	10.0	15.0	35.00	43.30	21.70	2.90	1.25	44.00	44.87	11.13	3.00	1.31
2.00	20.0	5.0	30.50	44.28	25.22	2.80	2.43	41.00	45.38	13.62	3.40	2.42
2.00	20.0	10.0	33.00	43.81	23.19	2.80	2.04	44.50	45.00	10.50	3.30	2.07
2.00	20.0	15.0	36.50	43.15	20.35	3.00	1.67	46.50	44.78	8.72	3.10	1.77

Table 24

DATA FOR THE CONSTRUCTION OF ISOKINETIC SUCROSE AND GLYCEROL GRADIENTS FOR MSE ROTOR 43127-102

Particle density g/ml	Temperature °C	Top concentration %(w/w)	Sucrose					Glycerol				
			C_h %(w/w)	Buffer 2x ml	H_2O ml	V_m ml	Velocity of 100 S particle at top speed ml/h	C_h %(w/w)	Buffer 2x ml	H_2O ml	V_m ml	Velocity of 100 S particle at top speed ml/h
1.20	5.0	5.0	22.50	45.63	31.87	6.25	4.26	33.00	46.03	20.97	8.25	4.35
1.20	5.0	10.0	26.00	44.96	29.04	6.75	3.19	36.50	45.63	17.87	8.00	3.46
1.20	5.0	15.0	28.50	44.48	27.02	6.50	2.28	39.50	45.29	15.21	7.75	2.73
1.20	10.0	5.0	22.50	45.68	31.82	6.00	4.97	32.50	46.16	21.34	7.75	5.08
1.20	10.0	10.0	26.00	45.01	28.99	6.50	3.75	36.00	45.76	18.24	7.50	4.07
1.20	10.0	15.0	28.50	44.53	26.97	6.25	2.70	40.00	45.32	14.68	7.75	3.24
1.20	20.0	5.0	24.00	45.51	30.49	6.50	6.55	33.50	46.21	20.29	7.50	6.73
1.20	20.0	10.0	26.00	45.13	28.87	6.00	5.00	37.50	45.76	16.74	7.50	5.46
1.20	20.0	15.0	29.50	44.47	26.03	6.50	3.66	41.00	45.38	13.62	7.50	4.40
1.30	5.0	5.0	25.50	45.05	29.45	6.25	4.41	36.50	45.63	17.87	8.25	4.45
1.30	5.0	10.0	29.00	44.38	26.62	6.75	3.46	38.50	45.40	16.10	7.50	3.63
1.30	5.0	15.0	31.50	43.90	24.60	6.50	2.61	42.00	45.01	12.99	7.50	2.95
1.30	10.0	5.0	26.00	45.01	28.99	6.25	5.15	38.00	45.54	16.46	8.50	5.20
1.30	10.0	10.0	29.00	44.44	26.56	6.50	4.06	39.50	45.37	15.13	7.50	4.26
1.30	10.0	15.0	31.50	43.96	24.54	6.25	3.09	42.50	45.04	12.46	7.25	3.49
1.30	20.0	5.0	25.50	45.22	29.28	5.75	6.78	38.00	45.71	16.29	7.75	6.87
1.30	20.0	10.0	29.50	44.47	26.03	6.25	5.40	40.50	45.43	14.07	7.25	5.71
1.30	20.0	15.0	33.50	43.71	22.79	7.00	4.17	43.50	45.11	11.39	7.00	4.73
1.40	5.0	5.0	27.00	44.76	28.24	6.25	4.49	39.00	45.35	15.65	8.50	4.50
1.40	5.0	10.0	31.00	44.00	25.00	7.00	3.59	41.00	45.12	13.88	7.75	3.71
1.40	5.0	15.0	33.00	43.62	23.38	6.50	2.78	44.00	44.79	11.21	7.50	3.06
1.40	10.0	5.0	27.50	44.72	27.78	6.25	5.24	38.50	45.48	16.02	8.00	5.25
1.40	10.0	10.0	29.50	44.34	26.16	6.00	4.21	42.00	45.09	12.91	7.75	4.36
1.40	10.0	15.0	33.50	43.58	22.92	6.50	3.29	43.50	44.93	11.57	7.00	3.61
1.40	20.0	5.0	28.50	44.66	26.84	6.25	6.89	40.00	45.49	14.51	7.75	6.94
1.40	20.0	10.0	30.50	44.28	25.22	6.00	5.60	42.50	45.21	12.29	7.25	5.83
1.40	20.0	15.0	34.00	43.62	22.38	6.25	4.43	44.50	45.00	10.50	6.75	4.89
1.50	5.0	5.0	28.00	44.57	27.43	6.25	4.54	39.50	45.29	15.21	8.25	4.53
1.50	5.0	10.0	30.00	44.19	25.81	6.00	3.67	41.50	45.07	13.43	7.50	3.76
1.50	5.0	15.0	33.50	43.52	22.98	6.25	2.89	44.50	44.73	10.77	7.25	3.12
1.50	10.0	5.0	28.50	44.53	26.97	6.25	5.29	40.00	45.32	14.68	8.00	5.29
1.50	10.0	10.0	30.50	44.15	25.35	6.00	4.31	42.50	45.04	12.46	7.50	4.42
1.50	10.0	15.0	34.00	43.49	22.51	6.25	3.41	44.00	44.87	11.13	6.75	3.69

Table 24 (continued)
DATA FOR THE CONSTRUCTION OF ISOKINETIC SUCROSE AND GLYCEROL GRADIENTS FOR MSE ROTOR 43127-102

Particle density g/ml	Temperature °C	Top concentration %(w/w)	Sucrose					Glycerol				
			C_h %(w/w)	Buffer 2x ml	H_2O ml	V_m ml	Velocity of 100 S particle at top speed ml/h	C_h %(w/w)	Buffer 2x ml	H_2O ml	V_m ml	Velocity of 100 S particle at top speed ml/h
1.50	20.0	5.0	29.00	44.56	26.44	6.00	6.96	41.50	45.32	13.18	7.75	6.99
1.50	20.0	10.0	31.50	44.09	24.41	6.00	5.72	43.00	45.16	11.84	7.00	5.91
1.50	20.0	15.0	35.00	43.43	21.57	6.25	4.59	45.00	44.94	10.06	6.50	4.99
1.60	5.0	5.0	28.00	44.57	27.43	6.00	4.57	39.50	45.29	15.21	8.00	4.55
1.60	5.0	10.0	30.00	44.19	25.81	5.75	3.72	41.50	45.07	13.43	7.25	3.79
1.60	5.0	15.0	34.50	43.34	22.16	6.50	2.95	44.50	44.73	10.77	7.00	3.16
1.60	10.0	5.0	28.00	44.63	27.37	5.75	5.33	40.00	45.32	14.68	7.75	5.31
1.60	10.0	10.0	32.50	43.77	23.73	6.50	4.37	42.50	45.04	12.46	7.25	4.46
1.60	10.0	15.0	35.00	43.30	21.70	6.50	3.49	44.50	44.82	10.68	6.75	3.74
1.60	20.0	5.0	29.00	44.56	26.44	5.75	7.01	41.50	45.32	13.18	7.50	7.01
1.60	20.0	10.0	33.00	43.81	23.19	6.25	5.81	43.00	45.16	11.84	6.75	5.96
1.60	20.0	15.0	35.50	43.34	21.16	6.25	4.69	45.50	44.89	9.61	6.50	5.06
1.70	5.0	5.0	28.50	44.48	27.02	6.00	4.59	39.50	45.29	15.21	7.75	4.56
1.70	5.0	10.0	30.50	44.10	25.40	5.75	3.76	42.00	45.01	12.99	7.25	3.82
1.70	5.0	15.0	34.50	43.34	22.16	6.25	3.00	45.00	44.68	10.32	7.00	3.20
1.70	10.0	5.0	28.50	44.53	26.97	5.75	5.35	40.50	45.26	14.24	7.75	5.33
1.70	10.0	10.0	31.00	44.06	24.94	5.75	4.41	44.00	44.87	11.13	7.50	4.49
1.70	10.0	15.0	35.00	43.30	21.70	6.25	3.54	45.00	44.76	10.24	6.75	3.78
1.70	20.0	5.0	31.00	44.18	24.82	6.25	7.04	42.00	45.27	12.73	7.50	7.03
1.70	20.0	10.0	33.50	43.71	22.79	6.25	5.86	44.50	45.00	10.50	7.00	5.99
1.70	20.0	15.0	36.00	43.24	20.76	6.25	4.76	46.00	44.83	9.17	6.50	5.11
1.80	5.0	5.0	29.00	44.38	26.62	6.00	4.61	40.00	45.23	14.77	7.75	4.57
1.80	5.0	10.0	31.50	43.90	24.60	6.00	3.79	42.50	44.95	12.55	7.25	3.84
1.80	5.0	15.0	35.00	43.24	21.76	6.25	3.04	44.50	44.73	10.77	6.75	3.22
1.80	10.0	5.0	29.50	44.34	26.16	6.00	5.37	41.00	45.20	13.80	7.75	5.34
1.80	10.0	10.0	32.00	43.87	24.13	6.00	4.45	43.50	44.93	11.57	7.25	4.51
1.80	10.0	15.0	35.50	43.20	21.30	6.25	3.58	45.50	44.71	9.79	6.75	3.80
1.80	20.0	5.0	30.00	44.37	25.63	5.75	7.07	42.50	45.21	12.29	7.50	7.05
1.80	20.0	10.0	32.50	43.90	23.60	5.75	5.91	44.00	45.05	10.95	6.75	6.02
1.80	20.0	15.0	36.50	43.15	20.35	6.25	4.82	46.50	44.78	8.72	6.50	5.14
1.90	5.0	5.0	28.00	44.57	27.43	5.50	4.62	41.00	45.12	13.88	8.00	4.58
1.90	5.0	10.0	32.50	43.71	23.79	6.25	3.81	42.00	45.01	12.99	7.00	3.85

Table 24 (continued)
DATA FOR THE CONSTRUCTION OF ISOKINETIC SUCROSE AND GLYCEROL GRADIENTS FOR MSE ROTOR 43127-102

Particle density g/ml	Temperature °C	Top concentration %(w/w)	Sucrose					Glycerol				
			C_h %(w/w)	Buffer 2x ml	H_2O ml	V_m ml	Velocity of 100 S particle at top speed ml/h	C_h %(w/w)	Buffer 2x ml	H_2O ml	V_m ml	Velocity of 100 S particle at top speed ml/h
1.90	5.0	15.0	35.00	43.24	21.76	6.25	3.07	45.50	44.62	9.88	7.00	3.24
1.90	10.0	5.0	30.50	44.15	25.35	6.25	5.39	41.50	45.15	13.35	7.75	5.35
1.90	10.0	10.0	33.00	43.68	23.32	6.25	4.47	44.00	44.87	11.13	7.25	4.52
1.90	10.0	15.0	35.50	43.20	21.30	6.25	3.62	45.00	44.76	10.24	6.50	3.82
1.90	20.0	5.0	31.00	44.18	24.82	6.00	7.09	43.00	45.16	11.84	7.50	7.06
1.90	20.0	5.0	33.50	43.71	22.79	6.00	5.94	44.50	45.00	10.50	6.75	6.04
1.90	20.0	10.0	35.50	43.34	21.16	5.75	4.86	47.00	44.73	8.27	6.50	5.17
1.90	20.0	15.0	29.50	44.29	26.21	6.00	4.63	40.50	45.18	14.32	7.75	4.59
2.00	5.0	5.0	32.00	43.81	24.19	6.00	3.83	43.00	44.90	12.10	7.25	3.86
2.00	5.0	10.0	34.00	43.43	22.57	5.75	3.09	45.00	44.68	10.32	6.75	3.25
2.00	5.0	15.0	29.50	44.34	26.16	5.75	5.40	41.00	45.20	13.80	7.50	5.36
2.00	5.0	5.0	32.50	43.77	23.73	6.00	4.49	43.50	44.93	11.57	7.00	4.54
2.00	10.0	10.0	36.00	43.11	20.89	6.25	3.64	46.00	44.65	9.35	6.75	3.84
2.00	10.0	15.0	32.00	43.99	24.01	6.25	7.10	43.50	45.11	11.39	7.50	7.07
2.00	20.0	5.0	32.00	43.99	24.01	6.25	5.97	45.50	44.89	9.61	7.00	6.06
2.00	20.0	10.0	33.00	43.81	23.19	5.75	4.90	47.00	44.73	8.27	6.50	5.19
2.00	20.0	15.0	37.00	43.06	19.94	6.25						

Table 25

DATA FOR THE CONSTRUCTION OF ISOKINETIC SUCROSE AND GLYCEROL GRADIENTS FOR MSE ROTOR 43127-103

Particle density g/ml	Temperature °C	Top concentration %(w/w)	Sucrose					Glycerol				
			C_h %(w/w)	Buffer 2x ml	H_2O ml	V_m ml	Velocity of 100 S particle at top speed ml/h	C_h %(w/w)	Buffer 2x ml	H_2O ml	V_m ml	Velocity of 100 S particle at top speed ml/h
1.20	5.0	5.0	22.00	45.73	32.27	6.50	2.04	31.50	46.20	22.30	8.25	2.09
1.20	5.0	10.0	24.50	45.24	30.26	6.25	1.53	35.00	45.80	19.20	8.00	1.66
1.20	5.0	15.0	27.50	44.67	27.83	6.25	1.09	37.50	45.52	16.98	7.50	1.31
1.20	10.0	5.0	21.50	45.87	32.63	6.00	2.39	31.50	46.27	22.23	8.00	2.44
1.20	10.0	10.0	24.00	45.39	30.61	5.75	1.80	34.50	45.93	19.57	7.50	1.95
1.20	10.0	15.0	28.50	44.53	26.97	6.75	1.30	39.50	45.37	15.13	8.00	1.55
1.20	20.0	5.0	23.50	45.60	30.90	6.75	3.15	32.50	46.32	21.18	7.75	3.23
1.20	20.0	10.0	25.50	45.22	29.28	6.25	2.40	36.00	45.93	18.07	7.50	2.62
1.20	20.0	15.0	28.00	44.75	27.25	6.00	1.76	38.50	45.65	15.85	7.00	2.12
1.30	5.0	5.0	25.00	45.15	29.85	6.50	2.12	34.00	45.92	20.08	8.00	2.14
1.30	5.0	10.0	28.50	44.48	27.02	7.00	1.66	37.00	45.57	17.43	7.50	1.74
1.30	5.0	15.0	31.00	44.00	25.00	6.75	1.26	39.50	45.29	15.21	7.00	1.42
1.30	10.0	5.0	23.50	45.49	31.01	5.75	2.47	35.50	45.82	18.68	8.25	2.50
1.30	10.0	10.0	28.00	44.63	27.37	6.50	1.95	38.50	45.48	16.02	7.75	2.05
1.30	10.0	15.0	31.00	44.06	24.94	6.50	1.48	41.00	45.20	13.80	7.25	1.68
1.30	20.0	5.0	25.00	45.32	29.68	6.00	3.26	36.00	45.93	18.07	7.75	3.30
1.30	20.0	10.0	28.50	44.66	26.84	6.25	2.60	39.50	45.54	14.96	7.50	2.74
1.30	20.0	15.0	31.00	44.18	24.82	6.00	2.00	42.50	45.21	12.29	7.25	2.27
1.40	5.0	5.0	26.50	44.86	28.64	6.50	2.16	35.00	45.80	19.20	7.75	2.16
1.40	5.0	10.0	28.00	44.57	27.43	6.00	1.72	39.50	45.29	15.21	7.75	1.78
1.40	5.0	15.0	32.50	43.71	23.79	6.75	1.34	40.50	45.18	14.32	6.75	1.47
1.40	10.0	5.0	27.00	44.82	28.18	6.50	2.52	36.00	45.76	18.24	7.75	2.52
1.40	10.0	10.0	29.00	44.44	26.56	6.25	2.02	41.00	45.20	13.80	8.00	2.09
1.40	10.0	15.0	32.50	43.77	23.73	6.50	1.58	42.00	45.09	12.91	7.00	1.74
1.40	20.0	5.0	26.00	45.13	28.87	5.75	3.31	37.50	45.76	16.74	7.50	3.33
1.40	20.0	10.0	30.00	44.37	25.63	6.25	2.69	41.50	45.32	13.18	7.50	2.80
1.40	20.0	15.0	33.00	43.81	23.19	6.25	2.13	43.00	45.16	11.84	6.75	2.35
1.50	5.0	5.0	27.50	44.67	27.83	6.50	2.18	36.00	45.69	18.31	7.75	2.17
1.50	5.0	10.0	29.50	44.29	26.21	6.25	1.76	40.50	45.18	14.32	7.75	1.81
1.50	5.0	15.0	32.50	43.71	23.79	6.25	1.39	41.50	45.07	13.43	6.75	1.50
1.50	10.0	5.0	26.00	45.01	28.99	5.75	2.54	37.00	45.65	17.35	6.75	2.54
1.50	10.0	10.0	30.00	44.25	25.75	6.25	2.07	41.50	45.15	13.35	7.75	2.12
1.50	10.0	15.0	33.00	43.68	23.32	6.25	1.64	42.50	45.04	12.46	6.75	1.77
1.50	20.0	5.0	27.00	44.94	28.06	5.75	3.34	38.50	45.65	15.85	7.50	3.36

Table 25 (continued)

DATA FOR THE CONSTRUCTION OF ISOKINETIC SUCROSE AND GLYCEROL GRADIENTS FOR MSE ROTOR 43127-103

Particle density g/ml	Temperature °C	Top concentration %(w/w)	Sucrose					Glycerol				
			C_h %(w/w)	Buffer 2x ml	H_2O ml	V_m ml	Velocity of 100 S particle at top speed ml/h	C_h %(w/w)	Buffer 2x ml	H_2O ml	V_m ml	Velocity of 100 S particle at top speed ml/h
1.50	20.0	10.0	31.00	44.18	24.82	6.25	2.75	42.00	45.27	12.73	7.25	2.84
1.50	20.0	15.0	34.00	43.62	22.38	6.25	2.20	44.00	45.05	10.95	6.75	2.40
1.60	5.0	5.0	27.50	44.67	27.83	6.25	2.19	37.00	45.57	17.43	7.75	2.18
1.60	5.0	10.0	29.50	44.29	26.21	6.00	1.79	40.50	45.18	14.32	7.50	1.82
1.60	5.0	15.0	33.50	43.52	22.98	6.50	1.42	41.50	45.07	13.43	6.50	1.52
1.60	10.0	5.0	28.00	44.63	27.37	6.25	2.56	37.00	45.65	17.35	7.50	2.55
1.60	10.0	10.0	29.50	44.34	26.16	5.75	2.10	41.50	45.15	13.35	7.50	2.14
1.60	10.0	15.0	34.00	43.49	22.51	6.50	1.67	43.00	44.98	12.02	6.75	1.80
1.60	20.0	5.0	28.50	44.66	26.84	6.00	3.37	38.50	45.65	15.85	7.25	3.37
1.60	20.0	10.0	30.50	44.28	25.22	5.75	2.79	42.00	45.27	12.73	7.00	2.86
1.60	20.0	15.0	34.50	43.52	21.98	6.25	2.25	44.00	45.05	10.95	6.50	2.43
1.70	5.0	5.0	28.00	44.57	27.43	6.25	2.20	37.50	45.52	16.98	7.75	2.19
1.70	5.0	10.0	30.00	44.19	25.81	6.00	1.81	41.00	45.12	13.88	7.50	1.83
1.70	5.0	15.0	33.50	43.52	22.98	6.25	1.44	44.00	44.79	11.21	7.25	1.54
1.70	10.0	5.0	28.50	44.53	26.97	6.25	2.57	38.00	45.54	16.46	7.50	2.56
1.70	10.0	10.0	30.50	44.15	25.35	6.00	2.12	42.00	45.09	12.91	7.50	2.15
1.70	10.0	15.0	34.00	43.49	22.51	6.25	1.70	43.50	44.93	11.57	6.75	1.81
1.70	20.0	5.0	29.00	44.56	26.44	6.00	3.38	39.00	45.60	15.40	7.25	3.38
1.70	20.0	10.0	31.00	44.18	24.82	5.75	2.82	43.50	45.11	11.39	7.25	2.88
1.70	20.0	15.0	35.00	43.43	21.57	6.25	2.29	44.50	45.00	10.50	6.50	2.45
1.80	5.0	5.0	26.50	44.86	28.64	5.50	2.21	37.00	45.57	17.43	7.50	2.20
1.80	5.0	10.0	31.00	44.00	25.00	6.25	1.82	41.50	45.07	13.43	7.50	1.84
1.80	5.0	15.0	34.50	43.34	22.16	6.50	1.46	43.00	44.90	12.10	6.75	1.55
1.80	10.0	5.0	27.00	44.82	28.18	5.50	2.58	38.00	45.54	16.46	7.50	2.56
1.80	10.0	10.0	31.50	43.96	24.54	6.25	2.14	42.50	45.04	12.46	7.50	2.16
1.80	10.0	15.0	35.00	43.30	21.70	6.50	1.72	44.00	44.87	11.13	6.75	1.83
1.80	20.0	5.0	29.50	44.47	26.03	6.00	3.39	39.50	45.54	14.96	7.25	3.39
1.80	20.0	10.0	32.00	43.99	24.01	6.00	2.84	43.00	45.16	11.84	7.00	2.89
1.80	20.0	15.0	36.00	43.24	20.76	6.50	2.32	45.00	44.94	10.06	6.50	2.47
1.90	5.0	5.0	28.00	44.57	27.43	6.00	2.22	37.50	45.52	16.98	7.50	2.20
1.90	5.0	10.0	30.00	44.19	25.81	5.75	1.83	41.00	45.12	13.88	7.25	1.85
1.90	5.0	15.0	34.00	43.43	22.57	6.25	1.47	44.00	44.79	11.21	7.00	1.56

Table 25 (continued)

DATA FOR THE CONSTRUCTION OF ISOKINETIC SUCROSE AND GLYCEROL GRADIENTS FOR MSE ROTOR 43127-103

Particle density g/ml	Temperature °C	Top concentration %(w/w)	Sucrose					Glycerol				
			C_h %(w/w)	Buffer 2x ml	H_2O ml	v_m ml	Velocity of 100 S particle at top speed ml/h	C_h %(w/w)	Buffer 2x ml	H_2O ml	v_m ml	Velocity of 100 S particle at top speed ml/h
1.90	10.0	5.0	28.00	44.63	27.37	5.75	2.59	38.50	45.48	16.02	7.50	2.57
1.90	10.0	10.0	30.50	44.15	25.35	5.75	2.15	43.00	44.98	12.02	7.50	2.17
1.90	10.0	15.0	34.50	43.39	22.11	6.25	1.74	43.50	44.93	11.57	6.50	1.84
1.90	20.0	5.0	29.00	44.56	26.44	5.75	3.40	39.00	45.60	15.40	7.00	3.39
1.90	20.0	10.0	33.00	43.81	23.19	6.25	2.85	43.50	45.11	11.39	7.00	2.90
1.90	20.0	15.0	35.00	43.43	21.57	6.00	2.34	44.50	45.00	10.50	6.25	2.48
2.00	5.0	5.0	27.00	44.76	28.24	5.50	2.22	38.00	45.46	16.54	7.50	2.20
2.00	5.0	10.0	31.50	43.90	24.60	6.25	1.84	42.00	45.01	12.99	7.50	1.86
2.00	5.0	15.0	33.00	43.62	23.38	5.75	1.48	43.50	44.84	11.66	6.75	1.56
2.00	10.0	5.0	27.50	44.72	27.78	5.50	2.59	38.00	45.54	16.46	7.25	2.57
2.00	10.0	10.0	32.00	43.87	24.13	6.25	2.16	42.50	45.04	12.46	7.25	2.18
2.00	10.0	15.0	33.50	43.58	22.92	5.75	1.75	44.50	44.82	10.68	6.75	1.84
2.00	20.0	5.0	29.50	44.47	26.03	5.75	3.41	39.50	45.54	14.96	7.00	3.40
2.00	20.0	10.0	32.50	43.90	23.60	6.00	2.87	43.50	45.11	11.39	7.00	2.91
2.00	20.0	15.0	36.50	43.15	20.35	6.50	2.35	45.50	44.89	9.61	6.50	2.49

Table 26

DATA FOR THE CONSTRUCTION OF ISOKINETIC SUCROSE AND GLYCEROL GRADIENTS FOR MSE ROTOR 43127-104

Particle density g/ml	Temperature °C	Top concentration %(w/w)	Sucrose					Glycerol				
			C_h %(w/w)	Buffer 2x ml	H_2O ml	V_m ml	Velocity of 100 S particle at top speed ml/h	C_h %(w/w)	Buffer 2x ml	H_2O ml	V_m ml	Velocity of 100 S particle at top speed ml/h
1.20	5.0	5.0	21.50	45.82	32.68	24.50	4.57	30.00	46.38	23.62	30.50	4.67
1.20	5.0	10.0	25.00	45.15	29.85	26.00	3.42	34.00	45.92	20.08	30.00	3.71
1.20	5.0	15.0	28.00	44.57	27.43	26.00	2.44	38.00	45.46	16.54	30.50	2.93
1.20	10.0	5.0	22.00	45.77	32.23	25.00	5.34	30.50	46.39	23.11	30.00	5.46
1.20	10.0	10.0	25.50	45.10	29.40	26.50	4.02	35.00	45.88	19.12	30.50	4.37
1.20	10.0	15.0	28.00	44.63	27.37	25.50	2.90	38.50	45.48	16.02	30.50	3.48
1.20	20.0	5.0	22.50	45.79	31.71	24.50	7.04	32.50	46.32	21.18	30.50	7.23
1.20	20.0	10.0	25.50	45.22	29.28	25.00	5.37	35.50	45.98	18.52	29.00	5.86
1.20	20.0	15.0	29.00	44.56	26.44	26.50	3.93	39.50	45.54	14.96	29.50	4.73
1.30	5.0	5.0	24.50	45.24	30.26	25.00	4.74	33.00	46.03	20.97	30.50	4.78
1.30	5.0	10.0	27.50	44.67	27.83	25.50	3.71	37.50	45.52	16.98	30.50	3.89
1.30	5.0	15.0	31.50	43.90	24.60	28.00	2.81	40.00	45.23	14.77	28.50	3.17
1.30	10.0	5.0	24.50	45.29	30.21	24.50	5.53	34.00	45.99	20.01	30.50	5.58
1.30	10.0	10.0	29.00	44.44	26.56	28.00	4.36	38.00	45.54	16.46	30.00	4.58
1.30	10.0	15.0	31.50	43.96	24.54	27.00	3.32	40.50	45.26	14.24	28.00	3.75
1.30	20.0	5.0	26.00	45.13	28.87	25.50	7.28	36.00	45.93	18.07	30.50	7.38
1.30	20.0	10.0	29.00	44.56	26.44	26.00	5.80	40.00	45.49	14.51	30.50	6.13
1.30	20.0	15.0	32.50	43.90	23.60	27.50	4.48	42.00	45.27	12.73	28.00	5.08
1.40	5.0	5.0	26.50	44.86	28.64	26.00	4.82	35.00	45.80	19.20	30.50	4.83
1.40	5.0	10.0	29.00	44.38	26.62	25.50	3.86	38.50	45.40	16.10	29.50	3.99
1.40	5.0	15.0	33.00	43.62	23.38	28.00	2.99	41.00	45.12	13.88	27.50	3.28
1.40	10.0	5.0	27.00	44.82	28.18	26.00	5.63	36.00	45.76	18.24	30.50	5.64
1.40	10.0	10.0	29.50	44.34	26.16	25.50	4.53	39.50	45.37	15.13	29.50	4.68
1.40	10.0	15.0	33.00	43.68	23.32	27.00	3.53	42.50	45.04	12.46	28.50	3.88
1.40	20.0	5.0	28.00	44.75	27.25	26.00	7.41	38.00	45.71	16.29	30.50	7.46
1.40	20.0	10.0	30.50	44.28	25.22	25.50	6.02	42.00	45.27	12.73	30.50	6.27
1.40	20.0	15.0	33.50	43.71	22.79	26.00	4.76	43.50	45.11	11.39	27.50	5.26
1.50	5.0	5.0	27.50	44.67	27.83	26.00	4.87	36.00	45.69	18.31	30.50	4.86
1.50	5.0	10.0	30.00	44.19	25.81	25.50	3.94	40.00	45.23	14.77	30.00	4.04
1.50	5.0	15.0	33.50	43.52	22.98	27.00	3.10	42.00	45.01	12.99	27.50	3.35
1.50	10.0	5.0	28.00	44.63	27.37	26.00	5.69	37.00	45.65	17.35	30.50	5.68
1.50	10.0	10.0	30.50	44.15	25.35	25.50	4.63	41.00	45.20	13.80	30.00	4.75
1.50	10.0	15.0	33.50	43.58	22.92	26.00	3.66	42.50	45.04	12.46	27.00	3.96

Table 26 (continued)

DATA FOR THE CONSTRUCTION OF ISOKINETIC SUCROSE AND GLYCEROL GRADIENTS FOR MSE ROTOR 43127-104

Particle density g/ml	Temperature °C	Top concentration %(w/w)	Sucrose					Glycerol				
			C_h %(w/w)	Buffer 2x ml	H_2O ml	V_m ml	Velocity of 100 S particle at top speed ml/h	C_h %(w/w)	Buffer 2x ml	H_2O ml	V_m ml	Velocity of 100 S particle at top speed ml/h
1.50	20.0	5.0	28.50	44.66	26.84	25.00	7.48	39.00	45.60	15.40	30.50	7.50
1.50	20.0	10.0	32.00	43.99	24.01	26.50	6.15	42.50	45.21	12.29	29.50	6.35
1.50	20.0	15.0	34.50	43.52	21.98	26.00	4.93	43.50	45.11	11.39	26.00	5.36
1.60	5.0	5.0	27.00	44.76	28.24	24.00	4.91	36.50	45.63	17.87	30.00	4.88
1.60	5.0	10.0	30.50	44.10	25.40	25.50	4.00	41.00	45.12	13.88	30.50	4.08
1.60	5.0	15.0	34.50	43.34	22.16	28.00	3.17	43.00	44.90	12.10	28.00	3.40
1.60	10.0	5.0	27.50	44.72	27.78	24.00	5.72	37.50	45.59	16.91	30.00	5.70
1.60	10.0	10.0	31.00	44.06	24.94	25.50	4.69	42.00	45.09	12.91	30.50	4.79
1.60	10.0	15.0	35.00	43.30	21.70	28.00	3.74	44.00	44.87	11.13	28.00	4.02
1.60	20.0	5.0	29.50	44.47	26.03	25.50	7.53	40.00	45.49	14.51	30.50	7.53
1.60	20.0	10.0	31.50	44.09	24.41	24.50	6.24	43.00	45.16	11.84	29.00	6.40
1.60	20.0	15.0	34.50	43.52	21.98	25.00	5.04	45.00	44.94	10.06	27.00	5.43
1.70	5.0	5.0	27.50	44.67	27.83	24.00	4.93	37.50	45.52	16.98	30.50	4.90
1.70	5.0	10.0	31.00	44.00	25.00	25.50	4.04	40.50	45.18	14.32	29.00	4.10
1.70	5.0	15.0	34.00	43.43	22.57	26.00	3.22	43.50	44.84	11.66	28.00	3.43
1.70	10.0	5.0	28.00	44.63	27.37	24.00	5.75	38.50	45.48	16.02	30.50	5.72
1.70	10.0	10.0	31.50	43.96	24.54	25.50	4.74	42.00	45.09	12.91	29.50	4.82
1.70	10.0	15.0	34.50	43.39	22.11	26.00	3.81	44.00	44.87	11.13	27.50	4.06
1.70	20.0	5.0	29.50	44.47	26.03	24.50	7.56	40.00	45.49	14.51	30.00	7.56
1.70	20.0	10.0	32.00	43.99	24.01	24.50	6.30	44.00	45.05	10.95	29.50	6.44
1.70	20.0	15.0	35.00	43.43	21.57	25.00	5.12	45.50	44.89	9.61	27.00	5.48
1.80	5.0	5.0	28.50	44.48	27.02	25.00	4.95	37.50	45.52	16.98	30.00	4.91
1.80	5.0	10.0	31.00	44.00	25.00	25.00	4.07	42.00	45.01	12.99	30.50	4.12
1.80	5.0	15.0	34.50	43.34	22.16	26.50	3.26	43.50	44.84	11.66	27.50	3.46
1.80	10.0	5.0	29.00	44.44	26.56	25.00	5.77	39.00	45.43	15.57	30.50	5.73
1.80	10.0	10.0	31.00	44.06	24.94	24.00	4.78	43.00	44.98	12.02	30.50	4.84
1.80	10.0	15.0	34.00	43.49	22.51	24.50	3.85	44.50	44.82	10.68	27.50	4.08
1.80	20.0	5.0	29.50	44.47	26.03	24.00	7.59	40.50	45.43	14.07	30.00	7.57
1.80	20.0	10.0	33.00	43.81	23.19	25.50	6.34	44.00	45.05	10.95	29.00	6.47
1.80	20.0	15.0	36.00	43.24	20.76	26.00	5.18	45.50	44.89	9.61	26.50	5.52
1.90	5.0	5.0	28.50	44.48	27.02	24.50	4.96	38.00	45.46	16.54	30.50	4.92
1.90	5.0	10.0	31.00	44.00	25.00	24.50	4.10	42.00	45.01	12.99	30.00	4.14
1.90	5.0	15.0	34.50	43.34	22.16	26.00	3.29	43.00	44.90	12.10	26.50	3.48

Table 26 (continued)

DATA FOR THE CONSTRUCTION OF ISOKINETIC SUCROSE AND GLYCEROL GRADIENTS FOR MSE ROTOR 43127-104

Particle density g/ml	Temperature °C	Top concentration %(w/w)	Sucrose					Glycerol				
			C_h %(w/w)	Buffer 2x ml	H_2O ml	V_m ml	Velocity of 100 S particle at top speed ml/h	C_h %(w/w)	Buffer 2x ml	H_2O ml	V_m ml	Velocity of 100 S particle at top speed ml/h
1.90	10.0	5.0	29.00	44.44	26.56	24.50	5.79	39.00	45.43	15.57	30.50	5.75
1.90	10.0	10.0	31.50	43.96	24.54	24.50	4.80	43.00	44.98	12.02	30.00	4.86
1.90	10.0	15.0	34.00	43.49	22.51	24.00	3.89	44.00	44.87	11.13	26.50	4.11
1.90	20.0	5.0	30.50	44.28	25.22	25.00	7.61	40.50	45.43	14.07	29.50	7.59
1.90	20.0	10.0	33.00	43.81	23.19	25.00	6.38	44.00	45.05	10.95	28.50	6.49
1.90	20.0	15.0	36.00	43.24	20.76	25.50	5.22	45.00	44.94	10.06	25.50	5.55
2.00	5.0	5.0	29.00	44.38	26.62	25.00	4.97	38.50	45.40	16.10	30.50	4.93
2.00	5.0	10.0	31.50	43.90	24.60	25.00	4.12	41.50	45.07	13.43	29.00	4.15
2.00	5.0	15.0	34.50	43.34	22.16	25.50	3.32	44.00	44.79	11.21	27.50	3.49
2.00	10.0	5.0	29.00	44.44	26.56	24.00	5.80	39.50	45.37	15.13	30.50	5.75
2.00	10.0	10.0	31.50	43.96	24.54	24.00	4.83	43.00	44.98	12.02	29.50	4.87
2.00	10.0	15.0	35.00	43.30	21.70	25.50	3.91	45.00	44.76	10.24	27.50	4.13
2.00	20.0	5.0	30.50	44.28	25.22	24.50	7.63	41.00	45.38	13.62	29.50	7.60
2.00	20.0	10.0	33.00	43.81	23.19	24.50	6.41	45.00	44.94	10.06	29.50	6.51
2.00	20.0	15.0	36.50	43.15	20.35	26.00	5.26	46.00	44.83	9.17	26.50	5.58

Table 27

DATA FOR THE CONSTRUCTION OF ISOKINETIC SUCROSE AND GLYCEROL GRADIENTS FOR MSE ROTOR 43127-106

Particle density g/ml	Temperature °C	Top concentration %(w/w)	Sucrose					Glycerol				
			C_h %(w/w)	Buffer 2x ml	H_2O ml	V_m ml	Velocity of 100 S particle at top speed ml/h	C_h %(w/w)	Buffer 2x ml	H_2O ml	V_m ml	Velocity of 100 S particle at top speed ml/h
1.20	5.0	5.0	23.00	45.53	31.47	61.00	6.08	32.00	46.15	21.85	73.00	6.21
1.20	5.0	10.0	25.50	45.05	29.45	59.00	4.56	35.00	45.80	19.20	69.00	4.94
1.20	5.0	15.0	29.00	44.38	26.62	64.00	3.25	39.00	45.35	15.65	71.00	3.90
1.20	10.0	5.0	23.00	45.58	31.42	59.00	7.10	32.50	46.16	21.34	72.00	7.27
1.20	10.0	10.0	26.00	45.01	28.99	60.00	5.35	36.00	45.76	18.24	70.00	5.81
1.20	10.0	15.0	29.50	44.34	26.16	65.00	3.85	39.50	45.37	15.13	70.00	4.62
1.20	20.0	5.0	24.00	45.51	30.49	60.00	9.36	34.50	46.10	19.40	73.00	9.61
1.20	20.0	10.0	26.50	45.03	28.47	59.00	7.15	37.00	45.82	17.18	68.00	7.80
1.20	20.0	15.0	30.00	44.37	25.63	64.00	5.22	40.50	45.43	14.07	68.00	6.29
1.30	5.0	5.0	26.00	44.96	29.04	61.00	6.31	35.50	45.74	18.76	73.00	6.36
1.30	5.0	10.0	28.50	44.48	27.02	60.00	4.94	39.00	45.35	15.65	71.00	5.18
1.30	5.0	15.0	32.00	43.81	24.19	63.00	3.74	41.00	45.12	13.88	65.00	4.21
1.30	10.0	5.0	26.00	45.01	28.99	59.00	7.36	36.00	45.76	18.24	72.00	7.43
1.30	10.0	10.0	29.50	44.34	26.16	63.00	5.80	39.50	45.37	15.13	70.00	6.09
1.30	10.0	15.0	32.50	43.77	23.73	64.00	4.42	42.00	45.09	12.91	66.00	4.98
1.30	20.0	5.0	27.50	44.85	27.65	62.00	9.69	38.00	45.71	16.29	72.00	9.82
1.30	20.0	10.0	30.00	44.37	25.63	61.00	7.72	41.00	45.38	13.62	69.00	8.16
1.30	20.0	15.0	33.00	43.81	23.19	62.00	5.96	43.00	45.16	11.84	64.00	6.76
1.40	5.0	5.0	27.00	44.76	28.24	58.00	6.42	37.00	45.57	17.43	72.00	6.43
1.40	5.0	10.0	30.50	44.10	25.40	62.00	5.13	41.00	45.12	13.88	72.00	5.30
1.40	5.0	15.0	33.50	43.52	22.98	63.00	3.98	43.00	44.90	12.10	66.00	4.37
1.40	10.0	5.0	27.50	44.72	27.78	58.00	7.49	38.00	45.54	16.46	72.00	7.51
1.40	10.0	10.0	31.00	44.06	24.94	62.00	6.02	42.00	45.09	12.91	72.00	6.23
1.40	10.0	15.0	34.00	43.49	22.51	63.00	4.70	43.50	44.93	11.57	65.00	5.16
1.40	20.0	5.0	28.50	44.66	26.84	58.00	9.85	40.00	45.49	14.51	72.00	9.92
1.40	20.0	10.0	32.00	43.99	24.01	62.00	8.01	43.00	45.16	11.84	69.00	8.33
1.40	20.0	15.0	34.50	43.52	21.98	61.00	6.33	44.00	45.05	10.95	61.00	6.99
1.50	5.0	5.0	28.00	44.57	27.43	58.00	6.48	38.50	45.40	16.10	73.00	6.47
1.50	5.0	10.0	31.00	44.00	25.00	60.00	5.25	41.00	45.12	13.88	68.00	5.37
1.50	5.0	15.0	34.00	43.43	22.57	61.00	4.12	43.00	44.90	12.10	63.00	4.46
1.50	10.0	5.0	28.50	44.53	26.97	58.00	7.56	39.50	45.37	15.13	73.00	7.56
1.50	10.0	10.0	32.00	43.87	24.13	62.00	6.15	42.00	45.09	12.91	68.00	6.31
1.50	10.0	15.0	35.00	43.30	21.70	63.00	4.87	44.00	44.87	11.13	63.00	5.27

Table 27 (continued)

DATA FOR THE CONSTRUCTION OF ISOKINETIC SUCROSE AND GLYCEROL GRADIENTS FOR MSE ROTOR 43127-106

Particle density g/ml	Temperature °C	Top concentration %(w/w)	Sucrose					Glycerol				
			C_h %(w/w)	Buffer 2x ml	H_2O ml	V_m ml	Velocity of 100 S particle at top speed ml/h	C_h %(w/w)	Buffer 2x ml	H_2O ml	V_m ml	Velocity of 100 S particle at top speed ml/h
1.50	20.0	5.0	30.00	44.37	25.63	60.00	9.95	41.00	45.38	13.62	71.00	9.98
1.50	20.0	10.0	33.00	43.81	23.19	62.00	8.18	44.00	45.05	10.95	68.00	8.44
1.50	20.0	15.0	35.50	43.34	21.16	61.00	6.55	45.00	44.94	10.06	61.00	7.13
1.60	5.0	5.0	29.00	44.38	26.62	59.00	6.53	39.00	45.35	15.65	72.00	6.50
1.60	5.0	10.0	31.50	43.90	24.60	59.00	5.32	42.00	45.01	12.99	69.00	5.42
1.60	5.0	15.0	35.00	43.24	21.76	63.00	4.22	44.00	44.79	11.21	64.00	4.52
1.60	10.0	5.0	29.00	44.44	26.56	57.00	7.61	40.00	45.32	14.68	72.00	7.59
1.60	10.0	10.0	32.50	43.77	23.73	61.00	6.24	43.00	44.98	12.02	69.00	6.37
1.60	10.0	15.0	35.00	43.30	21.70	60.00	4.98	44.00	44.87	11.13	61.00	5.34
1.60	20.0	5.0	30.50	44.28	25.22	59.00	10.02	42.00	45.27	12.73	71.00	10.02
1.60	20.0	10.0	34.00	43.62	22.38	63.00	8.30	44.50	45.00	10.50	67.00	8.51
1.60	20.0	15.0	36.00	43.24	20.76	60.00	6.70	45.50	44.89	9.61	60.00	7.23
1.70	5.0	5.0	29.00	44.38	26.62	57.00	6.56	39.50	45.29	15.21	72.00	6.52
1.70	5.0	10.0	32.00	43.81	24.19	59.00	5.38	42.00	45.01	12.99	67.00	5.46
1.70	5.0	15.0	35.50	43.15	21.35	63.00	4.29	44.00	44.79	11.21	62.00	4.57
1.70	10.0	5.0	30.00	44.25	25.75	59.00	7.65	41.00	45.20	13.80	73.00	7.61
1.70	10.0	10.0	32.50	43.77	23.73	59.00	6.31	43.00	44.98	12.02	67.00	6.41
1.70	10.0	15.0	36.00	43.11	20.89	63.00	5.06	44.50	44.82	10.68	61.00	5.40
1.70	20.0	5.0	31.00	44.18	24.82	58.00	10.06	42.50	45.21	12.29	71.00	10.05
1.70	20.0	10.0	33.50	43.71	22.79	58.00	8.38	45.00	44.94	10.06	67.00	8.57
1.70	20.0	15.0	36.50	43.15	20.35	60.00	6.81	46.00	44.83	9.17	60.00	7.30
1.80	5.0	5.0	29.50	44.29	26.21	58.00	6.58	40.00	45.23	14.77	72.00	6.53
1.80	5.0	10.0	32.50	43.71	23.79	60.00	5.42	43.00	44.90	12.10	69.00	5.48
1.80	5.0	15.0	36.00	43.05	20.95	64.00	4.34	44.00	44.79	11.21	61.00	4.60
1.80	10.0	5.0	30.00	44.25	25.75	57.00	7.68	41.50	45.15	13.35	73.00	7.63
1.80	10.0	10.0	33.50	43.58	22.92	62.00	6.35	44.00	44.87	11.13	69.00	6.44
1.80	10.0	15.0	36.00	43.11	20.89	61.00	5.12	45.00	44.76	10.24	61.00	5.43
1.80	20.0	5.0	31.00	44.18	24.82	57.00	10.10	43.00	45.16	11.84	71.00	10.08
1.80	20.0	10.0	34.00	43.62	22.38	59.00	8.44	44.50	45.00	10.50	64.00	8.60
1.80	20.0	15.0	37.00	43.06	19.94	61.00	6.89	46.00	44.83	9.17	59.00	7.35
1.90	5.0	5.0	29.50	44.29	26.21	57.00	6.60	40.50	45.18	14.32	73.00	6.54
1.90	5.0	10.0	33.00	43.62	23.38	61.00	5.45	43.00	44.90	12.10	68.00	5.50

Table 27 (continued)

DATA FOR THE CONSTRUCTION OF ISOKINETIC SUCROSE AND GLYCEROL GRADIENTS FOR MSE ROTOR 43127-106

Particle density g/ml	Temperature °C	Top concentration %(w/w)	Sucrose					Glycerol				
			C_η %(w/w)	Buffer 2x ml	H_2O ml	V_m ml	Velocity of 100 S particle at top speed ml/h	C_η %(w/w)	Buffer 2x ml	H_2O ml	V_m ml	Velocity of 100 S particle at top speed ml/h
1.90	5.0	15.0	36.00	43.05	20.95	63.00	4.38	44.50	44.73	10.77	62.00	4.63
1.90	10.0	5.0	30.50	44.15	25.35	58.00	7.70	41.50	45.15	13.35	72.00	7.64
1.90	10.0	10.0	33.00	43.68	23.32	58.00	6.39	44.00	44.87	11.13	68.00	6.46
1.90	10.0	15.0	36.00	43.11	20.89	60.00	5.17	45.50	44.71	9.79	62.00	5.46
1.90	20.0	5.0	31.50	44.09	24.41	57.00	10.13	43.00	45.16	11.84	70.00	10.09
1.90	20.0	10.0	34.50	43.52	21.98	60.00	8.49	45.50	44.89	9.61	66.00	8.63
1.90	20.0	15.0	37.50	42.96	19.54	62.00	6.95	47.00	44.73	8.27	61.00	7.39
2.00	5.0	5.0	30.00	44.19	25.81	58.00	6.62	41.00	45.12	13.88	73.00	6.55
2.00	5.0	10.0	33.00	43.62	23.38	60.00	5.48	43.00	44.90	12.10	67.00	5.52
2.00	5.0	15.0	36.00	43.05	20.95	62.00	4.41	44.50	44.73	10.77	61.00	4.65
2.00	10.0	5.0	30.50	44.15	25.35	57.00	7.72	42.00	45.09	12.91	73.00	7.65
2.00	10.0	10.0	34.00	43.49	22.51	62.00	6.42	44.00	44.87	11.13	67.00	6.48
2.00	10.0	15.0	36.00	43.11	20.89	59.00	5.21	45.50	44.71	9.79	61.00	5.49
2.00	20.0	5.0	32.00	43.99	24.01	58.00	10.15	43.50	45.11	11.39	70.00	10.11
2.00	20.0	10.0	34.50	43.52	21.98	59.00	8.53	45.50	44.89	9.61	65.00	8.66
2.00	20.0	15.0	37.50	42.96	19.54	61.00	7.00	46.50	44.78	8.72	59.00	7.42

Table 28

DATA FOR THE CONSTRUCTION OF ISOKINETIC SUCROSE AND GLYCEROL GRADIENTS FOR MSE ROTOR 43127-111

Particle density g/ml	Temperature °C	Top concentration %(w/w)	Sucrose					Glycerol				
			C_h %(w/w)	Buffer 2x ml	H_2O ml	V_m ml	Velocity of 100 S particle at top speed ml/h	C_h %(w/w)	Buffer 2x ml	H_2O ml	V_m ml	Velocity of 100 S particle at top speed ml/h
1.20	5.0	5.0	23.00	45.53	31.47	11.00	3.22	32.50	46.09	21.41	13.75	3.29
1.20	5.0	10.0	26.50	44.86	28.64	12.00	2.41	35.50	45.74	18.76	13.00	2.62
1.20	5.0	15.0	29.50	44.29	26.21	12.25	1.72	39.00	45.35	15.65	13.00	2.07
1.20	10.0	5.0	23.00	45.58	31.42	10.75	3.76	34.00	45.99	20.01	14.25	3.85
1.20	10.0	10.0	26.00	45.01	28.99	11.00	2.84	35.50	45.82	18.68	12.50	3.08
1.20	10.0	15.0	29.50	44.34	26.16	11.75	2.04	39.50	45.37	15.13	12.75	2.45
1.20	20.0	5.0	24.00	45.51	30.49	11.00	4.96	35.00	46.04	18.96	13.75	5.09
1.20	20.0	10.0	26.50	45.03	28.47	10.75	3.79	37.50	45.76	16.74	12.75	4.13
1.20	20.0	15.0	30.00	44.37	25.63	11.75	2.77	40.50	45.43	14.07	12.50	3.33
1.30	5.0	5.0	25.50	45.05	29.45	10.75	3.34	37.50	45.52	16.98	14.75	3.37
1.30	5.0	10.0	29.00	44.38	26.62	11.50	2.62	39.00	45.35	15.65	13.00	2.74
1.30	5.0	15.0	32.00	43.81	24.19	11.50	1.98	41.50	45.07	13.43	12.25	2.23
1.30	10.0	5.0	26.50	44.91	28.59	11.25	3.90	38.00	45.54	16.46	14.50	3.93
1.30	10.0	10.0	29.50	44.34	26.16	11.50	3.07	39.50	45.37	15.13	12.75	3.23
1.30	10.0	15.0	33.00	43.68	23.32	12.25	2.34	42.00	45.09	12.91	12.00	2.64
1.30	20.0	5.0	27.50	44.85	27.65	11.25	5.13	38.50	45.65	15.85	13.50	5.20
1.30	20.0	10.0	30.50	44.28	25.22	11.50	4.09	40.50	45.43	14.07	12.25	4.32
1.30	20.0	15.0	33.00	43.81	23.19	11.25	3.16	43.50	45.11	11.39	12.00	3.58
1.40	5.0	5.0	28.00	44.57	27.43	11.50	3.40	39.00	45.35	15.65	14.50	3.40
1.40	5.0	10.0	31.00	44.00	25.00	11.75	2.72	40.00	45.23	14.77	12.50	2.81
1.40	5.0	15.0	33.50	43.52	22.98	11.50	2.11	42.50	44.95	12.55	11.75	2.31
1.40	10.0	5.0	28.00	44.63	27.37	11.00	3.97	39.00	45.43	15.57	13.75	3.98
1.40	10.0	10.0	31.50	43.96	24.54	11.75	3.19	41.00	45.20	13.80	12.50	3.30
1.40	10.0	15.0	34.00	43.49	22.51	11.50	2.49	44.00	44.87	11.13	12.25	2.74
1.40	20.0	5.0	29.00	44.56	26.44	11.00	5.22	41.00	45.38	13.62	13.75	5.26
1.40	20.0	10.0	31.50	44.09	24.41	11.00	4.24	42.50	45.21	12.29	12.25	4.42
1.40	20.0	15.0	35.00	43.43	21.57	11.50	3.35	45.00	44.94	10.06	11.75	3.70
1.50	5.0	5.0	28.50	44.48	27.02	11.00	3.43	39.50	45.29	15.21	14.00	3.43
1.50	5.0	10.0	31.00	44.00	25.00	11.00	2.78	41.50	45.07	13.43	12.75	2.85
1.50	5.0	15.0	34.50	43.34	22.16	11.50	2.18	43.50	44.84	11.66	11.75	2.36
1.50	10.0	5.0	29.00	44.44	26.56	11.00	4.01	40.50	45.26	14.24	14.00	4.00
1.50	10.0	10.0	31.50	43.96	24.54	11.00	3.26	42.50	45.04	12.46	12.75	3.34
1.50	10.0	15.0	35.00	43.30	21.70	11.50	2.58	44.50	44.82	10.68	11.75	2.79

Table 28 (continued)

DATA FOR THE CONSTRUCTION OF ISOKINETIC SUCROSE AND GLYCEROL GRADIENTS FOR MSE ROTOR 43127-111

Particle density g/ml	Temperature °C	Top concentration %(w/w)	Sucrose					Glycerol				
			C_h %(w/w)	Buffer 2x ml	H_2O ml	V_m ml	Velocity of 100 S particle at top speed ml/h	C_h %(w/w)	Buffer 2x ml	H_2O ml	V_m ml	Velocity of 100 S particle at top speed ml/h
1.50	20.0	5.0	30.00	44.37	25.63	11.00	5.27	42.00	45.27	12.73	13.50	5.29
1.50	20.0	10.0	33.00	43.81	23.19	11.25	4.33	44.00	45.05	10.95	12.50	4.47
1.50	20.0	15.0	36.00	43.24	20.76	11.50	3.47	46.00	44.83	9.17	11.75	3.78
1.60	5.0	5.0	28.50	44.48	27.02	10.50	3.46	40.00	45.23	14.77	13.75	3.44
1.60	5.0	10.0	32.00	43.81	24.19	11.25	2.82	41.50	45.07	13.43	12.25	2.87
1.60	5.0	15.0	35.00	43.24	21.76	11.50	2.24	44.50	44.73	10.77	12.00	2.40
1.60	10.0	5.0	30.00	44.25	25.75	11.25	4.03	41.00	45.20	13.80	13.75	4.02
1.60	10.0	10.0	32.00	43.87	24.13	10.75	3.31	42.50	45.04	12.46	12.25	3.37
1.60	10.0	15.0	35.50	43.20	21.30	11.50	2.64	45.00	44.76	10.24	11.75	2.83
1.60	20.0	5.0	30.50	44.28	25.22	10.75	5.31	42.50	45.21	12.29	13.25	5.31
1.60	20.0	10.0	34.00	43.62	22.38	11.50	4.39	44.00	45.05	10.95	12.00	4.51
1.60	20.0	15.0	36.00	43.24	20.76	11.00	3.55	47.00	44.73	8.27	11.75	3.83
1.70	5.0	5.0	29.00	44.38	26.62	10.50	3.47	40.50	45.18	14.32	13.75	3.45
1.70	5.0	10.0	32.50	43.71	23.79	11.25	2.85	42.00	45.01	12.99	12.25	2.89
1.70	5.0	15.0	35.50	43.15	21.35	11.50	2.27	44.50	44.73	10.77	11.75	2.42
1.70	10.0	5.0	30.00	44.25	25.75	10.75	4.05	41.50	45.15	13.35	13.75	4.03
1.70	10.0	10.0	32.50	43.77	23.73	10.75	3.34	43.00	44.98	12.02	12.25	3.40
1.70	10.0	15.0	36.00	43.11	20.89	11.50	2.68	45.50	44.71	9.79	11.75	2.86
1.70	20.0	5.0	31.50	44.09	24.41	11.00	5.33	43.00	45.16	11.84	13.25	5.33
1.70	20.0	10.0	34.00	43.62	22.38	11.00	4.44	45.00	44.94	10.06	12.25	4.54
1.70	20.0	15.0	36.50	43.15	20.35	11.00	3.61	47.50	44.67	7.83	11.75	3.87
1.80	5.0	5.0	30.00	44.19	25.81	11.00	3.49	41.00	45.12	13.88	13.75	3.46
1.80	5.0	10.0	32.50	43.71	23.79	11.00	2.87	42.50	44.95	12.55	12.25	2.90
1.80	5.0	15.0	35.50	43.15	21.35	11.25	2.30	45.00	44.68	10.32	11.75	2.44
1.80	10.0	5.0	30.00	44.25	25.75	10.50	4.07	41.50	45.15	13.35	13.50	4.04
1.80	10.0	10.0	33.50	43.58	22.92	11.25	3.37	43.50	44.93	11.57	12.25	3.41
1.80	10.0	15.0	36.00	43.11	20.89	11.25	2.71	46.00	44.65	9.35	11.75	2.88
1.80	20.0	5.0	31.50	44.09	24.41	10.75	5.35	43.00	45.16	11.84	13.00	5.34
1.80	20.0	10.0	34.00	43.62	22.38	10.75	4.47	45.00	44.94	10.06	12.00	4.56
1.80	20.0	15.0	37.50	42.96	19.54	11.50	3.65	47.50	44.67	7.83	11.75	3.89
1.90	5.0	5.0	30.00	44.19	25.81	10.75	3.50	41.00	45.12	13.88	13.50	3.47
1.90	5.0	10.0	32.50	43.71	23.79	10.75	2.89	43.50	44.84	11.66	12.75	2.92

Table 28 (continued)

DATA FOR THE CONSTRUCTION OF ISOKINETIC SUCROSE AND GLYCEROL GRADIENTS FOR MSE ROTOR 43127-111

Particle density g/ml	Temperature °C	Top concentration %(w/w)	Sucrose					Glycerol				
			C_h %(w/w)	Buffer 2x ml	H_2O ml	V_m ml	Velocity of 100 S particle at top speed ml/h	C_h %(w/w)	Buffer 2x ml	H_2O ml	V_m ml	Velocity of 100 S particle at top speed ml/h
1.90	5.0	15.0	36.00	43.05	20.95	11.50	2.32	45.50	44.62	9.88	12.00	2.45
1.90	10.0	5.0	31.00	44.06	24.94	11.00	4.08	42.00	45.09	12.91	13.50	4.05
1.90	10.0	10.0	33.50	43.58	22.92	11.00	3.39	45.00	44.76	10.24	13.00	3.42
1.90	10.0	15.0	36.00	43.11	20.89	11.00	2.74	46.50	44.60	8.90	12.00	2.89
1.90	20.0	5.0	31.50	44.09	24.41	10.50	5.36	43.50	45.11	11.39	13.00	5.35
1.90	20.0	10.0	34.00	43.62	22.38	10.50	4.50	45.00	44.94	10.06	11.75	4.57
1.90	20.0	15.0	37.50	42.96	19.54	11.25	3.68	48.00	44.62	7.38	11.75	3.91
2.00	5.0	5.0	30.00	44.19	25.81	10.50	3.50	41.50	45.07	13.43	13.75	3.47
2.00	5.0	10.0	33.00	43.62	23.38	11.00	2.90	43.00	44.90	12.10	12.25	2.92
2.00	5.0	15.0	36.00	43.05	20.95	11.25	2.34	45.50	44.62	9.88	11.75	2.46
2.00	10.0	5.0	30.50	44.15	25.35	10.50	4.09	42.00	45.09	12.91	13.25	4.05
2.00	10.0	10.0	33.00	43.68	23.32	10.50	3.40	44.00	44.87	11.13	12.25	3.43
2.00	10.0	15.0	36.50	43.02	20.48	11.25	2.76	46.50	44.60	8.90	11.75	2.91
2.00	20.0	5.0	31.50	44.09	24.41	10.25	5.38	43.50	45.11	11.39	12.75	5.35
2.00	20.0	10.0	34.50	43.52	21.98	10.75	4.52	46.00	44.83	9.17	12.25	4.59
2.00	20.0	15.0	38.00	42.87	19.13	11.50	3.71	48.00	44.62	7.38	11.75	3.93

Table 29

DATA FOR THE CONSTRUCTION OF ISOKINETIC SUCROSE AND GLYCEROL GRADIENTS FOR MSE ROTOR 43127-115

Particle density g/ml	Temperature °C	Top concentration %(w/w)	Sucrose					Glycerol				
			C_h %(w/w)	Buffer 2x ml	H_2O ml	V_m ml	Velocity of 100 S particle at top speed ml/h	C_h %(w/w)	Buffer 2x ml	H_2O ml	V_m ml	Velocity of 100 S particle at top speed ml/h
1.20	5.0	5.0	21.00	45.92	33.08	4.75	3.19	32.00	46.15	21.85	6.75	3.26
1.20	5.0	10.0	25.00	45.15	29.85	5.25	2.39	34.00	45.92	20.08	6.00	2.59
1.20	5.0	15.0	28.00	44.57	27.43	5.25	1.70	39.00	45.35	15.65	6.50	2.04
1.20	10.0	5.0	22.00	45.77	32.23	5.00	3.72	32.00	46.22	21.78	6.50	3.81
1.20	10.0	10.0	25.00	45.20	29.80	5.00	2.81	36.00	45.76	18.24	6.50	3.05
1.20	10.0	15.0	27.50	44.72	27.78	4.75	2.02	39.00	45.43	15.57	6.25	2.42
1.20	20.0	5.0	21.50	45.98	32.52	4.50	4.91	34.50	46.10	19.40	6.75	5.04
1.20	20.0	10.0	25.50	45.22	29.28	5.00	3.75	36.00	45.93	18.07	6.00	4.09
1.20	20.0	15.0	28.50	44.66	26.84	5.00	2.74	39.00	45.60	15.40	5.75	3.30
1.30	5.0	5.0	25.00	45.15	29.85	5.25	3.30	34.50	45.86	19.64	6.50	3.33
1.30	5.0	10.0	28.50	44.48	27.02	5.50	2.59	38.00	45.46	16.54	6.25	2.71
1.30	5.0	15.0	31.50	43.90	24.60	5.50	1.96	41.00	45.12	13.88	6.00	2.21
1.30	10.0	5.0	25.50	45.10	29.40	5.25	3.86	35.50	45.82	18.68	6.50	3.89
1.30	10.0	10.0	27.00	44.82	28.18	4.75	3.04	38.00	45.54	16.46	6.00	3.19
1.30	10.0	15.0	30.00	44.25	25.75	4.75	2.31	41.00	45.20	13.80	5.75	2.61
1.30	20.0	5.0	25.00	45.32	29.68	4.75	5.08	36.50	45.87	17.63	6.25	5.15
1.30	20.0	10.0	28.50	44.66	26.84	5.00	4.05	40.50	45.43	14.07	6.25	4.27
1.30	20.0	15.0	32.00	43.99	24.01	5.25	3.12	42.50	45.21	12.29	5.75	3.54
1.40	5.0	5.0	26.00	44.96	29.04	5.00	3.36	37.00	45.57	17.43	6.75	3.37
1.40	5.0	10.0	30.00	44.19	25.81	5.50	2.69	40.50	45.18	14.32	6.50	2.78
1.40	5.0	15.0	31.00	44.00	25.00	4.75	2.08	41.00	45.12	13.88	5.50	2.29
1.40	10.0	5.0	27.00	44.82	28.18	5.25	3.92	37.00	45.65	17.35	6.50	3.93
1.40	10.0	10.0	28.50	44.53	26.97	4.75	3.16	41.50	45.15	13.35	6.50	3.26
1.40	10.0	15.0	31.50	43.96	24.54	4.75	2.46	43.50	44.93	11.57	6.00	2.71
1.40	20.0	5.0	28.00	44.75	27.25	5.25	5.16	38.50	45.65	15.85	6.25	5.20
1.40	20.0	10.0	30.00	44.37	25.63	5.00	4.20	42.50	45.21	12.29	6.25	4.37
1.40	20.0	15.0	33.00	43.81	23.19	5.00	3.32	43.50	45.11	11.39	5.50	3.66
1.50	5.0	5.0	27.00	44.76	28.24	5.00	3.40	37.50	45.52	16.98	6.50	3.39
1.50	5.0	10.0	29.00	44.38	26.62	4.75	2.75	40.00	45.23	14.77	6.00	2.82
1.50	5.0	15.0	32.50	43.71	23.79	5.00	2.16	42.00	45.01	12.99	5.50	2.34
1.50	10.0	5.0	27.50	44.72	27.78	5.00	3.96	37.50	45.59	16.91	6.25	3.96
1.50	10.0	10.0	29.50	44.34	26.16	4.75	3.23	41.00	45.20	13.80	6.00	3.31
1.50	10.0	15.0	33.00	43.68	23.32	5.00	2.55	42.00	45.09	12.91	5.25	2.76

Table 29 (continued)

DATA FOR THE CONSTRUCTION OF ISOKINETIC SUCROSE AND GLYCEROL GRADIENTS FOR MSE ROTOR 43127-115

Particle density g/ml	Temperature °C	Top concentration %(w/w)	Sucrose					Glycerol				
			C_h %(w/w)	Buffer 2x ml	H_2O ml	V_m ml	Velocity of 100 S particle at top speed ml/h	C_h %(w/w)	Buffer 2x ml	H_2O ml	V_m ml	Velocity of 100 S particle at top speed ml/h
1.50	20.0	5.0	28.50	44.66	26.84	5.00	5.21	38.50	45.65	15.85	6.00	5.23
1.50	20.0	10.0	30.50	44.28	25.22	4.75	4.29	43.00	45.16	11.84	6.00	4.42
1.50	20.0	15.0	34.00	43.62	22.38	5.00	3.44	44.50	45.00	10.50	5.50	3.74
1.60	5.0	5.0	27.50	44.67	27.83	5.00	3.42	38.00	45.46	16.54	6.50	3.40
1.60	5.0	10.0	31.50	43.90	24.60	5.50	2.79	41.50	45.07	13.43	6.25	2.84
1.60	5.0	15.0	32.50	43.71	23.79	4.75	2.21	43.50	44.84	11.66	5.75	2.37
1.60	10.0	5.0	28.00	44.63	27.37	5.00	3.99	38.50	45.48	16.02	6.25	3.98
1.60	10.0	10.0	31.50	43.96	24.54	5.25	3.27	42.50	45.04	12.46	6.25	3.34
1.60	10.0	15.0	33.00	43.68	23.32	4.75	2.61	43.50	44.93	11.57	5.50	2.80
1.60	20.0	5.0	28.50	44.66	26.84	4.75	5.25	39.50	45.54	14.96	6.00	5.25
1.60	20.0	10.0	32.50	43.90	23.60	5.25	4.35	43.50	45.11	11.39	6.00	4.46
1.60	20.0	15.0	34.50	43.52	21.98	5.00	3.51	44.50	45.00	10.50	5.25	3.79
1.70	5.0	5.0	28.00	44.57	27.43	5.00	3.44	38.00	45.46	16.54	6.25	3.42
1.70	5.0	10.0	30.00	44.19	25.81	4.75	2.82	42.00	45.01	12.99	6.25	2.86
1.70	5.0	15.0	33.50	43.52	22.98	5.00	2.25	44.00	44.79	11.21	5.75	2.39
1.70	10.0	5.0	28.00	44.63	27.37	4.75	4.01	38.00	45.54	16.46	6.00	3.99
1.70	10.0	10.0	30.50	44.15	25.35	4.75	3.30	41.50	45.15	13.35	5.75	3.36
1.70	10.0	15.0	34.00	43.49	22.51	5.00	2.65	44.00	44.87	11.13	5.50	2.83
1.70	20.0	5.0	29.00	44.56	26.44	4.75	5.27	40.00	45.59	14.51	6.00	5.27
1.70	20.0	10.0	33.00	43.81	23.19	5.25	4.39	43.50	45.11	11.39	5.75	4.49
1.70	20.0	15.0	35.00	43.43	21.57	5.00	3.57	45.00	44.94	10.06	5.25	3.82
1.80	5.0	5.0	28.50	44.48	27.02	5.00	3.45	38.50	45.40	16.10	6.25	3.42
1.80	5.0	10.0	31.00	44.00	25.00	5.00	2.84	42.50	44.95	12.55	6.25	2.87
1.80	5.0	15.0	34.50	43.34	22.16	5.25	2.27	43.50	44.84	11.66	5.50	2.41
1.80	10.0	5.0	29.00	44.44	26.56	5.00	4.02	38.50	45.48	16.02	6.00	4.00
1.80	10.0	10.0	31.50	43.96	24.54	5.00	3.33	43.50	44.93	11.57	6.25	3.37
1.80	10.0	15.0	35.00	43.30	21.70	5.25	2.68	43.50	44.93	11.57	5.25	2.85
1.80	20.0	5.0	29.50	44.47	26.03	4.75	5.29	39.50	45.54	14.96	5.75	5.28
1.80	20.0	10.0	32.00	43.99	24.01	4.75	4.42	44.00	45.05	10.95	5.75	4.51
1.80	20.0	15.0	36.00	43.24	20.76	5.25	3.61	45.50	44.89	9.61	5.25	3.85
1.90	5.0	5.0	28.00	44.57	27.43	4.75	3.46	38.50	45.40	16.10	6.25	3.43
1.90	5.0	10.0	32.00	43.81	24.19	5.25	2.86	42.00	45.01	12.99	6.00	2.88

Table 29 (continued)

DATA FOR THE CONSTRUCTION OF ISOKINETIC SUCROSE AND GLYCEROL GRADIENTS FOR MSE ROTOR 43127-115

Particle density g/ml	Temperature °C	Top concentration %(w/w)	Sucrose					Glycerol				
			C_h %(w/w)	Buffer 2x ml	H_2O ml	V_m ml	Velocity of 100 S particle at top speed ml/h	C_h %(w/w)	Buffer 2x ml	H_2O ml	V_m ml	Velocity of 100 S particle at top speed ml/h
1.90	5.0	15.0	34.00	43.43	22.57	5.00	2.30	43.00	44.90	12.10	5.25	2.42
1.90	10.0	5.0	28.50	44.53	26.97	4.75	4.04	38.50	45.48	16.02	6.00	4.01
1.90	10.0	10.0	32.50	43.77	23.73	5.25	3.35	43.00	44.98	12.02	6.00	3.39
1.90	10.0	15.0	34.50	43.39	22.11	5.00	2.71	43.00	44.98	12.02	5.00	2.86
1.90	20.0	5.0	29.00	44.56	26.44	4.50	5.31	40.00	45.49	14.51	5.75	5.29
1.90	20.0	10.0	33.00	43.81	23.19	5.00	4.45	44.00	45.05	10.95	5.75	4.53
1.90	20.0	15.0	35.00	43.43	21.57	4.75	3.64	44.50	45.00	10.50	5.00	3.87
2.00	5.0	5.0	29.00	44.38	26.62	5.00	3.47	38.00	45.46	16.54	6.00	3.43
2.00	5.0	10.0	31.50	43.90	24.60	5.00	2.87	43.00	44.90	12.10	6.25	2.89
2.00	5.0	15.0	35.50	43.15	21.35	5.50	2.31	44.00	44.79	11.21	5.50	2.44
2.00	10.0	5.0	29.50	44.34	26.16	5.00	4.04	39.00	45.43	15.57	6.00	4.01
2.00	10.0	10.0	32.00	43.87	24.13	5.00	3.36	43.50	44.93	11.57	6.00	3.40
2.00	10.0	15.0	35.50	43.20	21.30	5.25	2.73	44.00	44.87	11.13	5.25	2.88
2.00	20.0	5.0	30.00	44.37	25.63	4.75	5.32	40.50	45.43	14.07	5.75	5.30
2.00	20.0	10.0	32.50	43.90	23.60	4.75	4.47	44.50	45.00	10.50	5.75	4.54
2.00	20.0	15.0	36.50	43.15	20.35	5.25	3.67	46.00	44.83	9.17	5.25	3.89

Table 30

DATA FOR THE CONSTRUCTION OF ISOKINETIC SUCROSE AND GLYCEROL GRADIENTS FOR MSE ROTOR 43127-126

Particle density g/ml	Temperature °C	Top concentration %(w/w)	Sucrose					Glycerol				
			C_h %(w/w)	Buffer 2x ml	H_2O ml	V_m ml	Velocity of 100 S particle at top speed ml/h	C_h %(w/w)	Buffer 2x ml	H_2O ml	V_m ml	Velocity of 100 S particle at top speed ml/h
1.20	5.0	5.0	20.50	46.02	33.48	6.50	2.97	30.00	46.38	23.62	8.75	3.03
1.20	5.0	10.0	24.50	45.24	30.26	7.25	2.22	33.00	46.03	20.97	8.25	2.41
1.20	5.0	15.0	28.50	44.48	27.02	8.00	1.59	37.50	45.52	16.98	8.50	1.90
1.20	10.0	5.0	22.00	45.77	32.23	7.25	3.47	30.50	46.39	23.11	8.75	3.55
1.20	10.0	10.0	24.50	45.29	30.21	7.00	2.61	35.00	45.88	19.12	8.75	2.84
1.20	10.0	15.0	28.50	44.53	26.97	7.75	1.88	38.00	45.54	16.46	8.50	2.26
1.20	20.0	5.0	21.50	45.98	32.52	6.50	4.57	31.50	46.43	22.07	8.50	4.69
1.20	20.0	10.0	25.50	45.22	29.28	7.25	3.49	34.00	46.15	19.85	7.75	3.81
1.20	20.0	15.0	28.50	44.66	26.84	7.25	2.55	39.00	45.60	15.40	8.25	3.07
1.30	5.0	5.0	23.50	45.44	31.06	6.75	3.08	33.00	46.03	20.97	8.75	3.10
1.30	5.0	10.0	26.50	44.86	28.64	6.75	2.41	36.00	45.69	18.31	8.25	2.53
1.30	5.0	15.0	31.00	44.00	25.00	7.75	1.82	39.50	45.29	15.21	8.00	2.06
1.30	10.0	5.0	24.00	45.39	30.61	6.75	3.59	33.00	46.10	20.90	8.50	3.63
1.30	10.0	10.0	28.50	44.53	26.97	7.75	2.83	35.50	45.59	16.91	8.50	2.97
1.30	10.0	15.0	31.00	44.06	24.94	7.50	2.16	41.50	45.15	13.35	8.50	2.43
1.30	20.0	5.0	27.00	44.94	28.06	8.00	4.73	35.50	45.98	18.52	8.75	4.79
1.30	20.0	10.0	28.50	44.66	26.84	7.25	3.77	40.00	45.49	14.51	8.75	3.98
1.30	20.0	15.0	31.50	44.09	24.41	7.25	2.91	40.50	45.43	14.07	7.50	3.30
1.40	5.0	5.0	25.00	45.15	29.85	6.75	3.13	34.50	45.86	19.64	8.75	3.14
1.40	5.0	10.0	30.00	44.19	25.81	8.00	2.51	39.00	45.35	15.65	8.75	2.59
1.40	5.0	15.0	32.50	43.71	23.79	7.75	1.94	40.50	45.18	14.32	7.75	2.13
1.40	10.0	5.0	26.00	45.01	28.99	7.00	3.66	35.50	45.82	18.68	8.75	3.67
1.40	10.0	10.0	30.50	44.15	25.35	8.00	2.94	40.00	45.32	14.68	8.75	3.04
1.40	10.0	15.0	32.50	43.77	23.73	7.50	2.29	42.00	45.09	12.91	8.00	2.52
1.40	20.0	5.0	27.00	44.94	28.06	7.00	4.81	37.00	45.82	17.18	8.50	4.84
1.40	20.0	10.0	31.00	44.18	24.82	7.75	3.91	40.50	45.43	14.07	8.25	4.07
1.40	20.0	15.0	33.00	43.81	23.19	7.25	3.09	43.00	45.16	11.84	7.75	3.41
1.50	5.0	5.0	26.00	44.96	29.04	6.75	3.17	35.00	45.80	19.20	8.50	3.16
1.50	5.0	10.0	30.50	44.10	25.40	7.75	2.56	39.50	45.29	15.21	8.50	2.62
1.50	5.0	15.0	33.00	43.62	23.38	7.50	2.01	41.50	45.07	13.43	7.75	2.18
1.50	10.0	5.0	27.00	44.82	28.18	7.00	3.69	36.50	45.71	17.79	8.75	3.69
1.50	10.0	10.0	31.00	44.06	24.94	7.75	3.00	40.50	45.26	14.24	8.50	3.08
1.50	10.0	15.0	33.00	43.68	23.32	7.25	2.38	43.00	44.98	12.02	8.00	2.57

Table 30 (continued)

DATA FOR THE CONSTRUCTION OF ISOKINETIC SUCROSE AND GLYCEROL GRADIENTS FOR MSE ROTOR 43127-126

Particle density g/ml	Temperature °C	Top concentration %(w/w)	Sucrose					Glycerol				
			C_h %(w/w)	Buffer 2x ml	H_2O ml	V_m ml	Velocity of 100 S particle at top speed ml/h	C_h %(w/w)	Buffer 2x ml	H_2O ml	V_m ml	Velocity of 100 S particle at top speed ml/h
1.50	20.0	5.0	28.00	44.75	27.25	7.00	4.86	37.50	45.76	16.74	8.25	4.87
1.50	20.0	10.0	29.50	44.47	26.03	6.50	3.99	41.00	45.38	13.62	8.00	4.12
1.50	20.0	15.0	33.50	43.71	22.79	7.00	3.20	43.50	45.11	11.39	7.50	3.48
1.60	5.0	5.0	26.00	44.96	29.04	6.50	3.19	36.50	45.63	17.87	8.75	3.17
1.60	5.0	10.0	29.50	44.29	26.21	6.75	2.60	41.00	45.12	13.88	8.75	2.65
1.60	5.0	15.0	33.50	43.52	22.98	7.50	2.06	41.50	45.07	13.43	7.50	2.21
1.60	10.0	5.0	26.50	44.91	28.59	6.50	3.72	37.00	45.65	17.35	8.50	3.71
1.60	10.0	10.0	30.50	44.15	25.35	7.00	3.05	40.50	45.26	14.24	8.25	3.11
1.60	10.0	15.0	34.50	43.39	22.11	7.75	2.43	43.00	44.98	12.02	7.75	2.61
1.60	20.0	5.0	27.50	44.85	27.65	6.50	4.89	38.00	45.71	16.29	8.25	4.89
1.60	20.0	10.0	32.00	43.99	24.01	7.25	4.05	42.50	45.21	12.29	8.25	4.16
1.60	20.0	15.0	35.50	43.34	21.16	7.75	3.27	43.50	45.11	11.39	7.25	3.53
1.70	5.0	5.0	26.50	44.86	28.64	6.50	3.20	37.00	45.57	17.43	8.75	3.18
1.70	5.0	10.0	30.50	44.10	25.40	7.00	2.63	41.50	45.07	13.43	8.75	2.66
1.70	5.0	15.0	33.50	43.52	22.98	7.25	2.09	42.00	45.01	12.99	7.50	2.23
1.70	10.0	5.0	27.00	44.82	28.18	6.50	3.74	37.50	45.59	16.91	8.50	3.72
1.70	10.0	10.0	31.50	43.96	24.54	7.25	3.08	41.00	45.20	13.80	8.25	3.13
1.70	10.0	15.0	34.50	43.39	22.11	7.50	2.47	43.50	44.93	11.57	7.75	2.63
1.70	20.0	5.0	28.00	44.75	27.25	6.50	4.91	38.00	45.71	16.29	8.00	4.91
1.70	20.0	10.0	32.50	43.90	23.60	7.25	4.09	42.50	45.21	12.29	8.00	4.18
1.70	20.0	15.0	35.50	43.34	21.16	7.50	3.32	44.00	45.05	10.95	7.25	3.56
1.80	5.0	5.0	27.50	44.67	27.83	6.75	3.21	37.00	45.57	17.43	8.50	3.19
1.80	5.0	10.0	30.00	44.19	25.81	6.75	2.65	40.50	45.18	14.32	8.25	2.68
1.80	5.0	15.0	33.00	43.62	23.38	6.75	2.12	43.00	44.90	12.10	7.75	2.25
1.80	10.0	5.0	28.00	44.63	27.37	6.75	3.75	37.00	45.65	17.35	8.25	3.73
1.80	10.0	10.0	30.00	44.25	25.75	6.50	3.10	41.50	45.15	13.35	8.25	3.14
1.80	10.0	15.0	33.50	43.58	22.92	6.75	2.50	44.00	44.87	11.13	7.75	2.65
1.80	20.0	5.0	28.50	44.66	26.84	6.50	4.93	38.50	45.65	15.85	8.00	4.92
1.80	20.0	10.0	30.50	44.28	25.22	6.25	4.12	43.00	45.16	11.84	8.00	4.20
1.80	20.0	15.0	35.00	43.43	21.57	7.00	3.36	45.00	44.94	10.06	7.50	3.59
1.90	5.0	5.0	26.50	44.86	28.64	6.25	3.22	37.00	45.57	17.43	8.50	3.20
1.90	5.0	10.0	31.00	44.00	25.00	7.00	2.66	41.50	45.07	13.43	8.50	2.69

Table 30 (continued)

DATA FOR THE CONSTRUCTION OF ISOKINETIC SUCROSE AND GLYCEROL GRADIENTS FOR MSE ROTOR 43127-126

Particle density g/ml	Temperature °C	Top concentration %(w/w)	Sucrose					Glycerol				
			C_h %(w/w)	Buffer 2x ml	H_2O ml	V_m ml	Velocity of 100 S particle at top speed ml/h	C_h %(w/w)	Buffer 2x ml	H_2O ml	V_m ml	Velocity of 100 S particle at top speed ml/h
1.90	5.0	15.0	34.50	43.34	22.16	7.50	2.14	44.00	44.79	11.21	8.00	2.26
1.90	10.0	5.0	27.00	44.82	28.18	6.25	3.76	37.50	45.59	16.91	8.25	3.73
1.90	10.0	10.0	31.50	43.96	24.54	7.00	3.12	42.00	45.09	12.91	8.25	3.16
1.90	10.0	15.0	35.50	43.20	21.30	7.75	2.52	43.50	44.93	11.57	7.50	2.67
1.90	20.0	5.0	30.00	44.37	25.63	7.00	4.94	39.00	45.60	15.40	8.00	4.93
1.90	20.0	10.0	32.00	43.99	24.01	6.75	4.14	42.50	45.21	12.29	7.75	4.22
1.90	20.0	15.0	37.00	43.06	19.94	8.00	3.39	45.50	44.89	9.61	7.50	3.61
2.00	5.0	5.0	28.00	44.57	27.43	6.75	3.23	37.50	45.52	16.98	8.50	3.20
2.00	5.0	10.0	30.50	44.10	25.40	6.75	2.67	41.00	45.12	13.88	8.25	2.70
2.00	5.0	15.0	33.50	43.52	22.98	6.75	2.15	43.50	44.84	11.66	7.75	2.27
2.00	10.0	5.0	28.50	44.53	26.97	6.75	3.77	37.50	45.59	16.91	8.25	3.74
2.00	10.0	10.0	30.50	44.15	25.35	6.50	3.13	42.00	45.09	12.91	8.25	3.16
2.00	10.0	15.0	34.00	43.49	22.51	6.75	2.54	44.50	44.82	10.68	7.75	2.68
2.00	20.0	5.0	29.00	44.56	26.44	6.50	4.95	39.00	45.60	15.40	8.00	4.93
2.00	20.0	10.0	31.00	44.18	24.82	6.25	4.16	43.50	45.11	11.39	8.00	4.23
2.00	20.0	15.0	35.00	43.43	21.57	6.75	3.42	45.00	44.94	10.06	7.25	3.62

Table 31

DATA FOR THE CONSTRUCTION OF ISOKINETIC SUCROSE AND GLYCEROL GRADIENTS FOR MSE ROTOR 43127-501

Particle density g/ml	Temperature °C	Top concentration %(w/w)	Sucrose					Glycerol				
			C_h %(w/w)	Buffer 2x ml	H_2O ml	V_m ml	Velocity of 100 S particle at top speed ml/h	C_h %(w/w)	Buffer 2x ml	H_2O ml	V_m ml	Velocity of 100 S particle at top speed ml/h
1.20	5.0	5.0	23.00	45.53	31.47	13.75	1.56	33.00	46.03	20.97	17.50	1.60
1.20	5.0	10.0	26.00	44.96	29.04	14.25	1.17	35.50	45.74	18.76	16.25	1.27
1.20	5.0	15.0	29.50	44.29	26.21	15.25	.84	39.00	45.35	15.65	16.25	1.00
1.20	10.0	5.0	23.50	45.49	31.01	14.00	1.83	33.50	46.05	20.45	17.25	1.87
1.20	10.0	10.0	26.00	45.01	28.99	13.75	1.38	36.00	45.76	18.24	16.00	1.50
1.20	10.0	15.0	29.50	44.34	26.16	14.75	.99	39.50	45.37	15.13	16.00	1.19
1.20	20.0	5.0	24.00	45.51	30.49	13.75	2.41	35.00	46.04	18.96	17.00	2.47
1.20	20.0	10.0	26.50	45.03	28.47	13.50	1.84	37.00	45.82	17.18	15.50	2.01
1.20	20.0	15.0	30.00	44.37	25.63	14.50	1.34	40.50	45.43	14.07	15.50	1.62
1.30	5.0	5.0	26.50	44.86	28.64	14.50	1.62	36.50	45.63	17.87	17.50	1.63
1.30	5.0	10.0	29.00	44.38	26.62	14.25	1.27	39.00	45.35	15.65	16.25	1.33
1.30	5.0	15.0	32.50	43.71	23.79	15.00	.96	41.50	45.07	13.43	15.25	1.08
1.30	10.0	5.0	26.50	44.91	28.59	14.00	1.89	38.00	45.54	16.46	18.00	1.91
1.30	10.0	10.0	29.50	44.34	26.16	14.25	1.49	40.00	45.32	14.68	16.25	1.57
1.30	10.0	15.0	32.50	43.77	23.73	14.50	1.14	42.00	45.09	12.91	15.00	1.28
1.30	20.0	5.0	27.00	44.94	28.06	13.50	2.49	39.50	45.54	14.96	17.50	2.53
1.30	20.0	10.0	30.00	44.37	25.63	13.75	1.99	41.00	45.38	13.62	15.75	2.10
1.30	20.0	15.0	33.00	43.81	23.19	14.00	1.53	43.50	45.11	11.39	15.00	1.74
1.40	5.0	5.0	27.50	44.67	27.83	13.75	1.65	39.00	45.35	15.65	18.00	1.65
1.40	5.0	10.0	30.00	44.19	25.81	13.50	1.32	40.50	45.18	14.32	16.00	1.36
1.40	5.0	15.0	33.50	43.52	22.98	14.25	1.02	42.50	44.95	12.55	14.75	1.12
1.40	10.0	5.0	28.00	44.63	27.37	13.75	1.93	39.50	45.37	15.13	17.50	1.93
1.40	10.0	10.0	30.50	44.15	25.35	13.50	1.55	41.50	45.15	13.35	16.00	1.60
1.40	10.0	15.0	34.50	43.39	22.11	15.00	1.21	43.50	44.93	11.57	14.75	1.33
1.40	20.0	5.0	29.00	44.56	26.44	13.75	2.53	41.00	45.38	13.62	17.00	2.55
1.40	20.0	10.0	31.50	44.09	24.41	13.50	2.06	42.50	45.21	12.29	15.25	2.14
1.40	20.0	15.0	34.50	43.52	21.98	13.75	1.63	45.00	44.94	10.06	14.75	1.80
1.50	5.0	5.0	28.50	44.48	27.02	13.75	1.67	40.00	45.23	14.77	17.75	1.66
1.50	5.0	10.0	32.00	43.81	24.19	14.75	1.35	42.00	45.01	12.99	16.25	1.38
1.50	5.0	15.0	35.00	43.24	21.76	15.00	1.06	43.50	44.84	11.66	14.75	1.15
1.50	10.0	5.0	29.00	44.44	26.56	13.75	1.95	40.00	45.32	14.68	17.00	1.94
1.50	10.0	10.0	32.00	43.87	24.13	14.00	1.58	42.00	45.09	12.91	15.50	1.62
1.50	10.0	15.0	34.50	43.39	22.11	13.75	1.25	44.50	44.82	10.68	14.75	1.36

Table 31 (continued)

DATA FOR THE CONSTRUCTION OF ISOKINETIC SUCROSE AND GLYCEROL GRADIENTS FOR MSE ROTOR 43127-501

Particle density g/ml	Temperature °C	Top concentration %(w/w)	Sucrose					Glycerol				
			C_h %(w/w)	Buffer 2x ml	H_2O ml	V_m ml	Velocity of 100 S particle at top speed ml/h	C_h %(w/w)	Buffer 2x ml	H_2O ml	V_m ml	Velocity of 100 S particle at top speed ml/h
1.50	20.0	5.0	30.00	44.37	25.63	13.50	2.56	42.00	45.27	12.73	16.75	2.57
1.50	20.0	10.0	33.00	43.81	23.19	14.00	2.10	44.00	45.05	10.95	15.50	2.17
1.50	20.0	15.0	35.50	43.34	21.16	13.75	1.69	46.50	44.78	8.72	15.00	1.84
1.60	5.0	5.0	29.00	44.38	26.62	13.50	1.68	40.50	45.18	14.32	17.50	1.67
1.60	5.0	10.0	32.00	43.81	24.19	14.00	1.37	42.50	44.95	12.55	16.00	1.39
1.60	5.0	15.0	35.00	43.24	21.76	14.25	1.09	44.50	44.73	10.77	15.00	1.16
1.60	10.0	5.0	29.50	44.34	26.16	13.50	1.96	41.00	45.20	13.80	17.00	1.95
1.60	10.0	10.0	32.50	43.77	23.73	14.00	1.61	43.50	44.93	11.57	16.00	1.64
1.60	10.0	15.0	35.50	43.20	21.30	14.25	1.28	45.50	44.71	9.79	15.00	1.37
1.60	20.0	5.0	31.00	44.18	24.82	13.75	2.58	42.50	45.21	12.29	16.50	2.58
1.60	20.0	10.0	33.50	43.71	22.79	13.75	2.13	44.50	45.00	10.50	15.25	2.19
1.60	20.0	15.0	37.00	43.06	19.94	14.75	1.72	47.00	44.73	8.27	14.75	1.86
1.70	5.0	5.0	29.50	44.29	26.21	13.50	1.69	41.00	45.12	13.88	17.50	1.68
1.70	5.0	10.0	32.50	43.71	23.79	14.00	1.38	43.00	44.90	12.10	16.00	1.40
1.70	5.0	15.0	35.00	43.24	21.76	13.75	1.10	45.00	44.68	10.32	15.00	1.17
1.70	10.0	5.0	30.00	44.25	25.75	13.50	1.97	41.50	45.15	13.35	17.00	1.96
1.70	10.0	10.0	33.00	43.68	23.32	14.00	1.62	43.00	44.98	12.02	15.25	1.65
1.70	10.0	15.0	36.00	43.11	20.89	14.25	1.30	46.00	44.65	9.35	15.00	1.39
1.70	20.0	5.0	31.00	44.18	24.82	13.25	2.59	43.00	45.16	11.84	16.50	2.59
1.70	20.0	10.0	34.00	43.62	22.38	13.75	2.16	45.00	44.94	10.06	15.25	2.20
1.70	20.0	15.0	37.00	43.06	19.94	14.25	1.75	47.50	44.67	7.83	14.75	1.88
1.80	5.0	5.0	29.00	44.38	26.62	12.75	1.69	40.50	45.18	14.32	16.75	1.68
1.80	5.0	10.0	32.50	43.71	23.79	13.50	1.39	42.50	44.95	12.55	15.25	1.41
1.80	5.0	15.0	35.50	43.15	21.35	14.00	1.12	45.00	44.68	10.32	14.75	1.18
1.80	10.0	5.0	30.50	44.15	25.35	13.50	1.98	42.00	45.09	12.91	17.00	1.96
1.80	10.0	10.0	33.00	43.68	23.32	13.50	1.63	43.50	44.93	11.57	15.25	1.66
1.80	10.0	15.0	36.50	43.02	20.48	14.50	1.32	46.00	44.65	9.35	14.75	1.40
1.80	20.0	5.0	31.50	44.09	24.41	13.25	2.60	43.50	45.11	11.39	16.50	2.59
1.80	20.0	10.0	33.50	43.71	22.79	13.00	2.17	45.50	44.89	9.61	15.25	2.21
1.80	20.0	15.0	37.00	43.06	19.94	13.75	1.77	48.00	44.62	7.38	14.75	1.89
1.90	5.0	5.0	30.00	44.19	25.81	13.25	1.70	41.50	45.07	13.43	17.25	1.68
1.90	5.0	10.0	32.50	43.71	23.79	13.25	1.40	43.00	44.90	12.10	15.50	1.42

Table 31 (continued)

DATA FOR THE CONSTRUCTION OF ISOKINETIC SUCROSE AND GLYCEROL GRADIENTS FOR MSE ROTOR 43127-501

Particle density g/ml	Temperature °C	Top concentration %(w/w)	Sucrose					Glycerol				
			C_h %(w/w)	Buffer 2x ml	H_2O ml	V_m ml	Velocity of 100 S particle at top speed ml/h	C_h %(w/w)	Buffer 2x ml	H_2O ml	V_m ml	Velocity of 100 S particle at top speed ml/h
1.90	5.0	15.0	36.00	43.05	20.95	14.25	1.13	45.50	44.62	9.88	14.75	1.19
1.90	10.0	5.0	30.50	44.15	25.35	13.25	1.98	42.00	45.09	12.91	16.75	1.97
1.90	10.0	10.0	33.50	43.58	22.92	13.75	1.64	44.50	44.82	10.68	15.75	1.66
1.90	10.0	15.0	36.50	43.02	20.48	14.25	1.33	46.50	44.60	8.90	14.75	1.41
1.90	20.0	5.0	31.50	44.09	24.41	13.00	2.60	43.50	45.11	11.39	16.25	2.60
1.90	20.0	10.0	34.50	43.52	21.98	13.50	2.18	45.50	44.89	9.61	15.00	2.22
1.90	20.0	15.0	37.00	43.06	19.94	13.50	1.79	48.00	44.62	7.38	14.75	1.90
2.00	5.0	5.0	30.50	44.10	25.40	13.50	1.70	41.50	45.07	13.43	17.00	1.69
2.00	5.0	10.0	33.00	43.62	23.38	13.50	1.41	43.00	44.90	12.10	15.25	1.42
2.00	5.0	15.0	36.00	43.05	20.95	14.00	1.14	45.50	44.62	9.88	14.75	1.20
2.00	10.0	5.0	30.50	44.15	25.35	13.00	1.99	42.00	45.09	12.91	16.50	1.97
2.00	10.0	10.0	33.50	43.58	22.92	13.50	1.65	44.00	44.87	11.13	15.25	1.67
2.00	10.0	15.0	36.50	43.02	20.48	14.00	1.34	46.50	44.60	8.90	14.75	1.41
2.00	20.0	5.0	32.00	43.99	24.01	13.25	2.61	43.50	45.11	11.39	16.00	2.60
2.00	20.0	10.0	35.00	43.43	21.57	13.75	2.19	45.50	44.89	9.61	14.75	2.23
2.00	20.0	15.0	37.50	42.96	19.54	13.75	1.80	48.50	44.57	6.93	14.75	1.91

Table 32
DATA FOR THE CONSTRUCTION OF ISOKINETIC SUCROSE AND GLYCEROL GRADIENTS FOR MSE ROTOR 43127-503

Particle density g/ml	Temperature °C	Top concentration %(w/w)	Sucrose					Glycerol				
			C_h %(w/w)	Buffer 2x ml	H_2O ml	V_m ml	Velocity of 100 S particle at top speed ml/h	C_h %(w/w)	Buffer 2x ml	H_2O ml	V_m ml	Velocity of 100 S particle at top speed ml/h
1.20	5.0	5.0	21.50	45.82	32.68	42.00	5.41	29.50	46.44	24.06	50.50	5.53
1.20	5.0	10.0	24.50	45.24	30.26	42.50	4.05	34.00	45.92	20.08	51.50	4.39
1.20	5.0	15.0	27.50	44.67	27.83	42.50	2.89	37.50	45.52	16.98	50.50	3.47
1.20	10.0	5.0	22.00	45.77	32.23	42.50	6.32	30.50	46.39	23.11	51.50	6.46
1.20	10.0	10.0	25.00	45.20	29.80	43.00	4.76	34.00	45.99	20.01	49.50	5.17
1.20	10.0	15.0	28.00	44.63	27.37	43.00	3.43	38.00	45.54	16.46	50.00	4.11
1.20	20.0	5.0	22.50	45.79	31.71	42.00	8.32	32.00	46.37	21.63	51.00	8.55
1.20	20.0	10.0	25.50	45.22	29.28	42.50	6.36	35.50	45.98	18.52	49.50	6.94
1.20	20.0	15.0	28.00	44.75	27.25	40.50	4.65	38.50	45.65	15.85	47.50	5.60
1.30	5.0	5.0	24.00	45.34	30.66	41.50	5.61	33.00	46.03	20.97	51.50	5.65
1.30	5.0	10.0	27.50	44.67	27.83	43.50	4.39	37.00	45.57	17.43	51.00	4.61
1.30	5.0	15.0	30.50	44.10	25.40	43.50	3.32	40.00	45.23	14.77	48.50	3.74
1.30	10.0	5.0	24.50	45.29	30.21	41.50	6.54	33.50	46.05	20.45	51.00	6.60
1.30	10.0	10.0	27.50	44.72	27.78	42.00	5.16	37.50	45.59	16.91	50.00	5.42
1.30	10.0	15.0	30.50	44.15	25.35	42.00	3.93	40.00	45.32	14.68	46.50	4.43
1.30	20.0	5.0	25.50	45.22	29.28	42.00	8.61	35.50	45.98	18.52	51.50	8.73
1.30	20.0	10.0	28.00	44.75	27.25	41.00	6.87	39.50	45.54	14.96	50.50	7.25
1.30	20.0	15.0	31.50	44.09	24.41	43.00	5.30	41.50	45.32	13.18	46.50	6.01
1.40	5.0	5.0	25.50	45.05	29.45	41.50	5.71	34.50	45.86	19.64	51.00	5.71
1.40	5.0	10.0	29.00	44.38	26.62	43.50	4.56	39.00	45.35	15.65	51.50	4.71
1.40	5.0	15.0	31.50	43.90	24.60	42.00	3.54	41.00	45.12	13.88	47.00	3.88
1.40	10.0	5.0	26.00	45.01	28.99	41.50	6.66	35.50	45.82	18.68	51.50	6.68
1.40	10.0	10.0	29.00	44.44	26.56	42.00	5.35	40.00	45.32	14.68	51.50	5.54
1.40	10.0	15.0	32.00	43.87	24.13	42.00	4.18	42.50	45.04	12.46	48.50	4.59
1.40	20.0	5.0	27.00	44.94	28.06	41.50	8.76	37.50	45.76	16.74	51.50	8.82
1.40	20.0	10.0	30.00	44.37	25.63	42.50	7.12	41.00	45.38	13.62	49.50	7.41
1.40	20.0	15.0	33.00	43.81	23.19	42.50	5.63	43.50	45.11	11.39	47.00	6.22
1.50	5.0	5.0	26.50	44.86	28.64	41.50	5.76	35.50	45.74	18.76	51.00	5.75
1.50	5.0	10.0	29.50	44.29	26.21	42.00	4.67	39.50	45.29	15.21	50.00	4.78
1.50	5.0	15.0	32.50	43.71	23.79	42.50	3.67	42.00	45.01	12.99	47.00	3.97
1.50	10.0	5.0	27.00	44.82	28.18	41.50	6.73	36.50	45.71	17.79	51.00	6.72
1.50	10.0	10.0	30.00	44.25	25.75	42.50	5.47	41.00	45.20	13.80	51.50	5.61
1.50	10.0	15.0	33.00	43.68	23.32	42.50	4.33	42.50	45.04	12.46	46.00	4.69

Table 32 (continued)

DATA FOR THE CONSTRUCTION OF ISOKINETIC SUCROSE AND GLYCEROL GRADIENTS FOR MSE ROTOR 43127-503

Particle density g/ml	Temperature °C	Top concentration %(w/w)	Sucrose					Glycerol				
			C_h %(w/w)	Buffer 2x ml	H_2O ml	V_m ml	Velocity of 100 S particle at top speed ml/h	C_h %(w/w)	Buffer 2x ml	H_2O ml	V_m ml	Velocity of 100 S particle at top speed ml/h
1.50	20.0	5.0	28.00	44.75	27.25	41.50	8.85	38.00	45.71	16.29	50.00	8.88
1.50	20.0	10.0	31.00	44.18	24.82	42.00	7.27	42.00	45.27	12.73	49.00	7.51
1.50	20.0	15.0	34.00	43.62	22.38	43.00	5.83	44.00	45.05	10.95	45.50	6.34
1.60	5.0	5.0	27.00	44.76	28.24	41.00	5.80	36.50	45.63	17.87	51.50	5.78
1.60	5.0	10.0	30.50	44.10	25.40	43.50	4.73	40.50	45.18	14.32	50.50	4.82
1.60	5.0	15.0	33.00	43.62	23.38	42.50	3.75	42.50	44.95	12.55	46.50	4.02
1.60	10.0	5.0	27.50	44.72	27.78	41.00	6.77	37.50	45.59	16.91	51.50	6.75
1.60	10.0	10.0	31.00	44.06	24.94	43.50	5.55	41.50	45.15	13.35	50.50	5.66
1.60	10.0	15.0	33.50	43.58	22.92	42.00	4.43	43.00	44.98	12.02	45.50	4.75
1.60	20.0	5.0	28.00	44.75	27.25	39.50	8.91	38.50	45.65	15.85	49.00	8.91
1.60	20.0	10.0	31.50	44.09	24.41	41.50	7.38	43.00	45.16	11.84	49.50	7.57
1.60	20.0	15.0	34.50	43.52	21.98	42.50	5.96	44.50	45.00	10.50	45.00	6.43
1.70	5.0	5.0	27.50	44.67	27.83	41.00	5.83	37.00	45.57	17.43	51.50	5.79
1.70	5.0	10.0	30.50	44.10	25.40	42.00	4.78	41.00	45.12	13.88	50.50	4.85
1.70	5.0	15.0	33.50	43.52	22.98	42.50	3.81	43.00	44.90	12.10	46.50	4.06
1.70	10.0	5.0	28.00	44.63	27.37	41.00	6.80	38.00	45.54	16.46	51.50	6.77
1.70	10.0	10.0	31.00	44.06	24.94	42.00	5.61	42.00	45.09	12.91	50.50	5.70
1.70	10.0	15.0	34.00	43.49	22.51	42.50	4.50	43.50	44.93	11.57	45.50	4.80
1.70	20.0	5.0	29.00	44.56	26.44	40.50	8.95	39.00	45.60	15.40	49.00	8.94
1.70	20.0	10.0	32.00	43.99	24.01	41.50	7.45	43.50	45.11	11.39	49.50	7.62
1.70	20.0	15.0	35.00	43.43	21.57	42.50	6.05	45.00	44.94	10.06	45.00	6.49
1.80	5.0	5.0	27.50	44.67	27.83	40.00	5.85	37.50	45.52	16.98	51.50	5.81
1.80	5.0	10.0	30.50	44.10	25.40	41.00	4.82	41.00	45.12	13.88	49.50	4.88
1.80	5.0	15.0	33.50	43.52	22.98	41.50	3.86	42.50	44.95	12.55	44.50	4.09
1.80	10.0	5.0	28.00	44.63	27.37	40.00	6.83	38.00	45.54	16.46	50.50	6.78
1.80	10.0	10.0	31.00	44.06	24.94	41.00	5.65	42.50	45.04	12.46	50.50	5.73
1.80	10.0	15.0	34.00	43.49	22.51	41.50	4.56	44.50	44.82	10.68	47.00	4.83
1.80	20.0	5.0	29.00	44.56	26.44	39.50	8.98	39.00	45.60	15.40	48.00	8.96
1.80	20.0	10.0	32.50	43.90	23.60	42.00	7.50	43.50	45.11	11.39	48.50	7.65
1.80	20.0	15.0	35.50	43.34	21.16	43.00	6.13	45.50	44.89	9.61	45.00	6.53
1.90	5.0	5.0	28.00	44.57	27.43	40.50	5.87	37.50	45.52	16.98	51.00	5.82
1.90	5.0	10.0	31.00	44.00	25.00	41.50	4.85	41.50	45.07	13.43	50.00	4.89

Table 32 (continued)

DATA FOR THE CONSTRUCTION OF ISOKINETIC SUCROSE AND GLYCEROL GRADIENTS FOR MSE ROTOR 43127-503

Particle density g/ml	Temperature °C	Top concentration %(w/w)	Sucrose					Glycerol				
			C_h %(w/w)	Buffer 2x ml	H_2O ml	V_m ml	Velocity of 100 S particle at top speed ml/h	C_h %(w/w)	Buffer 2x ml	H_2O ml	V_m ml	Velocity of 100 S particle at top speed ml/h
1.90	5.0	15.0	34.00	43.43	22.57	42.50	3.90	43.00	44.90	12.10	45.00	4.11
1.90	10.0	5.0	28.50	44.53	26.97	40.50	6.85	38.00	45.54	16.46	49.50	6.80
1.90	10.0	10.0	31.50	43.96	24.54	41.50	5.68	42.50	45.04	12.46	50.00	5.75
1.90	10.0	15.0	34.50	43.39	22.11	42.50	4.60	44.00	44.87	11.13	45.00	4.86
1.90	20.0	5.0	29.50	44.47	26.03	40.00	9.00	39.50	45.54	14.96	48.00	8.97
1.90	20.0	10.0	33.00	43.81	23.19	42.50	7.55	44.00	45.05	10.95	48.50	7.68
1.90	20.0	15.0	35.50	43.34	21.16	42.00	6.18	45.50	44.89	9.61	44.50	6.57
2.00	5.0	5.0	28.00	44.57	27.43	40.00	5.88	38.00	45.46	16.54	51.50	5.83
2.00	5.0	10.0	31.00	44.00	25.00	41.00	4.87	42.00	45.01	12.99	50.50	4.91
2.00	5.0	15.0	34.50	43.34	22.16	43.50	3.92	43.00	44.90	12.10	44.50	4.13
2.00	10.0	5.0	29.00	44.44	26.56	41.00	6.86	38.50	45.48	16.02	50.00	6.81
2.00	10.0	10.0	31.50	43.96	24.54	41.00	5.71	43.00	44.98	12.02	50.50	5.76
2.00	10.0	15.0	34.50	43.39	22.11	41.50	4.63	44.00	44.87	11.13	44.50	4.88
2.00	20.0	5.0	29.50	44.47	26.03	39.50	9.02	39.50	45.54	14.96	47.50	8.99
2.00	20.0	10.0	32.50	43.90	23.60	40.50	7.58	44.00	45.05	10.95	48.00	7.70
2.00	20.0	15.0	35.50	43.34	21.16	41.50	6.22	46.00	44.83	9.17	45.00	6.60

REFERENCES

1. Brakke, *J. Am. Chem. Soc.,* 73, 1847 (1951).
2. Martin and Ames, *J. Biol. Chem.,* 236, 1372 (1961).
3. Fritsch, *Anal. Biochem.,* 55, 57 (1973).
4. Thomson and Mikuta, *Arch. Biochem. Biophys.,* 51, 487 (1954).
5. Strohmaier and Mussgay, *Z. Naturforsch.,* 146, 171 (1959).
6. De Duve, Berthet, and Beaufay, *Progr. Biophys. Biophys. Chem.,* 9, 326 (1959).
7. Nomura, Hall, and Spiegelman, *J. Mol. Biol.,* 2, 324 (1960).
8. Rosenbloom and Cox, *Biopolymers,* 4, 747 (1966).
9. Abelson and Thoman, *J. Mol. Biol.,* 18, 262 (1966).
10. Siegel and Monty, *Biochim. Biophys. Acta,* 112, 346 (1966).
11. Schumaker, *Adv. Biol. Med. Phys.,* 11, 245 (1967).
12. Dingman, *Anal. Biochem.,* 49, 124 (1972).
13. McEwen, *Anal. Biochem.,* 20, 114 (1967).
14. Brakke, *Virology,* 6, 96 (1958).
15. Noll, in *Techniques in Protein Biosynthesis,* Vol. 2, Campbell and Sargent, Eds., Pergamon Press, London, 1969, 101.
16. Reisner, Askey, and Aylmer, *Anal. Biochem.,* 46, 365 (1972).
17. Noll, *Nature,* 215, 350 (1967).
18. McCarthy, Stafford, and Brown, *Anal. Biochem.,* 24, 314 (1968).
19. Noll, *Anal. Biochem.,* 27, 130 (1969).
20. Barber, *Natl. Cancer Inst. Monogr.,* 21, 219 (1966).
21. van der Zeijst and Bult, *Eur. J. Biochem.,* 28, 463 (1972).
22. Gagen, *Biochemistry,* 5, 2553 (1966).
23. Schachman and Lauffer, *J. Am. Chem. Soc.,* 71, 536 (1949).
24. Schachman and Lauffer, *J. Am. Chem. Soc.,* 72, 4266 (1950).
25. Lauffer, Taylor, and Wunder, *Arch. Biochem. Biophys.,* 41, 453 (1952).
26. Wales and Williams, *J. Polym. Sci.,* 8, 449 (1952).
27. Lauffer and Bendet, in *Advances in Virus Research,* Vol. 2, Smith and Lauffer, Eds., Academic Press, New York, 1954, 241.
28. Katz and Schachman, *Biochim. Biophys. Acta,* 18, 28 (1955).
29. Lauffer and Taylor, *Arch. Biochem. Biophys.,* 42, 102 (1953).
30. Taylor, Epstein, and Lauffer, *J. Am. Chem. Soc.,* 77, 1270 (1955).
31. Hamilton, Cavalieri, and Peterman, *J. Biol. Chem.,* 237, 1155 (1961).
32. Brakke and van Pelt, *Anal. Biochem.,* 38, 56 (1970).

PHYSICAL CHEMICAL DATA FOR MIXED SOLVENTS USED IN LOW TEMPERATURE BIOCHEMISTRY

P. Douzou, G. Hui Bon Hoa, P. Maurel, and F. Travers

Binary and ternary mixed solvents made up of various proportions of water and organic solvents (methanol, ethylene glycol, propylene glycol, 2-methyl-2-4-pentanediol, glycerol, *N-N*-dimethylformamide, dimethylsulfoxide) have been selected to provide liquid fluid solutions at low temperatures (+20, −100°C) in order to work on biological systems (proteins, enzymes, nucleic acids) in static as well as in dynamic experiments. In the following tables, density (d), viscosity (η), dielectric constant (D), and protonic activity of several buffers (pa_H), measured in these solvents in the authors' laboratory, are given as a function of the temperature. The solvent composition refers to the volume percent of the several components. Freezing points (fp) given here are the temperatures for which a crystalline phase appears in the liquid; some mixed solvents showed supercooling (sc). The viscosity of the binary mixed solvents (water-methanol, water-ethylene glycol) and ternary solvents (water-methanol-ethylene glycol) has been determined in order to select a number of solvents reasonably fluid at low temperature for kinetic measurements on stopped flow apparatus.

Abbreviations used for organic solvents are: EGOH, ethylene glycol or 1-2-ethanediol; MetOH, methanol; MPD, 2-methyl-2-4-pentanediol; PrOH, propylene glycol or 1-2-propanediol; GlOH, glycerol or 1-2-3-propanetriol; DMSO, dimethylsulfoxide; DMF, *N-N*-dimethylformamide.

Table 1
DENSITY

	Temperature, °C								
	+20	+10	0	−10	−20	−30	−40	−50	−60
20% EGOH-80% H$_2$O	1.026	1.030	1.034	1.037	−	−	−	−	−
30% EGOH-70% H$_2$O	1.041	1.044	1.048	1.052	−	−	−	−	−
50% EGOH-50% H$_2$O	1.067	1.073	1.079	1.085	1.092	1.098	1.104	−	−
50% MetOH-50% H$_2$O	0.928	0.935	0.943	0.951	0.959	0.967	0.974	−	−
70% MetOH-30% H$_2$O	0.888	0.897	0.906	0.915	0.923	0.932	0.941	0.950	0.959
40% EGOH-20% MetOH-40% H$_2$O	1.025	1.032	1.038	1.044	1.051	1.057	1.064	1.070	1.077
25% EGOH-25% MetOH-50% H$_2$O	0.994	1.003	1.010	1.017	1.023	1.030	1.037	1.043	−
10% EGOH-50% MetOH-40% H$_2$O	0.934	0.942	0.950	0.958	0.967	0.975	0.983	0.992	1.000
10% EGOH-60% MetOH-30% H$_2$O	0.916	0.924	0.933	0.942	0.951	0.960	0.969	0.977	0.986
50% MPD-50% H$_2$O	0.990	0.996	1.003	1.009	1.016	1.022	1.029	−	−
50% PrOH-50% H$_2$O	1.039	1.045	1.051	1.057	1.062	1.068	1.074	1.079	1.085
50% GlOH-50% H$_2$O	1.136	1.142	1.147	1.153	1.158	1.163	1.168	1.174	1.179
50% DMSO-50% H$_2$O	1.077	1.084	1.091	1.097	1.104	1.110	1.117	1.124	1.130

From Travers, Douzou, Pederson, and Gunsalus, *Biochimie*, 57, 43 (1975).

Table 2
VISCOSITY[a,b] OF ETHYLENE GLYCOL-METHANOL-WATER MIXTURES[1]

	Temperature, °C					
	−10	−20	−30	−40	−50	−60
20% EGOH-80% H_2O	11	−	−	−	−	−
30% EGOH-70% H_2O	13	−	−	−	−	−
50% EGOH-50% H_2O	18	32	63	125	−	−
50% MetOH-50% H_2O	−	12	21	38	−	−
70% MetOH-30% H_2O	−	−	10	17	28	48
40% EGOH-20% MetOH-40% H_2O	15	26	48	91	195	420
25% EGOH-25% MetOH-50% H_2O	10	19	34	66	145	−
10% EGOH-50% MetOH-40% H_2O	−	11	19	35	69	140
10% EGOH-60% MetOH-30% H_2O	−	−	14	23	42	76

Compiled by P. Douzou, G. Hui Ban Hoa, P. Maurel, and F. Travers.

[a]Viscosity is in centipoises.
[b]Only values higher than 10 cps have been determined.

REFERENCE

1. Travers, Douzou, Pederson, and Gunsalus, *Biochimie,* in press.

DIELECTRIC CONSTANT AND FREEZING POINT

Table 3
ETHYLENE GLYCOL-WATER MIXTURES

% solvent	Temperature, °C													a^a	$b^a \times 10^3$	fp^b °C
	+20	+10	0	-10	-20	-30	-40	-50	-60	-70	-80	-90	-100			
0	80.4	84.2	88.1	—	—	—	—	—	—	—	—	—	—	1.945	2.00	0
10	77.7	81.4	85.3	—	—	—	—	—	—	—	—	—	—	1.931	2.02	-4
20	75.1	78.4	82.5	86.9	—	—	—	—	—	—	—	—	—	1.916	2.15	-10
30	72.0	75.7	79.5	84.0	—	—	—	—	—	—	—	—	—	1.912	2.20	-17
40	68.1	72.1	76.3	80.2	84.4	—	—	—	—	—	—	—	—	1.880	2.30	-26
50	64.5	68.4	72.4	76.5	80.7	85	89.3	—	—	—	—	—	—	1.860	2.35	-44
60	61.1	64.6	67.9	72.0	76.3	80.8	85.3	90.1	95.7	—	—	—	—	1.832	2.41	-69
70	56.9	60.0	63.4	67.5	71.3	75.3	79.8	84.5	89.5	94.6	100.0	106.1	112.1	1.803	2.47	<-100
80	53.0	55.6	58.8	62.3	66.2	70.0	74.2	78.5	83.2	88.1	93.0	—	—	1.770	2.50	-83
90	47.5	50.5	53.5	56.8	60.2	63.8	67.8	72.0	—	—	—	—	—	1.728	2.52	-50
100	41.9	44.7	47.6	50.3	—	—	—	—	—	—	—	—	—	1.675	2.54	-12.5

[a]The evolution of D as a function of temperature obeys Akerlof's law: log D = a – bt with t in °C.
[b]fp = freezing point.

From Travers and Douzou, *Biochimie, 56*, 509 (1974).

Table 4
METHANOL-WATER MIXTURES

% solvent	Temperature, °C													a^a	$b^a \times 10^3$	fp^b °C
	+20	+10	0	-10	-20	-30	-40	-50	-60	-70	-80	-90	-100			
0	80.4	84.2	88.1	–	–	–	–	–	–	–	–	–	–	1.945	2.00	0
40	63.8	67.7	71.9	75.6	79.5	83.5	87.9	–	–	–	–	–	–	1.855	2.20	-40
50	60.3	64.0	67.8	71.2	75.5	79.2	83.9	–	–	–	–	–	–	1.830	2.30	-49
60	55.1	58.7	62.5	66.0	70.5	73.8	78.2	82.2	86.7	–	–	–	–	1.790	2.50	-67
70	46.3	49.4	53.0	56.6	60.1	63.5	66.8	70.9	74.9	79.2	83.5	–	–	1.730	2.50	-85
80	43.7	46.4	49.5	52.3	55.4	58.6	61.9	65.7	69.3	73.5	77.8	82.8	88.4	1.695	2.45	<-100
100	33.6	35.4	37.9	40.6	42.7	45.4	48.3	51.3	54.6	58.0	62.0	66.5	–	1.580	2.65	-96

[a]The evolution of D as a function of temperature obeys Akerlof's law: $\log D = a - bt$ with t in °C.
[b]fp = freezing point.

From Travers and Douzou, *J. Phys. Chem.*, 74, 2243 (1970). With permission. Copyright by the American Chemical Society.

Table 5
2-METHYL-2-4-PENTANEDIOL-WATER MIXTURES

Temperature, °C

% solvent	+20	+10	0	-10	-20	-30	-40	-50	-60	-70	-80	-90	-100	a^a	$b^a \times 10^3$	fp^b °C
0	80.4	84.2	88.1	—	—	—	—	—	—	—	—	—	—	1.945	2.00	0
10	75.4	79.2	83.2	—	—	—	—	—	—	—	—	—	—	1.920	2.13	-1.5
20	70.8	74.7	78.8	—	—	—	—	—	—	—	—	—	—	1.895	2.33	-5
30	65.7	69.7	73.5	78.0	—	—	—	—	—	—	—	—	—	1.866	2.48	-10.5
40	59.6	63.7	67.3	71.3	75.0	—	—	—	—	—	—	—	—	1.829	2.50	-26
50	53.7	57	60.5	64.5	68.3	72.7	77.1	—	—	—	—	—	—	1.783	2.60	-48
60	47.7	50.6	53.5	56.9	60.5	64.1	68.1	72.0	76.5	81.2	86.2	91.7	97.5	1.728	2.59	sc
70	41.3	43.8	46.8	49.4	52.6	55.7	59.4	62.6	66.8	70.6	75.6	79.6	84.7	1.668	2.60	sc
80	35.6	37.8	40.3	42.9	45.3	48.4	51.6	54.9	58.9	62.1	66.5	70.9	75.1	1.605	2.71	sc
90	30.6	32.6	34.7	36.9	39.1	41.9	44.6	47.3	50.7	54.3	57.4	60.9	64.7	1.540	2.72	sc
100	26.3	28.0	29.9	31.9	31.0	36.2	38.9	41.0	44.4	46.9	50.3	53.5	57.4	1.476	2.80	sc

[a]The evolution of D as a function of temperature obeys Akerlof's law: $\log D = a - bt$ with t in °C.
[b]fp = freezing point; sc = supercooling.

From Travers and Douzou, *Biochimie*, 56, 509 (1974).

Table 6
1-2-PROPANEDIOL-WATER MIXTURES

% solvent	Temperature, °C													a^a	$b^a \times 10^3$	fp^b °C
	+20	+10	0	-10	-20	-30	-40	-50	-60	-70	-80	-90	-100			
0	80.4	84.2	88.1	—	—	—	—	—	—	—	—	—	—	1.945	2.00	0
10	76.2	79.8	83.6	—	—	—	—	—	—	—	—	—	—	1.922	2.02	-3.5
20	72.0	76	79.8	—	—	—	—	—	—	—	—	—	—	1.902	2.20	-8
30	68.0	71.7	76.6	80.0	—	—	—	—	—	—	—	—	—	1.879	2.38	-17
40	63.6	67.4	71.3	75.2	79.4	83.7	—	—	—	—	—	—	—	1.852	2.37	-38
50	58.3	61.6	65.4	69.3	73.3	77.5	81.8	86.7	92.0	97.6	103.5	108.1	114.8	1.816	2.47	sc
60	53.2	56.4	59.7	63.3	67.1	70.8	74.7	79.3	83.9	89.0	94.2	99.6	105.0	1.776	2.47	sc
70	47.6	50.4	53.5	56.6	60.1	63.7	67.6	71.6	76.1	80.6	85.7	90.6	96.2	1.728	2.55	sc
80	41.6	44.2	46.9	49.8	52.8	56.2	59.5	63.1	66.8	71.1	75.3	79.6	84.8	1.671	2.58	sc
90	36.2	38.5	40.9	43.5	46.2	49.0	52.3	55.6	59.1	62.8	66.8	70.6	75.5	1.612	2.66	sc
100	29.4	31.3	33.4	35.8	38.0	40.5	43.3	46.0	49.1	52.3	55.9	59.4	63.3	1.524	2.78	sc

[a] The evolution of D as a function of temperature obeys Akerlof's law: $\log D = a - bt$ with t in °C.
[b] fp = freezing point; sc = supercooling.

From Travers and Douzou, *Biochimie*, 56, 509 (1974).

Table 7
GLYCEROL-WATER MIXTURES

Temperature, °C

% solvent	+20	+10	0	-10	-20	-30	-40	-50	-60	-70	-80	-90	-100	a^a	$b^a \times 10^3$	fp^b °C
0	80.4	84.2	88.1	—	—	—	—	—	—	—	—	—	—	1.945	2.00	0
10	77.0	80.8	84.6	—	—	—	—	—	—	—	—	—	—	1.928	2.09	-3
20	74.3	77.9	81.6	—	—	—	—	—	—	—	—	—	—	1.913	2.10	-8
30	71.3	75.0	78.9	83.0	—	—	—	—	—	—	—	—	—	1.897	2.15	-14.5
40	68.6	72.0	75.7	80.0	84.5	—	—	—	—	—	—	—	—	1.881	2.23	-29
50	65.0	68.8	72.6	76.6	80.7	85.1	90.2	95.3	99.8	106.0	111.5	117.9	123.0	1.861	2.36	sc
60	61.5	65.1	68.7	72.8	77.0	81.3	86.1	91.0	96.2	102.5	108.5	114.0	121.1	1.837	2.43	sc
70	57.6	60.9	64.2	67.9	71.8	75.9	79.7	84.7	89.6	94.7	100.1	106.0	112.1	1.808	2.40	sc
80	53.5	56.4	59.6	63.0	66.5	70.1	73.9	78.2	82.3	87.2	91.8	97.2	103.0	1.775	2.35	sc
90	48.6	51.3	53.9	57.1	59.8	63.3	66.8	70.2	74.4	78.2	82.4	86.7	91.7	1.732	2.30	sc
100	43.4	45.7	48.2	50.5	—	—	—	—	—	—	—	—	—	1.683	2.24	18

[a]The evolution of D as a function of temperature obeys Akerlof's law: $\log D = a - bt$ with t in °C.
[b]fp = freezing point; sc = supercooling.

From Travers and Douzou, *Biochimie*, 56, 509 (1974).

Table 8
DIMETHYLSULFOXIDE-WATER MIXTURES

% solvent	Temperature, °C													a^a	$b^a \times 10^3$	fp^b °C
	+20	+10	0	-10	-20	-30	-40	-50	-60	-70	-80	-90	-100			
0	80.4	84.2	88.1	—	—	—	—	—	—	—	—	—	—	1.945	2.00	0
10	79.4	82.9	87.0	—	—	—	—	—	—	—	—	—	—	1.939	2.04	-3
20	78.8	82.7	86.5	90.7	—	—	—	—	—	—	—	—	—	1.937	2.03	-12
30	78.6	82.3	86.0	90.3	94.7	—	—	—	—	—	—	—	—	1.934	2.02	-19
40	77.2	80.8	85.1	89.1	93	97.3	101.2	—	—	—	—	—	—	1.928	2.01	-41
50	76.0	79.6	83.9	87.7	91.9	96.3	100.5	105.0	110.7	115.1	119.8	125.8	132.4	1.922	2.02	sc
60	73.6	77.4	81.0	83.9	88.7	92.6	96.4	101.0	105.3	110.9	115.1	122.0	127.9	1.908	1.98	sc
70	70.4	73.4	76.8	80.2	83.9	87.4	91.2	95.2	98.9	103.9	109.0	113.7	119.0	1.885	1.88	sc
80	65.4	68.0	71.0	74.1	77.1	80.0	—	—	—	—	—	—	—	1.851	1.75	-38
90	58.1	60.3	62.4	—	—	—	—	—	—	—	—	—	—	1.795	1.55	-7
100	45.0	—	—	—	—	—	—	—	—	—	—	—	—	1.703	1.48	18.5

[a] The evolution of D as a function of temperature obeys Akerlof's law: $\log D = a - bt$ with t in °C.
[b] fp = freezing point; sc = supercooling.

From Travers and Douzou, *Biochimie*, 56, 509 (1974).

Table 9
N-N-DIMETHYLFORMAMIDE-WATER MIXTURES

Temperature, °C

% solvent	+20	+10	0	-10	-20	-30	-40	-50	-60	-70	-80	-90	-100	a[a]	b[a] ×10³	fp[b] °C
0	80.4	84.2	88.1	—	—	—	—	—	—	—	—	—	—	1.945	2.00	0
10	78.2	82.3	86.7	—	—	—	—	—	—	—	—	—	—	1.938	2.20	-2.5
20	76.2	79.8	83.9	—	—	—	—	—	—	—	—	—	—	1.924	2.07	-7
30	73.9	76.9	81.1	84.6	—	—	—	—	—	—	—	—	—	1.908	1.96	-13
40	70.4	73.9	77.1	80.8	84.3	—	—	—	—	—	—	—	—	1.887	1.95	-25
50	66.5	69.9	72.9	76.2	79.4	83.1	86.8	—	—	—	—	—	—	1.863	1.92	-40
60	62.4	65.2	68.1	71.3	74.0	77.2	80.2	84.0	87.7	—	—	—	—	1.832	1.85	-62
70	57.7	60.1	62.6	65.3	68.0	70.8	73.8	76.8	79.8	83.4	86.9	—	—	1.797	1.78	-83
80	51.7	54.0	56.4	58.9	61.4	64.0	66.7	69.7	72.7	76.1	79.3	82.6	86.2	1.751	1.85	-100
90	45.1	47.3	49.6	51.8	54.3	56.9	59.5	62.5	65.2	68.7	72.0	75.6	—	1.696	2.00	-90
100	38.2	40.2	42.2	44.6	46.9	49.5	51.9	55.0	57.8	—	—	—	—	1.625	2.25	-62

[a]The evolution of D as a function of temperature obeys Akerlof's law: $\log D = a - bt$ with t in °C.
[b]fp = freezing point.

From Travers and Douzou, *Biochimie*, 56, 509 (1974).

Table 10
ETHYLENE GLYCOL-METHANOL-WATER MIXTURES

Temperature, °C

Solvent EGOH-MetOH-H₂O	+20	+10	0	-10	-20	-30	-40	-50	-60	a[a]	b[a] ×10³	fp[b] °C
40%-20%-40%	59.6	63.3	67	70.8	74.8	79.1	83.6	88.3	93.8	1.825	2.45	-71
25%-25%-50%	62.1	65.8	69.5	73.6	77.7	82.2	87.1	91.9	97.5	1.842	2.45	-50
10%-50%-40%	57.3	60.5	64.3	68.1	72.1	76.2	80.9	85.5	90.8	1.808	2.50	-69
10%-60%-30%	52.2	55.4	58.7	62.4	66.1	70.2	74.7	79.3	84.0	1.769	2.60	sc

[a]The evolution of D as a function of temperature obeys Akerlof's law: $\log D = a - bt$ with t in °C.
[b]fp = freezing point; sc = supercooling.

From Travers, Douzou, Pederson, and Gunsalus, *Biochimie*, 57, 43 (1975).

PROTONIC ACTIVITY OF SOME USUAL BUFFERS

Protonic activity in mixed solvents (determined by spectrophotometry of acid base indicator)[1] is defined as paH = $-\log a_H$, where a_H is the protonic activity in the mixed solvent. The reference state of the pa_H scale at any temperature is taken as that of the solvated proton in a solution of hydrochloric acid 10^{-2} M in the mixed solvent. In these conditions HCl is fully dissociated, and its activity coefficient $\gamma = a_H/[H^+]$ is calculated from the Debye-Huckel formula.

The absolute uncertainty on these measurements is ± 0.07. These pa_H scales in mixed solvents have been successfully tested by studying the activity of enzymic systems as a function of pa_H in these conditions.[2]

Table 11
PROTONIC ACTIVITY OF SOME USUAL BUFFERS[1]

| Buffers 10^{-2} M | pH water | Temperature,°C | | | | | |
| | | paH in 50% EGOH-50% H_2O | | | | | |
	+20	+20	0	−10	−20	−30	−40
Chloroacetate	2.35	2.90	2.95	2.95	3.00	3.05	3.05
	2.55	3.25	3.30	3.30	3.35	3.40	3.40
	2.85	3.55	3.60	3.60	3.65	3.70	3.70
Acetate	3.65	4.25	4.35	4.40	4.45	4.50	4.55
	4.10	4.65	4.75	4.80	4.85	4.90	4.95
	4.40	4.90	5.00	5.05	5.10	5.15	5.20
	4.55	5.05	5.15	5.20	5.25	5.30	5.35
	4.75	5.25	5.35	5.40	5.45	5.50	5.55
Cacodylate	5.4	5.90	6.00	6.05	6.10	6.15	6.25
	5.8	6.30	6.40	6.45	6.50	6.55	6.65
	6.2	6.65	6.75	6.80	6.85	6.90	7.00
	6.4	6.85	6.95	7.00	7.05	7.10	7.20
	7.0	7.30	7.40	7.45	7.50	7.55	7.65
Phosphate	6.0	6.60	6.75	6.80	6.90	7.00	7.10
	6.5	7.10	7.25	7.30	7.40	7.50	7.60
	7.0	7.60	7.75	7.80	7.90	8.00	8.10
	7.5	8.10	8.25	8.30	8.40	8.50	8.60
Tris	7.0	7.20	7.95	8.35	8.75	9.20	9.70
	7.5	7.65	8.40	8.80	9.20	9.65	10.15
	8.0	7.95	8.70	9.10	9.50	9.95	10.45
	8.5	8.45	9.20	9.60	10.00	10.45	10.95

Compiled by P. Douzou, G. Hui Bon Hoa, P. Maurel, and F. Travers.

REFERENCE

1. **Maurel, Hue Bon Hoa, and Douzou,** *J. Biol. Chem.,* 250, 4376 (1975).

Table 12
PROTONIC ACTIVITY OF SOME USUAL BUFFERS[1]

Temperature,°C

Buffers 10^{-2} M	pH water	paH in 50% MetOH-50% H_2O					
	+20	+20	0	−10	−20	−30	−40
Chloroacetate	2.35	2.90	2.95	2.95	3.00	3.05	3.10
	2.55	3.25	3.30	3.30	3.35	3.40	3.45
	2.85	3.60	3.65	3.65	3.70	3.75	3.80
Acetate	3.65	4.50	4.55	4.55	4.60	4.60	4.65
	4.10	4.85	4.90	4.90	4.95	4.95	5.00
	4.40	5.10	5.15	5.15	5.20	5.20	5.25
	4.55	5.30	5.35	5.35	5.40	5.40	5.45
	4.75	5.45	5.50	5.50	5.55	5.55	5.60
Cacodylate	5.4	6.30	6.30	6.35	6.35	6.40	6.40
	5.8	6.60	6.60	6.65	6.65	6.70	6.70
	6.2	6.95	6.95	7.00	7.00	7.05	7.05
	6.4	7.10	7.10	7.15	7.15	7.20	7.20
	7.0	7.55	7.55	7.60	7.60	7.65	7.65
Phosphate	6.0	7.20	7.25	7.25	7.30	7.30	7.35
	6.5	7.65	7.70	7.70	7.75	7.75	7.80
	7.0	8.10	8.15	8.15	8.20	8.20	8.25
	7.5	8.50	8.55	8.55	8.60	8.60	8.65
	8.0	8.90	8.95	8.95	9.00	9.00	9.05
Tris	7.0	7.25	7.80	8.15	8.55	8.90	9.35
	7.5	7.70	8.25	8.60	9.00	9.35	9.80
	8.0	7.95	8.50	8.85	9.25	9.60	10.05
	8.5	8.40	8.95	9.30	9.70	10.05	10.50
	9.0	8.85	9.40	9.75	10.15	10.50	10.95

Compiled by P. Douzou, G. Hui Bon Hoa, P. Maurel, and F. Travers.

REFERENCE

1. **Hui Bon Hoa and Douzou,** *J. Biol. Chem.,* 248, 4649 (1973).

Table 13
PROTONIC ACTIVITY OF SOME USUAL BUFFERS[1]

	pH water			paH in 70% MetOH-30% H_2O			
Buffers 10^{-2} M	**+20**	**+20**	**0**	**-10**	**-20**	**-30**	**-40**
Chloroacetate	2.35	3.20	3.20	3.20	3.20	3.20	3.20
	2.55	3.70	3.70	3.70	3.70	3.70	3.70
	2.85	4.05	4.05	4.05	4.05	4.05	4.05
Acetate	3.65	5.0	5.10	5.10	5.15	5.20	5.25
	4.10	5.35	5.45	5.45	5.50	5.55	5.60
	4.40	5.55	5.65	5.65	5.70	5.75	5.80
	4.55	5.75	5.85	5.85	5.90	5.95	6.00
	4.75	5.95	6.05	6.05	6.10	6.15	6.20
Cacodylate	5.4	6.65	6.70	6.70	6.75	6.80	6.80
	5.8	7.0	7.05	7.05	7.10	7.15	7.15
	6.2	7.35	7.40	7.40	7.45	7.50	7.50
	6.4	7.50	7.55	7.55	7.60	7.65	7.65
	7.0	8.0	8.05	8.05	8.10	8.15	8.15
Phosphate	6.0	7.75	7.85	7.90	7.95	8.00	8.10
	6.5	8.25	8.35	8.40	8.45	8.50	8.60
	7.0	8.60	8.70	8.75	8.80	8.85	8.95
	7.5	9.10	9.20	9.25	9.30	9.35	9.45
	8.0	9.45	9.55	9.60	9.65	9.70	9.80
Tris	7.0	7.30	7.95	8.40	8.80	9.25	9.80
	7.5	7.70	8.35	8.80	9.20	9.65	10.20
	8.0	8.00	8.65	9.10	9.50	9.95	10.50
	8.5	8.55	9.10	9.65	10.05	10.50	11.05
	9.0	9.10	9.65	10.20	10.60	11.05	11.60

Temperature,°C

Compiled by P. Douzou, G. Hui Bon Hoa, P. Maurel, and F. Travers.

REFERENCE

1. Hui Bon Hoa and Douzou, *J. Biol. Chem.,* 248, 4649 (1973).

Table 14
PROTONIC ACTIVITY OF SOME
USUAL BUFFERS[1]

Buffer 10^{-2} M	pH water	paH in 50% MPD–50% H_2O			
	+20	+20	0	−20	−25
Chloroacetate	2.2	2.55	2.55	2.55	2.55
	2.6	3.24	3.24	3.24	3.24
	3.0	3.76	3.76	3.76	3.76
	3.4	4.20	4.20	4.20	4.20
Acetate	3.8	4.55	4.59	4.65	4.66
	4.3	5.05	5.09	5.15	5.16
	4.8	5.56	5.60	5.66	5.67
	5.3	6.04	6.08	6.14	6.15
Cacodylate	5.5	6.08	6.09	6.12	6.12
	6.0	6.77	6.78	6.81	6.81
	6.5	7.34	7.35	7.38	7.38
	7.0	7.80	7.81	7.84	7.84
Phosphate	6.5	7.36	7.47	7.78	7.87
	7.0	8.05	8.16	8.47	8.56
	7.5	8.57	8.68	8.99	9.08
	8.0	8.90	9.01	9.32	9.41
Tris	8.0	8.10	8.64	9.39	9.60
	8.5	8.54	9.08	9.83	10.04
	9.0	8.90	9.44	10.19	10.40
	9.5	9.20	9.74	10.49	10.70
Carbonate	9.0	9.70	10.0	10.50	10.65
	9.5	10.67	10.97	11.47	11.62
	10.0	11.16	11.46	11.96	12.11
	10.5	11.61	11.91	12.41	12.56

Temperature,°C

Compiled by P. Douzou, G. Hui Bon Hoa, P. Maurel, and F. Travers.

REFERENCE

1. **Maurel, Hui Bon Hoa, and Douzou,** *J. Biol. Chem.,* 248, 4649 (1973).

Table 15
PROTONIC ACTIVITY OF SOME USUAL BUFFERS

		Temperature,°C					
	pH water	paH in 50% PrOH-50% H_2O					
Buffer 10^{-2} M	+20	+20	0	−10	−20	−30	−40
Chloroacetate	2.40	3.35	3.35	3.40	3.40	3.40	3.45
	2.60	3.55	3.55	3.55	3.60	3.60	3.60
	3.00	4.00	4.00	4.00	4.00	4.05	4.05
	3.40	4.40	4.40	4.40	4.40	4.45	4.45
Acetate	4.00	4.90	4.90	4.95	4.95	5.00	5.05
	4.50	5.35	5.35	5.40	5.40	5.45	5.45
	5.00	5.80	5.80	5.85	5.85	5.90	5.95
	5.60	6.30	6.35	6.35	6.40	6.40	6.45
Cacodylate	5.50	6.05	6.15	6.15	6.20	6.25	6.30
	6.00	6.55	6.60	6.65	6.70	6.75	6.80
	6.50	7.00	7.10	7.10	7.15	7.20	7.30
	7.00	7.45	7.50	7.55	7.60	7.65	7.70
Phosphate	6.50	7.30	7.40	7.45	7.55	7.60	7.70
	7.00	7.75	7.85	7.90	7.95	8.05	8.10
	7.50	8.10	8.20	8.25	8.35	8.40	8.45
	8.00	8.55	8.65	8.70	8.80	8.85	8.90
Tris	8.00	7.80	8.40	8.80	9.20	9.65	10.10
	8.50	8.20	8.80	9.20	9.55	10.00	10.45
	9.00	8.60	9.25	9.60	10.00	10.45	10.90
	9.50	8.90	9.55	9.90	10.30	10.75	11.15
Carbonate	9.50	9.70	10.00	10.15	10.35	10.55	10.75
	10.00	10.25	10.55	10.70	10.90	11.10	11.30
	10.50	10.60	10.90	11.10	11.25	11.45	11.65
	11.00	10.75	11.05	11.20	11.35	11.60	11.80

Table 16
PROTONIC ACTIVITY OF SOME USUAL BUFFERS

Buffer 10^{-2} M	pH water	paH in 50% GlOH-50% H_2O				
	Temperature,°C					
	+20	+20	0	-10	-20	-30
Chloroacetate	2.60	3.10	3.25	3.35	3.45	3.55
	3.00	3.50	3.65	3.75	3.85	3.95
	3.40	3.90	4.05	4.15	4.25	4.35
Acetate	3.80	4.25	4.45	4.55	4.70	4.80
	4.30	4.70	4.90	5.05	5.15	5.30
	4.80	5.15	5.35	5.45	5.60	5.75
	5.30	5.60	5.80	5.90	6.05	6.20
Cacodylate	5.50	5.70	5.90	6.05	6.15	6.30
	6.00	6.15	6.35	6.45	6.60	6.75
	6.50	6.65	6.85	7.00	7.10	7.25
	7.00	7.15	7.35	7.50	7.60	7.75
Phosphate	6.50	6.70	6.95	7.10	7.30	7.45
	7.00	7.15	7.45	7.60	7.75	7.90
	7.50	7.65	7.90	8.05	8.20	8.40
	8.00	8.10	8.30	8.50	8.65	8.80
Tris	8.00	8.05	8.90	9.35	9.80	10.35
	8.50	8.50	9.35	9.75	10.25	10.85
	9.00	8.85	9.70	10.15	10.65	11.20
	9.50	9.15	9.95	10.45	10.90	11.50
Carbonate	9.50	9.50	9.95	10.20	10.45	10.70
	10.00	9.95	10.40	10.60	10.90	11.20
	10.50	10.25	10.70	10.95	11.20	11.50
	11.00	10.45	10.90	11.10	11.40	11.70

From Maurel, Hui Bon Hoa, and Douzou, *J. Biol. Chem.,* 250, 1376 (1975). With permission.

Table 17
PROTONIC ACTIVITY OF SOME USUAL BUFFERS

	pH water	paH in 50% DMSO-50% H_2O					
Buffers $10^{-2}\ M$	+20	+20	0	-10	-20	-30	-40
Chloroacetate	2.60	3.05	3.15	3.20	3.30	3.40	3.45
	3.00	3.60	3.75	3.85	3.90	4.00	4.15
	3.40	3.95	4.10	4.20	4.25	4.35	4.45
Acetate	3.80	4.45	4.65	4.75	4.85	5.00	5.10
	4.30	4.90	5.10	5.20	5.30	5.45	5.60
	4.80	5.40	5.55	5.70	5.80	5.95	6.10
	5.30	5.75	5.95	6.05	6.20	6.30	6.45
Cacodylate	5.50	6.00	6.25	6.40	6.55	6.70	6.90
	6.00	6.45	6.70	6.85	7.00	7.15	7.35
	6.50	7.00	7.20	7.35	7.50	7.70	7.85
	7.00	7.45	7.70	7.85	8.00	8.15	8.35
Phosphate	6.50	7.75	8.10	8.30	8.50	8.80	9.00
	7.00	8.20	8.60	8.80	9.00	9.25	9.50
	7.50	8.60	8.95	9.20	9.40	9.65	9.90
	8.00	9.05	9.40	9.60	9.85	10.10	10.35
Tris	8.00	7.20	7.90	8.35	8.75	9.25	9.80
	8.50	7.55	8.25	8.70	9.15	9.65	10.15
	9.00	7.95	8.70	9.10	9.55	10.05	10.55
	9.50	8.25	9.00	9.40	9.85	10.35	10.85
Carbonate	9.50	10.75	11.30	11.60	11.90	12.30	12.65

From Maurel, Hui Bon Hoa, and Douzou, *J. Biol. Chem.*, 250, 1376 (1975). With permission.

Table 18
PROTONIC ACTIVITY OF SOME USUAL BUFFERS[1]

	pH water	Temperature,°C					
		paH in 50% DMF-50% H_2O					
Buffers 10^{-2} M	+20	+20	0	−10	−20	−30	−40
Chloroacetate	2.35	4.20	4.30	4.35	4.40	4.45	4.50
	2.55	4.40	4.50	4.55	4.60	4.65	4.70
	2.85	4.60	4.70	4.75	4.80	4.85	4.90
Acetate	3.65	5.90	6.05	6.15	6.25	6.35	6.50
	4.10	6.35	6.50	6.60	6.70	6.80	6.95
	4.40	6.55	6.70	6.80	6.90	7.0	7.15
	4.55	6.70	6.85	6.95	7.05	7.15	7.30
	4.75	6.90	7.05	7.15	7.25	7.35	7.50
Phosphate[a]	6.0	8.55	8.55	8.60	8.60	8.65	8.65
	6.5	8.90	8.90	8.95	8.95	9.00	9.00
	7.0	9.20	9.20	9.25	9.25	9.30	9.30
	7.5	9.35	9.35	9.40	9.40	9.45	9.45
	8.0	9.55	9.55	9.60	9.60	9.65	9.65
Tris	7.0	6.90	7.50	7.90	8.20	8.60	9.05
	7.5	7.35	7.95	8.35	8.65	9.05	9.50
	8.0	7.60	8.20	8.60	8.90	9.30	9.75
	8.5	8.20	8:80	9.20	9.50	9.90	10.35
	9.0	8.50	9.10	9.50	9.80	10.20	10.65

[a]Concentration is 10^{-3} M because of precipitation.

REFERENCE

1. **Hui Bon Hoa and Douzou**, *J. Biol. Chem.*, 248, 4649 (1973).

Table 19
PROTONIC ACTIVITY OF SOME USUAL BUFFERS

	pH water	paH in 40% EGOH-20% MetOH-40% H_2O Temperature,°C					
Buffer $10^{-2}\ M$	+20	+20	0	−10	−20	−30	−40
Chloroacetate	2.2	2.80	2.80	2.80	2.80	2.80	2.80
	2.6	3.40	3.40	3.40	3.40	3.40	3.40
	3.0	3.90	3.90	3.90	3.90	3.90	3.90
	3.4	4.25	4.25	4.25	4.25	4.25	4.25
Acetate	3.8	4.55	4.60	4.65	4.70	4.70	4.75
	4.3	5.10	5.15	5.20	5.25	5.25	5.30
	4.8	5.55	5.60	5.65	5.70	5.70	5.75
	5.3	6.10	6.15	6.20	6.25	6.25	6.30
Cacodylate	5.5	6.10	6.10	6.15	6.15	6.20	6.20
	6.0	6.70	6.70	6.75	6.75	6.80	6.80
	6.5	7.10	7.10	7.15	7.15	7.20	7.20
	7.0	7.60	7.60	7.65	7.65	7.70	7.70
Phosphate	6.5	7.45	7.45	7.45	7.45	7.50	7.50
	7.0	8.00	8.00	8.00	8.00	8.05	8.05
	7.5	8.45	8.45	8.45	8.45	8.50	8.50
	8.0	8.80	8.80	8.80	8.80	8.85	8.85
Tris	8.0	8.05	8.65	9.00	9.30	9.70	10.15
	8.5	8.65	9.25	9.60	9.90	10.30	10.75
	9.0	8.95	9.55	9.90	10.20	10.60	11.05
	9.5	9.20	9.80	10.15	10.45	10.85	11.30
Glycine	9.5	9.50	10.00	10.30	10.65	10.95	11.25
	10.0	9.95	10.45	10.75	11.10	11.40	11.70
	10.5	10.35	10.85	11.15	11.50	11.80	12.10
	11.0	10.60	11.10	11.40	11.75	12.05	12.35

From Travers, Douzou, Pederson, and Gunsalus, *Biochimie,* 57, 43 (1975).

Table 20
PROTONIC ACTIVITY OF SOME USUAL BUFFERS

Buffer 10^{-2} *M*	pH water	paH in 25% EGOH-25% MetOH-50% H_2O Temperature,°C					
	+20	+20	0	–10	–20	–30	–40
Chloroacetate	2.2	2.80	2.80	2.80	2.80	2.80	2.80
	2.6	3.35	3.35	3.35	3.35	3.35	3.35
	3.0	3.95	3.95	3.95	3.95	3.95	3.95
	3.4	4.35	4.35	4.35	4.35	4.35	4.35
Acetate	3.8	4.80	4.80	4.80	4.80	4.85	4.85
	4.3	5.30	5.30	5.30	5.30	5.35	5.35
	4.8	5.80	5.80	5.80	5.80	5.85	5.85
	5.3	6.30	6.30	6.30	6.30	6.35	6.35
Cacodylate	5.5	6.45	6.45	6.45	6.45	6.45	6.45
	6.0	7.10	7.10	7.10	7.10	7.10	7.10
	6.5	7.55	7.55	7.55	7.55	7.55	7.55
	7.0	8.00	8.00	8.00	8.00	8.00	8.00
Phosphate	6.5	7.85	7.90	7.90	7.95	8.00	8.05
	7.0	8.25	8.30	8.30	8.35	8.40	8.45
	7.5	8.80	8.85	8.85	8.90	8.95	9.00
	8.0	9.30	9.35	9.35	9.40	9.45	9.50
Tris	8.0	8.40	9.05	9.35	9.70	10.05	10.55
	8.5	8.80	9.45	9.75	10.10	10.45	10.95
	9.0	9.25	9.90	10.20	10.55	10.90	11.40
	9.5	9.50	10.15	10.45	10.80	11.15	11.65
Glycine	9.5	9.80	10.30	10.55	10.85	11.10	11.50
	10.0	10.30	10.80	11.05	11.35	11.60	12.00
	10.5	10.65	11.15	11.40	11.70	11.95	12.35
	11.0	11.00	11.50	11.75	12.05	12.30	12.70

From Travers, Douzou, Pederson, and Gunsalus, *Biochimie,* 57, 43 (1975).

Table 21
PROTONIC ACTIVITY OF SOME USUAL BUFFERS

| | pH water | | | Temperature,°C | | | |
| | | paH in 10% EGOH-60% MetOH-30% H_2O | | | | | |
Buffer $10^{-2}\ M$	+20	+20	0	–10	–20	–30	–40
Chloroacetate	2.2	3.15	3.15	3.15	3.15	3.20	3.20
	2.6	3.80	3.80	3.80	3.80	3.85	3.85
	3.0	4.35	4.35	4.35	4.35	4.40	4.40
	3.4	4.70	4.70	4.70	4.70	4.75	4.75
Acetate	3.8	5.15	5.15	5.20	5.20	5.30	5.30
	4.3	5.70	5.70	5.75	5.75	5.85	5.85
	4.8	6.20	6.20	6.25	6.25	6.35	6.35
	5.3	6.70	6.70	6.75	6.75	6.85	6.85
Cacodylate	5.5	6.65	6.65	6.65	6.65	6.75	6.75
	6.0	7.15	7.15	7.15	7.15	7.25	7.25
	6.5	7.60	7.60	7.60	7.60	7.70	7.70
	7.0	8.10	8.10	8.10	8.10	8.20	8.20
Phosphate	6.5	8.25	8.30	8.30	8.35	8.40	8.40
	7.0	8.75	8.80	8.80	8.85	8.90	8.90
	7.5	9.25	9.30	9.30	9.35	9.40	9.40
	8.0	9.60	9.65	9.65	9.70	9.75	9.75
Tris	8.0	7.90	8.50	8.80	9.20	9.60	10.05
	8.5	8.35	8.95	9.25	9.65	10.05	10.50
	9.0	8.85	9.45	9.75	10.15	10.55	11.00
	9.5	9.30	9.90	10.20	10.60	11.00	11.45
Glycine	9.5	9.45	10.00	10.30	10.75	11.25	12.05
	10.0	10.15	10.70	11.00	11.45	11.95	12.75
	10.5	10.55	11.10	11.40	11.85	12.35	13.15
	11.0	11.05	11.60	11.90	12.35	12.85	13.65

From Travers, Douzou, Pederson, and Gunsalus, *Biochimie,* 57, 43 (1975).

REFRACTIVE INDEX CHANGE WITH TEMPERATURE
FOR ETHYL ALCOHOL

V. S. Ananthanarayanan

$$n = 1.36250 - 10^{-6} [404(t-15) + 0.22(t-15)^2 + 0.0075(t-15)^3]$$

From Hall and Payne, *Physiol. Rev.,* 20, 249 (1922) as given by **Partington,** *Advanced Treatise on Physical Chemistry,* Vol. 4, John Wiley & Sons, New York, 1953, 31. With permission.

DENSITY CHANGE WITH TEMPERATURE OF
SOME ORGANIC SOLVENTS

The following table gives the parameters for the equation:

$$d_t = [d_s + 10^{-3} \alpha(t-t_s) + 10^{-6} \beta(t-t_s)^2 + 10^{-9} \gamma(t-t_s)^3] \pm 10^{-4} \Delta$$

where $t_s = 0°$ C. Where the density at $0°$ C has not been determined, a value has, in many cases, been arrived at by extrapolation. Such values are enclosed in parentheses and are given as a basis for calculation only.

Data given are for only a few solvents commonly used in physical chemical studies on biomolecules. See *International Critical Tables* for data on other solvents.

Formula	Name	d_s	α	β	γ	Δ	Range, °C
CCl_4	Carbon tetrachloride	1.63255	−1.9110	−0.690	−	2	0−40
$CHCl_3$	Chloroform	1.52643	−1.8563	−0.5309	−8.81	1	−53−55
CH_2O_2	Formic acid	1.2441	−1.221	0.126	−	20	0−40
$C_2H_2Cl_2O_2$	Dichloroacetic acid	(1.5919)	−1.375	−	−	20	0−100
$C_2H_4Cl_2$	Ethylene chloride	1.28248	−1.4217	−0.933	2.29	2	0−74
$C_2H_6O_2$	Glycol	1.1257	−0.5713	−2.766	10.9	10	0−136
$C_3H_8O_3$	Glycerol	1.2727	−0.5506	−1.016	1.270	10	0−280
C_6H_6	Benzene	(0.90005)	−1.0636	−0.0376	−2.213	2	11−72
C_6H_{12}	Cyclohexane	0.79707	−0.8879	−0.972	1.55	3	0−65
C_7H_8O	*m*-Cresol	(1.0495)	−0.7639	−0.471	−	3	9−153

From *International Critical Tables,* Vol. 3, Natl. Acad. Sci., Washington, D.C., 1928, 27. Reproduced with permission of the National Academy of Sciences.

For methyl alcohol the following equation reproduces the values of several authors at different temperatures, to 0.00002:

$$d_4^t = 0.80999 - 0.0_3 9253 t - 0.0_6 41 t^2$$

(where the subscript indicates the temperature of water to which the density is referred). Extrapolation by this equation to $60°$ gives values which are probably the best available.

For ethyl alcohol the following equation reproduces the available data to 0.0001 up $40°$ and to 0.0002 up to $80°$:

$$d_4^t = 0.80625 - 0.0_3 8461 t + 0.0_6 160 t^2 - 0.0_8 85 t^3$$

The densities of dimethyl formamide at different temperatures from 25 to $90°$ have been given by Gopal and Rizvi.[1]

REFERENCE

1. **Gopal and Rizvi,** *J. Ind. Chem. Soc.,* 43, 179 (1966).

Index

INDEX

A

Q

R